nature

The Living Record of Science
《自然》学科经典系列

总顾问：李政道（Tsung-Dao Lee）

英方总主编：Sir John Maddox
Sir Philip Campbell　　中方总主编：路甬祥

天文学的进程 II
PROGRESS IN ASTRONOMY II

（英汉对照）

主编：武向平

外语教学与研究出版社　·　麦克米伦教育　·　《自然》旗下期刊与服务集合

FOREIGN LANGUAGE TEACHING AND RESEARCH PRESS · MACMILLAN EDUCATION · NATURE PORTFOLIO

北京 BEIJING

图书在版编目（CIP）数据

天文学的进程 . II：英汉对照 / 武向平主编 . -- 北京：外语教学与研究出版社，2022.8

（《自然》学科经典系列 / 路甬祥等总主编）

ISBN 978-7-5213-3904-8

Ⅰ . ①天… Ⅱ . ①武… Ⅲ . ①天文学－文集－英、汉 Ⅳ . ①P1-53

中国版本图书馆 CIP 数据核字 (2022) 第 146176 号

出 版 人　王　芳
项目统筹　章思英
项目负责　刘晓楠　顾海成
责任编辑　刘晓楠
责任校对　王　菲　白小羽　夏洁媛
封面设计　孙莉明　高　蕾
版式设计　孙莉明
出版发行　外语教学与研究出版社
社　　址　北京市西三环北路 19 号（100089）
网　　址　http://www.fltrp.com
印　　刷　北京华联印刷有限公司
开　　本　787×1092　1/16
印　　张　47
版　　次　2022 年 9 月第 1 版　2022 年 9 月第 1 次印刷
书　　号　ISBN 978-7-5213-3904-8
定　　价　568.00 元

购书咨询：（010）88819926　电子邮箱：club@fltrp.com
外研书店：https://waiyants.tmall.com
凡印刷、装订质量问题，请联系我社印制部
联系电话：（010）61207896　电子邮箱：zhijian@fltrp.com
凡侵权、盗版书籍线索，请联系我社法律事务部
举报电话：（010）88817519　电子邮箱：banquan@fltrp.com
物料号：339040001

《自然》学科经典系列

（英汉对照）

总顾问：李政道（Tsung-Dao Lee）

英方总主编：Sir John Maddox
　　　　　　Sir Philip Campbell

中方总主编：路甬祥

英方编委：

Philip Ball

Arnout Jacobs

Magdalena Skipper

中方编委（以姓氏笔画为序）：

万立骏

朱道本

许智宏

武向平

赵忠贤

滕吉文

天 文 学 的 进 程

（英汉对照）

主编：武向平

审稿专家 （以姓氏笔画为序）

马宇蒨	毛淑德	冯珑珑	许　冰	杜爱民	李　然	肖伟科
吴学兵	何香涛	沈志侠	张华伟	陈　阳	陈含章	欧阳自远
季江徽	周礼勇	周济林	胡永云	夏俊卿	徐　栋	徐仁新
蒋世仰	臧伟呈	黎　卓				

翻译工作组稿人 （以姓氏笔画为序）

王晓蕾	王耀杨	刘　明	何　铭	沈乃澂	张　健	郭红锋
蔡则怡						

翻译人员 （以姓氏笔画为序）

冯翀	吕孟珍	刘霞	刘项琨	齐红艳	孙惠南	李海宁
肖莉	何钧	余恒	沈乃澂	武振宇	金世超	周杰
周旻辰	钱磊	梁恩思	谭秀慧			

校对人员 （以姓氏笔画为序）

于萌	马荣	马晨晨	王菲	王帅帅	王志云	王丽霞
王珊珊	元旭津	毛俊捷	化印	公晗	史未卿	白小羽
丛岚	冯翀	吉祥	吕秋莎	刘子怡	刘东亮	刘若青
刘项琨	闫妍	阮玉辉	李芳	李兆升	李盎然	杨茜
邱彩玉	何铭	何敏	何思源	邹伯夏	宋乔	张狄
张向东	张宜嘉	张瑞玉	陈云	陈露芸	范艳璇	罗小青
周小雅	周少贞	郑征	郑娇娇	宗伟凯	侯鉴璇	夏洁媛
顾海成	钱磊	郭红锋	郭晓博	黄小斌	黄元耕	黄雪嫚
崔天明	梁瑜	葛越	董静娟	蒋世仰	雷文欣	蔡迪
蔡则怡	蔡军茹					

Contents
目录

Volume II

(1973-2006)

Earth-Moon Mass Ratio from Mariner 9 Radio Tracking Data

S. K. Wong and S. J. Reinbold

Editor's Note

The ratio of the mass of the Earth to that of the Moon is used to predict the evolution of the Moon's orbit into the future. The ratio can be determined without any assumptions using radio tracking data from spacecraft. Here S. K. Wong and S. J. Reinbold do just that, using data from the Mariner 9 mission to Mars. They find the mass ratio to be 81.3007, very close to the currently accepted value.

THE navigation of the Mariner 9 spacecraft from Earth to Mars was performed using phase-coherent range and doppler tracking data recorded by the Jet Propulsion Laboratory (JPL) Deep Space Network. These data also determine the Earth–Moon mass ratio, which involves the following physics: as the Earth revolves about the centre of mass of the Earth–Moon system, a sinusoidal curve is impressed on the range and doppler tracking data with a frequency equal to the sidereal mean motion of the Moon. This signature is shown in Fig. 1, where a perturbation of 0.0003 was made in the mass ratio (μ^{-1} = mass of Earth over mass of Moon). This sinusoidal variation in the tracking data can be eliminated by finding a value for μ^{-1} that properly represents the amplitude of the barycentre motion of the Earth. The procedure is direct and for all practical purposes is completely uncoupled from other parameters used in reducing the tracking data.

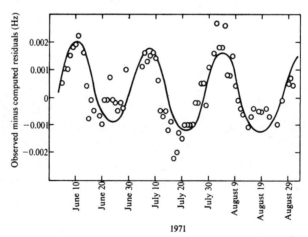

Fig. 1. The effect of Earth–Moon mass ratio on doppler residuals.

The mass ratio was determined from range and doppler data obtained over a period of

由水手9号射电跟踪数据得到的地月质量比

黄，赖因博尔德

编者按

地球质量与月球质量之比被用于预测月球轨道在未来的演化。不必进行任何假设，使用来自航天器的射电跟踪数据，我们就能够确定这个比值。本文中黄和赖因博尔德正是那样做的，他们使用的数据来自被派往火星的水手9号。他们得到的质量比是81.3007，非常接近现在普遍接受的值。

从地球飞向火星的水手9号航天器使用由喷气推进实验室（JPL）深空探测网记录的相位相干测距和多普勒跟踪数据进行导航。这些数据也确定了地月质量比，这涉及以下物理事实：地球围绕地月系统的质心转动，使得测距和多普勒跟踪数据成为正弦曲线，频率与月球相对于恒星的平均运动相同。这个特征如图1所示，其中对质量比作了0.0003的扰动（μ^{-1}为地球质量与月球质量之比）。通过找到恰当地反映地球质心运动幅度的μ^{-1}，就可以在跟踪数据中去除此正弦变化。这是一种直接的方法，在实际应用中完全和处理跟踪数据用到的其他参量无关。

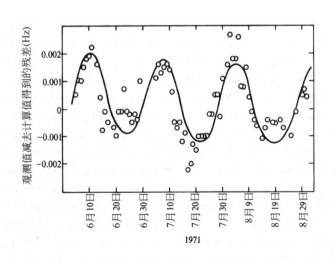

图 1. 地月质量比对多普勒残差的影响

地月质量比是基于15周的测距和多普勒数据（1971年6月5日到9月15日）

15 weeks (June 5 to September 15, 1971). The data coverage is shown in Table 1. We also show the statistics from the best determination. The data reduction was performed using the JPL Double Precision Orbit Determination Program[1], which uses a Cowell integrated trajectory and a batch least squares filter. In weighting the range data, we have taken extreme care to assure optimal data utilization without conflicting with the doppler data. The Mariner Mars 1969 results showed that such conflicts can cause significant perturbations in the estimated parameters.

Table 1. Tracking Data Statistics

Tracking station	Data type*	Number Of points†	Data interval, 1971 (UTC)	Mean residual‡	Root-mean-Square residual‡
Goldstone-Echo, Calif.	Doppler	1,069	6/5 08:37 to 9/14 04:59	0.000005 Hz	0.00144 Hz
	Range-Mark 1A	248	6/5 08:42 to 6/30 09:14	−17.73 RU	20.94 RU
	Range-Mu	31	7/12 05:40 to 9/6 02:49	−19.12 ns	134.55 ns
Goldstone-Mars, Calif.	Doppler	193	6/5 13:01 to 8/31 09:36	0.000002 Hz	0.00132 Hz
	Range-Tau	109	6/5 12:43 to 8/27 06:20	−33.67 ns	116.22 ns
Woomera, Australia	Doppler	1,638	6/5 01:15 to 9/12 17:56	−0.000012 Hz	0.00139 Hz
	Range-Mark 1A	594	6/5 01:09 to 7/13 17:40	5.28 RU	19.21 RU
Johannesburg, South Africa	Doppler	1,336	6/5 02:54 to 9/14 22:57	−0.000117 Hz	0.00140 Hz
	Range-Mark 1A	776	6/5 03:00 to 7/18 04:31	4.38 RU	20.60 RU
Cebreros, Spain	Doppler	387	6/26 00:43 to 9/12 00:36	0.000023 Hz	0.00188 Hz
	Range-Mark 1A	27	7/5 02:40 to 7/15 02:50	1.97 RU	21.04 RU

* Mark 1A = near-Earth ranging system; Tau and Mu = ranging systems using different ground hardware.
† Sample rate for doppler and range was 20 min.
‡ 1 Hz =65 mm/s; 1 RU (range unit) ≈ 1 m; 1 ns ≈ 0.15 m.

The Deep Space Network has three types of ranging systems: Mu, Tau, and Mark 1A. The Mu and Tau systems are capable of planetary distances, whereas the Mark 1A ranging system is limited to an effective one-way range of approximately 10^7 km. In weighting the range data the following factors were taken into consideration. First, assuming no external errors, the Mu and Tau systems are accurate to about 20 m. This includes system noise and transponder and ground equipment calibration errors. The Mark 1A ranging system is accurate to about 30 m. Second, because the radio signals travel through the ionosphere of the Earth and the interplanetary space plasma, there is a change in the radio signal path length. The group wave path length is increased, while the phase wave path length is decreased, corrupting both range and doppler measurements made from the radio signal. The charged particles of the Earth's ionosphere could account for as much as a 15 m error in range; the charged particles in the interplanetary medium (space plasma) could account for as much as a 22 m error in range (J. F. Jordan et al., paper presented at AAS/AISS Astrodynamics Conference, August 1970). Third, another possible cause of range error is the Z component of station location. This component is parallel to the Earth's spin axis. The computed range value is sensitive to incorrect Z values when the probe declination becomes large in absolute value. An equation relating the two is[3]

而确定的。表1给出了数据范围，同时我们也列出了由最佳测定结果得到的统计数据。使用了喷气推进实验室双精度轨道测定程序处理数据[1]，该程序使用了考埃尔轨道积分和批处理最小二乘滤波器。在考虑测距数据的权重时，我们非常谨慎，以确保使用最优的数据，避免和多普勒数据出现不一致。1969年的火星水手号结果显示，这样的不一致性会导致估计参数出现较大误差。

表 1. 跟踪数据的统计

跟踪站	数据类型 *	数据点数†	数据间隔，1971 年（UTC）	平均残差‡	均方根残差‡
戈尔德斯通－回波 加利福尼亚	多普勒	1,069	6/5 08:37 到 9/14 04:59	0.000005 Hz	0.00144 Hz
	测距 –Mark 1A	248	6/5 08:42 到 6/30 09:14	−17.73 RU	20.94 RU
	测距 –Mu	31	7/12 05:40 到 9/6 02:49	−19.12 ns	134.55ns
戈尔德斯通－火星 加利福尼亚	多普勒	193	6/5 13:01 到 8/31 09:36	0.000002 Hz	0.00132 Hz
	测距 –Tau	109	6/5 12:43 到 8/27 06:20	−33.67 ns	116.22ns
伍默拉 澳大利亚	多普勒	1,638	6/5 01:15 到 9/12 17:56	−0.000012 Hz	0.00139 Hz
	测距 –Mark 1A	594	6/5 01:09 到 7/13 17:40	5.28 RU	19.21 RU
约翰内斯堡 南非	多普勒	1,336	6/5 02:54 到 9/14 22:57	−0.000117 Hz	0.00140 Hz
	测距 –Mark 1A	776	6/5 03:00 到 7/18 04:31	4.38 RU	20.60 RU
塞夫雷罗斯 西班牙	多普勒	387	6/26 00:43 到 9/12 00:36	0.000023 Hz	0.00188Hz
	测距 –Mark 1A	27	7/5 02:40 到 7/15 02:50	1.97 RU	21.04RU

*Mark 1A 为近地测距系统；Tau 和 Mu 表示使用不同地面硬件设备的测距系统。
† 测距和多普勒位移数据的采样间隔是 20 分钟。
‡ 1 Hz=65 mm/s；1 RU（测距单位）≈ 1 m; 1 ns ≈ 0.15 m。

深空探测网有三类测距系统：Mu、Tau 和 Mark 1A。Mu 和 Tau 系统可以测量行星距离，而 Mark 1A 系统的单程有效范围仅限于大约 10^7 km。使用这些测距数据的权重需要考虑以下因素：第一，假设不存在外部误差，Mu、Tau 系统精确度为大约 20 m，这包括系统噪声、转发器和地面设备的定标误差。Mark 1A 测距系统精确度为大约 30 m。第二，由于射电信号穿过了地球的电离层和行星际空间的等离子体，所以其光程也会发生变化。群光程增加，而相光程减小，这影响了由射电信号得到的测距和多普勒测量结果。地球电离层中的带电粒子会造成多达 15 m 的测距误差；行星际介质中的带电粒子（空间等离子体）会造成多达 22 m 的测距误差（乔丹等人在美国宇航学会、美国航空航天协会主办的航天动力学会议上发表的文章，1970 年 8 月）。第三，造成测距误差的另一个可能的原因是站点位置中的 Z 分量，这个分量和地球的自转轴平行。当探测器赤纬的绝对值较大时，计算得出的测距值对不正确的 Z 值敏感，联系这两个量的方程为 [3]：

$$\Delta\rho = \Delta Z \sin \delta$$

Where ρ is the range datum and δ is the geocentric declination of the spacecraft.

Because previous space mission data did not yield significant information on the Z component of station locations, JPL analysts used the Z values obtained by the Smithsonian Astrophysical Observatory (SAO) in 1969. The change in the Z component was as much as 56 m. Assuming that the Z values from SAO may be in error by 30 m and the maximum absolute value in geocentric declination for Mariner Mars 1971 is 29.15°, the above equation would yield a range error of 14.6 m. The doppler data are insensitive to this Z-component error. Even though the Z component of a station location is not too well determined, the distances from the spin axis and the longitude are known to better than 3 and 5 m, respectively.

A number of solutions with different combinations of weights for each data type and different sets of estimated parameters were examined. The standard set of estimated parameters includes the probe position and velocity (6), solar pressure (3), attitude control leaks (3), station locations (15), and the Earth–Moon ratio (1). An *a priori* statistic of 0.0166 was applied to the mass ratio parameter. These solutions and their identifications are given as follows:

Case 1, Doppler only (doppler weight = 0.015 Hz) with standard estimated parameter set. Case 2, Range only (range weight = 100 m) with standard estimated parameter set. Case 3, Doppler and range (doppler weight = 0.015 Hz, range weight = 100 m) with standard estimated parameter set. Case 4, Doppler and range (doppler weight = 0.015 Hz, range weight = 100 m) with standard estimated parameter set plus Mars and Earth–Moon barycentre ephemeris parameters. Case 5, Doppler and range (doppler weight = 0.015 Hz, range weight = 50 m) with the estimated parameter set as in Case 4.

Solutions	μ^{-1}
Case 1	81.30068
Case 2	81.30067
Case 3	81.30067
Case 4	81.30067
Case 5	81.30068

All solutions yielded nearly the same mass ratio. Cases 1 and 2 give remarkable agreement on a mass ratio between the two data types. With such good agreement, the relative weight of the two data types becomes less significant. Cases 3 and 4 show that the lunar ephemeris error is probably too small to have an effect on the mass ratio estimate. Possible error sources are the periodic variations in the interplanetary medium. W. G. Melbourne has shown (12th Plenary Meeting of the Committee on Space Research, Prague, 1969) that a 28 day sinusoidal variation of solar flux of 0.1% could produce an error of about 0.001 in the mass ratio, but that it is not likely. Also, the agreement of mass ratios computed from the data gathered from several interplanetary spacecraft does not indicate

$$\Delta\rho = \Delta Z \sin\delta$$

其中，ρ 是测距值，δ 是航天器的赤纬。

由于之前的航天数据并未给出站点位置 Z 分量的重要信息，所以喷气推进实验室的分析者使用了由史密森天体物理台（SAO）1969 年得出的 Z 值。Z 值的变化达到了 56 m。假设史密森天体物理台得到的 Z 值存在 30 m 的误差，且 1971 年火星水手号赤纬的最大绝对值是 29.15°，则用上述方程可以得到一个 14.6 m 误差范围。多普勒数据对这个 Z 分量误差不敏感。尽管站点位置的 Z 分量没有得到充分确定，但其到自转轴的距离和经度可以分别在优于 3 m 和 5 m 的精确范围内确定。

在对每种数据类型取不同权重组合，并取不同的估计参数组后，可得到若干结果，本文对它们进行了检验。标准的估计参数组包括探测位置和速度（6）、太阳辐射压（3）、姿态控制的误差（3）、站点位置（15）和地月质量比（1）。将先验统计值 0.0166 作为质量比的初始参数。这些结果和相应的证认如下：

情况 1，基于标准估计参数组的多普勒位移（多普勒权重为 0.015 Hz）数据。情况 2，基于标准估计参数组的测距数据（测距权重为 100 m）。情况 3，基于标准估计参数组的多普勒位移和测距数据（多普勒权重为 0.015 Hz，测距权重为 100 m）。情况 4，基于标准估计参数组以及火星和地月质心的星历参数的多普勒位移和测距数据（多普勒权重为 0.015 Hz，测距权重为 100 m）。情况 5，基于和情况 4 相同的估计参数组的多普勒位移和测距数据（多普勒权重为 0.015 Hz，测距权重为 50 m）。

结果	μ^{-1}
情况 1	81.30068
情况 2	81.30067
情况 3	81.30067
情况 4	81.30067
情况 5	81.30068

所有的结果都得到了相近的质量比。情况 1、2 由两种类型的数据给出了相当一致的质量比。基于这么好的一致性，两组数据的相对权重就变得不太重要了。情况 3 和 4 表示月球星历表的误差对估算质量比结果的影响微不足道。误差的可能来源是行星际介质的周期性变化。墨尔本曾指出（空间研究委员会第 12 届全体会议，布拉格，1969 年）周期为 28 天的太阳流量正弦变化（幅度为太阳流量的 0.1%）会导致在质量比中出现 0.001 的误差，不过这好像不太可能。而且，由许多行星际航天

this sort of systematic error unless the phase of the flux variation is the same for each mission, which does not seem likely.

The results from the Mariner Mars 1971 data are given in Table 2 together with previous results obtained from Pioneers 8 and 9 and Mariners 2, 4, 5, 6 and 7. Values computed for Pioneers 8 and 9 and Mariner 2 were obtained from solutions using only doppler data. It is interesting to note that the Mariner 9 value and the mean of all spacecraft determined values of μ^{-1} are 81.3007. The deviations from the arithmetic mean of the Mariner and Pioneer values are tabulated in Table 1. Further, the mass ratio computed from the last five interplanetary spacecraft launched (Pioneers 8 and 9 and Mariners 6, 7 and 9) showed a spread of only 0.0004. The values for Pioneers 8 and 9 and Mariners 6, 7 and 9 are 81.3004, 81.3008, 81.3004, 81.3005 and 81.3007, respectively. This provides a good indication of the accuracy of μ^{-1}. The decrease in the fluctuation of the mass ratio value can be attributed to the improvement of data quality owing to changes in Deep Space Network tracking systems and the change in computer software from single precision to double precision.

Table 2. Estimates of the Earth–Moon Mass Ratio, μ^{-1}

Spacecraft	μ^{-1}	$\mu^{-1} - \overline{\mu^{-1}}$	Reference
Pioneer 8	81.3004±0.0001	−0.0003	5
Pioneer 9	81.3008±0.0001	0.0001	5
Mariner 2 (Venus)	81.3001±0.0013	−0.0006	6
Mariner 4 (Mars)	81.3015±0.0017	0.0008	7
Mariner 5 (Venus)	81.3013±0.0002	0.0006	5
Mariner 6 (Mars)	81.3004±0.0002	−0.0002	5
Mariner 7 (Mars)	81.3005±0.0002	−0.0002	5
Mariner 9 (Mars)	81.3007±0.0001	0.0000	

$\overline{\mu^{-1}}$ = arithmetic mean.

We thank W. L. Sjogren for discussion and review of this article. This research was supported by NASA.

(**241**, 111-112; 1973)

S. K. Wong and S. J. Reinbold
Jet Propulsion Laboratory, Pasadena, California 91103

Received September 11, 1972.

References:

1. Moyer, T., *Technical Report 32-1527* (Jet Propulsion Laboratory, Pasadena, 1971).

2. Mottinger, N. A., *et al.*, *Technical Memorandum 33-469* (Jet Propulsion Laboratory, Pasadena, 1970).

3. Anderson, J. D., thesis, University of California at Los Angeles (1967).

4. Null, G. W., Gordon, H. J., and Tito, D. A., *Technical Report 32-1108* (Jet Propulsion Laboratory, Pasadena, 1967).

器得到的数据计算出的质量比也是一致的，这表明没有这种系统误差，除非辐射流变化的相位对于每次航行都相同，但这是不大可能的。

　　表 2 中给出了 1971 年火星水手号的数据以及以前由先驱者 8 号、9 号和水手 2 号、4 号、5 号、6 号和 7 号得到的结果。对于先驱者 8 号、9 号和水手 2 号的结果，计算时只用了多普勒数据。有趣的是，水手 9 号得到的数值和所有飞行器得到的 μ^{-1} 平均值都是 81.3007。水手号和先驱者号的结果与算术平均值的偏差列在了表 1 里。而且，由最后 5 次（先驱者 8 号、9 号和水手 6 号、7 号和 9 号）发射的行星际航天器的数据计算出的质量比差别都在 0.0004 的范围内。先驱者 8 号、9 号和水手 6 号、7 号和 9 号的结果分别是 81.3004、81.3008、81.3004、81.3005 和 81.3007。这很好地表明了 μ^{-1} 的精度。质量比变化的减小主要归因于数据质量的提高，这都源于深空探测网络跟踪系统的改进以及在计算机软件数据类型中单精度变为了双精度。

表 2. 地月质量比估算结果，μ^{-1}

航天器	μ^{-1}	$\mu^{-1} - \overline{\mu^{-1}}$	参考资料
先驱者 8 号	81.3004 ± 0.0001	-0.0003	5
先驱者 9 号	81.3008 ± 0.0001	0.0001	5
水手 2 号（金星）	81.3001 ± 0.0013	-0.0006	6
水手 4 号（火星）	81.3015 ± 0.0017	0.0008	7
水手 5 号（金星）	81.3013 ± 0.0002	0.0006	5
水手 6 号（火星）	81.3004 ± 0.0002	-0.0002	5
水手 7 号（火星）	81.3005 ± 0.0002	-0.0002	5
水手 9 号（火星）	81.3007 ± 0.0001	0.0000	

$\overline{\mu^{-1}}$ 是算数平均值。

　　我们感谢肖格伦的讨论和对此文的审阅。该研究得到了美国国家航空航天局（NASA）的支持。

（冯翀 翻译；沈志侠 审稿）

The Search for Signals from Extraterrestrial Civilizations

J. C. G. Walker

Editor's Note

James Walker, an expert on the evolution of the terrestrial environment, here turns his gaze beyond the Earth's atmosphere to consider the feasibility of detecting radio signals from extraterrestrial civilizations. This notion had been pursued experimentally since 1960, when astronomer Frank Drake performed a radio-telescope search. Drake devised an equation for estimating the probability of intelligent civilizations on other worlds. Walker combines a related estimate for the number of habitable planets with the detection capabilities of telescopes to deduce how long observations might need to proceed before an "intelligent" signal is found. The result is dispiriting: even with an optimistic estimate of how many inhabitable planets produce technologically advanced civilizations, it could take over a thousand years to spot them.

Although the technology exists for exchanging radio messages with extraterrestrial civilizations, a successful search for such civilizations among the many stars that might support them could take more than a thousand years, even if most habitable planets are occupied by communicative civilizations.

THAT the technology exists for sending and receiving radio messages over interstellar distances is not in doubt[1-5] so that, if there are similar technological civilizations based on stars not too distant from the Sun, we can, in principle, communicate with them. A problem, however, is to determine which star out of a large number of candidates is the home of a potentially communicative civilization[6-8]. The subject of this paper is the search problem of interstellar communication.

Occurrence of Habitable Planets

Even the most optimistic estimates of the frequency of occurrence of potentially communicative civilizations suggest that a large number of stars will have to be searched before a civilization is encountered. If we let P_c be the probability that a given star has a communicative civilization, we may write

$$P_c = f P_{HP} \tag{1}$$

Where P_{HP} is the probability that the star has a habitable planet in orbit around it, and f is the fraction of habitable planets with communicative civilizations. This fraction involves

地外文明信号搜寻

沃克

编者按

本文中，一位地球环境演化方面的专家詹姆斯·沃克，将他的目光转向了地球大气之外，思考探测来自地外文明射电信号的可能性。这一理念自 1960 年开始实验性实施，当时天文学家弗兰克·德雷克用射电望远镜进行了一次搜寻。德雷克提出了一个估计其他星球中智慧文明出现概率的方程。沃克将对宜居行星数量的相关估计与望远镜的探测能力相结合，从而推导出需要多长时间的观测才能找到一个"智慧的"信号。结果令人沮丧：即使对宜居行星产生技术发达的文明的概率进行一个最乐观的估计，找到它们也可能需要超过一千年的时间。

尽管和地外文明交换射电信息的技术已经实现，但即使有通信能力的文明占据着大多数宜居行星，对能支持这样的文明的恒星的成功搜寻也可能将花费超过一千年。

毫无疑问，在星际距离上发送和接收射电信息的技术已经实现了 [1-5]。因此，如果太阳附近的其他恒星存在拥有类似技术的文明，原则上我们就可以与他们通信。然而，存在一个问题，就是如何在众多的候选体中确定哪颗恒星是这样的家园，即具有潜在的有通信能力的文明 [6-8]。本文的主题就是关于星际通信的搜寻问题。

宜居行星的发现

即使对潜在有通信能力的文明出现的概率做最乐观的估计，也必须搜寻大量的恒星才能找到一个地外文明。如果我们设 P_c 为一个给定恒星拥有通信能力的文明的概率，我们可以写出：

$$P_c = f P_{HP} \tag{1}$$

其中 P_{HP} 是围绕这颗恒星的轨道上有一颗宜居行星的概率，而 f 是此行星上存在有通

11

the probability that a communicative civilization will evolve on a habitable planet as well as the average lifetime of communicative civilizations[6,9-11].

It may be impossible to determine the value of f by other than empirical means, but P_{HP} is a quantity that can, in principle, be estimated from a knowledge of cosmogony and planetology. Dole[12], for example, has made a detailed estimate of the probabilities that planets on which Man could exist are in orbit about stars of different spectral classes. His considerations lead him to conclude that P_{HP} achieves a maximum value of 5.5% for stars of classes G0 to G4, and that 3.7% of stars in classes F2 to K1 have habitable planets; P_{HP} is zero for stars outside this range. Dole's estimates involve many assumptions, and improved values of P_{HP} will undoubtedly become available in time, but his values are adequate for present purposes. They show, first, that several tens of stars must be searched, even if f is of order unity and, second, that we can estimate P_{HP} for different stars and therefore can use this information to guide our search.

Strategy for the Search

One approach to the search problem is to assume that the other civilization will do most of the work, which implies that the search is limited to "supercivilizations" able to transmit detectable signals in all directions all the time[5,13]. We could not do such a thing[14], for the power requirement of an isotropic call signal detectable at a range of 100 light year is approximately equal to the world's present total power consumption[2,15]. We can signal over interstellar distances only by using a large radio telescope to concentrate the radiated energy into a narrow beam. It would be possible to use a number of transmitters to send continuous signals to an equal number of target stars, but for such a strategy to have a reasonable chance of success, the number of transmitters would have to exceed

$$P_c^{-1} = (f P_{HP})^{-1}$$

This number is larger than 18, using Dole's values of P_{HP}, and it may be very much larger, since f may be small.

Even if there were several interstellar transmitters—and there are not—a strategy of continuous transmission to a select group of stars would not, however, be optimal. So as f is unknown, the probability of success for the transmitting civilization is proportional to the number of stars called regularly. In order to call the largest possible number of stars for a given level of effort, transmitter time must be shared among different target stars. How often, then, should a signal be sent to a given star?

Von Hoerner[6] has analysed this problem in general terms, pointing out that it is necessary to develop an optimal search strategy for both transmitter and receiver, and then to assume that the target civilization will perform the same analysis and arrive at the same conclusion. I present here a strategy for which the probability of success can be evaluated, at least as a function of f.

信能力的文明的比例。这个比例包含有通信能力的文明在宜居行星上能进化出来的概率以及有通信能力的文明的平均寿命[6, 9-11]。

f值只能用经验性的方法来估计，但是原则上，P_{HP}却可以用天体演化学和行星学的知识来估算。例如，多尔[12]对不同光谱型的恒星周围存在人类能够生存的行星的概率进行了详细的估算。他得出结论：光谱型为G0到G4的恒星存在宜居行星的概率最大，P_{HP}达到5.5%；而光谱型F2到K1的恒星有3.7%的可能存在宜居行星；在其他类型的恒星中，P_{HP}为0。多尔的估计涉及很多假设，而且毫无疑问的是，随着时间发展P_{HP}值会被改进，但他得出的值对于现在的目的来说已经足够了。首先，它们表明即使f值为1，我们也必须搜寻至少数十颗恒星；其次，我们可以估算出不同恒星的P_{HP}，并用来指导我们的搜寻。

搜 寻 策 略

解决搜寻问题的一个方法是假设其他文明已经高度发展，这意味着搜寻仅限于"超级文明"，这些"超级文明"能够一直向各方向发射可探测的信号[5,13]。我们做不到这样[14]，因为发射一个各向同性的、在100光年的范围内可探测的呼叫信号的能量需求大约相当于现在世界总的能量消耗[2,15]。我们只能通过使用一台大型射电望远镜将辐射能量集中到一个狭窄的波束中在星际距离发送信号。用若干发射机发送持续的信号给相同数量的目标恒星是可能的，但是要让这个策略有合理的成功机会，发射机的数量必须超过

$$P_c^{-1} = (f P_{HP})^{-1}$$

采用多尔的P_{HP}值，这个数应该大于18，因为f可能会很小，所以这个数可能比18大得多。

即使有一些星际发射机——实际上没有——向挑选出的一组恒星持续发射信号的方法也不是最优策略。当f未知，成功向其他文明发送信号的概率与定期向其发送信号的恒星的数量成正比。为了通过一定程度的努力向尽可能多的恒星发送信号，发射机时间必须被不同目标恒星共享。那么，给一个特定恒星发送信号的频率应该是多少呢？

冯·赫尔纳[6]大体上分析了这个问题，他指出，有必要针对发射机和接收机建立一个优化的搜寻策略，他假设目标文明也会进行同样的分析并得到相同的结论。这里我提出一个策略，以此可以估计成功的概率，至少可以表示为f的一个函数。

An optimal search strategy should use all the information that we share with the target civilization. This includes the spectral classes of candidate stars and thus the values of P_{HP}; it includes the optimal spectral region in which to work, the region where unavoidable background noise is minimal[1-5] and the distances to the candidate stars.

This last quantity provides the only indication we have as to how often we should look at any given star. The natural repetition period[8] for a star at distance R is

$$T = \frac{2R}{c} \qquad (2)$$

where c is the velocity of light. It is the time for a contact signal to travel to the star and for a reply to return. I shall assume that the transmitter sends a contact signal every T years to every star within range having a non-zero P_{HP}. Excess transmitter capacity would be used to increase the range of the search rather than to provide more frequent contact signals.

Although there are a number of ideas about the wavelength to use for contact signals[1,5], a search in frequency cannot be eliminated entirely because extremely narrow bandwidths must be used for interstellar communication[2]. I shall not consider the frequency search explicitly, so for simplicity in the analysis I shall assign this task to the transmitter. Thus the contact signal to be sent out every T years will sweep slowly over the optimal spectral region, and the receiver may confine its search to a single frequency.

Probability of Success

With the transmitter strategy thus defined, it is possible to determine the optimal receiver strategy and evaluate the rate of success. From the point of view of the receiver, let us redefine P_c to be the probability that a given star has a civilization that sends a contact signal to the receiving star every T years at the wavelength on which the receiving civilization is listening. This redefinition introduces a corresponding change in the definition of the unknown fraction f, but no change in the known probability P_{HP}.

If, at the beginning of the search, the receiver devotes a period of time $\Delta\tau$ to listening to a given star, the probability that it will receive a call is

$$P_s = P_c \, \Delta\tau / T \qquad (3)$$

The rate of success in the search is therefore $f P_{HP}/T$ yr^{-1}, where P_{HP}/T depends on the spectral class and the distance of the target star and f is the same for all stars. Because the repetition period T increases linearly with distance to the target star, the success rate is highest for the closest stars. Using Dole's figures[17], we find a success rate for α Centauri of $1.3 \times 10^{-2} f$ yr^{-1}; for ε Eridani and τ Ceti the success rates are both about $1.5 \times 10^{-3} f$ yr^{-1}. For a G0 star at 100 light year, however, the success rate is $2.7 \times 10^{-4} f$ yr^{-1}.

一个最佳的搜寻策略是应该使用我们和目标文明共享的所有信息。这包括候选恒星的光谱型以及相关的 P_{HP} 值；也包括选取最佳的工作频段使得不可避免的背景噪声在此频段最小 [1-5]，还要考虑离候选恒星的距离。

最后这个计算结果将给出我们多久向指定恒星发射一次信息的唯一指标。对于一颗距离为 R 的恒星来说，自然重复周期 [8] 为

$$T = \frac{2R}{c} \tag{2}$$

其中 c 是光速，T 是一个通信信号传送到该恒星后再返回所需的时间。我假设信号发生器每 T 年向一非零 P_{HP} 区域内的所有恒星发送信号。额外的信号发送能力将用来增加搜索范围而不是更频繁地提供通信信号。

尽管对通信信号波长的选择存在很多观点 [1,5]，但是频率搜索不能完全舍弃，因为星际通信必须要在极窄的波段 [2]。我将不仔细考虑频率搜索，为简化分析，我把这个任务交给发射机。因此，每 T 年发送的通信信号将缓慢地扫过最优的频谱范围，同时接收机会把搜索限定在某一个单一频率上。

成功的概率

在这样定义的发射机策略下，就可能确定最优的接收机策略，同时也可以评估出成功的概率。从接收机的角度来讲，我们重新定义 P_c 为给定恒星拥有一个文明的概率，此文明每 T 年发送通信信号给可能的接收星，波长为接收信号的文明用于监听的波长。这种重新定义给未知参数 f 的定义带来相应的变化，但是不改变已知的概率 P_{HP}。

如果在搜索开始时，接收机使用时间周期 $\Delta\tau$ 来接收给定恒星的信号，则它接收到一个呼叫的概率为

$$P_s = P_c \,\Delta\tau/T \tag{3}$$

因此，搜索成功的概率就是 $f P_{HP} / T$ yr^{-1}，这里 P_{HP} / T 依赖于目标恒星的光谱型和距离，f 对所有恒星都相同。因为重复周期 T 随目标恒星的距离线性增加，所以对最近的恒星成功概率最高。使用多尔的图 [17]，我们发现对半人马座 α（南门二）的成功概率为 $1.3 \times 10^{-2} f$ yr^{-1}；对波江座（天苑四）和鲸鱼座（天仓五）成功概率均为 $1.5 \times 10^{-3} f$ yr^{-1}。对一颗 100 光年处的 G0 型恒星，成功概率为 $2.7 \times 10^{-4} f$ yr^{-1}。

But the receiver should not devote all its time to the closest star. This star may not be the home of a communicative civilization. After the receiver has devoted a large number n of randomly spaced listening periods $\Delta\tau$ to a given star, the probability that the receiver will have failed to receive a call, assuming that there is a transmitter associated with that star, is $\exp(-n\Delta\tau/T)$. The probability of success on the next look at the star is therefore

$$P_s(\tau) = f\,P_{HP}\,(\Delta\tau/T)\,\exp(-\tau/T) \tag{4}$$

where $\tau = n\Delta\tau$.

The optimal receiver strategy is now clear. The success rate is maximized if each listening period $\Delta\tau$ is devoted to the star for which $(P_{HP}/T)\,\exp(-\tau/T)$ is greatest. As the search progresses, the number of stars included in the search increases steadily, for each star within range the value of τ increases, and the instantaneous success rate grows steadily smaller. Each try, however, adds the greatest possible amount to the cumulative probability of success. How much time, on average, must elapse before success is achieved?

Suppose that there are N_i stars of spectral class i per unit volume and let $P_{HP}(i)$ be the probability that each of these stars has a habitable planet. After a total time t has been devoted to the search, following the strategy outlined above, the cumulative probability of success is

$$P_s = \frac{4}{3}\,\varepsilon f t \tag{5}$$

where ε is the instantaneous success rate at time t given by

$$\varepsilon = \left[\frac{\pi c^3}{32t}\sum_i N_i P_{HP}{}^4(i)\right]^{\frac{1}{4}} \tag{6}$$

From Dole's Table 18, $\sum_i N_i P_{HP}{}^4(i) = 1.24 \times 10^{-9}$ per cubic light year, so

$$\varepsilon = 3.3 \times 10^{-3}\,t^{-\frac{1}{4}}\,\text{yr}^{-1} \tag{7}$$

and
$$P_s = 4.4 \times 10^{-3}\,f t^{\frac{3}{4}} \tag{8}$$

where t is expressed in years.

On the average, contact will be achieved when $P_s = 1$ or after a search that has lasted

$$t_0 = 1{,}380 f^{-4/3}\text{yr} \tag{9}$$

Values of t_0 corresponding to several assumed values of f are shown in Table 1. Also shown is the average distance that separates communicative civilizations for these values of f.

但是接收机不应该把所有的时间都放在最近的恒星上。这颗恒星可能不是一个具有通信能力的文明的家园。在接收机投入了 n 个随机分布的接收周期 $\Delta\tau$ 给一颗特定恒星后，假定那里有一台发射机，那么这台接收机还未能接收一个呼叫的概率是 $\exp(-n\Delta\tau/T)$。因此，下一次搜索该星时成功的概率就是：

$$P_s(\tau) = f\,P_{HP}\,(\Delta\tau/T)\,\exp\,(-\tau/T) \tag{4}$$

其中 $\tau = n\Delta\tau$。

现在，最佳的接收策略已经清楚了。当投入到这颗恒星的每个接收周期 $\Delta\tau$ 使得 $(P_{HP}/T)\exp(-\tau/T)$ 最大时，成功概率最大。随着搜索的进行，搜索中包含的恒星数量稳步增长，搜索范围内每颗星的 τ 的增长，瞬时成功概率稳步减小。然而，每次尝试都会为累积的成功概率增加一个最大可能的量。问题是平均需要多少时间，才会成功搜索到地外文明的信号？

假定单位体积有 N_i 颗光谱型 i 的恒星，令 $P_{HP}(i)$ 为每一颗恒星有宜居行星的概率。在投入总时间 t 搜索后，按照上面提到的策略，累积的成功概率是

$$P_s = \frac{4}{3}\,\varepsilon ft \tag{5}$$

其中 ε 是在时刻 t 的瞬时成功率，由

$$\varepsilon = \left[\frac{\pi c^3}{32t}\sum_i N_i P_{HP}^4(i)\right]^{\frac{1}{4}} \tag{6}$$

给出。由多尔的表 18，$\sum_i N_i P_{HP}^4(i) = 1.24\times10^{-9}$ 每立方光年，因此

$$\varepsilon = 3.3 \times 10^{-3}\,t^{-\frac{1}{4}}\,\mathrm{yr}^{-1} \tag{7}$$

$$P_s = 4.4 \times 10^{-3}\,ft^{\frac{3}{4}} \tag{8}$$

其中 t 的单位为年。

平均来说，当 $P_s=1$ 或者一次搜索已经进行了

$$t_0 = 1{,}380\,f^{-4/3}\,\mathrm{yr} \tag{9}$$

后，我们就能够与地外文明建立联系。对应于一些假定的 f 值的 t_0 值列在表 1 中。另外也列出了对于这些 f 值，有通信能力的文明间的平均距离。

Table 1. Search Strategies for Various Distributions of Civilizations

Fraction of habitable planets occupied by communicative civilizations f	Average separation of communicative civilizations (light year)	Duration of search T_0(yr)
1	24	1.4×10^3
10^{-3}	240	1.4×10^7
10^{-6}	2,400	1.4×10^{11}

We see that even with optimistic assumptions concerning the frequency of occurrence of communicative civilizations, the time required for a successful search is long. Of course, t_0 is, strictly speaking, telescope time devoted to the search, not total elapsed time. The duration of the search is therefore inversely proportional to the number of receiving telescopes and could be shortened by a massive effort. Alternatively, it is possible that the transmitter strategy I have assumed is incorrect, and that more frequent calls would be optimal, say m calls every T years. In this case the duration of the search differs from the values in Table 1 by a factor of $1/m$, provided the receiving civilization knows the value of m.

The conclusion, therefore, is disappointing. If every habitable planet has a communicative civilization there might be 50 such civilizations within 100 light year of us[18]. We possess the technology to exchange messages with this multitude of other worlds, if only we can find them. Unless my assumed transmitter strategy is seriously in error, however, or unless habitable planets are substantially more abundant than Dole has concluded, the problem of finding the other worlds is overwhelming. These circumstances may limit us to a search for supercivilizations[16].

This research has been supported, in part, by a NASA grant.

(**241**, 379-381; 1973)

James C. G. Walker

Department of Geology and Geophysics, Yale University, New Haven, Connecticut 06520

Received January 5; revised September 11, 1972.

References:

1. Cocconi, G., and Morrison, P., *Nature*, **184**, 844 (1959).

2. Drake, F. D., *Sky and Telescope*, **19**, 140 (1959).

3. Webb, J. A., in *Institute for Radio Engineers Seventh National Communications Symposium Record: Communications- Bridge or Barrier*, **10** (1961).

4. Oliver, B. M., in *Interstellar Communication* (edit. by Cameron, A. G. W.), 294 (W. A. Benjamin, New York, 1963).

5. Shklovskii, I. S., and Sagan, C., *Intelligent Life in the Universe* (Holden-Day, San Francisco, 1966).

6. von Hoerner, S., *Science*, **134**, 1839 (1961).

7. Bracewell, R. N., *Nature*, **186**, 670 (1960).

8. Huang, S.-S., in *Interstellar Communication* (edit. by Cameron, A. G. W.), 201 (W. A. Benjamin, New York, 1963).

表 1. 对各种文明分布的搜索策略

有通信能力文明占据宜居行星的比例 f	有通信能力文明的平均间距（光年）	搜寻持续的时间 T_0（年）
1	24	1.4×10^3
10^{-3}	240	1.4×10^7
10^{-6}	2,400	1.4×10^{11}

我们看到，即使对可通信联系的文明出现的频率做最乐观的假设，成功的搜索所需的时间还是很长。当然，严格来说 t_0 是天文望远镜投入到搜索的时间，而不是流逝的总时间。搜索持续的时间反比于接收天文望远镜的数量，因此可以通过增加大量望远镜来缩短。另一种可能是，我所假设的发射机策略是不正确的，更加频繁的呼唤可能是最佳的，比如说每 T 年进行 m 次呼叫。在这种情况下，假如接收文明知道 m 的值，搜索的持续时间与表 1 中的值有一个 $1/m$ 因子的差别。

综上，结论有些令人失望，如果每个宜居的星球都有可通信联系的文明，那么在距离我们 100 光年内大约会有 50 个这样的文明[18]。我们掌握着同这众多的其他世界交换信息的技术，只要我们能找到他们。但是，除非我假设的发射机策略是严重错误的，或者宜居行星的数量比多尔推断的多，否则搜寻其他地外文明的困难几乎无法克服。这些情况可能会限制我们搜寻超级文明[16]。

本研究部分得到了美国国家航空航天局基金的支持。

（周旻辰 翻译；沈志侠 审稿）

19

9. Morrison, P., *Bull. Phil. Soc. Washington*, **16**, 58 (1962).

10. Pearman, J. P. T., in *Interstellar Communication* (edit. by Cameron, A. G. W.), 287 (W. A. Benjamin, New York, 1963).

11. Cameron, A. G. W., in *Interstellar Communication* (edit. by Cameron, A. G. W.), 309 (W. A. Benjamin, New York, 1963).

12. Dole, S. H., *Habitable Planets for Man*, second edition, Table 17 (American Elsevier, New York, 1970).

13. Kardashev, N. S., *Soviet Astronomy—A. J.*, **8**, 217 (1964).

14. Webb, J. A., in *Interstellar Communication* (edit. by Cameron, A. G. W.), 188 (W. A. Benjamin, New York, 1963).

15. Hubbert, M. K., in *Resources and Man* (Committee on Resources and Man of the National Academy of Sciences-National Research Council), 157 (W. H. Freeman, San Francisco, 1969).

16. *Extraterrestrial Civilizations* (edit. by Tovmasyan, G. M.) (translated by Israel Program for Scientific and Technical Translations, 1964).

17. Dole, *Habitable Planets for Man*, second edition, Table 22 (American Elsevier, New York, 1970).

18. Dole, *Habitable Planets for Man*, second edition, Table 19 (American Elsevier, New York, 1972).

On the Origin of Deuterium

F. Hoyle and W. A. Fowler

Editor's Note

Deuterium is hydrogen that contains a neutron in its nucleus. It is relatively abundant in the universe, most having been created in the first few minutes after the Big Bang. Here Fred Hoyle and William Fowler investigate several ways in which it could also be created in astrophysical situations. The shock waves associated with supernovae (exploding old stars), and cosmic rays hitting clouds of gas can both generate deuterium, but not enough to explain the observations. The arguments advanced here were an attempt to avoid invoking a Big Bang at all, which Hoyle spent the later years of his life opposing. The Big Bang is, however, now the generally accepted explanation for the origin and properties of the Universe.

The origin of deuterium has always been a problem for theories of stellar nucleosynthesis. A general solution is proposed and shown to be applicable under several astrophysical circumstances in the light of new observations of the Galactic abundance of deuterium.

CESARSKY, Moffet and Pasachoff[1] have recently observed an absorption feature in the spectrum of radiation from the Galactic centre at 327.38837±0.00001 MHz. They interpret this as arising from the ground-state hyperfine transition in deuterium near 91.6 cm which is analogous to the well-known 21 cm line of ordinary hydrogen. If they assume the feature observed to be due to noise they are able to set an upper limit

$$D/H < 5\times10^{-4}$$

while an analysis assuming the feature to be due to deuterium yields

$$3\times10^{-5} < D/H < 5\times10^{-4} \tag{1}$$

These results are to be compared with the terrestrial value of 1.5×10^{-4} and to the upper limit for the proto-solar value of 3×10^{-5} (refs. 2, 3).

Jefferts, Penzias and Wilson[4] report line emission from a cloud within the Orion Nebula at 144,828 MHz and attribute it to the $J = 2$ to $J = 1$ transition in DCN. In a separate investigation of the $J = 1$ to $J = 0$ transition at 72,414 MHz, the hyperfine components expected for DCN have also been found[5], setting the identification beyond reasonable doubt. The cloud in question probably has dimensions of the order of a light year and a mass of order $10^2\ M_\odot$, considerably less than the Orion Nebula itself. The H II region of Orion, which is the part seen optically, has mass $\sim10^3\ M_\odot$, whereas a larger scale molecular cloud, detected in the 2.6 mm radiation of the CO molecule, has been

22

氘的起源

氘是原子核内有一个中子的氢。它在宇宙中相对丰富，大部分是在宇宙大爆炸之后的几分钟内产生的。在本文中，弗雷德·霍伊尔和威廉·福勒研究了在天体物理条件下也可能形成氘的几种方式。虽然与超新星（老年恒星在演化末期的爆炸）有关的冲击波，以及宇宙射线撞击气体云都能产生氘，但这不足以解释观测结果。本文中提出的论点是从根本上避免援引宇宙大爆炸理论的一个尝试，霍伊尔在他的晚期生涯中一直反对这一理论。然而，宇宙大爆炸理论却是目前被人们普遍接受的对宇宙的起源和特性的解说。

氘的起源一直是恒星核合成理论中的一个问题。根据对银河系氘丰度的新观测，本文对氘起源问题提出了一个一般性的解答并表明这一解答在一些天体物理环境条件下具有可行性。

塞萨尔斯基、莫菲特和帕萨乔夫[1] 最近观测到在来自银河系中心的辐射光谱中的 327.38837 ± 0.00001 MHz 处具有吸收特征。他们把这解释为来自 91.6 cm 附近的氘基态的超精细跃迁，类似于众所周知的普通氢的 21 cm 线。如果假设观测到的特征是由噪声导致的，那么他们可以设定一个上限

$$D/H < 5 \times 10^{-4}$$

而另有一种分析，假设这一现象是由氘产生的，则

$$3 \times 10^{-5} < D/H < 5 \times 10^{-4} \tag{1}$$

这些结果将与地球上的 1.5×10^{-4} 和原始太阳的上限值 3×10^{-5} 相比较（参考文献 2、3）。

杰弗茨、彭齐亚斯和威尔逊[4] 报道了来自猎户座星云内部的一个云团在 144,828 MHz 处的发射线，并把它归因于 DCN 中从 $J=2$ 到 $J=1$ 的跃迁。在 72,414 MHz 处 $J=1$ 到 $J=0$ 跃迁的独立研究中也发现了预期的 DCN 的超精细跃迁组分 [5]，使得这一识别毫无异议。所涉及云团大小远小于猎户座星云本身，大小可能在 1 光年量级，质量处在 $10^2 M_\odot$ 量级。猎户座的 H II 区是可见光区，质量约为 $10^3 M_\odot$，然而据所罗门（个人交流）估计，在 CO 分子 2.6 mm 处辐射中探测到的较

estimated by Solomon (private communication) to have mass $\sim 10^5 \, M_{\odot}$.

The emission from the $J = 1$ to $J = 0$ transition is approximately as strong as that in $^1\text{H}^{12}\text{C}^{15}\text{N}$ and $^1\text{H}^{13}\text{C}^{14}\text{N}$, so

$$D/H = 6 \times 10^{-3} \qquad (2)$$

a remarkably high value. The ratio determined in this way applies to D and H in combination with CN, not to D and H in atomic form. Thus the D/H ratio in the interstellar gas could still be comparable with the terrestrial value, or with the value of 3×10^{-5} referred to above.

The position concerning deuterium has therefore changed from doubt concerning its widespread existence in the Galaxy to one in which it is a reasonable inference that D/H of order 10^{-4} occurs on a large scale, and the problem of the origin of deuterium now seems more urgent than it did before. Together with R. V. Wagoner[6], we found some years ago that significant quantities of D could arise in a low density Friedmann universe. Defining a parameter h from the relation (applicable after $e^{\pm}$-pair annihilation in the universe)

$$\rho_b = hT_9^3 \text{ g cm}^{-3} \qquad (3)$$

where ρ_b is the baryon mass density and T_9 is the radiation temperature measured in units of 10^9 K, we found D/H comparable with the terrestrial ratio of 1.5×10^{-4} when $h \simeq 6 \times 10^{-6}$. Setting $T_9 = 2.7 \times 10^{-9}$ K for the present temperature leads to $\rho_b \simeq 10^{-31}$ g ml.$^{-1}$ for the present baryon density, which is in good agreement with estimates which have been made of the average density of matter in galaxies by Oort[7] and Shapiro[8].

It is an unexpected feature of such a primordial mode of synthesis that the well-known cosmological parameter q_0 turns out to be small, and close to zero[6], instead of the value 0.5 required to "close" the universe. Many cosmological investigations, the formation of galaxies for example, are much more awkward in hyperbolic models ($q_0 \simeq 0$) than they are for $q_0 \geq 0.5$. One might seek to evade the resulting difficulties by arguing that, unlike an idealized Friedmann model, the actual universe is inhomogeneous. It might be supposed that the initial state of the universe was very patchy, with h falling below 10^{-5} in some places. On the other hand, the 2.7 K radiation background is exceedingly uniform, both locally and from one part of the sky to other distant parts of the sky. This uniformity, while not forbidding sufficiently fine scale inhomogeneities, would seem to us to constitute a warning against this line of argument.

It is also important in the new circumstances of the problem to reconsider possible astrophysical modes of origin for deuterium. Should it turn out that D/H is locally variable, astrophysical processes would be preferred to primordial synthesis, but if the D/H ratio is found to be universal then primordial synthesis would be preferred.

大尺度的分子云的质量约为 $10^5\ M_\odot$。

由 $J=1$ 到 $J=0$ 跃迁产生的辐射强度大约与 $^1H^{12}C^{15}N$ 和 $^1H^{13}C^{14}N$ 中的辐射强度相当，因此

$$D/H = 6 \times 10^{-3} \tag{2}$$

这是一个相当高的值。用这种方法确定的比值适用于与 CN 结合的 D 和 H，而对原子形式的 D 和 H 不适用。因此星际气体中的 D/H 比仍然有可能和地球上的值相当，或与上面提到的值 3×10^{-5} 相当。

于是对于氘的认识从怀疑其在银河系中的普遍存在性变为了下面这一合理的推断，即在大尺度上 D/H 处在 10^{-4} 的量级。而现在看起来有关氘起源的问题较之从前也更为紧迫了。几年前，我们和瓦戈纳一道发现在低密度的弗里德曼宇宙中存在大量的 D[6]。从关系式（在宇宙中的正负电子对湮灭后适用）

$$\rho_b = hT_9^3\ \text{g} \cdot \text{cm}^{-3} \tag{3}$$

中定义参数 h，其中 ρ_b 是重子质量密度，而 T_9 是以 10^9 K 为计量单位的辐射温度。我们发现在 $h \simeq 6 \times 10^{-6}$ 时，D/H 和地球上 1.5×10^{-4} 的值相当。设定当前温度为 $T_9 = 2.7 \times 10^{-9}$ K，可以得到现在的重子密度为 $\rho_b \simeq 10^{-31}$ gml^{-1}，这和奥尔特[7]以及夏皮罗[8]估计的星系中物质的平均密度具有很好的一致性。

为人熟知的宇宙学参数 q_0 最终被证明很小并且接近于零[6]，而不是"封闭"宇宙需要的 0.5，这是原初核合成模式所没有预料到的。许多宇宙学研究，如星系形成，在双曲模型中（$q_0 \simeq 0$）要比 $q_0 \geq 0.5$ 难处理得多。有人提出，与理想的弗里德曼模型不同，真实的宇宙是非均匀的，以此来规避由于该参数小导致的困局。可以假设宇宙的初始状态是非常不均匀的，在某些地方 h 小于 10^{-5}。另一方面，无论在局部天区还是从天空的一个区域到遥远的其他区域，2.7 K 的背景辐射值都非常均匀。这种均匀性，虽然没有完全排除很小尺度上不均匀性的存在，似乎预示着并不支持这一论点。

在新的困局下，重新考虑氘起源的可能的天体物理模式也是重要的。如果证明局部 D/H 具有变化，那么天体物理过程比原初核合成更为合理，但如果 D/H 比值具有均一性，则原初合成模式较合理。

Investigations going back to the middle nineteen fifties have repeatedly shown that D is best produced astrophysically under non-thermodynamic conditions. We imagine α particles projected at high speed into an ionized gas composed mainly of hydrogen. If the gas temperature has some moderate value, say 10^5 K for definiteness, nuclear reactions leading to D production can occur. D can be knocked out of an α particle by a spallation reaction, and neutrons knocked out of the ^{4}He can subsequently be captured by protons of the ambient gas. For example, α particles entering gas at speed $c/3$ have a kinetic energy relative to the gas of ~200 MeV, which is 50 MeV per nucleon, and at such a bombarding energy the total cross section for all the reactions leading to D production is about 5×10^{-26} cm^2, that is, 50 mb according to Audouze et al.[9]. Although the stopping cross section due to Coulomb scattering is greater than this, it is clear that a significant fraction of the α particles will produce a D nucleus. Because the D is thus formed within a comparatively low temperature gas it is not subject to subsequent breakup, except in rare cases where it happens to be hit by a further incoming α particle.

There are many ways in which this general idea can be used; for example, it can be applied to cosmic rays entering a cloud of gas. The production of D (and ^{3}He) through the spallation of ^{4}He was discussed a decade ago[10] in connexion with magnetic flares in the solar surface. The nuclear physics involved is largely independent of the acceleration mechanism. But processes involving cosmic rays are not capable of explaining large D concentrations of the kind that have now been reported. Cosmic rays are too wasteful of their energy in this respect. D production reaches the geometrical cross section and thus occurs most efficiently when the energy per nucleon is about 30 MeV. Energies as high as several GeV, which is where the main reservoir of cosmic ray energy lies, are not required. For greatest efficiency we must look therefore to processes involving speeds ~ $c/3$.

Such speeds have indeed been found for shock waves generated by stellar explosions[11]. The shock wave starts in the region of the explosion at a speed not much different from the speed of sound, which is always much less than c. As it travels outwards into the lower density regions of the envelope, however, the wave speeds up. It is therefore in the outer envelope that speeds of the required order have been reported in previous investigations.

The shock wave condition we have in mind can be applied much more generally than to a supernova. The basic requirement is for a supply of radiant energy (or of relativistic particles) to emerge from some local source into a diffuse outer envelope of gas containing ^{4}He. The larger the energy supply the better. For a supernova we expect ~10^{50} erg to be available, which is much less than a case reported recently by Appenzeller and Fricke[12-14]. Their work is concerned with objects having masses between 10^5 and 10^6 $M_\odot$—that is, masses of the order of the whole Orion Nebula. Under suitable conditions nuclear energy generation in a time scale of a few thousand seconds, yielding ~10^{56} erg, can lead to expansion and disruption of the whole object. Because most of the energy is taken up in the radiation field, it is possible that a bubble of radiation, say with total energy ~10^{55} erg, may work its way to the outer part of the object and may propagate thence into a surrounding diffuse cloud. Outbursts involving relativistic particles, also with energies of

自 20 世纪 50 年代中期开始的研究已经不断地表明，在天体物理中 D 在非热力学条件下最容易产生。我们想象 α 粒子以高速射入一团主要由氢组成的电离气体中。如果气体具有某一适中温度，比如确切的 10^5 K，导致 D 形成的核反应就可以发生。D 可以通过散裂反应从一个 α 粒子中被击出，从 ^{4}He 中击出的中子随后可以被周围气体中的质子俘获。例如，以 $c/3$ 的速度进入气体的 α 粒子相对气体的动能约为 200 MeV，也就是每个核子 50 MeV。根据奥杜兹等人的研究，在这样的轰击能量下，所有生成 D 反应的总截面大约是 5×10^{-26} cm^2，也就是 50 mb[9]。尽管库仑散射造成的阻止截面大于这一值，但显然，绝大部分的 α 粒子都将会产生 D 核。因为 D 是在相对低温的气体中形成的，除了一些很少见的情形（如：它刚好被另一个入射 α 粒子撞击），其不易继续分裂。

该一般性的想法在许多方面都可应用，例如，可应用于宇宙线进入一团气体云的情景。十年前，通过 ^{4}He 散裂产生的 D（和 ^{3}He）被认为与太阳表面的磁耀斑有关[10]。所涉及的核物理过程很大程度上独立于加速机制。但是与宇宙线有关的核物理过程不能解释目前已经报道的如此大丰度的 D。在这个方面，宇宙线太浪费能量了。D 的产量达到了几何截面，因此在每个核子能量大约为 30 MeV 时，产氘效率最高。这里并不需要高达几 GeV 的能量，也就是宇宙线能量的主要范围。为了达到最高的效率，我们必须考虑与速度约 $c/3$ 有关的过程。

事实上，这样的速度已经在恒星爆发产生的冲击波中被发现了[11]。冲击波在爆发区域以和声速差不多的速度开始，这一速度总是比 c 小很多。但是，当它向外传播进入低密度包层时，冲击波开始加速。因此，之前的研究中所报道的所需速度是在外包层中的速度。

我们已知的产生冲击的条件不仅在超新星中存在，在其他更普遍的情况下也存在。基本的要求是辐射能量的供应（或者相对论性粒子的供应），其形成于局域源进入弥散的含有 ^{4}He 的气体外包层过程中。能量供应越大越好。对于超新星，我们预计获得的能量约为 10^{50} erg，这比最近阿彭策勒和弗里克报告[12-14]的情形小很多。他们的工作涉及的天体质量在 $10^5 M_\odot$ 到 $10^6 M_\odot$ 之间，也就是整个猎户座星云质量所处的量级。在适宜条件下，数千秒内产生的核能约为 10^{56} erg，这一能量可以导致整个天体的膨胀和破裂爆炸。由于大部分能量被辐射场吸收，一个总能量约为 10^{55} erg 的辐射泡有可能成功地到达天体外部，并因此传播到周围弥漫的云中。出现在射电星系中的相对论性粒子爆发，也具有约 10^{55}~10^{56} erg 能量。事实上，来自射电源

$\sim 10^{55}$–10^{56} erg, occur in radio galaxies. Indeed, outbursts from radio sources involving energies up to $\sim 10^{60}$ erg have been considered.

The energy in question, whether radiant or in the form of relativistic particles, would escape from the source at speed c if it were not for the material of the surrounding envelope. But as the envelope becomes more tenuous the radiation is able to push material ahead of it at speeds which approach more and more to c. In such circumstances the material of the envelope is accelerated forward impulsively at a shock front, and nuclei present in it are subject to spallation if the shock becomes violent enough. Deuterium is then formed, either by direct spallation of ^{4}He or from neutrons coming from ^{4}He, the neutrons being captured subsequently by protons downstream of the shock.

The properties of shock waves are usually investigated through the equations of continuum mechanics. The complex physical processes taking place at the shock front are idealized by a discontinuity, much as an impulse to a body is idealized in classical mechanics. We define v_i, v_s as velocity components normal to the shock, with subscripts i, s referring to conditions upstream and downstream of the front respectively. For simplicity, taking hydromagnetic effects to be small, and taking a frame of reference in which the front is stationary, we have the following conservation relations across the front:

$$\rho_i v_i = \rho_s v_s$$
$$p_i + \rho_i v_i^2 = p_s + \rho_s v_s^2 \qquad (4)$$
$$u_i + \frac{p_i}{\rho_i} + \frac{1}{2}v_i^2 = u_s + \frac{p_s}{\rho_s} + \frac{1}{2}v_s^2$$

where p is the pressure and u is the internal energy per unit mass. Terms in p_i, u_i are small in our case. Omitting them, we can satisfy equations (4) with

$$\rho_s = \rho_i \frac{\gamma+1}{\gamma-1}, \quad v_s = v_i \frac{\gamma-1}{\gamma+1}, \quad p_s = \frac{2\rho_i v_i^2}{\gamma+1} \qquad (5)$$

where

$$(\gamma-1)\,\rho_s u_s = p_s \qquad (6)$$

γ is the ratio of specific heats downstream of the shock, and is close to 4/3 in a radiation dominated problem. Putting $\gamma = 4/3$ in equation (5) we get

$$\rho_s = 7\rho_i, \quad v_s = v_i/7, \quad p_s \simeq \frac{1}{3}aT_s^4 = \frac{6}{7}\rho_i v_i^2 \qquad (7)$$

The third of these equations can be regarded as determining v_i when T_s and ρ_i are given. The value to be used for T_s depends on the energy supply, while ρ_i is the density of the envelope upstream of the shock. From here on, we shall be concerned with situations in which T_s is high enough in relation to ρ_i for the resulting value of v_i to be of order $c/3$.

Material flowing through the front experiences a change of velocity $v_i - v_s = 6v_i/7$, which for $v_i = c/3$ is $2c/7$. If we now think of the material in terms of individual particles, instead of from the point of view of continuum mechanics, the particles experience a change

的能量高达约 10^{60} erg 的爆发也已经被考虑过。

如果没有周围包层的物质，无论是以辐射还是以相对论性粒子形式存在的能量都将以 c 的速度逃离源区。但是随着包层变得稀薄，辐射可以推动它前面的物质达到越来越接近 c 的速度。在这样的情况下，包层的物质在冲击波波前被有力地向前推动加速，如果冲击变得足够强烈，其中的原子核就易发生散裂。因此，氘要么由 ^{4}He 的直接散裂产生，要么由来自 ^{4}He 的中子被冲击波下游的质子俘获形成。

冲击波的性质通常用连续介质力学方程进行研究。在冲击波波前发生的复杂物理过程通过一个理想的不连续点来简化，正像在经典力学中对作用于一个物体的冲力的理想化一样。我们定义 v_i、v_s 为与冲击波正交的速度分量，下标 i、s 分别表示波前的上游和下游。为简单起见，假设磁流体动力学效应很小，采用波前为静止的参考系，我们得到以下跨越波前的守恒关系式：

$$\rho_i v_i = \rho_s v_s$$
$$p_i + \rho_i v_i^2 = p_s + \rho_s v_s^2 \tag{4}$$
$$u_i + \frac{p_i}{\rho_i} + \frac{1}{2}v_i^2 = u_s + \frac{p_s}{\rho_s} + \frac{1}{2}v_s^2$$

其中 p 是压强，u 是单位质量内能。对我们的情景而言，p_i、u_i 项值很小。忽略这些项，下列关系可以满足方程（4）

$$\rho_s = \rho_i \frac{\gamma+1}{\gamma-1}, \quad v_s = v_i \frac{\gamma-1}{\gamma+1}, \quad p_s = \frac{2\rho_i v_i^2}{\gamma+1} \tag{5}$$

其中

$$(\gamma-1)\rho_s u_s = p_s \tag{6}$$

γ 是冲击波下游的比热比，在辐射主导的问题里接近于 4/3。在方程（5）中令 $\gamma=4/3$，我们得到

$$\rho_s = 7\rho_i, \quad v_s = v_i/7, \quad p_s \simeq \frac{1}{3}aT_s^4 = \frac{6}{7}\rho_i v_i^2 \tag{7}$$

其中第三个方程可以看作在给定 T_s 和 ρ_i 的时候计算确定 v_i。T_s 的取值依赖于能量供应，而 ρ_i 是冲击波上游包层的密度。从这里开始，我们将考虑与 ρ_i 相联系的 T_s 值足够导致 v_i 值达到 $c/3$ 量级的情形。

流过波前的物质经历了一个 $v_i - v_s = 6v_i/7$ 的速度变化，在 $v_i = c/3$ 时变化值等于 $2c/7$。如果我们现在把物质想象为一些单独的粒子，而不是从连续介质力学的观点

in velocity of $2c/7$ as they pass from being upstream to being downstream of the shock. We have to think of the front as possessing a finite depth and of particles from upstream experiencing collisions as they pass through a finite shock zone. As a velocity of $2c/7$ is equivalent to a bombarding energy of ~40 MeV per nucleon we have a situation similar to that discussed above for α particles projected into a stationary gas. Thus ^{4}He present in material upstream of the shock will be subject to fragmentation as it passes through the front.

The present situation is actually more favourable to fragmentation than the case of α particles projected into an ambient gas, because all particles in the shock zone have come from upstream and have therefore experienced collisions. About half of the bombarding energy will be transferred to electrons, but otherwise the bombarding energy will be retained as random motions of protons and α particles within the shock zone, and the α particles will therefore be subject to breakup over the whole of the time they are within the shock zone.

Material may be considered to have flowed downstream of the shock when electrons have had sufficient time to radiate the kinetic energy they have acquired from the heavy particles. Radiation through bremsstrahlung occurs with cross section

$$\sigma_{\text{brems}} \simeq 4 \frac{e^2}{\hbar c} \left(\frac{e^2}{mc^2} \right)^2 \ln(2E) \tag{8}$$

where E is the electron energy in units of the rest mass. In our case the electron energies are ~20 MeV, and the logarithmic term in equation (8) is about 4, so the cross section is ~10 mb. This is about five times less than the spallation cross section. On the other hand, an electron of energy 20 MeV has a velocity about five times greater than a proton of the same energy. Hence

$$<\sigma v>_{\text{Bremsstrahlung}} \simeq <\sigma v>_{\text{Spallation}} \tag{9}$$

from which it follows that an appreciable fraction of the α particles must be fragmented by the time they pass downstream of the shock.

At this stage we have to distinguish two cases according to whether the gas density is high enough for neutrons from spallation to be captured by protons, or not. The former case is more efficient in its deuterium production by an order of magnitude. Considering this case, and taking the material of the initial cloud to have the usual helium concentration of ~0.25 by mass, and using say 40% for the fraction of the helium experiencing complete spallation, we arrive at a D concentration downstream of the shock of 0.1. Thus the mass density of D downstream of the shock is $0.1 \rho_s$. Using (7), and setting $v_i = c/3$, we have

$$\rho(D) = 0.1 \, \rho_s = \frac{7}{10} \rho_i = \frac{7}{10} \times \frac{7}{18} \frac{aT_s^4}{v_i^2}$$

$$= \frac{49}{20} \frac{aT_s^4}{c^2} = 2.7 \times 10^{-21} \, aT_s^4 \tag{10}$$

or
$$aT_s^4 / \rho(D) = 3.7 \times 10^{20} \text{ erg g}^{-1}$$

来看，这些粒子在从冲击波的上游到下游的过程中，其速度变化了 $2c/7$。我们必须认为波前具有有限深度，并且来自上游的粒子在通过有限冲击区时经历了碰撞。由于 $2c/7$ 的速度等效于每个核子约 40 MeV 的轰击能量，我们得到了与上面讨论过的 α 粒子射入静止气体类似的情景。因此，冲击波上游物质中的 ^4He 在通过波前的时候会碎裂。

事实上，由于冲击区中的所有粒子都来自上游并经历了碰撞，当前情形比 α 粒子射入周围气体的情形更有利于碎裂。虽然大约一半的轰击能量将被转移给电子，但除此之外的轰击能量将以冲击区内无规则运动的质子和 α 粒子形式存在，这些 α 粒子在冲击区内的整个时间段内都易发生分裂。

当电子有足够的时间将它们从重粒子处得到的动能辐射出去时，物质就可以被认为已经流到了激波的下游。发生韧致辐射的辐射截面为

$$\sigma_{韧致辐射} \simeq 4\frac{e^2}{\hbar c}\left(\frac{e^2}{mc^2}\right)^2 \ln(2E) \tag{8}$$

其中 E 是以静止质量为单位的电子能。在我们这个情形下，电子能量约为 20 MeV，方程（8）中对数项的值大约是 4，因此截面约为 10 mb。这大约是散裂截面的 1/5。另一方面，一个能量为 20 MeV 的电子的速度大约是具有同样能量的质子速度的 5 倍。因此

$$<\sigma v>_{韧致辐射} \simeq <\sigma v>_{散裂} \tag{9}$$

由此断定很大一部分 α 粒子在通过冲击区下游时必然会碎裂。

在这里，我们必须根据气体密度是否足够高到使得散裂产生的中子能被质子俘获两种情形来区分。前者氘的产生效率比后者高一个量级。鉴于此种情况，通常取初始云块物质中氦的质量丰度通常约为 0.25，假设其中的 40% 完全散裂，我们在冲击区下游得到的 D 的丰度是 0.1。因此冲击区下游 D 的质量密度是 $0.1\,\rho_s$。根据方程（7），令 $v_i = c/3$，我们得到

$$\rho(D)=0.1\,\rho_s=\frac{7}{10}\rho_i=\frac{7}{10}\times\frac{7}{18}\frac{aT_s^4}{v_i^2}$$

$$=\frac{49}{20}\frac{aT_s^4}{c^2}=2.7\times10^{-21}\,aT_s^4 \tag{10}$$

或者

$$aT_s^4/\rho(D)=3.7\times10^{20}\ \mathrm{erg\cdot g^{-1}}$$

This is a relation between energy supply and deuterium production. The energy "cost" in the radiation field to produce one gram of deuterium is seen to be $\sim 3.7 \times 10^{20}$ erg. So to produce an average D/H ratio equal to the proto-solar value of $\sim 3 \times 10^{-5}$ throughout a mass M requires $\sim 10^{16} M$ erg, from which we see that to obtain D/H $\sim 3 \times 10^{-5}$ throughout the Galaxy requires $\sim 3 \times 10^{60}$ erg. Because this value is within the range that can be contemplated it seems possible for the present process to give a galactic deuterium concentration comparable to the initial solar system value.

The present considerations require neutrons from the spallation of ^{4}He to be captured by protons. For this condition to be satisfied ρ_i must not be much less than $\sim 10^{-8}$ g cm^{-3}. Otherwise the efficiency of D production is reduced by an order of magnitude, and the energy requirement is increased correspondingly. In the rest of this article we shall take $\rho_i = 10^{-7}$ g cm^{-3} for definiteness, and will consider the rates which are then operative for various processes.

The first question to be asked is: what will be the order of the depth of the shock zone? Taking $\sim 3 \times 10^{-7}$ g cm^{-3} as the average density within the shock zone, and noting that $<\sigma v>_{\text{Bremsstrahlung}} \simeq 3 \times 10^{-16}$, we see that the electrons lose their energy in a time $\sim 2 \times 10^{-2}$ s. In this time material flowing at a mean speed of say $\sim v_i/2 = c/6$ travels a distance $\sim 10^8$ cm. This gives the order of the depth of the zone.

To prevent radiation downstream of the shock from simply streaming through the front it is necessary that the optical depth of material within the shock zone shall be greater than unity. For the numerical values considered in the previous paragraph there are about 30 g of material per unit area of the front. This is sufficient to dam back the radiation through Thomson scattering by the electrons. Before the radiation can penetrate the shock zone new material is then added from upstream. This circumstance does not depend on the particular numerical values used here. It depends essentially on the Thomson scattering for radiation being much larger than the bremsstrahlung cross section, $\sim 10^{-24}$ cm^2 compared with $\sim 10^{-26}$ cm^2.

Once bremsstrahlung transfers energy from the electrons to the radiation field other processes become involved, particularly Compton scattering. Because most of the bremsstrahlung energy from 20 MeV electrons consists of γ rays above 1 MeV, Compton scattering has a complex behaviour, certain scattering angles augmenting the radiation field, others transferring energy back to the electrons. A refined calculation would be necessary in order to consider such effects in detail. Here we shall simply take relation (9) to represent the order of magnitude of the relation of radiation losses to spallation. The conclusion from (9) is that an appreciable fraction of α particles are fragmented by the time they pass downstream of the shock.

Deuterium produced within the shock zone will itself be subject to spallation, as will be ^{3}He and T. Consequently we shall not regard D production as being due to the immediate spallation of ^{4}He, but as arising from neutrons released in the breakup of ^{4}He. The lifetime

这是能量供应和氘产量之间的关系式。在辐射场中，产生 1 g 氘需要"消耗"的能量约为 3.7×10^{20} erg。因此，在一个质量为 M 的天体中，为了使 D/H 达到原初太阳（约 3×10^{-5}）的平均值，需要约 $10^{16} M$ erg 的能量，由此我们发现，在整个银河系中，D/H 达到约 3×10^{-5} 所需能量约为 3×10^{60} erg。由于这个值在合理的范围内，看起来有可能通过这个过程产生和原初太阳系相当的星系的氘丰度。

目前的考虑要求 ^{4}He 散裂产生的中子被质子俘获，为满足这个条件，ρ_i 必然不能比约 10^{-8} g·cm^{-3} 小很多。否则 D 产生的效率会减小一个量级，同时能量需求会相应增加。为了明确，在本文余下的部分我们将取 $\rho_i = 10^{-7}$ g·cm^{-3}，并将考虑应用于各种过程的速率。

首要回答的问题是：冲击区的厚度在什么量级？取约 3×10^{-7} g·cm^{-3} 作为冲击区内的平均密度，并注意到 $<\sigma v>_{\text{韧致辐射}} \simeq 3 \times 10^{-16}$。我们发现电子在约 2×10^{-2} s 的时间内损失掉它们的能量。在这个时间内，流动物质的平均速度约为 $v_i/2 = c/6$，移动了约 10^8 cm 的距离。这给出了冲击区厚度的量级。

为防止冲击波下游的辐射简单地流过波前，冲击区域内物质的光深必须大于 1。对上一段中考虑的数值，单位面积的波前有大约 30 g 物质。这足以挡住由电子产生的汤姆逊散射的辐射。在辐射能够穿透冲击区之前，就有新的物质从上游补充进来。这种情况不依赖于这里用到的具体数值。对辐射来说，由于汤姆逊散射的截面远大于韧致辐射的截面（约 10^{-24} cm^2 对约 10^{-26} cm^2），光深在本质上由汤姆逊散射决定。

一旦韧致辐射把能量从电子传输到辐射场，其他过程就参与进来，特别是康普顿散射。由于 20 MeV 电子的韧致辐射大部分由能量高于 1 MeV 的 γ 射线组成，康普顿散射具有复杂的行为特征，某些散射角使辐射场扩张，其他散射角把能量返回给电子。为了详细地考虑这些效应，精确计算是必要的。这里我们将简单采用方程（9）表示与散裂有关的辐射损失程度。自关系式（9）得到的结论是，相当可观的一部分 α 粒子在通过冲击波下游时分裂了。

正如 ^{3}He 和 T 那样，在冲击波区域产生的氘自身也易发生散裂。因此，我们将不把 D 产生归因于 ^{4}He 的直接散裂，而是由 ^{4}He 分裂释放的中子生成。中子弱衰变

of the neutrons against weak decay ($\sim 10^3$ s) is very long compared to the time spent ($\sim 10^{-2}$ s) in the shock zone. So the neutrons move downstream to regions where thermodynamic conditions can be considered to be established. Putting $\rho_i = 10^{-7}$ g cm^{-3}, $v_i = c/3$ in (7) gives $T_s \simeq 7.6 \times 10^6$ K, which is far too low for deuterium formed by n+p $\rightarrow$ D+γ to be subject to spallation or to photodisintegration. The $<\sigma v>$ value for this reaction is 7×10^{-20} cm^3 s^{-1}, so for a hydrogen density $\simeq 7 \times 10^{-7}$ g cm^{-3} downstream of the shock it takes some 30 s for the neutrons to be captured by protons. Because this is much less than the neutron half-life, although amply long enough for the neutrons to flow downstream well clear of the shock, we conclude that essentially all neutrons go to form D.

To recapitulate to this point: Normal thermonuclear processes generate ^{4}He from hydrogen with very little production of deuterium. But if ^{4}He can be shaken loose into its constituent neutrons and protons in abnormal situations, such as those in the front of a high speed shock wave, conditions become favourable to D production provided the gas density is not so small that the neutrons decay before they are captured by protons. Conditions are then also favourable in that the temperature $T \simeq 7.6 \times 10^6$ K behind the front leads to a lifetime for D(p, γ) ^{3}He of $\sim 10^{10}$ s, which is much longer than the time required for a local exploding object to disperse. The D/H ratio can be locally higher than the terrestrial value by as much as 10^2, but after averaging with other exterior material the D/H value will be lowered. If the estimate of 3×10^{-5} for the proto-solar value is typical of the interstellar medium as a whole, then the process described here could be the origin of all the deuterium in the Galaxy. Accompanying the production of D in the shock front will be that of a smaller amount of ^{3}He, perhaps enough to yield the proto-solar values ($\sim 10^{-5}$) taken by Black[3] as typical. Further, because the explosion which produces the shock wave may well be due to nuclear energy generation converting a substantial fraction of hydrogen into helium, this process may also be the source of galactic ^{4}He. This brings us back (see pp. 23, 24 of ref. 6) full circle to the fact that conversion of hydrogen into helium in one part in four by mass yields the full energy of the background microwave radiation at 2.7 K and once again forces us to ask if there could have been a mechanism which provided the necessary thermalization. So far we have found no plausible affirmative answer to this question, but the coincidence of the numbers remains puzzling.

We thank Dr Klaus Fricke for discussions on the production of shock waves in the implosion–explosion of massive objects, and the authors of refs. 1, 4 and 5 for informing us of their work in advance of publication. This work was supported in part by the National Science Foundation.

(**241**, 384-386; 1973)

Fred Hoyle & William A. Fowler
California Institute of Technology, Pasadena, California

Received December 12, 1972.

寿命（约 10^3 s）与其在冲击波区内的停留时间（约 10^{-2} s）相比要长得多。所以中子向下游运动到可以被认为已形成热力学条件的区域。在（7）中令 $\rho_i=10^{-7}$ g·cm^{-3}，$v_i=c/3$ 得到 $T_s\simeq 7.6\times 10^6$ K，这一值远低于由 n+p → D+γ 产生氘的温度，以至于不易发生散裂或光致分裂。这个反应的 $<\sigma v>$ 值为 7×10^{-20} cm^3·s^{-1}，所以对于氢密度约为 7×10^{-7} g·cm^{-3} 的冲击区下游，质子需要大约 30 s 的时间将中子俘获。虽然这一时间足够让中子流到下游摆脱冲击，但由于其远小于中子的半衰期，我们认为在本质上所有中子都形成了 D。

总结起来说：由氢产生 ^{4}He 的一般热核过程几乎不生成氘。但是，如果在异常情况下，如在一个高速冲击波的前沿，^{4}He 能够被震动分解为组成它的中子和质子，在这种情况下，只要气体密度不小到中子在被俘获前就完全衰变，将会有利于 D 产生。随后的条件同样有利于 D 的形成，这是由于波前后面的温度为 $T\simeq 7.6\times 10^6$ K，导致 D（p,γ）^{3}He 的寿命约 10^{10} s，这比一个局域爆炸天体消散所需时间长得多。局部的 D/H 比可以比地球上的值高 10^2，但是和其他外部物质平均后，D/H 值会降低。如果总体上原初太阳的估计值 3×10^{-5} 是星际介质的特征值，那么这里描述的过程可能是银河系中所有氘的来源。伴随冲击波前锋中 D 的产生，将有少量的 ^{3}He 形成，这可能足以产生原初太阳中的值（约 10^{-5}），也就是布莱克[3]认为的特征值。另外，因为产生冲击波的爆炸有可能正是由于将很大一部分氢转化为氦产生的核能导致的，所以这个过程可能也是星系中 ^{4}He 的源头。这让我们回到（见参考文献 6 的 23、24 页）这样的事实，将氢转化为 1/4 质量的氦产生 2.7 K 的全部微波背景辐射的能量，并且再一次迫使我们提出问题：是否可能已经存在提供必要热化的机制。到目前为止，对这个问题，我们没有发现可信的确切答案，但是数字的巧合仍令人费解。

我们感谢克劳斯·弗里克博士对大质量天体在聚爆——裂爆中产生冲击波的讨论，也感谢参考文献 1、4 和 5 的作者们告知我们他们尚未发表的工作。这项工作部分得到了美国国家科学基金会的支持。

（钱磊 翻译；许冰 审稿）

References:

1. Cesarsky, D. A., Moffet, A. T., and Pasachoff, J. M., *Astrophys. J. Lett.* (in the press).

2. Reeves, H., Audouze, J., Fowler, W. A., and Schramm, D. N., *Astrophys. J.* (in the press).

3. Black, D. C., *Geochim. Cosmochim. Acta,* **36**, 347 (1972).

4. Jefferts, K. B., Penzias, A. A., and Wilson, R. W., *Astrophys. J. Lett.* (in the press).

5. Wilson, R. W., Penzias, A. A., Jefferts, K. B., and Solomon, P. M., *Astrophys. J. Lett.* (in the press).

6. Wagoner, R. V., Fowler, W. A., and Hoyle, F., *Astrophys. J.,* **148**, 3 (1967).

7. Oort, J. H., Solvay Conference on *Structure and Evolution of the Universe,* 163 (R. Stoops, Brussels, 1958).

8. Shapiro, S. L., *Astron. J.,* 76, 291 (1971).

9. Audouze, J., Epherre, M., and Reeves, H., *High Energy Nuclear Reactions in Astrophysics,* chap. 9 (Benjamin, New York, 1967).

10. Fowler, W. A., Greenstein, J. L., and Hoyle, F., *Geophys. J. Roy. Astron. Soc.,* **6**, 148 (1962).

11. Colgate, S. A., and White, R. H., *Astrophys. J.,* **143**, 626 (1966).

12. Appenzeller, I., and Fricke, K., *Astron. Astrophys.,* **12**, 488 (1971).

13. Appenzeller, I., and Fricke, K., *Astron. Astrophys.,* **18**, 10 (1972).

14. Appenzeller, I., and Fricke, K., *Astron. Astrophys.,* **21**, 285 (1972).

Discontinuous Change in Earth's Spin Rate following Great Solar Storm of August 1972

J. Gribbin and S. Plagemann

Editor's Note

Does solar activity influence how fast the Earth spins? The idea sounds unlikely, but in 1959, following one of the largest solar storms in recorded history, Andre Danjon reported an apparent sharp increase in the length of the day as measured by astronomical means. Here John Gribbin and Stephen Plagemann tested this controversial idea by taking advantage of an even larger solar storm in 1972. They found a similar effect: an abrupt change in the length of day correlated with an increase in cosmic rays from solar nuclear reactions. The authors speculated that these cosmic rays might influence the dynamics of the Earth's upper atmosphere, changing the Earth's rate of spin within a few days.

THE question of a link between changes in the Earth's spin rate and the activity of the Sun is of topical interest, and there is good evidence that the changing length of day is influenced by the mean level of solar activity[1,2]. The possibility of a one-to-one correlation between specific events on the Sun and specific changes in the length of day has remained more controversial, however, although there was a suggestion of such an effect associated with the great solar storm of 1959 (refs. 3–5). Specifically, Danjon suggested[3-5] that there was an increase in the length of day when the nucleonic component of solar cosmic rays increased; this was in addition to the usual steady increase in the length of day. Other observers questioned the reality of this effect (for a discussion of the controversy see ref. 6), and because the 1959 solar storm was the greatest recorded since the time of Galileo, there was no immediate hope of an independent test of Danjon's claim. In August 1972, however, an even greater disturbance occurred on the Sun[7-9]. It seemed to us that this might provide the ideal opportunity to resolve the controversy, and we have indeed found a discontinuous change in the length of day, and a change in the rate of change of the length of day (a glitch) immediately after that event. Changes in the length of day, and thus in the spin rate of the Earth, are revealed by regular measurements of Universal Time (UT) carried out at many observatories around the world. For our purpose, we are interested in UT2, the version of Universal Time with the effects of the Chandler Wobble and seasonal variations removed. The difference between Atomic Time (AT) and UT2 shows, on average, a monotonic increase as the Earth's spin slows down and the length of day increases.

In Fig. 1 we plot AT–UT2 for a period of one month on either side of the great solar storm of August 1972. The spot group associated with the flare activity built up from July 29, reaching a maximum size (covering 17° in longitude) on August 4. In Fig. 1, we have

1972年8月强太阳风暴过后地球
自转速度的不连续变化

格里宾，普拉格曼

编者按

太阳活动是否会对地球的自转速度产生影响？答案似乎是否定的，但是在1959年，有史以来最强的太阳风暴之一爆发过后，安德烈·丹戎报道了通过天文学方法，测量出日长（译者注：地球自转一周的时间）发生了显著增长。在本文中，通过研究1972年的一次更大规模的太阳风暴，约翰·格里宾和斯蒂芬·普拉格曼对这个有争议的观点进行了检验。他们发现一个相似的效应：日长发生突变，与太阳核反应发出的宇宙射线增长相关。作者推断这些宇宙射线可能会对地球高层大气的动力学过程产生影响，在几天内改变地球的自转速度。

地球自转速度的变化与太阳活动的关系是当前的热点问题，目前已有充分的证据证明，日长变化受太阳平均活动强度的影响[1,2]。然而，太阳上发生的特定事件与日长的特定变化之间存在一一对应的相互关系的可能性仍有许多争议。尽管已有学者指出，1959年的强太阳风暴曾对日长带来过这样的影响（参考文献3~5）。特别指出的是，丹戎曾经提出[3-5]当太阳宇宙射线的核子组分增加时日长就会增加，这一增加是附加在日长的稳定增加之外的。其他学者则对该效应的真实性持怀疑态度（关于该争议的讨论见参考文献6），而且由于1959年的强太阳风暴是自伽利略时代以来记录到的最强的一次，所以无法立即对丹戎的观点进行独立检验。然而，1972年8月太阳上发生了一次更强的扰动[7-9]。对我们来说，这可能为解决该争议提供了一次理想的机会，我们也确实发现了日长的不连续变化，以及在该事件之后日长变化速率立即发生了改变（即突变）。世界各地的多个天文台对世界时（UT）的常规测定都揭示了日长的变化，以及由此而带来的地球自转速度的变化。对我们来说，我们感兴趣的是UT2，即一种去除了钱德勒摆动以及季节性变化的影响的世界时。通常来说，随地球自转减慢即日长增加，原子时（AT）与UT2之间的差值显示为单调递增。

在图1中我们标出了1972年8月强太阳风暴爆发前后各一个月内，AT–UT2的变化情况。与耀斑活动相关的黑子群于7月29日开始活跃，在8月4日达最大规模（日面经度跨越了17°）。我们在图1中用箭头标志出了8月3日（JD 41532）的位

marked August 3 (JD 41532), the last day before the series of Forbush decreases which marked the solar activity[7], by an arrow. The expected change in spin rate and length of day is clearly visible; a similar plot using data for the six month period May to October 1972 shows (Fig. 2) that the jump is the largest such event recorded over that period. The change in slope indicated by the data of Fig. 2 is equally significant; this is qualitatively similar to the changes seen in pulsar spin rates during glitches, and we therefore borrow the term from the pulsar literature to describe this terrestrial process. At the time of the flare and sunspot activity, the slope flattens slightly, subsequently trending back towards a slope more typical of the data of the preceding three months over a period of several weeks.

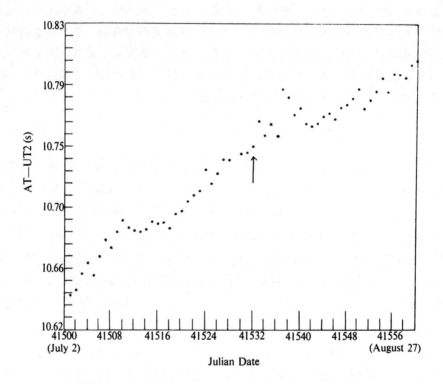

Fig. 1. Change in length of day for a period of one month on either side of date of great solar flare activity of August 1972. Arrow marks August 3, the last day before the flare activity. Large discontinuous change in AT–UT2 occurs on August 8.

These effects are not so dramatic that one would necessarily attribute them to an outside cause on the basis of these data alone, but they take on a greater significance in the light of out prediction, following Danjon, that just such a change should occur soon after a great solar flare. We are confident that the effect is real, and that the glitch was indeed caused by events associated with the solar activity of early August 1972.

置，这是福布什下降的前一天，福布什下降是太阳活动的标志[7]。从图中可以清楚地看到我们所预期的自转速度及日长的变化。我们还利用 1972 年 5~10 月这 6 个月的数据作了类似的图（图 2），如图所示，该跃变是这一时期内记录到的此类事件中最大的一次。图 2 数据所显示的斜率变化也同样显著，究其本质，它类似于脉冲星自转突快期间出现的自转速度的变化，因此我们借用描述脉冲星文献中的术语"突变"（glitch）来描述这一发生在地球上的情况。在太阳耀斑和太阳黑子活动期间，该斜率会略微变小，持续数周后，又回归至前三个月的数据所表现出的更加具有代表性的斜率上。

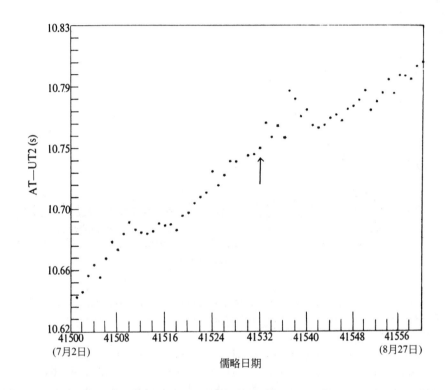

图 1. 1972 年 8 月强太阳耀斑活动发生日前后各一个月内日长的变化。箭头标示 8 月 3 日，是耀斑活动发生的前一天。AT–UT2 的较大不连续变化出现在 8 月 8 日。

　　由于自转与日长并未发生很大的变化，所以我们不能仅凭这些数据就将其归因于太阳风暴这一外部因素。但是据我们预测，也正如丹戎所言，这些变化应该有更重要的意义，因为它们紧随强太阳耀斑活动发生。我们确信，这种效应是真实存在的，并且突变确实是由与 1972 年 8 月初的太阳活动相关的事件所引起。

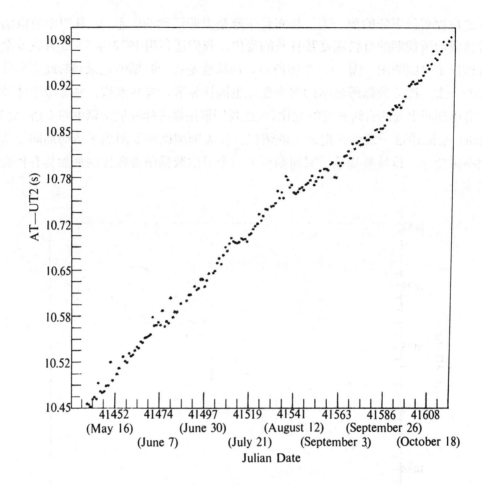

Fig. 2. As Fig. 1, for May to October 1972. The change in slope after the time of the great solar activity, and subsequent return towards the pre-flare slope, emphasize the importance of the event.

It is not difficult to envisage models which explain the delay of 5 days between commencement of the flare activity and the glitch. We will not discuss detailed mechanisms here, except to point out that solar phenomena are known to influence the large scale circulation of the Earth's atmosphere. For example, troughs in the circulation pattern at high latitudes are amplified when the level of solar cosmic rays reaching the Earth is high[10,11]. Like Schatzman[6], we believe that sudden variations in the length of day may be produced by meteorological phenomena induced by solar activity; in that case, it would be most unreasonable if it did not take a few days for these effects to show themselves in the AT−UT2 measurements.

We will discuss details of such a mechanism and further consequences of this discovery elsewhere. We thank the US Naval Observatory, Washington, for supplying the raw data used in the preparation of Figs. 1 and 2.

(**243**, 26-27; 1973)

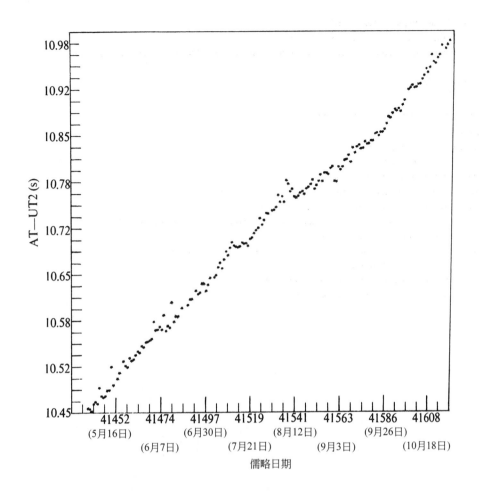

图 2. 1972 年 5~10 月 AT–UT2 的变化情况。强太阳活动后斜率发生变化，而随后又回归至耀斑发生前的斜率，这充分说明了该事件的重要性。

我们不难想出一个模型来解释为何耀斑活动开始到日长的突变发生之间有 5 天的滞后效应。在本文中我们不会讨论详细的机制问题，仅指出地球大气层的大尺度环流会受太阳活动的影响。例如，当到达地球的太阳宇宙射线水平较高时，高纬度地区环流模式中的低压槽就会加深 [10,11]。同沙茨曼 [6] 一样，我们认为日长的突然变化可能是由太阳活动导致的气象现象造成的。在这种情况下，AT–UT2 的测定结果中这类效应并未出现几天时间的滞后，那反而是最不合理的。

有关这一机制的详情以及该发现的进一步结果，我们将在其他文章中详细讨论。在准备绘制图 1 和图 2 的过程中，位于华盛顿的美国海军天文台向我们提供了原始数据，在此表示感谢。

（齐红艳 孙惠南 翻译；马宇蒨 审稿）

John Gribbin* and Stephen Plagemann†

*Nature, 4 Little Essex Street, London WC2R 3LF

†NASA Goddard Space Flight Center, Institute for Space Studies, 2880 Broadway, New York, NY 10025

Received March 26, 1973.

References:

1. Challinor, R. A., *Science*, **172**, 1022 (1971).

2. Gribbin, J., *Science*, **173**, 558 (1971).

3. Danjon, A., *CR Acad. Sci. Paris*, **254**, 2479 (1962).

4. Danjon, A., *CR Acad. Sci. Paris*, **254**, 3058 (1962).

5. Danjon, A., *Notes et Informations de l'Observatoire de Paris*, **8**, No. 7 (1962).

6. Schatzman, E., in *The Earth-Moon System* (edit. by Cameron, A. G. W., and Marsden, B. G.), 12 (Plenum, New York, 1966).

7. Pomerantz, M. A., and Duggal, S. P., *Nature*, **241**, 331 (1973).

8. Chupp, E. L., Forrest, D. J., Higbie, P. R., Surie, A. N., Tsai, C., and Dunphy, P. P., *Nature*, **241**, 333 (1973).

9. Mathews, T., and Lanzerotti, L. J., *Nature*, **241**, 335 (1973).

10. Macdonald, N. J., and Roberts, W. O., *J. Geophys. Res.*, **65**, 529 (1960).

11. Roberts, W. O., and Olsen, R. M., *J. Atmos. Sci.*, **30**, 135 (1973).

Cores of the Terrestrial Planets

K. E. Bullen

Editor's Note

In the early 1970s we were just beginning to understand the properties of the planets of the Solar System, as the first unmanned space missions sent back data. In particular, the densities of the terrestrial planets systematically decreased going from Mercury outwards to Mars. Keith Bullen here makes an early attempt to explain this by invoking a change in mineral composition for mixtures of iron and iron oxides. The idea only partially worked, failing to account for the cores of Mercury and the Moon. We now have much better data and a far more sophisticated understanding of the conditions in the early Solar System and how the planets were assembled, rendering Bullen's model obsolete.

The compositions of the cores of the terrestrial planets have been re-examined following a calculation by O. G. Sorokhtin that the iron oxide Fe_2O is stable at pressures reached in the Earth's core and his suggestion that the outer core may consist of Fe_2O. It is shown that the idea of an Fe_2O outer core in the Earth can be fairly well reconciled with a common overall composition for the planets Earth, Venus and Mars in a way that avoids the main objections to the earlier phase-transition theory.

THE presently favoured theory that the Earth's outer core consists predominantly of iron (possibly alloyed) requires[1,2] the overall compositions of the planets Earth, Venus and Mars to be markedly different. It is difficult to reconcile this result with accretion theories of the origin of the planets[1-3].

In an early attempt at reconciliation, Ramsey[4] and myself[5] independently advanced a "phase-transition theory" which treated the Earth's outer core as a high-pressure modification of the lower mantle material. This theory agreed with a common overall composition for the three planets and fitted well the then available observational data on the masses and radii of Venus and Mars and the polar flattening of Mars. The theory later met with difficulties.

Revisions of the planetary observational data reduced the quality of the original fit. The most serious surviving discrepancy is that the preferred observational estimate[6] of the radius of Venus is now 6,052±10 km, as against 6,270 km predicted[7] on the phase-transition theory.

The original Earth models A and B (refs 8, 9) used in deriving models of the other planets

46

类地行星的核

布伦

编者按

在 20 世纪 70 年代早期，当第一艘无人飞船传回数据时，人们才刚刚开始理解太阳系行星的性质，特别是类地行星的密度从水星向外到火星逐渐减小。本文中，基斯·布伦进行了一次早期尝试，他用铁和氧化铁混合物中矿物成分的变化解释这个现象。这个想法只是部分成立，它不能解释水星和月球的核。我们现在已经有好得多的数据和对早期太阳系的条件以及对行星是怎样形成的有着更细致的理解，布伦的模型过时了。

根据索罗赫京的计算，在地核的压强下氧化铁 Fe_2O 是稳定的，他提出类地行星的外核可能由 Fe_2O 构成，据此我们重新检验了类地行星的核的组成。这里证明了地球外核由 Fe_2O 构成的想法可以和地球、金星和火星共同的总体成分保持相当好的一致，避免了与早期提出的相变理论的主要矛盾。

现在普遍接受的理论认为地球外核主要是由铁（可能是合金）组成，这就要求[1,2]地球、金星和火星这些行星的总体成分有显著的不同。这和行星形成的吸积理论[1-3]难以协调。

在早期调和这一矛盾的尝试中，拉姆齐[4] 和我 [5] 都曾独立提出"相变理论"，这个理论认为地球外核为一种高压下变质了的下地幔物质。这个理论和其他三个行星的总体成分一致，而且和那时已有的金星及火星的质量和半径、火星的极向扁率的观测数据符合得很好。这理论后来遇到了困难。

对行星观测数据的修正降低了原先的拟合效果。最严重的偏差是：金星半径的最佳观测估计值[6]为 6,052 ± 10 km，和相变理论预测[7]的 6,270 km 不一致。

人们用地球模型 A 和 B（参考文献 8，9）来推导其他行星的情况。然而这两个原

47

have had to be amended (see refs 10, 11) to allow (among other things) for the reduction of the estimated Earth's central density[12] from about 18 to 13 g cm⁻³, and of the estimated moment of inertia coefficient[13] for the Earth from 0.3336 to 0.3309. Both revisions have slightly worsened the fit with Venus and Mars.

Geochemists have found difficulty, principally on the question of the packing of oxygen atoms in the lower mantle material, in reconciling the proposed phase transition with the large density jump (in the ratio 0.7 or more) at the Earth's mantle-core boundary, N say.

Transitory shock-wave experiments carried out in laboratories at high pressures have failed to supply positive evidence that a phase transition occurs at N.

Alternative Model

Having regard to various uncertainties and questions of experimental interpretation, I think it is unsound to assert, as do some investigators, that the difficulties have "disproved" the theory—indeed the theory still has some notable followers. But the case against it is sufficiently strong to make it desirable to seek alternatives.

As a possible alternative, I propose a theory which avoids the principal difficulties of the phase-transition theory while retaining the important feature that the pressure p_c at N is critically involved in the changes of property at N. The theory incorporates a suggestion of O. G. Sorokhtin (personal communication) that the Earth's outer core consists of Fe_2O. Sorokhtin has calculated that this oxide, which is unstable at ordinary pressures, becomes stable at the pressures in the Earth's core and has a density-pressure relation matching that in the Earth's outer core.

In other respects, I have deviated substantially from Sorokhtin. Whereas he attributes the occurrence of Fe_2O in the outer core to a breakdown of FeO into Fe_2O and oxygen, I have associated the occurrence of the Fe_2O with the equation $Fe_2O \rightleftharpoons FeO + Fe$, my reason being the extent of planetary fit that can be thereby achieved.

My proposal, which I refer to as the "Fe_2O theory", envisages a model family of planets with the following properties:

All planets of the family are composed of two primary materials—a basic mantle material X, and Fe_2O. (This is subject to qualification below.) For present purposes, the composition of X need not be specified, but X is likely to include some FeO.

In all the planets, the ratio of the mass of X to the mass M of the planet is the same.

In those planets which contain Fe_2O, the Fe_2O occurs as a distinct zone (the "outer" core) throughout which $p \geqslant p_c$, where p is the pressure.

初的地球模型被指出必须进行修正（参考文献 10，11）：（其中）要允许地球中心密度的估算值[12] 从大约 18 g·cm⁻³ 降到 13 g·cm⁻³，同时允许地球转动惯量系数的估算值[13] 从 0.3336 降到 0.3309。这两个修正都对金星和火星的拟合造成了轻微的不良影响。

地球化学家在协调所提出的相变理论与地球核幔边界(如 N 处)大的密度跳跃(比值为 0.7 或更大）时遇到了困难，困难主要在于氧原子在下地幔物质中的填充问题。

在实验室进行的高压下瞬变激波实验中，并未提供足以支持在 N 处发生相变的正面证据。

替 代 模 型

针对实验解释存在的各种问题和不确定性，我认为像一些研究学者们那样做以下断言——这些困难"反驳"了这个理论是不合适的，实际上这个理论还有一些著名的追随者。但是反对这个理论的理由也是足够强的，这使我们有必要去寻找替代模型。

作为一种可能的替代模型，我提出了一种新理论，避免了相变理论的主要困难，同时保留了其重要特点，即 N 处的压强 P_c 和 N 处的性质变化密切相关。该理论还包括了索罗赫京的意见（个人交流），他认为地球外核是由 Fe_2O 组成的。索罗赫京计算得出这种在一般压强下不稳定的氧化物在地核压强下变得稳定，并且有和地球外核相匹配的密度－压强关系。

在其他方面，我和索罗赫京的观点大相径庭。尽管他将外核中 Fe_2O 的产生归结为 FeO 分解为 Fe_2O 和氧原子，但我认为外核中 Fe_2O 的产生与方程 $Fe_2O \rightleftharpoons FeO+Fe$ 有关，我的理由在于这样可以达到和行星拟合的程度。

我这个称为"Fe_2O 理论"的提议假设了一个具有以下性质的行星模型族：

该族的行星都由两种主要物质组成，一种是基本的地幔物质 X，另一种是 Fe_2O。（这是受到以下这些限制的）基于现在的目的，X 的组成无须详细说明，但是其中可能包含一些 FeO。

在所有的行星中，X 的质量和行星质量 M 的比值是相同的。

在那些含有 Fe_2O 的行星中，Fe_2O 存在于 $p \geqslant p_c$ 的独立区域（地核的"外层"），其中 p 是压强。

In those planets where the first two properties would entail $p < p_c$ in an Fe_2O zone, some or all of the Fe_2O has broken down into FeO and Fe. This FeO, which I refer to as Y, forms part of the mantle and is additional to any FeO that may be part of X. The Fe falls to form an "inner" core.

Details of the Model

The second property secures a common overall composition for the family. The masses of Y and Fe in any planet are such as would combine precisely to form Fe_2O when $p \geqslant p_c$, and the ratio of the total mass of Fe, Fe_2O and Y to M is the same for the whole family.

The family has three subsets—H, J, K, say (Fig. 1). The subset H includes the smallest planets which have no Fe_2O zones and thus no outer cores; they have mantles composed of X and Y mixed, and inner cores of Fe. The subset K includes the largest planets which have mantles composed purely of X, outer cores of Fe_2O, and no inner cores. The subset J consists of intermediate planets which have mantles composed of X and some Y, outer cores of Fe_2O, and inner cores of Fe. For all the J planets, $\rho = \rho_c$ at the top of the outer core. The core-mantle mass ratios are the same for all H planets, and for all K planets (but the ratio is different for H and K).

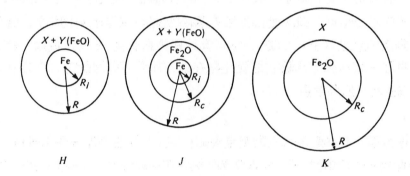

Fig. 1. Materials in the interiors of the three subsets (H, J, K) of terrestrial planets. The outermost zones, all containing the material X, are mantles. The Fe_2O zones are referred to as "outer cores", the Fe zones as "inner cores".

Earth and Venus would correspond to members of J, Mars of H. Unfortunately, no known planets correspond to K, so that an observational test of the subset K is not available. From evidence on the Earth, all Fe_2O zones (outer cores) would be expected to be fluid, and Fe zones (inner cores) solid.

The Fe_2O theory resembles the phase-transition theory in that all J planets have the same (critical) pressure p_c at the mantle-core boundaries, and all H planets lack outer cores. To a considerable extent, it is this feature which enables both theories to go close to fitting a common overall composition for Earth, Venus and Mars. Differences in detail arise because, in planets smaller than Earth, the Fe_2O theory entails larger iron (inner) cores than does the phase-transition theory.

在那些前两个性质限定了在 Fe_2O 区域中 $p<p_c$ 的行星中，部分或全部的 Fe_2O 已经被分解成 FeO 和 Fe。这些 FeO，我称其为 Y，组成地幔的一部分，并且可能作为 X 的一部分，即 FeO 的补充。Fe 则下落形成了"内"核。

模 型 细 节

第二个性质保证该模型族的行星有着相同的总体成分。在任何行星中 Y 和 Fe 的质量在 $p \geq p_c$ 时恰好形成 Fe_2O，而且 Fe、Fe_2O 和 Y 的质量之和与 M 的比值在整个模型族中都是相同的。

该模型族有三个子集——H、J、K，如（图1）。子集 H 包含那些最小的行星，它们没有 Fe_2O 区域，因而没有外核；它们的幔由 X 和 Y 混合组成，内核由 Fe 组成。子集 K 包含那些最大的行星，它们的幔完全由 X 组成，外核为 Fe_2O，没有内核。子集 J 包含中等的行星，其幔由 X 和一些 Y 组成，外核为 Fe_2O，内核由 Fe 组成。对于所有 J 类行星，在外核顶层 $\rho=\rho_c$。对于所有的 H 类以及所有 K 类行星，它们各自内部的核–幔的质量比相同（但 H 类和 K 类行星的核–幔质量比是不同的）。

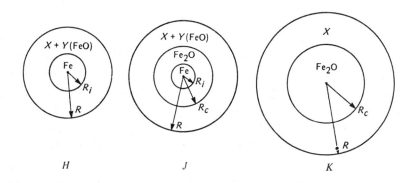

图 1. 三种类地行星模型族（H，J，K）内部的物质组成。最外层的地幔区域都含有 X 物质。Fe_2O 的区域称为"外核"，铁原子的区域称为"内核"。

地球和金星对应于 J 类成员，火星对应于 H 类成员。遗憾的是没有已知的行星属于 K 类，所以无法对 K 类行星进行观测检验。利用地球上的证据可以预期，所有的 Fe_2O 区域（外核）为液态，Fe 区域（内核）为固态。

这种 Fe_2O 理论和相变理论的类似之处在于，所有 J 类行星在核–幔边界有着相同的（临界）压强 P_c，并且所有的 H 类行星都没有外核。在相当大的程度上，这个特点让两种理论都较为接近对地球、金星和火星共有的总体成分的拟合。出现细节上的区别是因为，对于比地球小的行星，Fe_2O 理论所需的铁（内）核比相变理论的大。

Agreement with Observation

I have started from a simplified Earth model, taken to be a member of the subset J, in which the mean densities of the mantle, outer and inner cores are 4.5, 11 and 13 g cm^{-3}, respectively. I have provisionally neglected variations of density inside planetary zones and taken the ratio of the mean densities of X and Y as 0.7. For a first approximation, differences of compression between corresponding zones of the J planets have also been neglected. Possible temperature differences between different planetary interiors, volume changes due to chemical interaction, and so on, in the mixing of X and Y in mantles, and the presence of nickel in cores have been neglected as likely to affect the essential numerical detail only slightly.

With these simplifications, the postulates formally determine numerical details for all the J planets. In particular, a set of values of M, R (radius of the planet), R_c (radius of the outer core) and R_i (radius of the inner core) is determined.

The observational mass of Venus[14] is 4.87×10^{24} kg. For the J planet with this mass, the simplified Fe_2O theory gives R=6,010 km, R_c=2,710 km, R_i=2,110 km. Allowance for the neglected compression differences will increase this estimate of R because Venus, being smaller, is less compressed than the Earth. Thus the Fe_2O theory fits the observed radius (~6,050 km) of Venus as well as can be expected.

The value 600 km yielded for the thickness, T say, of the (presumed fluid) Venus outer core is also interesting. It may be contrasted with the value exceeding 900 km (ref. 7) on the phase-transition theory and substantially larger values on theories which assume a predominantly iron outer core. By yielding the lowest value of T, the Fe_2O theory accounts best for the failure to observe a significant magnetic field around Venus; the presumed seat of the Earth's magnetic field (the Earth's outer core) is 2,200 km thick.

The postulates also determine the ratio of the mass of the inner core to the mass M of any H planet as approximately 0.15. For a particular H planet of assigned mass, a first approximation to R is then formally yielded if values are available for the mean densities, ρ_1 and ρ_2 say, of the core and mantle. For Mars, M=0.642$\times10^{24}$ kg. For the H planet with this value of M, taking ρ_1, ρ_2=9.0, 3.7 g cm^{-3}, the simplified Fe_2O theory gives R=3,354 km; the yielded value of R is increased by 10 km if the trial value of ρ_1 is decreased by about 1.0 g cm^{-3}, or of ρ_2 by about 0.03 g cm^{-3}. Thus uncertainties in the appropriate values of ρ_1 and ρ_2 could possibly permit raising R to about 3,370 km, which is, however, less than the presently preferred observational value[3,15,16] of 3,388±5 km.

When hydrostatic conditions are assumed throughout Mars and ρ_1 and ρ_2 are treated as constant, the Fe_2O theory yields f =1/191 and δ=18 km, where f and δ are, respectively, the polar flattening and the excess of the equatorial over the polar radius. These results agree closely with the observed f and δ (see refs 3, 17). But allowance for density variation inside the mantle is likely to lower the calculated values to 1/200 (or even less) and 17 km.

与观测结果的吻合

我从一个简化的地球模型开始，假设它是子集 J 中的成员，其幔、外核和内核的平均密度分别是 $4.5 \ g \cdot cm^{-3}$、$11 \ g \cdot cm^{-3}$ 和 $13 \ g \cdot cm^{-3}$。我暂时忽略行星内部区域的密度变化，并将 X 和 Y 的平均密度比取为 0.7。作为一级近似，也忽略 J 型行星中相应区域压缩的差异。在 X 和 Y 混合的地幔中，行星内部不同地方可能的温度差异和由于化学相互作用导致的体积变化之类的因素，以及核中镍的存在都被忽略了，因为它们可能对主要数值细节的影响甚微。

有了这些简化，对于所有 J 类行星，这些假设从形式上确定了数值细节。特别是关于 M、R（行星的半径）、R_c（外核的半径）和 R_i（内核的半径）的一系列数值都被确定了。

金星的观测质量 [14] 为 $4.87 \times 10^{24} \ kg$。对于这个质量的 J 类行星，简化的 Fe_2O 理论得出 R=6,010 km、R_c=2,710 km 和 R_i=2,110 km。由于体积小的金星不如地球致密，模型中忽略了压缩率的不同，考虑压缩率的修正后将增加 R 的估算值。因此 Fe_2O 理论和金星观测半径（约 6,050 km）符合得和预计的一样好。

假设金星外核为液体，同样有趣的是，我们得到其外核的厚度值 T 为 600 km。这可能与相变理论得到的超过 900 km 的值（参考文献 7）以及假设主要由铁构成外核的理论得出的大很多的值不同。通过估算 T 的最小值，Fe_2O 理论能很好地解释为何没能观测到明显的围绕金星的磁场；而由地球磁场数据推测（地球外核）的厚度为 2,200 km。

这些假设也确定了任意 H 类行星内核质量与总质量 M 之比，大致为 0.15。对于一个特定质量的 H 类行星，如果核和幔的平均密度值，即 ρ_1 和 ρ_2 已知，就可以推出 R 的一级近似。对于火星，M=0.642×10^{24} kg。对于这样质量的 H 类行星，取 ρ_1=$9.0 \ g \cdot cm^{-3}$，ρ_2=$3.7 \ g \cdot cm^{-3}$，简化的 Fe_2O 理论给出 R=3,354 km，R 值会随着试验参数 ρ_1 降低 $1.0 \ g \cdot cm^{-3}$ 而增加 10 km，或者随着 ρ_2 降低 $0.03 \ g \cdot cm^{-3}$ 而增加 10 km。于是 ρ_1 和 ρ_2 合适值的不确定性，可能允许将 R 提高到 3,370 km。然而，这仍然小于现在的最佳观测值 [3,15,16] $3,388 \pm 5$ km。

当火星整体都采用了流体静力学条件的假设，并把 ρ_1 和 ρ_2 看作常数时，Fe_2O 理论得到的极向扁率和赤道半径超过极向半径的值分别为 f=1/191 和 δ=18 km。这些结果和观测到的 f 和 δ 值（见参考文献 3、17）很接近。但是考虑地幔内部的密度变化之后，计算值很可能分别下降至 1/200（或者更少）和 17 km。这些最新结果处

These last values are on the border[17] of the limits of error of the hydrostatic theory.

The phase-transition theory gives $R=3,390$ km and $f \approx 1/188$ (ref.18), assuming hydrostatic conditions. Thus the phase-transition theory gives the better fit with Mars unless the departures from hydrostatic conditions are considerable. This is principally because the Fe_2O theory requires a larger core, radius $R_i \sim 1,400$ km. It is likely that seismic records taken on Mars will in due course provide an observational estimate of R_i and so serve as a further test of the theories. (The gathering of seismic records on Venus is likely to be more difficult.)

Ironically, after the phase-transition theory had been shown to fit closely the then assumed observational estimate $R=3,390$ km for Mars, several investigators were disposed to reject that theory by insisting on observational values of 3,350 km or less[17]. A value of 3,350 km would fit the Fe_2O theory extremely well.

Assuming, as usual, that the mantle and iron core of Mars are solid (ref. 19) the Fe_2O theory, in common with the phase-transition theory, accounts satisfactorily for the failure to detect a magnetic field around Mars.

So the provisional calculations indicate that the Fe_2O theory fits the observational M and R for Venus extremely well, and M, R and f for Mars moderately well. The theory also gives the closest agreement of all theories with the observational evidence on magnetic fields.

Like all theories that are not extremely *ad hoc*, the Fe_2O theory does not fit the planet Mercury as Mercury now is, although the theory has an advantage over the phase-transition theory in requiring a substantially larger iron core. As I have previously suggested[20], Mercury, through its proximity to the Sun, has probably lost much of its primordial mantle by volatilization. The Fe_2O theory would suggest that the primordial Mercury (assumed to belong to the subset H) was larger than Mars.

The low density of the Moon precludes it from belonging to the family of planets considered here. The H planet with the Moon's mass would have an iron core of radius about 700 km, which is appreciably too large to fit evidence on the Moon's mean density and moment of inertia. Thus the Fe_2O theory here merely confirms that the Moon must have had an exceptional origin.

I hope to present later calculations which take detailed account of compression and density variation inside the planetary zones.

(**243**, 68-70; 1973)

K. E. Bullen
Department of Geophysics and Astronomy, University of British Columbia, Canada, and University of Sydney, NSW 2006

于流体静力学理论的误差限上 [17]。

在流体静力学条件的假设下，相变理论给出 $R=3,390$ km 和 $f \approx 1/188$（参考文献 18）。因此除非偏离流体静力学条件较大，否则相变理论与火星符合得更好。这主要是因为 Fe_2O 理论要求有一个较大的核，半径 R_i 约为 1,400 km。火星上的地震记录有可能将在适当的时候提供一个 R_i 的观测估计，成为对该理论的进一步检验。（搜集发生在金星上的地震记录可能更困难。）

具有讽刺意味的是，当相变理论被证明和当时的火星观测估计值 $R=3,390$ km 很接近时，一些学者打算拒绝那个理论，坚持认为观测值是 3,350 km 或者更小 [17]。3,350 km 这个数值和 Fe_2O 理论符合得极好。

Fe_2O 理论采用了和相变理论一样常见的假设，认为火星的幔和铁核都是固体（参考文献 19），这较好地解决了未能探测到火星周围磁场的问题。

所以暂时的计算结果表明 Fe_2O 理论给出的结果和金星的观测 M 和 R 符合得极好，与火星的 M、R 以及 f 拟合得较好。这个理论也是所有理论中和磁场的观测证据符合得最好的。

像所有非极端特殊的理论一样，Fe_2O 理论和现在的水星不符合，尽管这个理论相对相变理论而言，优势在于得到了更大的铁核。就像我以前提出的 [20]，水星由于接近太阳，可能已经通过蒸发损失了大量的原始地幔。Fe_2O 理论可以推论出原始水星（假定它属于 H 子集）比火星大。

月球的低密度将其排除在现在所讨论的这类行星家族之外。和月球同样质量的 H 类行星会有一个大约 700 km 半径的铁核，这大到不能与月球的平均密度和转动惯量相符合。因此 Fe_2O 理论只能确定月球肯定有一个特殊的起源。

我希望能够介绍详细考虑行星内部各区域的压缩率和密度变化细节的后续计算结果。

（冯翀 翻译；肖伟科 审稿）

Received January 29, 1973.

References:

1. Jeffreys, H., *Mon. Not. Roy. Astron. Soc., Geophys. Suppl.*, **4**, 62 (1937).

2. Bullen, K. E., *Rept. Austral. NZ Assoc. Adv. Sci.*, **23**, 25 (1937).

3. Cook, A. H., *Proc. Roy. Soc.*, A, **328**, 301 (1972).

4. Ramsey, W. H., *Mon. Not. Roy. Astron. Soc.*, **108**, 406 (1948).

5. Bullen, K. E., *Mon. Not. Roy. Astron. Soc.*, **109**, 457 (1949).

6. Dollfus, A., in *Surfaces and Interiors of Planets and Satellites,* 45 (Academic Press, London, 1970).

7. Bullen, K. E., *Mon. Not. Roy. Astron. Soc.*, **110**, 256 (1950).

8. Bullen, K. E., *Bull, Seismol. Soc. Amer.*, **32**, 19 (1942).

9. Bullen, K. E., *Mon. Not. Roy. Astron. Soc., Geophys. Suppl.*, **6**, 50 (1950).

10. Bullen, K. E., and Haddon, R. A., *Phys.Earth Planet. Int.*, **1**, 1 (1967).

11. Haddon, R. A., and Bullen, K. E., *Phys. Earth Planet. Int.*, **2**, 35 (1969).

12. Birch, F., *Geophys. J.*, **4**, 295 (1961).

13. Cook, A. H., *Space Sci. Rev.*, **2**, 355 (1963).

14. Anderson, J. D., *Jet Propulsion Lab. Tech. Rept.*, **32**, 816 (1967).

15. Bullen, K. E., *Nature,* **211**, 396 (1966).

16. Kliore, A., Cain, D. L., and Levy, G. S., in *Moon and Planets,* 226 (North-Holland, Amsterdam, 1967).

17. Bullen, K. E., *Mon. Not. Roy. Astron. Soc., Geophys. Suppl.*, **7**, 272 (1957).

18. Bullen, K. E., *Mon. Not. Roy. Astron. Soc.*, **109**, 689 (1949).

19. Bullen, K. E., *Mon. Not. Roy. Astron. Soc.* **133**, 229 (1966).

20. Bullen, K. E., *Nature,* **170**, 363 (1952).

Redshift of OQ 172

E. J. Wampler *et al.*

Editor's Notes

Quasars are among the most distant objects in the universe, which is demonstrated by the degree to which the light they emit is reddened. This paper reports the discovery of one of the more distant quasars. Quasars are now believed to be galaxies containing massive black holes which generate intense radiation by sucking in neighbouring matter. In the 1970s, however, it was still possible for people to argue that the redshift of quasars was a consequence of their great mass.

WE report the discovery of a second QSO with a redshift greater than 3. Spectra taken at the 120-inch at Lick Observatory[1] give $z=3.53$ for OQ172. Carswell and Strittmatter[2] have found the redshift of OH471 to be $z=3.40$. Our attention was drawn to the object by the work of Gent *et al.*[3] on Molonglo radio sources. Accurate radio coordinates for OQ172 were communicated to us ahead of publication by C. Hazard and H. Gent. They are:

$$\alpha \ (1950) \ 14 \ h \ 42 \ min \ 50.48 \ s$$
$$\delta \ (1950) \ 10° \ 11' \ 12.4''$$

A finding chart for OQ172 is given by Véron[4], who notes that it is a "blue stellar object". From our inspection of the Palomar Sky Survey plates we agree with Véron's assessment of the colour of OQ172, although Gent *et al.*[3] list it as stellar, that is an object with neutral colours. Radio fluxes for OQ172 are available from ref. 5, where the spectral index is given as −0.09 between 178 and 1,400 MHz. The tabulated fluxes at 1,400, 2,695 and 5,009 MHz are 2.42, 1.96 and 1.22 f.u., respectively[5]. The radio spectrum is therefore very flat and extends to high frequencies.

Figure 1 shows a spectrum of OQ172 obtained with the Cassegrain ITS system[6] at Lick Observatory. This spectrum is a composite formed by adding data from spectra obtained on different nights. Consequently the signal-to-noise ratio in the spectrum varies as a function of wavelength. It is particularly poor in the λ6000–λ6500 region where we have only one spectrum, to the red of λ7200 Å where OH emission features increase the radiation from the night sky, and shortward of 3500 Å where the atmospheric extinction becomes troublesome. In the spectral region between 4000 Å and 6000 Å the signal-to-noise ratio is good; most of the strong features present here are real and can be seen on the several individual spectra. The two strong emission features near λ5540 and λ7015 have been identified with Ly-α λ1216 and CIV λ1549. The measured wavelength of the emission peaks corresponds to a redshift of 3.56 for Ly-α and 3.53 for CIV. The very strong absorption in the blue wing of the feature identified as Ly-α has probably shifted

58

类星体OQ172的红移

编者按

类星体是宇宙中最遥远的天体之一，这一点可以用它们所发出光的红化大小来证明。本文报告了最遥远的类星体之一的发现，尽管现在人们认为类星体是含有大质量黑洞的星系，黑洞通过吞噬周围物质而产生强烈的辐射，不过在 20 世纪 70 年代，当时的人们仍可能会认为，类星体的红移是其巨大质量的结果。

我们报道的是第二颗红移超过 3 的类星体的发现。在利克天文台用 120 英寸望远镜得到的光谱[1]给出 OQ172 的红移为 $z = 3.53$。卡斯韦尔和斯特里特马特[2]发现 OH471 的红移为 $z = 3.40$。由于金特等人[3]对莫隆格勒射电源的研究，我们也开始关注这个天体。在哈泽德和金特的论文发表前，他们就将精确的射电坐标告诉了我们。它们是：

$$\alpha \ (1950) \ 14 \ h \ 42 \ min \ 50.48 \ s$$
$$\delta \ (1950) \ 10° \ 11' \ 12.4''$$

韦龙[4]提供了 OQ172 的证认图，他注意到那是一个"蓝色的恒星状天体"。尽管金特等人[3]将它列为恒星，但是根据我们在帕洛玛巡天底片中查阅的结果，我们赞同韦龙对 OQ172 颜色的评估，即它是一个具有中性颜色的天体。参考文献 5 中给出了 OQ172 的射电流量，并给出 178 MHz 和 1,400 MHz 之间的谱指数为 –0.09。1,400 MHz、2,695 MHz 和 5,009 MHz 处的流量分别为 2.42、1.96 和 1.22 流量单位[5]，因此射电频谱非常平坦并延伸到高频段。

图 1 显示了使用利克天文台的卡塞格林 ITS 系统获得的 OQ172 的光谱[6]。这条光谱是由不同夜晚得到的光谱数据叠加而成的，因此光谱中的信噪比是波长的函数。在 λ6,000~λ6,500 Å 光谱区，由于信噪比特别糟糕，我们只有一条光谱。在 λ7200 Å 的红端，OH 发射的辉光增加了夜空的辐射；而 3500 Å 处的短波端大气消光又成了棘手问题。在 4000 Å 到 6000 Å 之间的光谱区，信噪比很好；这里显示的大部分强谱线特征都是实际存在的，并且能在若干单独的光谱中看到。邻近 λ5540 Å 和 λ7015 Å 的两个强的发射特征已被确认为 Ly–α λ1216 和 CIV λ1549。对于 Ly–α，测量的发射线峰值波长对应 3.56 的红移，CIV 对应的红移为 3.53。被证认为 Ly–α 的蓝端线翼的极强吸收特征可能使观测波长移向真实峰位置的红端。为此，我们认为

the measured wavelength to the red of the true peak. For this reason we do not regard the slight difference in the measured redshifts of the lines as being significant. We found no other acceptable fit to the observed ratio of wavelengths that could not be rejected because of the absence of other strong lines in observed spectral regions. The presence of many strong absorption lines would also suggest a large redshift, say $z \gtrsim 2$. We therefore conclude that the emission line redshift is near 3.53. The position of several features assuming a redshift of 3.53 have been marked on Fig. 1.

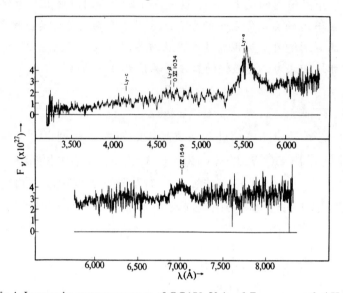

Fig. 1. Image tube scanner spectrum of OQ172. Units of $F\nu$ are erg cm^{-2} s^{-1} Hz^{-1}.

It is common for large redshift QSOs to have absorption features; OQ172 seems to have one of the richest absorption spectra known. The heavy absorption shortward of Ly-α increases the difficulty of identifying emission features in that spectral region. There is some indication of emission at the expected position of Ly-β and OVI but spectra with higher resolution will be needed to reduce the line blending and establish the continuum level before any such identification can be regarded as secure. In any case, there is no doubt that the continuum radiation remains strong far to the violet of the position of the Lyman edge in the emission line redshift system, a surprising result for an object with such strong absorption features. The strongest lines near the Ly-α emission peak have the correct separation to be identified with the CIV λ1549 doublet at $z=2.56$, raising the possibility that the absorption line systems have a much lower redshift than the emission line system. The Lyman continuum absorption associated with these features would then begin at the extreme violet end of our data where the signal-to-noise ratio is poor. It is also possible, in a gas with little turbulent motion, to have a substantially higher optical depth in Ly-α than in the Lyman continuum. One could therefore have strong, narrow, Ly-α absorption features without having appreciable absorption in the Lyman continuum. A programme to obtain higher resolution spectra of this object is now under way and should help resolve some of these questions.

观测到的谱线红移的微小差异并不重要。由于在观测的光谱范围内缺乏其他强线，我们没有找到其他不能被排除的、与观测到的波长比符合的拟合。很多强吸收谱线的存在也暗示着高红移，比如 $z \gtrsim 2$。我们由此得出结论，发射线红移大约为 3.53。图 1 中标注了假定红移为 3.53 时的一些特征谱线的位置。

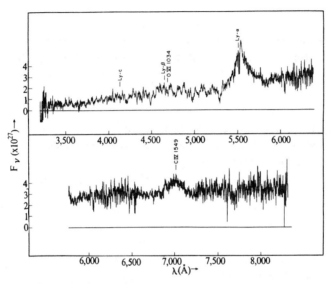

图 1. OQ172 的显像管扫描器（ITS）光谱。$F\nu$ 的单位为 $erg \cdot cm^{-2} \cdot s^{-1} \cdot Hz^{-1}$。

高红移的类星体通常具有吸收特征；OQ172 似乎是拥有已知最丰富吸收谱的类星体之一。Ly-α 短波端的强吸收增加了在该波段识别发射特征的难度。在 Ly-β 和 OVI 的预期位置存在某些发射迹象，但是需要更高分辨率的光谱以减少谱线的重叠并确定连续谱，这样的证认才是可靠的。但无论如何，在远离发射线红移系统的莱曼吸收边紫端的连续谱辐射无疑仍然是很强的，这对于一个有如此强吸收特征的天体来说是个惊人的结果。在 Ly-α 发射附近有合理间距的最强的吸收谱线被证认为 $z = 2.56$ 的 CIV λ1549 双线。这提高了一种可能性，即吸收线系的红移比发射线系低得多。伴随这些特征的莱曼连续谱吸收始于我们所得数据的极紫端，那里的信噪比很糟糕。在几乎不存在湍流运动的气体中，Ly-α 的光深可能比莱曼连续谱大得多。因此，在没有可观测的莱曼连续谱吸收的情况下，可以观测到强而窄的 Ly-α 吸收特征峰。获得该天体更高分辨率的光谱的工作正在进行，它将有助于解决这其中的一些问题。

OQ172 is very bright. It was estimated by Véron[4] to be magnitude 17.5. Our calibrated spectra, which should be accurate to ±20%, indicate that the V magnitude of the continuum is about 17.9. The strong continuum level, extending well into the ultraviolet, is in agreement with the blue classification given to OQ172 by Véron[4]. Although it is bright at short radio wavelengths it is too faint to appear in the 4C catalogue, otherwise it satisfies all the normal criteria for classification as a QSS. Other, similar objects presumably exist and could be discovered by the proper survey techniques without requiring highly accurate radio positions.

This research has been supported, in part, with grants from the National Science Foundation, the National Aeronautics and Space Administration and NATO.

(**243**; 336-337; 1973)

E. J. Wampler*, L. B. Robinson*, J. A. Baldwin* and E. M. Burbidge†
*Lick Observatory, Board of Studies in Astronomy and Astrophysics,University of California, Santa Cruz
†Royal Greenwich Observatory, Herstmonceux Castle

Received May 14, 1973.

References:

1. Fitch, L. T., Dixon, R. S., and Kraus, J. D., *Astron. J.*, 74, 612 (1969).

2. Carswell, R. F., and Strittmatter, P. A., *Nature*, 242, 394(1973).

3. Gent, H., Crowther, J. H., Adgie, R. L., Hoskins, D. G., Murdoch, H. S., Hazard, C., and Jauncey, D. L., *Nature*, 241, 261 (1973).

4. Véron, M. P., *Astron. Astrophys.*, 11, 1 (1971).

5. Witzel, A., Véron, P., and Véron, M. P., *Astron. Astrophys.*, 11, 171 (1971).

6. Robinson, L. B., and Wampler, E. J., *Publ. Astron. Soc. Pacific*, 84, 161 (1972).

OQ172 非常明亮。按照韦龙[4] 的估计，其亮度为 17.5 等。我们定标过的光谱可以精确到 ±20%，它显示连续谱的 V 星等大约为 17.9 等。充分延伸到紫外区域的强连续谱与韦龙将 OQ172 归类为蓝色[4] 一致。尽管它在短波长的射电波段很亮，但它在其他波长看还是太暗了，所以没有出现在 4C 星表中，否则它会符合归类为类星射电源的全部常规判据。在不要求高精度射电位置的情况下，利用恰当的巡天技术，可能发现还存在其他类似的天体。

此项研究部分得到了国家科学基金会、美国国家航空航天局、北大西洋公约组织的资助。

（王耀杨 翻译；吴学兵 审稿）

Black Hole Explosions?

S. W. Hawking

Editor's Note

It was realized more than two hundred years ago that there is a critical mass and radius beyond which light cannot escape the gravitational field of an object—such an object becomes a "black hole". This idea was rigorously validated by the theory of general relativity. Here Stephen Hawking shows that black holes have effective temperatures that are inversely related to their mass, and should therefore radiate photons and neutrinos from their event horizons—they are not fully "black". As this radiation proceeds, the black hole loses mass. Finally it emits large quantities of X-rays and gamma-rays, and disappears in an explosion. "Hawking radiation" from black holes is now widely expected, but has not yet been seen.

QUANTUM gravitational effects are usually ignored in calculations of the formation and evolution of black holes. The justification for this is that the radius of curvature of space-time outside the event horizon is very large compared to the Planck length $(G\hbar/c^3)^{1/2} \approx 10^{-33}$ cm, the length scale on which quantum fluctuations of the metric are expected to be of order unity. This means that the energy density of particles created by the gravitational field is small compared to the space-time curvature. Even though quantum effects may be small locally, they may still, however, add up to produce a significant effect over the lifetime of the Universe $\approx 10^{17}$ s which is very long compared to the Planck time $\approx 10^{-43}$ s. The purpose of this letter is to show that this indeed may be the case: it seems that any black hole will create and emit particles such as neutrinos or photons at just the rate that one would expect if the black hole was a body with a temperature of $(\kappa/2\pi)(\hbar/2k) \approx 10^{-6} (M_\odot/M)K$ where κ is the surface gravity of the black hole[1]. As a black hole emits this thermal radiation one would expect it to lose mass. This in turn would increase the surface gravity and so increase the rate of emission. The black hole would therefore have a finite life of the order of $10^{71} (M_\odot/M)^{-3}$ s. For a black hole of solar mass this is much longer than the age of the Universe. There might, however, be much smaller black holes which were formed by fluctuations in the early Universe[2]. Any such black hole of mass less than 10^{15} g would have evaporated by now. Near the end of its life the rate of emission would be very high and about 10^{30} erg would be released in the last 0.1 s. This is a fairly small explosion by astronomical standards but it is equivalent to about 1 million 1 Mton hydrogen bombs.

To see how this thermal emission arises, consider (for simplicity) a massless Hermitean scalar field ϕ which obeys the covariant wave equation $\phi_{;ab}g^{ab} = 0$ in an asymptotically flat space time containing a star which collapses to produce a black hole. The Heisenberg operator ϕ can be expressed as

64

黑洞爆炸？

霍金

编者按

人们在两百多年前就已经意识到，对于给定的半径有一个临界质量，当物体的质量超出临界质量时，其引力场就强到甚至连光都不能逃脱——这样的物体变成了"黑洞"。这一想法在广义相对论中得到了严格验证。本文中，斯蒂芬·霍金向我们表明黑洞有跟其质量成反比的等效温度，因此它必须从其事件视界向外辐射光子和中微子——即它们并不完全是"黑"的。黑洞在辐射过程中会损失质量，最终释放出大量的X射线和伽马射线，并在一次爆炸之后消失。虽然目前还没有直接观测到，但人们普遍相信黑洞存在"霍金辐射"。

在计算黑洞的形成和演化时，一般可忽略量子引力效应。这一点的合理性在于，在事件视界外的时空曲率半径远大于普朗克长度 $(G\hbar/c^3)^{1/2} \approx 10^{-33}$ cm，而在此尺度上预期度规的量子涨落是 1 的量级。这意味着由引力场产生的粒子的能量密度和时空曲率相比要小。虽然量子效应在局部很小，然而它们仍然可能在宇宙的寿命 $\approx 10^{17}$s 内积累产生重大的影响，这个时间远长于普朗克时间 $\approx 10^{-43}$ s。这篇快报的目的是说明，似乎任何黑洞都将以预期的速率产生和发射粒子，如中微子或光子，正如同黑洞是一个温度为 $(\kappa/2\pi)(\hbar/2k) \approx 10^{-6} (M_\odot/M)$ K 的物体所表现的那样，其中 κ 是黑洞的表面引力 [1]。当黑洞发射这类热辐射时，我们预期它将损失质量。这本身将增大它的表面引力，因而增大其发射速率。从此，黑洞将具有 $10^{71}(M_\odot/M)^{-3}$ s 量级的有限寿命。对于太阳质量的黑洞，这将比宇宙年龄更长。然而，可能存在许多较小的黑洞，它们是由早期宇宙中的涨落形成的 [2]。任何这类质量小于 10^{15} g 的黑洞到现在都应该蒸发殆尽了。在接近它生命终了时，其粒子发射速率将非常高，在最后 0.1 s 将释放约 10^{30} erg 能量。以天文学的标准来看，这是一个相当小的爆炸，但它相当于大约一百万个 100 万吨量级的氢弹爆炸。

为了解释黑洞热辐射是如何产生的，为简单起见，在一个包含由一颗恒星塌缩形成的一个黑洞的渐近平直时空中，考虑一个无质量的厄米标量场 ϕ，且它遵守协变波动方程 $\phi_{;ab}g^{ab} = 0$。海森堡算符 ϕ 可表示为：

$$\phi = \sum_i \{f_i a_i + \overline{f}_i a_i^+\}$$

where the f_i are a complete orthonormal family of complex valued solutions of the wave equation $f_{i;ab}g^{ab} = 0$ which are asymptotically ingoing and positive frequency—they contain only positive frequencies on past null infinity I^-[3,4,5]. The position-independent operators a_i and a_i^+ are interpreted as annihilation and creation operators respectively for incoming scalar particles. Thus the initial vacuum state, the state containing no incoming scalar particles, is defined by $a_i |0_-\rangle = 0$ for all i. The operator ϕ can also be expressed in terms of solutions which represent outgoing waves and waves crossing the event horizon:

$$\phi = \sum_i \{p_i b_i + \overline{p}_i b_i^+ + q_i c_i + \overline{q}_i c_i^+\}$$

where the p_i are solutions of the wave equation which are zero on the event horizon and are asymptotically outgoing, positive frequency waves (positive frequency on future null infinity I^+) and the q_i are solutions which contain no outgoing component (they are zero on I^+). For the present purposes it is not necessary that the q_i are positive frequency on the horizon even if that could be defined. Because fields of zero rest mass are completely determined by their values on I^-, the p_i and the q_i can be expressed as linear combinations of the f_i and the $\overline{f}_i$:

$$p_i = \sum_j \{\alpha_{ij} f_j + \beta_{ij} \overline{f}_j\} \text{ and so on}$$

The β_{ij} will not be zero because the time dependence of the metric during the collapse will cause a certain amount of mixing of positive and negative frequencies. Equating the two expressions for ϕ, one finds that the b_i, which are the annihilation operators for outgoing scalar particles, can be expressed as a linear combination of the ingoing annihilation and creation operators a_i and a_i^+

$$b_i = \sum_j \{\overline{\alpha}_{ij} a_j - \overline{\beta}_{ij} a_j^+\}$$

Thus when there are no incoming particles the expectation value of the number operator $b_i^+ b_i$ of the ith outgoing state is

$$\langle 0_- | b_i^+ b_i | 0_- \rangle = \sum_j |\beta_{ij}|^2$$

The number of particles created and emitted to infinity in a gravitational collapse can therefore be determined by calculating the coefficients β_{ij}. Consider a simple example in which the collapse is spherically symmetric. The angular dependence of the solution of the wave equation can then be expressed in terms of the spherical harmonics Y_{lm} and the dependence on retarded or advanced time u, v can be taken to have the form $\omega^{-1/2} \exp(i\omega u)$ (here the continuum normalisation is used). Outgoing solutions $p_{lm\omega}$ will now be expressed as an integral over incoming fields with the same l and m:

$$p_\omega = \int \{\alpha_{\omega\omega'} f_{\omega'} + \beta_{\omega\omega'} \overline{f}_{\omega'}\} d\omega'$$

$$\phi = \sum_i \{f_i a_i + \overline{f}_i a_i^+\}$$

其中，f_i 是波动方程 $f_{i;ab}g^{ab}=0$ 的一族渐近向内、频率为正且完备正交归一复数解，它们在过去类光无穷远 $I^{-[3,4,5]}$ 只含有正频率。对于入射的标量粒子，位置无关的算符 a_i 和 a_i^+ 分别解释为湮灭和产生算符。因此对于所有的 i，初始真空态，即不含有向内传播的标量粒子的态，可定义为 $a_i |0_-\rangle = 0$。算符 ϕ 也可以用代表向外的波和穿过事件视界的波的解表示：

$$\phi = \sum_i \{p_i b_i + \overline{p}_i b_i^+ + q_i c_i + \overline{q}_i c_i^+\}$$

其中，p_i 是波动方程的解，它们在视界上为零且是渐近向外的正频波（在未来类光无穷远 I^+ 为正频率），而 q_i 是不含向外成分的解（它们在 I^+ 为零）。就现在的目的而言，即使可以被定义，q_i 在视界处也不一定是正频的。因为零静止质量的场完全被它们在 I^- 的值确定，p_i 和 q_i 可以表示为 f_i 和 $\overline{f}_i$ 的线性组合：

$$p_i = \sum_j \{\alpha_{ij}f_j + \beta_{ij}\overline{f}_j\} \ 等$$

因为在塌缩期间度规的时间依赖性将导致一定量的正频和负频的混合，所以 β_{ij} 将不为零。令 ϕ 的两个表达式相等，可以发现向外传播标量粒子的湮灭算符 b_i 可以表示为向内湮灭和产生算符 a_i 和 a_i^+ 的线性叠加，即：

$$b_i = \sum_j \{\overline{\alpha}_{ij}a_j - \overline{\beta}_{ij}a_j^+\}$$

于是在没有向内态的粒子时，第 i 个向外态的粒子数算符 $b_i^+b_i$ 的期望值为：

$$\langle 0_- |b_i^+ b_i| 0_-\rangle = \sum_j |\beta_{ij}|^2$$

因此，在一次引力塌缩中产生并发射到无穷远的粒子的数目可以通过计算系数 β_{ij} 确定。考虑一个简单的例子，塌缩是球对称的。波动方程的解对角度的依赖可以用球谐函数 Y_{lm} 表示，对推迟或超前时间 u、v 的依赖可以取为 $\omega^{-1/2} \exp(i\omega u)$（这里使用了连续归一化）。向外的解 $p_{lm\omega}$ 现在可以表示为对相同 l 和 m 的向内的场的积分：

$$p_\omega = \int \{\alpha_{\omega\omega'}f_{\omega'} + \beta_{\omega\omega'}\overline{f}_{\omega'}\}d\omega'$$

(The *lm* suffixes have been dropped.) To calculate $\alpha_{\omega\omega'}$ and $\beta_{\omega\omega'}$ consider a wave which has a positive frequency ω on I^+ propagating backwards through spacetime with nothing crossing the event horizon. Part of this wave will be scattered by the curvature of the static Schwarzschild solution outside the black hole and will end up on I^- with the same frequency ω. This will give a $\delta(\omega-\omega')$ behaviour in $\alpha_{\omega\omega'}$. Another part of the wave will propagate backwards into the star, through the origin and out again onto I^-. These waves will have a very large blue shift and will reach I^- with asymptotic form

$$C\omega^{-1/2} \exp \{-i\omega\kappa^{-1} \log (v_0-v) + i\omega v\} \ for \ v < v_0$$

and zero for $v \geq v_0$, where v_0 is the last advanced time at which a particle can leave I^-, pass through the origin and escape to I^+. Taking Fourier transforms, one finds that for large ω', $\alpha_{\omega\omega'}$ and $\beta_{\omega\omega'}$ have the form:

$$\alpha_{\omega\omega'} \approx C \exp [i(\omega-\omega')v_0](\omega'/\omega)^{1/2} \cdot \Gamma(1-i\omega/\kappa) [-i(\omega-\omega')]^{-1+i\omega/\kappa}$$

$$\beta_{\omega\omega'} \approx C \exp [i(\omega+\omega')v_0](\omega'/\omega)^{1/2} \cdot \Gamma(1-i\omega/\kappa) [-i(\omega+\omega')]^{-1+i\omega/\kappa}$$

The total number of outgoing particles created in the frequency range $\omega \rightarrow \omega+d\omega$ is $d\omega \int_0^\infty |\beta_{\omega\omega'}|^2 d\omega'$. From the above expression it can be seen that this is infinite. By considering outgoing wave packets which are peaked at a frequency ω and at late retarded times one can see that this infinite number of particles corresponds to a steady rate of emission at late retarded times. One can estimate this rate in the following way. The part of the wave from I^+ which enters the star at late retarded times is almost the same as the part that would have crossed the past event horizon of the Schwarzschild solution had it existed. The probability flux in a wave packet peaked at ω is roughly proportional to $\int_{\omega_1}^{\omega_2'}\{|\alpha_{\omega\omega'}|^2 - |\beta_{\omega\omega'}|^2\} \, d\omega$ where $\omega_2' \gg \omega_1' \gg 0$. In the expressions given above for $\alpha_{\omega\omega'}$ and $\beta_{\omega\omega'}$ there is a logarithmic singularity in the factors $[-i(\omega-\omega')]^{-1+i\omega/\kappa}$ and $[-i(\omega+\omega')]^{-1+i\omega/\kappa}$. Value of the expressions on different sheets differ by factors of $\exp(2\pi n\omega\kappa^{-1})$. To obtain the correct ratio of $\alpha_{\omega\omega'}$ to $\beta_{\omega\omega'}$ one has to continue $[-i(\omega+\omega')]^{-1+i\omega/\kappa}$ in the upper half ω' plane round the singularity and then replace ω' by $-\omega'$. This means that, for large ω',

$$|\alpha_{\omega\omega'}| = \exp (\pi\omega/\kappa)|\beta_{\omega\omega'}|$$

From this it follows that the number of particles emitted in this wave packet mode is $(\exp(2\pi\omega/\kappa)-1)^{-1}$ times the number of particles that would have been absorbed from a similar wave packet incident on the black hole from I^-. But this is just the relation between absorption and emission cross sections that one would expect from a body with a temperature in geometric units of $\kappa/2\pi$. Similar results hold for massless fields of any integer spin. For half integer spin one again gets a similar result except that the emission cross section is $(\exp(2\pi\omega/\kappa)+1)^{-1}$ times the absorption cross section as one would expect for thermal emission of fermions. These results do not seem to depend on the assumption of exact spherical symmetry which merely simplifies the calculation.

这里略去了 lm 下标。为计算 $\alpha_{\omega\omega'}$ 和 $\beta_{\omega\omega'}$，考虑一个在 I^+ 为正频 ω，在时空中反向传播的不穿过视界的波。这个波的一部分将被黑洞外的静态史瓦西解的曲率散射并将在 I^- 上以相同的频率 ω 终止。这将导致 $\alpha_{\omega\omega'}$ 的 $\delta(\omega-\omega')$ 行为。这个波的另外一部分将向后传播到恒星中，通过原点然后向外再到 I^-。这些波将有非常大的蓝移，并将以渐近形式接近 I^-：

$$C\omega^{-1/2}\exp\{-i\omega\kappa^{-1}\log(v_0-v)+i\omega v\},\ \text{当}\ v<v_0\ \text{时}$$

当 $v\geqslant v_0$ 时，这个蓝移值为 0。其中，v_0 是最后的超前时间，此时的粒子尚可脱离 I^-，通过原点并逃向 I^+。作傅里叶变换可以发现对于大的 ω'，$\alpha_{\omega\omega'}$ 和 $\beta_{\omega\omega'}$ 有如下形式：

$$\alpha_{\omega\omega'}\approx C\exp[i(\omega-\omega')v_0](\omega'/\omega)^{1/2}\cdot\Gamma(1-i\omega/\kappa)[-i(\omega-\omega')]^{-1+i\omega/\kappa}$$

$$\beta_{\omega\omega'}\approx C\exp[i(\omega+\omega')v_0](\omega'/\omega)^{1/2}\cdot\Gamma(1-i\omega/\kappa)[-i(\omega+\omega')]^{-1+i\omega/\kappa}$$

在频率范围 $\omega\to\omega+d\omega$ 产生的向外粒子的总数为 $d\omega\int_0^\infty|\beta_{\omega\omega'}|^2d\omega'$，并由上面的表达式可以看出这个量是无限的。通过考虑在较晚的推迟时间的、峰值频率 ω 的向外波包，可以看到这个无限的粒子数对应于较晚推迟时间的一个稳态发射率。可以通过以下方法估计这个发射率。波中来自 I^+ 的在较晚推迟时间进入恒星的部分（如果这部分存在的话）和穿过史瓦西解的过去视界的部分几乎相同，峰值在 ω 处的波包中的概率流大致正比于 $\int_{\omega_1'}^{\omega_2'}\{|\alpha_{\omega\omega'}|^2-|\beta_{\omega\omega'}|^2\}d\omega$，其中，$\omega_2'\gg\omega_1'\gg0$。在上面给出的 $\alpha_{\omega\omega'}$ 和 $\beta_{\omega\omega'}$ 的表达式中，$[-i(\omega-\omega')]^{-1+i\omega/\kappa}$ 和 $[-i(\omega+\omega')]^{-1+i\omega/\kappa}$ 因子中有一个对数奇点。这个表达式在不同面上的值相差一个 $\exp(2\pi n\omega\kappa^{-1})$ 因子。为得到正确的 $\alpha_{\omega\omega'}$ 和 $\beta_{\omega\omega'}$ 的比值，必须在上半 ω' 平面围绕奇点对 $[-i(\omega+\omega')]^{-1+i\omega/\kappa}$ 进行延拓并将 ω' 换为 $-\omega'$。这意味着对于大的 ω'：

$$|\alpha_{\omega\omega'}|=\exp(\pi\omega/\kappa)|\beta_{\omega\omega'}|$$

由此可得，这个波包模式中发射的粒子数是从 I^- 入射到黑洞上的类似波包中已被吸收的粒子数的 $[\exp(2\pi\omega/\kappa)-1]^{-1}$ 倍，但这正是根据几何单位下温度为 $\kappa/2\pi$ 的物体所预期的吸收和发射截面之间的关系。类似结果对任何整数自旋的无质量场也同样成立。对半整数自旋，正如我们对费米子的热辐射所预期的那样，也能得到类似结果，只是发射截面是吸收截面的 $[\exp(2\pi\omega/\kappa)+1]^{-1}$ 倍，这些结果似乎并不依赖于只是为了简化计算而采取的精确球对称性假设。

Beckenstein[6] suggested on thermodynamic grounds that some multiple of κ should be regarded as the temperature of a black hole. He did not, however, suggest that a black hole could emit particles as well as absorb them. For this reason Bardeen, Carter and I considered that the thermodynamical similarity between κ and temperature was only an analogy. The present result seems to indicate, however, that there may be more to it than this. Of course this calculation ignores the back reaction of the particles on the metric, and quantum fluctuations on the metric. These might alter the picture.

Further details of this work will be published elsewhere. The author is very grateful to G. W. Gibbons for discussions and help.

(**248**, 30-31; 1974)

S. W. Hawking
Department of Applied Mathematics and Theoretical Physics and Institute of Astronomy, University of Cambridge

Received January 17, 1974.

References:

1. Bardeen, J. M., Carter, B., and Hawking, S. W., *Commun. math. Phys.*, **31**, 161–170 (1973).

2. Hawking, S. W., *Mon. Not. R. astr. Soc.*, **152**, 75-78 (1971).

3. Penrose, R., in *Relativity, Groups and Topology* (edit. by de Witt, C. M., and de Witt, B. S). Les Houches Summer School, 1963 (Gordon and Breach, New York, 1964).

4. Hawking, S. W., and Ellis, G. F. R., *The Large-Scale Structure of Space-Time* (Cambridge University Press, London 1973).

5. Hawking, S. W., in *Black Holes* (edit. by de Witt, C. M., and de Witt, B. S), Les Houches Summer School, 1972 (Gordon and Breach, New York, 1973).

6. Beckenstein, *J. D.*, *Phys. Rev.*, D7, 2333–2346 (1973).

　　贝肯斯坦[6]在热力学基础上提出，κ 的某个倍数应该被看作黑洞的温度。然而，他并未提出，黑洞可以像吸收粒子一样发射粒子。因此，巴丁、卡特和我曾经认为，κ 和温度之间的热力学的相似性只是一种类比。然而，目前的结果似乎表明，可能存在比这更多的内容。当然，这个计算忽略了粒子对度规的反作用以及度规本身的量子涨落，这些不排除会改变这个物理图像。

　　这项工作的进一步细节将在其他地方发表。作者非常感谢吉本斯给予的建议和帮助。

（沈乃澂 翻译；肖伟科 审稿）

Radio Sources with Superluminal Velocities

M. H. Cohen *et al.*

Editor's Note

As galaxies and quasars were observed using interferometric methods in the 1970s, it was found that the blobs of material that seemed to be travelling out from their centres were doing so at speeds apparently greater than the speed of light, at face value violating special relativity. This was used to argue by some astronomers that quasars actually were relatively near objects that had been ejected at high speeds from nearby galaxies. Here Michael Cohen and colleagues conclude that the evidence instead favours the genuinely "superluminal" picture. Astronomer Martin Rees had predicted that geometrical effects can account for this motion without violating relativity. However, some of these jets can't be explained this way, but require a more subtle model.

Radio data from four extragalactic sources, three quasars and one galaxy, show evidence for an apparent expansion faster than the speed of light. The data on these "superluminal" sources are reviewed, and their implications briefly discussed.

BRIGHTNESS distributions in four extragalactic radio sources have been seen to vary so rapidly that the apparent transverse velocity of expansion is greater than the velocity of light (assuming a cosmological origin for the redshift). The term superluminal will be used to describe this phenomenon. In this paper we review many of the observations of superluminal expansions, and also add some new material. Blandford, McKee, and Rees[1] have summarised theoretical ideas on the subject.

The observations have all been obtained with very-long-baseline interferometry (VLBI) systems using two to five radio telescopes, spaced over thousands of kilometres, to form multi-element interferometres. Various model-fitting and map-making procedures have been used to estimate parameters of the brightness distribution, such as diameter, separation and flux of components, position angle, and so on. In some cases the data have been sufficiently accurate and extensive to warrant many-parameter models, and detailed contour diagrams have been made. In general, however, the models are not unique. Some have been challenged[2,3]; but there still is general agreement among different workers on the main features of the sources. VLBI map-making is presently in a stage of rapid development; in particular there is a growing use of phase data, and more definitive results are expected in a few years.

超光速射电源

科恩等

编者按

20 世纪 70 年代，人们利用干涉测量法来观测星系和类星体时，发现从这些星系和类星体中心发出的物质团的运动速度似乎明显高于光速——从表面上看其数值违背了狭义相对论。一些天文学家据此认为类星体实际上更接近于从邻近星系高速喷射出来的物质。本文中迈克尔·科恩及其合作者们的结论是，这些证据反而支持了真正的"超光速"图像。天文学家马丁·瑞斯已经预言，在不违背相对论的情况下，用几何效应就能够解释这种运动。但是，一些这样的喷流不能通过这种方式解释清楚，而需要一种更加巧妙的模型。

来自四个河外源（三个类星体和一个星系）的射电数据表明，它们的视膨胀速度超过光速。本文对这些"超光速"源的数据进行了探讨，并简要讨论了它们的内在意义。

我们已经观察到四个河外射电源的亮度分布变化非常快，以至于膨胀的视横向速度超过了光速（假设是宇宙学红移）。以下用"超光速"这一术语来表征该现象。本文，我们探讨了很多例超光速膨胀的观测，同时也增添了一些新的观测资料。布兰福德、麦基和瑞斯 [1] 已经就这个问题总结了一些理论观点。

这些观测数据都是通过甚长基线干涉仪（VLBI）系统得到的。甚长基线干涉仪系统利用二至五个分布范围达几千公里的射电望远镜构成一个多天线干涉仪。我们使用多种模型拟合和成图的步骤来估算亮度分布的各种参数，如直径、位置角、成分源的角分离和流量等。对于一些源，我们已经掌握足够精确广泛的数据，可以去限制多参数模型，而且我们已经得到详细的等光度线图。不过大体上说，这些模型不是唯一的。一些模型已经遭到质疑 [2,3]，但是对于这些源的主要性质，不同研究者的意见总体上还是一致的。VLBI 成图目前正处于一个快速发展时期，特别是越来越多地使用相位数据，在几年内有望得到更加可靠的结果。

73

Four sources including three quasars (3C345, 3C273, and 3C279) and one radio galaxy (3C120) will be discussed. In each case we give simple angular measures and do not discuss the detailed models of brightness distribution. More complete discussions will be found in the references.

3C345 ($z = 0.595$)

Two observing groups[4,5] have independently shown that 3C345 could be accurately described as a simple double source in mid-1974. The two components were approximately Gaussian and contained most or all of the flux density at $\lambda = 2$, 2.8, and 3.8 cm. In these experiments three or four telescopes were used. The simple model was an excellent fit, and there was no requirement for, nor evidence of, a more complex structure. At most other epochs only two or three telescopes were used. The brightness distributions were not as well determined, of course, but a simple double fit the data and the separation, θ, was accurately determined in every case.

Figure 1 shows the separation of the centres of the two components as a function of time. The point at 6 cm is taken from a number of measurements made in 1968 and 1969 (ref. 8). The source was only slightly resolved at that early epoch and θ was determined from the best-fitting double. The position angle was assumed to be 105°, the value found more recently. The 3.8 cm point at 1971.1 (ref. 7) is similarly model dependent because the source is only partially resolved. The three points in 1976 are provisional because detailed model fitting is yet to be done; however, they are unlikely to change much because θ is well determined from the sharp minima of the visibility function.

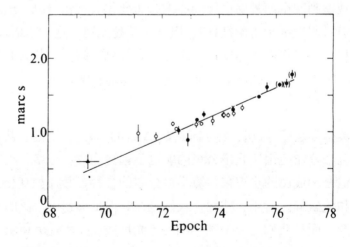

Fig. 1. Apparent separation of the two components of 3C345 as a function of time. The slope of the line is 0.17 marc s yr[-1]. The observations at 2 cm are taken from ref. 5; the 2.8-cm points are from refs 5 and 6, and the four points in 1972–73 are from private communications with N. Broten and M. Quigley (the points in 1976 are discussed in the text); the 3.8 cm points are from refs 4 and 7; and the 6-cm point is discussed in the text. ×, 2.0 cm; ●, 2.8 cm; ○, 3.8 cm; ▲, 6.0 cm.

我们将对四个源进行讨论，其中包括三个类星体（3C345、3C273 和 3C279）和一个射电星系（3C120）。对每个源我们只给出角度测量值，并不讨论亮度分布的详细模型。更完整的讨论见参考文献。

3C345（$z=0.595$）

两个观测小组 [4,5] 根据 1974 年中期的观测分别独立地指出 3C345 可以被准确地描述成简单的双源。两个子源近似高斯型，并包含 $\lambda=2$ cm、2.8 cm 和 3.8 cm 波段的绝大部分或是全部流量密度。这些观测使用了三至四台望远镜。简单模型就拟合得相当好，没有证据，也没必要去构造更复杂的模型。在其余观测时段只使用了二至三台望远镜。当然，亮度分布并没有被准确测定，但是用简单的双峰模型去拟合数据的话，每次观测都能得到准确的角分离 θ。

两个子源中心的角分离随时间变化如图 1 所示。6 cm 波段的数据点来自 1968 年和 1969 年的多次观测（参考文献 8）。在那样一个较早的时期，3C345 只能被略微分辨开，而角分离 θ 是通过最佳拟合的双峰模型得到的，其位置角被认定是 105°，这是最近才得到的。类似地，在 1971.1（参考文献 7）的 3.8 cm 波段的数据点也是依赖于模型得到的，因为此源只能部分地分辨开。1976 年的 3 个数据点是暂定的，因为具体的模型拟合还没有做。不过，它们不大可能将参数改变太多，因为根据可见度函数的强极小值可以很好地确定 θ。

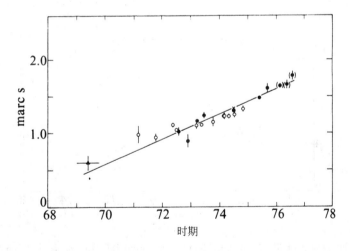

图 1. 3C345 的两个子源视角分离随着时间的变化。拟合直线的斜率是 0.17 marc s·yr⁻¹。2 cm 波段的观测来自参考文献 5；2.8 cm 波段的数据点来自参考文献 5 和 6，1972 年至 1973 年的四个数据点来自和布拉滕和奎格利的私人交流（文中讨论了 1976 年的数据点）；3.8 cm 波段的数据点来自参考文献 4 和 7；6 cm 波段的数据点在文中进行了讨论。×，2.0 cm；●，2.8 cm；○，3.8 cm；▲，6.0 cm。

The points in Fig. 1 show that the brightness distribution expanded by a factor of about 3 in 7 years, at the approximate rate 0.17 marc s per year. The expansion may not be uniform, however, for the slope seems to be flatter in 1972–73, and steeper in 1975–76. The difference in slope explains most of the discrepancy in the rates already reported in the literature[6,9]. The angular rate can be converted into an apparent transverse velocity at the source by multiplying by the cosmological distance and correcting for time dilation with a factor $(1+z)$.

$$\frac{v}{c} = \frac{\theta}{H_0 q_0^2 (1+z)} \left(q_0 z + (q_0-1)[(1+2q_0 z)^{\frac{1}{2}}-1] \right)$$

The line in Fig.1 corresponds to $v/c \approx 7$. ($H_0 = 55$ km s^{-1} Mpc^{-1} and $q_0 = 0.05$.)

The augular separation of the components of the double seems to be independent of wavelength over at least an octave. However, the detailed visibility curves show that the ratio of flux densities varies with wavelength; that is, the components have different spectra.

The position angle (PA) of the double was determined for all cases shown in Fig. 1 for which there were adequate data. There are no substantial differences with wavelength or time; the PA has been constant at $105° \pm 3°$.

Extrapolation of the rate shown in Fig. 1 to zero separation indicates that an "event" occurred in 1966. Figure 2 shows the total flux density of 3C345 at 3.8 cm. A major increase in flux density began in 1966; the flux doubled in two years and has since stayed high. The near coincidence of the times suggests that the outburst and the expansion have a common origin.

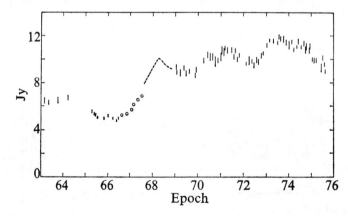

Fig. 2. Total flux density of 3C345 at $\lambda = 3.8$ cm. Points from 1963 to 1966.5 are taken from ref. 10; from 1966.6 to 1967.5, ref. 11; 1969.0 to 1971.6, ref 12; 1971.6 to 1974.4, ref 13; 1974.4 to 1975.6 from private communications with W. A. Dent, and the dashed line is interpolated between observations at 2.8 and 4.5 cm reported in ref. 14.

3C273 (z=0.158)

3C273 has a complex elongated structure (refs 3, 15 and 16 and W. D. Cotton *et al.*, in

图 1 中的数据点显示 7 年来亮度分布大约膨胀到了原来的 3 倍，每年近似膨胀 0.17 marc s。不过，膨胀可能不是匀速的，因为从斜率来看似乎 1972 年至 1973 年要平缓一些，而 1975 年至 1976 年要陡一些。斜率的不同可以解释文献 [6,9] 中报道过的大部分速率的差异。通过乘以宇宙学距离以及用因子（1+z）作时间膨胀的修正，可以将角速率转化为源的表观横向速度。

$$\frac{v}{c} = \frac{\theta}{H_0 q_0^2 (1+z)} \{q_0 z + (q_0-1)[(1+2q_0 z)^{\frac{1}{2}}-1]\}$$

图 1 中的直线对应于 $v/c \approx 7$。（$H_0 = 55$ km·s^{-1}·Mpc^{-1}，$q_0 = 0.05$。）

在至少一个倍频程内，两个子源的角分离似乎都和波长无关。不过，详细的可见度曲线显示子源的流量密度比随波长变化；也就是说，两个子源的谱形是不一样的。

我们利用图 1 中所有的观测结果得出双源的位置角（PA），这些观测数据是足够多的。位置角随着波长或时间没有明显的变化，一直保持在 105°±3°。

将图 1 中的速率外推到角分离为零的状况，表明"分裂事件"发生在 1966 年。图 2 显示了 3C345 在 3.8 cm 波段上的总流量密度。从 1966 年开始，流量密度有了大幅度的增加；流量在两年内成倍增长，此后一直很高。这些时间上的近乎重合表明爆发和膨胀的起源相同。

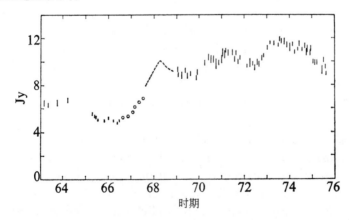

图 2. 3C345 在 λ=3.8 cm 波段的总流量密度。1963 到 1966.5 的数据点来自参考文献 10；1966.6 到 1967.5 的来自参考文献 11；1969.0 到 1971.6 的来自参考文献 12；1971.6 到 1974.4 的来自参考文献 13；1974.4 到 1975.6 的来自和登特的私人交流。虚线是根据参考文献 14 中的 2.8 cm 和 4.5 cm 波段观测得到的插值。

3C273（z=0.158）

3C273 具有复杂的伸长结构（参考文献 3、15、16 和科顿等人准备发表的论文）。

preparation). Its low declination (2°) means that the (u, v) coverage is poor and the models are more indeterminate than for high declination sources like 3C345. In consequence, different authors have proposed models of different character, some expanding with time[7,15], and others with stationary components whose intensities are suitably varied[3,17]. All observers agree that the overall size of the brightness distribution increased from 1971 to 1974, but the uniformity and nature of the increase has been argued.

Figure 6 of Schilizzi et al. (ref. 15) shows a succession of (u, v) diagrams with the observed lines of maxima and minima of the visibility function. The lines are not equally spaced; therefore, the source cannot be represented by a simple (that is, Gaussian) double. Furthermore, the relative spacings change with time, and successive sets of lines move toward the origin. This means that the shape changes with time, and that, in some measure, it is increasing in size. The visibility data shown by Schilizzi et al., however, do not define the brightness distribution uniquely.

In an attempt to escape difficulties caused by modeling with inadequate data, we have studied one simple measure of the overall size, the distance from the origin to the first minimum of the visibility function. Let w (in wavelengths) be this distance, then $\theta^* \equiv 1/(2w)$ is the angular separation of the components if the source is a simple double. In the more general case where the source consists of several isolated components strung out on a line, θ^* is a good measure of the overall angular scale. For example, in a wide range of triples which we studied numerically, θ^* varies from 1.0 to 0.65 times the overall size.

We plot θ^* at various epochs in Fig. 3. Many of these observations were taken on only one baseline, and the position angles were very poorly determined. So, to provide a consistent basis for calculating θ^*, we assumed PA = 64° in all cases. This is the typical value obtained when multi-baseline data are available, although the source is known to be nonlinear[3,16]. The error bars in Fig. 3 come from the uncertainty in estimating the locations of the minimum, and do not include any errors from an incorrectly assumed position angle. The arrows in mid-1972 signify lower and upper limits.

Further data exist at 2.8 cm for 1975 and 1976, but the character of the visibility function is different then from what it was earlier. The functions do not contain any clearly recognizable extrema (except at the origin), but drop rather smoothly to a "core" value. The brightness distributions may still contain "components", but all (or perhaps all but one) of them must be large enough to be individually resolved at the interferometer spacings which would otherwise produce maxima and minima. In contrast, at 6 cm the structure still had concentrations which could produce the observed minima. These recent results will be discussed separately.

The 2.8-, 3.8-, and 6.0-cm data all fit together very well up to 1975. We conclude that the basic shape was independent of wavelength over this range, although there were spectral differences in component intensities.

它的低赤纬（2°）意味着 (u, v) 覆盖很差，和 3C345 那样的高赤纬源相比其模型更加难以确定。因此不同作者提出的模型特性各不相同：有些随着时间膨胀[7,15]，而另一些则子源静止，但流量强度有些变化[3,17]。所有观测者都认同从 1971 年到 1974 年亮度分布总面积增大了，但是关于这一增大的均匀性和物理本质仍有争论。

斯基利齐等人的文章 (参考文献 15) 中的图 6 给出了连续的 (u, v) 图并表示出可见度函数极大和极小值的观测曲线。曲线不是等距离分布的，所以源不能用简单的双峰（高斯型）来表示。此外，相对间距随着时间改变，连续的线系朝着起点移动。这意味着形状随着时间改变，而且在某种程度上，它的尺度在增大。不过，由斯基利齐等人展示的可见度数据尚不能给出唯一的亮度分度。

为了避开用不充分的数据建立模型所带来的困难，我们研究了对源的整体尺寸的一次简单测量，其距离是从起点到可见度函数的第一个极小值。令 w（以波长形式）为距离，如果源为一个简单双峰源，那么 $\theta^* \equiv 1/(2w)$ 是子源的角分离。在更一般的情形下，源包含几个处在一条线上的独立子源，此时 θ^* 是对整体角尺度的一个很好的量度。例如，在我们进行数值分析的 3 倍范围内，θ^* 的变化范围在整体尺度的 1.0 倍到 0.65 倍之间。

我们在图 3 上标出了各个时期的 θ^*。大部分观测只建立在一个基线上，因此位置角很难确定。所以，为了给计算 θ^* 提供统一的基准，我们假定在所有情形下 PA=64°。这是在存在多基线数据时的典型值，尽管我们知道这个源是非线性的[3,16]。图 3 的误差棒来自估计最小值位置的不确定性，并没有包含由于位置角假设错误带来的误差。1972 年中期的箭头表示数据的下限和上限。

1975 年和 1976 年在 2.8 cm 波段存在更多的数据，但是那时可见度函数的特征与早期的有所不同。该函数不包含任何清晰可识别的极值（除了在起点处），而是略微平稳地下降到一个"核"值。亮度分布可能仍然含有"子源成分"，但是所有这些子源（或者可能有一个例外）必须足够大，以便在干涉仪间距中能够单独分辨出来，否则有可能产生极大和极小值。相反，在 6 cm 波段仍然出现集中点，从中可以得到观测的极小值。我们将对这些最近的观测结果分别进行讨论。

一直到 1975 年，2.8 cm、3.8 cm 和 6.0 cm 波段的数据都彼此符合得很好。我们断定在这期间亮度分布的基本形状与波长无关，尽管子源强度存在频谱差异。

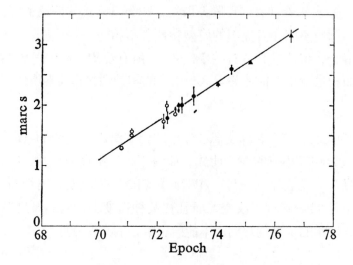

Fig. 3. "Size" of 3C273, where size $\theta^* = (2w)^{-1}$ and w is the distance to the first minimum of the visibility function. The slope of the line is 0.32 marc s yr^{-1}. The observations at 2.8 cm are taken from refs 15 and 16 and the points at 3.8 cm are taken from refs 7 and 17–19. The 6-cm points are our previously unpublished data. ● , 2.8 cm; ○ , 3.8 cm; ▲ , 6.0 cm.

The overall size of the brightness distribution increased rather steadily and more than doubled in 6 years. The word "expansion" may properly be used to describe this behavior. The points in Fig. 3 cluster around a line with slope $\theta^* = 0.32$ marc s per year ($v/c =$ 4.2). If the line is extrapolated it hits zero in 1966. The total flux density at $\lambda = 3.8$ cm is shown in Fig. 4, and, again, the "expansion" starts, roughly, near the beginning of a large sustained increase in flux density.

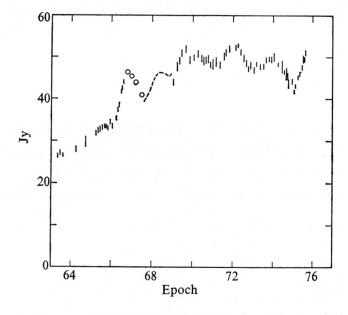

Fig. 4. Total flux density of 3C273 at $\lambda = 3.8$ cm. References for the points are as in Fig. 2.

80

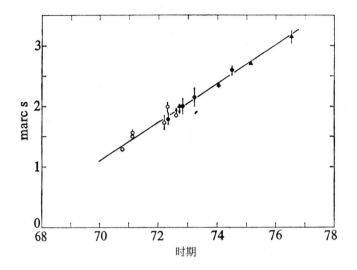

图 3. 3C273 的"尺度",其中尺度 $\theta^* = (2w)^{-1}$,w 是可见度函数的首个极小值距离。直线斜率为 0.32 marc s $\cdot$ yr^{-1}。2.8 cm 波段的观测来自参考文献 15 和 16,3.8 cm 波段的数据点来自参考文献 7、17、18 和 19。6 cm 波段的数据点是我们之前未发表的数据。●,2.8 cm;○,3.8 cm;▲,6.0 cm。

　　亮度分布的整体尺度增加得相当平稳,6 年间增加达两倍多。"膨胀"这个词能够恰当地描述这种现象。图 3 里的数据点聚集在一条斜率 $\theta^* = 0.32$ marc s $\cdot$ yr^{-1}($v/c = 4.2$)的直线周围。如果把线延长,它将在 1966 年到达零点。图 4 显示了在 $\lambda = 3.8$ cm 处的总流量密度,而且,"膨胀"再次大致在临近流量密度有持续大量增长的起始处开始。

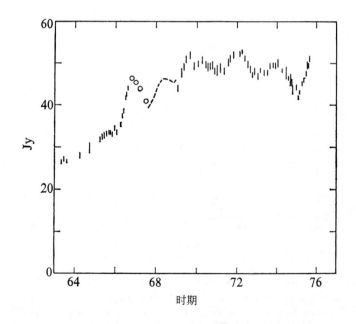

图 4. 3C273 在 $\lambda = 3.8$ cm 波段的总流量密度。数据点的参考文献来源和图 2 中的一样。

3C120 (z = 0.033)

3C120 varies rapidly in flux density and appearance, and has been observed frequently in an attempt to follow its variations[7,15,21-25]. Although many of the observations have been rather incomplete, in several cases very good data were obtained with 3 or 4 telescopes, and simple double models fit very well. The best of these cases comes from our observations at 2.8 cm in 1976.1, when 5 well defined maxima and minima were observed using telescopes at Green Bank, Fort Davis and Owens Valley. As with 3C345, a well-separated Gaussian double accurately fit all the visibility data. However, the compact structure in 3C120 never produced nearly all the total flux, the way it did in 3C345. The compact components of 3C120 are embedded in a larger, undetermined structure.

In all cases with good data the PA was near 65°. To determine a size in the poorer cases, we assumed that a double was always a reasonable representation of the source; and, when necessary, we also assumed PA = 65°.

The separation of the two components of the double is plotted in Fig. 5 for all values in the literature (with exceptions noted below) together with some of our previously unpublished data at 2.8 and 6.0 cm. The parentheses indicate cases where there was ambiguity; typically, the source was only weakly resolved, or the size determination was made from a single minimum on one baseline, or the first minimum was not directly observed.

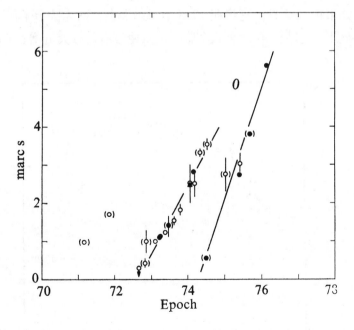

Fig. 5. Separation of the two components of 3C120, assuming a double at PA = 65° when necessary. The slopes of the lines are 1.8 and 2.9 marc s yr⁻¹. The early observations at 2.8 cm are from ref. 15, and the points from 1974.5 on are previously unpublished. The 3.8-cm data are from refs 7, 15, 21–25. The 6-cm point at 1974.5 is previously unpublished. ●, 2.8 cm; ○, 3.8cm; ▲, 6.0 cm.

3C120（$z=0.033$）

3C120 在流量密度和外形上变化得很快，为了跟踪它的变化，对它的观测较为频繁 [7,15,21-25]。尽管许多观测颇为不完整，但是有几个观测项目用三到四台望远镜得到了很好的数据，而且与简单的双峰模型拟合得非常好。这些观测中最好的一次是我们于 1976.1 在 2.8 cm 波段进行的观测，我们使用格林班克、戴维斯堡和欧文斯谷的望远镜观测到了 5 个确定得很好的极大和极小值。像 3C345 一样，分离得很好的高斯双峰模型精确拟合了所有的可见度数据。不过，3C120 中的致密结构不像 3C345 中的那样产生几乎全部总流量。3C120 中的致密成分处在一个更大的、未准确测量的结构里。

在所有数据质量良好的情况下，位置角接近于 65°。为了确定在数据质量较差情形下位置角的大小，我们假设双峰结构总是可以作为对这个源的合理描述，然而必要时，我们还是假定位置角为 65°。

文献中所有双峰两个子源的角分离数据（例外的情况在下面做了标注）和我们之前未发表的一些在 2.8 cm 和 6.0 cm 的数据都在图 5 上标出。圆括号代表存在模糊度的情况；通常源只能被微弱地分辨开，或是通过一条基线上的单个极小值来确定尺度，或是第一个极小值没有直接观测到。

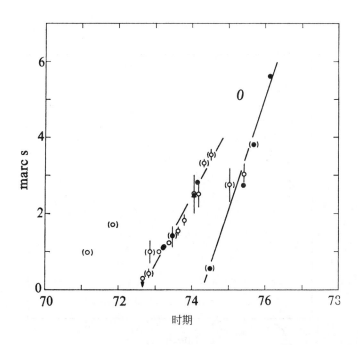

图 5. 3C120 两个子源的角分离，必要时假设其为位置角等于 65° 的双峰结构。两条直线斜率分别为 1.8 marc s·yr⁻¹ 和 2.9 marc s·yr⁻¹。2.8 cm 波段的早期观测来自参考文献 15，1974.5 之后的数据点以前未发表过。3.8 cm 波段的数据来自参考文献 7、15、21~25。1974.5 的 6 cm 波段的数据点以前未发表过。●，2.8 cm；○，3.8 cm；▲，6.0 cm。

We recognize two major epochs of expansion in Fig. 5, and have indicated them with straight lines whose slopes correspond to $v/c = 5$ and $v/c = 8$. At 1974.5 the 3.8 and 2.8 cm observations showed doubles of very different separation. No unique interpretation can be given for this, for the size and strength of the components of the doubles are very poorly determined. The observed double structure did not account for the total flux at either wavelength, so it is possible that both doubles existed at both wavelengths but were not recognized because of the differing resolutions. Thus, the close-spaced double seen at 2.8 cm might not have been recognized at 3.8 cm because its first minimum would have come at 1.9×10^8 wavelengths, whereas the maximum length of the baseline was only 1.0×10^8 wavelengths. Similarly, at 2.8 cm where the minimum length of the baseline was 1.6×10^8 wavelengths, components larger than about 1.5 marc s would have been resolved, and this double would have been missed. Spectral differences of the type apparently seen in 3C273 would increase the size of the outer components at 2.8 cm and accentuate this behaviour.

The points in 1971 were once interpreted as evidence for a superluminal expansion[21], but that result is ambiguous with only two points. Further data exist for early 1972 (refs 22, 23), but they are not plotted. The visibility was changing rapidly then, but remained very low. Therefore, the strong concentrations which must have existed in 1971 had dissipated, and by 1972.3 most of the flux came from regions greater than about 1 marc s in diameter.

During both epochs of expansion the PA was near 65° whenever a good model could be generated.

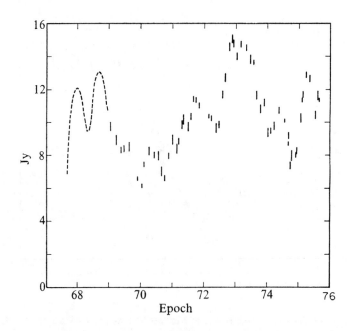

Fig. 6. Total flux density of 3C120 at $\lambda = 3.8$ cm. Points from 1969 to 1971.6 are taken from ref.12; 1971.6 to 1974.4, ref.13; 1974.4 to 1975.6, from private communications with W. A. Dent, and the dashed line is interpolated between observations at 2.8 and 4.5 cm reported in ref.14.

我们在图 5 中识别出两个主要的膨胀时期，并把它们用斜率对应于 $v/c=5$ 和 $v/c=8$ 的直线表示出来。在 1974.5，3.8 cm 和 2.8 cm 波段的观测显示双峰角分离明显不同。对此不能做出唯一的解释，因为双峰子源的大小、强度都确定得不好。在任一波段观测到的双峰成分都不能解释总流量，所以也可能是两个波段都存在两个双峰结构，但是因为两个波段的分辨率不同，因而没能在某一波段同时分辨出两个双峰结构来。因此，在 2.8 cm 波段看到的近距离间隔的双峰没能在 3.8 cm 波段识别出，可能由于它的第一个极小值在 1.9×10^8 个波长处，而 3.8 cm 的基线最大长度是 1.0×10^8 个波长。类似地，在 2.8 cm 波段基线的最小长度为 1.6×10^8 个波长，只有大于 1.5 marc s 的子源才能被分辨出，因而 3.8 cm 波段的双峰在 2.8 cm 波段却没有观测到。3C273 中明显看到的频谱差别可能在 2.8 cm 波段扩大了外层成分的尺度，因而突出了上面提到的这种情况。

1971 年的数据点一度被解释为是超光速膨胀的证据 [21]，不过只依据 2 个数据点，那样的结果是不明确的。1972 年早期有了更多的数据（参考文献 22、23），但还没有标绘。之后可见度变化很大，但仍然很低。所以，在 1971 年密集在一起的成分开始分离开来，到 1972.3，大部分流量就都来自直径大于 1 marc s 的区域。

在两个膨胀时期，只要模型拟合得很好，其位置角都在 65° 附近。

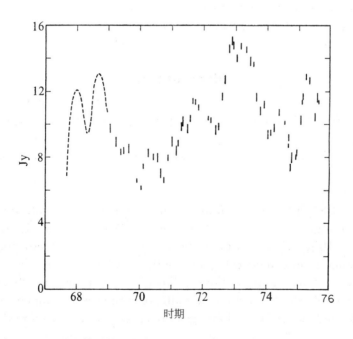

图 6. 3C120 在 λ=3.8 cm 波段的总流量密度。1969 到 1971.6 的数据点来自参考文献 12；1971.6 到 1974.4 的数据来自参考文献 13；1974.4 到 1975.6 的数据来自和登特的私人交流，虚线是根据参考文献 14 中的在 2.8 cm 和 4.5 cm 波段处观测得到的插值。

Figure 6 shows the total flux density of 3C120 at $\lambda = 3.8$ cm. This source varies more rapidly and has a greater fractional variation than 3C345 and 3C273. The largest outburst in Fig. 6 begins in mid-1972 and is coincident with the start of the first expansion shown in Fig. 5. The second expansion starts in early 1974 and does not appear to be correlated with any exceptionally large flux event.

3C279 ($z = 0.538$)

The available data on 3C279 are mainly at $\lambda = 3.8$ cm. from the period 1970–73 (refs 7, 17–20, 26). Most of this material is from only one baseline, and the interpretations are ambiguous.

The two earliest observations have been interpreted as an expanding double, with separation rate $\theta = 0.27$ marc s yr^{-1} ($v/c = 10$) (refs 7, 19). The rate is not well determined with only 2 points, but it is consistent with observations at 6 cm in early 1972 (ref. 17), which give the overall size as about 2 marc s. A minimum in the 13-cm visibility function on baselines from California to Australia[26] near the beginning of 1970 is also consistent with a simple double model based on the 3.8-cm observations, suggesting that the separation of the two components was similar over that range of wavelengths.

A sequence of observations at 3.8 cm in 1972 (ref. 17) can be interpreted as a new expansion event, since an unresolved weak core was seen in March and April 1972, and this became substantially resolved by November 1972. Another possibility is that all the 1972 data came from a triple source whose overall dimensions were constant, but whose component flux densities varied[17]. The published material is consistent with either model, and further data are necessary to discriminate between them.

Interpretations

It is of interest to ask what fraction of all sources are superluminal, and whether statistical arguments can be made for or against any of the customary interpretations. These are difficult questions because adequate surveys do not exist at the short centimeter wavelengths, and VLBI surveying is particularly incomplete. However, we can make some comments.

At $\lambda = 3.8$ cm, 3C345 is the weakest known superluminal source, with mean flux density $< S_{3.8} \sim 10$ Jy during the expansion. We know of 15 extragalactic sources which have been as strong as 10 Jy at 3.8 cm, at any epoch. Five of these are radio galaxies with most of the flux coming from extended regions. (3C123, 3C353, Vir A, Cen A, Cyg A). The remaining 10 sources are shown in Table 1. The fifth column in Table 1 gives the known range of flux density at $\lambda = 3.8$ cm. The sixth column gives a comment on the structure seen with VLBI, and the seventh column gives references for those sources not discussed in detail above.

图 6 显示了 3C120 在 λ=3.8 cm 的总流量密度。这个源比 3C345 和 3C273 变化更快，局部变化更大。图 6 中最强的爆发开始于 1972 年中期，和图 5 所示的第一次膨胀开始的时间相重合。第二次膨胀开始于 1974 年早期，并且看起来和任何超大流量事件都没有关系。

3C279（z=0.538）

1970 年到 1973 年期间 3C279 的数据主要集中在 λ=3.8 cm 波段（参考文献 7、17~20、26）。大部分数据是只基于一条基线得到的，而且不能给出明确的解释。

最早的两个观测被解释为是一个膨胀的双峰，分离速率为 θ = 0.27 marc s·yr[-1]（v/c=10）（参考文献 7、19）。由于只有两个数据点，因此这个速率不能很好地确定，但是它和 1972 年早期给出整体尺度约 2 marc s 的 6 cm 波段观测结果（参考文献 17）一致。1970 年初用从加利福尼亚到澳大利亚 [26] 的基线得到的 13 cm 波段可见度函数的最小值也和基于 3.8 cm 波段观测得到的简单双峰模型一致，这表明在这个波长范围内两个子源的分离相似。

1972 年在 3.8 cm 波段的一系列观测（参考文献 17）可以看成 3C279 发生一次新的膨胀事件，因为在 1972 年 3 月和 4 月看到了一个微弱的不可分辨的核，而它到 1972 年 11 月变得可以充分地分辨开来。还一个可能是 1972 年的所有数据都来自一个三源系统，它的整体尺度不变，但是子源的流量密度在变化 [17]。已经发表的数据跟任何一个模型都吻合，因此要区分这些模型就需要更多的数据。

结 果 分 析

超光速源在所有源中的比率以及是否存在统计上的论据来支持或者反对通常的解释，这些都是很有意义的问题。但是这些问题也都很难回答，因为在厘米短波上还没有足够多的巡天，并且 VLBI 巡天也特别不完整。不过，我们仍可以做一些评论。

在 λ=3.8 cm 波段，3C345 是已知超光速源中最弱的源，膨胀时的平均流量密度小于 $S_{3.8}$（$S_{3.8}$ 约为 10 Jy）。我们知道 15 个河外源任何时期在 3.8 cm 波段的流量都很大，达到 10 Jy。其中 5 个是射电星系，它们大部分的流量来自延展区域（3C123、3C353、Vir A、Cen A 和 Cyg A）。剩下的 10 个源列在表 1 中。表 1 的第 5 列给出了在 λ = 3.8 cm 波段源的流量密度的已知范围。第 6 列给出了对 VLBI 观测到的结构的意见，第 7 列给出前面没有具体讨论的源的参考文献。

Table 1. Strong compact extragalactic radio sources

Source	Name	Id	z	$S_{3.8}$(Jy)	VLBI	Refs
0316+41	3C84	G	0.018	20–58	slow variable	15, 24, 25, 27, 28
0355+50	NRAO150			7.3–15	stable double(?)	16, 24, 29
0430+05	3C120	G	0.032	6.4–15.1	superluminal	
0923+39	4C39.25	Q	0.699	8.5–11.8	stable double	5, 24, 29
1226+02	3C273	Q	0.158	28–53	superluminal	
1253–05	3C279	Q	0.538	11.2–17.5	superluminal	
1641+39	3C345	Q	0.595	8.8–11.9	superluminal	
2134+00		Q	1.94	11.0–13.4	slow variable	15, 24
2200+42	BLLac	G	0.07	4.8–14.2	rapid variable	16, 24, 29, 30
2251+15	3C454.3	Q	0.859	9.0–20.2	variable core	16, 24, 26, 29

3C84 has several components which vary in flux density but seem to have little motion. NRAO150 has been observed less than the others and its size is less than 0″.001 so that intercontinental observations are necessary for reasonable resolution.

4C39.25 has a well-defined double structure with constant separation, but the total flux density is variable. 2134+004 apparently is more complex than a simple double, and has slow weak variations in its brightness distribution. BL Lacertae has a double structure which is variable but keeps a constant position angle. The observed changes in BL Lac are not systematic as in the superluminal sources, but it may well be that shorter sampling intervals are required to see them, as this is the most variable source in Table 1. 3C454.3 has a core-halo structure. The variations in total flux density follow those of the unresolved core.

Nearly half the strong compact sources show a superluminal effect. This suggests that mechanisms which require us to be in an especially favorable position are unlikely to be at work. Luminous relativistically moving clouds are one such class, because to see them superluminally we have to be within a small solid angle[1]. However, the blue-shift of the approaching clouds also raises the possibility of a selection effect. Perhaps superluminal sources seem common partly because the blue shift raises their flux density above the minimum level being counted. To study this possibility we need to know the distribution of superluminal sources with flux density, which is unknown at present.

The superluminal sources all show a systematic expansion, and at least one of them shows two epochs of expansion with the same PA. This argues strongly against mechanisms which basically are random and allow both expansions and contractions. Gravitational lenses and some other propagation and opacity effects are of this type. So, too, would be a collection of separated sources which are independent, and flash at random times.

表 1. 河外致密强射电源

源	名字	Id	z	$S_{3.8}$ (Jy)	VLBI	参考文献
0316+41	3C84	G	0.018	20~58	缓慢变化	15、24、25、27、28
0355+50	NRAO150			7.3~15	稳定的双峰（？）	16、24、29
0430+05	3C120	G	0.032	6.4~15.1	超光速	
0923+39	4C39.25	Q	0.699	8.5~11.8	稳定的双峰	5、24、29
1226+02	3C273	Q	0.158	28~53	超光速	
1253-05	3C279	Q	0.538	11.2~17.5	超光速	
1641+39	3C345	Q	0.595	8.8~11.9	超光速	
2134+00			1.94	11.0~13.4	缓慢变化	15、24
2200+42	BLLac	G	0.07	4.8~14.2	快速变化	16、24、29、30
2251+15	3C454.3	Q	0.859	9.0~20.2	变化的核	16、24、26、29

3C84 有几个流量密度变化、但是看起来移动很小的子源成分。NRAO150 比其他源观测的少些，它的尺度小于 0″.001，所以需要通过洲际观测来加以分辨。

4C39.25 有非常确定的双峰结构，其分开速率不变但是总流量密度是变化的。2134+004 看起来明显比简单的双峰要复杂，而且它存在缓慢而微弱的亮度分布变化。蝎虎天体有一个变化的双峰结构，但是位置角不变。观测到的蝎虎天体的变化不像超光速源那样有系统性，但可能观测它们需要更短的取样间隔，因为它们是表 1 中变化最快的源。3C454.3 有一个核－晕结构。总流量密度随着不可分辨的核变化。

近半的致密强源表现出超光速效应。这意味着那些需要我们处于特别有利位置的机制不太可能再起作用。相对论运动的亮星云是这样的一类天体，为了以超光速观测到它们，我们必须处在一个很小的立体角内[1]。不过，云逼近时的蓝移也可能导致选择效应。也许超光速源看起来经常出现的部分原因是蓝移导致他们流量密度高于计数的最低水平。为了研究这种可能性，我们需要知道超光速源随流量密度的分布，这个分布目前还是未知的。

超光速源都显示出系统性膨胀，并且它们中至少有一个在两个时期内以相同的位置角膨胀。这与那些本质上是随机的，并且膨胀和收缩都可以存在的机制是严重相抵触的，诸如引力透镜和其他一些传播效应以及不透明度效应就是这种类型的机制。这样的话，我们应该收集到很多独立的、随机爆发的源。传播效应和不透明

Propagation and opacity effects also are unlikely because they predict that the separation would be wavelength dependent, contrary to observation.

The fact that the superluminal effect has been seen in a galaxy as well as in three quasars suggests that it cannot be used as evidence against a cosmological origin for the redshifts.

Thus what is now known can be summarized as follows. Nearly half the strong ($S_{3.8} \geq 10$ Jy) compact ($\theta < 0''.01$) sources show a superluminal expansion at centimetre wavelengths, and changes in overall size of up to factor 10 in periods of one to two years have been observed.

No systematic contractions have been seen. The separation of components in a source is basically independent of wavelength from 2 to 6 cm, although the components may have different spectra. In 3C273, and perhaps in 3C120, we may have observed an epoch where component evolution produced different sizes at different wavelengths.

The position angle is substantially constant during an expansion. In at least one case (3C120) there have been two distinct epochs of expansion. The position angle is the same for both.

The brightness distribution does not expand uniformly, but there is an overall trend as shown in the graphs. The epoch of zero separation, determined by extrapolation of the trend, may be close to the start of a major outburst in flux density.

The extensive data for 3C345 at 1974.5 and for 3C120 at 1976.2 strongly suggest a structure which can be represented as a simple double, with the two components being approximately equal. 3C273 is more complex. There are both simple (4C39.25) and complex (3C84) sources which do not show variations in the overall size of the brightness distribution although their total flux densities vary on time scales similar to those for the superluminal sources.

We thank W. A. Dent, and the members of the Canadian-British and the MIT-NASA VLBI groups, for giving us data in advance of its publication, and for useful comments. M. H. C. and J. D. R. are grateful to the Institute of Astronomy, Cambridge, for their hospitality during the period when part of this study was made.

This work was supported in part by grants from the US National Science Foundation. J. D. R. is a US National Science Foundation Graduate Fellow.

(**268**, 405-409; 1977)

度效应也是不可能的机制，因为它们预示着膨胀随波长而变，这与观测结果相反。

在一个星系和三个类星体中观测到超光速效应的事实表明超光速效应不能作为驳斥红移的宇宙学起源的证据。

因此，我们将目前已知的内容总结如下。近半数的强（$S_{3.8} \geq 10$ Jy）致密（$\theta < 0''.01$）源在厘米波段上表现出超光速膨胀，并且在 1 到 2 年的时间内观测到整体尺度增大了 10 倍。

我们没有观察到系统性的收缩。在一个源中，虽然子源会有不相同的谱形，但是子源的分离程度在 2 cm 到 6 cm 的波段范围内基本上与波长无关。在 3C273 中（也可能在 3C120 中）我们也许已经观测到，在一段时间内，子源的演化在不同波长上会导致不同的尺度。

在一个膨胀过程中位置角非常稳定。至少在这样一个源（3C120）中存在明显的两个膨胀时期，而位置角都是一样的。

虽然亮度分布的膨胀并不均匀，但如图所示，存在一个整体的趋势。从外推得出的零角分离时间可能接近于一次流量密度大爆发的起始点。

3C345 在 1974.5 的大量数据和 3C120 在 1976.2 的大量数据有力地表明，这两个源都可用简单的双峰结构来表示，其两个子源近似相同。3C273 要更复杂些。另外还存在一个简单的源（4C39.25）和一个复杂的源（3C84），尽管它们的总流量密度像那些超光速源一样随着时间尺度变化，但它们在亮度分布的整体尺度上都没有显示出变化。

我们感谢登特以及加拿大 – 大不列颠和麻省理工学院 – 美国国家航空航天局的甚长基线干涉仪小组的成员将尚未发表的数据提供给我们，并且提出有益的意见。科恩和罗姆尼向剑桥天文研究所在部分研究工作进行期间提供的热情帮助表示感谢。

本项工作得到美国国家科学基金会的部分资助。罗姆尼是美国国家科学基金会研究员。

（肖莉 翻译；何香涛 审稿）

M. H. Cohen*, K. I. Kellermann†, D. B. Shaffer†, R. P. Linfield*, A. T. Moffet*, J. D. Romney*, G. A. Seielstad*, I. I. K. Pauliny-Toth‡, E. Preuss‡, A. Witzel‡, R. T. Schilizzi§ and B. J. Geldzahler‖
*Owens Valley Radio Observatory, California Institute of Technology, Pasadena, California 91125
†National Radio Astronomy Observatory, Green Bank, West Virginia 24944
‡Max-Planck-Institut fur Radioastronomie, Bonn 1, Federal Republic of Germany
§Netherlands Foundation for Radio Astronomy, Radiosterrenwacht, Dwingeloo, The Netherlands
‖University of Pennsylvania, Philadelphia, Pennsylvania 19174

References:

1. Blandford, R. D., McKee, C. F. & Rees, M. J. *Nature* 267, 211 (1977).

2. Fort, D. N. *Astrophys. J. Lett.* 207, L155 (1976).

3. Legg, T. H. *et al. Astrophys. J.* 211, 21 (1977).

4. Wittels, J. J. *et al. Astron. J.* 81, 933 (1976).

5. Shaffer, D. B. *et al. Astrophys. J.* (in the press).

6. Cohen, M. H. *et al. Astrophys. J. Lett.* 206, L1 (1976).

7. Cohen, M. H. *et al. Astrophys. J.* 170, 207 (1971).

8. Kellermann, K. I. *et al. Astrophys. J.* 169, 1 (1971).

9. Wittels, J. J. *et al. Astrophys. J. Lett.* 206, L75 (1976).

10. Aller, H. D. & Haddock, F. T. *Astrophys. J.* 147, 833 (1967).

11. Kellermann, K. I. & Pauliny-Toth, I. I. K. *Ann. Rev. Astron. Astrophys.* 6, 431 (1968).

12. Dent, W. A. & Kojoian, G. *Astron. J.* 77, 819 (1972).

13. Dent, W. A. & Kapitsky, J. E. *Astron. J.* 81, 1053 (1976).

14. Medd, W. J., Andrew, B. H., Harvey, G. A. & Locke, J. L. *Mem. R. astron. Soc.* 77, 109 (1972).

15. Schilizzi, R. T. *et al. Astrophys. J.* 201, 263 (1975).

16. Kellermann, K. I. *et al. Astrophys. J.* 211, 658 (1977).

17. Kellermann, K. I. *et al. Astrophys. J. Lett.* 189, L19 (1974).

18. Knight, C. A. *et al. Science* 172, 52 (1971).

19. Whitney, A. R. *et al. Science* 173, 225 (1971).

20. Niell, A. E., Kellermann, K. I., Clark, B. G. & Shaffer, D. B. *Astrophys. J. Lett.* 197, L109 (1975).

21. Shaffer, D. B., Cohen, M. H., Jauncey, D. L. & Kellermann, K. I. *Astrophys. J. Lett.* 173, L147 (1972).

22. Shapiro, I. I. *et al. Astrophys. J. Lett.* 183, L47 (1973).

23. Kellermann, K. I. *et al. Astrophys. J. Lett.* 183, L51 (1973).

24. Wittels, J. J. *et al. Astrophys. J.* 196, 13 (1975).

25. Hutton, L. K. thesis, Univ. Maryland (1976).

26. Gubbay, J. *et al. Astrophys. J.*(in the press).

27. Legg, T. H. *et al. Nature* 244, 18 (1973).

28. Pauliny-Toth, I. I. K. *et al. Nature* 259, 17 (1976).

29. Shaffer, D. B. *et al. Astrophys. J.* 201, 256 (1975).

30. Clark, B. G. *et al. Astrophys. J. Lett.* 182, L57 (1973).

93

0957+561 A, B: Twin Quasistellar Objects or Gravitational Lens?

D. Walsh *et al.*

Editor's Note

Within the framework of general relativity, space is curved by the presence of mass. It was realized in the 1920s and 1930s that this implies that light from distant galaxies may be bent by intervening mass into a ring, or that the distortion could produce several images of the same galaxy—an effect called gravitational lensing. Here Dennis Walsh at Jodrell Bank in England and his coworkers report the first such example. During a spectroscopic study of two blue star-like objects separated by about 6 arcseconds, they realized that the objects were quasars with essentially identical spectra. They concluded that they might be seeing the effect of gravitational lensing by an intervening galaxy. Many such cases are now known.

0957+561 A, B are two QSOs of mag 17 with 5.7 arc s separation at redshift 1.405. Their spectra leave little doubt that they are associated. Difficulties arise in describing them as two distinct objects and the possibility that they are two images of the same object formed by a gravitational lens is discussed.

SPECTROSCOPIC observations have been in progress for several years on QSO candidates using a survey of radio sources made at 966 MHz with the MkIA telescope at Jodrell Bank. Many of the identifications have been published by Cohen *et al.*[1] with interferometric positions accurate to ~2 arc s and a further list has been prepared by Porcas *et al.*[2]. The latter list consists of sources that were either too extended or too confused for accurate interferometric positions to be measured, and these were observed with the pencil-beam of the 300 ft telescope at NRAO, Green Bank at λ 6 cm and λ 11 cm. This gave positions with typical accuracy 5–10 arc s and the identifications are estimated as ~80% reliable.

The list of Porcas *et al.* includes the source 0957+561 which has within its field a close pair of blue stellar objects, separated by ~6 arc s, which are suggested as candidate identifications. Their positions and red and blue magnitudes, m_R and m_B, estimated from the Palomar Observatory Sky Survey (POSS) are given in Table 1 and a finding chart is given in Fig. 1. Since the images on the POSS overlap, the magnitude estimates may be of lower accuracy than normal, but they are very nearly equal and object A is definitely bluer than object B. The mean position of the two objects is 17 arc s from the radio position, so the identification is necessarily tentative.

0957+561A、B：双类星体还是引力透镜？

沃尔什等

编者按

在广义相对论框架下，空间由于质量的存在而弯曲。在二十世纪二三十年代，人们意识到，这意味着来自遥远星系的光可能会被处于中间的质量扭曲成一个环，或者这种扭曲会使同一个星系产生多个像——这被称为引力透镜效应。本文中英格兰焦德雷尔班克天文台的丹尼斯·沃尔什和他的同事们报告了第一个这样的例子。在对两个相距6角秒的蓝色恒星状天体的光谱研究中，他们意识到这两个天体是具有本质上完全相同光谱的类星体。他们由此得出结论，他们可能看到了由中间星系形成的引力透镜效应。现在我们已经知道了很多这种例子。

0957+561A、B是两颗17等、红移1.405、相距5.7 arc s的类星体。从它们的光谱上可以毫无疑问地看出两者是相关联的。将它们描述为两个不同的天体存在一定困难，本文将讨论它们由同一个天体在引力透镜效应下形成两个像的可能性。

焦德雷尔班克天文台的MkIA望远镜在966 MHz对射电源进行过巡天，近年来对类星体候选体的分光观测在不断发展。许多得到证认的天体已经由科恩等人[1]发表，其干涉测量的位置精确到约2 arc s，波卡斯等人[2]已经准备了进一步的观测列表。后一个列表包括由于本身太过延展或者干涉测量位置不够精确的天体，这些天体是使用格林班克美国国家射电天文台口径300英尺的望远镜在波长λ为6 cm和11 cm上用笔型波束探测到的。它给出典型精度5~10 arc s的位置，而据估计这些被证认的天体约80%是可靠的。

波卡斯等人的列表中包括源0957+561，在该源的区域内有一对密近的蓝色恒星状天体，两者距离约6 arc s，它们被认为是类星体认证中的候选天体。它们的位置及红、蓝星等（m_R、m_B）都是由帕洛玛巡天POSS的结果估算给出的，列于表1；图1给出了证认图。由于其图像位于帕洛玛巡天的重叠部分，所以估计所得星等精度可能低于正常情况，但两者仍非常接近，而且天体A肯定比天体B更蓝。两个天体距射电源的平均距离是17 arc s，所以这个证认必然是尝试性的。

Table 1. Positions and magnitudes of 0957+561 A, B

Object	RA	Dec (1950.0)	M_R	M_B
0957+561A	09 57 57.3	+56 08 22.9	17.0	16.7
0957+561B	09 57 57.4	+56 08 16.9	17.0	17.0

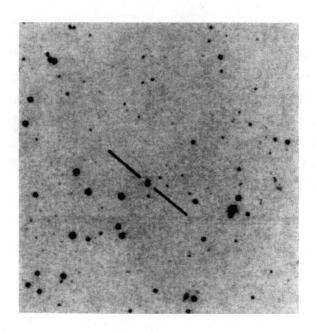

Fig. 1. Finding chart for the QSOs 0957+561 A and B. The chart is 8.5 arc min square with the top right hand corner north preceding and is from the E print of the POSS.

Observations

The two objects 0957+561 A, B were observed on 29 March 1979 at the 2.1 m telescope of the Kitt Peak National Observatory (KPNO) using the intensified image dissector scanner (IIDS). Sky subtraction was used with circular apertures separated by 99.4 arc s. Some observational parameters are given in Table 2. The spectral range was divided into 1,024 data bins, each bin 3.5 Å wide, and the spectral resolution was 16 Å. After 20-min integration on each object it was clear that both were QSOs with almost identical spectra and redshifts of ~1.40 on the basis of strong emission lines identified as C IV λ1549 and C III] λ1909. Further observations were made on 29 March and on subsequent nights as detailed in Table 2. By offsetting to observe empty sky a few arc seconds from one object on both 29 and 30 March it was confirmed that any contamination of the spectrum of one object by light from the other was negligible. On 1 April the spectral range was altered slightly by tilting the grating to cover the anticipated redshifted wavelength of Mg II λ2798 which was just beyond the limiting wavelength on previous nights.

表 1. 0957+561A、B 的位置及星等

天体	赤经	赤纬（1950.0）	红星等	蓝星等
0957+561A	09 57 57.3	+56 08 22.9	17.0	16.7
0957+561B	09 57 57.4	+56 08 16.9	17.0	17.0

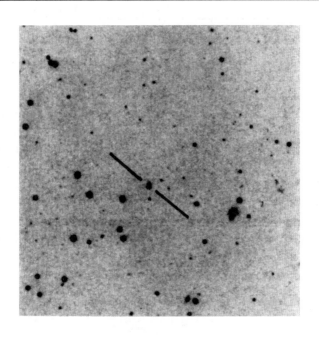

图 1. 类星体 0957+561A 和 B 的证认图。该图为 8.5（arc min），右上角方向为北，来自帕洛玛巡天（POSS）中的 E 色图片。

观　测

0957+561A、B 两个天体是在 1979 年 3 月 29 日，由基特峰国家天文台（KPNO）的 2.1 m 望远镜观测的，所用的仪器是加强型析像扫描器（IIDS）。天光去除是用相距 99.4 arc s 的圆形光孔得到的。一些观测参数列于表 2。整个光谱范围被分成了 1024 个数据段，其中每段的宽度为 3.5 Å，光谱分辨率为 16 Å。在对每个天体积分 20 分钟后，可以明显看出两者是有着几乎相同光谱的类星体，基于被认证为 C IV λ1549 和 C III] λ1909 的强发射线，可知红移大概是 1.40。进一步的观测在 3 月 29 日及之后的夜里进行，具体细节列于表 2。在 29 日和 30 日的晚上都对距离天体几角秒的无星天空进行了观测并作为校正，结果证实，一个天体对另外一个天体光谱产生的污染可以忽略不计。4 月 1 日，通过将光栅倾斜，光谱范围出现了细微的改变，以便覆盖预期 Mg II λ2798 红移后的波长，这条谱线在之前的夜里刚好在波长极限之外。

Table 2. List of IIDS observations with 2.1 m telescope

Date (1979)	Aperture (arc s)	Seeing (arc s)	Spectral range (Å)	Integration time, each object (min)
29 March	3.4	4	3,200–6,700	40
30 March	1.8	1	3,200–6,700	20
1 April	3.4	3	3,500–7,000	60

The spectra obtained on 1 April are shown in Fig. 2. Data on observed spectral lines are given in Table 3. These were taken from the spectra using the interactive picture processing system (IPPS) which makes a linear interpolation between two selected continuum points and calculates the centroid and equivalent width of the emission above the interpolated line. Data from all three nights were used in compiling Table 3; that on 1 April had double the signal-to-noise ratio of the other two nights and was weighted accordingly. The O IV] λ1402 line is outside the spectral range of Fig. 2 but was present in data taken on the other two nights. Although we believe that Mg II λ2798 is detected in the data of Fig. 2 for 0957+561B,and He II λ1640 is also detected taking into account all three nights' data, the low signal-to-noise ratio and poorly defined continuum prevent us deriving useful observed wavelengths or equivalent widths.

Table 3. Wavelengths, equivalent widths (EW) and derived redshifts from IIDS 2.1 m observations

λ_{em}		O IV] 1402	C IV 1549	He II 1640	C III] 1909	Mg II 2798
A	λ_{obs}(Å)	3373	3729.5	3938	4584.5	6739
	EW(Å)	24	68	11	54	28
	z(vacuum)	1.407	1.4082	1.402	1.4026	1.409
B	λ_{obs} (Å)	3376	3728.7	Present	4582.6	Present
	EW (Å)	26	70	—	55	—
	z (vacuum)	1.409	1.4077	—	1.4016	—

The data on the C IV λ1549 and C III] λ1909 lines are much more accurate than those on the other lines and we believe the r.m.s. errors in the observed wavelengths of the centroids of these lines are not greater than 3 Å while the r.m.s. errors in the equivalent widths are estimated to be 7 Å. Within the limits of observational error, the corresponding lines in each object are identical in observed wavelength and equivalent width. For each object there is a difference in the redshift derived from the C IV and C III] lines which is significantly greater than the combined r.m.s. error in each. This may be associated with the problem of giving a precise meaning to the redshift of a broad line of somewhat

表 2. 2.1 m 望远镜的加强型析像扫描器（IIDS）观测结果列表

日期 （1979）	孔径 （arc s）	视宁度 （arc s）	光谱范围 （Å）	每个天体的积分时间 （min）
3 月 29 日	3.4	4	3,200~6,700	40
3 月 30 日	1.8	1	3,200~6,700	20
4 月 1 日	3.4	3	3,500~7,000	60

图 2 中给出了 4 月 1 日得到的光谱。观测的谱线数据列于表 3。这些由光谱得到的数据使用交互式图像处理系统（IPPS）进行处理，该系统会对选定的连续两点进行线性插值，并计算该区域的中心及发射线高出插值线的等值宽度。三个夜晚的数据合并组成了表 3；其中 4 月 1 日的信噪比要比另外两天的信噪比高一倍，因此计算中进行了相应的加权。其中 O IV] λ1420 谱线在图 2 的光谱区域外，但却出现在了另外两个晚上得到的数据中。尽管我们相信，在图 2 的 0957+561B 数据中探测到了 Mg II λ2798 谱线，并且也探测到了 He II λ1640 谱线 (考虑三个晚上的数据)，但是较低的信噪比及难以确定的连续谱使我们无法得到有用的观测波长或等值宽度。

表 3. 由 2.1 m 望远镜加强型析像扫描器所得波长、等值宽度 (EW) 及推算出的红移

λ_{em}	O IV] 1402	C IV 1549	He II 1640	C III] 1909	Mg II 2798
A { λ_{obs} (Å) EW (Å) z (真空)	3373 24 1.407	3729.5 68 1.4082	3938 11 1.402	4584.5 54 1.4026	6739 28 1.409
B { λ_{obs} (Å) EW (Å) z (真空)	3376 26 1.409	3728.7 70 1.4077	存在 – –	4582.6 55 1.4016	存在 – –

C IV λ1549 和 C III] λ1909 谱线数据比其它谱线数据精确很多，我们相信这些谱线中心观测波长的均方根误差不会大于 3Å，同时等值宽度的均方根误差估计为 7Å。在观测误差范围内，每个天体相对应谱线的观测波长和等值宽度相同。每个天体由 C IV 和 C III] 谱线得出的红移值有所不同，这种差别比它们各自的综合均方根误差大很多。这可能与精确测量一条不规则宽线的红移会存在问题有关。由 CIV 和 C III] 发射线得到的红移平均值对 A 是 1.4054，对 B 是 1.4047，这个差别在测量误

irregular shape. The mean values of the redshift from the C IV and C III] emission lines are 1.4054 for A and 1.4047 for B, the difference being within the errors of measurement.

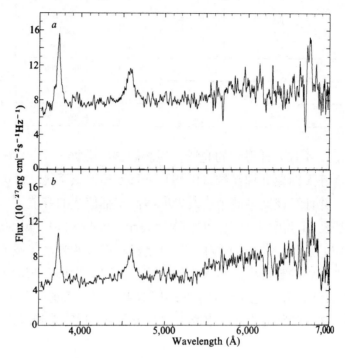

Fig. 2. IIDS scans of 0957+561 A(*a*) and B(*b*). The data are smoothed over 10 Å and the spectral resolution is 16 Å.

Although no attempt was made to carry out accurate spectrophotometry, some characteristics of the continua seem fairly well defined. Below about 5,300 Å they appear to have identical shapes, with QSO A brighter than B by 0.35 mag. Above 5,300 Å, however, the flux from B rises more steeply than that from A and they are equal at ~6,500 Å. These results are consistent with the magnitude estimates of Table 1.

The pair of QSOs provides unusual opportunity to investigate the origin of absorption lines in QSO spectra, a matter which is still in dispute. Accordingly, spectra having a resolution of about 2 Å were obtained of both QSOs on 30 March using the image tube spectrograph attached to the University of Arizona 2.3 m telescope. As in the observations described above, the seeing during the observations was sufficiently good for contamination of the spectrum of one QSO by the light from the other to be negligible. A portion of the tracings of the two plates covering the C IV emission line region is shown in Fig. 3. The absorption lines which have been identified are indicated on the figure, and the measured wavelengths (using a Grant measuring engine) are presented in Table 4, with the corresponding redshifts. The wavelengths of the C IV emission lines given in Table 4 were measured from the tracings by smoothing over the noise and finding the centre of symmetry for each line. Comparison with Table 3 shows that the agreement in wavelength for the C IV emission lines between the two sets of observations is within the errors of measurement.

差范围内。

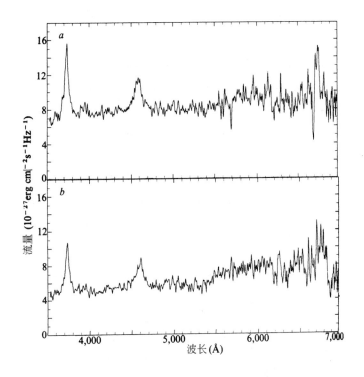

图 2．加强型析像扫描器对类星体 0957+561 A（*a*）、B（*b*）的扫描结果。这些数据以 10 Å 进行平滑，
光谱的分辨率为 16 Å。

尽管没有尝试进行精确的分光光度测量，但是一些连续谱的特征看起来还是相
当清楚的。在大约 5,300 Å 之内两者有着相同的形状，类星体 A 比 B 亮 0.35 个星
等；然而在 5,300 Å 之外，B 的流量比 A 的流量增加更快，在约 6,500 Å 处二者相等。
这些结果与表 1 中的星等估计值相吻合。

这对类星体为研究类星体光谱中吸收线的起源提供了不同寻常的机会，这是个
仍有争议的问题。相应地，在 3 月 30 日使用连接在亚利桑那大学 2.3 m 望远镜上的
显像管摄谱仪得到了这两个类星体的分辨率大约为 2 Å 的光谱。与以上描述的观测
情况一样，观测期间大气视宁度足够好，一个天体的光谱受到另外一个天体的光的
污染可以忽略。图 3 中给出了这两个底片中覆盖了 C IV 发射线区域的光谱。已经被
认证的吸收线在图中标出，观测所得波长（使用格兰特测量引擎）及相对应的红移
值列于表 4。表 4 中列出的 C IV 发射线波长值是根据光谱对噪声进行平滑处理并确
定了每条谱线的对称中心后得到的。和表 3 比较可以看出，在测量误差内，两组观
测中所得的 C IV 发射线的波长一致。

Table 4. Wavelengths, identifications, and derived redshifts from image-tube spectra, 2.3 m observations

Object	0957+561A		0957+561B	
Identification	λ_{air}	z (vacuum)	λ_{air}	z (vacuum)
—	3536.4	—	—	—
Si II 1526	3648.2	1.3903	(defect)	
C IV 1548	3699.9	1.3905	3700.1	1.3906
C IV 1550	3705.9	1.3904	3707.4:	1.3914:
C IV 1549(em)	3728.9:	1.4078:	3732.2:	1.4100:
—	3835.1	—	—	—
Fe II 1608	3844.0	1.3905	—	—
Al II 1670	3992.9	1.3905	3993.6	1.3909

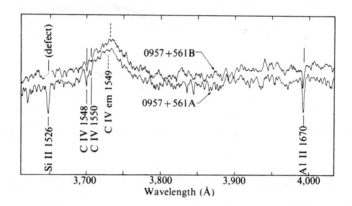

Fig. 3. Microdensitometer tracings of portions of the spectra of 0957+561 A and B. Original dispersion of the plates was 47 Å mm^{-1}. The solid lines mark the position of absorption features in the two QSOs and the dashed lines mark the adopted centres of the C IV emission line.

Low ionisation absorption systems (ones with Si II and Al II strengths > C IV strengths) are clearly present at z_{abs} =1.390 in both QSOs. Even in the low resolution IIDS spectrum of QSO A there is clear evidence for Fe II λ2383 and Mg II λ2798 absorption. Fe II λλ2600 and 2344 are possibly also present. Weak and possibly real absorption lines also appear in the image tube spectrum at λ3536.1 and λ3835.1 of QSO A. The features at λ3835.1 and λ3844.0 have a separation close to that of the Mg II doublet (at redshift 0.372). However, λ3844 is already identified with Fe II λ1608 in the 1.390 system so that the evidence for Mg II at 0.372 is not convincing. In QSO B, the absorption lines seem to be weaker than in QSO A on the basis of both the plate and IIDS data and none are seen in the low resolution spectrum. Unfortunately, a dust speck on the mask used to suppress image tube noise obliterated the Si II line in the spectrum of this object.

表 4. 由 2.3 m 望远镜观测得到的波长、证认结果及由图像管光谱计算得出的红移值

天体	0957+561A		0957+561B	
证认结果	λ_{air}	z（真空）	λ_{air}	z（真空）
—	3536.4	—	—	—
Si II 1526	3648.2	1.3903	（坏点）	
C IV 1548	3699.9	1.3905	3700.1	1.3906
C IV 1550	3705.9	1.3904	3707.4:	1.3914:
C IV 1549（发射）	3728.9:	1.4078:	3732.2:	1.4100:
—	3835.1	—	—	—
Fe II 1608	3844.0	1.3905		
Al II 1670	3992.9	1.3905	3993.6	1.3909

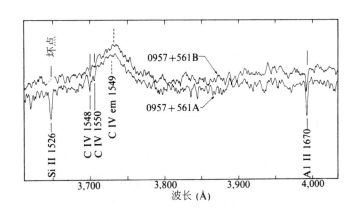

图 3．用显微光密度计对 0957+561A 和 B 测得的光谱中的部分光谱轮廓。各底片原始的色散为 47 Å·mm⁻¹。实线标记了两个类星体的吸收线中特征谱线所在位置，虚线标记了所采用的 C IV 发射线的中心位置。

在 z_{abs}=1.390 处，两个类星体中都明显地出现了低电离吸收系统（其中 Si II 和 Al II 谱线的强度 > C IV 谱线的强度）。即使在类星体 A 的低分辨率加强型析像扫描器（IIDS）光谱中，也可明显发现 Fe II λ2383 和 Mg II λ2798 吸收线存在的证据。Fe II λλ2600 和 2344 也有可能存在。微弱但可能是真实的吸收线也出现在了从类星体 A 的显像管摄谱仪光谱中 λ3536.1 和 λ3835.1 处。在 λ3835.1 和 λ3844.0 处的谱线间距和 Mg II 双线间距（红移为 0.372）接近。然而在 z_{abs}=1.390 的系统中已经将 λ3844 证认为 Fe II λ1608，所以在红移 0.372 处的 Mg II 的证据也变得不可信了。无论是从底片还是从加强型析像扫描器数据都可以看出，类星体 B 中的吸收线都要比类星体 A 的吸收线弱；在较低分辨率的光谱中没有看到这些吸收线。遗憾的是，在遮罩上原本用来限制显像管噪声的灰尘斑点，使该天体光谱中的 Si II 谱线彻底消失了。

The difference between the two absorption redshifts amounts to a velocity difference ΔV_{abs} $(B - A)$ of only about +45 km s^{-1}. However, in addition to the errors in estimating the line centres, somewhat larger errors occur in the zero point of the wavelength scales from plate to plate amounting typically to 100 km s^{-1}. As a result the difference between the absorption line redshifts in QSO A and QSO B cannot be considered significant.

The image tube data on the C IV emission lines give a velocity difference $\Delta V_{em}(B - A)$ of +265 km s^{-1}. This is also subject to the zero point error, but the major source of error is the uncertainty of ~1.5 Å (=120 km s^{-1}) in estimating the position of each line centre. The IIDS data on the C IV and C III] lines permit two independent estimates of the velocity difference leading to a mean $\Delta V_{em}(B - A) = -95$ km s^{-1} with an error slightly larger than for the image tube data. Combining both sets of data, the resulting ΔV_{em} $(B - A)$ is +120± 150 km s^{-1}. Again, the difference between the emission redshifts in the two QSOs cannot be considered significant.

The differences $z_{em}-z_{abs}$ for each QSO based on Table 4 are not affected by the zero point error. They correspond to relative velocities of 2,170 km s^{-1} and 2,400 km s^{-1} for A and B respectively. The relative velocities each have an error of ~120 km s^{-1} due to the uncertainty in the emission line centres. Thus the difference in relative velocities of 230 km s^{-1} seems somewhat larger than the measuring error.

Therefore, either the absorption redshifts, or the emission redshifts may be equal, but possibly not both.

Finally, a plate was obtained on 2 April with the University of Arizona 1.5 m telescope. The seeing was relatively poor (~2.5 arc s), but the two images were well resolved and their measured separation was 5.7 arc s.

Discussion

The great similarity in the spectral characteristics of these two QSOs which have the same redshift and which are separated by only 6 arc s seems to constitute overwhelming evidence that the two are physically associated, regardless of the nature of their redshifts, and we do not think that a useful *a posteriori* statistical test of this assertion can be carried out. In the rest of the discussion, however, we shall assume the QSO redshifts are cosmological. The same similarities further suggest that we may be dealing with a single source which has been split into two images by a gravitational lens. We shall consider this possibility after examining the more conventional explanation involving two distinct QSOs.

In the conventional interpretation of two adjacent QSOs we must either regard it as a coincidence that the emission spectra are so nearly the same, or assume that the initial conditions, age and environment influencing the development of the QSOs have been so similar that they have evolved nearly identically. For $q_0=0$ and $H_0=50$ km s^{-1} Mpc^{-1}

两条吸收线红移值之差对应的速度差 $\Delta V_{abs}(B-A)$ 仅有大约 +45 km·s^{-1}。然而，除了在估计这些谱线中心时的误差外，在某种程度上各底片之间确定波长的零点也会产生更大的误差，其典型量级为 100 km·s^{-1}。因此类星体 A 和 B 中吸收线对应的红移之差并不显著。

C IV 发射线的显像管数据给出的速度差 $\Delta V_{em}(B-A)$ 为 +265 km·s^{-1}。这也与零点确定的误差相关，但误差的主要来源是在估计每条谱线中心位置时的不确定性，大约为 1.5 Å（=120 km·s^{-1}）。根据加强型析像扫描器对 C IV 和 C III] 光谱线给出的数据，可以产生两个独立的速度差估计值，两者给出的平均值为 $\Delta V_{em}(B-A)=$ −95 km·s^{-1}，这个结果的误差要比显像管数据的误差稍微大一些。结合两组观测数据得到，$\Delta V_{em}(B-A)=+120\pm150$ km·s^{-1}。因此，这两个类星体存在的发射线红移之差也不显著。

基于表 4 的每个类星体的差值 $z_{em}-z_{abs}$ 不受零点误差影响。它们对应于类星体 A 和 B 的相对速度分别为 2,170 km·s^{-1}、2,400 km·s^{-1}。由于在确定发射线中心时的不确定，每个相对速度都存在大约 120 km·s^{-1} 的误差。因此相对速度 230 km·s^{-1} 的差值看起来要比测量误差大。

因此，两个天体的吸收线对应的红移值或者发射线对应的红移值有可能会相等，但可能不会都相等。

最后，我们使用亚利桑那大学口径 1.5 m 的望远镜在 4 月 2 日获得了一张底片。其大气视宁度相对较差（约 2.5 arc s），但两个星像很好地被区别开了，测得它们之间的距离为 5.7 arc s。

讨 论

这两个有着相同红移值且间距只有 6 arc s 的类星体在光谱上的巨大相似性成了两个天体有物理联系的压倒性证据，不管它们红移的本质是什么。我们认为没有一种归纳性的统计检验能有效地予以解释。然而在接下来的讨论中，我们将假设类星体的红移是宇宙学红移。同样的相似之处进一步表明，我们处理的可能是一个射电源，由于引力透镜而分裂为两个像。在检验更传统的涉及两个分开的类星体的解释之后，我们将考虑这个可能性。

在两个相邻类星体的传统解释中，我们必须把发射线如此近乎相同看作是巧合，或者假设两者的初始条件、年龄及影响演化的环境非常相似，导致两者的演化几乎相同。对于 $q_0=0$、$H_0=50$ km·s^{-1}·Mpc^{-1}，对应 $\theta=5.7$ arc s 的投影线距离为 68.5 kpc。

the projected linear separation corresponding to $\theta = 5.7$ arc s is 68.5 kpc. The difference between emission line velocities is well within the dispersion in velocities found by Stockton[3] between QSOs and associated galaxies, and the masses implied by orbital motion are of the order of $10^{11} M_\odot$ (because of the errors in ΔV, this is more like an upper limit).

The conventional interpretation of the sources as two QSOs requires additional coincidences to explain the absorption line systems regardless of the mechanism invoked to explain the absorption. Weymann et al.[4] have described three classes into which absorption systems found in QSOs in this redshift range may be placed. The first class involves ejection of material from the QSO. If the ejection of the two systems were caused by the two QSOs separately it would be an additional coincidence that the ejection velocities were so similar. If the lines arose from radial ejection by one of the QSOs, then the nearly identical redshift of the two absorption systems (the difference between which we take to be ≤ 150 km s^{-1}) requires a rather small angle between the direction of motion of the ejected cloud and the line of sight to the second QSO against which the cloud is projected. This implies a distance of the ejected material from the ejecting QSO of ~ 185 kpc. This in turn implies exceedingly large masses and energies for the ejected material for reasonable covering factors. This argument is very similar to that made by Wolfe et al.[5] for the 21 cm absorption in 3C286.

The second class of absorption involves intervening clouds associated with a cluster in which the QSOs are embedded. The velocity differences between the emission and absorption systems of A, B are typical of this class, but we must either ascribe the agreement in redshift of the two absorption systems to chance or assume that the two absorption systems are part of a common halo associated with a galaxy in the same cluster. An unusual feature if this last alternative is true is that the ionisation state is very low for this class. In the survey of Weymann et al. only one of about 20 absorption systems was a low ionisation system similar to those in A, B. The third class of absorption involves cosmologically distant intervening material. Neither the agreement in redshift of the systems in A and B nor their low ionization is then especially remarkable, but we must then ascribe to chance the fact that the intervening material happens to be at a redshift so near to the emission redshift.

We now consider the possibility that a gravitational lens is operating. The theory of gravitational imaging in a cosmological context has been considered elsewhere (see ref. 6 and refs therein) and we simply quote the main results of applying this theory. The following are the relevant parameters involved in considering the gravitational lens hypothesis: the angular separation of the images, the shape of the images and their sizes, and the amplification of the two images. There is no evidence on the plate taken on 2 April or on the POSS for any departure of the images from stellar images. The magnitude difference between A and B (Table 1) is $\sim$0.3. mag and this is confirmed by our observations.

发射线速度差是在斯托克顿 [3] 得到的类星体和寄主星系速度弥散的范围之内的，由轨道运动对应得到的质量在 $10^{11}M_\odot$ 的量级上（由于存在 ΔV 的误差，所以这更像一个上限）。

就算不考虑用于解释吸收的机制，将这些源看作两个类星体的传统解释也要求额外的巧合事件来解释吸收线系统。魏曼等人 [4] 对这个红移范围内类星体可能的吸收系统提出了三种情况。第一种情况涉及类星体中的物质喷射。如果这两个系统中的喷射是由这两个类星体分别造成的，那么喷射速度如此相似将是一个额外的巧合。如果这些光谱是由其中的一个类星体的径向喷射产生的，则对于这样两个红移很接近的吸收线系统（这两个系统差值我们设定为 $\leq 150\ \mathrm{km \cdot s^{-1}}$）需要满足：喷射云的运动方向和第二个类星体的视线方向只存在一个很小的夹角，喷射云对着它运动。这表明被喷射物质和喷出物质的类星体的距离约是 185 kpc。这反过来表明，对于合理的覆盖因子，喷射物质有极大的质量和能量。这个论据和沃尔夫等人 [5] 由 3C286 中的 21 cm 吸收线提出的观点很接近。

第二类吸收涉及类星体所在的星系团内的中间云。在这种情况中，类星体 A 和 B 的吸收线和发射线系统存在典型速度差，但我们不得不将两个吸收线系统红移的一致性归因于巧合，或者假设这两个吸收线系统是来自同一个星系团中的同一个星系晕。如果是后一种情况，那么将会有一个非常罕见的特征，即对于这类情况而言电离态非常低。在魏曼等人的巡天中，在大约二十个吸收线系统中只有一个系统是和类星体 A、B 类似的低电离系统。第三类吸收系统的情况涉及了宇宙学距离的中间物质。A 和 B 中红移的一致性以及它们较低的电离度都不是特别值得注意的，但是我们必须将中间物质红移碰巧处于和发射线十分接近的位置归为巧合。

我们现在考虑引力透镜效应发生的可能性。在宇宙学框架内引力透镜成像理论在很多地方被考虑过（见参考文献 6 以及其参考文献），在此我们只是简单引用这个理论的主要结论。以下是在考虑引力透镜假说中涉及的相关参数：像的角分离、像的形状和它们的大小以及这两个像的放大率。在 4 月 2 日得到的底片或者帕洛玛巡天（POSS）得到的观测结果中都没有证据显示图像偏离恒星图像。类星体 A 和 B 的星等差值（表 1）约为 0.3 星等，这已经被我们的观测所证实。

The 0.3 mag difference between the two components requires that the amplification of QSO light is ~4 for the brighter image, and thus implies a normal luminosity for the QSO. (This is also suggested by the absence of a strong narrow component in the C IV emission which might be expected if the source were a strongly amplified Seyfert nucleus.) The maximum angular size of the lens is only ~8 times that of the object, so we should not expect to resolve it on the sky.[6]

If the matter responsible for the gravitational imaging is far from the QSO, then, from simple euclidean space calculations we estimate that at redshift z_L its mass must be $\sim 10^{13} z_L M_\odot$, and require that it be contained in a radius $\lesssim 30$ kpc. If a galaxy is the cause, then a lower limit of $z_L \sim 0.1$ is likely from its absence on our plate material. However, the centre of such a galaxy must be within ~0.5 arc s of the direct line between the QSO and the observer. The chance of finding such an alignment with a massive elliptical galaxy obtained by folding in our mass requirement with Schechter's[7] luminosity function (with a mass-to-light ratio of 30) is roughly 10^{-5}, although the precise number depends quite strongly on the magnitude differences and angular separations allowed. Thus, while such coincidences must be very rare, it is not out of the question that we should have one example in the ~1,000 QSOs known.

An apparent objection arises from the difference in the shapes of the continua between the two QSOs. It is possible that differential reddening along the two light paths may be responsible. Note that the observed break at 5,300 Å corresponds to an emitted wavelength of 2,200 Å in the rest system of the QSOs. This is the wavelength of a well known resonance in interstellar extinction by dust in our Galaxy, and a model can be constructed to explain the observed continuum ratio incorporating the 2,200 Å feature at the redshift of the QSOs. This would imply that the intrinsic flux from B exceeds that from A.

Further observations would shed light on the gravitational lens hypothesis. If the flux from the object is variable, the light curves of the two images should be similar but with a relative time delay due to the difference in path lengths. The lag depends on the details of the geometry, but with the parameters discussed above would be expected to be of the order of months to years. Determination of the radio structure would also clearly be of great value.

We thank S. Tapia and Barbara Schaefer for technical assistance, Geoff Burbidge for his comments, and the KPNO staff for their help. R. F. C. thanks the SRC for support and R. J. W. acknowledges support from NSF grant AST 77-23055. D. W. and R. F. C. are visiting astronomers, Kitt Peak National Observatory, which is operated by the Association of Universities for Research in Astronomy, Inc., under contract with the NSF.

Since submission of this article we have heard that on 19, 20, and 21 April the two QSOs were observed by N. Carleton, F. Chaffee and M. Davis (of the Smithsonian Astrophysical Observatory) and R. J. W. using the SAO photon-counting reticon spectrograph attached to the SAO–UA multiple mirror telescope. The observations covered the range 5,900–

对更明亮的像来说，两个部分之间 0.3 个星等的差要求类星体的光增强约 4 倍，这表明该类星体的光度是正常的。（如果这个射电源是一个强烈放大的塞弗特活动星系核，在 CIV 发射谱中可能会出现一个强而窄的发射线。）由于引力透镜的最大角度只有天体分开角度的 8 倍左右，所以我们不能期待天空中会有这种图像 [6]。

由简单的欧氏空间计算，如果造成引力透镜成像的物质远离类星体，那么我们估计在红移 z_L 处其质量必须达到大约 $10^{13} z_L M_\odot$，并且要求其半径 ≤ 30 kpc。假设存在一个这样的星系，鉴于它在底片上毫无显示，可以得出 z_L 的下限大约是 0.1。然而，这样一个星系的中心必须处于从观测者到类星体之间直线角度的 0.5 arc s 左右以内。既要保证与这样一个大质量椭圆星系排成一线，又要满足我们根据谢克特 [7] 光度函数（其中质光比为 30）给出的质量要求，概率大约是 10^{-5}，不过具体数字强烈依赖于容许的星等差和角距。因此，尽管这样的巧合必然非常少见，但我们还是有可能在大约 1,000 个已知的类星体中找到一个例子。

一个明显的反驳来自两个类星体连续谱形状的不同。这有可能是因为两者光线路径有不同的红化。请注意，在波长 5,300 Å 观测到的中断在类星体静止系统中对应于波长 2,200 Å。这是我们银河系中由于尘埃的星际消光形成的一条著名谱线，可以建立一个模型，解释观测到的类星体红移处包括 2,200 Å 特征的连续谱比值。这可能暗示着类星体 B 的内禀流量超过类星体 A。

进一步的观测将辨明引力透镜假设。如果来自天体的流量随时间变化，那么由于路径长度的差异，这两个图像的光变曲线应该相似但是存在一个相对的时间延迟。延迟依赖于几何细节，但是以上文讨论的参数可以预计是几个月到几年的量级。显然射电结构的测定也具有巨大的价值。

我们感谢塔皮亚和芭芭拉·谢弗给予技术支持，杰夫·伯比奇给出了宝贵意见，以及美国基特峰国家天文台全体工作人员的帮助。卡斯韦尔感谢科学研究理事会给予的支持，魏曼感谢国家科学基金会基金 AST 77-23055 的资助。沃尔什和卡斯韦尔是访问天文学家，基特峰国家天文台是与国家科学基金会签约、由大学天文研究协会联合运行的。

自提交本文以来，我们了解到，在 4 月 19 日、20 日和 21 日卡尔顿、查菲和戴维斯（于史密森天体物理台）观测了这两个类星体；魏曼使用了安装在 SAO–UA 的多镜面望远镜上的 SAO 光子计数摄谱仪对它们进行了观测。此次观测覆盖的波长范围从

7,100 Å with a resolution of 4 Å FWHM. Details will be reported elsewhere, but the main results are: (1) to within the measuring errors the Mg II emission lines have the same profiles and observed equivalent widths (85 and 76±12 Å for A and B respectively) and the same redshift (1.4136±0.0015 for both). (2) Absorption lines due to Fe II $\lambda\lambda2586$, 2599, Mg II $\lambda\lambda2795$, 2802 and Mg I $\lambda2852$ are present in both objects but are somewhat stronger in A. The mean heliocentric redshifts of the two absorption systems are 1.3915 for A and 1.3914 for B. A cross-correlation analysis confirms that the difference in the two adsorption redshifts is remarkably small and corresponds to a velocity difference of 7 ± 10 km s^{-1}. These observations strengthen the case for a gravitational lens.

(**279**, 381-384;1979)

D. Walsh[*], R. F. Carswell[†] and R. J. Weymann[‡]

[*] University of Manchester, Nuffield Radio Astronomy Laboratories, Jodrell Bank, Macclesfield, Cheshire, UK

[†] Institute of Astronomy, Cambridge, UK

[‡] Steward Observatory, University of Arizona, Tucson, Arizona 85721

Received 25 April; accepted 8 May 1979.

References:

1. Cohen, A. M., Porcas, R. W., Browne, I. W. A., Daintree, E. J. & Walsh, D. *Mem. R. astr. Soc.* **84**, 1 (1977)

2. Porcas, R. W. *et al. Mon. Not. R. astr. Soc.* (submitted).

3. Stockton, A. N. *Astrophys. J.* **223**, 747 (1978).

4. Weymann, R. J., Williams, R. E., Peterson, B. M. & Turnshek, D. A. *Astrophys. J.* (submitted).

5. Wolfe, A. M., Broderick, J. J., Condon, J. J. & Johnston, K. J. *Astrophys. J. Lett.* **208**, L47 (1976).

6. Sanitt, N. *Nature* **234**, 199 (1971).

7. Schechter, P. *Astrophys. J.* **203**, 297 (1976).

5,900 Å 到 7,100 Å，分辨率为半峰全宽 4 Å。细节将在他处报道，但是主要结果如下：（1）在测量误差以内，Mg II 发射线有相同的谱线轮廓和观测等值宽度（类星体 A、B 分别对应为 85 Å 和 76±12 Å）以及相同的红移（两者都是 1.4136±0.0015）。(2) Fe II λλ2586、2599，Mg II λλ2795、2802 和 Mg I λ2852 的吸收线存在于两个天体中，但类星体 A 中更强。两组吸收系统的日心平均红移分别为 1.3915（A）和 1.3914（B）。交叉相关分析确认，这两个吸收线红移之差非常小，对应于速度差 7±10 km·s⁻¹。这些观测结果进一步强化了引力透镜观点。

（冯翀 翻译；何香涛 审稿）

A Millisecond Pulsar

D. C. Backer *et al.*

Editor's Note

Pulsars are rapidly rotating neutron stars that emit regular radio pulses. They generally slow down as they age, as rotational energy is radiated away in the pulses. So it was a surprise when Donald Backer and coworkers discovered that a repeating radio source with no evidence of being "young" was actually a pulsar with a very short period (time between pulses) of about 1.6 milliseconds. More "millisecond pulsars" were discovered subsequently, and it is now believed that they are "recycled" pulsars: old neutron stars that accrete gas from a close companion star. The angular momentum of the gas "spins up" the old star. Millisecond pulsars have been used to test predictions of general relativity, such as gravitational radiation.

The radio properties of 4C21.53 have been an enigma for many years. First, the object displays interplanetary scintillations (IPS) at 81 MHz, indicating structure smaller than 1 arc s, despite its low galactic latitude $(-0.3°)$[1]. IPS modulation is rare at low latitudes because of interstellar angular broadening. Second, the source has an extremely steep $(\sim \nu^{-2})$ spectrum at decametric wavelengths[2]. This combination of properties suggested that 4C21.53 was either an undetected pulsar or a member of some new class of objects. This puzzle may be resolved by the discovery and related observations of a fast pulsar, 1937+214, with a period of 1.558 ms in the constellation Vulpecula only a few degrees from the direction to the original pulsar, 1919+21. The existence of such a fast pulsar with no evidence either of a new formation event or of present energy losses raises new questions about the origin and evolution of pulsars.

A literature search in 1979 led to the suggestion that the steep-spectrum, IPS source was superposed on a flat-spectrum $(\nu^{-0.1})$ object with a diameter of $\sim$60 arc s located to the west of the 4C position by one interferometer lobe $(-31.6$ s$)$. Lobe identification errors in the 4C catalogue can occur in regions of confusion such as the galactic plane.

The superposition of two source components, one very compact with a steep spectrum and the other extended with a flat spectrum, was reminiscent of the radio properties of the Crab nebula and its pulsar in the pre-pulsar era[3]. The conjecture was made that the compact component in 4C21.53W was a pulsar as yet undetected due to pulse broadening of its radiation over one period by interstellar scattering. However, the IPS measurement placed an upper limit on interstellar scattering which, in turn, placed a limit on the pulse broadening. The conclusion was that only a very short period pulsar, $P \leqslant 10$ ms, would have been missed in metre wavelength searches owing to this effect. Searches for such a short

一颗毫秒脉冲星

巴克尔等

编者按

脉冲星是快速旋转的、发出规则射电脉冲的中子星。随着时间推移，转动能被脉冲辐射带走，通常脉冲星旋转会越来越慢。因此，当唐纳德·巴克尔和同事们发现一颗没有"年轻"证据的重复射电源其实就是一颗脉冲星，其自转周期（脉冲之间的间隔）很短，大约只有约 1.6 ms 的时候，人们感到震惊。随着更多的"毫秒脉冲星"被发现，现在它们被认为是"再加速"脉冲星：从其密近伴星吸积气体的年老中子星。这些气体的角动量使得那颗年老中子星自转"加速"。毫秒脉冲星已被用于检验广义相对论的预言，如引力辐射。

多年来，4C21.53 的射电性质一直是个谜。首先，尽管这个源的银纬低（−0.3°）[1]，但是它在 81MHz 表现出行星际闪烁(IPS)，表明它的结构应小于 1 arc s。由于星际角度展宽，在低纬度 IPS 调制很少见。其次，这个源在十米波段的频谱极其陡（~v^{-2}）[2]。结合这两种性质我们可以看出，如果 4C21.53 不是未被探测到的脉冲星，就是某一类新型天体中的成员。一颗快转脉冲星 1937+214 的发现及相关观测可能解开这个谜。这颗快转脉冲星位于狐狸座，自转周期为 1.558 ms，距离最初发现的脉冲星 1919+21 只有几度。存在如此快速转动的脉冲星，同时又没有与新的形成过程或当前能量损失相关的证据，这引发了关于脉冲星起源和演化的新疑问。

通过查找 1979 年的文献得知：这个陡谱的 IPS 源叠加在一个平谱源（$v^{-0.1}$）上。平谱源直径约为 60 arc s，位于其 4C 位置以西一个干涉仪波瓣大小（−31.6 arc s）处。4C 源表中的波瓣证认错误会产生于在像银道面这样易混淆的区域。

源的两个成分叠加在一起：一个是具有陡谱的致密源，另一个是具有平谱的延展源。这让人想到在脉冲星发现之前，蟹状星云和其中心脉冲星的射电属性[3]。据推测 4C21.53W 中的致密成分是一颗脉冲星，但星际散射使其在一个周期内的脉冲辐射展宽，从而尚未被探测到。不过，IPS 测量给出的星际散射上限，可以反过来限定脉冲展宽。结论是只有周期非常短的脉冲星（$P \leqslant 10$ ms）才可能由于这种效应在米波段脉冲星搜寻时被遗漏。在厘米波段，脉冲展宽效应将大为减弱。1979 年阿

period pulsar at centimeter wavelengths, where pulse broadening would be much reduced, were conducted at Arecibo Observatory and at Owens Valley Radio Observatory in 1979 without success.

After the 1979 pulsar searches, Erickson (personal communication) located a steep-spectrum compact source, 4C21.53E, east of the 4C position by one 4C interferometer lobe (+31.6 s). This observation provided evidence against the superposition hypothesis. Furthermore, Very Large Array (VLA) observations at 5 GHz by one of us (D.C.B.) showed that 4C21.53E was a compact double source with separation of 0.8 arc s.

Interest in the extended western object, 4C21.53W, returned when decametric observations at the Clark Lake Radio Observatory showed that both 4C21.53E and 4C21.53W had very steep spectra below 100 MHz (ref. 4). In addition, the Clark Lake observations indicated that the western source showed IPS at 34 MHz. The inferred brightness temperature exceeded $>10^{12}$ K.

Observations of 4C21.53W with the Westerbork Synthesis Radio Telescope (WSRT) at 609 MHz in January 1982 confirmed a suspected position discrepancy based on a Culgoora measurement at 80 MHz (ref. 5) and a Bonn measurement at 5,000 MHz (ref. 6) (Fig. 1). We suspected that the Culgoora position was dominated by the steep-spectrum component and that the Bonn position was dominated by the flat-spectrum, extended component. The division of 4C21.53W into two components, evident in Fig. 1., confirmed our suspicion. The southern and northern components were named 4C21.53W(com) or 1937+214 and 4C21.53W(ext) or 1937+215, respectively. A brief observation at the VLA confirmed the position and steep spectrum of 1937+214. A map of 4C21.53W at 1,415 MHz from a 12-h observation with the WSRT in August 1982 clearly resolved the compact and extended source component (Fig. 2). Recent WSRT observations and 1979 total power observations from Arecibo are summarised in Table 1. The spectra are decomposed into compact and extended object contributions. The Bonn measurement[6] is included for completeness. Erickson's decametric observations of this source are reported elsewhere[7].

雷西博天文台和欧文斯谷射电天文台在厘米波段搜寻过这样的短周期脉冲星，但都没有成功。

在 1979 年的脉冲星搜寻后，埃里克森（个人交流）在 4C 位置以东一个 4C 干涉阵波束（+31.6 s）处发现了一个致密陡谱源 4C21.53E。该观测提供的证据反驳了上述叠加猜想。此外，我们中的一位成员（巴克尔）利用甚大阵（VLA）在 5 GHz 发现 4C21.53E 是一个间隔 0.8 arc s 的致密双源。

当克拉克湖射电天文台在十米波段的观测显示 4C21.53W 和 4C21.53E 的频率在 100 MHz 以下具有非常陡的谱（参考文献 4）时，人们的兴趣回归到西边的延展源 4C21.53W 上。另外，克拉克湖射电天文台的观测显示西边的源在 34 MHz 处有 IPS 并且测算的亮温度超过 10^{12} K。

1982 年 1 月，韦斯特博克综合孔径射电望远镜（WSRT）在 609 MHz 对 4C21.53W 的观测证实了根据 80 MHz 卡尔哥拉的测量（参考文献 5）和 5,000MHz 波恩的测量（参考文献 6）所推测的位置偏差（图 1）。我们怀疑卡尔哥拉测量的位置由陡谱成分主导，波恩测量的位置由延展平谱成分主导。图 1 中 4C21.53W 分裂成为两个源，这证实了我们的想法。南边和北边的成分分别被命名为 4C21.53W（com）或 1937+214 和 4C21.53W（ext）或 1937+215。VLA 的短期观测确认了 1937+214 的位置和陡谱。WSRT 于 1982 年 8 月在 1,415 MHz 对 4C21.53W 进行 12 小时观测后的合成图清晰地分辨出致密和延展源成分（图 2）。表 1 总结了最近的 WSRT 观测和 1979 年阿雷西博总强度观测。频谱被分解成致密源和延展源两种成分的贡献。为了完整起见，我们把波恩的测量 [6] 也包括进来。埃里克森在十米波段对这个源的观测发表在其他文章中 [7]。

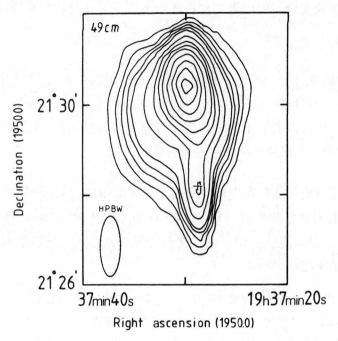

Fig. 1. Image of extended (north) and compact (south, +) components of 4C21.53W from a 12-h synthesis with the WSRT at 608.5 MHz on 15 January 1982. The synthesized beamwidth shown at the lower left is 31.3×80.4 arc s in RA and Dec, respectively.

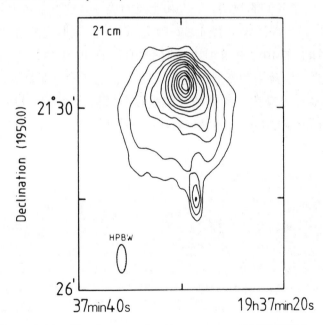

Fig. 2. Image of extended (north) and compact (south) components of 4C21.53W from a 12-h synthesis observation with the WSRT at 1,415 MHz on 8 August 1982. The synthesized beamwidth shown at the lower left is 13.3×37.0 in RA and Dec respectively.

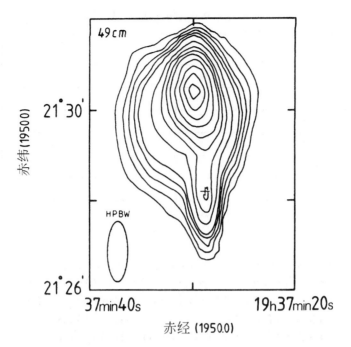

图 1. 1982 年 1 月 15 日 WSRT 在 608.5 MHz 频段上 12 小时合成后得到的 4C21.53W 延展(北)和致密(南，+）成分图。合成的波束宽度示于左下角，为 31.3 arc s × 80.4 arc s（分别在赤经和赤纬方向）。

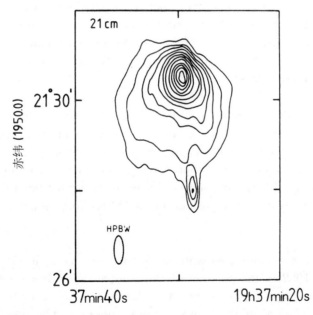

图 2. 1982 年 8 月 8 日 WSRT 在 1,415 MHz 频段上 12 小时观测后合成的 4C21.53W 延展(北)和致密(南)成分图。合成的波束宽度示于左下角，为 13.3 arc s × 37.0 arc s（分别在赤经和赤纬方向）。

Table 1. Flux densities of 4C21.53W

Frequency (MHz)	Instrument	Total flux density (Jy)	Compact flux density (Jy)	Compact polarization (%)	Extended flux density (Jy)
430	Arecibo	1.60±0.30	~0.3*	—	~1.3±0.30
609	WSRT	1.15±0.05	0.130	28±2	1.02±0.05
1,380	Arecibo	0.96†±0.05	~0.02*	—	0.94±0.05
1,415	WSRT	0.93±0.02	0.017	15±2	0.91±0.02
2,380	Arecibo	0.90†±0.05	—	—	0.90±0.05
5,000	Bonn[6]	0.91†	—	—	0.92

* Estimated from compact source spectral index between 609 and 1,415 MHz
† Corrected for beam sizes.

The spectrum of the extended source is approximately that of an H II region, $v^{-0.1}$. The H 166α recombination line was detected at the Arecibo Observatory at the position of the extended source in November 1982. The line temperature is ~1% of the continuum temperature. The velocity of the line is +2 km s^{-1} and its FWHM is ~25 km s^{-1}. These properties are comparable with those of an ordinary H II region. The small velocity does not distinguish between kinematic distances of 0 and 8.5 kpc.

The proximity of these two components of 4C21.53W on the sky suggests a possible physical connection. The morphology of the 21-cm continuum maps indicates that the exciting star may be displaced 43 arc s northwards from the centre of the H II region defined by the near-circular, low-level contours. On the other hand, the compact source is displaced 117 arc s southward from the centre. We will return later to the possible connection between these objects.

Interest in further pulsar searches was rekindled following the discovery of a steep-spectrum component in 4C21.53W at decametric wavelengths[4]. Momentum for the search increased when the 609-MHz WSRT map was available. A pulsar search sensitive to periods >4 ms was carried out by Boriakoss (personal communication) at the Arecibo Observatory in March 1982 without success. Detection in September 1982 of strong linear polarization at the compact source position in the WSRT maps made a pulsar detection a near certainty.

A new pulsar search was conducted with the 305-m antenna at the Arecibo Observatory on 25 September 1982 at the position of the compact 15 mJy component detected in the 1,400-MHz synthesis observations. Two harmonics of a millisecond periodicity, 1.558 ms, were discovered at the compact source position. The signal was present for only 3 min of a 7 min sample, and was not seen on a following day at either 1,400 or 2,380 MHz.

表 1 4C21.53W 的流量密度

频率（MHz）	设备	总流量密度（Jy）	致密成分流量密度（Jy）	致密成分偏振%	延展成分流量密度（Jy）
430	阿雷西博射电望远镜	1.60±0.30	~0.3*	—	~1.3±0.30
609	WSRT	1.15±0.05	0.130	28±2	1.02±0.05
1,380	阿雷西博射电望远镜	0.96†±0.05	~0.02*	—	0.94±0.05
1,415	WSRT	0.93±0.02	0.017	15±2	0.91±0.02
2,380	阿雷西博射电望远镜	0.90†±0.05	—	—	0.90±0.05
5,000	波恩射电望远镜 [6]	0.91†	—	—	0.92

* 从致密源 609 MHz 到 1,415 MHz 的谱指数估计得到
† 考虑波束大小后的修正值

延展源的频谱接近于典型的 H II 区的频谱（$v^{-0.1}$）。阿雷西博天文台 1982 年 11 月在延展源位置观测到 H 166α 复合线。谱线温度大约是连续谱温度的 1%。谱线对应的速度为 +2 km·s^{-1}，其半高全宽（FWHM）约为 25 km·s^{-1}。这些性质和普通 H II 区类似。这样小的速度不能区分开 0 kpc 和 8.5 kpc（译者注：kpc 表示千秒差距，1 秒差距 =3.08568×10^{16} 米）的运动学距离。

4C21.53W 的两个成分在天球上靠得很近，这表明它们之间可能存在物理关联。21 cm 连续谱成像显示，激发星可能向北偏离 H II 区中心 43 arc s，而 H II 区中心由接近圆形、低值等流量线确定。另一方面，致密源的位置在中心以南 117 arc s。我们之后再来讨论这些天体之间可能的联系。

进一步寻找脉冲星的兴趣随着在十米波段上 4C21.53W 陡谱子源的发现被重新激起 [4]。609 MHz WSRT 射电图使得搜寻脉冲星的势头大大增加。博里亚科斯（个人交流）1982 年 3 月在阿雷西博天文台进行了一次对周期 >4 ms 脉冲星敏感的搜寻，但没有成功。1982 年 9 月用 WSRT 射电图上在致密源位置观测到强的线偏振，几乎可以确定这个观测探测到的就是一个脉冲星。

1982 年 9 月 25 日利用阿雷西博天文台 305 m 天线，我们在 1,400 MHz 对在综合孔径天线探测到的 15 mJy 的致密子源位置处，展开了新的脉冲星搜寻。在致密源位置发现了周期为 1.558 ms 的两个谐波。但是信号在 7 min 的测量数据中只出现了 3 min，而且第二天在 1,400 MHz 和 2,380 MHz 频段都没观测到。

In November 1982 the pulsar search was intensified at Arecibo. In addition, we planned a search for interstellar scintillation (ISS) based on the possibility that the compact object was small, but not pulsing. Deep ISS modulation was detected at 1,400 MHz at the position of the compact source (Fig. 3). The amplitude was consistent with the flux density found in synthesis observations. The frequency and time correlation lengths, roughly 2 MHz and 5 min, respectively suggested a relatively small dispersion measure to the object, <100 electrons pc cm^{-3}. This observation indicated immediately that the compact source 4C21.53W (com) or 1937+214, was extremely small as only pulsars have shown ISS previously. In addition, the modulation bandwidth and time scale were consistent with the single detection in September. On the following day the millisecond pulsar was confirmed.

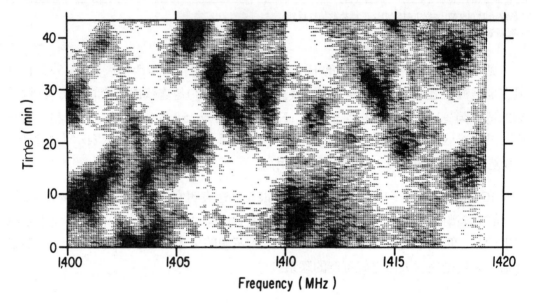

Fig. 3. Interstellar scintillation observation of compact component in 4C21.53W from Arecibo observations on 6 November 1982. Individual spectra were obtained for two polarizations in two 10-MHz bands every 30 s using a 252-channel, 1-bit autocorrelator. Spectra from the two polarization channels were summed and smoothed to a resolution of 100 kHz. The peak intensity in this dynamic spectrum is ~50 mJy.

The waveform of the pulsar contains a main pulse and an interpulse of comparable intensity separated by ~180° (Fig. 4). This morphology repeats precisely the main pulse/interpulse morphology of the Crab pulsar. The Crab pulsar has an additional "precursor" component preceding its main pulse at metre wavelengths. The full widths at half intensity of the pulse components in Fig. 4 are <125 μs, or 8% of the period. The pulses are readily detected with the Arecibo telescope at 1,400 and 430 MHz with a fast signal averager and integrations of a few hundred pulses when positioned on a peak of the ISS modulation pattern (Fig. 3). The waveforms at both frequencies are similar.

1982 年 11 月在阿雷西博天文台我们进一步加强了脉冲星搜寻。另外，基于存在致密源很小又无脉冲的可能，我们计划进行星际闪烁（ISS）搜寻。在 1,400 MHz 频段致密源位置处探测到显度 ISS 调制现象（图 3），其幅度和综合孔径观测到的流量密度一致。频率和时间相关长度分别约为 2 MHz 和 5 min，这表明该天体色散量相对较小，电子柱密度 < 100 pc · cm⁻³。这一观测表明致密源 4C21.53W（com）或 1937+214 的尺度极小，而之前只有脉冲星表现出 ISS 效应。另外，调制带宽和时标与 9 月份的单次观测一致。接下来一天的观测便确认了这颗毫秒脉冲星。

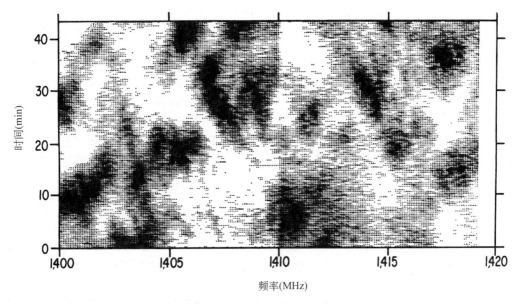

图 3. 1982 年 11 月 6 日阿雷西博观测中对 4C21.53W 致密成分的星际闪烁观测结果。10 MHz 带宽的两路独立偏振信号通过 252 通道，利用 1 比特自相关器每 30 s 采集一次。两个偏振通道的谱加起来，作 100 kHz 分辨率的平均。该动态谱的峰值强度为 ~50 mJy。

这颗脉冲星的波形包含一个主脉冲和一个中间脉冲，两者强度相当，分开约 180°（图 4）。这一形态非常类似蟹状星云脉冲星的主脉冲和中间脉冲的形态。蟹状星云脉冲星在米波波段还有一个早于主脉冲的额外"前导"成分。图 4 中脉冲成分的半高宽均小于 125 μs，或者说是周期的 8%。当阿雷西博望远镜对准 ISS 调制模式（图 3）的峰值处时，在 1,400 MHz 和 430 MHz 利用快速信号平均器对几百个脉冲积分就可以稳定地探测到脉冲。这两个波段的波形是相似的。

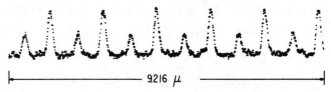

$\longmapsto$ 9,216 μ $\longmapsto$

Fig. 4. Waveform of the millisecond pulsar from a signal averager oscilloscope display. Sample spacing is 9 μs. The full trace is roughly six periods, 9,216 μs. The integration lasted ~75 s. The waveform consists of a main pulse and an interpulse separated by nearly 180° of rotational phase. Errors in timing the signal averager and a 20 μs RC time constant are responsible for most of the pulse width.

The first observations have resulted in the following parameters: RA (1950.0) 19 h 37 min 28.72 s, Dec (1950.0) 21° 28′ 01.3″, Barycentric period 0.001 557 807 (JD 244 5282), Dispersion measure 75 electrons pc cm^{-3}. The accuracy of all values is a few parts in the last decimal place. The distance is estimated as 2,500 pc using an average n_e of 0.03 cm^{-3}. There is no evidence for binary motion in timing measurements during the second week in November. Timing observations have been initiated to determine the period derivative at the Arecibo Observatory. A previous estimate of the period derivative[8], based on a comparison of measured periods in September and November, has not been substantiated by timing data in November. We suspect that the September measurement was corrupted by sampling errors. Analysis of the November observations results in an upper limit to P of 10^{-15} s per s.

Our original hypothesis that 4C21.53W was a fast pulsar superposed on an extended synchrotron-emitting nebula was only half correct. Now we are faced with a more profound enigma: the existence of a pulsar rotating near the maximum rate possible for a neutron star with a surprising lack of evidence of energetic activity in the vicinity of the pulsar.

The present rotation rate, 642 Hz, is very near the maximum rate of 2,000 Hz where centrifugal forces balance gravitational forces at the surface of a 1 $M_\odot$ neutron star with 10 km radius. Matter on the equator would have a velocity of 0.13c. Changes in the equilibrium figure of the pulsar resulting from energy losses to magnetic dipole and gravitational quadrupole radiation (discussed below) may occur as abrupt starquakes if there is such a close balance between gravitational and centrifugal forces. The balance could be tipped in favour of gravity if the pulsar is denser than $\Omega^2/G \sim 2\times10^{14}$ g cm^3. The amplitude and frequency of starquakes, observable in pulse arrival time measurements, may be able to distinguish between standard (1 $M_\odot$) and high density models for the millisecond pulsar.

The present rotational energy content is 7×10^{51} erg for a neutron star moment of inertia 10^{45} g cm^2. This energy is comparable to the entire mechanical energy output of a supernova event. A higher density star, as suggested above, could reduce the present energy content if its moment of inertia were smaller than 10^{45} g cm^2. Both the rapid spin and the large energy content are indicative of a young object as energy losses to magnetic dipole and gravitational quadrupole radiation are strong functions of the rotation rate, Ω^4 and Ω^6, respectively[9]. The minimum energy loss for this pulsar will be the observed radio emission which amounts to 3×10^{30} erg s^{-1} assuming a beam solid angle of 1 sr.

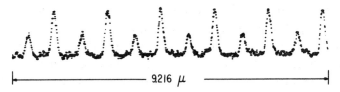

图 4. 信号平均示波器显示的毫秒脉冲星波形。取样间隔是 9 μs。包含约 6 个周期，共 9,216 μs。积分持续约 75 s。波形包含分开接近 180° 旋转相位的一个主脉冲和一个间脉冲。脉冲宽度主要来源于信号平均器的测时误差和 20 μs 的 RC 时间常数。

首次观测得到如下参数：赤经（1950.0）19 时 37 分 28.72 秒，赤纬（1950.0）21° 28′ 01.3″，太阳系质心系周期 0.001 557 807 s（儒略日（JD）244 5282），色散量为 75 个电子秒差距每立方厘米。所有值的精确度达到小数点最后一位。利用 n_e 为 0.03 cm^{-3} 的平均电子密度可以估算出距离为 2,500 pc。11 月第二周的测时观测没有显示此脉冲星存在于双星系统的迹象。为了得到周期变化率，我们在阿雷西博天文台进行了多次测时观测。之前通过比较 9 月和 11 月测得周期给出的周期变化率[8] 没有被 11 月的测时数据所肯定。我们怀疑 9 月份的观测因取样不当而损坏。通过分析 11 月的观测数据能够得到周期变化率的上限为每秒 10^{-15} s。

我们最初关于 4C21.53W 是快转脉冲星叠加在延展的同步辐射星云的猜想只对了一半。现在我们面临着一个意义更深远的谜团：脉冲星以接近中子星极限的速度旋转，令人吃惊的是周围却没有剧烈活动的证据。

目前 642 Hz 的自转速率非常接近 2,000 Hz 的最大自转速率（离心力和引力在 1 个太阳质量、半径 10 km 的中子星表面达到平衡时中子星的转速）。赤道物质的速度可达到 0.13c。假如引力和离心力接近这一平衡，那么因磁偶极和引力四极辐射（下面将讨论到）能量损失所致脉冲星平衡状态的改变可能以突发星震的形式发生。如果脉冲星密度大于 $\Omega^2/G \sim 2 \times 10^{14}$ g·cm^3，引力在平衡中将占优势。可由脉冲到达时间测量得到的星震幅度和频率也许能鉴别毫秒脉冲星的标准模型（1 $M_\odot$）和高密度模型。

转动惯量为 10^{45} g·cm^2 的中子星目前总转动能为 7×10^{51} erg。这个能量相当于超新星事件的全部机械能输出。对于上面提及的这种更高密度星，如果其转动惯量小于 10^{45} g·cm^2，那么星体的转动能将有所减少。因为磁偶极和引力四极辐射均强烈依赖于星体自转频率，分别遵从 Ω^4 和 Ω^6 [9]，所以快速自转率和高能量都暗示着年轻天体。假设辐射束的立体角为 1 sr，这颗脉冲星的最小能量损失即为观测到的射电辐射，共计 3×10^{30} erg·s^{-1}。

The age of this pulsar is puzzling. The maximum spin rate for a neutron star is ~2,000 Hz. A model for the Crab pulsar by Ostriker and Gunn[9] predicts a period decay from this rate to 100 Hz in 1 yr due to gravitational quadrupole radiation. We do not observe such a rapid decay. Furthermore, there is a surprising absence of evidence of any debris from a recent neutron star formation event. Our radio maps show no synchrotron-emitting nebula in the vicinity of the pulsar. Einstein observations place a limit of 1.5×10^6 K for the surface temperature of a neutron star at the pulsar position for the indicated distance of 2 kpc, and exclude the possibility of a synchrotron nebula (D. J. Helfand, personal communication). These X-ray limits do not allow for possible heavy extinction. There is no source at the pulsar position in the COS B catalogue[10]. Lick Observatory observations reveal a 20 mag optical star at the position of the pulsar[11]. In direct analogy to the Crab pulsar, we expect this object will show optical pulsations.

We conclude that despite the large spin and rotational energy, this pulsar is not young. Evidently it has found a way to preserve a large fraction of its original energy. Minimal energy loss requires low values for the perpendicular magnetic dipole moment and the gravitational quadrupole moment. The first binary pulsar, 1913+16, is an example of a rapidly rotating neutron star (17 Hz) with a very low moments based on period derivative measurements. Observations of the spin decay will determine the dominant energy loss mechanism.

Two factors lead us to suggest that the pulsar and the H II region are related: (1) The two objects are in the same area of sky in a region of relatively low radio confusion. (2) The brightest part of the H II region and the pulsar are displaced to opposite sides of the near circular, low-level contours of the H II region.

We propose that the pulsar and the exciting star of the H II region were formerly members of a binary system. One of the components evolved and went through a quiet neutron-star formation stage. Formation of neutron stars in binary systems is also required to explain X-ray and radio pulsar binaries. A large and asymmetric energy and momentum transfer to the neutron star from the other component must have provided the escape velocity necessary to disrupt the binary orbit. This transfer also provides a means for creating a massive, high density object. The present separation of the pulsar from the centre for the HII region suggests an epoch for the disruption event of 7,800 yr ago and a distance of 2 kpc if we assume a typical pulsar transverse velocity of 150 km s^{-1}. Proper motion of the pulsar would be southward in declination of 0.015 arc s yr^{-1}.

We thank the staff of Arecibo Observatory for support and our colleagues for many exciting discussions. This research is supported by grants from the NSF and the Netherlands Foundation for Radio Astronomy. The Arecibo Observatory is part of the National Astronomy and Ionosphere Centre which is operated by Cornell University under contract with NSF.

(**300**, 615-618; 1982)

这颗脉冲星的年龄还是个谜。对中子星最大的自转率约为 2,000 Hz，奥斯特里克和冈恩 [9] 的蟹状星云脉冲星模型预言：由于引力四极辐射，将发生周期衰减，在一年内从这个自转率降到 100 Hz。但是我们没有观测到这种快速衰减，更没有发现由近期形成中子星而产生的任何遗迹的证据。我们的射电成图显示脉冲星附近没有同步辐射的星云。在脉冲星的位置、推测距离 2 kpc 处，爱因斯坦卫星的观测给出中子星表面温度上限为 1.5×10^6 K，并排除了同步辐射星云的可能性（赫尔方，个人交流）。这些 X 射线方面的限制没有考虑可能的很强的消光。在 COS B 星表 [10] 里没找到位于这颗脉冲星位置的源。利克天文台在这颗脉冲星位置观测到一颗 20 星等的星 [11]。直接类比蟹状星云脉冲星，我们预期这一天体将显示光学脉动。

我们得出结论，尽管这颗脉冲星的自转速率和转动能都很大，但它并不年轻。显然它找到了一种保存大部分原初能量的方式。最小的能量损失要求垂直磁偶极矩和引力四极矩都很小。基于对周期变化率的测量，第一颗脉冲双星 1913+16 就是一颗磁矩和质量矩很低的快转自旋中子星（17 Hz）的例子。自转衰减的观测将确定主导的能量损失机制。

我们基于两个因素认为脉冲星和 H II 区成协：(1)这两个天体位于同一天区，那里的射电混淆度相对较低；(2)H II 区最亮部分和脉冲星反向偏离 H II 区的中心（由近圆形的低值等流量线而定）。

我们认为脉冲星和 H II 区的激发星曾经是一个双星系统。双星成员之一演化并历经了一个并不剧烈的中子星形成阶段。我们也需要中子星在双星系统中的形成来解释 X 射线和射电脉冲双星。从伴星到中子星的巨大而不对称的能量和角动量转移一定能够提供足够的逃逸速度来瓦解双星轨道。这一转移也为产生大质量、高密度天体提供了途径。如果我们取 150 km · s^{-1} 作为脉冲星的典型横向速度，目前脉冲星和 H II 区中心的间距表明：在 7,800 年前，2 kpc 距离处曾发生一起双星轨道瓦解事件。脉冲星应该沿着赤纬向南自行，速度为 0.015 arc s · yr^{-1}。

我们对阿雷西博天文台员工的支持以及与我们同事许多令人兴奋的讨论表示感谢。本研究由国家科学基金和荷兰射电天文学基金会资助。阿雷西博天文台隶属于国家科学基金资助、康奈尔大学管理的国家天文与电离层研究中心。

（肖莉 周旻辰 翻译；徐仁新 审稿）

D. C. Backer*, Shrinivas R. Kulkarni*, Carl Heiles*, M. M. Davis[†] and W. M. Goss[‡]

* Radio Astronomy Laboratory and Astronomy Department, University of California, Berkeley, California 94720, USA
[†] National Astronomy and Ionosphere Center, Arecibo, Puerto Rico
[‡] Kapteyn Laboratorium, Groningen, The Netherlands

Received 22 November; accepted 25 November 1982.

References:

1. Readhead, A. C. S. & Hewish, A. *Nature* **236**, 440 (1972).

2. Rickard, J. J. & Cronyn, W. *Astrophys. J.* **228**, 755 (1979).

3. Hewish, A. & Okoye, S. E. *Nature* **207**, 55 (1965).

4. Erickson, W. C., *Bull. Am. Astr. Soc.* **12**, 799 (1980).

5. Slee, O. B., *Austr. J. Phys. Suppl.* **36**, 1 (1977).

6. Altenhoff, W. J., Downes, D., Goad, L., Maxwell, A. & Rinehart, R. *Astr. Astrophys. Suppl.* **1**, 419 (1979).

7. Erickson, W. C. *Astrophys. J.* (in the press).

8. Backer, D. C., Kulkarni, S. R., Heiles, C., Davis, M. M. & Goss, W. M. *IAU Circ.* No. 3743 (1982).

9. Ostriker, J. P. & Gunn, J. E. *Astrophys. J.* **157**, 1395 (1969).

10. Swanenberg, B. N. *et al. Astrophys. J. Lett.*, **243**, L69 (1981).

11. Djorgovski, S. *Nature* **300**, 618-619 (1982).

Evidence for a Common Central-engine Mechanism in all Extragalactic Radio Sources

S. Rawlings and R. Saunders

Editor's Note

Quasars are the most extreme example of "active galactic nuclei" (AGN). All such objects generate a lot of energy within a small space, but their observed characteristics vary widely. While it was generally believed that the intense energy output is caused by massive black holes at the centres of these galaxies, direct evidence was lacking. Steve Rawlings and Richard Saunders of Cambridge University here describe a relationship between the total energy of AGN and the luminosity carried by narrow emission lines. They concluded that all AGN have the same origin, with a black hole "central engine". This rapidly became known as the "standard model" and is the currently accepted picture.

Extragalactic radio sources produce radio waves and narrow emission lines by very different physical processes: synchrotron radio emission arises from lobes filled with magnetized plasma extending over scales of kiloparsecs to megaparsecs, fed by the total kinetic power, Q, of jets driven by a central engine, whereas narrow-line luminosity L_{NLR} arises from gas, typically concentrated in the inner few kiloparsecs, that has been photoionized by a nuclear source. We report here the discovery of a close relationship between Q and L_{NLR}—an approximate proportionality which extends over four orders of magnitude from low-Q radio sources with relaxed structures to high-Q, radio-luminous classical double-lobe radio galaxies. Objects with broad Balmer lines follow the same trend as those without, showing that quasar-like photoionizing sources are ubiquitous but not always obvious. Moreover, all radio-source central engines channel at least as much power into the jets as is radiated by accretion: this high efficiency implies that the engine is a massive spinning black hole which both powers the jets and controls the accretion rate.

WE have recently found a positive correlation between the narrow-line luminosities and the radio luminosities of an unbiased sample of low-redshift ($z < 0.5$) Fanaroff and Riley[1] (hereafter FR) class II radio galaxies[2], providing quantitative evidence that the production of nuclear optical line emission and the production of large-scale radio emission are somehow intrinsically linked. The radio luminosity, L_{rad}, of an extended extragalactic radio source is much smaller than the bulk kinetic power Q that goes into increasing the energy (stored in fields and particles) of the growing lobes and pushing back the external medium[3]. Because Q is a much more direct measure than L_{rad} of what powers the radio source, and as some scatter in the relationship between L_{rad} and L_{NLR} may be induced simply by the effects of the lobe environment, we investigate here the observational relation between Q and L_{NLR}.

所有河外星系射电源有共同中央引擎机制的证据

罗林斯，桑德斯

编者按

类星体是"活动星系核"中最极端的例子，它们在狭小的空间内产生巨大的能量，但是观测特性变化很大。人们普遍认为其剧烈的能量产出是这些星系中心的大质量黑洞造成的，但却缺乏直接的证据。本文中，剑桥大学的史蒂夫·罗林斯和理查德·桑德斯描述了活动星系核的总能量与窄发射线光度之间的关系。他们的结论是：所有的活动星系核有相同的起源，那就是拥有一个黑洞"中央引擎"。很快该理论即作为"标准模型"为人所知，并且也是现在人们接受的图景。

河外射电源通过很不相同的物理过程产生射电波和窄发射线：同步射电辐射产生于延伸数千秒差距到数百万秒差距的充满磁化等离子体的瓣状结构，由中央引擎驱动的喷流的总动能功率(Q)提供能量；而窄线光度L_{NLR}则来自被核心源光致电离的气体，这些气体一般聚集在靠内的几千秒差距范围内。在此我们报告一个新发现，即喷流功率Q与窄发射谱线的光度L_{NLR}近似成正比关系；对从疏松结构的低喷流功率射电源到高喷流功率的经典双瓣强射电星系，此关系在超过4个量级的范围内均成立。有、无宽巴耳末线的天体具有相同的趋势，表明类似类星体的光致电离源是普遍存在的，但并非总是明显的。此外，所有射电源中央引擎为喷射提供了至少和吸积盘辐射一样多的功率：这么高的效率意味着该中央引擎是质量巨大的旋转黑洞，该黑洞既为喷流提供功率又控制着吸积率。

我们最近发现，低红移（$z<0.5$）FRII（FR，法纳罗夫与赖利的缩写[1]）射电星系[2]的一个无偏样本的射电光度与窄线的光度存在正相关，这提供了定量证据表明星系核光学谱线辐射的产生与大尺度射电辐射的产生有着某种内在联系。一个延展的河外射电源的射电光度为L_{rad}，总动能功率Q注入射电瓣里（存储在场和粒子中）使瓣的能量增加，不断膨胀，并推回外部介质，而L_{rad}比Q要小得多[3]。因为对向射电源提供能量而言，与L_{rad}相比，Q是一种更加直接的测量量；在L_{rad}与L_{NLR}的关系中的一些弥散可能只是瓣的环境效应引起的，因此我们在此只研究观测到的Q与L_{NLR}的关系。

For a total lobe energy E, an efficiency η that allows for work done on the external medium, and a lobe age T, the power $Q = E/T\eta$. We first take E as the equipartition energy[4]. (The main assumption here is that the electrons and the magnetic field make an equal contribution to the total energy density, which corresponds closely to the minimum value required to produce the observed synchrotron emission; this energy density is integrated over the volume of the radio source to produce the equipartition energy.) We take η to be 0.5, a value suggested by identifying compact hotspots as the working surfaces of relativistic jets[5]. We can thus evaluate Q for a source if T is known. Spectral ages are available for a number of radio galaxies; these are derived using standard formulae that relate the curvature of radio spectra to the energy losses incurred by relativistic electrons because of their own synchrotron radiation and because they scatter microwave-background photons by the inverse Compton process[6]. Where spectral ages are unavailable, we can estimate the age as follows. For a lobe of length $D/2$, width W and growth speed $V = D/2T$, we can relate the equipartition pressure p_{eq} in the lobe to the confining density ρ by $\rho V^2 = kD^2W^{-2}p_{eq}$; k reflects how the lobes are confined. D, W and p_{eq} are all straightforward to measure. For $k = 25$ there is good agreement between the estimated and spectral ages of the 16 FRII radio galaxies for which both have been reliably measured[7], and we therefore adopt this value. Sufficiently accurate estimates of ρ can be made from X-ray and optical data[5].

We have evaluated Q for the 39 FRII radio galaxies in our unbiased sample of ref. 2; details of the assumptions made and the data used are given in Table 1. In Fig. 1 we plot Q against L_{NLR}; the method used to calculate L_{NLR} from spectrophotometric data is described in the caption. There is a strong positive correlation extending over roughly two orders of magnitude in each variable, which is significantly better than the correlation between L_{rad} and L_{NLR} (ref. 2). Our unbiased sample was drawn from the complete 3CR sample[8], but our selection criteria excluded three classes of radio source: (1) those of FR class I; (2) those associated with quasars rather than galaxies; (3) all objects at $z > 0.5$. To include these in our investigation, we also plot those 3CR objects in each class for which there are suitable data to calculate Q and L_{NLR}. For the FRIs, values of E and spectral ages were used as above; for the quasars and high-z radio galaxies, growth speeds of $0.1c$ were assumed based on existing values of T for similar objects (for example, ref. 6). With the addition of these objects, Q and L_{NLR} are correlated over approximately four orders of magnitude in each variable with scatter of only one order of magnitude; the best-fit line $Q \propto L_{NLR}^{0.9\pm0.2}$ implies that the relationship is close to a proportionality.

当瓣总能量为 E，对外部介质做功效率为 η，瓣的年龄为 T 时，功率 $Q = E/T\eta$。我们首先把 E 看作均分能量 [4]。（这里主要的假设是，电子和磁场对于总能量密度的贡献相等，该能量密度值接近产生观测到的同步辐射所需的最小值；该能量密度通过对整个射电源体积进行积分从而得到均分能量）我们令 $\eta = 0.5$，该值是把致密热斑确认为相对论喷流的工作面得到的 [5]。因此，如果 T 是已知的，那么我们就可以估计出一个射电源的功率 Q。若干射电星系的光谱年龄是已知的，这是通过标准方程得到的，这些方程给出了射电光谱的曲率与相对论电子的能量损失的关系，而这些电子的能量损失是由于它们自身的同步辐射以及对微波背景辐射光子的逆康普顿散射 [6]。对于光谱年龄未知的情况，我们可以按照如下方式估计年龄：对于长度为 $D/2$，宽度为 W，膨胀速度为 $V = D/2T$ 的瓣，我们可以通过公式 $\rho V^2 = kD^2W^{-2}p_{eq}$ 将瓣中的均分压强 p_{eq} 与约束密度 ρ 联系起来，其中 k 反映瓣的约束程度，D、W 和 p_{eq} 都可以直接测量得到。对于 $k = 25$ 的情况，对 16 个 FRII 型射电星系，根据上式估计的年龄与光谱年龄有很好的一致性 [7]，这些年龄都是经过可靠测量的，因此我们将 k 取值为 25。从 X 射线和光学数据中可以足够精确地估计 ρ 值 [5]。

在参考文献 2 我们的无偏样本中，我们已经估算出 39 个 FRII 型射电星系的 Q 值；其中所作假设和采用数据详见表 1。图 1 给出了 Q 与 L_{NLR} 的关系图，并且在图注中描述了利用分光光度数据计算 L_{NLR} 的方法。在每个变量的大致跨越两个量级的范围内，Q 与 L_{NLR} 存在强的正相关，明显高于 L_{rad} 与 L_{NLR} 之间的相关性（文献 2）。我们的无偏样本取自完整的 3CR 样本 [8]，但是我们的挑选标准排除了三类射电源：（1）FRI 型射电源；（2）与类星体相伴而不是与星系相伴的射电源；（3）所有 $z > 0.5$ 的天体。为了在我们的研究中包含这些射电源，我们也绘出了具有适合计算 Q 与 L_{NLR} 的数据的以上三类 3CR 天体。对于 FRI 型射电源，均分能量 E 与光谱年龄的取值将如上使用；对于类星体和高红移的射电星系，根据类似天体已有的 T 值可以假定其增长速度为 $0.1c$（参考文献 6）。加入以上天体之后，在每个变量跨越大约四个量级的范围内，Q 与 L_{NLR} 相关，而且弥散仅为一个量级。最佳拟合曲线 $Q \propto L_{NLR}^{0.9\pm0.2}$ 表明二者的关系接近正比。

Table 1. Data for calculation of jet power

a, Unbiased sample: 39 FRII radio galaxies with $z < 0.5$

Name	T (yr)	n_e (m^{-3})	Q (W)	Name	T (yr)	n_e (m^{-3})	Q (W)
3C33		5×10^1	9×10^{37}	3C284	3×10^7		1×10^{38}
3C35	1×10^8		1×10^{37}	3C285	4×10^7		1×10^{37}
3C42		5×10^2	3×10^{38}	3C295		1×10^4	7×10^{38}
3C46		1×10^2	4×10^{38}	3C300	3×10^7		1×10^{38}
3C67		1×10^4	2×10^{38}	3C303		5×10^2	5×10^{37}
3C79	6×10^6		6×10^{38}	3C319	3×10^7		4×10^{37}
3C98		5×10^1	3×10^{37}	3C321		1×10^1	1×10^{38}
3C109		5×10^1	6×10^{38}	3C326		1×10^0	1×10^{38}
4C14.11		1×10^3	5×10^{37}	3C341		1×10^1	2×10^{39}
3C123		1×10^3	5×10^{38}	3C349		5×10^1	2×10^{38}
3C173.1		1×10^2	5×10^{38}	3C381		1×10^2	2×10^{38}
3C184.1	3×10^7		4×10^{37}	3C382		5×10^1	3×10^{37}
DA240		5×10^0	1×10^{37}	3C388		1×10^3	3×10^{37}
3C192		5×10^1	5×10^{37}	3C390.3		1×10^2	5×10^{37}
3C219	6×10^7		1×10^{38}	3C401		1×10^3	1×10^{38}
3C223		1×10^1	2×10^{38}	3C436	2×10^7		9×10^{37}
4C73.08	3×10^7		6×10^{37}	3C438		5×10^3	3×10^{38}
3C234	5×10^6		3×10^{38}	3C452	3×10^7		6×10^{37}
3C236	8×10^7		1×10^{38}	3C457		1×10^1	1×10^{39}
3C244.1	7×10^6		3×10^{38}				

b, 3CR FRIs with 178-MHz radio luminosity $< 2\times10^{25}$ W Hz^{-1}sr^{-1}

Name	T (yr)	n_e (m^{-3})	Q (W)	Name	T (yr)	n_e (m^{-3})	Q (W)
3C31	4×10^8		8×10^{35}	3C338	3×10^7		1×10^{36}
3C66B	3×10^7		2×10^{36}	NGC6251	1×10^8		5×10^{36}
3C76.1	8×10^7		1×10^{36}	3C465	2×10^8		5×10^{36}
3C264	1×10^7		3×10^{36}				

Equipartition energy was calculated assuming no energy in protons, minimum and maximum frequencies of 10 MHz and 10 GHz, and a volume-filling factor of one. Spectral ages T, where available, were taken from the literature (refs 6, 7, 14 for the FRIIs and refs 15–21 for the FRIs); confining electron number densities n_e were estimated, where required, using the available X-ray and optical data[5,7] and were related to gas density ρ by $\rho = 1.4 n_e m_p$ (m_p is the proton rest mass). High-z FRII radio sources studied were 9 3CR quasars with $z < 1$ (3C47, 3C175, 3C196, 3C215, 3C249.1, 3C263, 3C275.1, 3C334, 3C351) and 24 3CR radio galaxies (3C6.1, 3C22, 3C34, 3C55, 3C172, 3C175.1, 3C184, 3C217, 3C226, 3C228, 3C247, 3C263.1, 3C265, 3C268.1, 3C277.2, 3C280, 3C289, 3C330, 3C337, 3C340, 3C343.1, 3C352, 3C427.1, 3C441). The references for the radio data for all objects are given in ref. 8. We take the Hubble constant H_0 to be 50 km s^{-1} Mpc^{-1}, and the cosmological density parameter Ω_0 and the cosmological constant Λ to be zero.

表 1. 喷流功率的计算数据

a, 无偏样本：39 个 FRII 射电星系 (z<0.5)

名称	T (yr)	$n_e(\mathrm{m^{-3}})$	Q (W)	名称	T (yr)	$n_e(\mathrm{m^{-3}})$	Q (W)
3C33		5×10^1	9×10^{37}	3C284	3×10^7		1×10^{38}
3C35	1×10^8		1×10^{37}	3C285	4×10^7		1×10^{37}
3C42		5×10^2	3×10^{38}	3C295		1×10^4	7×10^{38}
3C46		1×10^2	4×10^{38}	3C300	3×10^7		1×10^{38}
3C67		1×10^4	2×10^{38}	3C303		5×10^2	5×10^{37}
3C79	6×10^6		6×10^{38}	3C319	3×10^7		4×10^{37}
3C98		5×10^1	3×10^{37}	3C321		1×10^1	1×10^{38}
3C109		5×10^1	6×10^{38}	3C326		1×10^0	1×10^{38}
4C14.11		1×10^3	5×10^{37}	3C341		1×10^1	2×10^{39}
3C123		1×10^3	5×10^{38}	3C349		5×10^1	2×10^{38}
3C173.1		1×10^2	5×10^{38}	3C381		1×10^2	2×10^{38}
3C184.1	3×10^7		4×10^{37}	3C382		5×10^1	3×10^{37}
DA240		5×10^0	1×10^{37}	3C388		1×10^3	3×10^{37}
3C192		5×10^1	5×10^{37}	3C390.3		1×10^2	5×10^{37}
3C219	6×10^7		1×10^{38}	3C401		1×10^3	1×10^{38}
3C223		1×10^1	2×10^{38}	3C436	2×10^7		9×10^{37}
4C73.08	3×10^7		6×10^{37}	3C438		5×10^3	3×10^{38}
3C234	5×10^6		3×10^{38}	3C452	3×10^7		6×10^{37}
3C236	8×10^7		1×10^{38}	3C457		1×10^1	1×10^{39}
3C244.1	7×10^6		3×10^{38}				

b, 3CR FRI 类射电星系, 其 178 MHz 的射电光度 $< 2\times10^{25}$ $\mathrm{W\cdot Hz^{-1}\cdot sr^{-1}}$

名称	T (yr)	$n_e(\mathrm{m^{-3}})$	Q (W)	名称	T (yr)	$n_e(\mathrm{m^{-3}})$	Q (W)
3C31	4×10^8		8×10^{35}	3C338	3×10^7		1×10^{36}
3C66B	3×10^7		2×10^{36}	NGC6251	1×10^8		5×10^{36}
3C76.1	8×10^7		1×10^{36}	3C465	2×10^8		5×10^{36}
3C264	1×10^7		3×10^{36}				

在均分能量的计算中，假定质子没有能量，最小和最大频率分别为 10 MHz、10 GHz，体积填充因子为 1。表中光谱年龄 T（如果有的话）皆取自参考文献（关于 FRII 型的文献参见 6、7、14，关于 FRI 型的文献参见 15~21）；在需要的地方，束缚电子数密度 n_e 可以通过可用的 X 射线和光学数据进行估算 [5,7]，并且通过公式 $\rho = 1.4n_e m_p$ 与气体密度 ρ 联系起来（m_p 是质子的静止质量）。这里研究的高红移 FRII 型射电源是 9 个 $z<1$ 的 3CR 类星体（分别为：3C47、3C175、3C196、3C215、3C249.1、3C263、3C275.1、3C334、3C351）和 24 个 3CR 射电星系（3C6.1、3C22、3C34、3C55、3C172、3C175.1、3C184、3C217、3C226、3C228、3C247、3C263.1、3C265、3C268.1、3C277.2、3C280、3C289、3C330、3C337、3C340、3C343.1、3C352、3C427.1、3C441）。以上所有天体的射电数据的参考文献均在参考文献 8 中给出。我们采用的哈勃常数 H_0 为 50 $\mathrm{km\cdot s^{-1}\cdot Mpc^{-1}}$，宇宙学密度参数 Ω_0 和宇宙学常量 Λ 均为 0。

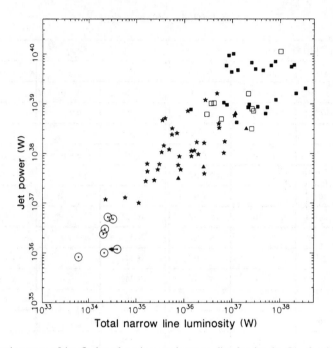

Fig. 1. Bulk kinetic power of jet Q plotted against total narrow-line luminosity L_{NLR} for the radio sources in Table 1. $\odot$, FRI sources; $\star$, FRII radio galaxies in our unbiased sample of objects with $z < 0.5$; $\blacktriangle$, broad-line radio galaxies; $\blacksquare$, radio galaxies with $0.5 < z < 1$; $\square$, quasars with $z < 1$. L_{NLR} has been measured for all objects in our unbiased sample; there are no missing points or upper limits. For the FRIs, we have used the 7 with 178-MHz radio luminosity $< 2\times10^{25}\,\mathrm{W\,Hz^{-1}\,sr^{-1}}$ for which suitable data are available; there are 7 other FRIs for which there are no data but these have similar structures, sizes and radio luminosities to those used here and bias is unlikely. For the high-z FRII radio sources, there are suitable data for 24 of the 28 possible radio galaxies and 9 of the 12 possible quasars. The spectrophotometric data for most of the objects were taken from refs 2, 11, 22, with additional data for 3C76.1 (ref. 23; an upper limit), 3C196 (ref. 24), 3C275.1 (ref. 25), the new identification for 3C326 (ref. 26), 3C334 (ref. 27), 3C338 (refs 8, 28). We measured a flux of Hα + [N II] of $7.5\times10^{-18}\,\mathrm{W\,m^{-2}}$ for NGC6251 during the observing run described in ref. 29. For 3C31, 3C66B and 3C465, we used the data in ref. 30, calibrated against the stellar light of our NGC6251 spectrum (using an aperture correction), and obtained Hα + [N II] fluxes of 5.5, 11.0 and $7.5\times10^{-18}\,\mathrm{W\,m^{-2}}$, respectively. As these spectra imply a low ionization parameter we have assumed for these objects that $L_{H\alpha\,+\,[N\,II]} \approx L_{[O\,II]}$. L_{NLR} represents the total luminosity of all optical narrow lines and Lyα, and is calculated[7] as $3\times(3\times L_{[O\,II]} + 1.5\times L_{[O\,III]})$, where we have made use of the close relationships between the fluxes of low- and high-ionization lines with [O II] (3,727 Å) and [O III] (5,007 Å), respectively, and have adopted a median value for the ratio of total recombination-line radiation to total forbidden-line radiation. If only one oxygen line was measured we estimated the other, where possible, using measurements of other lines of similar ionization, otherwise we used the relation in ref. 29 at $z < 0.5$ and $L_{[O\,III]} = 4\times L_{[O\,II]}$ at $z > 0.5$ (ref. 22); $L_{[N\,II]}$ ([N II] at 6,548 and 6,583 Å) was sometimes substituted for $L_{[O\,II]}$.

A striking feature of Fig. 1 is that quasars and broad-line radio galaxies follow the general trend defined by the narrow-line radio galaxies (NLRGs), despite the vast differences in their continua. This is important evidence supporting models in which NLRGs have central quasars hidden from us by obscuration but not hidden from their associated narrow-line regions[9,10]. We note that quasars may appear to have higher L_{NLR} than NLRGs with similar L_{rad} (ref. 11) because objects likely to be classified as quasars will be biased

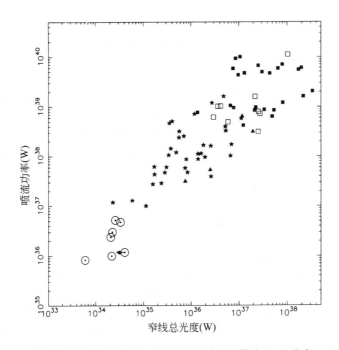

图 1. 给出了表 1 中射电源的喷流总动能功率 Q 与窄线总光度 L_{NLR} 的关系。⊙代表 FRI 型射电源；★代表我们的 $z<0.5$ 的天体的无偏样本中的 FRII 型射电星系；▲代表宽线射电星系；■代表 $0.5 < z < 1$ 的射电星系；□代表 $z < 1$ 的类星体。在我们的无偏样本中，已测量了所有天体的 L_{NLR} 值，不存在丢失的点或上限。对于 FRI 型射电源，我们采用了 178 MHz 波段射电光度小于 2×10^{25} $W \cdot Hz^{-1} \cdot sr^{-1}$ 的 7 个射电源，它们有合适的数据；有另外 7 个 FRI 型射电源没有数据，但它们与以上采用的 7 个具有类似的结构、尺寸和射电光度，因此不太可能有偏差。对于高红移的 FRII 型射电源，28 个可能的射电星系中有 24 个有合适的数据，12 个可能的类星体中有 9 个。大多数研究对象的分光光度数据来源于参考文献 2、11 和 22，还有一些数据来自 3C76.1（参考文献 23，上限值）、3C196（参考文献 24）、3C275.1（参考文献 25）、新证认的 3C326（参考文献 26）、3C334（参考文献 27）和 3C338（参考文献 8 和 28）。在参考文献 29 描述的观测过程中，我们测量了 NGC6251 的 Hα+[N II] 流量，为 7.5×10^{-18} $W \cdot m^{-2}$。对于 3C31、3C66B 和 3C465，我们采用了参考文献 30 中的数据，根据我们测量的 NGC6251 恒星光谱（使用了孔径校正）进行了校准，获得的 Hα+[N II] 流量分别为 5.5×10^{-18} $W \cdot m^{-2}$、11.0×10^{-18} $W \cdot m^{-2}$、7.5×10^{-18} $W \cdot m^{-2}$。由于这些光谱意味着较低的电离参数，我们假定 $L_{Hα + [N II]} \approx L_{[O II]}$。$L_{NLR}$ 代表了所有光学窄线和 Lyα 弧的总光度，可以通过 $3 \times (3 \times L_{[O II]} + 1.5 \times L_{[O III]})$ 计算得到 [7]，上式利用了 [O III]（5,007 Å）和 [O II]（3,727 Å）的高低电离谱线流量的紧密关系，并且使用了总复合线辐射和总禁线辐射比率的中值。如果只测量了一条氧的谱线，那么我们就在可能的情况下通过测量类似电离态的其他谱线估计另一条谱线；否则，我们将在 $z<0.5$ 时采用参考文献 29 中的关系式，在 $z>0.5$ 时参考文献 22 中的公式 $L_{[O III]} = 4 \times L_{[O II]}$，在有些情况下 $L_{[O II]}$ 可以被 $L_{[N II]}$（[N II] 位于 6,548 Å 和 6,583 Å 的谱线）代替。

　　图 1 中值得注意的是，类星体和宽线射电星系遵循由窄线射电星系（NLRGs）确定的大致趋势，尽管它们的连续谱差异巨大。这一点是支持如下模型的重要证据：NLRGs 存在中心类星体，被遮挡而不为我们所见，但是可从与它们相伴的窄线区发现 [9,10]。我们注意到，与具有相似的 L_{rad} 的 NLRGs 相比，类星体看起来可能具有更高的 L_{NLR}（参考文献 11），因为可能被归为类星体的天体倾向于具有较高的光致电离

to higher photoionizing luminosity L_{phot} and thus higher L_{NLR}, and the combination of Q–L_{NLR} proportionality and its scatter will contrive to give quasars apparently higher L_{NLR} at constant Q and thus similar L_{rad}.

We now consider accretion rates, first in the low-Q ($Q \leqslant 10^{38}$ W) objects of Fig. 1. These often contain huge stored energies E ($\sim 10^{53}$ J, equivalent to 5×10^5 $M_\odot$) and must therefore contain central masses $\geqslant 10^7$ $M_\odot$, with corresponding Eddington luminosities $L_{Edd} \geqslant 10^{38}$ W. For L_{phot} to exceed such values given the observed $L_{NLR} \leqslant 10^{36}$ W, covering factors $\kappa \leqslant 0.01$ are required, smaller than those observed in quasars (we assume equal power in photon flux per frequency decade); further, their low ionization parameters would require extreme gas densities in the narrow-line region (electron density $n_e \gg 10^8$ m^{-3}). Thus the accretion rate must be below the Eddington limit in low-Q objects.

In high-Q objects, where the energy stored in lobes is similar but L_{phot} much higher than in low-Q objects, super-Eddington accretion cannot be similarly ruled out. Any change from sub- to super-Eddington accretion as a function of Q is achieved without a discontinuity in Fig. 1, however, as may be possible with the recent model of Bell[12]. Alternatively, in all objects with radio jets, the accretion may be sub-Eddington and the central masses correspondingly large. The fraction of the power from the nucleus that is channelled into jets supports this hypothesis. Figure 1 shows that this fraction is roughly the same for all objects and indeed even for $\kappa \approx 0.01$, the radiated and bulk kinetic power are roughly equal. Thus any process whereby accretion drives the jets directly would be implausibly efficient. (Both equipartition energy E and spectral age T have the same dependence on the volume-filling factor, which is the only factor that allows E to be an overestimate of the true lobe energy. Thus, as a decrease in η also increases Q, the values of Q we use are truly minimum ones, making an even better case for high efficiency.) We are led to a model in which the radiation arises from accretion, but the jet power is produced by a machine, which, to satisfy Fig. 1, must also *control* the accretion rate. A possible model involves jets that are driven by the extraction of rotational energy from a spinning black hole, ringed by an ion-supported torus[13]. Whatever the model, it must explain the two key features of the ratio of the outputs in radiated and bulk kinetic form seen in Fig. 1: (1) why it is at most of the order of unity; (2) why it is independent of the total power over four orders of magnitude. Both features may prove particularly challenging to models that exclude central massive black holes.

Finally, we emphasize that the objects we studied were known *a priori* to contain large-scale radio jets. Many objects have high narrow-line luminosities but negligible large-scale radio emission, such as the radio-quiet quasars. These do not obey the relation of Fig. 1. This implies that the formation of radio jets requires specific physical attributes independent of the accretion rate (such as that the central black hole be large and spinning), or that the bulk outflow is formed but then disrupted within the host galaxy. Either way, the radio-quiet and radio-loud objects are physically distinct.

(**349**, 138-140; 1991)

光度 L_{phot}，因而有较高的 L_{NLR}，而 Q 与 L_{NLR} 的正比性以及它的弥散将使得类星体在 Q 值恒定以及相似 L_{rad} 的条件下具有明显较高的 L_{NLR}。

现在我们来考虑图 1 中低喷流功率 Q（$Q \leqslant 10^{38}$ W）天体的吸积率。这其中经常包含巨大的储能 E（约 10^{53} J，等价质量为 5×10^5 $M_\odot$），因此必须包含有中心质量 $\geqslant 10^7$ $M_\odot$，相应的爱丁顿光度 $L_{Edd} \geqslant 10^{38}$ W。对于超过该值的光致电离光度 L_{phot}，考虑到观测到的 $L_{NLR} \leqslant 10^{36}$ W，覆盖因子需要满足 $\kappa \leqslant 0.01$，该值比类星体中观测到的要小（我们假定每 10 倍频程的光子流量具有相同的功率）；而且，它们的低电离参数在窄线区将要求极端的气体密度（电子密度 $n_e \gg 10^8$ m^{-3}）的存在。因此，吸积率必须小于低 Q 值天体的爱丁顿极限。

在高喷流功率的天体中，存贮在瓣中的能量相似，但是其 L_{phot} 比低 Q 值天体高得多，"超爱丁顿吸积"不能类似地被排除。如图 1 所示，吸积率从亚爱丁顿到超爱丁顿吸积范围内是 Q 的连续函数，然而，这一点在最近的贝尔（Bell）模型中也是可能的 [12]。或者存在另一种可能，在所有的有射电喷流的天体中，吸积也许都是"亚爱丁顿吸积"，中心质量也相应地很大。核心区功率传送给喷流的比例支持这个假说。从图 1 可以看出，对于所有的天体而言，这个比例大体是相同的（即使对于 $\kappa \approx 0.01$ 也是如此）；辐射功率与总动能功率是大体相等的。因此，任何借助吸积直接驱动喷流的过程都异常有效。（均分能量 E 和光谱年龄 T 对体积填充因子有相同的依赖关系，该填充因子是允许 E 的估计值高于实际的瓣能量的唯一因子。因此，因为 η 的减小会增加 Q，我们使用的 Q 值实际上是最小值，使得较高效率的情况更好）。因此我们有以下的模型：辐射来自吸积，但是喷流功率由某种机制产生，该机制为了满足图 1 必须也能够**控制**吸积率。一个可能的模型涉及喷流被由旋转黑洞抽出的旋转能量所驱动，黑洞外环绕了一个离子支撑的环 [13]。但无论模型是什么，它必须能够解释图 1 中辐射能与总动能比值的两个关键特征：（1）为什么比值正好是 1 的量级；（2）为什么比值大小与总功率大小（在超过四个量级范围上）无关。这两个特征对于那些排除中心大质量黑洞的模型都构成了相当大的挑战。

最后，我们强调，我们研究的天体事先已知包含大尺度的射电喷流。许多天体具有高窄线光度却有可以忽略的大尺度射电辐射，例如射电宁静类星体。这些天体不遵循图 1 中的关系。这意味着，射电喷流的形成需要独立于吸积率的特殊物理属性（比如中心黑洞大且在旋转），或者尽管形成大规模外流但是很快在宿主星系中分散开。不过，无论是哪种情况，射电宁静天体和射电噪天体在物理上是有区别的。

（金世超 翻译；蒋世仰 审稿）

Steve Rawlings and Richard Saunders

Mullard Radio Astronomy Observatory, Cavendish Laboratory, Madingley Road, Cambridge CB3 OHE, UK

Received 5 September; accepted 8 November 1990.

References:

1. Fanaroff, B. L. & Riley, J. M. *Mon. Not. R. astr. Soc.* **167**, 31P-35P (1974).

2. Rawlings, S., Saunders, R., Eales, S. A. & Mackay, C. D. *Mon. Not. R. astr. Soc.* **240**, 701-722 (1989).

3. Longair, M. S., Ryle, M. & Scheuer, P. A. G. *Mon. Not. R. astr. Soc.* **164**, 243-270 (1973).

4. Miley, G. K. *A. Rev. Astr. Astrophys.* **18**, 165-218 (1980).

5. Rawlings, S. in *The Interstellar Medium In External Galaxies* (eds Hollenbach, D. J. & Thronson, H. A.) 188-190 (NASA Conf. Publ. 3084, 1990).

6. Leahy, J. P., Muxlow, T. W. B. & Stephens, P. M. *Mon. Not. R. astr. Soc.* **239**, 401-440 (1989).

7. Rawlings, S. thesis, Univ. of Cambridge (1988).

8. Laing, R. A., Riley, J. M. & Longair, M. S. *Mon. Not. R. astr. Soc.* **204**, 151-187 (1983).

9. Scheuer, P. A. G. in *Superluminal Radio Sources* (eds Zensus J. A. & Pearson T. J.) 104-113 (Cambridge University Press, 1987).

10. Barthel, P. D. *Astrophys. J.* **336**, 606-611 (1989).

11. Jackson, N. & Browne, I. W. A. *Nature* **343**, 43-45 (1990).

12. Bell, A. R. *Nature* **345**, 136-138 (1990).

13. Rees, M. J., Begelman, M. C., Blandford, R. D. & Phinney, E. S. *Nature* **295**, 17-21 (1982).

14. Alexander, P. & Leahy, J. P. *Mon. Not. R. astr. Soc.* **225**, 1-26 (1987).

15. Laycock, S. C. thesis, Univ. of Cambridge (1987).

16. Northover, K. J. E. *Mon. Not. R. astr. Soc.* **165**, 369-379 (1973).

17. Macklin, J. T. *Mon. Not. R. astr. Soc.* **203**, 147-155 (1983).

18. Bridle, A. H. & Vallée, J. P. **86**, 1165-1174 (1981).

19. Burns, J. O., Schwendeman, E. & White, R. A. *Astrophys. J.* **271**, 575-585 (1983).

20. Saunders, R., Baldwin, J. E., Pooley, G. G. & Warner, P. J. *Mon. Not. R. astr. Soc.* **197**, 287-300 (1981).

21. Leahy, J. P. *Mon. Not. R. astr. Soc.* **208**, 323-345 (1984).

22. McCarthy, P. J. thesis, Univ. of California at Berkeley (1989).

23. Yee, H. K. C. & Oke, J. B. *Astrophys. J.* **226**, 753-769 (1978).

24. Fabian, A. C., Crawford, C. S., Johnstone, R. M., Allington-Smith, J. R. & Hewett, P. C. *Mon. Not. R. astr. Soc.* **235**, 13P-18P (1988).

25. Hintzen, P. & Stocke, J. *Astrophys. J.* **308**, 540-545 (1986).

26. Rawlings, S., Saunders, R., Miller, P., Jones, M. E. & Eales, S. A. *Mon. Not. R. astr. Soc.* **246**, 21P-23P (1990).

27. Steiner, J. E. *Astrophys. J.* **250**, 469-477 (1981).

28. Gunn, J. E., Stryker, L. L. & Tinsley, B. M. *Astrophys. J.* **249**, 48-67 (1981).

29. Saunders, R., Baldwin, J. E., Rawlings, S., Warner, P. J. & Miller, L. *Mon. Not. R. astr. Soc.* **238**, 777-790 (1989).

30. Yee, H. K. C. & De Robertis, M. M. in *Active Galactic Nuclei* (eds Osterbrock, D. E. & Miller, J. S.) 457-459 (Kluwer, Dordrecht, 1989).

Acknowledgements. We thank P. Miller for help with compiling data on 3CR objects and P. Hughes for comments. SR is supported by a Research Fellowship at St John's College Cambridge.

A Planetary System around the Millisecond Pulsar PSR1257+12

A. Wolszczan and D. A. Frail

Editor's Note

Millisecond pulsars spin so rapidly that it is possible to make extremely precise measurements of their rotation rate—they are very good clocks. Here Alex Wolszczan and Dale Frail analyse the timing and pulse characteristics of the pulsar PSR 1257+12, and find that these suggest it is orbited by two planets, of masses 2.8 and 3.4 times the mass of the Earth. These were the first "extrasolar" planets (those around stars other than the Sun) to be discovered. But their environment does not seem conducive to life, and so there was still much excitement when the first planet around a Sun-like star was discovered only a few years later.

Millisecond radio pulsars, which are old ($\sim 10^9$ yr), rapidly rotating neutron stars believed to be spun up by accretion of matter from their stellar companions, are usually found in binary systems with other degenerate stars[1]. Using the 305-m Arecibo radiotelescope to make precise timing measurements of pulses from the recently discovered 6.2-ms pulsar PSR1257+12 (ref. 2), we demonstrate that, rather than being associated with a stellar object, the pulsar is orbited by two or more planet-sized bodies. The planets detected so far have masses of at least 2.8 $M_\oplus$ and 3.4 $M_\oplus$, where $M_\oplus$ is the mass of the Earth. Their respective distances from the pulsar are 0.47 AU and 0.36 AU, and they move in almost circular orbits with periods of 98.2 and 66.6 days. Observations indicate that at least one more planet may be present in this system. The detection of a planetary system around a nearby (~ 500 pc), old neutron star, together with the recent report on a planetary companion to the pulsar PSR1829–10 (ref. 3) raises the tantalizing possibility that a non-negligible fraction of neutron stars observable as radio pulsars may be orbited by planet-like bodies.

THE 6.2-ms pulsar PSR1257+12 (Fig. 1) was discovered during the search at high galactic latitudes for millisecond pulsars conducted in February 1990 with the 305-m Arecibo radiotelescope at a frequency of 430 MHz (ref. 2). The characteristics of this survey and the details of data analysis are described elsewhere[4]. The confirming observations made on 5 July 1990 have been followed by routine pulse timing measurements of the new pulsar. A total of 4,040 pulse time-of-arrival (TOA) observations have been accumulated so far, with the Arecibo radiotelescope, the 40-MHz, three-level correlation spectrometer and the Princeton Mark III pulsar processor at 430 MHz and 1,400 MHz. A typical uncertainty in the TOAs derived from 1-min pulse integrations is ~ 15 μs.

毫秒脉冲星PSR1257+12的行星系统

沃尔兹森，弗雷尔

编者按

毫秒脉冲星自转速度非常快，我们可以极其精确地测定它们的自转速率——它们是非常准的时钟。本文亚历山大·沃尔兹森和戴尔·弗雷尔分析了脉冲星 PSR1257+12 的测时和脉冲特征，发现这些特征表明有分别为 2.8 倍地球质量和 3.4 倍地球质量的两个行星环绕其转动。这是最早发现的"系外"行星（围绕除太阳外的其他恒星的行星），但是它们的环境似乎不利于生命存在，所以仅仅几年后当人们发现第一颗围绕类太阳恒星公转的行星时仍感到十分兴奋。

射电毫秒脉冲星是年老的（约 10^9 年），通过吸积伴星物质而自转加速的快速旋转中子星，通常发现于与另外的简并星组成的双星系统中 [1]。我们使用直径 305 米的阿雷西博射电望远镜精确测量最近发现的 6.2 毫秒脉冲星 PSR1257 + 12 （参考文献 2）的脉冲到达时间，表明这颗脉冲星为两颗甚至更多颗行星大小的天体所环绕运动，而不是跟一颗恒星相伴。目前探测到的这两颗行星质量至少为 2.8 $M_\oplus$ 和 3.4 $M_\oplus$，这里 $M_\oplus$ 为地球质量。它们各自距离脉冲星 0.47 AU 和 0.36 AU，在周期为 98.2 天和 66.6 天的近圆轨道上运行。观测显示这个系统可能至少还存在一颗行星。在附近的（约 500 秒差距）年老中子星周围探测到的一个行星系统，以及最近报道的脉冲星 PSR1829–10 的行星伴星（参考文献 3），促使人们思考这样一种可能：相当一部分表现为射电脉冲星的中子星周围可能被类行星天体环绕。

6.2 毫秒脉冲星 PSR1257 + 12 （图 1）是于 1990 年 2 月用 305 米阿雷西博射电望远镜在 430 MHz 频段上对高银纬区域搜寻毫秒脉冲星时发现的（参考文献 2）。这次巡天的特点和数据分析细节另文介绍 [4]。1990 年 7 月 5 日确认这颗新脉冲星后，我们对它做了一系列常规脉冲到达时间测量。目前我们使用阿雷西博射电望远镜、40 MHz 的三级相关频谱仪以及在 430 MHz 和 1,400 MHz 频段的普林斯顿标记 III 脉冲星处理器，总共积累了 4,040 次脉冲到达时间（TOA）观测。1 分钟的脉冲积分得到的典型 TOA 误差约为 15μs。

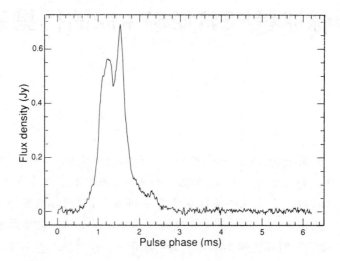

Fig. 1. The average pulse profile of PSR1257+12 at 430 MHz. The effective time resolution is ~12 μs.

The standard analysis of the timing data has been carried out using the model fitting program TEMPO[5] and the Center for Astrophysics Solar System ephemeris PEP740R. With a growing time span of the TOA measurements, it had gradually become clear that the TOAs showed an unusual variability superimposed on an annual sinusoidal pattern caused by a small (~1′) error in the assumed pulsar position. To separate these effects unambiguously, a timing-independent, interferometric position of PSR1257+12 was measured with the Very Large Array (VLA) in its A-array configuration on 19 July and again on 18 September 1991. The ~0.1″ accuracy of the resulting pulsar position was achieved by referencing the fringe phase to a point-source calibrator 1.7° away.

A least-squares fit of a simple model, which involved the pulsar's rotational period, P, and its derivative, $\dot{P}$, as free parameters and the fixed VLA position (Table 1), to the timing data spanning the period of 486 days resulted in post-fit residuals shown in Fig. 2a. The residuals, which measure the difference between the predicted and the actual TOAs, show a quasiperiodic "wandering" over the entire pulsar period. A closer examination of this effect has revealed that it was caused by two strict periodicities of 66.6 and 98.2 days in the pulse arrival times. This is further demonstrated in Fig. 2b and c, which shows post-fit residuals after fitting each of the two periods separately to the above data, assuming simple keplerian binary models involving a low-mass binary companion to the pulsar. Evidently, fitting for one of the assumed binary periods leaves the other one as a post-fit residual, implying that the pulse arrival times of PSR1257+12 are indeed affected by two independent periodicities. Further detailed analysis has shown that the periodicities are independent of radio frequency and that other millisecond pulsars routinely observed at Arecibo with the same data acquisition equipment show no such effect in their timing residuals.

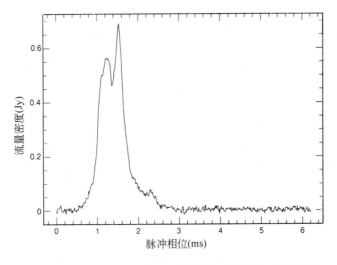

图 1. PSR1257 + 12 在 430 MHz 上的平均脉冲轮廓。有效时间分辨率约为 12μs。

我们使用模型拟合程序 TEMPO[5] 和天体物理太阳系中心历表 PEP740R 对测时数据进行标准分析。随着 TOA 测量时间跨度的增大，我们逐渐清楚，TOAs 显示了由于脉冲星假设位置的小误差（约 1′）导致的周年正弦形态上叠加了一个不寻常的变化。为了明确地区分这些效应，1991 年 7 月 19 日和 9 月 18 日，我们利用甚大阵（VLA）以 A 阵形测量了 PSR1257 + 12 的干涉仪位置，这与测时无关。以 1.7° 外的点源校准源的条纹相位为参考，我们得到的脉冲星位置坐标精度达到了约 0.1″。

用一个包括脉冲星自转周期 P，及其导数 $\dot{P}$ 作为自由参数的简单模型，加上确定的 VLA 位置（表 1），对跨度 486 天的测时数据做最小二乘法拟合，得到图 2a 中显示的拟合残差。残差是用来测量模型预测值和实际 TOAs 之间的差，这个源的残差表明在整个脉冲星周期内存在一个准周期"游荡"。对这个效应的进一步研究显示，这是脉冲到达时间中两个严格的 66.6 天和 98.2 天的周期性导致的。假设一个简单开普勒双星模型，其中脉冲星有一个小质量伴星，对以上两个周期的每个周期的数据分别进行拟合，拟合后残差分别如图 2b 和图 2c 所示，进一步证明了这两个周期性的存在。很明显，对假设的两个双星周期性中的一个进行拟合，得到的拟合后残差正好显示出另一个周期性。这表明 PSR1257 + 12 的脉冲到达时间实际上受到两个相互独立的周期性影响。进一步的仔细分析显示周期性和射电频率无关，在阿雷西博利用同样数据接收仪器对其他毫秒脉冲星进行常规观测，它们的测时残差没有显示这种效应。

Table 1. Parameters of the PSR1257 + 12 system

Pulsar parameters		
Rotational period, P	$0.00621853193177 \pm 0.00000000000001$ s	
Period derivative, $\dot{P}$	$1.21 \times 10^{-19} \pm 2 \times 10^{-21}$ s s^{-1}	
Right ascension (B1950.0, VLA)	12 h 57 min 33.131 s $\pm$ 0.015	
Right ascension (B1950.0, timing)	12 h 57 min 33.126 s $\pm$ 0.003	
Declination (B1950.0, VLA)	$12° 57' 05.9'' \pm 0.1$	
Declination (B1950.0, timing)	$12° 57' 06.60'' \pm 0.02$	
Epoch	JD 2448088.9	
Dispersion measure	10.18 ± 0.01 pc cm^{-3}	
Flux density (430 MHz)	20 ± 5 mJy	
Flux density (1,400 MHz)	1.0 ± 0.2 mJy	
Surface magnetic field, B	8.8×10^8 G	
Characteristic age, τ_c	0.8×10^9 yr	
Keplerian orbital parameters		
Projected semimajor axis, $a_1 \sin i$	1.31 ± 0.01 light ms	1.41 ± 0.01 light ms
Eccentricity, e	0.022 ± 0.007	0.020 ± 0.006
Epoch of periastron, T_o	JD 2448105.3 $\pm$ 1.0	JD 2447998.6 $\pm$ 1.0
Orbital period, P_b	5751011.0 ± 800.0 s	8487388.0 ± 1800.0 s
Longitude of periastron, ω	$252° \pm 20°$	$107° \pm 20°$
Parameters of the planetary system		
Planet mass, $m_{2,3}$ ($M_\oplus$)	$3.4/\sin i$	$2.8/\sin i$
Distance from the pulsar, d (AU)	0.36	0.47
Orbital period, P_b (days)	66.6	98.2

Millisecond pulsars are extremely stable rotators. Systematic timing observations of objects like the 1.5-ms pulsar 1937+21 (ref. 6) have not revealed any timing noise, quasiperiodic TOA variations or "glitches" at the level often found in the population of younger pulsars and believed to be related to neutron star seismology[7]. The frequency independence of the amplitude of TOA variations in PSR1257+12 rules out propagation phenomena, such as those detected in eclipsing binary pulsars[8,9], as a possible source of the observed periodicities. The integrated pulse profiles of PSR1257+12 do not show any morphological changes that might indicate the presence of a free precession of the pulsar spin axis[10,11] or any magnetospheric phenomena at the level that could lead to periodic TOA variations of the measured magnitude. Consequently, the most plausible remaining alternative is that PSR1257+12 has two low-mass companions and that their orbital motion is responsible for the observed TOA variations.

表 1. PSR1257+12 系统的参数

脉冲星参数		
自转周期，P	0.00621853193177 ± 0.00000000000001 s	
周期导数，$\dot{P}$	$1.21 \times 10^{-19} \pm 2 \times 10^{-21}$ s · s^{-1}	
赤经（B1950.0，VLA）	12 h 57 min 33.131 s ± 0.015	
赤经（B1950.0，测时）	12 h 57 min 33.126 s ± 0.003	
赤纬（B1950.0，VLA）	12°57′05.9″ ± 0.1	
赤纬（B1950.0，测时）	12°57′06.60″ ± 0.02	
历元	JD 2448088.9	
色散量	10.18 ± 0.01 pc · cm^{-3}	
流量密度（430 MHz）	20 ± 5 mJy	
流量密度（1,400 MHz）	1.0 ± 0.2 mJy	
表面磁场强度，B	8.8×10^8 G	
特征年龄，τ_c	0.8×10^9 年	
开普勒轨道参数		
投影半长轴，$a_1 \sin i$	1.31 ± 0.01 light ms	1.41 ± 0.01 light ms
偏心率，e	0.022 ± 0.007	0.020 ± 0.006
近星点历元，T_0	JD 2448105.3 ± 1.0	JD 2447998.6 ± 1.0
轨道周期，P_b	5751011.0 ± 800.0 s	8487388.0 ± 1800.0 s
近星点经度，ω	252° ± 20°	107° ± 20°
行星系统参数		
行星质量，$m_{2,3}$（$M_{\oplus}$）	3.4 / sin i	2.8 / sin i
与脉冲星的距离，d（AU）	0.36	0.47
轨道周期，P_b（天数）	66.6	98.2

　　毫秒脉冲星是极为稳定的转子。对 1.5 毫秒脉冲星 1937+21（参考文献 6）这样的天体进行系统的测时观测，并没有发现任何测时噪声、准周期 TOA 变化和经常在年轻脉冲星族中发现的自转突变（自转突变被认为同中子星星震学相关[7]）。PSR1257＋12 的 TOA 变化幅度和频率无关，这点排除了传播效应作为观测到周期性原因的可能性，例如那些在掩食脉冲双星中探测到的传播效应[8,9]。PSR1257＋12 的积分脉冲轮廓没有显示任何一种形态变化能够表明脉冲星自转轴存在一个自由进动[10,11]，或者一些磁层现象大到能够产生所测量到的幅度的 TOA 周期变化。因此，最有可能的情况是，PSR1257＋12 具有两个小质量伙伴星，它们的轨道运动导致了观测到的 TOA 变化。

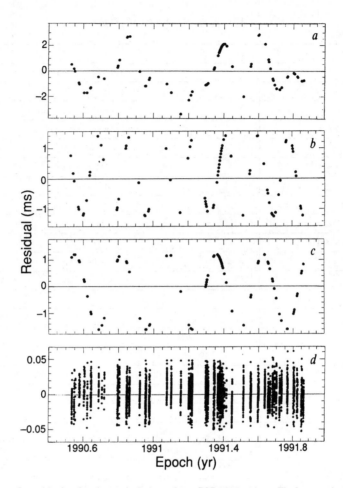

Fig. 2. The post-fit residuals of pulse arrival times from PSR1257+12. *a*, Fit for rotational parameters only (VLA position of the pulsar fixed); *b*, fit for a 98.2-day keplerian orbit (leaves a 66.6-day periodicity as residual); *c*, fit for a 66.6-day keplerian orbit (leaves a 98.2-day periodicity as residual); *d*, fit for all parameters of *a–c*.

To analyse this exciting possibility further, we have modified the code of the TEMPO program to accommodate a timing model including multiple, noninteracting keplerian orbits. The result of a fit of the model including the rotational and positional parameters of the pulsar as well as the two, five-parameter keplerian orbits to the entire data set is shown in Fig. 2*d*. The post-fit r.m.s. residual of the resultant timing model is ~18 μs (comparable to the individual TOA uncertainties) and the remaining residuals contain very little systematic variation. The model parameters of the pulsar and its assumed two companions are listed in Table 1. The derived parameters have been obtained from standard considerations involving very low-mass objects in keplerian orbits around a $1.4M_\odot$ neutron star. The pulsar period variations predicted by this model are displayed in Fig. 3 together with the observed period values derived from timing data. Note that the presence of the planet-sized bodies orbiting the pulsar results in apparent period variations of only ± 15 ps. This is caused by the pulsar's spatial motions which are characterized by the projected maximum velocity and displacement amplitudes of ±0.7 m s^{-1} and ±900 km, respectively.

146

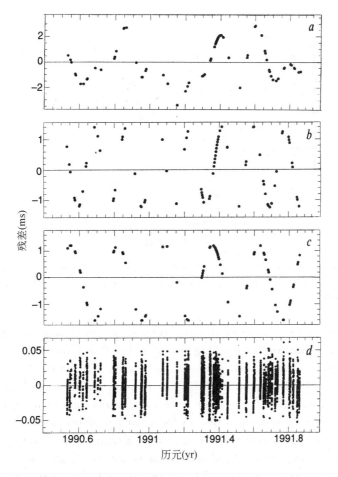

图 2. PSR1257+12 脉冲到达时间的拟合后残差。a，仅对自转参数的拟合（脉冲星的 VLA 坐标确定）；b，对 98.2 天开普勒轨道的拟合（残差的周期为 66.6 天）；c，对 66.6 天开普勒轨道的拟合（残差的周期为 98.2 天）；d，用 a~c 所有参数拟合。

为了进一步分析这种令人兴奋的可能性，我们修改了 TEMPO 程序的代码来适应包含多个非相互作用的开普勒轨道的测时模型。用包含脉冲星的旋转和位置参数以及其他两个伴星的 5 个开普勒轨道参数的模型拟合整个数据集，结果见图 2d。合成测时模型的拟合后均方根残差约为 18 μs（相当于单个 TOA 误差），剩下的残差几乎不存在系统变化。脉冲星和假设的两个伴星的模型参数见表 1。导出参数是通过有非常小质量天体绕 $1.4M_\odot$ 的中子星做开普勒轨道运动的标准考量得到的。这个模型预言的脉冲星周期变化和由测时数据推出的观测周期值都示于图 3。注意，脉冲星周围环绕的行星尺度天体的存在仅产生 ±15 ps 的可见的周期变化。这是脉冲星的空间运动导致的，空间运动的特征为最大投影速度和位移幅度分别为 ±0.7 m·s⁻¹ 和 ±900 km。

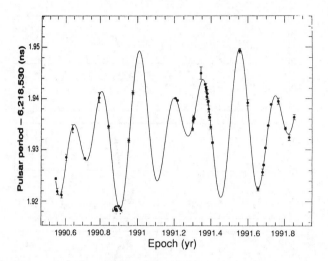

Fig. 3. Period variations of PSR1257+12. Each period measurement is based on observations made on at least two consecutive days. The solid line denotes changes in period predicted by a two-planet model of the 1257+12 system.

The high quality of the two-companion model fit to the pulsar timing data covering several orbital cycles provides compelling evidence that PSR1257+12 possesses a planetary system consisting of at least two planet-sized bodies. The possibility of more planets around PSR1257+12 is indicated by a 0.7″ discrepancy between the VLA and the timing positions, which is considerably greater than the conservative error estimates (Table 1). This discrepancy may arise from a bias in the timing position caused by a presence of a third planet with orbital period close to one year. Because the currently measured value of the second period derivative, $\ddot{P}$, of the pulsar is not significant, the effect of any outer planets that could be present in the 1257+12 system would be entirely absorbed in $\dot{P}$.

The characteristics of the 1257+12 planets are not unlike those of the inner Solar System. Both planets circle the pulsar at distances similar to that of Mercury in its orbit around the Sun. Assuming a random distribution of the inclinations, i, of orbital planes, there is a 50% probability that $i \geqslant 60°$, so that the median values of the permissible masses are $3.2 M_\oplus$ and $3.9 M_\oplus$, respectively (see Table 1). Interestingly, the ratio of orbital periods, 1.476 ±0.001, is close to a 3/2 orbital resonance of the type often encountered in the Solar System, either between planetary satellites, between Jupiter and some asteroids, or even Neptune and Pluto[12]. Also, the similarity of the measured eccentricities combined with the ~180° separation of the pericentres of the orbits can be easily understood in terms of secular perturbations of the orbital elements of the two planets (see, for example ref. 13). Finally, a hypothetical energy flux at the planets' distance from the pulsar (~0.4 AU) due to the neutron star's rotational energy loss ($\dot{E} = I\omega\dot{\omega}$, where I is the moment of inertia and $\omega = 2\pi/P$) is ~4 × 10^7 erg cm^{-2} s^{-1} . This is ~30 times the solar constant and corresponds to a black-body temperature of ~670 K, which is similar to the measured dayside surface temperature of Mercury (see, for example, ref. 14).

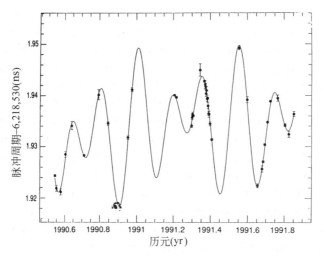

图 3. PSR1257+12 周期的变化。每一周期测量值都基于至少连续两天的观测。实线表示 1257+12 的 2 个行星系统的模型所预期的周期变化。

两个伴星模型对覆盖几个轨道周期的脉冲星测时数据的高质量拟合强有力地证明了 PSR1257+12 是由至少两颗行星尺度天体组成的行星系统。更多天体环绕 PSR1257 +12 的可能性蕴含在 VLA 和测时位置 0.7″ 的差异中，这比保守的误差估计要大很多（表 1）。这个差异可能是由于轨道周期接近一年的第三颗行星的存在导致的测时位置偏差。因为目前该脉冲星的周期二阶导数 $\ddot{P}$ 的测量值不是很显著，1257＋12 系统中存在外行星的效应可整个被 $\ddot{P}$ 吸收。

PSR1257+12 的行星特征与太阳系内行星相似。两个行星都以类似水星绕太阳的轨道距离环绕着脉冲星运动。假定轨道平面倾角 i 随机分布，$i \geqslant 60°$ 的概率为 50%，所以两行星可能的质量中值分别为 3.2 $M_{\oplus}$ 和 3.9 $M_{\oplus}$（见表 1）。有趣的是，两个轨道周期的比率为 1.476±0.001，接近于太阳系中常遇到的 3/2 轨道共振类型，或者是行星的卫星之间，或者是木星和一些小行星之间，甚至是海王星和冥王星之间 [12]。并且，对于测量到的两个轨道相似偏心率和约 180° 的近心点间隔，考虑到两个行星轨道成分的长期扰动，这都是可以很容易理解的（参见文献 13）。最后，假设的由于中子星自转动能损失（$\dot{E} = I\omega\dot{\omega}$，这里 I 是转动惯量，$\omega=2\pi/P$）导致的行星距离脉冲星（约 0.4 AU）处的能流约为 4×10^7 erg·cm^{-2}·s^{-1}，这约为太阳常数的 30 倍，对应的黑体温度约为 670 K，与测量到的水星光面的表面温度相似（参见文献 14）。

The possibility of the existence of planet-sized bodies orbiting neutron stars has been contemplated in the past[15,16]. The most recent evidence has been presented by Bailes et al[3], who have detected a ~6-month periodicity in the timing residuals of a relatively young pulsar, PSR1829–10, which is interpreted in terms of the orbital motion of a $\sim 10 M_\oplus$ companion. A number of mechanisms have been proposed to explain a planet around PSR1829–10 (refs 17–21). In the case of PSR1257+12, its old age and a low surface magnetic field (Table 1) are typical of neutron stars which are believed to evolve in low-mass binary systems and are spun up to the observed millisecond periods by accretion of matter from their stellar companions[1]. As it is not very likely that any primordial planets would survive this kind of evolution[22], the observed 1257+12 system probably consists of "second generation" planets created at or after the end of the pulsar's binary history. Because a second supernova explosion is not expected to occur in a low-mass binary, such planets could form a stable system of bodies orbiting an old neutron star. Another important evolutionary constraint is provided by the observed very low eccentricities ($e \approx 0.02$) of the planetary orbits and the near-resonance ratio of the orbital periods. These characteristics suggest that the planets were created from some form of accretion disk that would naturally provide means to circularize the orbits and to bring them close to a 3/2 resonance[12]. Consequently, it seems that any plausible mechanism for the creation of a planetary system around a millisecond pulsar must provide a way to remove its stellar companion in a manner different from the disruption of a binary system caused by a supernova explosion and to retain enough circumpulsar matter to form planet-sized objects.

Within the constraints provided by the observational evidence, it is tempting to postulate that the 1257+12 system simply represents one of the possible outcomes of a neutron star evolution in a low-mass binary system[1]. As two examples of companion-star evaporation by millisecond pulsars have already been detected[7,8], this mechanism seems to account naturally for the absence of such a companion to PSR1257+12. In this case, one way to supply the material for planet creation could be a disk formed out of the ablated stellar matter ejected from the binary system[23,34]. The existence of such an "outer disk" around the eclipsing binary pulsar PSR1957+20 could explain the observed orbital decay of this system[25].

The results described here strongly suggest that one of the nearby galactic millisecond pulsars, PSR1257+12, is accompanied by a system of two or more planet-sized bodies. The pulse timing method (similar to the analysis of single-line spectroscopic binaries) was used to detect a ±0.7 m s⁻¹ pulsar "wobble" caused by orbital motions of the planets. In fact, planetary masses smaller than that of the Moon could easily be detected with this technique. At present, this kind of accuracy is entirely inaccessible to the optical methods of planetary system detection[26]. The existence of both the 1257+12 planetary system and the object orbiting the pulsar PSR1829–10 seems to suggest that planets can form under a variety of conditions. This notion and the possibility of a non-negligible frequency of occurrence of planets around neutron stars, if confirmed, will have far-reaching consequences for our understanding of the formation and evolution of planetary systems

过去研究人员一直设想中子星周围具有行星尺度的天体环绕的可能性 [15,16]。最近的证据来自拜莱斯等人的研究 [3]，他们在一颗相对年轻的脉冲星 PSR1829–10 的测时残差中探测到一个约 6 个月的周期性，这可以用一颗质量约为 $10 \, M_⊕$ 的伴星的轨道运动来解释。研究者提出了一系列尝试解释一颗行星环绕 PSR1829–10 的机制（参考文献 17~21）。对于 PSR1257+12，它较老的年龄和较低的表面磁场强度（表 1）是一类中子星的典型特征，这类中子星被认为是在小质量双星系统中演化，并通过吸积伴星的物质而自转加速到现在我们观测到的毫秒周期 [1]。由于原初行星不大可能在这样的演化中存活下来 [22]，观测到的 1257+12 系统可能是由在脉冲星的双星历史末期或之后产生的"二代"行星组成。因为在小质量双星中不大可能发生第二次超新星爆发，这样的行星才可能形成环绕一颗年老中子星运动的稳定天体系统。另外一个重要的演化限制是观测到的行星轨道的极小的偏心率（e 约等于 0.02）和轨道周期的近共振比率。这些特征表明行星是从某种吸积盘产生的，这种吸积盘能够自然提供圆化轨道的途径，从而让它们接近于 3/2 共振 [12]。因此，任何生成环绕毫秒脉冲星的行星系统的可能机制都必须提供一种移除恒星伴星的途径，这种途径不同于超新星爆发所引起的双星系统的瓦解，并且能够留住足够多的环绕脉冲星的物质来形成行星尺度的天体。

在所提供的观测证据的限制之内，我们倾向于推测 1257+12 系统仅代表中子星在小质量双星系统中演化的一个可能结果 [1]。随着研究人员探测到两颗毫秒脉冲星蒸发伴星的事例 [7,8]，这个机制似乎自然地解释了为什么 PSR1257+12 缺少这样一颗伴星。在这种情形下，生成行星物质的一个来源可能是双星系统喷出的融化恒星物质所形成的一个物质盘 [23,24]。如果在食脉冲双星 PSR1957+20 周围存在这种"外盘"，就能够解释这个系统观测到的轨道衰减 [25]。

上述结果强有力地表明，PSR1257+12——一颗邻近的河内毫秒脉冲星，由两个或更多行星尺度的天体系统环绕。用脉冲测时方法（类似于单谱线分光双星分析）已经探测到一个由行星轨道运动引起的 $\pm 0.7 \, \mathrm{m \cdot s^{-1}}$ 的脉冲星"摇晃"。事实上，这种技术可以轻易探测到质量小于月球的行星。目前，这种精度在探测行星系统的光学方法中尚未达到 [26]。1257+12 行星系统和环绕脉冲星 PSR1829–10 的天体似乎表明行星能在多种环境下形成。这个想法以及以不可忽略的频率出现围绕中子星的行星这样的可能性一旦被证实，将对我们理解行星系统的形成和演化以及未来搜寻太

and for future strategies of searches for planets outside the Solar System.

(**355**, 145-147; 1992)

A. Wolszczan[*] and D. A. Frail[†]

[*] National Astronomy and Ionosphere Center, Arecibo Observatory, Arecibo, Puerto Rico 00613, USA

[†] National Radio Astronomy Observatory, Socorro, New Mexico 87801, USA

Received 21 November; accepted 9 December 1991.

References:

1. Bhattacharya, D. & van den Heuvel, E. P. J. *Phys. Rep.* **203**, 1-124 (1991).

2. Wolszczan, A. *IAU Circ.* No. 5073 (1990).

3. Bailes, M., Lyne, A. G. & Shemar, S. L. *Nature* **352**, 311-313 (1991).

4. Wolszczan, A. *Nature* **350**, 688-690 (1991).

5. Taylor, J. H. & Weisberg, J. M. *Astrophys. J.* **345**, 434-450 (1989).

6. Davis, M. M., Taylor, J. H., Weisberg, J. M. & Backer, D. C. *Nature* **315**, 547-550 (1985).

7. Alpar, M. A., Nandkumar, R. & Pines, D. *Astrophys. J.* **311**, 197-213 (1986).

8. Fruchter, A. S., Stinebring, D. R. & Taylor, J. H. *Nature* **333**, 237-239 (1988).

9. Lyne, A. G. *et al. Nature* **347**, 650-652 (1990).

10. Shaham, J. *Astrophys. J.* **214**, 251-260 (1977).

11. Nelson, R. W., Finn, L. S. & Wasserman, I. *Astrophys. J.* **348**, 226-231 (1990).

12. Peale, S. J. *Ann. Rev. Astr. Astrophys.* **14**, 215-246 (1976).

13. Dermott, S. F. & Nicholson, P. D. *Nature* **319**, 115-120 (1986).

14. Vilas, F. in *Mercury* (eds Vilas, F., Chapman, C. R. & Matthews, M. S.) 59-76 (University of Arizona Press, Tucson, 1988).

15. Richards, D. W., Pettengill, G. H., Counselman, C. C. III & Rankin, J. M. *Astrophys. J.* **160**, L1-6 (1970).

16. Demiański, M. & Prószyński, M. *Nature* **282**, 383-385 (1979).

17. Podsiadlowski, Ph., Pringle, J. E. & Rees, M. J. *Nature* **352**, 783-784 (1991).

18. Fabian, A. C. & Podsiadlowski, Ph. *Nature* **353**, 801-801 (1991).

19. Lin, D. N. C., Woosley, S. E. & Bodenheimer, P. H. *Nature* **353**, 827-829 (1991).

20. Krolik, J. H. *Nature* **353**, 829-831 (1991).

21. Wasserman, I., Cordes, J. M., Finn, L. S. & Nelson, R. W. *Cornell Univ. preprint* (1991).

22. Nakano, T. *Mon. Not. R. astr. Soc.* **224**, 107-130 (1987).

23. Shu, F. H., Lubow, S. H. & Anderson, L. *Astrophys. J.* **29**, 223-241 (1979).

24. Rudak, B. & Paczyński, B. *Acta Astr.* **31**, 13-24 (1981).

25. Ryba, M. F. & Taylor, J. H. *Astrophys. J.* **380**, 557-563 (1991).

26. Black, D. C. *Space Sci. Rev.* **25**, 35-81 (1980).

Acknowledgements. We thank D. Backer, P. Nicholson, B. Paczyński, F. Rasio, S. Shapiro, J. Taylor and S. Teukolsky for discussions. We also thank J. Taylor for making available the Princeton Mark III processor. Arecibo Observatory is part of the National Astronomy and Ionosphere Center, which is operated by Cornell University under contract with the NSF. The VLA is operated by the NRAO under cooperative agreement with the NSF. D.A.F. is supported by an NSERC postdoctoral fellowship and an NRAO Jansky fellowship.

阳系外行星的策略产生深远的影响。

（肖莉 翻译；徐仁新 审稿）

Discovery of the Candidate Kuiper Belt Object 1992 QB₁

D. Jewitt and J. Luu

Editor's Note

In 1951 it was hypothesized that the outer solar system (beyond Neptune's orbit) contains a population of icy bodies, in the so-called Kuiper Belt. Many searches were made, which revealed a couple of bodies beyond Saturn's orbit. By the early 1990s, telescopes were being equipped with semiconductor-based cameras that were hundreds or thousands of times more sensitive than earlier technology, such as photographic film. Here David Jewitt and Jane Luu use such a camera to find the first Kuiper Belt object. More objects were rapidly found, and by 2000 it was clear that the "planet" Pluto was really just another Kuiper Belt object. Pluto was demoted from planetary status in 2006.

The apparent emptiness of the outer Solar System has been a long-standing puzzle for astronomers, as it contrasts markedly with the abundance of asteroids and short-period comets found closer to the Sun. One explanation for this might be that the orbits of distant objects are intrinsically short-lived, perhaps owing to the gravitational influence of the giant planets. Another possibility is that such objects are very faint, and thus they might easily go undetected. An early survey[1] designed to detect distant objects culminated with the discovery of Pluto. More recently, similar surveys yielded the comet-like objects 2060 Chiron[2] and 5145 Pholus[3] beyond the orbit of Saturn. Here we report the discovery of a new object, 1992 QB₁ moving beyond the orbit of Neptune. We suggest that this may represent the first detection of a member of the Kuiper belt[4,5], the hypothesized population of objects beyond Neptune and a possible source of the short-period comets[6-8].

OUR observations are part of a deep-imaging survey[9] of the ecliptic, made with the University of Hawaii 2.2-m telescope on Mauna Kea. The survey uses Tektronix 1,024 × 1,024 pixel and 2,048 × 2,048 pixel charge-coupled devices (CCDs) at the $f/10$ Cassegrain focus. Both CCDs have anti-reflection coatings which yield quantum efficiencies of ~90% at wavelength $\lambda \approx 7,000$ Å (K. Jim, personal communication). Survey observations are obtained in sets of four images per field with a total timebase of 2 or more hours. Each image is exposed for 900 s while autoguiding at sidereal rate. Because objects in the outer Solar System have small proper motions, our survey was optimized to detect slowly moving objects (SMOs). The angular motions of SMOs are sufficiently small that little trailing-loss results from sidereal tracking. This strategy is found to provide optimum

柯伊伯带天体候选体 1992 QB$_1$ 的发现

朱维特，刘丽杏

编者按

1951 年，人们猜测外太阳系（太阳系内海王星轨道之外的空间）在后来被称为"柯伊伯带"的区域内有很多冰冷的天体。人们进行了很多搜寻，但仅在土星轨道之外找到了一些天体。到了 20 世纪 90 年代早期，人们开始在望远镜上安装基于半导体的相机，其灵敏度比照相胶片等早期技术提高了成百上千倍。大卫·朱维特和刘丽杏使用这样的相机找到了第一个柯伊伯带天体。很快更多天体被发现了。到 2000 年，人们清楚地意识到"行星"冥王星其实只是另一个柯伊伯带天体而已。冥王星于 2006 年被排除行星范围。

外太阳系表观上的空旷一直都是困扰天文学家的疑团，这与在更靠近太阳的地方发现的大量小行星及短周期彗星形成了鲜明的对比。对此，一种解释是，或许远处天体的轨道寿命本身就较短，寿命短可能是因为巨型行星的引力作用。另一种可能的解释是，这类天体非常暗淡，因此可能很难被探测到。以探测太阳系远距离天体为目的的一次早期巡天[1]发现了冥王星的存在。后来，类似的巡天在土星轨道之外发现了类彗天体 2060 喀戎[2] 和 5145 福鲁斯[3]。在此我们报告运行于海王星轨道之外的新天体 1992 QB$_1$ 的发现。我们认为这可能是第一次探测到的柯伊伯带成员[4,5]，柯伊伯带即人们猜测中的海王星之外的一族天体，也是短周期彗星的一个可能来源[6-8]。

我们的观测是利用位于冒纳凯阿火山的夏威夷大学 2.2 米望远镜进行的黄道深空成像巡天[9]项目的一部分。该观测使用 $f/10$ 的卡塞格林焦点处的美国泰克 1,024 × 1,024 像素和 2,048 × 2,048 像素电荷耦合元件（CCD）。以上两种 CCD 都有减反射镀膜，这使得波长 7,000 Å 处的量子效率达到 90% 左右（吉姆，个人交流）。在实际观测中对每个视场都拍了四幅图片，总的时间是 2 小时或者更长。在以恒星周天转速自动跟踪时，每幅图像的曝光时间是 900 s。由于外太阳系天体具有微小的自行，所以我们的巡天优先探测那些缓慢运动天体（SMOs）。这些 SMO 由于自身的角运动足够小，几乎没有跟踪造成的拖尾损失。人们发现这种观测方法对于预期的 SMO 的线性的、相

155

sensitivity to the linear, correlated motion expected of slowly moving objects. By restricting observations to stellar images of full width at half maximum (FWHM) ⩽ 1.0 arcsec, and to moon-less skies, we obtain limiting magnitudes $m_R \sim 25$. To date, a sky area of 0.7 square degrees has been imaged to this depth. We are extending our coverage to 1 square degree.

We confine our observations to the opposition direction, where the angular motion of distant objects is retrograde and primarily due to the Earth's motion. At opposition, the parallactic angular motion $\dot{\theta}$ (arcsec per hour) is given by $\dot{\theta} = 148[(1-R^{-1/2})/(R-1)]$, where R (in astronomical units, AU) is the heliocentric distance[9]. Thus, a measurement of $\dot{\theta}$ yields R directly. Our observations are restricted to the spring and autumnal equinoxes to benefit further from the large galactic latitude of the opposition point, and from the resultant low density of background stars. In fact, field galaxies pose a worse contamination problem than do field stars in our observations.

The object 1992 QB$_1$ was detected in real time on UT 1992 August 30 (ref. 10). The discovery images are shown in Fig. 1, where the faster motion of a nearby, unnumbered main-belt asteroid emphasizes the remarkably small angular motion (and hence large distance) of QB$_1$. The slow motion of the SMO (2.6 arcsec h^{-1} west and 1.1 arcsec h^{-1} south) was confirmed on UT August 31 and September 01. Absence of detectable diurnal parallax confirmed that the object was not a near-Earth asteroid whose proper motion fortuitously resembled that of a SMO. Positions of 1992 QB$_1$ were obtained with reference to stars in the Hubble Guide Star Catalogue. Measurements relative to these stars suggested an absolute astrometric accuracy of ±1 arcsec, and the relative positions were precise to about ±0.3 arcsec. Astrometric positions were communicated to B. Marsden at the Center for Astrophysics. Follow-up astrometry from the 2.2-m telescope was obtained on UT 1992 September 25, and other observers have reported astrometric measurements in the following three months[11-13]. Observations from 1992 August 30 to December 25 yielded a current heliocentric distance $R \approx 41$ AU, and were used to calculate the orbital parameters listed in Table 1 (ref. 13).

Table 1. Preliminary orbital parameters

Semi-major axis	a	44.4 AU
Eccentricity	e	0.11
Inclination	i	2.2°
Orbital period	P	296 years
Perihelion date	T	AD 2023
Perihelion distance	q	39.6
Aphelion distance	Q	49.1

Based on astrometry in the interval 1992 August 30 to December 25. Orbit solution is by Marsden[13].

关的运动最为灵敏。通过限制观测星像的半峰全宽（FWHM），在不大于 1.0 arcsec、无月夜的条件下，我们得到的极限星等为 $m_R \sim 25$。迄今，我们对 0.7 平方度的天区进行了这种深度的成像。我们正在将覆盖范围扩大到 1 平方度。

我们将观测限制在冲日方向，在那里，遥远天体的角运动主要是由于地球运动导致的逆行。在冲日点，视差角运动 $\dot{\theta}$(arcsec·h^{-1}) 为 $\dot{\theta} = 148[(1-R^{-1/2})/(R-1)]$，其中 R（单位为天文单位，AU）是与太阳的距离 [9]。因此，由 $\dot{\theta}$ 的测量就可以直接算出 R。我们的观测限制在了春分点和秋分点附近，进一步得益于冲日点的高银纬以及此处背景星的低密度。事实上，在我们的观测中，场星系造成的污染比场星造成的污染严重。

1992 QB₁ 是在世界时 1992 年 8 月 30 日实时发现的（参考文献 10）。图 1 显示了发现 1992 QB₁ 时的图像，图中其附近较快速运动的未编号的主带小行星凸现了 QB₁ 天体非常小的角运动（是由于距离遥远）。该 SMO 的缓慢运动（以 2.6 arcsec·h^{-1} 向西、1.1 arcsec·h^{-1} 向南运动）在世界时 8 月 31 日和 9 月 1 日得到了证实。基于没有发现周日视差可以确定，该天体不是一颗运动恰好类似于 SMO 的近地小行星。1992 QB₁ 的位置是参考哈勃导星星表的恒星得到的。相对于这些星的测量，给出的天体测量的绝对精度为 ±1 arcsec，相对位置的精度约为 ±0.3 arcsec。天体测量位置的结果转达给了天体物理中心的马斯登。世界时 1992 年 9 月 25 日我们用 2.2 米望远镜进行了后续天体测量，其他观测者在接下来的三个月内报告了天体测量结果 [11-13]。从 1992 年 8 月 30 日到 12 月 25 日的观测结果得出了现在的日心距 $R \approx 41$ AU，并且被用来计算轨道参数，相应的轨道参数列于表 1 中（参考文献 13）。

表 1. 轨道参数初步结果

半长轴	a	44.4 AU
偏心率	e	0.11
轨道倾角	i	2.2°
轨道周期	P	296 年
过近日点日期	T	AD 2023
近日点距离	q	39.6
远日点距离	Q	49.1

基于 1992 年 8 月 30 日到 12 月 25 日的天体测量。轨道解算由马斯登完成 [13]。

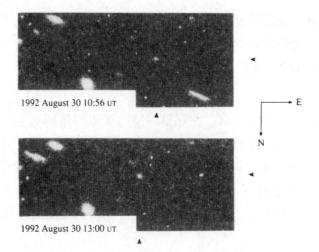

Fig. 1. Discovery images of 1992 QB₁ (marked by arrows), obtained UT 1992 August 30 at the University of Hawaii 2.2-m telescope, using a Mould R filter (central wavelength λ_c = 6,500 Å, FWHM $\Delta\lambda$ = 1,250Å). The images show regions of a 2,048 × 2,048 pixel Tektronix CCD (each pixel subtends 0.22 arcsec). Stellar images are about 0.8 arcsec FWHM. The elongated object to the lower right in the top image is a main-belt asteroid: it appears in the top left of the bottom image and demonstrates the extraordinarily slow motion of 1992 QB₁. The field shown is ~90 arcsec in width.

At distance R = 41 AU, geocentric distance Δ = 40 AU and phase angle α = 0.5°, the apparent red magnitude m_R = 22.8 ± 0.2 on August 30 corresponds to absolute magnitude $\overline{H_R}$ = 6.6 ± 0.2. For comparison, the absolute magnitude of the distant comet 2060 Chiron in its faint state is $\overline{H_R}$ ≈ 6.3 ± 0.1 (ref. 14). Lacking a measurement of the albedo, we cannot determine the size of 1992 QB₁. However, an albedo p_R = 0.04, similar to that of a comet nucleus, suggests a diameter d ≈ 250 km, roughly one-eighth the size of Pluto. The Kron–Cousins colours of 1992 QB₁ were measured on UT 1992 August 30 and September 01. Our best estimates are $m_V - m_R$ = 0.6 ± 0.1 and $m_R - m_I$ = 1.0 ± 0.2, to be compared with solar colours $m_V - m_R$ = 0.32 and $m_R - m_I$ = 0.4, respectively. Here m_V, m_R and m_I are the stellar magnitudes measured with V (5,500 Å), R (6,500 Å) and I (8,000 Å) filters. The $m_R - m_I$ colour is based on a single I filter image. Nevertheless, it seems that 1992 QB₁ is substantially redder than sunlight, inconsistent with a surface of pure ice but consistent with dirty ice, or one contaminated with organic compounds. For comparison, the corresponding colours of the distant object 5145 Pholus are $m_V - m_R$ ≈ 0.7 and $m_R - m_I$ ≈ 0.7 (ref. 15), and all other known cometary nuclei are substantially less red[16]. Both Pholus and 1992 QB₁ may retain primitive organic mantles produced by prolonged cosmic-ray irradiation. Such mantles would be disrupted or buried by rubble mantles on active, near-Sun comets[15,16].

Figure 2 shows the surface brightness profile measured from images taken UT 1992 September 01. The profile of a nearby field star is shown for comparison. Figure 2 supplies no evidence for a resolved coma down to surface brightness 29 mag per square arcsec, at a distance of 1.5 arcsec from the centre of the image. The absence of resolved coma allows us to place a model-dependent limit on the mass loss rate from 1992 QB₁, using a profile-

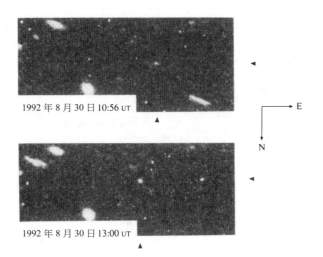

图 1. 1992 QB₁（用箭头标出）的发现图。于世界时 1992 年 8 月 30 日用夏威夷大学 2.2 米望远镜得到，所用的是 R 滤光片（中心波长为 $\lambda_c = 6{,}500$ Å，半峰全宽为 $\Delta\lambda = 1{,}250$ Å）。这幅图像显示 $2{,}048 \times 2{,}048$ 像素美国泰克 CCD（每个像素对应于 0.22 arcsec）的观测区域。恒星图像半峰全宽大约为 0.8 arcsec。上图中右下角被拉长的天体是主带小行星：它在下面一幅图中出现在了左上角，这说明了 1992 QB₁ 极其缓慢的运动。该区域宽度约为 90 arcsec。

在日心距 $R = 41$ AU，地心距离 $\Delta = 40$ AU 和相位角为 $\alpha = 0.5°$ 的情况下，8 月 30 日的红视星等 $m_R = 22.8 \pm 0.2$，对应绝对星等 $\overline{H_R} = 6.6 \pm 0.2$。作为比较，遥远彗星 2060 喀戎在暗弱态的绝对星等为 $\overline{H_R} \approx 6.3 \pm 0.1$（参考文献 14）。由于缺乏对星体反照率的测量，我们无法确定 1992 QB₁ 的大小。不过假设其具有类似于彗核的反照率 $p_R = 0.04$，则可以估算其直径 $d \approx 250$ km，大约为冥王星大小的八分之一。1992 QB₁ 的克朗－卡普斯色指数是在世界时 1992 年 8 月 30 日到 9 月 1 日之间测量的。我们的最佳估计是 $m_V - m_R = 0.6 \pm 0.1$ 和 $m_R - m_I = 1.0 \pm 0.2$，相较而言太阳的值是 $m_V - m_R = 0.32$ 和 $m_R - m_I = 0.4$。这里 m_V、m_R 和 m_I 分别是用 V(5,500 Å)、R(6,500 Å) 和 I(8,000 Å) 波段的滤光片测量的恒星星等。$m_R - m_I$ 的色指数基于单幅 I 波段图像。似乎 1992 QB₁ 比太阳光红很多，和纯净冰的表面的性质不一致，而和脏冰或者受了有机化合物污染的冰的表面性质一致。相较而言，遥远天体 5145 福鲁斯相对应的色指数值为 $m_V - m_R \approx 0.7$ 和 $m_R - m_I \approx 0.7$（参考文献 15），而所有其他已知彗核都没这么红 [16]。福鲁斯和 1992 QB₁ 可能保留了长期宇宙线的辐射产生的原始有机幔。在活动的近日彗星上，这样的幔很可能会被破坏或被碎石幔掩盖 [15,16]。

图 2 显示了世界时 1992 年 9 月 1 日拍摄的图像中测得的天体面亮度轮廓。另一颗临近场星的亮度分布被用于比较。在距图像中心 1.5 arcsec 处，暗至面亮度 29 mag·arcsec⁻²，图 2 中没有可分辨的彗发的迹象。没有可分辨的彗发让我们可以利用一种轮廓拟合方法 [17]，给出依赖模型的 1992 QB₁ 的质量损失率的上限。这个上限是 $dm/dt < 0.7$ kg·s⁻¹，故 1992 QB₁ 的活动程度至多是靠近太阳的哈雷型彗星活动

fitting method[17]. This limit is $dm/dt < 0.7$ kg s^{-1}, so 1992 QB$_1$ is at least 10^4 times less active than a Halley-class, near-Sun comet and probably less active than 2060 Chiron $(dm/dt \approx 1$ kg s$^{-1})$[18]. Weaker activity cannot be constrained by existing observations, however.

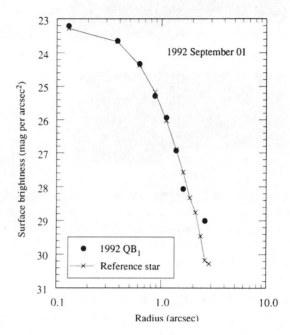

Fig. 2. Surface brightness profile of 1992 QB$_1$, measured UT 1992 September 01 with 0.8 arcsec FWHM images. The profile was computed from four separate R-filter images of total integration time 3,600 s. A scaled profile of a nearby field star is shown for reference. The image of 1992 QB$_1$ is consistent with a point source down to surface brightness ~29 magnitudes per square arcsec.

As our survey is still in progress, we will not discuss in detail the statistics implied by the detection of 1992 QB$_1$. It is, however, interesting to make a crude estimate of the Kuiper belt population. We note that 1992 QB$_1$ was detected in a survey that has so far covered 0.7 square degree. The implied surface density of similar objects is of order 1 per square degree. The inclination of 1992 QB$_1$ is 2 degrees (Table 1), so the angular width of the belt is at least 4 degrees, and is presumably much larger. A lower limit to the projected area of the Kuiper belt is $360 \times 4 = 1{,}440$ square degree, and the number of similar objects is thus $N \geqslant 1{,}440$. A minimal belt mass, based on an assumed diameter of 250 km and density of 10^3 kg m^{-3} for each object, is $M \geqslant 1 \times 10^{22}$ kg $\approx 2 \times 10^{-3}$ M_{Earth} (about 1 Pluto mass). This is compatible with the upper limit to the belt mass inferred from dynamical observations of P/Halley, $M \approx 1$ M_{Earth} (refs 19–21), and with the minimal mass needed to supply the observed flux of short period comets, $M \approx 0.02$ M_{Earth} (ref. 6). The estimated mass must be augmented to account for Kuiper belt comets too small or too distant to be detected in the present survey, but this depends on the (poorly known) mass and spatial distribution of cometary nuclei. M is a lower limit to the true mass.

The faintness of 1992 QB$_1$ makes it unlikely that earlier observations will be identified, so limiting the accuracy of the derived orbital parameters. The astrometric timebase used to

程度的 1/10,000，可能比 2060 喀戎 $(\mathrm{d}m/\mathrm{d}t \approx 1\ \mathrm{kg}\cdot\mathrm{s}^{-1})^{[18]}$ 的活动程度低。然而，更弱的活动性以现在的观测水平就无法限制了。

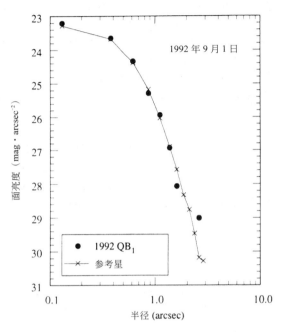

图 2. 在世界时 1992 年 9 月 1 日得到的半峰全宽为 0.8 arcsec 的图像中测量的 1992 QB₁ 面亮度轮廓。该图是由四块独立的 R 滤光片总积分时间为 3,600 s 的图像计算而来的。一颗标度过面亮度轮廓的临近场星也显示出来作为参考。1992 QB₁ 的图像和面亮度暗至 29 mag·arcsec⁻² 的点源一致。

由于我们的巡天还在进行中，所以我们不详细讨论 1992 QB₁ 探测的统计结果。然而对柯伊伯带天体数量进行粗略的估计还是很有意义的。我们注意到 1992 QB₁ 是在覆盖了 0.7 平方度的巡天中被发现的。这暗示了类似天体的表面密度为每平方度 1 个的量级。1992 QB₁ 的轨道倾角是 2°（表 1），所以柯伊伯带的角宽度至少有 4°，或者更宽。对于柯伊伯带的投影面积的下限估计是 $360 \times 4 = 1,440$ 平方度，因此类似天体的数量是 $N \geqslant 1,440$。基于直径为 250 km 和密度为 $10^{3}\ \mathrm{kg}\cdot\mathrm{m}^{-3}$ 的假定，可得柯伊伯带的最小质量为 $M \geqslant 1 \times 10^{22}\ \mathrm{kg} \approx 2 \times 10^{-3}\ M_{\mathrm{Earth}}$（大约一个冥王星的质量）。这与对哈雷彗星（P/Halley）动态观测推得的柯伊伯带质量上限数据一致，即 $M \approx 1\ M_{\mathrm{Earth}}$（参考文献 19~21）；并且与对应观测到的短周期彗星流量所需最小质量也是吻合的，即 $M \approx 0.02\ M_{\mathrm{Earth}}$（参考文献 6）。对总质量的估计应该对目前无法被观测到的太小或太远的柯伊伯带彗星而作出补偿，但这种补偿又依赖于目前尚知之甚少的彗核的质量和空间分布。M 是真实质量的下限。

由于 1992 QB₁ 暗弱，所以在之前的观测中不大可能被证认出来，故而限制了推导的轨道参数的精度。得到表 1 中参数的天体测量时基只是轨道周期的千分之

obtain the parameters in Table 1 is only 10^{-3} of the orbit period. Although these parameters are not finalized, it is likely that the perihelion lies beyond the orbits of the gas giant planets (the expected perihelion $q \approx 40$ AU, compared with the Neptune aphelion $Q_N = 30.3$ AU)[13]. It thus seems that 1992 QB₁ escapes strong planetary perturbations and is, in this sense, more primitive than its presumed cousins at smaller distances, 2060 Chiron and 5145 Pholus. Theory suggests that, although many orbits between the giant planets are unstable on timescales short compared with the age of the Solar System, orbits beyond Neptune may be stable for longer times[22-27]. In particular, low-eccentricity orbits with semi-major axes of ~44 AU (Table 1) have lifetimes in excess of 10^9 yr (ref. 27), supporting the idea that 1992 QB₁ is a Kuiper belt comet.

Note added in proof: On UT 1993 March 28 we detected a second slow moving object (*IAU Circ.* 5370, 29 March 1993). This object, 1993 FW, has the same magnitude and angular motion as 1992 QB₁, and is presumably also resident in the Kuiper belt. Our detection of 1993 FW supports the Kuiper belt population estimate given above.

(**362**, 730-732; 1993)

David Jewitt[*] & Jane Luu[†]

[*] Institute for Astronomy, University of Hawaii, 2680 Woodlawn Drive, Honolulu, Hawaii 96822, USA
[†] Department of Astronomy, 601 Campbell Hall, University of California at Berkeley, Berkeley, California 94720, USA

Received 13 January; accepted 30 March 1993.

References:

1. Tombaugh, C. W. in *Planets and Satellites* (eds Kuiper G. P. & Middlehurst, B. M.) 12-30 (Univ. Chicago Press, Chicago, 1961).

2. Kowal, C., Liller, W. & Marsden, B. in *Dynamics of the Solar System. Proc. IAU Symp.* No. 81 (ed. Duncombe R.) 245-249 (Reidel, Dordrecht, 1979).

3. Scottie, J. V. *IAU Circ.* No. 5434 (1992).

4. Edgworth, K. E. *Mon. Not. R. astr. Soc.* **109**, 600-609 (1949).

5. Kuiper, G. P. in *Astrophysics* (ed. Hynek, J. A.) 357-424 (New York, McGraw-Hill. 1951).

6. Duncan, M., Quinn, T. & Tremaine, S. *Astrophys. J.* **328**, L69-L73 (1988).

7. Whipple, F. *Proc. Natl. Acad. Sci.* **51**, 711-717 (1964).

8. Fernandez, J. A. *Mon. Not. R. astr. Soc.* **192**, 481-491 (1980).

9. Luu, J. X. & Jewitt, D. C. *Astr. J.* **95**, 1256-1262 (1988).

10. Jewitt, D. C. & Luu, J. X. *IAU Circ.* No. 5611 (1992).

11. McNaught, R. & Steel, D. *Minor Planet Circ.* 20878 (1992).

12. Hainault, O. & Elst, E. *Minor Planet Circ.* 20902 (1992).

13. Marsden, B. G. *IAU Circ.* No. 5684 (1992).

14. Luu, J. X. & Jewitt, D. C. *Astr. J.* **100**, 913-932 (1990).

15. Mueller, B., Tholen, D., Hartmann, W. & Cruikshank, D. *Icarus* **97**, 150-154 (1992).

16. Jewitt, D. C. *Cometary Photometry* in *Comets in the Post-Halley Era* (ed. Newburn, R.) 19-65 (Kluwer, Dordrecht, 1990).

17. Luu, J. X. & Jewitt, D. C. *Icarus* **97**, 276-287 (1992).

18. Luu, J. X. & Jewitt, D. C. *Astr. J.* **100**, 913-932 (1990).

19. Hamid, S. E., Marsden, B. & Whipple, F. *Astr. J.* **73**, 727-729 (1968).

20. Yeomans, D. K. *Proc. 20th ESLAB Symp.* 419-425 (ESA SP-250, Heidelberg, 1986).

21. Hogg, D. W., Quinlan, G. D. & Tremaine, S. *Astr. J.* **101**, 2274-2286 (1991).

22. Torbett, M. V. *Astr. J.* **98**, 1477-1481 (1989).

23. Torbett, M. V. & Smoluchowski, R. *Nature* **345**, 49-51 (1990).

一。尽管这些参数还没有最终确定，但该彗星的近日点可能位于气态巨行星轨道之外（预期的近日点距 $q \approx 40$ AU，相比较海王星远日点距 $Q_N = 30.3$ AU）[13]。所以看起来 1992 QB₁ 逃过强烈行星摄动，从这种意义上来说，相对距离更近的 2060 喀戎和 5145 福鲁斯，它更为原始。理论认为，尽管在与太阳系年龄相比较短的时标上许多在巨型行星之间的轨道不稳定，但在更长的时标 [22-27] 上海王星之外的轨道可能是稳定的。特别是，半长轴 ~44 AU（表 1）的低偏心率轨道的寿命超过 10^9 年（参考文献 27），这支持 1992 QB₁ 是一颗柯伊伯带彗星的观点。

附加说明： 在世界时 1993 年 3 月 28 日我们探测到了第二个缓慢运动的天体（国际天文学联合会快报 *IAU Circ.* 5370，1993 年 3 月 29 日）。1993 FW 这个天体和 1992 QB₁ 有着相同的星等和角运动，它也很有可能是柯伊伯带中的天体。我们探测到 1993 FW 支持了上文中对柯伊伯带天体数量的估计。

（冯翀 翻译；周礼勇 审稿）

24. Quinn, T., Tremaine, S. & Duncan, M. *Astrophys. J.* **355**, 667-679 (1990).

25. Duncan, M. & Quinn, T. in *Protostars and Planets III* (Tucson, in the press).

26. Holman, M. J. & Wisdom, J. *Astr. J.* (in the press).

27. Levison, H. & Duncan, M. *Astrophys. J.* (in the press).

Acknowledgements. We thank B. Marsden for discussions about 1992 QB$_1$. G. Luppino for CCD cameras, the Institute for Astronomy TAC for consistent allocation of time to this project, and A. Pickles for obtaining supporting astrometric images. The Planetary Astronomy program of NASA provides financial support at the UH 2.2-m telescope. J.X.L. is in receipt of a Hubble Fellowship.

Possible Gravitational Microlensing of a Star in the Large Magellanic Cloud

C. Alcock *et al.*

Editor's Note

The existence of dark matter—invisible matter that exceeds the amount of visible matter in the universe by a factor of 5–6—is supported by several observations, not least that it is required to bind galaxies together. But almost nothing is known about what it consists of. A possibility explored in the early 1990s was that it is in the form of small bodies collectively known as "massive compact halo objects", or MACHOs, because much of the dark matter in galaxies seems to be in their outer spherical "halos". Such bodies should produce a "microlensing" effect: the relativistic bending of light (gravitational lensing) by bodies smaller than galaxies. This paper is one of two to report the first microlensing signals ever seen. But it now appears that MACHOs are too rare to account for dark matter.

There is now abundant evidence for the presence of large quantities of unseen matter surrounding normal galaxies, including our own[1,2]. The nature of this "dark matter" is unknown, except that it cannot be made of normal stars, dust or gas, as they would be easily detected. Exotic particles such as axions, massive neutrinos or other weakly interacting massive particles (collectively known as WIMPs) have been proposed[3,4], but have yet to be detected. A less exotic alternative is normal matter in the form of bodies with masses ranging from that of a large planet to a few solar masses. Such objects, known collectively as massive compact halo objects[5] (MACHOs), might be brown dwarfs or "jupiters" (bodies too small to produce their own energy by fusion), neutron stars, old white dwarfs or black holes. Paczynski[6] suggested that MACHOs might act as gravitational microlenses, temporarily amplifying the apparent brightness of background stars in nearby galaxies. We are conducting a microlensing experiment to determine whether the dark matter halo of our Galaxy is made up of MACHOs. Here we report a candidate for such a microlensing event, detected by monitoring the light curves of 1.8 million stars in the Large Magellanic Cloud for one year. The light curve shows no variation for most of the year of data taking, and an upward excursion lasting over 1 month, with a maximum increase of ~2 mag. The most probable lens mass, inferred from the duration of the candidate lensing event, is ~0.1 solar mass.

THE MACHO Project[7,8] uses the gravitational microlens signature to search for evidence of MACHOs in the Galactic halo, which is thought to be at least three times as massive as the visible disk[2]. (Two other groups are attempting a similar search[9,10].)

大麦哲伦云的一颗恒星可能存在微引力透镜效应

阿尔科克等

编者按

种种证据表明，宇宙中存在暗物质，尤其是暗物质是束缚星系所必需的。这种不可见物质的含量是宇宙中可见物质含量的 5~6 倍。然而我们对其物质组成却几乎一无所知。20 世纪 90 年代初，由于发现星系中大部分暗物质似乎存在于星系晕的外层，人们提出了一种可能性，即认为暗物质是以一种被统称为"晕族大质量致密天体 (MACHO)"的小天体的形式存在的。这样的小天体可以产生微引力透镜效应：比星系小的天体产生的对光的相对论性偏折（引力透镜效应）。这篇文章是发现首例微引力透镜信号的两篇文章之一（另一篇文章请见本书第 178 页）。然而现在看来 MACHO 的数目似乎太少了，不足以解释暗物质。

现在有充分的证据表明，在正常星系周围存在着大量看不见的物质，包括我们的银河系 [1,2]。我们不了解这种"暗物质"的性质，但它不可能由普通的恒星、尘埃或气体组成，否则它们就会被轻易地探测到。人们已经提出若干种奇异粒子模型，如轴子、有质量中微子或者其他弱相互作用大质量粒子（简称为 WIMP）[3,4]，但目前还没有探测到其中任何一种。另一种不太奇异的可能性是各种天体中的正常物质，质量范围从一个大行星到几个太阳。这类天体统称为晕族大质量致密天体 [5]（MACHO），可能是褐矮星、类木星天体（这类天体质量太小，不能通过自身核聚变产生能量）、中子星、老年的白矮星或者黑洞。帕金斯基提出 [6]，MACHO 可能会产生微引力透镜现象，暂时放大近邻星系中背景星的视亮度。我们正在进行一项微引力透镜实验，检验我们银河系的暗物质晕是否由 MACHO 构成。我们在本文中报告了通过监测大麦哲伦云的 180 万颗恒星的光变曲线一年而探测到的一个微引力透镜候选事件。该事件中恒星的光变曲线存在持续超过一个月的向上偏移，且最大增量大约为 2 mag（星等），但是在这一年中其他绝大部分数据采集期间里没有变化。从候选透镜事件持续的时间可以推断，最可能的透镜天体质量大约相当于 0.1 倍太阳质量。

MACHO 项目 [7,8] 利用微引力透镜信号来搜寻银晕中存在 MACHO 的证据，一般认为银晕的质量至少是可见银盘的 3 倍 [2]（另外两个研究组也在进行类似的搜寻 [9,10]）。如果我们银河系的大部分暗物质都集中在 MACHO 中，则大麦哲伦云

If most of our Galaxy's dark matter resides in MACHOs, the "optical depth" for microlensing towards the Large Magellanic Cloud (LMC) is about 5×10^{-7} (independent of the mass function of MACHOs), so that at any given time about one star in two million will be microlensed with an amplification factor $A > 1.34$ (ref. 5). Our survey takes advantage of the transverse motion of MACHOs relative to the line-of-sight from the observer to a background star. This motion causes a transient, time-symmetric and achromatic brightening that is quite unlike any known variable star phenomena, with a characteristic timescale $t = 2r_E/v_\perp$ where r_E is the Einstein ring radius and $v_\perp$ is the MACHO velocity transverse to the line-of-sight. For typical halo models the time $t \sim 100 \sqrt{M_{macho}/M_\odot}$ days[5] (where $M_\odot$ is the mass of the Sun). The amplification can be large, but these events are extremely rare; for this reason our survey was designed to follow > ten million stars over several years.

The survey employs a dedicated 1.27-m telescope at Mount Stromlo. A field-of-view of 0.5 square degrees is achieved by operating at the prime focus. The optics include a dichroic beam-splitter which allows simultaneous imaging in a "red" beam (6,300–7,600 Å) and a "blue" beam (4,500–6,300 Å). Two large charge-coupled device (CCD) cameras[11] are employed at the two foci; each contain a 2×2 mosaic of $2,048 \times 2,048$ pixel Loral CCD imagers. The 15-μm pixel size corresponds to 0.63 arcsec on the sky. The images are read out through a 16-channel system, and written into dual ported memory in the data acquisition computer. Our primary target stars are in the LMC. We also monitor stars in the Galactic bulge and the Small Magellanic Cloud. As of 15 September 1993, over 12,000 images have been taken with the system.

The data are reduced with a crowded-field photometry routine known as Sodophot, derived from Dophot[12]. First, one image of each field that was obtained in good seeing is reduced in a manner similar to Dophot to produce a "template" catalogue of star positions and magnitudes. Normally, bright stars are matched with the template and used to determine an analytic point spread function (PSF) and a coordinate transformation. Photometric fitting is then performed on each template star in descending order of brightness, with the PSF for all other stars subtracted from the frame. When a star is found to vary significantly, it and its neighbours undergo a second iteration of fitting. The output consists of magnitudes and errors for the two colours, and six additional useful parameters (such as the χ^2 of the PSF fit and crowding information). These are used to flag questionable measurements, that arise from cosmic ray events in the CCDs, bad pixels and so on.

These photometric data are subjected to an automatic time-series analysis which uses a set of optimal filters to search for microlensing candidates and variable stars (which we have detected in abundance[13]). For each microlensing candidate a light curve is fitted, and the final selection is done automatically using criteria (for example, signal-to-noise, quality of fit, wave-length independence of the light curve and colour of the star) that were established empirically using Monte Carlo addition of fake events into real light curves.

This analysis has been done on four fields near the centre of the LMC, containing 1.8

168

（LMC）方向的微引力透镜的"光深"大约为 5×10^{-7}（与 MACHO 的质量函数无关）。这样在任意给定时刻下，200 万颗恒星中就大约有一颗可以被微引力透镜放大，放大因子为 $A > 1.34$（参考文献 5）。我们的巡天观测借助于 MACHO 相对于观察者到背景星视线的横向运动。这种运动可以引起一种暂现的、时间对称且无色差的增亮现象，这种现象不同于任何已知的变星现象，其特征时标为 $t = 2r_E/v_\perp$，其中 r_E 为爱因斯坦环半径，$v_\perp$ 为 MACHO 相对视线的横向运动速度。对于典型的晕模型，时间 t 大约为 $100\sqrt{M_{macho}/M_\odot}$ 天[5]，其中 $M_\odot$ 为太阳质量。这种放大效应可以很强，但是这类事件却极为罕见。因此，我们就需要在数年内对超过 1,000 万颗恒星进行巡天观测。

这次巡天观测使用了位于澳洲斯特朗洛山的 1.27 m 专用望远镜，该望远镜在主焦点工作状态下的视场为 0.5 平方度。其光学系统包括一个双色分束器，可以同时用红波段光束（6,300~7,600 Å）和蓝波段光束（4,500~6,300 Å）分别成像。在两条光束的焦点上分别放置电荷耦合器件（CCD）照相机[11]，每台相机含有一个 2×2 的 Loral 成像器（2,048 × 2,048 像素）。像素大小为 15 μm，对应于天空中的 0.63 arcsec。图像数据通过一个 16 通道系统读取，然后写入数据采集计算机中的双端存储器中。我们主要的目标恒星位于大麦哲伦云中，我们也监测银河系核球与小麦哲伦云中的恒星。截止到 1993 年 9 月 15 日，该系统已经获得超过 12,000 张图像。

这些数据通过密集星场测光程序 Sodophot 进行处理，该程序源自 Dophot[12]。首先，对每个星场中视宁度较好的图像通过和 Dophot 相似的方式进行处理，从而生成关于恒星位置及星等的"模板"星表。一般情况下，将较亮的恒星与模板相匹配，这样可以用来确定其解析的点扩散函数（简称为 PSF）和坐标转换。在对其他所有恒星用 PSF 修正之后，按照模版恒星亮度递减的顺序，我们对每颗模板恒星进行光度拟合。当发现一颗恒星的亮度变化显著时，则对该星及其邻星进行第二次迭代拟合。输出结果包括两种颜色的相应星等和误差，以及其他六种有用的参数（如 PSF 拟合的 χ^2 和星场密集分布信息），这些参数被用来标记有问题的测量数据，诸如 CCD 中的宇宙线事件、坏像元，等等。

随后，我们对这些测光数据进行一种自动时间序列分析，该分析通过使用一组最优滤波器来寻找微透镜候选体和变星（我们已经探测到很多[13]）。对于每一个微引力透镜候选体，我们进行光变曲线拟合，然后自动地根据判断标准（例如，信噪比、拟合品质、与光变曲线无关的波长以及恒星的颜色）作出最后筛选。这些判断标准是经验性地用蒙特卡罗方法在真实的光变曲线中加入伪事件得出的。

这种分析被应用于靠近大麦哲伦云中心附近的四个区域。这些区域包含了

million stars, with approximately 250 observations for each star. The candidate event reported here occurs in the light curve of a star at coordinates $\alpha = 05$ h 14 min 44.5 s, $\delta = -68° \, 48' \, 00''$ (J2000). (A finding chart is available on request from C.A.). The star has median magnitudes $V \sim 19.6$, $R \sim 19.0$, consistent with a clump giant (metal-rich helium core burning star) in the LMC. These magnitudes are estimated using colour transformations from our filters to V and R that have been derived from observations of standard stars.

Our photometry for this star, from July 1992 to July 1993, is shown in Fig. 1, and the candidate event is shown on an expanded scale in Fig. 2, along with the colour light curve. The colour changes by < 0.1 mag as it brightens and fades (the candidate "event"). A mosaic, showing portions of some of the CCD images used, is shown in Fig. 3, with the relevant star at the centre. The integrated number of PSF photoelectrons detected above the sky background in the template image is $\sim 10^4$, for a 300 s exposure. The increase in counts during the peak is highly significant, as is clear from the figures. Also shown in Fig. 2 is a fit to the theoretical microlensing light curve (see ref. 6). The four parameters fit are (1) the baseline flux, (2) the maximum amplification $A_{\max} = 6.86 \pm 0.11$, (3) the duration $t = 33.9 \pm 0.26$ d, (4) the centroid in time 433.55 ± 0.04 d. The quoted errors are formal fit errors. Using the PSF fit uncertainties as determined by the photometry program, the best-fit microlensing curve gives a χ^2 per degree of freedom of 1.6 (for 443 d.f.).

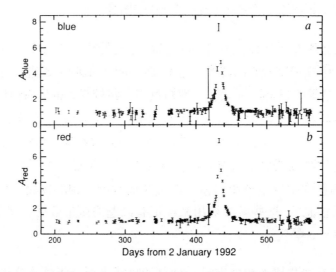

Fig. 1. The observed light curve with estimated $\pm 1\sigma$ errors. *a*, Shows A_{blue}, the flux (in linear units) divided by the median observed flux, in the blue passband. *b*, Is the same, for the red passband.

180 万颗恒星，每颗恒星大约进行了 250 次观测。这里报告候选事件产生于坐标为 $\alpha = 05\,h\,14\,min\,44.5\,s$，$\delta = -68°\,48'\,00''$（J2000）的恒星的光变曲线中（如果需要可以从阿尔科克处得到认证图）。该恒星的中位星等 $V\sim19.6$、$R\sim19.0$，与大麦哲伦云中的团簇巨星（富含金属且正经历氦核燃烧的恒星）相符。这些星等是通过我们的观测波段与 V、R 波段的颜色转换关系估计出来的，所采用的颜色转换关系是通过对标准星的观测导出的。

我们从 1992 年 7 月至 1993 年 7 月对该星的测光结果，如图 1 所示；候选事件的放大的光变曲线以及颜色光变曲线如图 2 所示。随着它的明暗变化（候选"事件"），其颜色变化小于 0.1 mag。图 3 给出了部分分析使用的 CCD 照片图，其中心位置为我们研究的恒星。对于 300 s 的曝光时间，我们在模板照片中探测到高于天空背景的 PSF 光电子的积分数为 $\sim10^4$。从图中我们可以清楚地看到峰值期间的计数增长非常明显。我们在图 2 中也相应给出了理论微透镜光变曲线的拟合结果[6]。四个拟合参数分别为：(1) 基线流量，(2) 最大放大率 $A_{max} = 6.86 \pm 0.11$，(3) 持续时间 $t = 33.9 \pm 0.26\,d$，(4) 放大峰值时刻，第 $433.55 \pm 0.04\,d$。引用的误差是标准的拟合误差。使用由测光程序给出的 PSF 拟合不确定度，最佳拟合微引力透镜曲线给出每个自由度的 χ^2 为 1.6（对于 443 个自由度）。

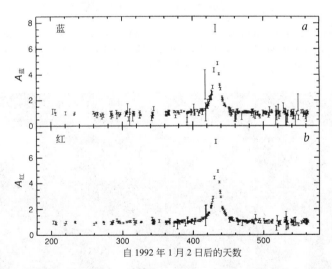

图 1. 观测到的光变曲线，估计误差为 $\pm1\sigma$。a，给出了蓝光波段的放大率 $A_{蓝}$，即流量（线性坐标）除以观测中值流量（蓝光波段）。b，与 a 一样，但针对红光波段。

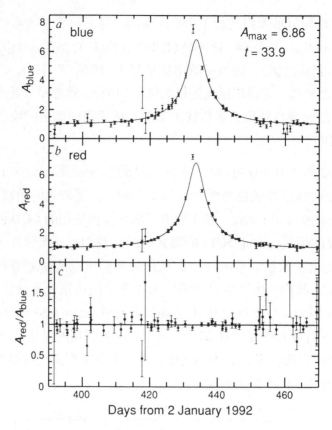

Fig. 2. As in Fig. 1, with an expanded scale around the candidate event. The smooth curve shows the best-fit theoretical microlensing model, fitted simultaneously to both *c* is the colour light curve, showing the ratio of red to blue flux, normalized so that the median is unity.

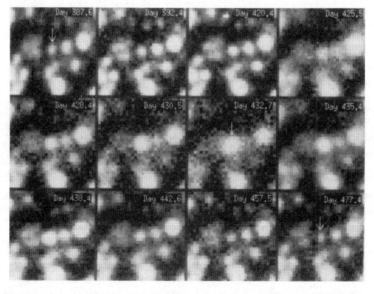

Fig. 3. Selected red CCD frames centred on the microlens candidate, showing observations before, during and after the event. The numbers on each frame indicate the days after 2 January 1992.

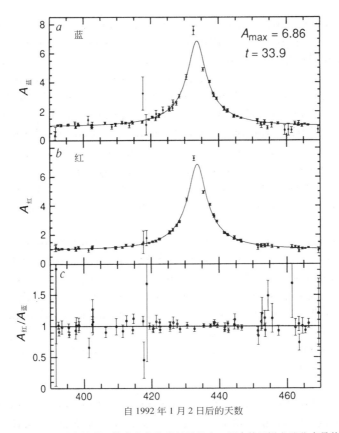

图 2. 如图 1，该图只是将候选事件附近的光变曲线进行了放大。图中的平滑曲线代表最佳拟合的理论微引力透镜模型，拟合过程同时拟合红光和蓝光两个波段的数据。*c* 图为颜色光变曲线图，给出的是红光和蓝光波段流量的比值。这里流量均已归一化，从而使得各自波段的中位值为 1。

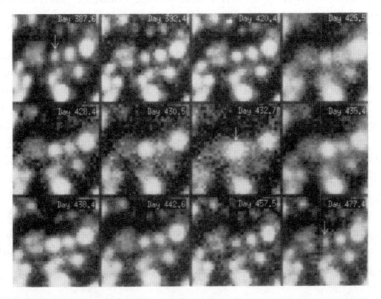

图 3. 筛选以微引力透镜候选天体为中心的红光波段的部分 CCD 图片，分别给出了该事件发生前、发生中和发生后的观测图像。每张图片上的数字表示 1992 年 1 月 2 日后的天数。

A number of features of the candidate event are consistent with gravitational microlensing: the light curve is achromatic within measurement error, and it has the expected symmetrical shape. If this is a genuine microlensing event, the mass of the deflector can be estimated. Because the duration depends upon the lens mass, the relative velocity transverse to the line-of-sight and the distance to the lens (none of which are known), the lens mass cannot be uniquely determined from the duration. But by using a model of the mass and velocity distributions of halo dark matter, one can find the relative probability that a MACHO of mass M_{macho} gave rise to the event. Thus, if this is genuine microlensing, Fig. 9 of ref. 5 implies the most likely mass is $\sim$0.12 $M_{\odot}$, with masses of 0.03 $M_{\odot}$ and 0.5 $M_{\odot}$ being roughly half as likely. However, this method does not properly take into account our detection efficiencies, and should be considered only a rough estimate.

The mass range given above includes brown dwarfs and main sequence stars. Any microlensing star is very unlikely to be a red dwarf of the Galactic stellar halo, because one can show that the optical depth τ_s for microlensing by main sequence stars of the stellar halo is very low. Even if the mass function of the stellar halo rises as steeply as $dN/dM \propto M^{-4}$, as suggested recently[14] (here N is the number of stars per unit stellar mass interval), τ_s is still a few hundred times smaller than the 5×10^{-7} optical depth estimated for MACHO microlensing. The chance of finding such a stellar microlensing event among our 1.8 million stars is therefore very small.

The prospects for direct observation of a lensing object are not favourable. Even a star of 0.5 $M_{\odot}$, for example, would have $V \sim 24$, and for many years would be within a small fraction of an arcsecond of the much brighter LMC star.

We emphasize that the observed stellar brightening could be due to some previously unknown source of intrinsic stellar variability. The fit discrepancy near the peak is not yet understood; a more refined analysis of the data is under way. We do not yet have a spectrum of the star. A crucial test of the hypothesis that we are seeing gravitational microlensing by MACHOs in the galactic halo will be the detection of other candidates. So far, we have analysed only $\sim$15% of our first year's frames and we plan to continue observations until 1996; this should allow us to determine if microlensing is really the cause. Additional events should show the theoretical distribution of maxima, and should be representative of both the colour–magnitude diagram and the spatial structure of the LMC. No repeats should be seen in any given star. (While this paper was in preparation, we were informed by J. Rich (personal communication) of the candidate events reported by the EROS collaboration. Note that the two groups use different definitions of characteristic time.)

If such candidates do result from microlensing we should be able to determine the contribution of MACHOs to the dark matter in the Galactic halo. The results presented here encourage us to believe this will happen.

(**365**, 621-623; 1993)

该候选事件的许多特征与微引力透镜现象相符：在测量误差范围内，光变曲线是无色差的，而且具有预期的对称形状。如果这是一个真实的微引力透镜事件，则可以估算出偏折天体的质量。因为持续时间依赖于透镜质量、横穿视线的相对速度以及与透镜天体的距离（这些量都是未知的），因此透镜质量不能由持续时间唯一确定。但是通过使用晕暗物质的质量和速度分布模型，我们可以计算出一颗质量为 M_{macho} 的 MACHO 导致该事件的相对概率。因此，如果这是真实的微引力透镜事件，从参考文献 5 中的图 9 可推出其最可能的质量约为 0.12 $M_\odot$，而质量分别为 0.03 $M_\odot$ 和 0.5 $M_\odot$ 的概率大致降低一半。然而，以上方法并没有很好地将我们的探测效率考虑进去，因此只能认为是一种粗略的估计。

上面给出的质量范围涵盖了褐矮星和主序星。任何形成微引力透镜事件的恒星都不太可能是银河系恒星晕中的红矮星，因为可以发现恒星晕中的主序星形成的微引力透镜事件的光深 τ_s 都是非常小的。即使根据最近提出的按照 $dN/dM \propto M^{-4}$ 变化 [14] 的恒星晕的质量函数（这里的 N 代表每单位恒星质量间隔的恒星数量），τ_s 仍然比 MACHO 微引力透镜事件的光深估计值（5×10^{-7}）小数百倍。因此，在我们观测的 180 万颗恒星中发现这样一个恒星微引力透镜事件的概率是很小的。

而直接观测到透镜天体的前景也并不乐观。比如即使对于一颗质量为 0.5 $M_\odot$ 的恒星，其星等 $V \sim 24$，在许多年间它可能还是处在大麦哲伦云中比它亮得多的一颗恒星附近若干分之一角秒范围之内。

在这里我们需要强调一下，观测到的恒星变亮还可能是由一些以前未知的恒星内禀变化导致的。峰值附近的拟合偏差还不能很好地给出解释，我们正在对这些数据进行更加细致的分析。我们还没有这颗恒星的光谱。关于观测到银晕中 MACHO 形成微引力透镜事件的假设，关键的检验是能否进一步探测到其他候选天体。迄今，我们仅分析了第一年图像的大约 15%，我们打算将观测持续到 1996 年，这将帮助我们确定造成这一现象的真正原因是不是微引力透镜事件。其他事件将揭示出最大放大倍数的理论分布，并可以反映大麦哲伦云的颜色－星等图和空间结构，在任何给定的恒星中都不应该看到重复事件。（就在本文准备过程中，里奇通知我们，EROS 合作研究小组也报道了一些候选事件（个人交流）。但是请注意，两个研究小组采用的特征时间定义不同。）

如果这类候选事件的确源于微引力透镜现象，那么我们应该可以确定 MACHO 对于银晕暗物质的贡献。本文结果使我们相信上述推论可以被确认。

（金世超 翻译；何香涛 审稿）

C. Alcock[*†], C. W. Akerlof[†¶], R. A. Allsman[*], T. S. Axelrod[*], D. P. Bennett[*†], S. Chan[‡], K. H. Cook[*†], K. C. Freeman[‡], K. Griest[†‖], S. L. Marshall[†§], H-S. Park[*], S. Perlmutter[†], B. A. Peterson[‡], M. R. Pratt[†§], P. J. Quinn[‡], A. W. Rodgers[‡], C. W. Stubbs[†§] & W. Sutherland[†]

[*] Lawrence Livermore National Laboratory, Livermore, California 94550, USA

[†] Center for Particle Astrophysics, University of California, Berkeley, California 94720, USA

[‡] Mt Stromlo and Siding Spring Observatories, Australian National University, Weston, ACT 2611, Australia

[§] Department of Physics, University of California, Santa Barbara, California 93106, USA

[‖] Department of Physics, University of California, San Diego, California 92039, USA

[¶] Department of Physics, University of Michigan, Ann Arbor, Michigan 48109, USA

Received 22 September; accepted 30 September 1993.

References:

1. Trimble, V. *A. Rev. Astr. Astrophys.* **25**, 425-472 (1987).

2. Fich, M. & Tremaine, S. *A. Rev. Astr. Astrophys.* **29**, 409-445 (1991).

3. Primack, J. R., Seckel, D. & Sadoulet, B. *A. Rev. Nucl. Part. Sci.* **B38**, 751-807 (1988).

4. Kolb, E. W. & Turner, M. S. *The Early Universe* (Addison Wesley, New York, 1990).

5. Griest, K. *Astrophys. J.* **366**, 412-421 (1991).

6. Paczynski, B. *Astrophys. J.* **304**, 1-5 (1986).

7. Bennett, D. *et al. Ann. N.Y. Acad. Sci.* **688**, 612-618 (1993).

8. Alcock, C. *et al. Astr. Soc. Pacif. Conf. Ser.* **34**, 193-202 (1992).

9. Magneville, C. *Ann. N.Y. Acad. Sci.* **688**, 619-625 (1993).

10. Udalski, A. *et al. Ann. N.Y. Acad. Sci.* **688**, 626-631 (1993).

11. Stubbs, C. W. *et al.* in *Charge-coupled Devices and Solid State Optical Sensors III* (ed. Blouke, M.) *Proc. of the SPIE* **1900**, 192-204 (1993).

12. Schechter, P. L., Mateo, M. L. & Saha, A. *Publ. Astron. Soc. Pac.* **105**, 1342-1353 (1993).

13. Cook, K. H. *et al. Bull. Am. Astr. Soc.* **24**, 1179 (1993).

14. Richer, H. B. & Fahlman, G. G. *Nature* **358**, 383-386 (1992).

Acknowledgements. We are grateful for the support given our project by the technical staff at the Mt Stromlo Observatory. Work performed at LLNL is supported by the DOE. Work performed by the Center for Particle Astrophysics on the UC campuses is supported in part by the Office of Science and Technology Centers of the NSF. Work performed at MSSSO is supported by the Bilateral Science and Technology Program of the Australian Department of Industry, Technology and Commerce. K.G. acknowledges a DOE OJI grant, and C.W.S. thanks the Sloan Foundation for their support.

Evidence for Gravitational Microlensing by Dark Objects in the Galactic Halo

E. Aubourg *et al.*

Editor's Note

The unseen "dark matter" that is needed to account for the gravitational cohesion of rotating spiral galaxies is one of the most profound current mysteries of science. This matter apparently exceeds its visible counterpart by a factor of 5–6, yet there is still no real understanding of what it consists of. In this paper a team of French astronomers explore the possibility that it could be constituted of dark, massive compact bodies in the spherical "halos" of galaxies like our own, which are called MACHOs. Such objects would occasionally be expected to bend and amplify the light from stars beyond, an effect called microlensing. The paper is one of two that report the first detection of microlensing by MACHOs, and it concludes that the objects are probably 0.01–1 times the mass of our sun.

The flat rotation curves of spiral galaxies, including our own, indicate that they are surrounded by unseen haloes of "dark matter"[1,2]. In the absence of a massive halo, stars and gas in the outer portions of a galaxy would orbit the centre more slowly, just as the outer planets in the Solar System circle the Sun more slowly than the inner ones. So far, however, there has been no direct observational evidence for the dark matter, or its characteristics. Paczyński[3] suggested that dark bodies in the halo of our Galaxy can be detected when they act as gravitational "microlenses", amplifying the light from stars in nearby galaxies. The duration of such an event depends on the mass, distance and velocity of the dark object. We have been monitoring the brightness of three million stars in the Large Magellanic Cloud for over three years, and here report the detection of two possible microlensing events. The brightening of the stars was symmetrical in time, achromatic and not repeated during the monitoring period. The timescales of the two events are about thirty days and imply that the masses of the lensing objects lie between a few hundredths and one solar mass. The number of events observed is consistent with the number expected if the halo is dominated by objects with masses in this range.

THE "EROS" (Expérience de Recherche d'Objets Sombres) collaboration is searching for microlensing events using the European Southern Observatory at La Silla, Chile[4,5]. We have two complementary programmes. The first uses $5° \times 5°$ Schmidt plates of the Large Magellanic Cloud (LMC) that allow us to monitor about eight million stars with a sampling rate of no more than two measurements per night. This makes the programme primarily sensitive to lens masses in the range $10^{-4}\ M_\odot < M < 1\ M_\odot$ (where $M_\odot$ is the

利用微引力透镜效应发现银晕中
存在暗天体的证据

为了解释转动的旋涡星系的引力束缚问题（旋转曲线），我们需要引入看不见的"暗物质"，这是当今科学最重大的未解之谜之一。这种暗物质的含量是可见物质含量的 5~6 倍，但对其具体的物质组成我们尚无真正的了解。在这篇文章中，一组法国天文学家探索了暗物质由与我们银河系类似星系的球状晕中的大质量致密暗天体（这些天体被称为 MACHO）组成的可能性。预期这些天体偶尔会使背后恒星发出的光偏折、变亮，这种效应称为微引力透镜效应。这篇文章是报告发现首例 MACHO 微引力透镜信号的两个工作之一，其结果表明这些小天体的质量可能只是我们太阳质量的 0.01~1 倍。

包括我们自己银河系在内的旋涡星系有着平坦的自转曲线，这暗示着它们都被不可见的"暗物质"[1,2] 晕包围着。如果没有这样大质量晕的存在，星系靠外部分的恒星以及气体将绕中心转得更慢，正如太阳系带外行星绕太阳转动的速度要比带内行星慢一样。然而，迄今为止还没有关于暗物质及其性质的直接观测证据。帕金斯基 [3] 提出，可以通过我们星系晕中暗天体作为引力"微透镜"使临近星系中恒星发出的放大来探测它们。这种微引力透镜事件所持续的时间和暗天体的质量、距离及速度有关。我们已经对大麦哲伦云中 300 万颗恒星的亮度进行了长达三年的监测，在此报告两个可能的微引力透镜事件。这些恒星的亮度变化在时间上对称，在不同波段一致并且在观测期间没有重复。这两次微引力透镜事件的时标约为 30 天，表明透镜天体的质量介于百分之几到一倍太阳质量之间。观测到的微引力透镜事件数量与假设晕由这种质量范围的天体主导所预期的数量相符。

经验性探测暗天体（EROS）合作项目利用位于智利拉西亚的欧南台来寻找微引力透镜事件 [4,5]。我们有两个互补的计划。第一个计划是使用大麦哲伦云（LMC）的 $5° \times 5°$ 施密特底片，这可以使我们以每晚不超过两次观测的采样率对约 800 万颗恒星进行监测。这使得该项目易于探测质量在 $10^{-4} M_\odot < M < 1 M_\odot$（其中 $M_\odot$ 是太阳质量）范围之内的透镜天体，对应的透镜事件的平均持续时间在 $1\,\mathrm{d} < \tau < 100\,\mathrm{d}$ 之间。

solar mass), corresponding to mean lensing durations in the range $1\ \mathrm{d} < \tau < 100\ \mathrm{d}$. The probability that a given star in the LMC is amplified by more than 0.3 magnitudes at a given time is calculated to be $\sim 0.5 \times 10^{-6}$ (refs 3, 6). For a deflector of mass M the typical timescale for the amplification is $\tau = 70\sqrt{M/M_\odot}$ d. The light curve of such an event should be symmetric in time, achromatic, and the event should not be repeated. Over the period 1990–93, a total of 304 Schmidt plates of the LMC were taken for us at La Silla with red or blue filters. Exposure times were typically one hour, permitting us to monitor stars down to the twentieth magnitude with a mean photometric precision of about 15% (r.m.s.). The transparency of the plates is digitized in 10 μm (0.67 arcsec) steps by the "MAMA" (Machine Automatique a Mesurer pour l'Astronomie) at the Observatoire de Paris[7]. The relation between transparency and star luminosity has been established using charge-coupled device (CCD) images scattered through the Schmidt-plate field.

The second programme uses a CCD camera consisting of a mosaic of sixteen 579×400 pixel Thomson THX 31157 CCDs covering about $1° \times 0.4°$. It is mounted on a 40-cm reflector (f/10) refurbished with the help of the Observatoire de Haute Provence. We have used this to observe one field in the bar of the LMC from December 1991 to March 1992 and from August 1992 to March 1993. As of March 1993, a total of 8,100 exposures had been taken with red and blue filters. About 100,000 stars are seen on each image, with a mean photometric precision of about 6%. Compared to the Schmidt-plate programme, the number of stars is a factor 80 smaller but the rapid sampling time (an image pair every 22 minutes) makes the CCD programme sensitive to deflector masses in the range $10^{-7}\,M_\odot < M < 10^{-3}\,M_\odot$, corresponding to event durations in the range $1\ \mathrm{h} < \tau < 3\ \mathrm{d}$.

After preliminary processing (digitizing the Schmidt plates and flat-fielding the CCD images), the data reduction for both programmes follows basically the same procedure. First, one reference image for each colour was constructed by combining ten plates or 50 CCD images taken with good atmospheric conditions. We used a star finding algorithm to establish a star catalogue for each reference image. Next, each image is aligned with the reference using bright, isolated stars. The positions of the stars on the reference image then serve as input to a photometric fitting programme to determine the luminosity of each catalogue star on the new image. The image is then aligned "photometrically" with the reference by requiring that the mean luminosity of stars in a given luminosity band equal the mean luminosity in the catalogue. (The small number of intrinsically variable stars in the catalogue does not affect this procedure.) Successive images then add one point to the blue or red light curve of each star in the catalogue.

After data reduction, each light curve is tested for the presence of time variations using a variety of algorithms. For microlensing-like events, we use a simple algorithm that scans curves for sequences of measurements that are significantly above the mean value. The light curve is selected as a microlensing candidate if it exhibits one and only one such sequence simultaneously in both colours. The precise value of the threshold for acceptance is chosen using estimates of measurement errors so that random fluctuations

根据计算, 大麦哲伦云中某一恒星在指定时间变亮超过 0.3 星等的概率约为 0.5×10^{-6} (参考文献 3, 参考文献 6)。对于一个质量为 M 的天体所产生的光偏折, 其光度变亮的典型时标为 $\tau = 70\sqrt{M/M_\odot}$ 天。这种微引力透镜事件的光变曲线应该在时间上具有对称性、在不同波段一致并且不可重复。1990~1993 年间, 我们在拉西亚天文台用红或蓝滤光片共得到 304 块大麦哲伦云的施密特底片。典型曝光时间为一小时, 这使我们可以在平均测光精度为 15%(rms, 均方根) 的条件下监测到暗至 20 星等的恒星。这些施密特底片的透明度在巴黎天文台[7] 被天文自动测量仪器(MAMA) 以 10 μm(0.67 arcsec) 为步长进行了数字化。利用整个施密特底片视场的电荷耦合器件(CCD) 图像, 我们得到了透明度和恒星光度的关系。

第二个项目是采用由 16 个 579 × 400 像素的 Thomson THX 31157 型 CCD 拼接组成的 CCD 照相机, 覆盖大约 1° × 0.4° 范围的天区。该照相机被安装在了由上普罗旺斯天文台参与修复的 40 cm 口径的反射望远镜(f/10) 上。用此 CCD 照相机, 我们从 1991 年 12 月到 1992 年 3 月及 1992 年 8 月到 1993 年 3 月之间观测了大麦哲伦云棒的一个天区。至 1993 年 3 月, 我们使用红和蓝滤光片总共得到 8,100 次曝光。在平均测光精度约为 6% 的条件下, 每张图像上约可看见 100,000 颗恒星。这和前面提到的施密特底片项目相比, 虽然恒星的数量是施密特底片项目中恒星数量的 1/80, 但是较短的采样时间(每幅图像平均只要 22 分钟) 使得这个 CCD 项目易于探测质量在 $10^{-7} M_\odot < M < 10^{-3} M_\odot$ 范围之内的偏折天体, 对应透镜事件的平均持续时间在 $1 \text{ h} < \tau < 3 \text{ d}$ 之间。

在初期处理(施密特底片信号的数字化和 CCD 图像的平场处理) 后, 两个项目的数据处理都遵循基本相同的步骤。首先, 通过结合良好大气条件下拍摄的 10 幅施密特底片或者 50 幅 CCD 图像, 我们得到了每种颜色下的参考图像。利用一种寻找恒星的算法, 我们为每幅参考图像建立一个星表。随后, 通过亮的孤立星将每一幅图像和参考图像对齐。这些参考图像上恒星的位置随即作为一个测光拟合程序的输入值, 以确定新图像星表中每颗恒星的光度。然后, 这些图像在“测光”上和参考图像对齐, 要求在指定的某一光度范围内恒星的平均光度和星表中的平均光度相同。(星表中少数内禀变星不会对此步骤产生影响。) 于是接连观测的图像就在星表中每颗恒星的蓝或红的光变曲线上增加了一个点。

在数据处理之后, 我们可以对每条光变曲线采用多种算法测试其时变性。对于类微引力透镜事件, 我们采用一种简单的算法, 即扫描光变曲线中明显超过平均值的测量序列。如果有且只有一个这种序列可以在蓝和红波段同时出现, 则此光变曲线被选为微引力透镜候选事件。我们通过估计误差范围来选择接受阈值的准确数值, 从而排除非变星的随机涨落。每幅图像的测光误差可以近似看作是星等的函数, 但

of intrinsically stable stars are not accepted. The photometric errors are estimated as a function of magnitude for each image but may still vary from one star to another by 20% according to the star's environment. We estimate the efficiency of the cuts to accept real microlensing events using Monte Carlo-generated lensing events, produced with the observed photometric resolution and observing sequence. We superimposed the Monte Carlo signature of microlensing events onto both observed (flat) light curves and simulated light curves. These curves were then subjected to the same algorithms as the real data. For simulated events with peak amplifications > 1.34 inserted into the data during the period of the observing seasons, the efficiency ranges from ~25% for events with timescales of 6 days to 50% for events with timescales of 30 days. These numbers differ from 100% because of the sampling period and the photometric resolution for faint stars.

Measured curves passing the above selection criteria were then inspected visually and subjected to further analysis to determine their compatibility with the microlensing hypothesis. At this stage, we found it necessary to eliminate only those light curves that exhibit variations on a timescale comparable to that of the total observing period. Such events cannot be tested for the presence of subsequent variations. Remaining events are fitted for the theoretical microlensing light curve. The parameters of the fit are the off-lensing luminosity, the maximum amplification, the time of maximum amplification, and the timescale of the microlensing. The light curves of the two colours are fitted separately (to test for wavelength independence) and then simultaneously.

We found no candidates in an analysis performed on the 1991–92 CCD data (20% of the total data). For this data, we expected about three candidates if the halo is entirely comprised of dark objects in the range 10^{-7} to $10^{-5}\,M_\odot$.

A preliminary analysis of 40% of the Schmidt-plate data has revealed two events that are consistent with the microlensing hypothesis. The light curves are shown in Figs 1 and 2, with the event characteristics listed in Table 1. They are the only curves so far analysed that show one significant amplification event with no further variations. (No curves have been found that show two or more examples of microlensing-like behaviour.) The curves are consistent with the theoretical curve; χ^2s are good within the estimated 20% uncertainty in the photometric errors. The events are wavelength independent at the 10% level, that is within errors. The amplification is near the median amplitude expected ($\delta m = 1.0$) for detectable events with these time-scales. Neighbouring stars show no variations over the whole observing period. The off-lensing magnitudes of the two stars are near the average for stars in our catalogue. Candidate 2 is on the main sequence while candidate 1 is between the main sequence and giant branch of the colour–magnitude diagram. Only the observation of further events will tell us if there is an accumulation of events in a given region of the diagram indicating variable-star phenomena.

是仍可能因为恒星的环境不同而在不同恒星间变化20%。我们用蒙特卡罗方法模拟得到的透镜事件（用观测的测光分辨率和观测序列产生）估计阈值对于接受真实微引力透镜事件的效率。我们在观测到的（平坦的）光变曲线和模拟的光变曲线上叠加微引力透镜事件的蒙特卡罗特征信号。这些曲线将在之后同真实数据一样用同样的算法进行分析。对于在观测时间段的数据中插入峰值放大率大于1.34的模拟透镜信号的模拟事件，探测效率的范围从时标为6天事件的25%到时标为30天事件的50%。而对于暗星而言，由于不同的采样周期以及测光分辨率，探测效率可以相差100%。

对满足上述选择判据的观测到的光变曲线，我们随即再通过人眼检验以及进一步的分析，以确定它们是否符合微引力透镜假设。在此阶段，我们发现只需剔除那些光度变化时标和整个观测周期差不多的光变曲线。这样的事件不能被用于检验之后的变化。剩下的事件将利用微引力透镜的理论光变曲线进行拟合。拟合参数包括未发生透镜现象时的光度、放大率的最大值、放大率达到最大的时刻以及微引力透镜事件的时标。两种颜色的光变曲线被分别（以此来检验波长的独立性）拟合，然后再同时拟合。

根据分析，我们没有在1991~1992年的CCD数据（总数据量的20%）中发现候选透镜事件。对于这些数据，如果银晕整体上均由范围在$10^{-7} M_\odot < M < 10^{-5} M_\odot$之间的暗天体组成，那么我们预期发现大约3个候选透镜事件。

对施密特底片40%的数据的初步分析显示，有两个候选事件符合微引力透镜假设。光变曲线如图1、图2所示，相关特征参数列于表1中。它们是迄今分析发现仅有的、存在明显光度放大且没有其他变化的光变曲线。（目前没有发现任何一个光变曲线表现出两次或者更多的类微引力透镜现象。）这些曲线和理论曲线一致；考虑测光误差中大约20%的不确定性，我们得到了很好的χ^2。在误差之内，在10%的水平上事件不依赖于波长。其放大率接近这类时标下可探测到的事件的预期（放大率）中位值（$\delta m = 1.0$）。在整个观测期间邻近的恒星没有变化。这两颗恒星未发生透镜事件时的星等值接近我们星表中恒星星等的平均值。在颜色-星等图中，候选天体1位于主序带和巨星支之间，而候选天体2则位于主序带上。只有对更多事件的观测才可以判断颜色-星等图上代表变星现象的区域是否存在事件的积累。

Table 1. Characteristics of the two microlensing candidates

	Candidate 1	Candidate 2
Coordinates of star (J2000)	α = 5 h 26 m 36 s δ = −70° 57′37″	α = 5 h 06 m 05 s δ = −65° 58′34″
b magnitude	19.3 ± 0.2	19.3 ± 0.2
b–r	0.3 ± 0.2	0.0 ± 0.2
Date of maximum amplification	1 February 1992	29 December1990
Event duration (τ in days)	27 ± 2	30 ± 3
Maximum amplification in magnitudes (blue filter)	1.0 ± 0.1	1.1 ± 0.2
Maximum amplification in magnitudes (red filter)	1.0 ± 0.1	1.3 ± 0.2
Maximum amplification in magnitudes (combined fit)	1.0 ± 0.1	1.2 ± 0.2
χ^2 (combined fit)	192 for 248 d.o.f.	167 for 131 d.o.f.

Errors in the magnitudes include fitting errors as well as systematic uncertainties in the magnitude-plate transparency relation. The χ^2 were calculated using the estimated errors which are known only to an estimated precision of 20%. The timescale τ is the time taken by the dark object to cross an angle corresponding to one "Einstein radius" (ref. 3). This definition differs from the traditional one which is the time the dark object remains within the Einstein ring. The traditional definition is meaningless for lensing events with a minimum impact parameter greater than the Einstein radius.

If the events are interpreted as arising from microlensing, the lensing objects would have a mass between a few × 10^{-2} and 1 $M_\odot$. This range is based on simulations with the standard isothermal halo yielding the observed flat rotation curve out to the LMC (refs 2, 6). The range is wide because, for fixed mass, the lensing time varies due to the uncertain distance and speed of the lensing object. The allowed masses include those expected for brown dwarfs and dim main sequence stars. If they are located in the Galactic halo, they would not be detectable. If the halo consists entirely of such objects, we estimate that we would have observed about six events if the dominant mass is 10^{-2} $M_\odot$ and about one event if it is 1 $M_\odot$, in agreement with the observed number.

Obviously, this interpretation of the events must be confirmed by further observations of the stars in question and, especially, by the discovery of further events, which would permit us to study their statistical properties. In particular, the distribution of amplification magnitudes must be shown to be consistent with the microlensing hypothesis and the microlensed stars must be distributed throughout the observed colour–magnitude diagram and throughout the LMC. Until this is done, it is not possible to rule out the possibility that we are dealing with a new type of variable star. During the preparation of this paper we learned that a similar microlensing event has been observed by the "MACHO" collaboration (C. Alcock, personal communication).

表 1. 两个微引力透镜候选事件的特征

	候选事件 1	候选事件 2
恒星的坐标 (J2000)	$\alpha = 5\,h\,26\,m\,36\,s$ $\delta = -70°\,57'\,37''$	$\alpha = 5\,h\,06\,m\,05\,s$ $\delta = -65°\,58'\,34''$
蓝星等	19.3 ± 0.2	19.3 ± 0.2
蓝星等 – 红星等	0.3 ± 0.2	0.0 ± 0.2
亮度增益最大的日期	1992 年 2 月 1 日	1990 年 12 月 29 日
现象时标 (τ 以天为单位)	27 ± 2	30 ± 3
星等增益的最大值 (蓝滤光片)	1.0 ± 0.1	1.1 ± 0.2
星等增益的最大值 (红滤光片)	1.0 ± 0.1	1.3 ± 0.2
星等增益的最大值 (综合拟合)	1.0 ± 0.1	1.2 ± 0.2
χ^2 (综合拟合)	192/248 自由度	167/131 自由度

星等误差中既包括拟合误差也包括星等 – 施密特底片透明度关系的系统不确定性。χ^2 是用估计的误差计算的，估计的误差精度只有 20%。时标 τ 是指暗天体穿过一个"爱因斯坦半径"(参考文献 3) 对应角度所用的时间。这个定义与传统定义不同，传统定义的时标是暗天体处在爱因斯坦环内的时间。对于最小影响参数大于爱因斯坦半径的透镜事件，这一传统的时标没有意义。

如果这些事件被解释为由微引力透镜事件导致，那么可以估计透镜天体的质量在数倍 $10^{-2}\,M_\odot$ 到 $1\,M_\odot$ 之间。这个质量范围基于标准等温晕模型下的模拟，该模型可以给出银晕外围直到大麦哲伦云范围之内的平坦自转曲线 (参考文献 2, 6)。这个范围较宽是因为对于给定透镜质量，透镜现象持续时间会由于距离以及透镜天体速度的不同而不同。这些可能的质量涵盖了褐矮星以及暗弱的主序星。如果它们位于银晕中，那么它们是不可被直接探测到的。假设银晕完全由这种天体组成，我们估计如果主要是质量为 $10^{-2}\,M_\odot$ 的透镜天体，会发现 6 个事件；如果质量为 $1\,M_\odot$ 的透镜天体主导，会发现 1 个事件，和观测到的数量一致。

显然，这些事情的解释必须由对所研究恒星的进一步观测证实，特别是发现更多的事件，可以让我们研究它们的统计性质。特别地，放大星等的分布应该和微引力透镜假设的理论预言相符合，被引力透镜放大的恒星应该遍布观测到的颜色 – 星等图以及大麦哲伦云。在做到这一步之前，我们没法排除所研究的天体是一种新型变星的可能性。在准备这篇论文期间，我们得知"MACHO"合作组 (阿尔科克，个人交流) 观测到一个类似的微引力透镜事件。

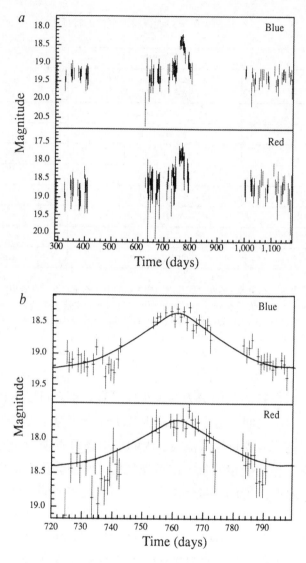

Fig. 1. *a*, The measured magnitudes for candidate 1 as a function of time. The time is counted from 1 January 1990. The error bars correspond to the estimated 1σ errors. *b*, The light-curve of candidate 1 on an expanded scale. The curve shows the best fit for the microlensing hypothesis. The parameters of the best fit are shown in Table 1.

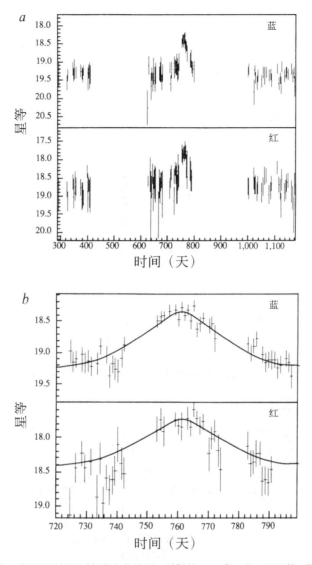

图 1. a，候选事件 1 的测量星等随时间的变化关系。时间从 1990 年 1 月 1 日开始。误差棒是估算得到的 1σ 误差值。b，在扩展刻度坐标下的候选事件 1 的光变曲线。该曲线显示观测结果与微引力透镜假说吻合得很好。最佳拟合的参数列于表 1 中。

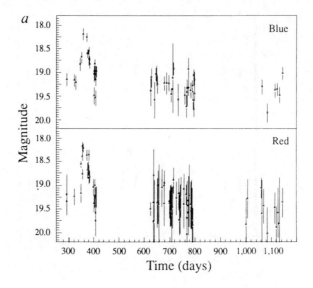

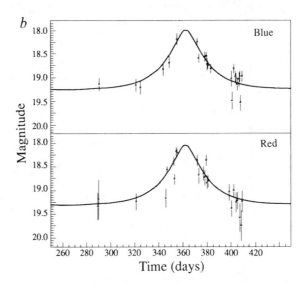

Fig. 2. As Fig.1, for candidate 2

The EROS collaboration is continuing to collect and analyse data and it is expected that further results will be presented within a year's time.

Note added in proof: We have located in our data the star involved in the candidate microlensing presented by the MACHO collaboration (C. Alcock *et al.,* this issue). Because the star is very faint in the blue band, we have reliable measurements only in the red. (This fact explains the rejection of the event in our analysis.) We confirm in this colour the existence and characteristics of the MACHO event. No other significant luminosity variations are seen in the 1990–91 and 1991–92 seasons, reinforcing the microlensing interpretation of the event.

(**365**, 623-625; 1993)

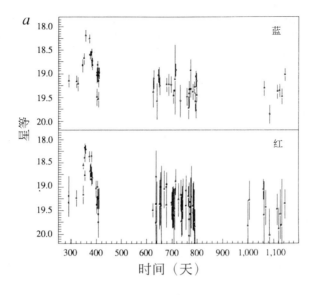

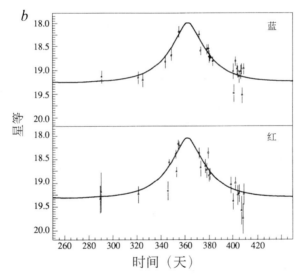

图 2. 对应于图 1，观测对象为候选事件 2

　EROS 合作组将继续收集和分析数据，预计一年之内将会有更多新的结果。

　附加说明：在我们的数据中找到了上文中提到的 MACHO 合作组（阿尔科克等人，请见本刊中另一篇文章）提供的候选微引力透镜天体。因为这颗恒星在蓝波段非常暗，我们只能在红波段得到可靠的测量结果。(这就是我们在分析时没有筛选出该事件的原因。)我们在该颜色下验证了 MACHO 所产生的透镜事件的存在性和特征。在 1990~1991 年和 1991~1992 年期间，并未发现该目标恒星存在其他明显的光度变化，这也进一步证明了可以用微引力透镜事件解释该例子。

（冯翀 翻译；臧伟呈 毛淑德 审稿）

E. Aubourg[*], P. Bareyre[*], S. Bréhin[*], M. Gros[*], M. Lachièze-Rey[*], B. Laurent[*], E. Lesquoy[*], C. Magneville[*], A. Milsztajn[*], L. Moscoso[*], F. Queinnec[*], J. Rich[*], M. Spiro[*], L. Vigroux[*], S. Zylberajch[*], R. Ansari[†], F. Cavalier[†], M. Moniez[†], J.-P. Beaulieu[‡], R. Ferlet[‡], Ph. Grison[‡], A. Vidal-Madjar[‡], J. Guibert[§], O. Moreau[§], F. Tajahmady[§], E. Maurice[||], L. Prévôt[||] & C. Gry[¶]

[*] DAPNIA, Centre d'Études de Saclay, 91191 Gif-sur-Yvette, France

[†] Laboratoire de l'Accélérateur Linéaire, Centre d'Orsay, 91405 Orsay, France

[‡] Institut d'Astrophysique de Paris, 98 bis Boulevard Arago, 75014 Paris, France

[§] Centre d'Analyse des Images de l'Institut National des Sciences de l'Univers, Observatoire de Paris, 61 Avenue de l'Observatoire, 75014 Paris, France

[||] Observatoire de Marseille, 2 Place Le Verrier, 13248 Marseille 04, France

[¶] Laboratoire d'Astronomie Spatiale de Marseille, Traverse du Siphon, Les Trois Lucs, 13120 Marseille, France

Received 22 September; accepted 30 September 1993.

References:

1. Trimble, V. A. Rev. Astr. Astrophys. 25, 425-472 (1987).

2. Primack, J. R., Seckel, D. & Sadoulet, B. A. Rev. Nucl. Part. Sci. 38, 751-807 (1988).

3. Paczyński, B. Astrophys. J. 304, 1-5 (1986).

4. Aubourg, E. et al. Messenger 72, 20-27 (1993).

5. Aubourg, E. thesis, Univ. Paris (1992).

6. Griest, K. et al. Astrophys. J. 372, L79-82 (1991).

7. Berger, J. et al. Astr. Astrophys. Suppl. Ser. B7, 389 (1991).

8. Alcock, C. et al. Nature 365, 621-623 (1993).

Acknowledgements. We thank C. Alcock for discussions at the beginning of this project. We thank A. Bijaoui, Ph. Veron, the staff at the Observatoire de Haute Provence, the ESO staff at La Silla, and the technical staff of the collaborating laboratories for their advice and help. We thank D. Bennett of the MACHO collaboration for help and interesting discussions. This work is based on observations at the European Southern Observatory, La Silla, Chile, and is funded by DSM-CEA, IN2P3-CNRS, INSU-CNRS, with support from ESO.

A Search for Life on Earth from the Galileo Spacecraft

C. Sagan *et al.*

This paper is a canny mixture of prescience and opportunism. By using a planned fly-by of the Earth by NASA's Galileo spacecraft in 1990 to investigate our planet as though it were an unknown, potentially habitable world, Carl Sagan and co-workers set the scene for observations of extrasolar planets in the early twenty-first century. The paper takes the inventive approach of treating the Earth as a hitherto unexplored planetary environment, asking whether, in such a situation, we could identify definitive evidence of life. The researchers use the Galileo observations of the chemistry of Earth's atmosphere, images of the planetary surface, and detection of radio-wave emissions, to infer that the presence of water-based life, probably intelligent, seems highly likely.

In its December 1990 fly-by of Earth, the Galileo spacecraft found evidence of abundant gaseous oxygen, a widely distributed surface pigment with a sharp absorption edge in the red part of the visible spectrum, and atmospheric methane in extreme thermodynamic disequilibrium; together, these are strongly suggestive of life on Earth. Moreover, the presence of narrow-band, pulsed, amplitude-modulated radio transmission seems uniquely attributable to intelligence. These observations constitute a control experiment for the search for extraterrestrial life by modern interplanetary spacecraft.

AT ranges varying from ~100 km to ~100,000 km, spacecraft have now flown by more than 60 planets, satellites, comets and asteroids. They have been equipped variously with imaging systems, photometric and spectrometric instruments extending from ultraviolet to kilometre wavelengths, magnetometers and charged-particle detectors. In none of these encounters has compelling, or even strongly suggestive, evidence for extraterrestrial life been found. For the Moon, Venus and Mars, orbiter and lander observations confirm the conclusion from fly-by spacecraft. Still, extraterrestrial life, if it exists, might be quite unlike the forms of life with which we are familiar, or present only marginally. The most elementary test of these techniques—the detection of life on Earth by such an instrumented fly-by spacecraft—had, until recently, never been attempted.

Galileo is a single-launch Jupiter orbiter and entry probe currently in interplanetary space and scheduled to arrive in the Jupiter system in December 1995. It could not be sent directly to Jupiter; instead, the mission incorporated two close gravitational assists at the

用伽利略木星探测器探测地球生命

萨根等

编者按

本文是预知结果与取巧验证的巧妙结合。通过利用美国国家航空航天局伽利略探测器 1990 年按计划掠过地球的机会，把我们的行星当成未知的、潜在的宜居世界进行研究，卡尔·萨根与合作者们为二十一世纪早期对太阳系外行星的观测设定了情景。本文将地球当作从未探索过的行星环境，以此创新方法研究在这样的情形下我们是否可以确定地证认生命的存在。这些研究者使用伽利略探测器对地球大气化学、地球表面图像和无线电波辐射进行探测，推断地球存在水基生命，看起来很有可能是有智慧的。

伽利略号探测器在 1990 年 12 月从地球掠过时，发现了丰富的气态氧存在的证据：一种广泛分布在地球表面、在可见光谱红端有尖锐吸收边缘的色素，以及大气中 CH_4 处于极端热力学非平衡态。所有这些都有力地证明地球存在生命。此外，观测到的窄带的、脉冲的、振幅调制的无线电发射信号似乎只能归结为智慧文明。这些观测结果构成了现代行星际探测器探索地球外生命的对照实验。

现在，各种探测器已经在 100~100,000 km 的距离上掠过了 60 多颗行星、卫星、彗星和小行星。它们都装备了各种各样从紫外到千米波的成像系统、光度计和光谱仪，以及磁强计和带电粒子探测器。在这些交会中没有发现令人信服的甚或是强有力的生命存在的证据。对于月球、金星和火星，轨道飞行器和着陆器的观测已经证实了飞掠探测器得出的结论。然而，如果有外星生命的话，其形式很可能不是我们熟知的那样，或者是很少量的存在。应用以上这些技术进行最初步的检测——用装备有这些技术设备的飞掠探测器探测地球上的生命——直到最近还从来没有人尝试过。

伽利略号是单向发射的木星轨道飞行器和进入行星际空间的探测器，计划在 1995 年 12 月抵达木星系。它不能直接向木星发射，取而代之的是，它的任务包括两次紧密的引力加速，一次在地球，另一次在金星。这大大延长了转运时间，但

Earth and one at Venus. This greatly lengthened the transit time, but it also permitted close observations of the Earth. The Galileo instruments were not designed for an Earth encounter mission, so an appropriate control experiment has fortuitously been arranged: a search for life on Earth with a typical modern planetary probe.

Closest approach to Earth in the 8 December 1990 encounter was 960 km over the Caribbean Sea. The Earth was approached from its night side, so all data in reflected light were acquired post-encounter. Evidence relevant to the search for life was obtained with the near-infrared mapping spectrometer (NIMS), the ultra-violet spectrometer (UVS), the solid-state imaging system (SSI), and the plasma wave spectrometer (PWS). These instruments are described in refs 1–4, respectively. Because of the high encounter velocity, most NIMS and SSI data were obtained 2 hours to 3.5 days after closest approach.

In what follows, we do not assume properties of life otherwise known on Earth, but instead attempt to derive our conclusions from Galileo data and first principles alone. We compare these conclusions with ground-based and low-altitude measurements that we describe, collectively, as the "ground-truth Earth". A necessary but not sufficient condition for the presence of life is a marked departure from thermodynamic equilibrium[5-8]. Once candidate disequilibria are identified, alternative explanations must be eliminated. Life is the hypothesis of last resort. As an analogy, mountains are in mechanical disequilibrium, given the Earth's erosional environment, but orogenic processes build mountains faster than wind, water and plate tectonics destroy them. Ozone is in substantial thermodynamic disequilibrium in the Earth's atmosphere, but this is driven by solar ultraviolet (UV) photochemistry. Bursts of high-brightness-temperature kilometre-wave radiation occur in the Earth's auroral zones, but are pumped by the interaction of the solar wind with the Earth's magnetosphere. No biological explanations of these disequilibrium phenomena need to be sought. As we will show, however, Galileo found such profound departures from equilibrium that the presence of life seems the most probable cause.

Chemistry

NIMS spectral and radiometric measurements indicate the presence of water in several forms. Spectra of Antarctica[9] show distinctive features due to condensed water, which are broadened and shifted to longer wavelengths compared to gas-phase features. The radiometric temperatures ($\sim-30\,°C$) indicate ice, and analysis of the spectra[9] give snow grain sizes of 50–200 μm. This ice and snow cover occurs on continental scale around the South Pole.

At higher latitudes are extensive areas with higher, nearly uniform temperatures just at, or slightly above, the melting point of water. In particular, the radiometric temperatures measured at 5 μm, uncorrected for atmospheric absorption, are $\sim-3\,°C$ at high southern latitudes, increasing to 8–18 °C at midlatitudes. The average 1 μm albedo of these extensive areas is $\sim4\%$, much smaller than the albedos of snow, clouds and rocky surfaces, but

同时也让我们得以仔细观测地球。伽利略号探测器并不是为探测地球而设计的，所以临时安排了一个合适的对照试验：利用典型的现代行星探测器搜寻地球上的生命。

1990 年 12 月 8 日的交会中距地球最近时是在加勒比海面上空 960 km 处。探测器进入到地球的夜面，所以所有反射光数据在交会之后获得。搜寻和生命相关的证据分别通过近红外成像光谱仪（NIMS）、紫外光谱仪（UVS）、固体成像系统（SSI）和等离子体波谱仪（PWS）获得。在参考文献 1~4 中分别介绍了这些探测设备。由于交会时速度大，大多数的 NIMS 和 SSI 近地点探测数据在过最近点之后的 2 小时到 3.5 天得到。

接下来，我们不去假设地球上已知生命的特征，相反，我们尝试从伽利略号提供的数据和第一性原理中得出结论。我们将这些结论同我们所描述的以地面状况为基础和对近地表的测量结果——称之为“实况地球”相比较。对于生命的存在而言，一个必要不充分条件就是显著偏离热力学平衡[5-8]。一旦候选体的不平衡被证实，那么其他的解释就必须被排除。有生命是最终才采用的假设。作为类比，考虑到地球的侵蚀性环境，山体处于力学的非平衡态，但是造山过程比风、水和板块构造对山体的破坏快。在地球大气层中，O_3 实际上处于热力学非平衡态，但这是受到太阳紫外线光化学的作用所致。高亮温度的千米波辐射爆发在地球的极光带，但却是受太阳风和地球磁层的相互作用激发的。这些不平衡现象不需要生物学解释。然而正如我们将要展示的，伽利略探测器发现了如此深刻的偏离平衡现象，生命的存在是解释这种现象最可能的原因。

化　学

NIMS 频谱和辐射测量结果说明了水以数种形式存在。南极的频谱[9]显示冷凝水造成的独有的特征，这些冷凝水的频谱相比于它的气相特征，波谱被展宽并且向长波方向移动。辐射温度（约 −30 ℃）表明存在冰，频谱分析[9]还显示存在 50~200 μm 大小的米雪。在规模上，这些冰雪覆盖了南极附近的大陆。

高纬度地区是一片广泛具有更高的、差不多处于或稍稍高于水的熔点的均一温度的地区。特别是，不考虑大气吸收的修正，在南半球高纬度地区测到的 5 μm 处的辐射温度大约是 −3 ℃。这个温度在中纬度地区增加到 8~18 ℃。这些广阔地区 1 μm 处的平均反照率约是 4%，这比雪、云和岩石表面的反照率小得多，但是却和包括

consistent with the low diffuse reflectance of dielectric liquid surfaces, including water. In many of the NIMS images, greatly enhanced specular reflection is observed[10]—implying the existence of large areas that are macroscopically smooth and homogenous (that is, not granular) and most easily explained by the presence of liquid surfaces of oceanic dimensions. Evidence of gas-phase H_2O is found over the entire planet. Representative NIMS infrared spectra in the 0.7–1.0 μm range, and in the 2.4–5.2 μm range, are shown in Fig. 1. They were obtained over a fairly clear area in the eastern Pacific, north of Borneo. Analysis[10] of the vibration–rotation bands of water vapour gives typical abundances of ~1,000 parts per million (p.p.m.) or ~0.6 g cm^{-2}. The observed high humidities, found over most of the planet, along with the preceding discussion, imply that the oceans are composed of liquid water.

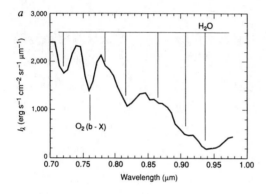

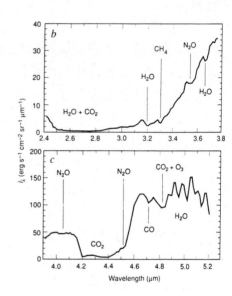

Fig. 1. a, Galileo long-wavelength-visible and near-infrared spectra of the Earth over a relatively cloud-free region of the Pacific Ocean, north of Borneo. The incidence and emission angles are 77° and 57° respectively. The (b $^3\Sigma_g^+$–X$^3\Sigma_g^-$) O—O band of O_2 at 0.76 μm is evident, along with a number of H_2O features. Using several cloud-free regions of varying airmass, we estimate an O_2 vertical column density of 1.5 km-amagat ± 25%. b and c, Infrared spectra of the Earth in the 2.4–5.2 μm region. The strong v_3 CO_2 band is seen at the 4.3 μm, and water vapour bands are found, but not indicated, in the 3.0 μm region. The v_3 band of nitrous oxide, N_2O, is apparent at the edge of the CO_2 band near 4.5 μm, and N_2O combination bands are also seen near 4.0 μm. The methane (0010) vibrational transition is evident at 3.31 μm. A crude estimate[10] of the CH_4 and N_2O column abundances is, for both species, of the order of 1 cm-amagate (≡ 1 cm path at STP).

Spectral data aside, from the albedo and heliocentric distance of the Earth alone it follows that the equilibrium temperature of the planet is only 20 °C or so below the freezing point of water—so that even a modest greenhouse effect would bring temperatures high enough that water, a cosmically abundant molecule, could exist in all three of its phases. Abundant surface liquid water is seen nowhere else in the contemporary Solar System. The dielectric constant, solvation properties, heat capacity, and temperature range of the liquid state are among the nonparochial reasons that water seems an ideal medium for life[11].

水在内的介电液体表面较低的漫反射率一致。在很多的 NIMS 图像中，观测到了大大增强的镜面反射[10]——这表明存在着大面积宏观上平滑且均匀的区域（也就是说，不是颗粒状的），用存在着海洋规模的液体表面来解释这个观测结果是最容易的。在整颗行星的所有地区都能找到气态水存在的证据。图 1 显示了非常具有代表性的 0.7~1.0 μm 波段和 2.4~5.2 μm 波段的 NIMS 近红外光谱。光谱是在婆罗洲北部的东太平洋上非常晴朗的地区获得的。对水蒸气的振动-转动谱带的分析[10] 得出了其丰度的典型值约是 1,000 ppm（ppm：百万分之一）或 0.6 g·cm^{-2}。在地球大部分地区观测到的高湿度，加上之前的讨论，说明海洋是由液态水组成的。

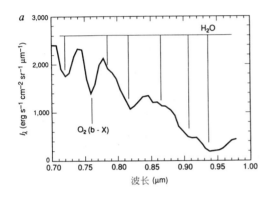

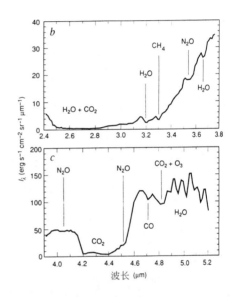

图 1. a，伽利略探测器在婆罗洲北部的太平洋上空相对无云地区探测到地球长波可见光和近红外光谱。入射角和出射角分别是 77° 和 57°。0.76 μm 处 O_2 的（b'$\sum_g^+$-X$^3\sum_g^-$）O—O 带明显可见，同时伴有一些水的特征线。通过使用几个不同大气质量的无云地区的光谱数据，我们估计了一个 1.5 km 阿马加 ±25% 的 O_2 垂直柱密度。b 和 c，2.4~5.2 μm 波段范围内的地球红外光谱。在 4.3 μm 处观测到强的 CO_2 的 υ_3 带，并且在 3.0 μm 带上发现了水蒸气带，但是没有标明。N_2O 的 υ_3 带位于 CO_2 带边缘，在 4.5 μm 附近出现；N_2O 组合带也在 4.0 μm 附近被发现了。CH_4（0010）的振动跃迁在 3.31 μm 处很明显。CH_4 和 N_2O 柱丰度的粗略估计[10] 都是 1 cm 阿马加（恒等于标准温度和压强下 1 cm 路径）。

不使用频谱数据，单从反照率和地球的日心距就可知道地球的平衡温度只比水的凝固点低 20 ℃ 左右。所以即使是最微弱的温室效应都会让温度升到足够高，以使一种宇宙中丰富的分子——水能够以它的三种相存在。在目前的太阳系中任何其他地方都没有看到如此大量水。液态水的介电常数、溶解特性、热容和温度范围都有力地说明，水看起来是生命理想的媒介[11]。

Figure 1 is notable for the presence of the A band of molecular oxygen at 0.76 μm. This transition is spin-forbidden, and the strength of the feature indicates ~200 g cm^{-2} of O_2 (Table 1). So large an abundance of O_2 is also unique among all the worlds in the Solar System. Oxygen can be generated by the UV photodissociation of water and the subsequent Jeans escape of H to space. But can the accumulation of so much O_2 over geological time be understood?

Table 1. Constituents of the Earth's atmosphere (volume mixing ratios)

Molecule	Standard abundance (ground-truth Earth)	Galileo value*	Thermodynamic equilibrium value	
			Estimate 1†	Estimate 2‡
N_2	0.78		0.78	
O_2	0.21	0.19 ± 0.05	0.21§	
H_2O	0.03–0.001	0.01–0.001	0.03–0.001	
Ar	9×10^{-3}		9×10^{-3}	
CO_2	3.5×10^{-4}	$5 \pm 2.5 \times 10^{-4}$	3.5×10^{-4}	
CH_4	1.6×10^{-6}	$3 \pm 1.5 \times 10^{-6}$	$< 10^{-35}$	10^{-145}
N_2O	3×10^{-7}	$\sim 10^{-6}$	2×10^{-20}	2×10^{-19}
O_3	10^{-7}–10^{-8}	$> 10^{-8}$	6×10^{-32}	3×10^{-30}

* Galileo values for O_2, CH_4 and N_2O from NIMS data; O_3 estimate from UVS data.
† From ref. 16 (P, 1 bar; T, 280 K).
‡ From ref. 17 (P, 1 bar; T, 298 K).
§ The observed value; it is in thermodynamic equilibrium only if the under-oxidized state of the Earth's crust is neglected.

Certainly, Venus and Mars—where atmospheric water vapour is being UV photodissociated—display very low O_2 abundances, < 1 and ~10^{-3} p.p.m., respectively[12,13], despite large quantities of water lost through photodissociation and atmospheric escape of H. But a real understanding of the Earth's steady-state O_2 abundance requires at the least knowledge of the oxidation state of the surface, surface erosion rates, and temperatures at the tropopause, the mesopause and the exobase. Galileo did not provide this data set. Even if it had, we could not be confident that present circumstances are typical of Earth history. An upper bound on the UV generation rate of O_2 is set by the photon-limited water photolysis rate, ~10^{13} cm^{-2} s^{-1} on the ground-truth Earth[14]—which yields the present O_2 abundance in < 10^5 yr. If instead we use the ground-truth present H escape flux[14] (which is H_2O diffusion-limited) and assume that all the H atoms are H_2O-derived, the present O_2 abundance would require several times the age of the Earth to accumulate. If there is substantial oxidation of the crust, these timescales will be longer. But the lack of impact craters in Galileo imagery, and pervasive wind and water on Earth suggest continuing exposure of fresh, oxidizable regolith. Accordingly, oxidation of the Earth's crust should be more extensive than on Venus and Mars, and yet these planets have much less atmospheric O_2. Galileo's observations of O_2 thus at least raise our suspicions about the presence of life. If detailed modelling showed solar UV to be insufficient, a process seems needed whereby the much more abundant, but

在图 1 中可以明显看出在 0.76 μm 处存在着 A 带的氧分子。这种跃迁是自旋禁阻的，这种特性的强度表明存在大约 200 g·cm^{-2} 的 O_2（表 1）。如此大量的 O_2 在太阳系的所有行星中也是很独特的。O_2 可以通过水的紫外（UV）光致离解和接下来 H 向太空的金斯逃逸作用而产生。但是在地质时期内积累起这么多的 O_2 能说得通吗？

表 1. 地球大气的组成（体积混合比）

分子	标准丰度（实况地球）	伽利略值 *	热力学平衡值	
			估值 1†	估值 2‡
N_2	0.78		0.78	
O_2	0.21	0.19 ± 0.05	0.21§	
H_2O	0.03～0.001	0.01～0.001	0.03～0.001	
Ar	9×10^{-3}		9×10^{-3}	
CO_2	3.5×10^{-4}	$(5 \pm 2.5) \times 10^{-4}$	3.5×10^{-4}	
CH_4	1.6×10^{-6}	$(3 \pm 1.5) \times 10^{-6}$	$< 10^{-35}$	10^{-145}
N_2O	3×10^{-7}	$\sim 10^{-6}$	2×10^{-20}	2×10^{-19}
O_3	10^{-7}～10^{-8}	$> 10^{-8}$	6×10^{-32}	3×10^{-30}

* O_2、CH_4 和 N_2O 的伽利略值来自 NIMS 数据；O_3 的伽利略值由 UVS 数据估计得出。
† 来自参考文献 16（压强为 1 bar，温度为 280 K）。
‡ 来自参考文献 17（压强为 1 bar，温度为 298 K）。
§ 观测值；只有在忽略地壳的氧化状态时才处于热力学平衡。

当然，尽管金星和火星大气中的水蒸气通过紫外光致离解以及 H 逃逸出大气层而大量损失，但仍显示出非常低的 O_2 含量，分别为 < 1 ppm 和约 10^{-3} ppm[12,13]。但对地球的稳态 O_2 丰度的真正理解要求至少了解表面的氧化状态、表面侵蚀率以及对流层顶、中间层顶和外大气层底的温度。伽利略号没有提供这些数据。即使有这些数据，我们也不能确定当前的环境就能代表地球的历史。O_2 的紫外产生速率的上限是由水的光解速率限定的，而水的光分解速率受到光子限制，在实况地球中约为 10^{13} cm^{-2}·s^{-1}[14]——按照这样的速率会在小于 10^5 年的时间之内得到当前的 O_2 丰度。相反，如果我们使用实况地球 H 逃逸流量[14]（这受到水扩散的限制），并且假设所有的 H 原子都是 H_2O 生成的，那么就需要用地球年龄很多倍的时间才能积累到目前的 O_2 丰度。如果存在大量的地壳氧化作用，那么时标将会更长。但是在伽利略号获得的影像中缺乏陨击坑，以及地球上普遍存在的风和水表明有新鲜的、可被氧化的土壤持续暴露出来。相应地，地壳的氧化作用比在金星和火星规模更大，然而这些行星上却只有比地球少得多的 O_2。因此伽利略号关于 O_2 的观测至少引起了我们对于生命存在的猜测。如果详细的模拟显示太阳紫外光子对于光致离解过程是不足的，那么就需要一个过

much less energetic visible light photons are used in series for H_2O photodissociation (at least two visible photons would be needed per photodissociation event). Apparently, only biological systems can accomplish this.

In Fig. 1*b* and *c*, the presence of carbon dioxide, methane, nitrous oxide and ozone are indicated. All are greenhouse gasses; together with water vapour, they mainly account for the discrepancy between the equilibrium and measured radiometric temperatures of the Earth's surface.

From straightforward photochemical theory it follows[15] that one consequence of the high oxygen abundance is a stratospheric ozone layer opaque in the middle UV. Even a quick look at the Galileo UVS raw data shows the strong Hartley bands of O_3 at wavelengths greater than 0.21 μm in the spectrum of the Earth but not of Venus. (O_3 absorption is also found in NIMS infrared spectra.) The presence of disequilibrium O_3 is not in itself a sign of life, because UV photochemistry is a disequilibrium process, but the photochemical abundance of O_3 is related approximately logarithmically to the O_2 abundance, so the observed quantity of O_3 does imply a substantial abundance of O_2. Therefore a train of argument may exist[15] from abundant O_3 to abundant O_2 to life. The Hartley bands provide substantial optical depth at wavelengths less than ~0.3 μm, suggesting that in addition to the oceans (where life could be protected at depth from UV radiation), life on the land is also possible provided structural molecular bond strengths exceed ~50 kcal mol^{-1}. In this category fall a large number of organic functional groups, including C—C (70–130 kcal mol^{-1}), C—H (80–110), C—O (~90) and C—N (80–100). Corresponding multiple bonds are even stronger. Of course this is not, by itself, an argument for organic biochemistry on Earth.

The agreement of ground-truth atmospheric abundances with those determined by Galileo is reasonably good (Table 1). Also shown in Table 1 are two independent estimates[16,17], made 25 years apart, of the thermodynamic equilibrium abundances expected of minor constituents in the observed excess of O_2. Such calculations are characteristically performed by minimizing the free energy of the system while simultaneously satisfying the equilibrium constants of all known reactions of all known reactants. This approach is less vulnerable to unknown reaction pathways or erroneous rate constants than is absolute reaction rate kinetics. But it is an idealization at best, because thermodynamic equilibrium may not be achieved in the age of the Solar System, and because there are prominent nonequilibrium processes including photochemistry and biology. Ozone is present at more than 20 orders of magnitude above its thermodynamic expectation value, but, as noted above, this disequilibrium is not considered evidence for life because a UV photochemical pathway from O_2 to this O_3 abundance is well known.

The circumstance of methane is different. It is oxidized quickly to CO_2 and H_2O, and at thermodynamic equilibrium there should not be a single methane molecule in the Earth's atmosphere[16,17]. The disparity between observation and thermodynamic equilibrium is about 140 orders of magnitude. Clearly there is some mechanism that pumps CH_4 into the Earth's atmosphere so rapidly that substantial steady-state abundances accrue before

程，将更多但能量小得多的可见光光子用于一系列水的光致离解过程（每次光致离解事件至少需要两个可见光光子）。很显然，只有生物过程才能满足这个条件。

在图 1*b* 和 1*c* 中，显示存在 CO_2、CH_4、N_2O 和 O_3。它们都是温室气体，跟水蒸气混合在一起。它们是导致地球表面平衡温度和实际观测到的辐射温度之间的偏差的主要原因。

直接从光化学理论来看[15]，高氧含量的一个结果是存在对中紫外波段不透明的平流臭氧层。 即使快速浏览一下伽利略号的 UVS 获取的原始数据也能发现地球光谱中波长比 0.21 μm 长的强 O_3 哈脱莱吸收带，这在金星光谱中是不存在的。(NIMS 红外光谱中同样发现了 O_3 的吸收谱线。) O_3 的非平衡态的存在本身不是生命的迹象，因为紫外光化学是一个非平衡过程，但是 O_3 的光化学丰度是与 O_2 丰度近似成对数关系的，所以观测到的 O_3 量确实表明有可观的 O_2 丰度。这样，可能存在一系列从丰富的 O_3 到丰富的 O_2 再到生命的论证[15]。哈脱莱吸收带使得波长小于 0.3 μm 处有相当大的光深，这表明除了海洋（在海洋深处生命可以免受紫外线的侵害），这片土地上的生命也可能提供键能超过 50 kcal·mol^{-1} 的分子键。在这类分子键中有一大批有机的官能团，包括 C—C(70~130 kcal·mol^{-1})，C—H(80~110 kcal·mol^{-1})，C—O(约 90 kcal·mol^{-1}) 和 C—N(80~100 kcal·mol^{-1})。相应的多重键更强。当然，就其本身而言，这并不是地球上有机生物化学的论据。

伽利略号测量的值和实况地球大气层的丰度一致性相当好（表 1）。在表 1 中也可以看到两个相隔 25 年、相对独立的、对过量 O_2 中所预期的微量成分的热动力学平衡丰度的估计[16,17]。这样的计算一般是通过在使系统自由能最小并同时满足所有已知反应物的所有已知的反应的平衡常量而进行的。这种方法对未知的反应路径或者错误的速率常数来说，要比绝对的反应速率动力学更稳妥。但是这最多是一个理想化的方法，因为热力学平衡在太阳系的年龄内可能无法实现，还因为事实上存在着包括光化学和生物学在内的明显的不平衡过程。现在的 O_3 丰度要比热动力学预期值高 20 个数量级。但是正如上面所说，这种不平衡并不是生命存在的证据，因为存在一个广为人知的从 O_2 到 O_3 丰度的紫外光化学途径。

CH_4 的情况不同。它能很快被氧化成 CO_2 和 H_2O，而且在热力学平衡态，地球大气层中不应该存在单独的 CH_4 分子[16,17]。观测和热力学平衡值大约差 140 个数量级。很明显，存在着某些机制将 CH_4 快速地注入地球的大气层，以至于在其氧化能够跟上节奏之前就获得巨大的稳态丰度。一直以来就存在一种观点，在富氧环境中

oxidation can keep pace. It has long been suggested that an extreme disequilibrium abundance of a reduced gas such as CH_4 in an O_2-rich atmosphere could be evidence for life on Earth[6-8,16]. The total emission flux of CH_4 into the atmosphere of the ground-truth Earth from all sources[18] required to sustain the steady state abundance is 500 ± 100 Tg yr^{-1} (where 1 Tg is 10^{12}g), or $\sim 10^{-4}$ g cm^{-2} yr^{-1}, an oxidation of all atmospheric CH_4 roughly every decade. Conceivably, methane could be injected into the Earth's atmosphere by other means (for example, the decomposition of prebiological organic matter) but so large a rate of injection, corresponding to ~ 1 bar of CH_4 after only 10^7 yr, seems highly implausible. Oxidation of such organic matter has certainly gone to completion on, for example, Venus. In fact, on the ground-truth Earth the contribution of volcanos, fumaroles, and earthquakes to atmospheric CH_4 is negligible[18]. About half the annual CH_4 injection is thought to arise from natural systems (methane bacteria and so on) and approximately half is anthropogenic; rice cultivation, biomass burning and flatulence from domesticated ruminants are among the principal sources[18]. All these sources are biological, including those deriving from fossil fuels. Thus the disequilibrium abundance of CH_4 in the Earth's oxidizing atmosphere is found by ground-based observations to be indeed caused by biology, as naive consideration of the Galileo data would suggest. It is also evidence that organic chemistry plays some role in life on Earth.

The presence of nitrous oxide (N_2O) in the atmosphere at high disequilibrium abundances is another indicator of biological processes. Nitrous oxide is partly lost to photodissociation, with an atmospheric lifetime[19] of about 50 years. Although there are non-biological mechanisms for producing N_2O (for example lightning), the major source of N_2O on the ground-truth Earth is nitrogen-fixing bacteria and algae that convert soil and oceanic nitrate (NO_3^-) to N_2 and N_2O.

Imaging

A typical Galileo image of the Earth shows continents, oceans, the Antarctic polar cap and a highly time-variable configuration of clouds. The six bands used in SSI global images can be combined in various ways to visualize specific spectral contrasts in the clouds or on the surface. Of the many possibilities, three categories of band combinations prove most informative. First is (RED, GRN, VIO) (equivalent wavelengths 0.670, 0.558 and 0.407 μm respectively), which gives an approximately natural colour view (Fig. 2a). Large expanses of blue ocean and apparent coastlines are present, and close examination of the images shows a region of specular reflection in ocean but not on land. Clouds cover much of the land surface, but in clear areas extreme albedo contrasts are seen. The lighter areas have a colour compatible with mineral soils, while a greenish tint can be perceived in the darkest areas.

类似 CH_4 这种还原气体的极端不平衡丰度可以作为地球上存在生命的证据[6-8,16]。要保持稳态丰度，所有排放源产生的到实况地球大气层的 CH_4 排放总量[18]需要保持在 500 ± 100 Tg·yr^{-1}（1 Tg 等于 10^{12} g），或者 $\sim 10^{-4}$ g·cm^{-2}·yr^{-1}，这大约是每十年大气层中氧化的 CH_4 总量。可以想象，CH_4 可以通过其他方式注入大气层（比如，通过生命起源前的有机物质的分解），但是仅仅经过 10^7 年就能产生约 1 bar 的 CH_4，这所需要的 CH_4 注入速率是很大的，这似乎非常不可能。这种有机物质的氧化在金星这样的星体上已经结束了。实际上，实况地球上的火山口、喷气孔和地震发生时向大气中喷发出来的 CH_4 是可以忽略的[18]。大约每年一半的 CH_4 的注入量都是通过自然系统增加的（CH_4 细菌等等），还有大概一半是人造的：水稻的种植、生物质的燃烧和驯化的反刍动物的肠胃气都是主要的来源[18]。所有这些来源都是生物的，包括那些化石燃料燃烧产生的。这样，通过地球表面观测到的地球氧化大气层中的 CH_4 丰度的不平衡现象被证实是通过生物产生的，正如对伽利略探测器的数据的简单考虑所示。这也表明有机化学在地球生命中发挥了一定的作用。

大气层中处于高度非平衡丰度的 N_2O 的存在是生物过程的另外一个迹象。在 50 年大气生命周期内[19] N_2O 一部分被光致离解。虽然，存在着生成 N_2O 的非生物学机制（比如闪电），但是实况地球上 N_2O 的主要来源还是能将土壤和海洋中 NO_3^- 转换成 N_2 和 N_2O 的固氮细菌和藻类。

成　像

一幅典型的伽利略探测器得到的图像显示了大陆、海洋、南极极地冰冠以及时变的云。在 SSI 的全球成像中使用的六个波段可以通过各种方法结合起来，将云层中或者地球表面不同的光谱可视化。在这么多的可能性中，三种波段结合方法被证明最有信息量。首先是（红，绿，紫）（波长分别是 0.670 μm、0.558 μm 和 0.407 μm），这三个波段近似给出了自然色景象（图 2a）。图中呈现了广阔的蓝海和明显的海岸线。仔细检查图像发现了一个海洋中产生镜面反射的区域，但是在陆上没有发现镜面反射的区域。云覆盖了绝大部分陆地，但是在晴朗区域可以观察到极端的星体反照率的反差。浅色区域色彩与矿质土壤一致，同时在颜色最深的区域可以看到绿色色调。

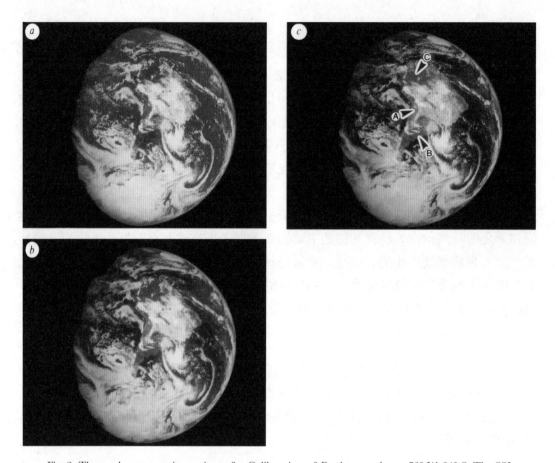

Fig. 2. Three colour-composite versions of a Galileo view of Earth centred over 72° W, 34° S. The SSI bands are identified by the following mnemonics and mean wavelengths in micrometres: VIO (0.407), GRN (0.558), RED (0.670), 0.73 (0.734/N), 0.76 (0.760/N), 0.89 (0.887/N), 1.0 (0.984). N indicates a narrow-band filter (the others are broadband), and effective wavelengths and radiometric calibration for data presented here are for the SSI camera with the clear lens cap on, which was the case for the encounters described here. Of these seven bands, six (all except 0.89) were used for global imaging of the Earth at 20–30 km resolution. (Galileo also imaged Australia and Antarctica at higher resolution (~one to a few km per pixel), but these imaging sequences did not use bands beyond 0.73 μm). a, Approximately natural-colour view constructed from (RED, GRN, VIO) filters; b, colour-composite produced from mapping the (1.0 μm, RED, GRN) bands to red, green and blue; c, colour-composite produced from closely spaced (0.76 μm, 0.73 μm, RED) bands, which uniquely separate rock and soil (grey tones) from areas that have unusual, very steep spectral slopes near 0.7 μm, possibly indicative of photosynthetic pigments in autotrophic terrestrial life. Areas whose spectra are shown in Fig. 3 are marked with A, B, C on this composite.

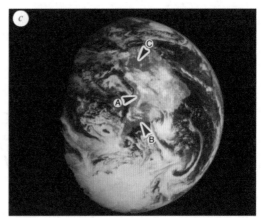

图 2. 伽利略探测器观测到的以西经 72°和南纬 34°为中心的三组地球色彩合成视觉图。SSI 不同波段通过以下的助记符号和以微米为单位的平均波长标示：紫 (0.407)、绿 (0.558)、红 (0.670)、0.73 (0.734/N)、0.76 (0.760/N)、0.89 (0.887/N)、1.0 (0.984)。N 代表一个窄带滤光片 (其他代表宽带)，这里呈现的数据的有效波长和辐射定标是针对开启无色的镜头盖的 SSI 成像相机，这就是我们在这里描述的相遇的情形。在这 7 个波段中，有 6 个 (0.89 除外) 是用来以 20~30 km 分辨率拍摄地球全球图像的。(伽利略探测器还以更高的分辨率来拍摄澳大利亚和南极的图像 (每像素一到几千米)，但是这些拍摄结果没有使用超过 0.73 μm 的波段。) a，从 (红，绿，紫) 滤光片合成的接近自然光的视图；b，将 (1.0 μm，红，绿) 波段映射为红、绿和蓝得到的颜色合成；c，由密近排布的 (0.76 μm，0.73 μm，红) 波段得到的颜色合成，这就能区分来自不寻常的、在 0.7 μm 附近有非常陡的谱斜率的区域与岩石和土壤 (灰色调)，这也许就可以表明自养型陆地生命的光合作用的色素。图 3 显示的光谱对应的地区在这张合成图中标注为 A、B、C。

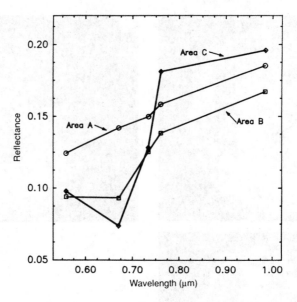

Fig. 3. Representative spectra from three areas on the land surface (see Fig. 2c). A gently sloping spectrum (circles, Area A) is consistent with any of several types of rock or soil. An intermediate spectrum (squares, Area B) shows some evidence of an absorption band near 0.67 μm (RED). Substantial areas on the surface have an unusual spectrum (diamonds, Area C) with a strong absorption in the RED band and a steep band edge just beyond 0.7 μm. This spectrum is inconsistent with all likely rock and soil types, and is plausibly associated with photosynthetic pigments (see text).

Second, we may translate (1.0, RED, GRN) (equivalent wavelength of 1.0 is 0.984 μm) bands to red, green and blue, respectively, in a false-colour image (Fig. 2b). This version reveals two kinds of sharp spectral contrasts. First, the cyan colour of the bright polar cap is caused by absorption in the 1.0 μm band, which confirms its H_2O-ice composition, and second, some land areas appear bright red, indicating a high albedo at 1.0 μm along with a low albedo in the visible spectrum (especially RED). Even without additional spectral information, a GRN/RED/1.0 signature so extreme is inconsistent with most dry or hydrated rocks or soils that might be expected on the surface of a terrestrial planet.

Using the 0.73 and 0.76 μm bands deepens the mystery. Figure 3 shows average mean spectra for three classes of land surface. (VIO is not plotted because of strong, differential atmospheric scattering corrections that are not fully modelled). The spectrum of area A (circles) has a gentle slope that rises in brightness uniformly from GRN to 1.0 μm, consistent with a variety of dark rock or mineral-soil surfaces. The spectrum of area B, an intermediate-albedo area, shows an absorption centred at RED. Although such areas are common on the surface, no common igneous, altered igneous, or sedimentary rock/soil surface displays such a signature[20]. The spectrum of area C is from a large low-albedo area in the northern part of the continent. Its very unusual spectrum has a strong absorption in the visible spectrum and a very steep spectral slope from 0.67 (RED) to 0.76 μm.

The third category of colour composite uses three closely spaced bands (0.76, 0.73, RED), and emphasizes strong spectral slopes over this narrow wavelength range (0.67–0.76 μm)

206

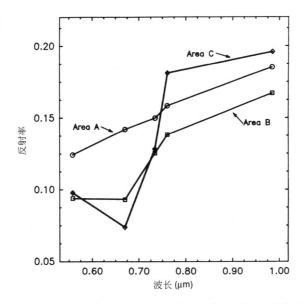

图 3. 地球表面三个地区的代表性光谱（见图 2c）。任何几个类型的岩石或者土壤都跟一条斜率缓和的光谱（圆圈，A 区域）一致。中间的光谱（方形，B 区域）显示 0.67 μm（红色）附近的吸收带的一些证据。地球表面的大量区域具有不寻常的光谱（菱形，C 区域），这条光谱在红端有强吸收带且在超出 0.7 μm处有较陡的带边缘。这条光谱与所有可能的岩石和土壤都不一致，可能和光合作用的色素有关。

其次，在一幅伪彩色图像中（图 2b）我们可以将（1.0，红，绿）（1.0 对应的等效波长为 0.984 μm）分别转换为红色、绿色和蓝色。这幅图像揭示出了两种截然不同的光谱。第一种，青色代表的闪亮的极盖是由在 1.0 μm 波段吸收引起的，这证实了它的成分就是水结成的冰。第二种，一些陆地地区显示出亮红色，表明 1.0 μm 处的高反照率伴随着可见光光谱（尤其是红端）中的较低的反照率。即使是没有其他的光谱信息，一个如此极端的绿 / 红 / 1.0 的特征与预计的类地行星表面的大部分干燥或者含水的岩石或者土壤也不一致。

0.73 μm 和 0.76 μm 波段更加深了这种神秘感。图 3 展示出了地表的三种典型的平均谱。（紫色没有绘制出来，因为其强烈、与众不同的大气散射修正不能完全地建模）。A 区域内的光谱（圆圈）有一条从绿色到 1.0 μm 亮度均匀上升的斜线，这与一些暗色岩石和矿质土表面一致。B 区域是一个中等反照率的区域，其光谱显示出以红端为中心的吸收带。虽然这些区域在地球表面很常见，但是不常见的火成岩、变相的火成岩或者沉积岩 / 土壤表面却不显示这样的特征[20]。C 区域的光谱来自北部大陆大片的低反照率地区。它非同寻常的光谱在可见光波段有一个强吸收，且在 0.67 μm（红色）到 0.76 μm 之间有非常陡的光谱斜率。

第三种色彩组合使用了三个紧密分布的波段（0.76，0.73，红），并突出了光谱

(Fig. 2*c*). Areas that have gently sloped spectra consistent with plausible rocks and soils appear grey here, whereas areas dominated by the unusual surface material with a strong absorption edge near 0.7 μm appear in greenish-yellow/orange hues. (The only other substantial contrast are the magenta areas in clouds, due to the 0.73 μm water-vapour band.)

These spectra cannot by uniquely interpreted; however, because they are inconsistent with any known rock or soil types on terrestrial planets of iron silicate surface composition, with or without aqueous alteration[20], unusual materials must be considered. The possibility naturally arises that the strong RED absorption is the signature of a light-harvesting pigment in a photosynthetic system—already suggested, as discussed above, as one possible explanation of the large atmospheric O_2 abundance. Plant life might have to be widespread in order to sustain this O_2 abundance in the face of likely loss by oxidation of the crust. Substantial areas of the surface seem to be covered by this pigment, which on the ground-truth planet is of course chlorophyll *a* and *b*. Photosynthesis might also be driven by oceanic microbes and plants, but the strong red-to-infrared absorption of water makes their pigments much more difficult to detect.

Morphology and Topographic Resolution

During this fly-by, Galileo's highest resolution systematic imaging of Earth was obtained at 1–2 km per pixel. A two-part mosaic of the ~2,900 by 4,000 km continent of Australia was produced: the eastern half of the continent was imaged mostly with four-band (VIO, GRN, RED, 0.73) coverage at ~1 km per pixel, and the western half with three-band (VIO, GRN, RED) coverage. A mosaic of Antarctica was also produced in three bands (VIO, GRN, RED) at ~2 km per pixel (at the pole).

A later six-band global imaging sequence at ~26 km per pixel showed Australia to be dominated by large central and western deserts, but with some 1.0 μm-bright areas (presumably plant life) concentrated toward the eastern and northern coasts, while Antarctica was found to be almost entirely covered by H_2O ice (Fig. 2*b*). In ~1 km per pixel Galileo mosaics of the Australian continent, no reworking of the surface into geometric patterns, or other compelling indications of artefacts of a technical civilization could be discerned. Although we find some large-scale albedo boundaries in Australian coastal regions that can be associated *a posteriori* with agriculture, in our judgement they are not sufficiently distinctive to be, by themselves, indicative of intelligent life. Kilston *et al.*[21] have demonstrated from a large array of daytime orbital imagery that very few images of the ground-truth Earth at resolution ~1 km reveal evidence of life.

In a study using daytime satellite imaging at 0.1 km resolution, C.S. and Wallace[22] concluded that the chance of finding convincing artifacts in a random frame was only ~10^{-2}. They further estimated, as a function of resolution, the threshold (in terms of fraction of the ground-truth surface image) for detection of a civilization at the current

(0.67~0.76 μm 这个狭窄波长范围内的) 中明显的倾斜 (图 2c)。具有缓和斜率的光谱的区域可能是岩石和土壤,在这里表现为灰色,但是那些被不寻常的表面物质覆盖的地区在 0.7 μm 附近有强烈的吸收边缘,在这里表现为青黄色或者橘色。(根据 0.73 μm 处的水蒸气波段,唯一一个有巨大差别的就是洋红色的有云的区域。)

我们无法对这些光谱进行唯一的解读。因为无论是否有水侵,它们都无法与类地行星表面上任何由铁硅酸盐组成的已知的岩石或者是土壤类型相符合 [20],所以应该考虑特殊物质。自然而然地想到,红端强烈的吸收可能是光合系统中集光色素的信号——正如上面所讨论的,这可以作为大气层中大量 O_2 存在的一个可能解释。为了维持这个 O_2 丰度,植被可能需要广泛分布以抵消由于地壳氧化而导致的 O_2 消耗。大量区域的表面似乎被这种色素覆盖着,当然这在地球上就是叶绿素 a 和 b。光合作用很可能也是受到海洋微生物和植物的驱动,但是水在红色到红外范围的强烈吸收使得它们的色素难以被探测到。

地貌和地形的分辨率

在这次飞掠过程中,伽利略号获得了关于地球的每像素 1~2 km 的高分辨率图像。通过拼接方法将两部分拼接起来,约 2,900 km × 4,000 km 的澳大利亚大陆图像就产生了:大陆东部主要用了四个波段(紫,绿,红,0.73),每像素覆盖约 1 km;西部地区用了三个波段(紫,绿,红)。南极地区也用三个波段(紫,绿,红)以每像素约 2 km (在极点)合成了拼接图像。

之后的每像素约 26 km 的六波段全景成像序列显示,澳大利亚中部和西部是大片的沙漠,但是有一些 1.0 μm 亮的地区(可能是植物)主要集中在东部和北部海岸。而几乎整个南极洲都被水冰覆盖着(图 2b)。在澳大利亚大陆的每像素 1 km 的伽利略拼接图像中,没有发现对表面改造形成几何图案,也没有发现技术文明的其他有力证据。虽然我们在澳大利亚海岸地区发现了一些大规模反照边缘,根据后验这跟农业耕作有关,但是根据我们的判断这些东西本身还不足以成为智慧生命活动的指示。基尔斯顿等人 [21] 已经通过一系列日间轨道成像证明在约 1 km 的分辨率范围内几乎没有什么实况地球的成像能够成为生命存在的证据。

在一个用 0.1 km 分辨率的日间卫星图像研究中,萨根和华莱士 [22] 得出结论,在一个随机区域内发现人类存在痕迹的可能性只有约 0.01。他们估计了一个阈值(根据实况地球的表面图像部分),这个阈值是关于分辨率的函数,它检测了在目前人类

human level of development (type A) and a hypothetical civilization that reworks its planetary surface to a significantly greater degree (type B). Their Fig. 29 predicts that even with complete surface coverage, detection of a type A civilization would require ~2-km resolution, and type B, ~10-km resolution. Galileo's Australia mosaic represents 2.3% of the whole surface of the Earth imaged at ~1 km per pixel, and the Antarctic mosaic adds another 4% at ~2 km per pixel. Unluckily, these are among the most sparsely settled regions of the ground-truth planet. Imaging about 1% of the surface at this resolution would detect a type B civilization, but imaging nearly all the surface would be required to detect one of type A[22]. A type B civilization is either wholly absent, or present in other continents than Australia and Antarctica. Extensive imaging of the Earth at resolution better than 0.1 km would have readily detected signs of life[23]. The foregoing analysis is based on visible and near-infrared Galileo imaging in reflected sunlight. No usable imaging data were obtained from the night hemisphere of the planet.

Radio Emission

Ground-truth television and radar transmissions are predicted to generate a non-thermal radio emission spectrum so strong and striking as to announce the presence of intelligent life not just over interplanetary, but over interstellar distances[24,25]. With a rough knowledge of the composition and structure of the Earth's atmosphere and the solar UV spectrum, it is a straightforward matter[26] to derive the order of magnitude of N_e, the ionospheric electron density in the E and F layers, yielding a plasma frequency f_p ~1–5 MHz. The higher value is the maximum for the sunlit hemisphere. Radio emissions from a technical civilization at the Earth's surface should be detectable only at $f > f_p$.

During the Galileo fly-by, the plasma wave instrument detected radio signals, plausibly escaping through the nightside ionosphere from ground-based radio transmitters. Of all Galileo science measurements, these signals provide the only indication of intelligent, technological life on Earth. They are illustrated in the PWS frequency–time spectrograms in Fig. 4. The transmitter signals can be seen near the top of the spectrograms in Fig. 4a, starting at about 18:00 UT and extending to about 20:25 UT, shortly before closest approach (C/A). These signals were detected only during the inbound, nightside pass, and not on the outbound, dayside, pass. Also seen in Fig. 4a are a series of Type III solar radio bursts[27] and several bursts of auroral kilometric radiation[28]. The narrow-band emission line labelled f_p is an electrostatic oscillation excited at the local electron plasma frequency by thermal fluctuations in the ionospheric plasma. The sharp peak in f_p near closest approach is caused by the local peak in N_e as the spacecraft passed through the ionosphere at an altitude of ~950 km.

技术水平条件下的文明（A 类）和一个假想的很大程度上可以重塑行星表面的文明（B 类）。他们的图 29 预测，即使表面完全被覆盖，探测 A 类文明也需要约 2 km 的分辨率，探测 B 类文明需要 ~10 km 的分辨率。伽利略探测器对澳大利亚的拼接图像每像素 1 km，仅代表着地球表面图像的 2.3%，而南极的拼接图像增加了 4% 的地球表面图像，分辨率为每像素 2 km。非常不幸的是，这两个地区都是这个星球上最人烟稀少的地区之一。用这种方法对表面的 1% 进行成像就会探测到 B 类文明的存在，但是要探测一个 A 类文明需要对全部表面进行成像[22]。B 类文明或者是完全不存在的，或者是在除了澳大利亚和南极以外的大陆存在。比 0.1 km 分辨率更高的大面积成像方法将会更容易探测到生命的迹象[23]。前述的分析建立在反射太阳光的可见光和近红外的伽利略图像基础上。未获得这颗行星的夜半球可用的成像数据。

无线电辐射

我们预测实况地球电视和雷达传输可以产生强烈的非热无线电辐射，这足以证明生命不仅仅存在于行星际的距离上也存在于恒星际的距离上[24,25]。通过对地球大气层和太阳紫外光谱组成和结构的粗略了解，可以直接[26]得到在 E 层和 F 层的电离层的电子密度 N_e 的数量级，对应大约 1~5 MHz 的等离子体频率。较大的 N_e 值是日照半球的最大值。地球表面技术文明产生的无线电辐射只有在 $f > f_p$ 范围内才能被探测到。

在伽利略号飞掠期间，等离子体波谱仪探测到了无线电信号，可能是从夜面地面无线电发射器发射出来的。在所有伽利略号科学测量中，这些信号是唯一表明了地球上存在着智慧技术生命的指标。它们在图 4 中用等离子体波谱仪（PWS）频率–时间谱图表示。发射器的信号可以在图 4a 谱图的顶部看见，大约从 18:00 UT 到 20:25 UT，仅仅稍微早于最接近时刻（C/A）。这些信号只在经过夜面时探测到，在日间离开时没有探测到。在这张图中我们还可以看到一系列 III 型太阳射电暴[27]以及几次极光千米波辐射暴发[28]。标记为 f_p 的窄带发射线是电离层等离子体的热涨落以局域电子–等离子体频率激发的静电振荡。在最接近时刻附近的 f_p 处的峰是由探测器在大约 950 km 高度穿过电离层时的当地 N_e 的峰值导致的。

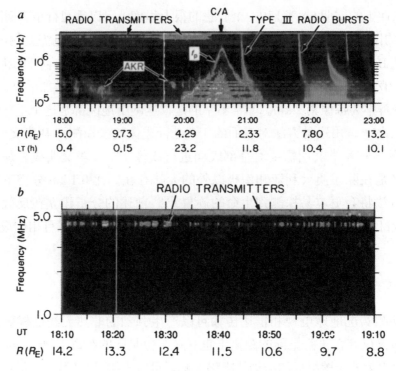

Fig. 4. A frequency–time spectrogram of the radio signals detected by the Galileo plasma wave instrument. The intensities are coded in the sequence blue–green–yellow–red, with blue lowest and red highest. Several natural sources of radio emission are shown in *a*, including auroral kilometric radiation (AKR). Modulated emission at $f > 4\,\text{MHz}$ is shown with an expanded time scale in *b*. Modulated patterns of this type are characteristic of the transmission of information, and would be highly unusual for a naturally occurring radio source. (UT, universal time; R is distance of Galileo from Earth in units of Earth's radius, R_E; LT, local time.)

The identification of the narrow-band emissions near 4–5 MHz as ground-based radio transmitters is based on studies by LaBelle *et al.*[29] and Keller[30] of the ground-truth planet; satellite observations clearly showed the signals originating from surface transmitters. That a similar conclusion could be drawn from the aforementioned Galileo fly-by data alone, without any previous knowledge of the existence of radio transmitters on Earth, is plausible but not beyond doubt. The rapid increase in the signal strength as the spacecraft approaches the Earth clearly indicates a near-Earth origin. The fact that the signals are only observed on the nightside inbound pass, where the ionosphere propagation cutoff frequency is sufficiently low to allow the signals to escape through the ionosphere, but not on the dayside where the cutoff prevents escape, strongly indicates a source beneath the ionosphere. Taking into account the expected exponential decrease in ionospheric N_e with increasing altitude, the theoretical maximum $f_\text{p} \sim 5$ MHz is in good agreement with the peak local plasma frequency (~ 2 MHz) observed at C/A, which was over the sunlit hemisphere at 16:30 LT. (These data might also be consistent with a class of nocturnal transmitting stations orbiting the Earth above the ionosphere, but the conclusion of intelligent life would be unchanged.)

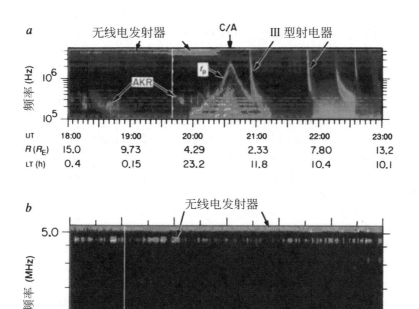

图 4. 伽利略号等离子体波谱仪捕捉到的无线电信号的频率–时间谱图。强度用蓝–绿–黄–红来表示。蓝色最低，红色最高。图 *a* 中显示了几种自然源的无线电发射情况，其中包括极光千米波辐射（AKR）。图 *b* 中显示了一个扩展的时间段内 $f > 4\,\text{MHz}$ 的调制辐射。这种类型的调制波形是信息传递的特征体现，对一个自然发生的电磁波资源来说是非常罕见的。（UT 代表世界时间；R 代表以地球半径（R_E）为单位计算的伽利略探测器距离地球的距离；LT 代表地方时）。

对 4~5 MHz 附近的地面无线电发射器的窄带发射的识别是基于拉贝尔等人[29] 和凯勒[30] 关于行星表面实况的研究。卫星观测清晰地展示了由地面发射器发射的信号。在没有关于地球无线电发射器存在的任何先验知识的情况下，从前面提到的伽利略号数据中可以总结出一个类似的结论，它看似合理却不是毋庸置疑的。当探测器接近地球的时候，信号强度的快速增长清晰地表明这个信号是源自近地地区。事实是信号仅仅在经过夜面时在电离层的传播截止频率足够低到让信号逃离电离层时才能被观察到，而不是在太阳照射的截止频率让信号无法逃脱的日面，这有力地表明在电离层下有一个发射源。考虑到在电离层中 N_e 随着高度的增加而指数降低，f_p 的理论最大值（约 5 MHz）和 16:30 LT 的日照半球上 C/A 处观测到的等离子体频率峰值（约 2 MHz）是一致的。（这些数据可能和在电离层之上绕地球旋转的夜间传输站的数据一致，但是智慧生命存在的结论可能不会改变。）

More difficult to estimate from first principles is the nightside f_p, because it depends on various losses and secondary ionization processes that cannot easily be evaluated from Galileo data alone. But because the primary (UV) ionization source is absent on the nightside, it is clear that N_e there will be much lower than on the dayside. From remote observations of Venus and Mars, it is entirely plausible that the nightside f_p should be depressed by factors of three to ten, relative to dayside values, permitting the escape of radio signals through the nightside ionosphere at 4–5 MHz, and providing a consistent overall interpretation of the day–night asymmetry in the observed radio signal. Ground-truth measurements of the night-time ionosphere show that $f_p < 4$ MHz is quite common, particularly at high latitudes[31].

The signals are confined mainly to two or three distinct channels near the top of the spectrogram (Fig. 4). The fact that the central frequencies of these signals remain constant over periods of hours strongly suggests an artificial origin. Naturally generated radio emissions almost always display significant long-term frequency drifts. Even more definitive is the existence of pulse-like amplitude modulation. When the spectrum in Fig. 4a is expanded (Fig. 4b), the individual narrow-band components can be seen to have a complex modulation pattern. Although the time resolution of the instrument (18.67 s) is inadequate to decode the modulation, such modulation patterns are never observed for naturally occurring radio emissions and implies the transmission of information. On the basis of these observations, a strong case can be made that the signals are generated by an intelligent form of life on Earth.

Conclusions

From the Galileo fly-by, an observer otherwise unfamiliar with the Earth would be able to draw the following conclusions: The planet is covered with large amounts of water present as vapour, as snow and ice, and as oceans. If any biota exists, it is plausibly water-based. There is so much O_2 in the atmosphere as to cast doubt on the proposition that UV photodissociation of water vapour and the escape of hydrogen provide an adequate source. An alternative explanation is biologically mediated photodissociation of water by visible light as the first step in photosynthesis. An unusual red-absorbing pigment that may serve this purpose, corresponding to no plausible mineral, is found widely on land. Methane is detected at ~1 p.p.m., some 140 orders of magnitude higher than the thermodynamic equilibrium value in this oxygen-rich atmosphere. Only biological processes are likely to generate so large a disparity. But how plausible a world covered with carbon-fixing photosynthetic organism, using H_2O as the electron donor and generating a massive (and poisonous) O_2 atmosphere, might be to observers from a very different world is an open question.

Several per cent of the land surface was imaged at a resolution of a few kilometres—entirely in Australia and Antarctica. No unambiguous sign of technological geometrization was found. Narrow-band pulsed amplitude-modulated radio transmission above the

从第一性原理上更难估计的是夜面的 f_p，因为它依赖于各种各样的损失和二次电离过程，它们不能简单地仅仅从伽利略数据中推算出来。但是因为主要的（紫外，UV）电离源不在夜面，所以很清楚地知道夜面的 N_e 比日面要低很多。从对金星和火星的远距离观测可知，完全可信的是夜面的 f_p 为日面值的 $\frac{1}{10} \sim \frac{1}{3}$，以允许无线信号通过夜面电离层以 4~5 MHz 逃逸，并且提供一致的、完整的关于日面和夜面观测到的无线信号不对称的解释。夜间电离层地面实况测量表明 $f_p < 4$ MHz 是非常常见的情况，特别是在高纬度[31]。

信号主要被限制在谱图顶部附近的两、三个不同的通道之内（图 4）。这些信号的中心频率保持几个小时不变的事实表明存在一个人工源。自然产生的无线电辐射几乎总是呈现明显的长期的频率漂移。更具决定性的是存在着像脉冲信号的调幅。当图 4a 中的频谱被扩展（图 4b），可以看到单个窄带成分都有一个复杂的调制模式。尽管该仪器的时间分辨率（18.67 s）不足以解调，但这种调制方式从未在自然产生的无线电辐射中看到过，这暗示了信息的传输。在这些观测的基础上，可以做出这样一个结论：信号是由地球上一种智慧生命形式产生的。

结　论

从伽利略号飞掠可知，不熟悉地球的观察者将会得出如下的结论：这颗行星是由大量水覆盖的，这些水以水蒸气、雪、冰和海洋等形式存在。如果有任何生物群存在，那么它们几乎都是水基的。在大气中有太多的 O_2，这使我们对水蒸气紫外光致离解和 H 的逃逸提供足够氧气源的观点提出了疑问。一个可能的解释是通过生物调控的可见光光解水作为光合作用的第一步。一种不同寻常的吸收红光的色素广泛地存在于陆地上，它可能满足这一要求，与此一致的是没有发现合理的矿物。探测到的 CH_4 浓度约为百万分之一，在这样一个 O_2 丰富的大气层中，这个值比热力学平衡值高出大约 140 个数量级。只有生物过程能够产生如此大的差异。但是从一个完全不同的世界来观察，一个由固碳的光合生物覆盖，将 H_2O 作为电子给体，并产生大量的（有毒的）O_2 层的世界有多大的可能性，这是一个开放的问题。

有部分陆地表面是以几千米的分辨率成像的——这部分陆地都是在澳大利亚和南极洲。没有发现由于技术导致的几何图形的清晰特征。在等离子体频率之上的窄

plasma frequency strongly suggests the presence of a technological civilization. Most of the evidence uncovered by Galileo would have been discovered by a similar fly-by spacecraft as long ago as about 2 billion (10^9) years. In contrast, modulated narrow-band radio transmissions could not have been detected before this century.

The identification of molecules profoundly out of thermodynamic equilibrium, unexplained by any non-biological process; widespread pigments that cannot be understood by geochemical processes; and modulated radio emission are together evidence of life on Earth without any *a priori* assumptions about its chemistry; 200 g cm^{-2} O_2 is at least suggestive of biology. Negative results in spacecraft exploration of other planets have thus much wider application than merely to "life as we know it". Of course putative ecosystems that are only weakly coupled to the surface environment (for example, subsurface[32,33]) are not excluded by such observations.

Surface oceans of liquid water are unique in the Solar System; conceivably this is part of the reason that Earth is the only planet in the Solar System with abundant surface life. Similar spectroscopic methods, although without resolution of the disk, may be useful in examining planets of other stars for indigenous life, as has been suggested for O_2 by Owen[34]. Large filled-aperture or interferometric telescopes intended for investigating extrasolar planets, especially those of terrestrial mass, might incorporate into their design the wavelength range and spectral resolution that has proved useful in the present work. In the radio search for extraterrestrial intelligence (SETI), optimum transmission through interstellar space and plausible planetary atmospheres lies in the 1–3 GHz range, not the MHz range used here.

The Galileo mission constitutes an apparently unique control experiment on the ability of fly-by spacecraft to detect life at various stages of evolutionary development on other worlds in the Solar System. Although a similar opportunity arises in the summer 1999 Earth fly-by of the ESA/NASA Cassini spacecraft on its way to Saturn, there are, because of funding constraints, no plans to observe the Earth with Cassini. Although a great deal more exploration remains to be done before such conclusions can be considered secure, our results are consistent with the hypothesis that widespread biological activity now exists, of all the worlds in this Solar System, only on Earth.

(**365**, 715-721; 1993)

Carl Sagan[*], **W. Reid Thompson**[*], **Robert Carlson**[†], **Donald Gurnett**[‡] **& Charles Hord**[§]

[*] Laboratory for Planetary Studies, Cornell University, Ithaca, New York 14853, USA

[†] Atmospheric and Cometary Sciences Section, Jet Propulsion Laboratory, Pasadena, California 91109, USA

[‡] Department of Physics and Astronomy, University of Iowa, Iowa City, Iowa 52242-1479, USA

[§] Laboratory for Atmospheric and Space Physics, University of Colorado, Boulder, Colorado 80309, USA

Received 17 February; accepted 14 September 1993.

带脉冲调制的无线电传输强烈暗示存在一个技术文明。大部分伽利略号发现的证据可能已经在大约 20 亿年前被一个类似的飞掠探测器发现了。相反,调制的窄带无线电传输在这个世纪之前未探测到。

对热力学极度不平衡且无法用非生物过程解释的分子的认证、无法用地球化学过程理解的广泛分布的色素以及调制的无线电辐射,这些都是在没有任何有关其化学过程的先验假设时在地球上存在生命的证据。200 g·cm^{-2} 的 O_2 是适合生物生存的最低标准。在探测器探索其他星球时否定的研究结果具有更广阔的应用空间而不仅仅是应用于"我们所知道的生命形式"。当然,这样的观察并没有排除公认的与地表环境弱耦合的生态系统(比如,地下 [32,33])。

在太阳系的各层次天体中液态水海洋的表面是独特的。可以相信,这是地球是太阳系中唯一一个表面有丰富生命的行星的原因。虽然还没有行星盘的分辨率,但相似的光谱学方法在检验其他行星是否具有原住生命时具有参考价值,就像曾经由欧文 [34] 对 O_2 的建议那样。为搜索系外行星特别是和地球质量类似的行星而使用连续孔径或者干涉望远镜时,应该将对于当前研究有用的波长范围和谱分辨率包含在内。在用无线电进行地外文明探索时,适宜在星际空间或者是可能存在的行星大气层传播的无线电的频率在 1~3 GHz 波段,而不是在这里所用的 MHz 波段。

伽利略探测器任务构成了一个独特的对于飞掠型探测器探测太阳系中其他世界各种不同的演化阶段的对照实验。虽然在 1999 年的夏季,由 ESA/NASA 从地球发射的卡西尼号土星探测器在它去往土星的旅程中有一个类似的机会。但因为预算的限制,没有为卡西尼号土星探测器安排观测地球的计划。尽管在保证结论准确无误之前还需进行更多的探索,但我们的结果仍然和假设一致,那就是在太阳系中大范围的生物活动只存在于地球上。

(刘霞 翻译;欧阳自远 审稿)

References:

1. Carlson, R. W. *et al. Space Sci. Rev.* **60**, 457-502 (1992).

2. Hord, C. W. *et al. Space Sci. Rev.* **60**, 503-530 (1992).

3. Belton, M. J. S. *et al. Space Sci. Rev.* **60**, 413-455 (1992).

4. Gurnett, D. A. *et al. Space Sci. Rev.* **60**, 341-355 (1992).

5. Lederberg, J. *Nature* **207**, 9-13 (1965).

6. Lovelock, J. *Nature* **207**, 568-570 (1965).

7. Lovelock, J. E. *Proc. R. Soc.* B**189**, 167-181 (1975).

8. Sagan, C. *Proc. R. Soc.* B**189**, 143-166 (1975).

9. Carlson, R. W., Arakelian, T. & Smythe, W. D. *Antarct. J. U.S.* (in the press).

10. Drossart, P. J. *et al. Planet. Space Sci.* (submitted).

11. Henderson, L. J. *The Fitness of the Environment: An Inquiry into the Biological Significance of the Properties of Matter* (Peter Smith, Gloucester, Mass., 1913).

12. von Zahn, U., Kumar, R. S., Neimann, H. & Prinn, R. in *Venus* Ch. 13 (eds Hunten, D. M., Colin, L., Donahue, T. M. & Moroz, V. I.) (Univ. of Arizona Press, Tucson, 1983).

13. Owen, T. in *Mars* Ch. 25 (eds Keiffer, H. H., Jakosky, B. M., Snyder, C. W. & Matthews, M. S.) (Univ. of Arizona Press, Tucson, 1992).

14. Walker, J. G. C. *Evolution of the Atmosphere* (Macmillan, New York, 1977).

15. Léger, A., Pirre, M. & Marceau, F. J. *Astr. Astrophys.* (in the press).

16. Lippincott, E. R., Eck, R. V., Dayhoff, M. O. & Sagan, C. *Astrophys. J.* **147**, 753-764 (1967).

17. Chameides, W. L. & Davis, D. D. *Chem. Engng. News* **60**, 38-52 (1992).

18. Hogan, K. B., Hoffman, J. S. & Thompson, A. M. *Nature* **354**, 181-182 (1991).

19. Lewis, J. S. & Prinn, R. G. *Planets and Their Atmospheres* (Academic, New York, 1984).

20. Bowker, D. E., Davis, R. E., Myrick, D. L., Stacy, K. & Jones, W. T. Ref. Publ. No. 1139 (NASA, Washington, 1985).

21. Kilston, S. D., Drummond, R. R. & Sagan, C. *Icarus* **5**, 79-98 (1966).

22. Sagan, C. & Wallace, D. *Icarus* **15**, 515-554 (1971).

23. Sagan, C. *et al.* in *Biology and the Exploration of Mars* Ch. 9 (eds Pittendrigh, C. S., Vishniac, W. & Pearman, J. P. T.) (Natl. Acad. Sci. Washington, 1966).

24. Shklovskii, I. S. & Sagan, C. *Intelligent Life in the Universe* (Holden Day, San Francisco, 1966).

25. Sullivan, W. T., Brown, S. & Wetherill, C. *Science* **199**, 377-388 (1978).

26. Chapman, S. *Proc. Phys. Soc.* **43**, 483-501 (1931).

27. Fainberg, J. & Stone, R. G. *Space Sci. Rev.* **16**, 145-188 (1974).

28. Gurnett, D. A. *J. Geophys. Res.* **79**, 4227-4238 (1974).

29. LaBeile, J., Trumann, R. A., Boehm, M. H. & Gewecke, K. *Radio Sci.* **24**, 725-737 (1989).

30. Keller, A. thesis, Univ. Iowa (1990).

31. Hines, C. O., Paghis, I., Hartz, T. R. & Fejer, J. A. *Physics of the Earth's Upper Atmosphere* (Prentice Hall, Englewood Cliffs, 1965).

32. Lederberg, J. & Sagan, C. *Proc. Natl. Acad. Sci. U.S.A.* **48**, 1473-1475 (1962).

33. Gold, T. *Proc. Natl. Acad. Sci. U.S.A.* **89**, 6045-6049 (1992).

34. Owen, T. in *Strategies for the* Search *for Life in the Universe* (ed. Papagiannis, M.) 177-185 (Reidel, Dordrecht, 1980).

Acknowledgements. We are grateful to W. O'Neil, F. Fanale, T. Johnson, C. Chapman, M. Belton and other Galileo colleagues, as well as W. Sullivan, for encouragement and support; and to J. Lederberg, A. Léger, J. Lovelock, A. McEwen, T. Owen and J. Tarter for comments. This research was supported by the Galileo Project Office, Jet Propulsion Laboratory, NASA and by a grant from NASA's Exobiology Program.

Deuterium Abundance and Background Radiation Temperature in High-redshift Primordial Clouds

A. Songaila *et al.*

Editor's Note

Atoms of deuterium (heavy hydrogen) contain one proton and one neutron. Deuterium is depleted in our Milky Way galaxy, as it has been burned in stars whose gas is then returned to the interstellar medium. Here Antoinette Songaila and colleagues measure the deuterium abundance in a cloud of gas at a redshift of about 3.3, meaning that it is being seen only about 2 billion years after the Big Bang. They found that deuterium in this cloud is much more abundant than it is today in the Milky Way. This is consistent with current models of element formation (nucleosynthesis) associated with the first few minutes after the Big Bang, and so acts to verify our understanding of that process.

Measurements of the deuterium abundance in the early Universe provide a sensitive test of the "standard" Big Bang cosmology. The probable detection of deuterium absorption by a gas cloud between us and a distant quasar suggests an abundance much greater than estimated from observations in the Milky Way, and consistent with the amount of presently observed luminous matter comprising all the baryons in the Universe. The same spectra imply a cosmic background temperature for the early Universe which is consistent with our expectations from standard Big Bang cosmology.

THE primary observational supports of Big Bang cosmology are the spectrum and isotropy of the cosmic microwave background radiation (CMBR)[1-4] and the measurement of light-element abundances predicted by standard Big Bang nucleosynthesis (SBBN)[5-10]. Given a present-day CMBR temperature, SBBN predicts the composition of material emerging from the first few minutes of the Big Bang as a function of the universal density of baryons. Measurements of the relative primordial abundances of H, D, ^{3}He, ^{4}He, and ^{7}Li test the consistency of SBBN—whether there is any value of the baryon density for which the theory predicts all of these abundances correctly. They also provide a measurement of the baryon density, $\Omega_b h^2$ (in units where the cosmic critical density $\Omega = 1$ and the Hubble constant $H_0 = 100h$ km s^{-1} Mpc^{-1}), which can be compared with other data—the density of stars, gas and mass in the Universe.

Of the light elements, only H and ^{4}He have been measured outside the Galaxy. ^{7}Li has been measured in Galactic metal-poor old halo stars, yielding what is presumed to be a

220

高红移原初气体云中的氘丰度
和背景辐射温度

编者按

氘原子(重氢)包含一个质子和一个中子。在我们的银河系中，氘几乎消耗殆尽，这是由于在恒星核反应中会发生氘燃烧，而其产生的气体则会回到星际介质当中去。这里安托瓦内特·桑盖拉和同事们在一个红移大约3.3左右，也就是大爆炸仅仅20亿年之后的气体云中测量了氘丰度。他们发现该气体云中的氘丰度远高于今天银河系的氘丰度。这个结果与当前描述大爆炸后几分钟之内元素形成(核合成)的模型十分吻合，因此这也进一步证实了我们对这一过程的理解。

对早期宇宙中氘丰度的测量可以很灵敏地检验"标准"大爆炸宇宙论。对遥远类星体和我们之间的一块气体云进行观测，我们探测到可能的氘吸收线，这表明气体云中的氘丰度远高于从银河系观测所得到的估计值，并且与目前观测得到的包含宇宙中所有重子发光物质的含量一致。相同的光谱还暗示早期宇宙的背景温度与标准大爆炸宇宙论的预言一致。

支持大爆炸宇宙论的主要观测证据包括宇宙微波背景辐射(CMBR)[1-4]的光谱和各向同性，以及标准大爆炸核合成模型(SBBN)[5-10]预言的轻元素丰度测量。给定一个现在的宇宙微波背景辐射温度，标准大爆炸核合成理论可以预言出大爆炸之初的几分钟内合成的物质组分随宇宙重子密度的变化关系。测量氢(H)、氘(D)、氦3(^{3}He)、氦4(^{4}He)以及锂7(^{7}Li)的相对原始丰度可以检验大爆炸核合成理论的自洽性——是否存在一个使得所有丰度预言都正确的重子密度。这些相对丰度还可以用于测量宇宙重子密度，$\Omega_b h^2$(以宇宙临界密度 $\Omega = 1$，哈勃常数 $H_0 = 100h$ km · s^{-1} · Mpc^{-1} 为单位)，这个量可以与其他观测数据相比较——比如，宇宙中恒星密度、气体密度和质量密度。

在所有的轻元素中，只有 H 和 ^{4}He 的丰度在银河系以外被测量过。人们在银河系内年老的贫金属晕族星中也对 ^{7}Li 的丰度进行了测量，这样测得的丰度据推测接近

nearly primordial abundance, although this interpretation has been questioned[10]. As yet, deuterium and ^{3}He have been observed only within the highly evolved and chemically active disk of our Galaxy. Moreover, because D is the most fragile of isotopes, it is easily destroyed in stars, where some of it is converted into ^{3}He and heavier elements, and only some of these daughter products are returned to the interstellar medium. The primordial abundances of these elements have been estimated by modelling the chemical evolution of the Galaxy, a complex process involving the cycling of interstellar gas through generations of stars[11]. Therefore, in practice the use of these elements in testing the SBBN model has been limited by the need to use evolutionary corrections to the observed abundances. Here we aim at a precise measurement of the ratio of the abundance of deuterium to hydrogen (D/H) in a chemically unevolved environment, bypassing the need for such corrections. The measurement is also confirmation of the predicted universality of the abundances over great distances—a new test of the cosmological principle. Among the light elements, only the ^{4}He/H ratio has previously been measured in more than one galaxy, and no ratios have been measured at cosmological distances.

This measurement also affects the SBBN estimate of Ω_b. The most precisely and reliably known primordial abundance is the helium mass fraction, estimated[12] from seven nearby metal-poor galaxies to be $Y_p = 0.228 \pm 0.005$. The baryon density implied by this estimate alone is rather low, $\Omega_b h^2 = 0.005$, barely more than the number of baryons we know to be present in luminous stars[13,14] or high-redshift quasar absorbers[15] ($\Omega_b \approx 2.9 \pm 0.6 \times 10^{-3}$). With such a low value for $\Omega_b h^2$ however, SBBN predicts a very high value for deuterium, $D/H \approx 2 \times 10^{-4}$, ~10 times the value of D/H observed in the interstellar medium. Chemical evolution models predict that with such a high D/H, the presolar nebula should be much more enriched in ^{3}He than allowed by observations of the solar wind[11]. The best chemical evolution model fits, and the constraint of the $(D+^3He)/H$ abundance from Solar System measurements, both suggest a primordial D/H in the range $1.9 \times 10^{-5} < D/H < 6.8 \times 10^{-5}$ (for example, ref. 11).

For this reason, the preferred canonical range of baryon density believed[9-11] to yield the best consistency with all the light elements is $\Omega_b h^2 \approx 0.010$–$0.015$. Although this formally predicts Y_p between 2σ and 3σ (where σ is standard deviation) above the observed value, systematic uncertainties in the Y_p determination may be large enough to overpower the formal errors[13]. If the true value of $\Omega_b h^2$ does lie in this range, most of the baryons are in some dark, as yet unobserved, form.

Truly primordial abundances can be measured by studying the absorption spectra of high-redshift quasars. Light from quasars encounters many foreground gas clouds at various redshifts, leading to absorption lines of a variety of species which provide a sensitive probe of cloud composition. This technique offers an opportunity to discover directly the value of D/H in a variety of very distant and fairly chemically unevolved environments, where much less time has elapsed since the Big Bang and where one can verify that little stellar enrichment has occurred, hence little deuterium destruction. Furthermore, some high-column-density quasar absorption clouds—certain of the so-called "Lyman limit

原始丰度，尽管有人对这一解释表示怀疑[10]。迄今，我们也只在高度演化且化学活跃的银盘内对氘和 ³He 进行过观测。不仅如此，由于氘是氢最不稳定的同位素，因此它在恒星中很容易被破坏，变成 ³He 或者更重的元素，然后这些产物中只有一部分回到星际介质中。这些元素的原始丰度可以利用星系化学演化模型来进行估算，但须考虑星际气体在各代恒星间复杂的循环过程[11]。因此，实际利用这些元素对标准大爆炸核合成模型进行的检验受限于必须考虑对观测丰度做演化修正。在这里我们的目标是在无化学演化环境中对氘氢丰度比(D/H)进行精确测量，无需这些修正。这个测量还可以用来确认丰度值在很大距离范围内的普适性——这是一种对宇宙学原理的新检验。在所有轻元素中，只有 ⁴He/H 比在不止一个星系中被测量过，但从未在宇宙学尺度上被测量过。

另外，这个测量的结果还会影响标准大爆炸核合成理论对 Ω_b 的估计。原始丰度的研究中最精确和可信的测量来自对邻近七个贫金属星系的氦丰度测量[12]，$Y_p = 0.228 \pm 0.005$。由此估算的重子密度很低，$\Omega_b h^2 = 0.005$，只比目前已知的在高光度恒星[13,14] 或者高红移类星体吸收天体[15] 中的重子含量($\Omega_b \approx (2.9 \pm 0.6) \times 10^{-3}$)稍高一点。然而在如此低的重子密度的情况下，标准大爆炸核合成理论却能预言出非常高的氘丰度，$D/H \approx 2 \times 10^{-4}$，大约是星际介质中 D/H 观测结果的 10 倍。化学演化模型预言，在如此高的 D/H 值下，太阳前星云中 ³He 的含量要比太阳风允许的观测结果高得多[11]。最好的化学演化模型的拟合结果以及太阳系内 (D+³He)/H 丰度测量结果的限制，都倾向于 D/H 值在 $1.9 \times 10^{-5} < D/H < 6.8 \times 10^{-5}$ 的范围内(例如，参考文献 11)。

由于这个原因，人们普遍相信重子密度的范围[9-11] 是 $\Omega_b h^2 \approx 0.010 \sim 0.015$，这样与所有轻元素丰度观测一致性最好。尽管这样预言的 Y_p 会在 2σ 到 3σ 水平上高于观测值(σ 是标准差)，但是 Y_p 的系统误差可能大到超过前者的误差[13]。如果重子密度的真实值确实处在这个范围之内，那么大部分重子必然是以某种暗的、尚未被观测到的形式存在。

事实上，真实的原始丰度可以通过观测高红移类星体的吸收光谱来进行测量。高红移类星体的光线在传播过程中会穿过位于不同红移处的前景气体云，在光谱的不同位置产生对应不同元素的大量吸收线，因此这些吸收线可以用来精确探测气体云的具体元素组成。这种技术给了我们一个直接观测那些遥远的、在基本未发生化学演化环境中的 D/H 值的机会，这些环境处于大爆炸之后很短的时间内，且可以证实其所对应的恒星元素增丰效应还很不显著，因此氘也几乎未被破坏。不仅如此，有些高柱密度的类星体吸收云，即所谓的莱曼系限系统(LLS)，它们对于氢原子的

systems" (LLS), which are optically thick to photoelectric absorption by hydrogen—are nearly ideal sites for measuring D/H accurately[16]. The H column density of a cloud can be measured quite precisely from the optical depth at the Lyman continuum limit or from line profile fitting of multiple lines of the series. A simple model of a multicomponent gas cloud, consisting of a series of velocity components, each one characterized by a velocity, a column density and spectral "b-parameter" ($b \equiv (2kT/m)^{1/2}$ where m is the atomic mass and T the temperature) can thereby be overconstrained by simultaneously fitting many lines of the Lyman series. Deuterium is measured by its isotopically shifted absorption lines, displaced towards the blue from the hydrogen lines by an apparent velocity $v = cm_e/2m_p = 82$ km s^{-1} (where m_e and m_p are the masses of the electron and proton, respectively). The much smaller D column density is nevertheless detectable and accurately measured in the lowest-order lines.

This important programme has been attempted for years by us and others on 4-m class telescopes without notable success, because the faintness of the high-redshift quasars (that is, high enough redshift to see the Lyman series from the ground) makes high signal-to-noise spectroscopy too difficult at the required resolution ~ 10 km s^{-1}—to constrain the cloud model reliably. In addition, a certain amount of good fortune is required to locate suitably clean absorbers where the deuterium line is not contaminated by other intervening hydrogen lines.

We have now finally obtained data of the required quality, using the recently introduced high spectral resolution spectrograph on the Keck 10-m telescope (Mauna Kea, Hawaii), and report here the result of the first measurements, based on one night of data. We find absorption consistent with a detection D/H $\approx (1.9\text{--}2.5) \times 10^{-4}$ in a chemically unevolved cloud in the line of sight to the quasar Q0014+813; because in any single instance we cannot rule out the possibility of a chance H contamination at exactly the D velocity offset, this result should be considered as an upper limit until further observations of other systems are made. As more measurements are made, we would expect to see a consistent minimum value of D/H emerging as the true, primordial value.

The same spectra also allow a test of another fundamental prediction of relativistic cosmology—namely, that the temperature of the cosmic microwave background radiation T_{CMBR} should evolve with redshift, z, as $T_{CMBR} \propto (1+z)$—which is a more direct constraint than other arguments[17] against alternative models for the origin of the background radiation. Recent experiments have provided extraordinarily precise confirmation of the isotropy of the cosmic microwave background radiation and of its black-body spectrum; the current value of T_{CMBR} is now extremely accurately measured, at 2.73 K, both directly[1,2] and from cyanogen excitation in the interstellar medium[18]. The value of T_{CMBR} at high z can similarly be measured through fine-structure splitting of atomic carbon lines in high-z quasar absorption line systems[19,20]. Although attempts have been made to use this technique on C II and C I lines[20,21], the results have again been limited by the difficulty of obtaining very high signal-to-noise spectra on these very faint ($V = 16\text{--}17$, where V is visual magnitude) quasars. We present below such a measurement of C II 1,334 Å in the Lyman limit absorber in quasar

光电吸收是光学厚的，是精确测量 D/H 值近乎完美的理想场所[16]。从莱曼连续谱系限的光深，或者通过莱曼系多条吸收线的轮廓拟合，我们可以精确测量气体云中氢元素的柱密度。对于一个包含了多种速度组成、简单的多成分气体云模型，我们可以用速度、柱密度和谱参数 $b(b \equiv (2kT/m)^{1/2})$ 来描述其中的各个成分，其中 m 是原子质量，T 是温度，这些量均可以通过同时拟合多条吸收线来进行很好的限制。对氘的测量是通过其由于同位素效应而移动的谱线进行的，该谱线比氢线稍朝蓝端移动，相应的视速度为 $v = cm_e/2m_p = 82 \text{ km} \cdot \text{s}^{-1}$（其中 m_e 和 m_p 分别是电子和质子质量）。虽然氘的柱密度很小，但还是可观测的，人们已经通过最低阶的谱线对其进行了精确的测量。

多年来我们和其他人都曾经用 4 米级的望远镜尝试进行这个重要的项目，但是并没有取得明显的成功。这是因为暗淡的高红移类星体（其红移需高到足以利用地面望远镜接收莱曼系光子），在所需的分辨率下（约为 $10 \text{ km} \cdot \text{s}^{-1}$）很难得到高信噪比的光谱，即很难可靠地限制气体云模型。另外当然还需要一点好运气，这样才能找到不受其他氢吸收线污染的干净的氘吸收线。

使用 10 米凯克望远镜（夏威夷冒纳凯阿火山）上最新引进的高分辨率光谱仪，我们最终获得了所需分辨率的数据。在这里我们报告基于一个晚上观测数据所得到的首次测量结果。我们在类星体 Q0014+813 的视线方向上一个无化学演化的气体云中发现了对应 $D/H \approx (1.9 \sim 2.5) \times 10^{-4}$ 的吸收线。因为对任何一个单独的事例我们都不能排除正好在对应于氘谱线速度的地方存在氢污染的可能性，我们观测到其他系统之前，这个值应该被理解为一个上限。随着更多观测的进行，我们期待可以得到自洽的 D/H 的最小值，这将给出真正的原始丰度。

这些光谱还可以用来检验另一个相对论宇宙学的基本预言——宇宙微波背景辐射的温度 T_{CMBR} 随红移 z 的演化规律，即满足 $T_{CMBR} \propto (1+z)$——相对于其他论证方法[17] 来说，该检验可以对不同微波背景辐射起源理论给出更直接的限制。最近的实验以极高的精度确认了宇宙微波背景辐射的各向同性和黑体谱特征，且通过直接观测[1,2] 或者从星际介质中的氰激发[18] 得到了迄今为止对微波背景辐射温度最准确的测量结果，T_{CMBR} 约为 2.73 K。而高红移处的 T_{CMBR} 也可以类似地通过高红移类星体吸收线系统中因碳原子精细结构而导致的谱线分裂来进行测量[19,20]。尽管此前已有人利用 C II 和 C I 线进行相关尝试[20,21]，但是结果依旧受限于难以获得这些暗淡类星体高信噪比的光谱（$V = 16 \sim 17$，V 为目视星等）。下面，我们给出对类星体 Q0636+68 莱曼系限吸收体中的 C II 1,334 Å 谱线的测量结果。这个结果对 $z = 2.9092$

Q0636+68, which places a new tighter limit of 13.5 K on T_{CMBR} at $z = 2.9092$, corresponding to a redshifted value $T_{CMBR}/(1+z) < 3.5$ K measured at wavelength $611/(1+z)$ μm.

Abundance Measurements

Observations of the two quasars, Q0636+68 ($V \approx 16.5$) and Q0014+813 ($V = 16.9$), were made on 11 November 1993 with the Keck 10-m telescope, as four 40-min exposures and one 50-min exposure for Q0636+68 (3.5 h total), and six 40-min exposures for Q0014+813 (4.0 h total). A $1.2'' \times 3.8''$ slit was used. Each observation covered the range 3,500–5,900 Å at a resolution $R = 36,000$ and, at the longer wavelengths of interest here, sky- and read-noise and detector dark current were negligible, so that the signal-to-noise ratio was determined only by source counts.

The two quasars have strong LLS, at $z = 2.909$ in Q0636, and $z = 2.813$ in Q0014 (ref. 22). Both are metal-line systems and have very complex structure, stretching over several hundred kilometres per second in velocity. In particular, there is contamination by high-negative-velocity hydrogen at the expected position of deuterium, making measurement of D/H essentially impossible. In this they resemble certain interstellar lines of sight[23] which are too complicated to be useful for D/H measurement.

Q0014 also has a much weaker LLS at $z = 3.3201$ corresponding to the modestly high hydrogen atom column density ($N_H \approx 10^{17}$ atoms cm^{-2}) metal-poor cloud extensively studied by Chaffee and collaborators[24,25]. This cloud may contain some elements heavier than helium, as indicated by the possible presence of Si III 1,206 Å and C III 977 Å absorption lines. Although the lines occur at the proper wavelengths, they lie in a region with many Lyman-α absorption lines, and Chaffee et al.[25] argue that these cannot be attributed to Si and C because the line strengths are not consistent with the expected intensities. Irrespective of this point, the metallicity in the Chaffee cloud is very low. Even in the unlikely case that the cloud were predominantly neutral, the C II and C I line limits ($< 10^{12}$ cm^{-2} from our data) would place the metallicity at less than one-fortieth of the solar value, whereas if C III or C IV were the dominant ionization state, the hydrogen would be highly ionized, whether the cloud were photoionized or collisionally ionized[25,26] and in this case both Chaffee et al.[25] and Donahue and Shull[27] concluded that the metallicity was less than, or about, $10^{-3.5}$ of solar.

Chaffee et al. found that the neutral hydrogen in this cloud was well fitted by two components, each with $N_H \approx 5 \times 10^{16}$ cm^{-2} and $b \sim 28$ km s^{-1}, and with a separation of N_H 110 km s^{-1}. With our higher signal-to-noise, higher resolution data, we have reanalysed the neutral hydrogen model, using the Lyman break at a wavelength of 3,939 Å (Fig. 1) to obtain the combined neutral hydrogen column density. We compared the flux in the 40 Å below the break to that between 3,960 Å and 4,000 Å, and evaluated the ratio of the maximum peaks (0.58), the fifth maximum peaks (0.55), the medians (0.58) and the means (0.60). The maximum peak

的宇宙微波背景辐射温度给出了更好的限制，其温度约为 13.5 K，对应在波长为 611/(1+z) μm 处测得 $T_{CMBR}/(1+z) < 3.5$ K。

丰 度 测 量

我们在 1993 年 11 月 11 日使用凯克 10 米望远镜对两个类星体 Q0636+68 ($V = 16.5$) 和 Q0014+813 ($V = 16.9$) 进行了观测，其中包括对 Q0636+68 的 4 段 40 分钟和 1 段 50 分钟的曝光（总共 3.5 小时），对 Q0014+813 的 6 段 40 分钟的曝光（总共 4 小时）。在整个观测过程中，我们使用了 $1.2'' \times 3.8''$ 的光缝。每一次观测覆盖的波长为 3,500~5,900 Å，分辨率为 $R = 36,000$。同时，在感兴趣的较长波段区间，由于天空噪声、读噪声和探测器暗流噪声均可以忽略，因此信噪比仅取决于源计数。

这两个类星体都有很强的莱曼系限系统，其中 Q0636 的莱曼系限系统出现在红移为 $z = 2.909$ 处，而 Q0014 的莱曼系限系统则出现在红移为 $z = 2.813$ 处（参考文献 22）。这两者都是金属谱线系统，并具有很复杂的结构，其速度展宽可以超过每秒几百千米。特别的是，在预期氘谱线的位置上存在高负速度氢所带来的污染，这使得对 D/H 的精确测量几乎变得不可能。这与某些视线方向的星际介质[23] 相似，太复杂以至于无法用于对 D/H 的测量。

Q0014 在红移 $z = 3.3201$ 处还有一个弱得多的莱曼系限系统，对应一个具有中等高氢原子柱密度（每平方厘米约 10^{17} 个原子）的贫金属气体云。该气体云已被查菲及其合作者们广泛地研究过[24,25]。从可能存在的 Si III 1,206 Å 和 C III 977 Å 吸收线的迹象判断，这块气体云可能包含比氢更重的元素。尽管这些吸收线出现在相应的位置，但是它们所在的波长范围内还存在很多莱曼 α 吸收线。并且由于这些吸收线的强度与预期的强度不符，查菲等人[25] 认为这些并不是碳和硅的吸收线。不考虑这一点，这块查菲气体云的金属丰度很低。即便这块气体云处于不太可能的中性状态，C II 和 C I 线给出的限制（我们的数据显示 $< 10^{12}$ cm^{-2}）也表明该气体云的金属丰度小于太阳的四十分之一。然而如果电离的 C III 和 C IV 是主要的状态，那么不管气体云是光致电离还是碰撞电离[25,26]，氢都将是高度电离的。在这种情况下查菲等人[25]、多纳休和沙尔[27] 都推测该气体云的金属丰度大约仅为太阳的 $10^{-3.5}$ 或更低。

查菲等人发现这块气体云中的中性氢可以用两部分来进行很好的拟合，每一部分对应：$N_H \approx 5 \times 10^{16}$ cm^{-2}，$b \sim 28$ km·s^{-1}，两部分 N_H 之间相差 110 km·s^{-1}。利用所得的高信噪比、高分辨率的数据我们重新分析这个中性氢模型，用波长为 3,939 Å 处的莱曼断裂（图 1）去计算整体的中性氢柱密度。我们比较了莱曼断裂波长以下 40 Å 处以及 3,960 Å 到 4,000 Å 处的流量，并计算了最高峰（0.58）、第五最高峰（0.55）、中位数（0.58）和平均值（0.60）之间的比值。这里最高峰的比值给出了氢柱密度的最

ratios constitute the best estimate, with the mean serving as an upper bound as it is raised by the presence of the Lyman series in the upper wavelength range. Together they give an optical depth $\tau = 0.55 \pm 0.05$, corresponding to $N_H = 9.5 \times 10^{16} \pm 0.8 \times 10^{16}\,\mathrm{cm}^{-2}$.

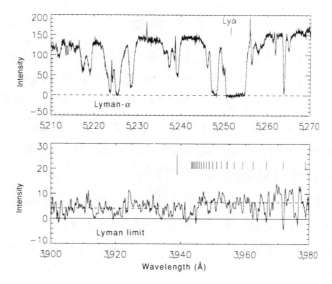

Fig. 1. The region of Lyman-α (5,250 Å; top) and the Lyman break (3,939 Å; bottom) in the $z = 3.32$ Lyman limit system (LLS) in the quasar Q0014+813. Six CCD (charge-coupled device) frames were flat-fielded with a quartz lamp exposure, spatially registered and added with a filter to remove cosmic rays; one-dimensional spectra were formed by subtracting a heavily smoothed sky and then summing the spatial points along the slit corresponding to a particular spectral pixel. The spatial summation included all points where the flux exceeded 10% of the maximum on the source. The spectra were precisely wavelength-calibrated using a Th–Ar lamp and approximately flux-calibrated (F_v) using observations of the white dwarf star Feige 110. The resulting signal-to-noise ratio was ~87 per resolution element at Lyman-α and much less (~15 per resolution element) at the Lyman break, because of the rapid drop-off in both detector efficiency and the cross-disperser blaze function at and below 4,000 Å. The positions of Ly-α(top) and Lyman 9 to Lyman 30 and the Lyman break (bottom) are shown as vertical lines. The dashed lines indicate the median values below and above the Lyman break.

To determine the b-values and relative strengths of the strongest components, which are the key to measuring D/H, it is desirable to work with the unsaturated very-high-order Lyman series lines. The structure of this region is, however, complex, being deep in the Lyman forest, and fitting of individual lines is somewhat subjective (compare Fig. 1). To alleviate this problem we formed the medians in velocity space of the lines Lyman 7 to Lyman 13, and Lyman 14 to Lyman 20 (Fig. 2). This eliminated contaminating lines, leaving only the relatively invariant Lyman lines themselves. In the Lyman 14–20 composite, only the strong component at $-17.5\,\mathrm{km\,s}^{-1}$ is clearly seen. The uncertainty in the central velocity is ~$5\,\mathrm{km\,s}^{-1}$. Fitting this with a single cloud, we find a b-value of $21\,\mathrm{km\,s}^{-1}$ with an acceptable range of 18–$28\,\mathrm{km\,s}^{-1}$. If we assume that the line approximates Lyman 17, because the oscillator strengths vary only slowly in the high Lyman series, we find $N_H = 5.5 \times 10^{16}\,\mathrm{cm}^{-2}$. The primary uncertainty in N_H is the placement of the continuum, and the value could range from $8.7 \times 10^{16}\,\mathrm{cm}^{-2}$ to $4.0 \times 10^{16}\,\mathrm{cm}^{-2}$. The second component at $80\,\mathrm{km\,s}^{-1}$ shows up clearly in the Lyman 7–13 composite. This component is best fitted with two clouds, each

佳估计，而考虑到莱曼系出现在波长范围的长波段，平均值的比值可作为上限。最终结果表明，光深 $\tau = 0.55 \pm 0.05$，对应 $N_H = 9.5 \times 10^{16} \pm 0.8 \times 10^{16}$ cm^{-2}。

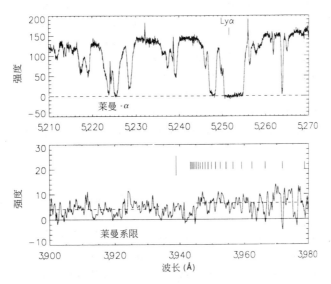

图1. 类星体 Q0014+813 在 $z = 3.32$ 处莱曼系限系统中对应的莱曼 α（5,250 Å，上图）和莱曼断裂（3,939 Å，下图）的光谱。六幅 CCD（电荷耦合器件）观测图像，通过石英灯进行平场校正、并进行空间校正后，使用一定的滤波函数滤去宇宙线干扰。减去高度平滑的天空背景后，沿着光缝将对应于某一光谱像素的空间点相叠加后得出一维光谱。空间叠加包含了所有超过源最大值10%的点。利用钍-氩灯，我们对光谱的波长进行了精确定标，并用白矮星 Feige 110 的观测进行近似流量定标（F_ν）。最终得到莱曼 α 处单位可分辨单元的信噪比约为87，而莱曼断裂处的信噪比要低很多（对应单位可分辨单元的信噪比约为15），这是因为在 4,000 Å 及以下，传感器效率和横向色散器闪耀函数都将迅速降低。莱曼 α（上图），莱曼9到莱曼30和莱曼断裂（下图）的位置均已用竖线标出。而虚线则标出高于或低于莱曼断裂的中值。

为了确定 b 值以及最强成分的相对强度这些测量 D/H 的关键量，我们必须处理未饱和的很高阶的莱曼系谱线。这些谱线所对应的区域处在莱曼森林中，结构非常复杂，因此要拟合单一谱线就多少受主观因素影响（比较图1）。为了解决这个问题，我们在速度空间上画出莱曼7到莱曼13，和莱曼14到莱曼20的中值（图2）。这样做可以排除掉干扰的谱线，只留下相对不变的莱曼线系。对于莱曼14~20的合成谱，只有在 -17.5 km·s^{-1} 的强成分可以清晰地被辨认。中心速度的误差为 5 km·s^{-1}。若用单一气体云模型来拟合，我们发现 b 值为 21 km·s^{-1}，可接受的范围是 18~28 km·s^{-1}。如果我们假设这条谱线近似是莱曼17，考虑到振子强度在高阶莱曼系的变化非常缓慢，我们可以得到 $N_H = 5.5 \times 10^{16}$ cm^{-2}。这里 N_H 的主要误差来源是连续谱的精确定位，因此 N_H 的可取值范围为 8.7×10^{16} cm^{-2} 到 4.0×10^{16} cm^{-2}。在莱曼7~13的成分中，我们还清晰地发现了 80 km·s^{-1} 处的第二个成分。这个成

with $b = 16\,\mathrm{km\,s^{-1}}$ and velocities of $80\,\mathrm{km\,s^{-1}}$ and $113\,\mathrm{km\,s^{-1}}$. The cloud column densities are $N_H = 1.5 \times 10^{16}\,\mathrm{cm^{-2}}$ and $5.5 \times 10^{15}\,\mathrm{cm^{-2}}$ if we adopt a fit with the oscillator strength of Lyman 10. As we move up the Lyman series, weaker components with $N_H \sim 10^{15}\mathrm{cm^{-2}}$ are seen, also with b-values around $21\,\mathrm{km\,s^{-1}}$. The final cloud model is summarized in Table 1 and the fit to the Lyman series is shown in Fig. 2. The total column density of the model fit is $7.9 \times 10^{16}\,\mathrm{cm^{-2}}$, which is slightly lower than the value we have estimated from the Lyman break. This could reflect the uncertainty in the strongest component caused by the continuum fit, or possibly some additional material not well described by static thermal slabs; either way, the near agreement argues against significant systematic errors in estimating N_H. For the purposes of estimating D/H, we shall assume that cloud 1 has N_H in the range $(5.5\text{–}7.3) \times 10^{16}\,\mathrm{cm^{-2}}$. This is consistent with the range of $N_H = (2.3\text{–}8.7) \times 10^{16}\,\mathrm{cm^{-2}}$ found by Chaffee et al.[24].

The Chaffee cloud is very well suited for a determination of the D/H abundance ratio because the high-column-density component lies at the blue end of the complex. In Fig. 3 we show the predictions of the cloud model of Table 1 for Dα as a function of the value of D/H assuming that $b_{\mathrm{deuterium}} = b_{\mathrm{hydrogen}} / \sqrt{2}$ —that is, thermal broadening. The feature seen in the blue wing of Lyman-α is precisely at the expected position of deuterium and, if interpreted as deuterium, gives a best fit of D/H $= 2.5 \times 10^{-4}$ for $N_H = 5.5 \times 10^{16}\,\mathrm{cm^{-2}}$. If the larger value of N_H obtained from fitting the break were used, this becomes 1.9×10^{-4}.

Table 1. Best cloud model ($z = 3.32015$)

$v\,(\mathrm{km\,s^{-1}})$	$b\,(\mathrm{km\,s^{-1}})$	$N\,(\mathrm{cm^{-2}})$
−17.5	21.0	5.5×10^{16}
13.5	24.0	1.8×10^{15}
80.0	16.0	1.5×10^{16}
113.0	16.0	5.5×10^{15}
155.0	21.5	6.5×10^{14}

v, Central velocity; b, Doppler parameter; N, column density.

分用两块气体云模型拟合得最好，每块气体云都有 $b = 16 \ \mathrm{km \cdot s^{-1}}$，且速度分别为 $80 \ \mathrm{km \cdot s^{-1}}$ 和 $113 \ \mathrm{km \cdot s^{-1}}$。如果用莱曼 10 的振子强度来进行拟合，那么气体云的柱密度分别为 $1.5 \times 10^{16} \ \mathrm{cm^{-2}}$ 和 $5.5 \times 10^{15} \ \mathrm{cm^{-2}}$。如果我们转向更高阶的莱曼线，将可以看到柱密度 N_H 约为 $10^{15} \ \mathrm{cm^{-2}}$ 的相对更弱的成分，同样对应 b 值约为 $21 \ \mathrm{km \cdot s^{-1}}$。在表 1 中我们总结了最终的气体云模型，并把对莱曼线系的拟合结果显示在图 2 中。最终模型拟合得到的总柱密度为 $7.9 \times 10^{16} \ \mathrm{cm^{-2}}$，这比直接用莱曼断裂所估计的值稍小。这可能是因为连续谱拟合带来最强成分的不确定性，也可能还存在一些不能用静态热板块模型进行很好描述的因素。但无论如何，这个接近吻合的数值表明在估计 N_H 时并没有明显的系统误差。为了估计 D/H，我们将假设第一块气体云的柱密度为 $(5.5 \sim 7.3) \times 10^{16} \ \mathrm{cm^{-2}}$，这和查菲等人的结果 $(2.3 \sim 8.7) \times 10^{16} \ \mathrm{cm^{-2}}$ 相吻合 [24]。

查菲气体云非常适合用来计算 D/H 丰度比，因为其中的高柱密度成分处于复合体的蓝端。在图 3 中，假设热致宽 $b_D = b_H / \sqrt{2}$，我们基于表 1 中给出的气体云模型预言给出了 Dα 随 D/H 变化的函数关系。可以看到，莱曼 α 的蓝端刚好位于氘的位置，如果解释为氘，那么在 $N_H = 5.5 \times 10^{16} \ \mathrm{cm^{-2}}$ 的条件下，有最佳拟合值 D/H $= 2.5 \times 10^{-4}$。如果换成使用拟合莱曼断裂所得到的更大的 N_H 时，D/H 的最佳拟合值变为 1.9×10^{-4}。

表 1. 最佳气体云模型 $(z = 3.32015)$

$v(\mathrm{km \cdot s^{-1}})$	$b(\mathrm{km \cdot s^{-1}})$	$N(\mathrm{cm^{-2}})$
−17.5	21.0	5.5×10^{16}
13.5	24.0	1.8×10^{15}
80.0	16.0	1.5×10^{16}
113.0	16.0	5.5×10^{15}
155.0	21.5	6.5×10^{14}

v，中心速度；b，多普勒参数；N，柱密度。

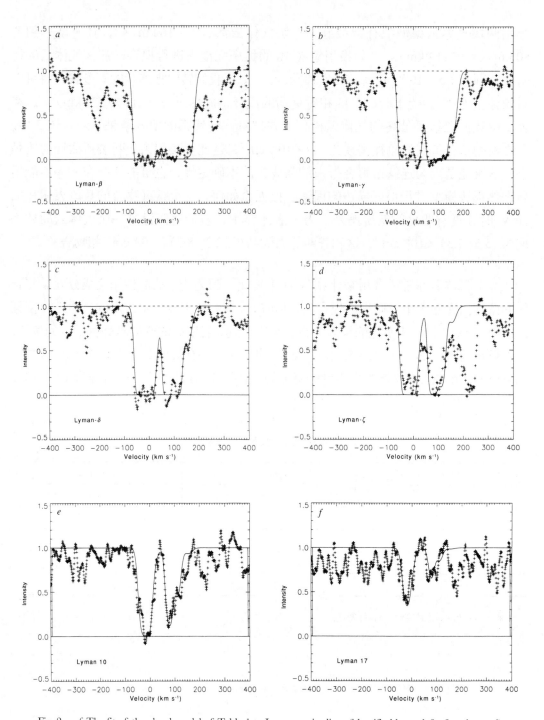

Fig. 2. *a–f*, The fit of the cloud model of Table 1 to Lyman series lines (identified lower left of each panel) of the *z* = 3.32015 LLS in Q0014+813. The profile labelled "Lyman 10" is a median in velocity space of the Lyman 7 to Lyman 13 profiles, and "Lyman 17" is a composite of Lyman 14 to Lyman 20. Crosses mark the data values and the solid line the model fit.

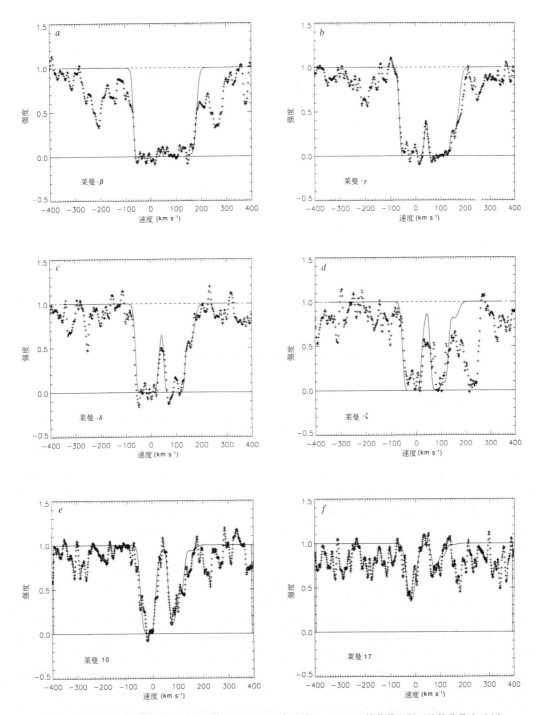

图 2. a~f, 采用表 1 气体云模型对类星体 Q0014+813 中红移 $z = 3.32015$ 的莱曼系限系统的莱曼线系(在每张小图左下注明)所得到的拟合结果。标注"莱曼 10"的轮廓线是莱曼 7 到莱曼 13 轮廓线速度空间的中值,标注"莱曼 17"的轮廓线是莱曼 14 到莱曼 20 的合成谱。十字是数据点,实线是模型拟合。

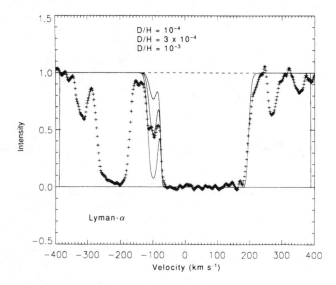

Fig. 3. The prediction of the cloud model of Table 1 for deuterium-α absorption in Q0014+813. Crosses mark data values, and the three solid lines show model fits for D/H = 10^{-4} (top), 3×10^{-4} (middle) and 10^{-3} (bottom).

Implications for the Big Bang

In the case of a single absorber, we cannot rule out the possibility that an errant hydrogen cloud, with a column density 2.5×10^{-4} that of the main cloud, happens to be floating at the velocity shift where we expect to find deuterium; note in particular that the signal to noise is insufficient to distinguish the narrower thermal broadening expected for D ($b_{\text{deuterium}} = b_{\text{hydrogen}} / \sqrt{2}$). But we can place limits on the likelihood of such a chance interpolation. The redshift of the Chaffee cloud lies close to the redshift of the quasar (3.41) and the number density of Lyman-α forest lines increases steeply with decreasing wavelength in this region. The Ly-α line lies at the red end of echelle order 68, where there are 12 Ly-α forest lines in the observed 75 Å which would be strong enough to contaminate the deuterium region, whereas in order 67, which is 80 Å to the red, the number has dropped to 4 in 75 Å. A rate of 12 lines per 75 Å would amount to a probability of 3% that a Ly-α line might be observed at random within ± 5 km s^{-1} of the deuterium position. Although this is quite small, until similar abundances are regularly found in other absorbers, it is necessary to regard our measurement as an upper limit on the deuterium abundance. Viewed in this light, the limit is encouragingly close to the values of D/H predicted from Galactic chemical evolution models, and suggests that, as further clouds are studied, we will be able to refine the measurement of D/H to significantly constrain SBBN and chemical evolution models. The quality of the data is sufficient to detect even much lower abundances, so that a small statistical survey of similar absorbers should remove the current ambiguity due to possible contamination.

It is also possible that this might prove to be the best measurement to date of the primordial D/H, and as such it is interesting to examine its potential consequences for cosmological theory. A value of D/H as high as 2.5×10^{-4} is acceptable within SBBN, is just consistent

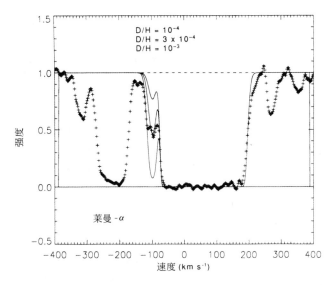

图 3. 表 1 气体云模型对类星体 Q0014+813 中的氘-α 吸收所做的预言。十字是数据点，三条实线是对 D/H 比分别为 10^{-4}（上）、3×10^{-4}（中）和 10^{-3}（下）的模型拟合。

大爆炸的含义

在只观测到一个吸收体的情况下，我们不能排除这是一块柱密度为主气体云的 2.5×10^{-4} 倍漂浮的氢气体云且正好以产生我们期望观测到氘线的速度在漂移的可能性。特别值得注意的是，（这一数据的）信噪比还不足以有效地区分具有较窄热致宽的氘（$b_D = b_H / \sqrt{2}$）。但是我们可以对产生这种现象的概率加以限制。查菲气体云红移和类星体红移（3.41）很接近，而莱曼 α 森林谱线的数密度在这个区域随着波长的减小而迅速增长。莱曼 α 谱线位于中阶梯光栅第 68 级的红端，在观测的 75 Å 波长内有 12 条莱曼 α 森林谱线，可以强到足以污染氘的区域。而在第 67 级，也就是距红端大约 80 Å 处，莱曼 α 谱线的数目在 75 Å 内下降到 4 条。每 75 Å 内的 12 条谱线造成莱曼 α 谱线随机出现在氘线附近 5 km·s^{-1} 内的可能性是 3%。尽管很小，但在更多的吸收体中发现相似的结果前，我们还是应该把我们测得的氘丰度仅仅视为是一个上限。这样看来，这个极限很接近星系化学演化模型预言的 D/H 值。这表明，随着更多的气体云研究，我们将可以不断改进 D/H 测量的精度，并用来更好地限制标准大爆炸核合成和化学演化模型。数据的质量将足以用来探测更低的丰度，因此一个小的寻找类似吸收体的统计巡天观测应该可以很好地排除目前可能的污染所造成的不确定性。

与此同时，这可能也是迄今为止对原初 D/H 值的最佳测量，因此很有必要进一步讨论该结果对宇宙学理论的潜在影响。首先，高达 2.5×10^{-4} 的 D/H 值在标准大爆炸核合成理论中是可接受的，且这个值与 ^{7}Li 的结果相吻合[10]，甚至与 ^{4}He 符合

with ^{7}Li measurements[10], and is even favoured by ^{4}He (ref. 28). It has the consequence of reducing Ω_b a factor of three below its canonical value, to $0.005h^{-2}$ (refs 9, 10), which roughly agrees[13-15] with the inventory of luminous and gaseous matter in the Universe. Because evolution after the Big Bang is a net destroyer of deuterium, it is not surprising to find relatively high D/H in a relatively unevolved primordial cloud.

There are however difficulties reconciling such a high value of D/H with the much lower values found in the interstellar medium[29] (of the order of 10^{-5}), and especially with the Solar System abundance of ^{3}He. Although ^{3}He can be both created and destroyed in different types of stars, models including a normal mix of stars predict[11] that Galactic chemical evolution leads to at best only a slight decrease with time in the sum $(D+^3He)/H$, so that the $^3He/^4He$ ratio measured locally in the solar wind seems to imply a low presolar value of $(D+^3He)/H \approx 4 \times 10^{-5}$, contradicting our D/H measurement. It is of course possible that the chemical evolution models are inadequate; for example, it could be that the presolar nebula at the time of its formation contained abundance atypical of the Galaxy. (The possibility of such spatial fluctuations in $^3He/H$ is suggested by interstellar measurements of $^3He/H$ using hyperfine transitions[30], which show a range of values for $^3He/H$ from $\lesssim 1 \times 10^{-5}$ to 15×10^{-5}.) These uncertainties make the use of local ^{3}He observations ambiguous for cosmological arguments, and motivate further measurements of D/H at high redshift, which should give a much cleaner test of SBBN.

Demonstration of a low value of Ω_b would have a wide impact on other aspects of cosmological theory. For example, many models of galaxy formation predict that the baryon mass fraction in galaxy clusters ought to be Ω_b, and a low value of Ω_b exacerbates their conflict with the observed high baryon fraction ($> 0.05h^{-3/2}$) of the Coma cluster[31]. A low baryon density would also conflict with massive spiral galaxy halos being made primarily of baryonic objects, a prediction which can be tested by gravitational microlensing experiments[32,33]. If we are indeed seeing almost all of the baryons already, dynamical evidence points even more strongly to halos made of non-baryonic dark matter.

Cosmic Background Temperature

The evolution of the microwave background temperature with redshift is a little-doubted prediction of relativistic cosmology which is nevertheless worth testing precisely. For example, this test complements other arguments[17,34] constraining alternatives to the Big Bang origin of the background radiation. At the redshifts of our absorbers, we can observe transitions from the ground state of singly ionized atomic carbon C II, as well as from the fine structure split state C II* which lies 64 cm^{-1} above the ground state. At these redshifts, the frequency difference is only about three times the peak frequency of the CMBR, so C II* is predicted to have a non-negligible population just due to the presence of the microwave background. We measure the relative populations of C II and C II* from the strengths of ultra-violet absorption lines from the two states.

得更好(参考文献 28)。但是该结果却可以使 Ω_b 减小到人们公认值的三分之一,约 0.005h^{-2}(参考文献 9,10),这与宇宙中发光物质和气体物质的总量大致相当[13-15]。由于大爆炸结束后氘就只消耗而不产生,因此在一个无演化的原初气体云中发现较高的 D/H 值并不奇怪。

然而,要调和这一高 D/H 值与在星际介质中测得的低得多的值[29](约 10^{-5})之间的矛盾,尤其是与太阳系中 ^{3}He 丰度间的矛盾却非常困难。尽管 ^{3}He 在不同的恒星中既可以产生也可以减少,但是具有正常的恒星组合的模型预言[11],星系的化学演化最多只能造成(D+^{3}He)/H 随时间缓慢减少,因此在太阳风中测得的 ^{3}He/^{4}He 暗示原始太阳可能具有较小的(D+^{3}He)/H 值,约 4×10^{-5},这与我们测得的 D/H 值矛盾。当然也有可能是星系化学演化模型不完善导致的,例如,太阳前星云在它诞生之时包含的丰度不是银河系典型的星系丰度。(用超精细跃迁[30]测得的星际 ^{3}He/H 值揭示了 ^{3}He/H 值的这种空间涨落的可能性,对应值的范围是 $\leqslant 1 \times 10^{-5}$ 到 15×10^{-5}。) 这使得我们利用本地测得的 ^{3}He 值讨论宇宙学问题时存在不确定性,因此我们需要观测得到高红移的 D/H 值,这样可以给出对标准大爆炸核合成更准确的检验。

如果低的 Ω_b 被证实,那将对于宇宙学很多其他方面产生重要的影响。例如,很多星系形成模型预言星系团中的重子质量百分比应该就是 Ω_b,而较小的 Ω_b 将加剧它们与观测的后发星系团中高重子比例($> 0.05h^{-3/2}$)之间的矛盾[31]。低重子密度也和大质量旋涡星系晕主要由重子物质构成的预言相矛盾,这个预言可以用微引力透镜实验进行检验[32,33]。如果我们确实已经看到了所有的重子物质,那么动力学证据则更加强烈地证明星系晕是由非重子暗物质组成。

宇宙背景温度

虽然宇宙微波背景温度随红移的演化是相对论宇宙学的一个很少被怀疑的预言,然而精确的检验还是非常有必要的。例如,这个检验可以与其他研究手段一起用于限制关于背景辐射的各种非大爆炸起源的模型[17,34]。在我们研究的吸收体的红移处,我们可以观测到一次电离碳原子 C II 的基态跃迁,也可以看到在基态之上 64 cm^{-1} 的精细结构分裂状态 C II* 的跃迁。在这些红移处,这个频率差仅仅是宇宙微波背景辐射峰值频率的 3 倍,因此正因为存在微波背景,模型预言存在不可忽略数量的 C II*。我们利用这两个态的紫外吸收线强度测量了 C II 和 C II* 的相对数。

We have observed the C II line at 1,334.532 Å corresponding to the LLS at $z = 2.909$ found by Sargent *el al.*[22] in the quasar Q0636+68. The C I multiplets are too weak in this system to be useful for this test while N II lies within the Lyman forest. As Fig. 4 shows, the C II line consists of three strong, and a number of weaker, components stretching over more than $\pm 100\,km\,s^{-1}$, the strongest of which has a redshift, $z = 2.9034$ which we have chosen to be the zero velocity. Comparison with other lines such as Si II leaves no doubt that all the components seen are C II lines, and confirms the accuracy of the wavelength scale.

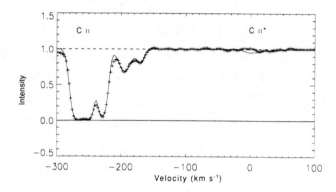

Fig. 4. C II 1,334.532 Å and C II* 1,335.708 Å absorption in the $z = 2.9034$ LLS in Q0636+68. Crosses show the data values and the positions of C II and C II* are labelled. The solid line shows a model in which the column density of C II* is 1% of C II, to show where C II* would appear if it were present. The rest equivalent width of the three strongest C II components is 248 mÅ, corresponding to a minimum column density $N_{(CII)} = 1.5 \times 10^{14}\,cm^{-2}$. More precisely, profile-fitting gives $N_{(CII)} = 4.6 \times 10^{14}\,cm^{-2}$. The equivalent width of C II* for all three main components is $\leqslant 0.4\,mÅ \pm 0.6\,mÅ$, corresponding to a 2σ upper limit of $10^{12}\,cm^{-2}$ in this line.

Profile-fitting the three strongest components gives a column density, $N_{(CII)} = 4.6 \times 10^{14}\,cm^{-2}$. Considerable additional material could be hidden in strongly saturated components, but this is the minimum value needed to give an accurate fit. By contrast, we do not see the line at 1,335.708 Å from the excited fine-structure level C II*. We have determined the noise level in this region empirically by fitting the column densities at random positions at neighbouring ("wrong") wavelengths and find that the equivalent width of C II* for all three main components is $0.4\,mÅ \pm 0.6\,mÅ$, corresponding to a 2σ upper limit of $10^{12}\,cm^{-2}$ in this line. The relative populations of the two states then imply a 2σ limit, $T_{CMBR} < 13.5\,K$, which should be compared with the prediction of 10.66 K at this redshift; or, representing the temperature as $T_{CMBR} = T_0(1+z)^{\alpha}$, we find that the 2σ upper limit on α is 1.15. This is a firm upper bound: saturation effects could only raise the column density of the C II (and so lower the temperature limit) as would any additional excitation of the C II* line, other than the CMBR. This is the tightest limit to date on T_{CMBR} at high redshift; the previous best limit of Meyer *et al.*[21] gave $T_{CMBR} < 16\,K$ at $z = 1.776$, or $\alpha < 1.73$.

It seems that the fine-structure lines will provide a sensitive new test of relativistic cosmology. The extremely weak excitation of the fine structure levels in this absorber suggests that the population of C II* is dominated just by the cosmic background, so that we are likely to be able to actually determine T_{CMBR} with improved observations. Because the dependence

在类星体 Q0636+68 的吸收线系统中，我们观测到对应 1,334.532 Å 的 CⅡ 线，这将对应在 $z = 2.909$ 处萨金特等人发现的莱曼系限系统[22]。CⅠ 的多重线因强度太弱无法用于检验，而 NⅡ 则位于莱曼森林内。如图 4 所示，CⅡ 线包含了 3 个强的和一些弱的成分，速度展宽大于 $\pm 100\ km \cdot s^{-1}$，其中最强的成分在红移 $z = 2.9034$ 处，我们将其取为零速度。通过和 SiⅡ 等其他线进行比较，结果证实这些吸收线毫无疑问是 CⅡ 线，这也进一步确定了波长尺度的准确性。

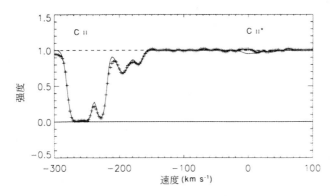

图 4. 类星体 Q0636+68 中对红移 $z = 2.9034$ 的莱曼系限系统中的 CⅡ 1,334.532 Å 与 CⅡ*1,335.708 Å 吸收线的观测。十字是数据点，CⅡ 与 CⅡ* 的位置标在图上。实线是 CⅡ* 柱密度为 CⅡ 的 1% 的模型，这是为了显示如果 CⅡ* 存在的话它应该存在的位置。三条最强的 CⅡ 谱线成分的静止等值宽度为 248 mÅ，对应于最小柱密度 $N_{(CⅡ)} = 1.5 \times 10^{14}\ cm^{-2}$，轮廓拟合给出了更精确的柱密度 $N_{(CⅡ)} = 4.6 \times 10^{14}\ cm^{-2}$。CⅡ* 三个主要成分的等值宽度 $\leqslant 0.4\ mÅ \pm 0.6\ mÅ$，对应于这条线柱密度 $10^{12}\ cm^{-2}$ 的 2σ 上限。

对三个最强成分的轮廓进行拟合，结果表明其柱密度 $N_{(CⅡ)} = 4.6 \times 10^{14}\ cm^{-2}$。当然，在强饱和的成分中可能还藏有相当多的额外物质，但这个数值是给出准确拟合的下限。相对而言，我们没有看到 1,335.708 Å 的处于激发态的 CⅡ* 的谱线。我们通过拟合附近（"错误"）波长的随机位置的柱密度来估算噪声水平，结果发现三个主要成分的 CⅡ* 的等值宽度为 $0.4\ mÅ \pm 0.6\ mÅ$，在 2σ 水平上对应于一个上限为 $10^{12}\ cm^{-2}$ 的柱密度。这两个状态的相对数量比表明在 2σ 水平上，$T_{CMBR} < 13.5\ K$，而对应理论给出该红移处的预言为 10.66 K。如果将温度写成 $T_{CMBR} = T_0 (1+z)^{\alpha}$ 的形式，我们发现 α 的 2σ 上限是 1.15。这是一个很可靠的上限：因为饱和效应只会提供不同于宇宙微波背景辐射的额外激发机制来产生更多的激发态 CⅡ* 线，从而只能提高 CⅡ 的柱密度（而降低温度上限）。这是迄今为止对高红移 T_{CMBR} 的最强限制；之前由迈尔等人做出的限制[21]给出在 $z = 1.776$，或者说 $\alpha < 1.73$ 时，$T_{CMBR} < 16\ K$。

看上去精细结构谱线可以提供一个精确检验相对论宇宙学的新方法。吸收体中处于精细结构能级激发态的数量非常少，这表明 CⅡ* 的量主要由微波背景辐射决定，因此可以通过改进对 CⅡ* 的观测精度来更好地测量 T_{CMBR}。因为等值宽度对温度的依

of equivalent width on temperature is only logarithmic, rather precise measurements of the temperature should be obtainable. In the present case, an increase in signal-to-noise ratio of a factor of 5 (less would be needed at higher redshift) must result in a significant detection of $C II^*$ if the relativistic prediction is correct. Note that systems at different redshifts yield measurements of the background at various present-day frequencies $64/(1+z)\,cm^{-1}$, and so also constrain the shape of the spectrum. Because this technique for measuring the temperature always yields an upper limit, a single well-constrained measurement lower than predicted would have catastrophic repercussions for standard Big Bang theory, whereas a firm measurement of the temperature at the expected level would rule out other possible explanations for the background radiation.

(368, 599-604; 1994)

A. Songaila[*†], **L. L. Cowie**[*†], **C. J. Hogan**[‡] **& M. Rugers**[‡]

[*] Institute for Astronomy, University of Hawaii, 2680 Woodlawn Drive, Honolulu, Hawaii 96822, USA

[‡] Department of Astronomy, University of Washington, Seattle, Washington 98195, USA

Received 12 January; accepted 10 March 1994.

References:

1. Mather, J. C. *et al. Astrophys. J.* **354**, L37-L40 (1990).

2. Mather, J. C. *et al. Astrophys. J.* **420**, 439-444 (1994).

3. Smoot, G. F. *et al. Astrophys. J.* **396**, L1-L5 (1992).

4. White, M., Scott, D. & Silk, J. *A. Rev. Ast. Astrophys.* (in the press).

5. Peebles, P. J. E. *Astrophys. J.* **146**, 542-552 (1966).

6. Wagoner, R., Fowler, W. & Hoyle, F. *Astrophys. J.* **148**, 3-49 (1967).

7. Reeves, H., Audouze, J., Fowler, W. & Schramm, D. N. *Astrophys. J.* **179**, 909-930 (1973).

8. Epstein, R., Lattimer, J. & Schramm, D. N. *Nature* **263**, 198-202 (1976).

9. Walker, T. P., Steigman, G., Schramm, D. N., Olive, K. A. & Kang, H. S. *Astrophys. J.* **376**, 51-69 (1991).

10. Smith, M. S., Kawano, L. H. & Malaney, R. A. *Astrophys. J. Suppl. Ser.* **85**, 219-247 (1993).

11. Steigman, G. & Tosi, M. *Astrophys. J.* **401**, 150-156 (1992).

12. Pagel, B. E. J., Simonson, E. A., Terlevich, R. J. & Edmunds, M. G. *Mon. Not. R. astr. Soc.* **255**, 325-345 (1992).

13. Pagel, B. E. J. *Phys. Scripta* **T36**, 7-15 (1991).

14. Binney, J. & Tremaine, S. *Galactic Dynamics* (Princeton Univ. Press, 1987).

15. Wolfe, A. M. *Ann. N.Y. Acad. Sci.* **688**, 281-296 (1993).

16. Webb, J. K., Carswell, R. F., Irwin, M. J. & Penston, M. V. *Mon. Not. R. astr. Soc.* **250**, 657-665 (1991).

17. Peebles, P. J. E., Schramm, D. N., Turner, E. I. & Kron, R. G. *Nature* **352**, 769-776 (1991).

18. Roth, K. C., Meyer, D. M. & Hawkins, I. *Astrophys. J.* **413**, L67-L71 (1993).

19. Bahcall, J. N. & Wolf, R. A. *Astrophys. J.* **152**, 701-729 (1968).

20. Bahcall, J. N., Joss, P. C. & Lynds, R. *Astrophys. J.* **182**, L95-L98 (1973).

21. Meyer, D. M., Black, J. H., Chaffee, F. H., Foltz, C. & York, D. G. *Astrophys. J.* **308**, L37-L41 (1986).

22. Sargent, W. L. W., Steidel, C. C. & Boksenberg, A. *Astrophys. J. Suppl. Ser.* **69**, 703-761 (1989).

23. Cowie, L. L., Laurent, C., Vidal-Madjar, A. & York, D. G. *Astrophys. J.* **229**, L81-L85 (1979).

24. Chaffee, F. H., Foltz, C. B., Röser, H.-J., Weymann, R. J. & Latham, D. W. *Astrophys. J.* **292**, 362-370 (1985).

25. Chaffee, F. H., Foltz, C. B., Bechtold, J. & Weymann, R. J. *Astrophys. J.* **301**, 116-123 (1986).

26. Shapiro, P. R. & Moore, R. T. *Astrophys. J.* **207**, 460-483 (1976).

27. Donahue, M. & Shull, J. M. *Astrophys. J.* **383**, 511-523 (1991).

28. Vangioni-Flam, E. & Audouze, J. *Astr. Astrophys.* **193**, 81-86 (1988).

29. Linsky, J. L. *et al. Astrophys. J.* **402**, 694-709 (1993).

30. Bania, T. M., Rood, R. T. & Wilson, T. L. *Astrophys. J.* **323**, 30-43 (1987).

赖满足对数关系，因此可以期待得到更加精确的温度测量结果。以目前的情况为例，如果相对论的预言是正确的，那么增加至 5 倍的信噪比将足以探测到显著的 C II* 吸收（在更高红移处则并不要求这么高的信噪比）。注意到测量不同红移处的系统对应的是今天光谱中不同的频率，$64/(1+z)$ cm^{-1}，因此通过对这些系统的观测将可以用来限制谱的形状。因为这种对温度的测量总是给出上限，如果探测到的数值低于预言的结果，那么将对标准大爆炸理论带来灾难性的后果。然而，如果确定温度在所预期的区间内则可排除对背景辐射起源的其他解释。

（周杰 翻译；张华伟 审稿）

31. White, S. D. M., Navarro, J. F., Evrard, A. E. & Frenk, C. S. *Nature* **366**, 429-433 (1993).

32. Alcock, C. *et al. Nature* **365**, 621-623 (1993).

33. Aubourg, E. *et al.* **365**, 623-625 (1993).

34. Wright, E. L. *et al. Astrophys. J.* **420**, 450-456 (1994).

Acknowledgements. We are grateful to the designers and builders of the Keck 10-m telescope and of HIRES. We particularly thank M. Keane and B. Shaeffer for their help in obtaining the observations. This work was supported at the University of Hawaii by the State of Hawaii and by NASA, and at the University of Washington by the US NSF and NASA.

Distance to the Virgo Cluster Galaxy M100 from Hubble Space Telescope Observations of Cepheids

W. L. Freedman *et al.*

Editor's Note

In astronomy it is essential to know the distance to any object, because only then can the observed brightness at any wavelength be converted to the true luminosity. Variable stars called Cepheids are central to working out this "distance ladder", because their luminosity is closely related to their period and can thus be determined independently. Here Wendy Freedman and colleagues use the Hubble Space Telescope to determine the Cepheid distance to the galaxy M100, and thereby determine an accurate value for the Hubble constant, which measures the expansion rate of the Universe. This was one of the key design goals for the HST. The initial value of ~80 km s^{-1} Mpc^{-1} was refined over the following decade.

Accurate distances to galaxies are critical for determining the present expansion rate of the Universe or Hubble constant (H_0). An important step in resolving the current uncertainty in H_0 is the measurement of the distance to the Virgo cluster of galaxies. New observations using the Hubble Space Telescope yield a distance of 17.1 ± 1.8 Mpc to the Virgo cluster galaxy M100. This distance leads to a value of $H_0 = 80 \pm 17$ km s^{-1} Mpc^{-1}. A comparable value of H_0 is also derived from the Coma cluster using independent estimates of its distance ratio relative to the Virgo cluster.

WITHIN the framework of general relativity, the evolution of the Universe can be specified by the Friedmann equation which relates the expansion rate H, to the mean mass density ρ, the curvature k, and a possible additional term, called the cosmological constant Λ (identified with the gravitational effects of the vacuum energy density). In a uniform and isotropic Universe the relative expansion velocity v is proportional to the relative distance r such that $v = H \times r$. Thus a determination of the present-day value of the Hubble constant H_0 determines both the expansion timescale and the size scale of the Universe. The Hubble constant also provides constraints on the density of baryons produced in the Big Bang, the amount of dark matter, and how structure formed in the early Universe[1,2].

Despite 65 years of study, the value of the Hubble constant has remained in dispute. Although the measurement of relative velocities of galaxies is straightforward, the measurement of accurate distances has always been more difficult. For distances out to ~5 Mpc, there is now general agreement[3,4] to a level of better than ~ $\pm 10\%$. However, for

由哈勃空间望远镜观测造父变星到室女星系团成员星系 M100 的距离

弗里德曼等

编者按

在天文学中，知道任意天体的距离是重要的，因为只有这样才能将任意波长处观测到的亮度转换为真实的光度。被称作造父变星的变星对得到"距离阶梯"尤为重要，因为它们的光度和它们的光变周期紧密相关，因而它们的光度可以独立地测定。在这篇文章中，温迪·弗里德曼和他的同事们利用哈勃空间望远镜测定到 M100 星系的造父距离，并从而测定哈勃常数（衡量宇宙膨胀率的参数）的精确值。确定哈勃常数是哈勃空间望远镜的主要设计目标之一。它最初的值 ~80 km·s^{-1}·Mpc^{-1} 在之后的几十年中得到了修正。

测定到星系的精确距离对于测定目前的宇宙膨胀率或哈勃常数（H_0）是至关重要的。解决目前 H_0 不确定性的重要一步是测量到室女星系团的距离。使用哈勃空间望远镜新的观测得到的到室女星系团成员星系 M100 的距离为 17.1±1.8 Mpc。由这一距离得到 $H_0 = 80\pm17$ km·s^{-1}·Mpc^{-1}。利用后发星系团相对室女星系团距离比进行独立的测定，也得到了相近的 H_0 值。

在广义相对论的框架内，宇宙的演化可以用弗里德曼方程描述，这一方程将膨胀率 H 与平均质量密度 ρ、曲率 k 以及一个称作宇宙学常数 Λ（等同于真空能量密度的引力作用）的可能的附加项联系起来。在一个均匀且各向同性的宇宙中，相对膨胀速度 v 正比于相对距离 r，即 $v = H \times r$。因此，测定哈勃常数的当前值 H_0 就同时得到了宇宙的膨胀时标和尺度。哈勃常数也对大爆炸时期产生的重子物质密度、暗物质的数量以及宇宙早期结构的形成都具有约束作用 [1,2]。

尽管经过了 65 年的研究，但是哈勃常数的值仍然存在争议。尽管测量星系间的相对速度是直接的，然而测量星系间的精确距离却困难得多。目前普遍认为，对于大约 5 Mpc 的距离的测定，测量的精度高于 ±10%[3,4]。然而，对于像到室女星系团这么遥远的目标的距离，测量结果存在很大的分歧 [5,6]，其测量值大约在 15 Mpc 到

distances as great as the Virgo cluster there is a significant discrepancy[5,6], with quoted values ranging from about 15 to 24 Mpc. Not only is there a range of published distances, but there is a tendency for the values to cluster at the extremes, giving rise to the so-called "short" and "long" distance scales.

The most accurate means of measuring the distances to galaxies has proved to be the application of a relationship between the period and the luminosity for a class of supergiant variable stars known as classical Cepheids. In fact, Cepheids were used by Edwin Hubble to demonstrate the extragalactic nature of the spiral nebulae[7,8]. They are relatively young, massive stars with luminosities ~1,000 to 100,000 times brighter than the Sun ($-2 < M_V < -7$ mag). They are well calibrated; they are easy to identify in external galaxies because of their variability and brightness; and they have measured dispersions in the V- and I-band period–luminosity (P–L) relationships amounting to only ± 0.25 and ± 0.20 mag, respectively[3]. In addition, the underlying physical basis for the Cepheid P–L relation is well understood.

Unfortunately, Cepheids are not luminous enough to be observed out to distances where galaxies are participating in the free expansion of the Universe. The motions of individual galaxies can be perturbed by interactions with nearby neighbouring galaxies; in addition, galaxies can participate in large-scale flows[9,10]. Hence, to measure the pure Hubble flow, other (secondary) distance techniques must be used to extend the extragalactic distance scale beyond the observable range of the Cepheids, to recession velocities of a few thousand km s^{-1} where peculiar velocities (which can amount to several hundred km s^{-1}) are small in comparison to the expansion velocity.

The Virgo cluster of galaxies is close enough to be studied in detail; and yet it is far enough away that it is of cosmological interest. It is rich in both spiral and elliptical galaxies and therefore it has played a critical role in the extragalactic distance scale and determination of the Hubble constant. Thus obtaining a Cepheid distance to the Virgo cluster represents a crucial step in resolving the current uncertainty in H_0 and resolving the dichotomy in distance estimates. A variety of other techniques for measuring distances have been applied to the Virgo cluster; hence, an accurate direct measure of the distance to this cluster can be used to calibrate (or set the zero point for) other secondary distance indicators.

The Key Project

The goal of the Hubble Space Telescope Key Project on the extragalactic distance scale is to provide a measure of the Hubble constant accurate to 10%. This aim is non-trivial given that the history of previous attempts to measure the extragalactic distance scale is replete with examples where large systematic errors were eventually revealed. Hence, the determination of accurate distances requires careful attention to eliminating potential sources of systematic error.

24 Mpc 之间。已经发表的距离不仅范围很大，而且到星系团的距离还存在着两种极端趋势，即所谓"短"的和"长"的距离尺度。

已经证实的测量星系团距离最精确的方法是应用被称作经典造父变星的一类超巨星变星的周光关系。事实上，造父变星曾被埃德温·哈勃用来解释旋涡星云位于银河系之外 [7,8]。它们是相对年轻的大质量恒星，其光度是太阳的约 1,000 倍到 10,000 倍（-2 mag $< M_V < -7$ mag）。它们已经被很好地定标过；由于其光变和亮度，它们很容易在河外星系中被识别出来；它们在 V 波段与 I 波段的周光关系中测量出的弥散度分别只有 ± 0.25 mag 和 ± 0.20 mag[3]。另外，我们对于造父变星周光关系的物理机制也已经有了较好的了解。

遗憾的是，造父变星由于亮度不够，在宇宙自由膨胀的星系中，还不能被观测到。个别星系的运动可能在与邻近星系的相互作用中受到干扰；另外，星系可能参与大尺度的流动 [9,10]。因此，为了测量纯粹的哈勃流，必须使用其他的（次级的）距离测量技术，以便将河外星系的距离尺度扩展到造父变星的可观测范围之外，在这一区域退行速度为每秒几千千米，而本动速度（大小只有每秒几百千米）与宇宙的膨胀速度相比则很小。

室女星系团距离我们足够近，可以详细地加以研究；而且从宇宙学研究的角度来看，它也足够远。该星系团富含旋涡星系和椭圆星系，这使得它在河外星系的距离尺度和哈勃常数测定上都扮演着至关重要的角色。因此，获得室女星系团的造父距离，对于解决当前 H_0 测定的不确定性以及距离估计上的分歧，都是关键的一步。鉴于各种测定距离的其他方法都已经用到了室女星系团；因此，精确地直接测量到这个星系团的距离可以用来定标（或者设置零点）所有其他的次级示距天体。

关 键 项 目

哈勃空间望远镜关于河外距离尺度的关键项目主要目标是将哈勃常数的测量精度提高到 10%。考虑到之前河外距离尺度测量的历史中充斥的许多事例（最终都被证实存在着很大的系统误差），就会发现这一目标是意义非凡的。因此，精确距离的测量要求仔细地去消除潜在的各种系统误差。

The strategy adopted by the Key Project team on the extragalactic distance scale is threefold. The first goal is to discover Cepheids, and thereby measure accurate primary distances to spiral galaxies located in the field and in small groups that are suitable for the calibration of several independent secondary methods. (These secondary methods include: the Tully–Fisher relationship[11,12], the surface-brightness fluctuation method[13], the planetary nebula luminosity function[14], the expanding photosphere method applied to type II supernovae[15,16], and the measurement of the luminosities of type Ia supernovae[6,17-19].) The second objective is to provide a check on potential systematic errors in the Cepheid distance scale through independent distance estimates to the nearby galaxies, M31, M33 and the Large Magellanic Cloud (LMC) and, in addition, to undertake an empirical test of the sensitivity of the zero point of the Cepheid P–L relationship to heavy-element abundances. The third and most challenging observational goal is to make Cepheid measurements of distances to three spiral galaxies in the Virgo cluster and two members of the Fornax cluster.

The Distance to M100

Our first observations aimed at finding Cepheids in a Virgo cluster galaxy were made in a two-month period beginning in April 1994 with a sequence of 12 1-hour V-band exposures of a field ~2 arcmin east of the nucleus of M100. The observing strategy was designed to provide well-sampled light curves for Cepheids having periods ranging from 10 to 60 days. In addition, I-band exposures were taken back-to-back with 4 of the V observations. We present here sample light curves for the Cepheids discovered, their V-band and I-band P–L relationships, and a preliminary estimate of the distance to M100. Details of the data reduction and analysis, tabulation of the photometry, and identification of the variables will be presented elsewhere (L.F. et al., manuscript in preparation; R.H. et al., manuscript in preparation). Three independent calibrations of the photometric zero points were made; the final resulting uncertainties in the V and I zero points amount to ±0.05 and ±0.04 mag, respectively. A description of the sampling strategy used for the optimal discovery of variables, and of the method used for identification of variables, determination of mean magnitudes, reddening, and distance have been published elsewhere[20].

V-band light curves for twelve M100 Cepheids are illustrated in Fig. 1. The total sample of Cepheids has a range of periods from 20 to 65 days, and mean V magnitudes ranging from 25.0 to 26.5 mag. It is evident from the small scatter in the light curves that the quality of the photometry being obtained by the Hubble Space Telescope (HST) is excellent. Internal estimates of the random errors for single data points are found to be ±0.14 mag at $V \approx 25$ mag and ±0.17 mag at $V \approx 26$ mag. The errors in the mean magnitudes for the 12 epochs amount to ±0.04 and ±0.05 mag at these same magnitude levels.

关键项目的研究小组在河外距离尺度问题上采取的策略包括三个目标。第一个目标是发现造父变星，在此基础上测量到位于星场中以及位于小星系群内的旋涡星系的精确初级距离，这一结果适合于定标几种相互独立的次级距离测量方法。(这些次级距离测量方法包括：塔利 – 费希尔关系[11,12]、面亮度起伏方法[13]、行星状星云光度函数法[14]、在 II 型超新星上使用的膨胀光球法[15,16]，以及 Ia 型超新星光度测量法[6,17-19]。)第二个目标是通过邻近星系 M31、M33 以及大麦哲伦云（LMC）距离的独立测算检验造父距离的潜在系统误差。此外，还可以经验性地检验造父变星周光关系的零点对重元素丰度的敏感性。第三个也是最具挑战性的观测目标是利用造父变星测量出室女星系团内的三个旋涡星系以及天炉星系团内两个成员星系的距离。

到 M100 的距离

我们的第一个观测目标是寻找室女星系团内的造父变星，通过对 M100 核心区域偏东约 2 arcmin 的视场进行 12 幅为一序列，每幅曝光时间为 1 小时的 V 波段测光，这一观测过程从 1994 年 4 月开始延续了两个月。采取这样的观测策略是为了给光变周期处于 10 天到 60 天的造父变星提供较好的光变曲线采样点。另外，紧接着 4 组 V 波段观测，我们进行了 I 波段的观测。这里，我们给出了所发现的造父变星的样本光变曲线，它们在 V 波段和 I 波段的周光关系，以及初步估计的到 M100 的距离。数据处理与分析的细节，测光列表以及变星的认证将另文发表（温迪·弗里德曼等，稿件准备中；罗伯特·希尔等，稿件准备中）。我们使用了三种独立的测光零点定标；V 波段和 I 波段零点最终结果的不确定性分别为 ±0.05 mag 和 ±0.04 mag。用于最优地发现变星的采样策略，以及认证变星、确定平均星等、红化值以及测量距离的方法都已经在其他地方发表[20]。

在图 1 中展示了 M100 内 12 颗造父变星在 V 波段的光变曲线。所有造父变星样本的光变周期从 20 天到 65 天，平均 V 星等的范围从 25.0 mag 到 26.5 mag。由光变曲线很小的散射证明哈勃空间望远镜的测光质量是非常好的。单一数据点的随机误差的自身估计在 $V \approx 25$ mag 时为 ±0.14 mag，在 $V \approx 26$ mag 时为 ±0.17 mag。对于 12 次观测得到的平均星等的误差在相同星等条件下分别为 ±0.04 mag 以及 ±0.05 mag。

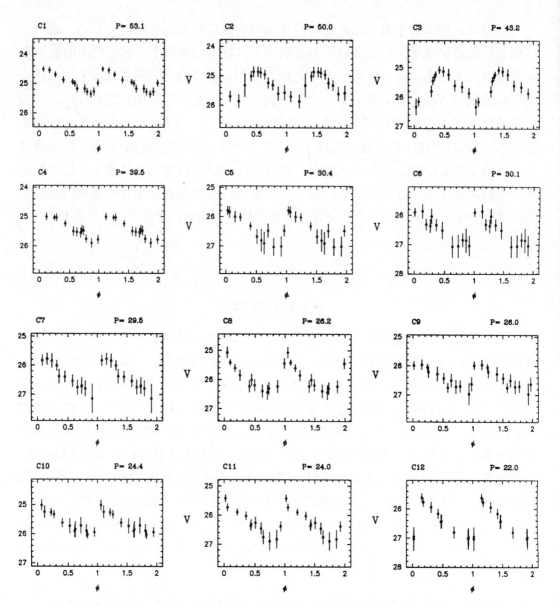

Fig. 1. Sample light curves for twelve Cepheids in M100. The periods of the stars illustrated (~20 to 50 days) are typical of the total sample discovered in M100. In addition to their periods, Cepheids are characterized by their distinctive light curves showing a rapid rise to maximum light, followed by a slower linear decline to minimum light. Cepheids pulsate because their atmospheres are not in hydrostatic equilibrium. The instability occurs as a result of a change in the opacity of a helium ionization zone with temperature. As singly ionized helium becomes doubly ionized, the opacity increases with increasing temperature. This changing opacity acts as a valve so that when the atmosphere is compressed and at higher temperature, it becomes more opaque to radiation, and the thermal energy increases. Subsequently, it expands and cools, becoming more transparent. As the pressure on the atmosphere decreases it then collapses and the cycle repeats.

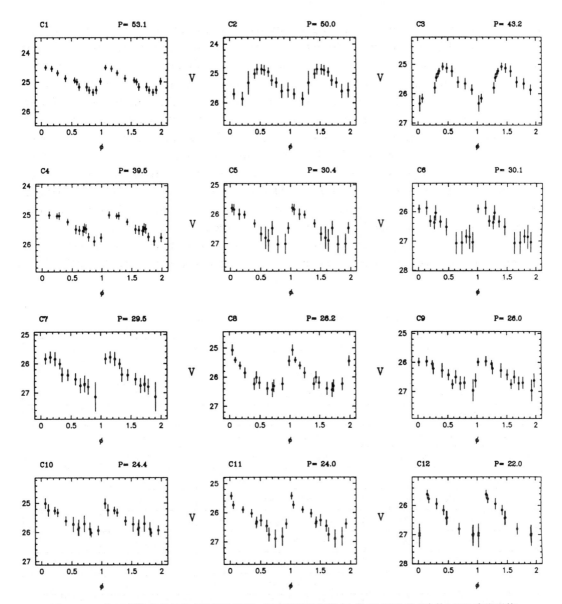

图 1. M100 中 12 颗造父变星光变曲线的示例。图中注明的周期(大约 20 天到 50 天)是 M100 内发现的造父变星总样本中的典型值。除了周期性之外,造父变星的特性主要来自它们独特的光变曲线:快速增亮到最大光度,然后缓慢、线性地降低到最小光度。造父变星的脉动是由于它们的大气并未处于流体静力平衡状态。当氦电离区的不透明度随着温度变化时不稳定性就会出现。当一次电离的氦被二次电离时,不透明度会随着温度的增加而增加。变化的不透明度的作用就像阀门一样,当大气被压缩并处于更高温度时,它变得对辐射更加不透明,大气的热能增加。接下来,大气开始膨胀并且冷却,变得更加透明。当作用于大气的压力下降时,大气又开始收缩,新的循环又开始。

In Fig. 2, we present the time-averaged, intensity-weighted V-band and I-band P–L relations for 20 Cepheids in M100 (white circles). We compare the brightness of the M100 Cepheids with those in the LMC (black circles) to obtain a distance ratio (M100/ LMC). In logarithmic units of magnitudes, the so-called "distance modulus" (μ) is given by $5 \times \log_{10}$ (distance in parsecs) -5.0. A correction for the dimming effect due to interstellar dust is obtained by fitting the apparent V and I moduli to a standard interstellar extinction law. The apparent I-band modulus to M100 is measured to be $\mu_I = 31.25$ mag; the corresponding V P–L relation for the same sample of Cepheids yields $\mu_V = 31.31$ mag, implying a mean extinction of $A_V = 0.15 \pm 0.17$ for the M100 Cepheid sample. The true (reddening-corrected) distance modulus to M100 is determined to be 31.16 ± 0.20 mag, corresponding to a distance of 17.1 ± 1.8 Mpc. A detailed listing of the errors for this determination is given in Table 1. The random errors in the apparent moduli amount to $< 4\%$ in distance. The uncertainty in the true distance is dominated by systematics due primarily to the adopted LMC distance and correction for reddening, giving a total uncertainty of $\pm 10\%$.

Table 1. Error budget in the M100 distance scale

Source of uncertainty	Type of uncertainty	Error (mag.)	Cumulative error
Distance to M100			
[A] LMC Distance	Independent estimates	± 0.10	
[B] WFPC2 V-band zero point	Ground-based / on-orbit calibration	± 0.05	
[C] WFPC2 I-band zero point	Ground-based / on-orbit calibration	± 0.04	
[D] Extinction-corrected Cepheid modulus	[B] and [C] are uncorrelated	± 0.17	
[E] Charge transfer efficiency	4% across entire chip	± 0.01	
[F] Cepheid V-band modulus	(V-band P–L dispersion) / $\sqrt{N}$	± 0.06	
[G] Cepheid I-band modulus	(I-band P–L dispersion) / $\sqrt{N}$	± 0.07	
[H] Cepheid true modulus	[F] and [G] are correlated	± 0.03	
[I] = [A] + [D] + [E] + [H]	Combined in quadrature		± 0.20
Distance to Virgo			
[J] Velocity of Virgo cluster	± 80 km s^{-1}	± 0.11	
[K] Back-to-front geometry of Virgo	± 3 Mpc	± 0.35	
[L] = [I] + [J] + [K]	Combined in quadrature $H_0 = 82 \pm 17$ km s^{-1} Mpc^{-1}		± 0.42
Distance to Coma			
[M] Velocity of Coma cluster	± 100 km s^{-1}	± 0.02	
[N] Virgo / Coma Relative distance	Secondary indicators	± 0.10	
[O] = [I] + [K] + [M] + [N]	Combined in quadrature $H_0 = 77 \pm 16$ km s^{-1} Mpc^{-1}		± 0.42

A quantitative overview of the propagation of errors associated with our new data set and its application to the extragalactic distance scale. Errors in the photometric zero points and statistical uncertainties associated with the

在图 2 中，我们给出了 M100 内的 20 颗造父变星的时间平均的和强度加权的 V 波段与 I 波段的周光关系（白圆圈）。我们将 M100 内造父变星的亮度与大麦哲伦云内的造父变星（黑圆圈）相比较以得到距离比（M100/LMC）。以星等这种对数性单位，被称作"距离模数"的 μ 可以表述为 $5 \times \log_{10}$（以秒差距为单位的距离）-5.0。通过将 V 波段与 I 波段视星等的模数与标准的星际消光律拟合可以得到由于星际尘埃造成的消光影响的修正量。测量得到的到 M100 的 I 波段视距离模数为 $\mu_I = 31.25$ mag；对于同一造父变星样本，由相应的 V 波段的周光关系求得 $\mu_V = 31.31$ mag，这表明对于 M100 内的造父变星样本，平均消光 $A_V = 0.15 \pm 0.17$。真实的（红化改正的）到 M100 的距离模数为 31.16 ± 0.20 mag，对应的距离为 17.1 ± 1.8 Mpc。表 1 中详尽地列出了这一测定结果的误差来源。视距离模数的随机误差造成了 $< 4\%$ 的距离误差。真实距离的不确定性由所采用的大麦哲伦云的距离以及红化改正主导，其总的不确定性为 $\pm 10\%$。

表 1. M100 距离尺度中的误差估计

误差来源	误差类型	误差（mag）	累计误差
到 M100 的距离			
[A] LMC 距离	独立估计	±0.10	
[B] WFPC2 V 波段零点	地基 / 在轨定标	±0.05	
[C] WFPC2 I 波段零点	地基 / 在轨定标	±0.04	
[D] 消光改正的造父变星模数	[B] 和 [C] 不相关	±0.17	
[E] 电荷转移效率	整个探测器为 4%	±0.01	
[F] 造父变星 V 波段模数	（V 波段周光关系弥散）/ $\sqrt{N}$	±0.06	
[G] 造父变星 I 波段模数	（I 波段周光关系弥散）/ $\sqrt{N}$	±0.07	
[H] 造父变星真实的模数	[F] 和 [G] 是相关的	±0.03	
[I] = [A]+[D]+[E]+[H]	正交合成		±0.20
到室女星系团的距离			
[J] 室女星系团的速度	±80 km·s^{-1}	±0.11	
[K] 室女星系团前后的几何形状	±3 Mpc	±0.35	
[L] = [I]+[J]+[K]	正交合成 $H_0 = 82 \pm 17$ km·s^{-1}·Mpc^{-1}		±0.42
到后发星系团的距离			
[M] 后发星系团的速度	±100 km·s^{-1}	±0.02	
[N] 室女与后发星系的相对距离	次级示距天体	±0.10	
[O] = [I]+[K]+[M]+[N]	正交合成 $H_0 = 77 \pm 16$ km·s^{-1}·Mpc^{-1}		±0.42

与我们新的数据相关的误差传递的定量概述及其在河外星系距离尺度中的应用。测光零点的误差和与平均视距离模数相关的统计不确定性都传递进了真实距离模数的测量当中。就测光定标中的独立误差而言，假设各项是完全不相

mean apparent distance moduli both propagate into the determination of a true distance modulus. In the case of independent errors in the photometric calibration the terms are assumed to be fully uncorrelated giving

$$\varepsilon_{\mu_0}^U = R_{VI} \, (\varepsilon_I^2 + \varepsilon_I^2)^{1/2}$$

where $R_{VI} = A_V / E(V-I) = 2.6$ is the ratio of total-to-selective absorption. For the individual wavelength-dependent moduli the uncertainties are correlated in sign and magnitude because in this case the Cepheid samples defining the means are identical, thereby giving

$$\varepsilon_{\mu_0}^C = R_{VI} \, (\varepsilon_V - \varepsilon_I)$$

As can readily be seen, uncertainties in the photometric zero points dominate the error budget for the calculated distance to M100; with additional ground-based observing this error can be significantly reduced.

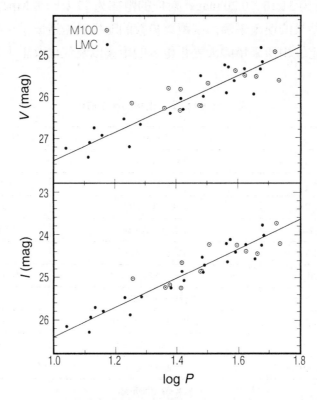

Fig. 2. Composite period-luminosity relations (*V*-band upper panel; *I*-band lower panel) for Cepheids in M100 (white circles) and Cepheids in the LMC (black circles, shifted to the distance of M100). Apparent *V* and *I* distance moduli for M100 are obtained by minimizing the residuals in the combined *P–L* relations for the two galaxies, and determining the relative offset with respect to the calibrating LMC sample. Consistent with previous studies[3,20], only high signal-to-noise variables (those having an average of their absolute deviations from the mean exceeding 1.5 times the mean error) are plotted, whereas only stars with log*P* < 1.8 are included in the fit. The difference in the *V* and *I* apparent moduli for M100 is assumed to be due to interstellar dust present both in M100 and our own galaxy. Correcting for this effect yields a reddening-corrected (true) distance to M100 of 17.1 ± 1.8 Mpc.

This first Cepheid distance to M100 falls toward the low end of the published range of distance values to the Virgo cluster. How does this distance compare with previous distance estimates to M100 itself? We discuss four recent measurements of the distance to M100. The first

关的，则可以得到：

$$\varepsilon_{\mu_0}^U = R_{VI}(\varepsilon_V^2 + \varepsilon_I^2)^{1/2}$$

其中 $R_{VI} = A_V/E(V-I) = 2.6$ 是整体吸收和选择性吸收的比。就目前的情况来说，确立这一方式的造父变星的样本是相同的，所以对于单独的波长依赖的模数而言，不确定性在符号和数量上都是相关的，因此有

$$\varepsilon_{\mu_0}^C = R_{VI}(\varepsilon_V - \varepsilon_I)$$

容易看出，测光零点的不确定性主导了计算得到的到 M100 的距离结果的误差区间，使用附加的地基观测，这一误差可以显著减小。

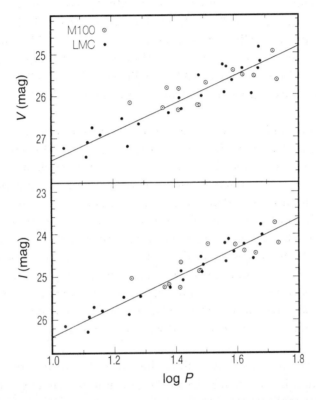

图 2. 合成的造父变星周光关系图（上图为 V 波段，下图为 I 波段），其中 M100 的造父变星标为白圆圈，大麦哲伦云内的造父变星标为黑圆圈并且移动到 M100 的距离处。M100 在 V 波段和 I 波段的视距离模数可以通过将这两个星系合并的周光关系的残差最小化而得到，并且得到相对于定标的大麦哲伦云造父变星样本的相对偏差。同已有的研究一致的是 [3,20]，图中只画出了具有高信噪比的变星样本（它们相对于平均星等的绝对偏差大于 1.5 倍的平均误差），只有周期 $\log P < 1.8$ 的恒星用到拟合中。假设 M100 在 V 波段和 I 波段视距离模数结果的不同是由 M100 内以及我们所在星系内都存在的星际尘埃造成的。对这一影响的改正可以得到 M100 红化改正了的（真实的）距离 17.1 ± 1.8 Mpc。

这第一个到 M100 的造父距离处在已经发表的到室女星系团距离范围的低值端。如何将这个距离与之前估计的到 M100 的距离进行比较呢？我们讨论四个最近

measurement is based on a comparison of the observed diameters of the ScI galaxies. Assuming that M100 has the same diameter as M101, Sandage[21] concluded that M100 is located at a distance of 27.7 Mpc. Ten years earlier, de Vaucouleurs[22] discussed the relative positions of 37 spirals in the Virgo cluster based on a variety of secondary distance indicators available at that time. He found a distance to M100 of 11.8 Mpc, which places M100 on the near side of the Virgo cluster which he estimated to be at a distance of 15.0 Mpc. A third estimate of the distance to M100 is obtained from the expanding photosphere method[15,16] applied to the type II supernova SN1979c. The most recent application of this method yields a distance of 15 ± 4 Mpc. As discussed further (J.R.M. *et al.*, manuscript in preparation), the distances to the four galaxies now available using both Cepheids and type II supernovae all show good agreement. Fourth, M100 is contained, in the Virgo galaxy sample studied by Pierce and Tully[23]. On the basis of the Tully–Fisher (TF) relationship, these authors conclude that M100 lies at a distance of 18.4 ± 2.2 Mpc, which they claim is within 1σ of their mean distance to the Virgo cluster (15.6 ± 1.5 Mpc). A somewhat closer distance of 14.5 ± 2.7 Mpc is given in Table 3 by Pierce[24]. The M100 Tully–Fisher and type II supernova estimates are in good agreement with the distance derived by Cepheids.

Where does M100 lie with respect to the centre of the Virgo cluster? An answer to this question is needed to determine H_0 from the present data. M100 is projected on the sky $3.9°$ from the centre of the Virgo cluster as defined by the giant elliptical galaxy M87. The angular extent of the core is about $6°$ radius, corresponding to 1.8 Mpc at a distance of 17.1 Mpc. The full front-to-back depth of the elliptical galaxy component of the cluster has been estimated to be 8 Mpc (ref. 25). This range would imply an r.m.s. variation in depth of ± 2 Mpc for the elliptical galaxies, if the radial distribution is gaussian. However, the distribution in depth of the spiral galaxies in the cluster appears to be larger, with a full range of ~ 13 Mpc (ref. 23), and this distribution is more complex than a simple gaussian model. With these uncertainties in mind we conservatively assign an error of $\pm 20\%$ in the mean distance of the Virgo cluster. The confidence limits that can be placed on this Virgo distance will be discussed in more details (J.R.M. *et al.*, manuscript in preparation).

By contrast, is there any reason to believe that M100 lies in the foreground? Sandage and Bedke[26] present a list of 75 galaxies in the direction of the Virgo cluster that resolve most easily into individual stars. On the basis of their resolution, 45 of these galaxies are classified as "excellent", "good" or "good-to-fair", whereas M100 itself is classified as "fair". It then seems unlikely that M100 is significantly closer than most of the Virgo spiral galaxies discussed by Sandage because resolution is distance dependent. Nevertheless, the line-of-sight position of M100 with respect to the core of the Virgo cluster is the major source of uncertainty in what follows.

There is a second source of uncertainty in estimating H_0 from the Virgo cluster even if the distance were accurately known. The cluster is sufficiently far that its Hubble recession velocity is larger than its peculiar velocity. However, there are many difficulties in determining the Hubble velocity: our own Local Group is infalling toward the Virgo cluster, and the presence of foreground plus background galaxies complicates the determination of the

的关于 M100 距离的测量。第一个测量是基于比较观测到的 ScI 型星系的直径。假设 M100 与 M101 有相同的直径,桑德奇[21] 得出结论,M100 位于 27.7 Mpc 距离处。十年前,德·沃库勒尔[22] 基于当时已知的各种次级示距天体,讨论了室女星系团内 37 个旋涡星系的相对位置。他发现到 M100 的距离为 11.8 Mpc,这一距离表明 M100 位于室女星系团的近端,他估计室女星系团的距离为 15.0 Mpc。到 M100 距离的第三种估计是通过将膨胀光球方法应用到 II 型超新星 SN1979c 得到的[15,16]。使用这一方法得到的最新的结果为 15±4 Mpc。进一步讨论表明(杰里米·莫尔德等,稿件准备中),由造父变星方法和 II 型超新星方法测定的到这四个星系的距离非常一致。第四,M100 包含在皮尔斯与塔利所研究的室女星系团的星系样本中[23]。基于塔利–费希尔关系,这些作者得出,M100 位于 18.4±2.2 Mpc 距离处,并认为这一结果在 1σ 的误差范围内与室女星系团的平均距离 15.6±1.5 Mpc 是一致的。在皮尔斯的表格 3 中也列出了更接近的距离 14.5±2.7 Mpc[24]。由塔利–费希尔关系以及 II 型超新星得到的 M100 的距离与使用造父变星得到的结果是非常一致的。

相对于室女星系团的中心,M100 位于什么位置呢?回答这一问题对于由目前的数据确定 H_0 的值是必需的。M100 到由巨椭圆星系 M87 确定的室女星系团中心的角距离为 3.9°。核心区域的角半径大约为 6°,在 17.1 Mpc 的距离上相当于 1.8 Mpc。星系团内的椭圆星系的纵深范围大约为 8 Mpc[25]。这个范围表明,对于椭圆星系而言,如果其径向分布是高斯形式的,那么纵深的均方根误差(rms)应该是 ±2 Mpc。然而,星系团内旋涡星系的纵深分布显示出更大的范围,大致为 13 Mpc[23],这种分布比简单的高斯模型更为复杂。基于以上不确定性的考虑,我们保守地估计室女星系团平均距离的误差为 ±20%。关于室女星系团距离的置信度限制将在其他地方进行详细讨论(杰里米·莫尔德等,稿件准备中)。

相比之下,有什么理由相信 M100 位于室女星系团的前方呢?桑德奇与贝德克[26] 列出了在室女星系团方向最容易分辨出单颗恒星的 75 个星系。根据分辨率,其中 45 个星系被分类成"极清楚"、"很清楚"或"大致清楚",而 M100 本身被分类成"清楚"。由于分辨率是依赖于距离的,M100 不太可能明显比室女星系团内其他桑德奇讨论过的大部分旋涡星系更靠近我们。然而,M100 相对于室女星系团核心的视线方向的位置是下面将讨论的不确定性的主要来源。

即使准确地知道距离,在用室女星系团估计 H_0 时,仍然存在着第二个不确定因素。这个星系团已经足够远,使得它的哈勃退行速度大于它的本动速度。然而,在确定哈勃速度的过程中仍然存在很多困难:我们自身所在的本星系群正朝向室女星系团沉降,前景与背景星系的存在使测定室女星系团速度本身复杂化了。本星系群

Virgo cluster velocity itself. The range of corrections for the Local Group centroid, the adopted infall velocity and the random velocity can lead to Virgo recession velocities[27] from 1,200 to 1,600 km s^{-1}. Adopting the recession velocity of $1,404 \pm 80$ km s^{-1} preferred by Huchra[27] and a Virgo cluster distance of 17.1 Mpc yields a value of $H_0 = 82 \pm 17$ km s^{-1} Mpc^{-1}. The cumulative errors entering this estimate are listed in Table 1. (However, we note that adopting the velocity[6,31] of $1,179 \pm 17$ km s^{-1} results in $H_0 = 69 \pm 14$ km s^{-1} Mpc^{-1} for the same distance.)

A means of avoiding the uncertainty in the Virgo cluster recession velocity in determining H_0 is to make use of the well measured distance ratio between the Virgo and Coma clusters[28]. The Coma cluster is located about 6 times more distant than the Virgo cluster. Its peculiar velocity is $\lesssim 80$ km s^{-1} (ref. 29) and it has a recession velocity[9,30,31] of $7,200 \pm 100$ km s^{-1}. Hence its peculiar velocity contributes a very small fractional uncertainty to the cosmological velocity. The Virgo–Coma cluster distance ratio has been measured using a variety of techniques (for example, the Tully–Fisher relationship, the D_n–Σ relation, and type Ia supernovae). There is excellent agreement among a number of different authors[4,32] resulting in a mean relative Coma–Virgo distance modulus of $3.71 \pm \lesssim 0.10$ mag. Adopting a distance of 17.1 Mpc for the Virgo cluster yields a distance to the Coma cluster of 94 Mpc. The corresponding value of the Hubble constant is $H_0 = 7,200/94 = 77 \pm 16$ km s^{-1} Mpc^{-1}. This value agrees well with values determined for the Virgo cluster velocities given above. Thus a value of the Hubble constant $\sim 80 \pm 17$ km s^{-1} Mpc^{-1} is indicated extending out to a distance of ~ 100 Mpc.

Systematic Errors

Could there by a systematic error in the local calibrators, possibly associated with the zero point of the Cepheid distance scale? This possibility seems very unlikely given that the distances to nearby galaxies have been measured using several different methods completely independent of the Cepheids; for example, using RR Lyrae variables[33,34], using the photoionized disk around SN1987A[35], the expanding photosphere method applied to SN1987A[15], and using the luminosity of the tip of the red giant branch[36]. Although as recently as a decade ago the distances to the local calibrators were in disagreement at a level of a factor of two, newer data obtained from CCDs and near-infrared arrays have led to a convergence of the distances to local galaxies with differences amounting to less than 0.3 mag, or 15% in distance[37].

Are we measuring the true global value of H_0? Given the observational evidence for large-scale velocity flows[9,38], and theoretical models for structure formation (such as cold dark matter), it is possible that measurement of H_0 locally does not yield a value representative of that on a larger global scale[39,40]. However, there are several direct lines of evidence that do not support this hypothesis. At the distance of the Coma cluster such fluctuations are predicted to be less than 5–10%, based both on models and considerations of the observed local bulk motion with respect to the microwave background[39]. At least three distance indicators presently extend well beyond the distance of the Coma cluster and yield internally consistent values of H_0 both locally and out to redshifts $> 14,000$ km s^{-1}. These are the type

258

中心位置的修正幅度，所采用的降落速度和随机速度可能导致室女座星系团的退行速度 [27] 为 1,200 km·s^{-1} 到 1,600 km·s^{-1}。采用修兹劳 [27] 提出的室女星系团的退行速度 1,404 ± 80 km·s^{-1} 以及 17.1 Mpc 的距离，得出哈勃常数 H_0 = 82 ± 17 km·s^{-1}·Mpc^{-1}。估算中的累计误差列在表 1 当中。（然而，我们注意到，如果采用 1,179 ± 17 km·s^{-1} 的速度 [6,31]，对于同样的距离将得出 H_0 = 69 ± 14 km·s^{-1}·Mpc^{-1}。）

避免测定 H_0 过程中室女星系团退行速度不确定性的一种方法是：利用已经准确测定的室女星系团与后发星系团的距离比 [28]。后发星系团位于大约 6 倍于室女星系团的距离上。它的本动速度 ≤ 80 km·s^{-1} [29]，退行速度 [9,30,31] 为 7,200 ± 100 km·s^{-1}。因此，其本动速度对宇宙学速度只贡献了非常小的不确定性。室女–后发星系团的距离比已经使用各种技术（例如，塔利–费希尔关系、D_n–$\sum$关系以及 Ia 型超新星）测量过。由许多作者得到的结果是很一致的 [4,32]，后发–室女星系团之间的相对距离模数的平均值为 3.71 ± ≤ 0.10 mag。如果采用 17.1 Mpc 作为室女星系团的距离，那么到后发星系团的距离为 94 Mpc。相应的哈勃常数的值 H_0 = 7,200/94 = 77 ± 16 km·s^{-1}·Mpc^{-1}。这一结果与前面采用室女星系团的速度测定的结果是很一致的。因此，延伸到约 100 Mpc 的距离之外，哈勃常数大约为 80 ± 17 km·s^{-1}·Mpc^{-1}。

系 统 误 差

局域定标天体里的系统误差会不会与造父距离标尺的零点有关联呢？这种可能性是非常低的，因为测量邻近星系所使用的各种方法完全不依赖于造父变星；例如使用天琴 RR 型变星 [33,34]，使用超新星 SN1987A 周围的光电离盘 [35]，将膨胀光球方法用于 SN1987A [15]，以及使用红巨星支上端的光度 [36]。尽管近十年以来，到局域定标天体的距离相差了一倍。然而，由 CCD 和近红外探测阵列得到的较新的数据表明到近邻星系的距离的误差量已经减小到小于 0.3 mag，或者按距离小于 15% [37]。

我们测量的是 H_0 真实的全局值吗？尽管给出了大尺度速度流的观测证据 [9,38] 以及结构形成的理论模型（例如冷暗物质），局部区域里测量的 H_0 不一定能代表更大的宇宙学尺度上的值 [39,40]。然而，存在几条直接的证据链不支持这样的假设。基于理论模型以及考虑观测到的相对于微波背景的局部团块运动 [39]，在后发星系团的距离处预测这样的扰动小于 5%~10%。目前，至少有三种示距天体很好地扩展到了后发星系团的距离之外，并且对于近邻的以及红移大于 14,000 km·s^{-1} 的地方都得到了自洽的 H_0 值。这些示距天体是 II 型超新星 [15,16]、Ia 型超新星 [6] 以及最亮的星系团成

II supernovae[15,16], type Ia supernovae[6], and brightest cluster members[41,42]. Moreover, the Hubble diagram for the brightest cluster members of Sandage and Hardy[42] is well-defined to velocities $> 40,000$ km s^{-1} and appears to remain linear out to $\sim 100,000$ km s^{-1}. The zero-point calibration of these various methods is still a subject of debate (and the goal of the Key Project) but, for example, it is concluded[41] that the variation of H_0 over the redshift range $0.01 < z < 0.05$ is constrained to be within $\pm 7\%$ r.m.s.

Implications

The HST measurement of the Cepheid distance to M100 enables us to place constraints on the range of plausible values of H_0 and the expansion age. The current limits on H_0 are illustrated in Fig. 3. Also shown in Fig. 3 are the expansion ages for various values of (H_0, Ω_0) corresponding to 14 ± 2 Gyr. These limits are broadly consistent with estimates of the ages of globular clusters from stellar evolution theory[43-45], estimates of the age of the Galactic disk based on white dwarf cooling times[46,47], and radioactive dating of elements[48,49]. All of these age estimates generally span a full range between 10 to 18 Gyr, consistent with the 1σ error limits quoted above. A value of $H_0 = 80 \pm 17$ km s^{-1} Mpc^{-1} is consistent with a low-density $(0.1 < \Omega < 0.3)$ Universe and $T_0 = 12$ Gyr.

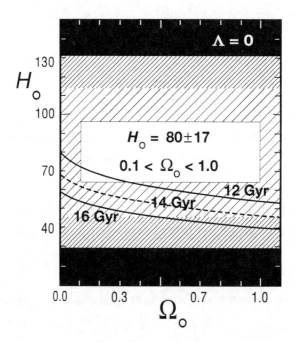

Fig. 3. Current limits on H_0 (in km s^{-1} Mpc^{-1}) and the age and density of the Universe. Expansion ages of 14 ± 2 Gyr, corresponding to globular cluster ages, are illustrated as a function of H_0 and the density parameter Ω for models with the cosmological constant $\Lambda = 0$. The 2σ, 3σ and 4σ confidence intervals on the Hubble constant presented here are shown by the lightly hatched, densely hatched and black horizontal areas, respectively. Broad limits on Ω are illustrated at 0.1 and 1.0.

员 [41,42]。而且，对于桑德奇与哈迪的最亮的星系团成员星系 [42]，其哈勃图可以很好地延伸到速度大于 40,000 km·s^{-1}，一直到 ~100,000 km·s^{-1} 仍能保持线性关系。以上各种方法的零点定标仍然是有争议的（包括关键项目的目标）。然而，举例来说，我们可以得出的结论是 [41]，将红移限定在 $0.01 < z < 0.05$ 的范围内，H_0 的变化范围限定在了均方根 ±7% 的范围内。

启　示

使用哈勃空间望远镜测定的到 M100 的造父距离使得我们可以对哈勃常数 H_0 以及膨胀年龄的合理值的范围进行限制。图 3 中给出了当前对 H_0 的限制。图 3 中也显示了对应于膨胀年龄为 140 亿±20 亿年的各种值（H_0, Ω_0）。这些年龄限制与基于恒星演化理论的球状星团的年龄 [43-45]，基于白矮星冷却时间的银盘的年龄 [46,47]，以及元素的放射性计年 [48,49] 都是一致的。所有这些年龄估计大体上覆盖了 100 亿年到 180 亿年间的范围，在上述的 1σ 误差范围内是一致的。$H_0 = 80 \pm 17$ km·s^{-1}·Mpc^{-1} 的结果同一个低密度（$0.1 < \Omega < 0.3$）以及年龄 $T_0 = 120$ 亿年的宇宙一致。

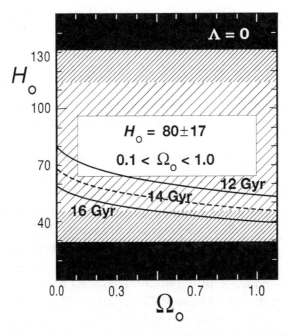

图 3. 当前对 H_0（单位为 km·s^{-1}·Mpc^{-1}）以及宇宙年龄和密度的限制。对具有宇宙学常数 $\Lambda = 0$ 的模型，图中画出了对应于球状星团的年龄，140 亿±20 亿年的膨胀年龄作为 H_0 以及密度参数 Ω 的函数的变化趋势。哈勃常数 2 倍、3 倍、4 倍标准误差的置信区间在图中分别显示为稀疏的阴影线、致密的阴影线以及黑色的水平区域。图中标明了位于 0.1 到 1.0 之间的对于 Ω 宽松的限制。

The Einstein–de Sitter cosmological model (often referred to as the standard model) has a mean mass density equal to the critical density ($\Omega = 1$) and a cosmological constant, $\Lambda = 0$. In the standard model the expansion age is given by $T_0 = 2/3\ H_0^{-1}$. (For models with $\Lambda > 0$ the Universe undergoes an accelerating expansion resulting in ages that are older than $\Lambda = 0$ models[50].) For $H_0 = 80\ \text{km s}^{-1}\ \text{Mpc}^{-1}$ the standard model gives an expansion age of 8 Gyr, well below the other age estimates cited above. This "age conflict" suggests that either the standard cosmological model needs to be revised, or present theories (or observations) bearing on stellar and galactic evolution may need to be re-examined.

Concerns and Future Plans

The weakest point in the present analysis is of course that we have measured a distance to only one galaxy. However, there has recently been reported[51] the discovery of three Cepheid candidates in the Virgo cluster spiral galaxy NGC4571 based on observations taken at the Canada–France–Hawaii Telescope. It is concluded from these data that NGC4571 lies at a distance of 14.9 ± 1.2 Mpc, yielding a Hubble constant $H_0 = 87 \pm 7\ \text{km s}^{-1}\ \text{Mpc}^{-1}$. Clearly, however, the distances to more galaxies are required to define the mean Virgo cluster distance. We do not wish to mislead the reader into believing that the problem of determining H_0 has been solved. It has not. Our analysis has shown that the remaining uncertainty in the value of the Hubble constant is dominated by systematic errors, which are difficult to quantify with a sample of only one galaxy. An accuracy of 10% or better in the value of H_0 will only be reached when we have measured Cepheid distances to a larger sample of galaxies so that the magnitude of these systematic errors can be assessed directly. However, the independent estimate of the distance to NGC4571, the consistency of the type II supernovae and Tully–Fisher distance scales with the direct Cepheid measurement to M100, and the agreement of the distance to the Virgo cluster based on elliptical galaxies[14,25], lead us to conclude that the evidence at this time favours a value of $H_0 \approx 80 \pm 17\ \text{km s}^{-1}\ \text{Mpc}^{-1}$.

The two remaining years of the Key Project will be critical. We have been awarded time and are currently obtaining data to measure Cepheid distances to a total of ~20 calibrating spirals both in the field and in small groups (for example, Leo I and Coma I), and including two additional galaxies in each of the Virgo and Fornax clusters. A direct comparison of the Tully–Fisher, surface brightness fluctuation, planetary nebula luminosity function, and type I and II supernovae (and other) distance scales can then be made. In parallel, independent horizontal-branch distances to M33 and M31 will be measured, and main-sequence fitting undertaken for LMC clusters, allowing further checks on the zero point of the Cepheid P–L relation. These data will provide a direct means of assessing the systematic errors in the current extragalactic distance scale at each stage of its application, and allow a determination of the Hubble constant to unprecedented accuracy. The results of this paper suggest that a Hubble constant, accurate to 10%, and measured through secondary techniques out to scales of hundreds of Mpc, is now a realizable goal.

(**371**, 757-762; 1994)

爱因斯坦 – 德西特宇宙模型（通常称作标准模型）的平均物质密度等于临界密度（$\Omega = 1$），宇宙学常数 $\Lambda = 0$。在标准模型中膨胀年龄由 $T_0 = 2/3\,H_0^{-1}$ 决定。（对于 $\Lambda > 0$ 的模型，宇宙在加速膨胀从而导致了比 $\Lambda = 0$ 的模型更老的年龄[50]。）对于 $H_0 = 80\ \mathrm{km \cdot s^{-1} \cdot Mpc^{-1}}$，标准模型给出的膨胀年龄为 80 亿年，远远小于上述其他年龄的估计值。这一"年龄矛盾"表明要么标准宇宙模型需要修正，要么目前关于恒星和星系演化的理论（或者观测）需要重新检验。

各种考虑及未来的计划

在目前分析当中最为薄弱的一点是我们只测量了到一个星系的距离。然而，最近已经有报告指出基于加拿大 – 法国 – 夏威夷望远镜对室女星系团内的旋涡星系 NGC4571 的观测发现了三个造父变星的候选体[51]。从这些数据得出的结论是 NGC4571 位于距离 14.9 ± 1.2 Mpc 处，得出的哈勃常数 $H_0 = 87 \pm 7\ \mathrm{km \cdot s^{-1} \cdot Mpc^{-1}}$。然而，很明显，确定室女星系团的平均距离需要确定到更多星系的距离。我们并不想误导读者相信 H_0 测定的问题已经解决了。这一问题还没有解决。我们的分析表明哈勃常数的值仍然存在的不确定性主要由系统误差主导，而这些系统误差对于只有一个星系的样本而言很难给出定量结果。只有当测出更大的星系样本的造父距离时，H_0 值的精度会达到 10%，甚至更好，而系统误差的大小也可以直接估计。然而，对于 NGC4571 距离的独立测量，使用 Ⅱ 型超新星以及塔利 – 费希尔关系得到的距离尺度与直接使用造父变星测定的到 M100 的距离的一致。以及与基于椭圆星系测得的到室女星系团距离一致[14,25]，使得我们得出以下结论：目前的证据表明哈勃常数最为可取的结果是 $H_0 \approx 80 \pm 17\ \mathrm{km \cdot s^{-1} \cdot Mpc^{-1}}$。

剩下的两年对关键项目将是至关重要的。我们已经获得了观测时间，并且目前正在获取数据，以便测量总数大约为 20 个定标的旋涡星系的造父距离。这些星系有的位于星场中，有的处在小的星系群（举例来说如狮子座 Ⅰ 星系和后发星系团 Ⅰ），还有另外两个星系分别在室女星系团和天炉星系团。这样一来，塔利 – 费希尔关系、面亮度起伏、行星状星云光度函数以及 Ⅰ 型和 Ⅱ 型超新星（以及其他方法）得出的距离就可以进行直接的比较。与此同时，也将测量到 M33 和 M31 独立的水平支星距离，以及对大麦哲伦云内星团做主序星拟合，使得可以对造父变星的周光关系零点进行进一步检测。这些数据将为当前河外星系距离尺度在每一应用阶段系统误差的估计提供一个直接的方法，并且能够使哈勃常数的测定达到空前的精度。本文的结果表明，目前，哈勃常数的值精度达到 10%，并且通过各种次级技术测量到几百兆秒差距的距离是可实现的目标。

（武振宇 翻译；何香涛 审稿）

Wendy L. Freedman[*], Barry F. Madore[†], Jeremy R. Mould[‡], Robert Hill[*], Laura Ferrarese[§¶], Robert C. Kennicutt Jr[‖], Abhijit Saha[¶], Peter B. Stetson[#], John A. Graham[**], Holland Ford[§¶], John G. Hoessel[††], John Huchra[‡‡], Shaun M. Hughes[§§] & Garth D. Illingworth[‖‖]

[*] Carnegie Observatories, 813 Santa Barbara Street, Pasadena, California 91101, USA

[†] NASA/IPAC Extragalactic Database, Infrared Processing and Analysis Center, Jet Propulsion Laboratory, California Institute of Technology, Pasadena, California 91125, USA

[‡] Mount Stromlo and Siding Spring Observatories, Private Bag, Weston Creek Post Office ACT 2611, Sydney, Australia

[§] Department of Physics and Astronomy, Bloomberg 501, Johns Hopkins University, 3400 North Charles Street, Baltimore, Maryland 21218, USA

[‖] Steward Observatory, University of Arizona, Tucson, Arizona 85721, USA

[¶] Space Telescope Science Institute, Homewood Campus, Baltimore, Maryland 21218, USA

[#] Dominion Astrophysical Observatory, 5071 West Saanich Road, Victoria, British Columbia, V8X 4M6, Canada

[**] Department of Terrestrial Magnetism, Carnegie Institution of Washington, 5241 Broad Branch Road North West, Washington DC 20015, USA

[††] Department of Astronomy, University of Wisconsin, 475 North Charter Street, Madison, Wisconsin 53706, USA

[‡‡] Harvard-Smithsonian, Center for Astrophysics, 60 Garden Street, Cambridge, Massachusetts 02138, USA

[§§] Royal Greenwich Observatory, Madingley Road, Cambridge CB3 0HA, UK

[‖‖] Lick Observatory, University of California, Santa Cruz, California 95064, USA

Received 20 September; accepted 26 September 1994.

References:

1. Fukugita, M., Hogan, C. & Peebles, P. J. E. *Nature* **366**, 309-312 (1993).

2. Kolb, E. W. & Turner, M. S. *The Early Universe* (Addison-Wesley, Redwood City, 1990).

3. Madore, B. F. & Freedman, W. L. *Publ. Astron. Soc. Pac.* **103**, 933-957 (1991).

4. van den Bergh, S. *Publ. Astron. Soc. Pac.* **104**, 861-883 (1992).

5. Jacoby, G. H. *et al. Publ. Astron. Soc. Pac.* **104**, 599-662 (1992).

6. Sandage, A. R. & Tammann, G. A. *Astrophys. J.* **415**, 1-9 (1993).

7. Hubble, E. P. *Astrophys. J.* **62**, 409-443 (1925).

8. Hubble, E. P. *Astrophys. J.* **63**, 236-274 (1926).

9. Aaronson, M. *et al. Astrophys. J.* **302**, 536-563 (1986).

10. Dressler, A. *et al. Astrophys. J.* **313**, L37-L42 (1987).

11. Tully, R. B. & Fisher, J. R. *Astr. Astrophys.* **54**, 661-673 (1977).

12. Pierce, M. J. & Tully, R. B. *Astrophys. J.* **387**, 47-55 (1992).

13. Tonry, J. & Schneider, D. P. *Astr. J.* **96**, 807-815 (1988).

14. Jacoby, J., Ciardullo, R. & Ford, H. C. *Astrophys. J.* **356**, 332-349 (1990).

15. Schmidt, B. P., Kirshner, R. P. & Eastman, R. G. *Astrophys. J.* **395**, 366-386 (1992).

16. Schmidt, B. P., Kirshner, R. P. & Eastman, R. G. *Astrophys. J.* **432**, 42-48 (1994).

17. Branch, D. &. Miller, D. L. *Astrophys. J.* **405**, L5-L8 (1993).

18. Phillips, M. M. *Astrophys. J.* **413**, L105-L108 (1993).

19. Saha, A. *et al. Astrophys. J.* **425**, 14-34 (1994).

20. Freedman, W. L. *et al. Astrophys. J.* **427**, 628-655 (1994).

21. Sandage, A. R. *Astrophys. J.* **402**, 3-14 (1993).

22. de Vaucouleurs, G. *Astrophys. J.* **253**, 520-525 (1982).

23. Pierce, M. J. & Tully, R. B. *Astrophys. J.* **330**, 579-595 (1988).

24. Pierce, M. J. *Astrophys. J.* **430**, 53-62 (1994).

25. Tonry, J. L., Ajhar, E. A. & Luppino, G. A. *Astr. J.* **100**, 1416-1423 (1990).

26. Sandage, A. R. & Bedke, J. *Atlas of Galaxies Useful For Measuring the Cosmological Distance Scale*, SP-496 (NASA, 1988).

27. Huchra, J. in *Extragalactic Distance Scale* (eds van den Bergh, S. & Pritchet, C. J.) *Publ. Astron. Soc. Pac. Conf. Series* **4**, 257-280 (1988).

28. Tammann, G. A. & Sandage, A. R. *Astrophys. J.* **294**, 81-95 (1985).

29. Han, M. & Mould, J. R. *Astrophys. J.* **396**, 453-459 (1992).

30. Fukugita, M. *et al. Astrophys. J.* **376**, 8-22 (1991).

31. Jergen, H. & Tammann, G. A. *Astr. Astrophys.* **276**, 1-8 (1993).

264

32. de Vaucouleurs, G. *Astrophys. J.* **415**, 10-32 (1993).

33. Pritchet, C. J. & van den Bergh, S. *Astrophys. J.* **316**, 517-529 (1987).

34. Saha, A., Freedman, W. L., Hoessel, J. G. & Mossman, A. E. *Astr. J.* **104**, 1072-1085 (1992).

35. Panagia, N. *et al. Astrophys. J.* **380**, L23-L26 (1991).

36. Lee, M. G., Freedman, W. L. & Madore, B. F. *Astrophys. J.* **417**, 553-559 (1993).

37. Freedman, W. L. & Madore, B. F. in *New Perspectives on Stellar Pulsation and Pulsating Stars* (eds Nemec, J. & Matthews, J.) (Cambridge Univ. Press, 1993).

38. Lynden-Bell, D., Lahav, O. & Burstein, D. *Mon. Not. R. astr. Soc.* **241**, 325-345 (1989).

39. Turner, E., Cen, R. & Ostriker, J. P. *Astr. J.* **103**, 1427-1437 (1992).

40. Bartlett, J. G. *et al. Science* (submitted).

41. Lauer, T. & Postman, M. *Astrophys. J.* **400**, L47-L50 (1992).

42. Sandage, A. R. & Hardy, E. *Astrophys. J.* **183**, 743-757 (1973).

43. Renzini, A. in *Observational Tests of Inflation* (eds Shanks, T. *et al.*), (Kluwer, Boston, 1991).

44. Sandage, A. R. *Astr. J.* **106**, 719-725 (1993).

45. Shi, X., Schramm, D. N., Dearborn, D. S. P. & Truran, J. W. *Science* (submitted).

46. Winget, D. E. *et al. Astrophys. J.* **315**, L77-L81 (1987).

47. Pitts, E. & Taylor, R. J. *Mon. Not. R. astr. Soc.* **255**, 557-560 (1992).

48. Fowler, W. in *14th Texas Symp. on Relativistic Astrophysics* (ed Fenyores, E. J.) 68-78 (N.Y. Acad. Sci., New York, 1989).

49. Clayton, D. D. in *14th Texas Symp. on Relativistic Astrophysics* (ed. Fenyores, E. J.). 79- 89 (N.Y. Acad. Sci., New York, 1989).

50. Carroll, S. M., Press, W. J. & Turner, E. L. *A. Rev. Astr. Astrophys.* **30**, 499-542 (1992).

51. Pierce, M. J. *et al. Nature* **371**, 385-389 (1994).

Acknowledgements. We sincerely thank the many scientists, engineers and astronauts who contributed to HST, making these observations possible. This research was supported by NASA, through a grant from the Space Telescope Science Institute (which is operated by the Association of Universities for Research in Astronomy), and by the NSF. This work has benefitted from the use of the NASA/IPAC Extragalactic Database (NED).

Gravitationally Redshifted Emission Implying an Accretion Disk and Massive Black Hole in the Active Galaxy MCG–6–30–15

Y. Tanaka *et al.*

Editor's Note

Although today the presence of supermassive black holes at the centres of most galaxies is accepted almost universally by astronomers, in 1995 the situation was not yet resolved. One of the key predictions from the combination of general relativity and black-hole physics was that spectral emission lines from gas near the event horizon should show a characteristic width, and a shape redshifted by the strong gravity there. Here Yasuo Tanaka and colleagues report the first such gravitationally redshifted line (from highly ionized iron) in the galaxy MCG–6–30–15. Several other examples have since been found. Combined with the orbits of stars near the centre of the Milky Way, they make the case for supermassive black holes at galactic centres essentially airtight.

Active galactic nuclei and quasars are probably powered by the accretion of gas onto a supermassive black hole at the centre of the host galaxy[1], but direct confirmation of the presence of a black hole is hard to obtain. As the gas nears the event horizon, its velocity should approach the speed of light; the resulting relativistic effects, and a gravitational redshift arising from the proximity to the black hole, should be observable, allowing us to test specific predictions of the models with the observations. Here we report the detection of these relativistic effects in an X-ray emission line (the $K\alpha$ line) from ionized iron in the galaxy MCG–6–30–15. The line is extremely broad, corresponding to a velocity of $\sim 100,000$ km s^{-1} and asymmetric, with most of the line flux being redshifted. These features indicate that the line most probably arises in a region between three and ten Schwarzschild radii from the centre, so that we are observing the innermost region of the accretion disk.

RECENT optical and radio observations[2-4] of gas motions show a large, central concentration of mass (probably a black hole) in a number of galaxies. But this gas lies beyond $\sim 30,000$ Schwarzschild radii so the characteristic signatures of the black hole itself—relativistic effects due to its strong gravitational field—were not observed. X-rays are produced much closer to the black hole and provide better access to these effects, particularly if a spectral line is emitted for which energy shifts can be clearly measured.

An iron $K\alpha$ emission line close to 6.4 keV and an additional, hard-continuum component are imprinted on the primary, power-law, X-ray continuum of most Seyfert 1 active galactic nuclei (AGN)[5,6]. They arise via fluorescence and back-scattering ("reflection") from optically

活动星系 MCG-6-30-15 的引力红移辐射暗示吸积盘和大质量黑洞的存在

田中靖郎等

编者按

虽然现在天文学家们普遍接受了"在绝大部分星系的中心都存在超大质量黑洞"这一观点，但是在 1995 年的时候，这个问题还没有得到解决。在结合了广义相对论和黑洞物理的理论中，一个关键的预言是：来自黑洞视界附近的气体的光谱发射线，应该有一个特征展宽，以及经过强引力红移后的形状。在这篇文章中，田中靖郎和他的合作者们报道了在 MCG-6-30-15 星系中探测到的第一条引力红移谱线（来自高电离铁的）。至今为止，已经有几个其他这样的例子存在。结合银河系中心附近的恒星运动轨道，这些观测基本证实了星系中心存在超大质量黑洞的观点。

活动星系核与类星体很可能由位于寄主星系中心的超大质量黑洞的气体吸积提供能量[1]，但是直接确认黑洞的存在却非常困难。随着气体靠近视界，其速度也会接近光速；这时就会观察到由于靠近黑洞而产生的相对论效应和引力红移，为我们观测检验模型的具体预言创造了条件。这里我们将报道对星系 MCG-6-30-15 中铁离子 X 射线的发射线（Kα 线）的相对论效应的探测。该发射线非常宽，对应速度约为 100,000 km·s^{-1}，并且具有非对称结构，其中的大部分谱线流量发生红移。这些特征表明，这些谱线很可能来自距离黑洞中心 3~10 倍施瓦西半径的区域；因此，事实上我们观察的是吸积盘最内区。

最近对气体运动的光学和射电观测[2-4]表明，在许多星系中存在大质量的中心聚集（可能是黑洞）。但是这些气体位于大约 30,000 倍施瓦西半径之外，因此并没有观测到黑洞自身独有的特征（由于强引力场产生的相对论效应）。X 射线产生于更靠近黑洞的地方，因此也提供了更好的途径研究这些效应，尤其是当有一条谱线发射出来，并且可以清楚地测出能量改变的情况下。

大多数赛弗特 1 型（Seyfert 1）活动星系核（AGN）的主要 X 射线幂律连续谱上[5,6]，会叠加一条约 6.4 keV 的铁 Kα 发射线，以及一个额外的硬连续谱成分。这些谱线产

thick material which covers about half the sky seen by the primary continuum source[7,8]. The data do not allow the geometry of the material to be determined unambiguously, but are consistent with an irradiated accretion disk[9-12]. The emission lines from such a system are expected to be very broad, as most fluorescence occurs in a regime where Doppler and gravitational shifts are large[9,12,13]. Clear evidence for significant broadening—at a level compatible with accretion-disk models—has now come from new data (from the ASCA satellite) on several AGN, including MCG–6–30–15[14,15]. The emission lines should also have a characteristic profile. The strongest lines are produced in a face-on geometry[10], where gravitational and transverse Doppler effects dominate, producing a profile skewed to the red. For edge-on disks the line is weaker, and the blue side of the profile (the blue "wing") more prominent[9,16].

MCG–6–30–15 was observed by ASCA[17] on 23 July 1994 for approximately four days, with both the CCD (charge-coupled device) detectors (SIS) and gas-scintillation proportional counters (GIS) in operation. Here we present only the SIS data, which have the better energy resolution; the GIS and SIS data are entirely consistent. A complicating factor in the analysis of the integrated SIS spectra is the presence of features from highly ionized gas in the line of sight, the so-called warm absorber[18], which is now well established in this source[14,19,20]. Initially, then, we fitted the spectra in the range 0.4–10 keV with a model consisting of a power law emitter and photoionized absorber. This shows that the absorption only affects the spectrum below 2 keV, and therefore in analysing the line data we have restricted the energy range to 2.5–10 keV. Previous data from the Ginga satellite also showed[6] strong evidence for a reflection continuum component accompanying the iron-emission line. We have assumed a face-on slab subtending 2π sr at the X-ray source and calculated the reflection spectrum using the model of Lightman and White[8]. As expected from previous observations[6], the spectrum and residuals for the 3–10 keV continuum fit show a well defined excess in the residuals at ~6 keV (Fig. 1), most probably an iron Kα line, but much broader than the instrument resolution for a single narrow line. We emphasize that the statistical quality of the data is sufficient to allow the precise determination of the continuum above and below the line; the derived line profile depends little on the uncertainty in the continuum shape, or the normalization of the reflection continuum.

It is clear from the residuals (Fig. 1) that the line profile is not gaussian. It consists of a relatively narrow core around 6.4 keV with a very broad wing extending to the red. We have modelled the line with two gaussians, one each for the core and wing (see Fig. 1 legend for details). This fit allows us to determine the best-fit model for the continuum. The deviations of the data from this continuum model represent the line profile, which is shown in Fig. 2. The data have been rebinned to approximately the instrumental resolution and are shown in incident (photon) flux. The profile is clearly extremely broad and asymmetric. Most of the flux is strongly redshifted from the rest energy of even the lowest-energy Kα lines (6.4 keV), by as much as $\Delta E / E \approx 0.3$. The full width at zero intensity corresponds to a velocity of the order of $100,000 \text{ km s}^{-1}$.

生于光学厚物质中的荧光辐射和反向散射（"反射"），这些光学厚物质大约覆盖了主要连续谱源的半个天区[7,8]。虽然从这些谱线数据中尚不能清楚地断定这些物质的几何形态，但是它们和一个被辐射的吸积盘相符合[9-12]。由于大多数荧光辐射发生在多普勒效应和引力偏移较强的物理条件下，因此这类系统的发射线将会很宽[9,12,13]。包括 MCG-6-30-15 在内的若干活动星系核的最新数据（数据源自宇宙学和天体物理学高新卫星，ASCA）也提供了显著展宽的明确证据（在与吸积盘模型相符合的水平上）[14,15]。这些发射线也应该具有特征轮廓。强发射线产生自正向（朝向我们观测者）几何[10]，在正向几何中，由引力效应和横向多普勒效应主导，使得谱线轮廓偏向红端。而侧向吸积盘的发射线较弱，谱线轮廓的蓝端（蓝翼）更加突出[9,16]。

从 1994 年 7 月 23 日起，ASCA 卫星对 MCG-6-30-15 进行了大约四天的观测[17]，观测中采用了电荷耦合器件（CCD）探测器（SIS）与气体闪烁正比计数器（GIS）。在这里我们只给出 SIS 数据，该数据具有较好的能量分辨率；GIS 和 SIS 数据是完全一致的。在积分 SIS 光谱的分析过程中，一个使问题复杂化的因素是存在位于视线方向的高度电离气体的特征，即温吸收体[18]，如今这种温吸收体在这个源中已经得到了清楚的认识[14,19,20]。最初，我们用包含幂律发射体和光电离吸收体的模型对 0.4~10 keV 范围内的光谱进行拟合。这种拟合表明这种吸收只影响能量小于 2 keV 的光谱，因此在分析谱线数据的时候，我们将能量限制在 2.5~10 keV 范围内。银河号卫星以前提供的数据也给出很强的证据[6]，显示铁的发射线总伴有一个反射连续谱成分。我们假设在 X 射线源处正向平面所对应着的立体角为 2π sr，然后采用莱特曼和怀特的模型[8]计算反射光谱。正如以前观测所预期的那样[6]，对 3~10 keV 连续谱拟合给出的光谱和残差都表明，在 6 keV 附近的拟合残差明显过大（如图 1），这说明发射线很可能是铁 $K\alpha$ 线，但是它比仪器对单条窄线能分辨的宽度要宽。这里我们强调一下，这些数据的统计质量足以精确地确定该谱线能量附近的连续谱。推导出来的谱线轮廓几乎不依赖于连续谱形状的不确定度或反射连续谱的归一化。

从残差（图 1）中可以清楚地看出该谱线的轮廓并不是高斯形的；该轮廓包括一个位于 6.4 keV 附近的相对窄的核和一个延伸到红端的非常宽的翼。我们采用两个高斯函数来拟合该谱线，分别代表核成分和翼成分（详见图 1 的图例）。这种拟合让我们得以确定连续谱的最佳拟合模型。观测数据与连续谱模型的差异代表了该谱线的轮廓，如图 2 所示，其中的数据已经调整到接近仪器的分辨率，并表示为入射（光子）流量的形式。从图中可知谱线轮廓非常宽且不对称；大部分流量即使相对于静止能量最低的 $K\alpha$ 线（6.4 keV）也发生了很大的红移，偏移量高达 $\Delta E / E \approx 0.3$；零强度全宽对应的速度量级为 100,000 km·s^{-1}。

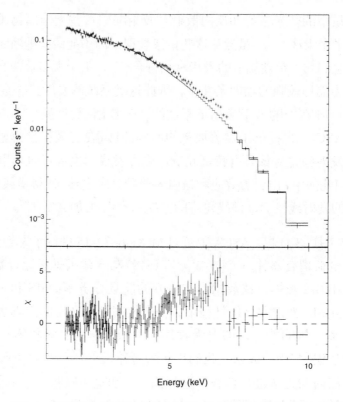

Fig. 1. X-ray spectrum of MCG–6–30–15, as observed by the ASCA satellite using the SIS detectors.
Top panel, observed spectrum (crosses) and fitted "power-law plus continuum reflection" model (stepped
line; the model has the same energy bins as the data). This fit excludes data between 5 and 7 keV. Bottom
panel, data-minus-model residuals. The emission line is clearly visible with an asymmetry to the red.
Parametrizing the observed emission line with a single gaussian improved the fit by $\Delta\chi^2 = 82$ with a
centroid energy of $E_K = 5.92^{+0.16}_{-0.15}$ keV, width $\sigma = 0.74^{+0.24}_{-0.18}$ keV and equivalent width EW $= 330^{+180}_{-120}$ eV. The
power-law continuum has photon index $\Gamma = 2.05 \pm 0.07$. Adding an additional gaussian improved the fit
further, with $\Delta\chi^2 = 39$, to $\chi^2 = 656.0$ (675 degrees of freedom). The double-gaussian parametrization consists of
a relatively narrow ($\sigma = 0.18$ keV) component at 6.4 keV and a broader ($\sigma = 0.64$ keV) wing at ~5.5 keV. The
red wing carries more flux than the core, with equivalent widths of 200 eV and 120 eV respectively.

From the large redshift, we can immediately reject the hypothesis that Compton scattering
produces the observed line profile[21]. Such scattering requires that energy is transferred
from the photons to the electrons and thus that $kT < 1.6$ keV in any scattering plasma
($4kT <$ photon energy). The energy shift per scattering is small (~80 eV) and a high
Thomson optical depth (~5) is required for the scattering plasma to produce the total shift
in the line centroid. Such a plasma would also Compton scatter the continuum photons to
lower energies, creating a break at around 20 keV. No such break is observed[22].

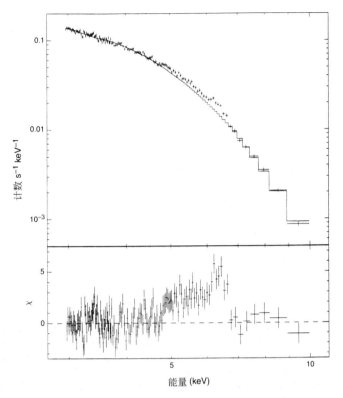

图 1. MCG-6-30-15 的 X 射线光谱，由 ASCA 卫星使用 SIS 探测器观测得到。上部分的图给出了实际观测能量谱（十字标记）和"幂律加反射连续谱"的模型拟合曲线（阶梯线；模型和数据有着相同的能量间隔）。该拟合不包括 5~7 keV 之间的数据。下部分的图是将上部分中的观测数据与相应拟合值相减后得到的残差，显然发射线不对称且向红端发生偏移。观测到的发射线可以通过用单个高斯函数对观测到的发射线进行参数化来改进谱线的拟合，所使用的拟合参数为：$\Delta\chi^2 = 82$，平均能量 $E_{\rm K} = 5.92^{+0.16}_{-0.15}$ keV，宽度 $\sigma = 0.74^{+0.24}_{-0.18}$ keV，等值宽度 EW $= 330^{+180}_{-120}$ eV。幂律连续谱的光子谱指数 $\Gamma = 2.05 \pm 0.07$。引入另一高斯函数可以进一步改进以上拟合，相应参数为：$\Delta\chi^2 = 39$，$\chi^2 = 656.0$（675 个自由度）。双高斯参量拟合包括位于 6.4 keV 的较窄成分（$\sigma = 0.18$ keV）和位于约 5.5 keV 附近较宽的翼成分（$\sigma = 0.64$ keV）。其中红翼部分比中心部分的流量更大，二者等值宽度分别为 200 eV 和 120 eV。

　　从这个较高的红移，我们可以立刻排除以下假说：观测到的谱线轮廓是由康普顿散射产生的[21]。这种散射需要能量从光子传递给电子，因此对于任何发生散射的等离子体（$4kT <$ 光子能量），其 $kT < 1.6$ keV。每次散射的能量偏移是非常小的（~80 eV），且为了实现谱线中心的整体偏移需要较高的汤姆孙光深（~5）。这种等离子体同时也会通过康普顿散射使连续谱的光子能量变低，从而在 20 keV 附近的位置发生截断。然而我们并没有观察到这种截断[22]。

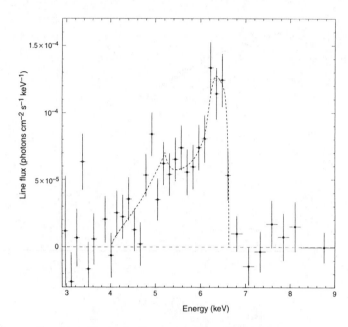

Fig. 2. The line profile of iron Kα in the X-ray emission from MCG–6–30–15. The data have been rebinned to approximately the instrumental resolution. The emission line is very broad, with full width at zero intensity of ∼100,000 km s^{-1}. There is a marked asymmetry at energies lower than the rest-energy of the emission line (6.35 keV at the source redshift of 0.008). The most plausible mechanisms which can produce this extensive red wing are transverse Doppler and gravitational shifts close to a central black hole. The dotted line shows the best-fit line profile from the model of Fabian et al.[9], an externally-illuminated accretion disk orbiting a Schwarzschild black hole.

The Doppler width of the line implies velocities of the order of $0.3c$ (where c is the velocity of light), indicating highly relativistic motions. The line flux drops sharply above the rest energy of 6.4 keV, which excludes any velocity distribution where substantial amounts of the material is moving towards us. This would cause enhancement of the blue wing, which is not observed. Symmetrical outflows and spherical or quasi-spherical distributions of orbiting clouds therefore cannot account for the observed profile. The observation of broad redshifted lines in several objects[15], but no strong blue-shifted lines, argues against any asymmetrical-outflow hypothesis in which the flow is directed away from us (some objects should then have the flow directed towards us). It is therefore most probable that the material is moving primarily transversely at a large fraction of the speed of light with the width of the line being produced by transverse Doppler and gravitational redshifts close to a central black hole. The profile seems to be remarkably similar to that predicted for a face-on accretion disk[9,12].

To determine whether our line profile can be accounted for by a black hole and accretion disk, we initially employed a model which assumes a disk orbiting a Schwarzschild black hole[9]. The free parameters in this model are the inclination of the disk, i, its inner and outer radii, R_i and R_o and the index of the emissivity function, α; the line emissivity is assumed to vary as $R^{-\alpha}$. Initially, we assumed a rest energy of 6.4 keV appropriate to species less ionized than Fe XVI (more highly ionized species are discussed later). The model gives $\chi^2 = 655.7$ (676 degrees of freedom), very similar to the double-gaussian model, but with

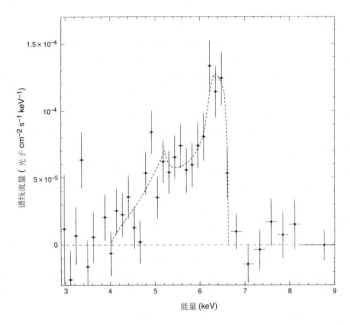

图 2. MCG−6−30−15 星系 X 射线发射中铁 Kα 发射线的谱线轮廓。图中的数据已经重新调整到接近仪器分辨率。图中发射线非常的宽,零强度全宽大约为 100,000 km·s^{-1}。在低于发射线静止能量(6.35 keV,对应源红移量为 0.008)的范围内,谱线轮廓具有明显的非对称性。产生这个高强度红翼轮廓的最合理机制是横向多普勒效应和中心黑洞附近的引力偏移。图中的点线给出了谱线轮廓的最佳拟合,该拟合结果由法比安等人 [9] 的模型(一个围着施瓦西黑洞运行且外部被照亮的吸积盘)得到。

由谱线的多普勒宽度可推出对应速度大约在 $0.3c$ 的量级(其中 c 为光速),这意味为高度相对论性运动。当能量超过 6.4 keV 的静止能量时,谱线流量迅速下降,从而排除了大量物质朝向我们运动引起的任何速度分布。因为朝向我们的运动将引起谱线的蓝翼加强,然而迄今尚未观察到这一现象。因此,对称外向流以及做轨道运动的气体云的球状或准球状分布不可能解释观察到的轮廓。在一些天体中都观测到了宽的红移谱线而没有强的蓝移谱线 [15],这与任何非对称外向流假说相矛盾,这些假说认为流向是远离我们的(那么一些天体就应该有朝向我们的喷流)。因此,最可能的情况是,这些物质主要以接近光速的速度横向运动,谱线宽度由多普勒效应和中心黑洞附近的引力红移造成。该轮廓看起来和正向吸积盘的预言非常相似 [9,12]。

为了确定我们的谱线轮廓是否可以用黑洞和吸积盘来解释,我们最初采用以下模型:假设存在一个环绕施瓦西黑洞的吸积盘 [9]。该模型中的自由参量是:吸积盘的倾角,用 i 表示;其内外半径,分别用 R_i 和 R_o 表示;发射率函数的指数,用 α 表示;假定谱线发射率按 R^α 规律变化。最初,我们将静止能量设定为 6.4 keV,该值适合电离程度比 Fe XVI 弱的核素(电离程度更高的核素在后面进行讨论)。模型给出 $\chi^2 = 655.7$(676 个自由度),与双高斯模型非常相似,但是少了一个自由参数。最

one fewer free parameter. The best-fitting value is $\alpha = 0.7$, but is not well determined. For a centrally illuminated disk, we expect a rather flat emissivity profile in the central regions, $\alpha \approx 1$ increasing to $\alpha \approx 3$ in the outer disk[10]. We have fixed α at these two values, which both fall within the 68% confidence region in the fit, to determine the other parameters. For $\alpha = 1$, we find $i = 30.2°^{+1.5}_{-2.7}$, $R_i = 3.4^{+3.0}_{-0.4} R_s$ and $R_o = 7.4^{+2.5}_{-0.8} R_s$, with equivalent width EW = 390^{+90}_{-130} eV. For $\alpha = 3$, the parameters are $i = 29.7°^{+2.9}_{-3.9}$, $R_i = 3.7^{+1.8}_{-0.7} R_s$, $R_o = 10.0^{+12.5}_{-3.2} R_s$ and EW = 380^{+100}_{-110} eV, very similar to those for $\alpha = 1$. Note that the minimum value for R_i in the Schwarzschild geometry is $3R_s$, so the lower error bar on this value is not statistical. In the Kerr metric of a spinning black hole the last stable orbit is closer to the central hole. We have also tested this condition[19], for which the parameters, $i = 26.8°^{+2.1}_{-1.0}$, $R_i = 4.7^{+0.83}_{-0.9} R_s$, $R_o = 16.7^{+\infty}_{-7.6} R_s$, $\alpha = 4.5^{+3.9}_{-3.7}$ and EW = 300^{+70}_{-50} eV, and χ^2 are very similar to those of the Schwarzschild black hole. The uncertainties in the parameter values depend on the assumptions inherent in the models. For example, the value of R_o does not necessarily represent the true outer radius of the accretion disk, but is simply that radius beyond which no significant iron Kα line emission occurs. The rather low value derived in our fits implies that the emission line is produced very near to the central black hole and that the primary X-rays originate close to the surface of the inner disk.

An iron-line equivalent width of 250–400 eV is a factor of ~2–3 higher than the predictions of simple disk reflection models[10,11]. There are several possible explanations for this. First, a modest increase of a factor ~1.5–3 in the iron abundance relative to the lighter elements would produce a line flux compatible with that reported here[10,23], the primary uncertainty being the assumed cosmic abundances. Another possibility is that there is another source of line emission. For example, there may be a contribution to the emission line from a molecular torus[24-26] surrounding the nucleus at a distance of ~1 pc. Because of this large distance we expect the Kα line to be narrow compared to the instrumental resolution and centred at 6.4 keV in the rest frame. The addition of a narrow (dispersion $\sigma = 10$ eV) gaussian at 6.4 keV to our "power-law plus reflection continuum" model gives an equivalent width of only ~45 eV. This is a conservative upper limit to the torus contribution, because it allows no emission at 6.4 keV from the disk line. A large contribution from a torus is therefore unlikely, and in any case does not change our conclusions regarding the remainder of the line. A final possibility is that the material is highly ionized[27,28]. An increase in the effective fluorescence yield and reduced photoelectric opacity between 6 and 7 keV would then result in a higher line flux, provided that a large proportion of the iron is in the helium-like and hydrogen-like states. Such a scheme is compatible with our data; the increase in rest energy to 6.7–6.9 keV can be compensated for by a smaller inclination angle and/or concentrating the emission closer to the black hole, thereby increasing the gravitational shift.

Given the strong evidence for the black-hole/accretion-disk model in MCG–6–30–15, we infer that the broad lines observed in other AGN (even though less well-resolved) are produced by the same process. Modelling these profiles can reveal extraordinary detail about the central regions of the AGN, as illustrated by the disk line fit above. We already have indications that the X-rays are produced very close to the surface of the disk, and close to the black hole. Observations of variability of the emission line, and its time-lag

佳拟合值是 $\alpha = 0.7$，但是还没有很好地确定。对于一个中心受照的吸积盘，在中心区域将存在相当平坦的发射轮廓，并从内部的 $\alpha \approx 1$ 逐渐增加到外部的 $\alpha \approx 3$ [10]。下面我们采用 $\alpha \approx 1$ 与 $\alpha \approx 3$ 两个取值，这两个值的拟合都落在 68% 的置信区域内。对于 $\alpha = 1$ 的情况，我们发现 $i = 30.2°^{+1.5}_{-2.7}$，$R_i = 3.4^{+3.0}_{-0.4} R_s$ 以及 $R_o = 7.4^{+2.5}_{-0.8} R_s$，等值宽度为 EW $= 390^{+90}_{-130}$ eV。对于 $\alpha = 3$ 的情况，相应参数分别为 $i = 29.7°^{+2.9}_{-3.9}$，$R_i = 3.7^{+1.8}_{-0.7} R_s$ 以及 $R_o = 10.0^{+12.5}_{-3.2} R_s$，等值宽度为 EW $= 380^{+100}_{-110}$ eV，与 $\alpha = 1$ 情况的取值相近。注意到在施瓦西几何中，R_i 的最小值为 $3R_s$，因此该值的下误差棒不具有统计性。在一个自旋转黑洞的克尔度规中，最内的稳定轨道更为靠近黑洞中心。我们也检验了这一情况 [19]，其参数为 $i = 26.8°^{+2.1}_{-1.0}$、$R_i = 4.7^{+0.83}_{-0.9} R_s$、$R_o = 16.7^{+\infty}_{-7.6} R_s$、$\alpha = 4.5^{+3.9}_{-3.7}$、EW $= 300^{+70}_{-50}$ eV 以及 χ^2，与施瓦西黑洞的相应参数非常相近。这些参数的不确定度依赖于模型的内在假设。比如，R_o 的值不一定代表实际的吸积盘外半径，而只是简单地表示该半径以外就没有明显的铁 $K\alpha$ 谱线发射。我们通过拟合得到的半径参数值非常的小，这意味着发射线产生于非常靠近黑洞中心的地方，主要的 X 射线来自吸积盘内区的表面附近。

250~400 eV 的铁线等值宽度是简单吸积盘反射模型预言的 2~3 倍 [10,11]。对于这一现象存在以下几种可能的解释。一种可能是，如果铁元素的丰度是更轻元素丰度的 1.5~3 倍，其谱线流量就会与文中的值相符 [10,23]；其主要的不确定度来自假定的宇宙丰度。另一种可能是，存在另一个线发射源，比如约 1 pc 的距离处围绕核心运动的分子环 [24-26] 可能也会贡献部分发射线。由于该距离较大，$K\alpha$ 线相对仪器分辨率应该较窄，且在静止参考架中的中心能量位于 6.4 keV。在"幂律加反射连续谱"模型中添加一个位于 6.4 keV 的窄高斯分布（弥散度 $\sigma = 10$ eV），则可以给出仅约 45 eV 的等值宽度。这是分子环贡献的保守上限，因为该模型允许吸积盘可以不发射 6.4 keV 的谱线。因此，环面不可能有非常大的贡献，并且任何情况都不可能改变我们关于剩余谱线的结论。最后一种可能性是，物质高度电离 [27,28]。如果一大部分铁处于类氦和类氢状态，在 6~7 keV 之间有效荧光产额的增加和光电不透明度的减小将造成更高的谱线流量。这种机制与我们的数据也是一致的；其中静止能量增加到 6.7~6.9 keV 可以通过以下方式得到补偿，即减小倾角和（或）将辐射集中到更靠近黑洞的位置，从而增加了引力偏移。

鉴于 MCG–6–30–15 中存在黑洞 / 吸积盘模型的有力证据，我们推断在其他活动星系核（即使分辨率稍差）所观察到的宽线也是由于相同的过程产生的。对这些轮廓的模型拟合可以进一步揭示活动星系核中心区域的更多细节，如上面提到的盘谱线拟合。通过模型拟合，我们已经获得以下迹象，即 X 射线在非常靠近吸积盘表面

with respect to variations in the intensity of the continuum emission may allow us to
measure the black hole mass and spin[9,29].

(**375**, 659-661; 1995)

Y. Tanaka[*†], K. Nandra[‡], A. C. Fabian[‡], H. Inoue[*], C. Otani[*], T. Dotanl[*], K. Hayashida[§], K. Iwasawa[‖], T. Kll[*], H. Kunieda[‖], F. Maklno[*] & M. Matsuoka[¶]

[*] Institute of Space and Astronautical Science, 3-1-1 Yoshinodai, Sagamihara, Kanagawa 229, Japan
[†] Max-Planck-lnstitut fur Extraterrestrische Physik, D-85740 Garching, Germany
[‡] Institute of Astronomy, Madingley Road, Cambridge CB3 OHA, UK
[§] Osaka University, Machikaneyama-cho 1-1, Osaka, Japan
[‖] Department of Astrophysics, Nagoya University, Chikusa-ku, Nagoya 464-01, Japan
[¶] RIKEN, Institute of Physical and Chemical Research, Hirosawa, Wako, Saitama 351-01, Japan

Received 2 March; accepted 23 May 1995.

References:

1. Rees, M. J. A. *Rev. Astr. Astrophys.* **22**, 471-506 (1984).

2. Ford, H. C. *et al. Astrophys. J.* L27-L30 (1994).

3. Harmes, R. J. *et al. Astrophys. J.* L35-L38 (1994).

4. Miyoshi, M. *et al. Nature* **373**, 127-129 (1995).

5. Pounds, K. A., Nandra, K., Stewart, G. C., George, I. M. & Fabian, A. C. *Nature* **344**, 132- 133 (1990).

6. Nandra, K. & Pounds, K. A. *Mon. Not. R. astr. Soc.* **268**, 405-429 (1994).

7. Guilbert, P. W. & Rees, M. J. *Mon. Not. R. astr. Soc.* **233**, 475-484 (1994).

8. Lightman, A. P. & White, T. R. *Astrophys. J.* **335**, 57-66 (1988).

9. Fabian, A. C., Rees, M. J., Stella, L. & White, N. E. *Mon. Not. R. astr. Soc.* **238**, 729-736 (1989).

10. George, I. M. & Fabian, A. C. *Mon. Not. R. astr. Soc.* **249**, 352-367 (1991).

11. Matt, G., Perola, G. C. & Piro, L. *Astr. Astrophys.* **245**, 75 (1991).

12. Matt, G., Perola, G. C., Piro, L. & Stella, L. *Astr. Astrophys.* **257**, 63 (1992).

13. Laor, A. *Astrophys. J.* **376**, 90-94 (1991).

14. Fabian, A. C. *et al. Publ. Astron. Soc. Jpn* **46**, L59-L64 (1994).

15. Mushotzky, R. F. *et al. Mon. Not. R. astr. Soc.* **272**, L9-L12 (1995).

16. Chen, K., Halpern, J. P. & Filippenko, A. V. *Astrophys. J.* **339**, 742-751 (1989).

17. Tanaka, Y., Inoue, H. & Holt, S. S. *Publ. Astron. Soc. Jpn* **46**, L37-L41 (1994).

18. Halpern, J. P. *Astrophys. J.* **281**, 90-94 (1984).

19. Nandra, K., Pounds, K. A. & Stewart, G. C. *Mon. Not. R. astr. Soc.* **242**, 660-668 (1990).

20. Nandra, K. & Pounds, K. A. *Nature* **359**, 215-216 (1992).

21. Czerny, B., Zbyszewska, M. & Raine, D. J. in *Iron Line Diagnostics in X-ray Sources* (ed. Treves, A.) 226-229 (Springer, Berlin, 1991).

22. Zdziarski, A. A., Johnson, W. N., Done, C., Smith, D. & McNaron-Brown, K. *Astrophys. J.* **438**, L63-L66 (1995).

23. Reynolds, C. S., Fabian, A. C. & Inoue, H. *Mon. Not. R. astr. Soc.* (submitted).

24. Krolik, J. H. & Kallman, T. R. *Astrophys. J.* **320**, L5-L8 (1987).

25. Ghisellini, G., Haardt, F. & Matt, G. *Mon. Not. R. astr. Soc.* **267**, 743-754 (1994).

26. Krolik, J. H., Madau, P. & Zycki, P. *Astrophys. J.* **420**, L57-L61 (1994).

27. Ross, R. R. & Fabian, A. C. *Mon. Not. R. astr. Soc.* **261**, 74-82 (1993).

28. Matt, G., Fabian, A. C. & Ross, R. *Mon. Not. R. astr. Soc.* **262**, 179-186 (1993).

29. Stella, L. *Nature* **344**, 747-749 (1990).

Acknowledgements. We thank A. Laor for use of his disk fine model. Y.T., K.N. and A.C.F. were supported by an Alexander von Humboldt Research Award, PPARC, and the Royal Society and British Council, respectively.

并且非常接近黑洞的地方产生。对发射线的变化，以及对其相对连续谱发射强度变化的时滞的观测，让我们有可能测量黑洞的质量和自转 [9,29]。

（金世超 翻译；吴学兵 审稿）

A Jupiter-mass Companion to a Solar-type Star

M. Mayor and D. Queloz

Editor's Note

Although "extrasolar" planets had several years earlier been discovered around pulsars, the exotic environment precluded any possibility of these worlds hosting "life as we know it". Michel Mayor and Didier Queloz, along with four other groups, had been searching for the signature of planets orbiting solar-like stars for over a decade. Numerous claims had been reported, all subsequently shown to be wrong. But here Mayor and Queloz report the reliable signature of a Jupiter-mass planet in the spectra of the star 51 Pegasi. The planet was orbiting very close to its parent star—0.05 astronomical units (AU—the Sun-Earth distance). Many such planets have now been discovered, and they are generically called "hot Jupiters".

The presence of a Jupiter-mass companion to the star 51 Pegasi is inferred from observations of periodic variations in the star's radial velocity. The companion lies only about eight million kilometres from the star, which would be well inside the orbit of Mercury in our Solar System. This object might be a gas-giant planet that has migrated to this location through orbital evolution, or from the radiative stripping of a brown dwarf.

FOR more than ten years, several groups have been examining the radial velocities of dozens of stars, in an attempt to identify orbital motions induced by the presence of heavy planetary companions[1-5]. The precision of spectrographs optimized for Doppler studies and currently in use is limited to about $15\,\mathrm{m\,s^{-1}}$. As the reflex motion of the Sun due to Jupiter is $13\,\mathrm{m\,s^{-1}}$, all current searches are limited to the detection of objects with at least the mass of Jupiter (M_J). So far, all precise Doppler surveys have failed to detect any jovian planets or brown dwarfs.

Since April 1994 we have monitored the radial velocity of 142 G and K dwarf stars with a precision of $13\,\mathrm{m\,s^{-1}}$. The stars in our survey are selected for their apparent constant radial velocity (at lower precision) from a larger sample of stars monitored for 15 years[6,7]. After 18 months of measurements, a small number of stars show significant velocity variations. Although most candidates require additional measurements, we report here the discovery of a companion with a minimum mass of 0.5 M_J, orbiting at 0.05 AU around the solar-type star 51 Peg. Constraints originating from the observed rotational velocity of 51 Peg and from its low chromospheric emission give an upper limit of 2 M_J for the mass of the companion. Alternative explanations to the observed radial velocity variation (pulsation or spot rotation) are unlikely.

类太阳恒星的一个类木伴星

马约尔，奎洛兹

编者按

尽管系外行星在很多年前就被发现围绕着脉冲星公转，其特殊的环境却排除了这些地方存在任何已知生命形式的可能性。米歇尔·马约尔和迪迪埃·奎洛兹以及其他四个团队，在超过十年的时间里一直在搜寻绕类太阳恒星运转的行星的特征。很多的断言都被报道过了，但随后都被证实是不对的。本文中，马约尔和奎洛兹报道了在飞马座 51 恒星光谱中发现的近似木星质量的行星存在的可靠特征。行星以非常近的距离——0.05 个天文单位（天文单位 AU：地球到太阳的距离）——绕它的母星公转。许多类似的行星现在已经被发现，它们一般都被称为"热木星"。

从飞马座 51 恒星观测到的视向速度周期性变化可以推断出其存在一个木星质量量级的伴星。该伴星和主星的距离只有 8,000,000 km，等于完全在太阳系中的水星轨道内。这个可能为气态巨行星的天体大概是在轨道演化过程中移居到此位置的，或者来自褐矮星的辐射剥离。

早在十多年前，就有一些小组致力于观测许多恒星的视向速度，试图确认由于存在巨型行星伴星[1-5]而产生的轨道运动。光谱仪的精度因为多普勒运动的研究而提高，现在可用的精度已提高至大约 15 m·s^{-1}。太阳相对于木星的反应运动的精度是 13 m·s^{-1}，现阶段的所有研究都局限于被探测的天体至少要有木星的质量（M_J）。迄今为止，所有精确的多普勒巡天观测都还未探测到任何类木行星及褐矮星。

自从 1994 年 4 月，我们已经监测了 142 个 G 型矮星和 K 型矮星精度为 13 m·s^{-1} 的视向速度。我们观测的恒星都是从 15 年来监测得到的大的恒星样本[6,7]中，根据其不变的可见视向速度（较低精度下）挑选出的。在 18 个月的测量后，少数恒星表现出了显著的速度变化。虽然许多候选天体还需要进行更多的观测，但是我们先在此公布质量下限是 0.5 M_J，距飞马座 51 恒星的轨道距离为 0.05 AU 的伴星的相关发现。源于观测到的飞马座 51 恒星的自转速度和低层色球辐射的限制，可以得到这个伴星的质量上限是 2 M_J。关于观测到的视向速度变化的其他解释（脉动或黑子自转）是不太可能的。

The very small distance between the companion and 51 Peg is certainly not predicted by current models of giant planet formation[8]. As the temperature of the companion is above 1,300 K, this object seems to be dangerously close to the Jeans thermal evaporation limit. Moreover, non-thermal evaporation effects are known to be dominant[9] over thermal ones. This jovian-mass companion may therefore be the result of the stripping of a very-low-mass brown dwarf.

The short-period orbital motion of 51 Peg also displays a long-period perturbation, which may be the signature of a second low-mass companion orbiting at larger distance.

Discovery of Jupiter-mass Companion(s)

Our measurements are made with the new fibre-fed echelle spectrograph ELODIE of the Haute-Provence Observatory, France[10]. This instrument permits measurements of radial velocity with an accuracy of about $13\ \mathrm{m\,s^{-1}}$ of stars up to 9 mag in an exposure time of < 30 min. The radial velocity is computed with a cross-correlation technique that concentrates the Doppler information of about 5,000 stellar absorption lines. The position of the cross-correlation function (Fig. 1) is used to compute the radial velocity. The width of the cross-correlation function is related to the star's rotational velocity. The very high radial-velocity accuracy achieved is a result of the scrambling effect of the fibres, as well as monitoring by a calibration lamp of instrumental variations during exposure.

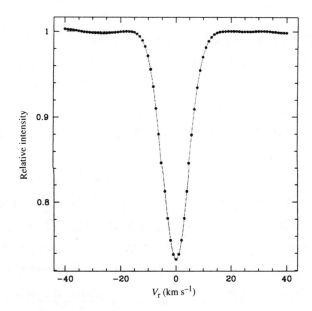

Fig. 1. Typical cross-correlation function used to measure the radial velocity. This function represents a mean of the spectral lines of the star. The location of the gaussian function fitted (solid line) is a precise measurement of the Doppler shift.

现在的巨型行星形成模型 [8] 显然不能预测该伴星和飞马座 51 恒星之间有这么近的距离。由于这个伴星的温度在 1,300 K 以上，它可能非常接近金斯热蒸发极限。而且，还有已知的非热蒸发作用主要影响 [9] 着这些热天体。因此这个木星质量量级的伴星可能是一个小质量褐矮星剥离后的结果。

飞马座 51 恒星存在着短周期的轨道运动，还伴随着长周期的扰动。这也许表示在更远的距离处，还有第二个小质量的伴星在做轨道运动。

木星质量量级伴星的发现

在法国的上普罗旺斯天文台 [10]，我们利用新型的光纤反馈中阶梯光栅摄谱仪 ELODIE 进行测量。这个设备在小于 30 分钟的曝光时间内，对亮度达 9 mag 的恒星，可以测出精度达 13 m · s^{-1} 的视向速度。这些视向速度是基于一种互相关法，利用约 5,000 条恒星吸收线的多普勒信息计算得到的。互相关函数（图 1）的位置被用来计算视向速度。互相关函数的宽度和恒星的自转速度有关。高精度的视向速度是来源于光纤的加扰效应，以及在曝光过程中定标灯对仪器变化的监测。

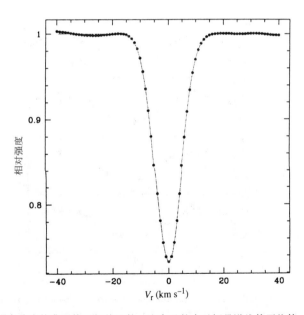

图 1. 被用来计算视向速度的典型的互相关函数。这个函数表示恒星谱线的平均值。拟合出的高斯函数（实线）的位置精确地表征出了多普勒频移。

The first observations of 51 Peg started in September 1994. In January 1995 a first 4.23-days orbit was computed and confirmed by intensive observations during eight consecutive nights in July 1995 and eight in September 1995. Nevertheless, a $24 \, \mathrm{m \, s^{-1}}$ scatter of the orbital solution was measured. As this is incompatible with the accuracy of ELODIE measurements, we adjusted an orbit to four sets of measurements carried out at four different epochs with only the γ-velocity as a free parameter (see Fig. 2). The γ-velocity in Fig. 3 shows a significant variation that cannot be the result of instrumental drift in the spectrograph. This slow perturbation of the short-period orbit is probably the signature of a second low-mass companion.

Table 1. Orbital parameters of 51 Peg

P	4.2293 ± 0.0011 d
T	$2,449,797.773 \pm 0.036$
e	0 (fixed)
K_1	$0.059 \pm 0.003 \, \mathrm{km \, s^{-1}}$
$a_1 \sin i$	$(34 \pm 2) \, 10^5 \, \mathrm{m}$
$f_1(m)$	$(0.91 \pm 0.15) \, 10^{-10} \, M_{\odot}$
N	35 measurements
$(O-C)$	$13 \, \mathrm{m \, s^{-1}}$

P, period; T, epoch of the maximum velocity; e, eccentricity; K_1 half-amplitude of the velocity variation; $a_1 \sin i$, where a_1 is the orbital radius; $f_1(m)$, mass function; N, number of observations; $(O-C)$, r.m.s. residual.

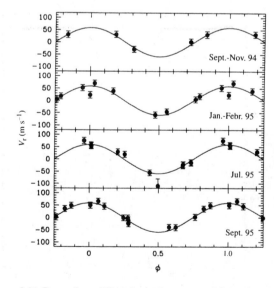

Fig. 2. Orbital motion of 51 Peg at four different epochs corrected from the γ-velocity. The solid line represents the orbital motion fitted on each time span with only the γ-velocity as a free parameter and with the other fixed parameters taken from Table 1.

1994 年 9 月，对飞马座 51 恒星进行了第一次观测。在 1995 年 1 月首次计算出了周期为 4.23 天的轨道，该计算值分别于 1995 年 7 月和 1995 年 9 月通过连续 8 天的密集观测结果所证实。但是，精度为 24 m·s⁻¹ 的轨道离散值还是被测量到了。这是和 ELODIE 仪器的精度不相合的，所以我们分别在四个不同时间，对同一个轨道采用了四套测量方法以进行调整，其中只有 γ 速度被定为自由参量（见图 2）。在图 3 中，γ 速度表现出的明显变化不可能来自摄谱仪仪器本身的零点漂移。这些短周期轨道运动中的小扰动可能是第二个小质量伴星存在的特征。

表 1. 飞马座 51 恒星的轨道参数

P	4.2293 ± 0.0011 d
T	$2,449,797.773 \pm 0.036$
e	0（固定的）
K_1	0.059 ± 0.003 km·s⁻¹
$a_1 \sin i$	$(34 \pm 2)10^5$ m
$f_1(m)$	$(0.91 \pm 0.15)10^{-10}\ M_\odot$
N	35 次测量
$(O-C)$	13 m·s⁻¹

P，周期；T，最大速度出现时间；e，偏心率；K_1，速度变化的半振幅；$a_1 \sin i$，其中 a_1 是轨道半径；$f_1(m)$，质量函数；N，观测次数；$(O-C)$，均方根残差。

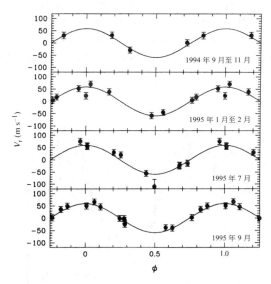

图 2. 在四个不同时期，修正了 γ 速度的飞马座 51 恒星的轨道运动。实线表示了分别在每个时间跨度内，仅取 γ 速度为自由参量，其他参量取自表 1 时的轨道运动拟合图。

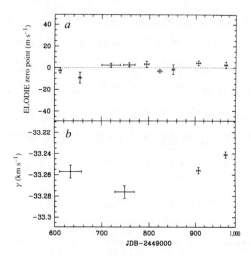

Fig. 3. *a*, ELODIE zero point computed from 87 stars of the sample having more than two measurements and showing no velocity variation. No instrumental zero point drift is detected. *b*, Variation of the γ-velocity of 51 Peg computed from the orbital fits displayed in Fig. 2. Considering the long-term stability of ELODIE this perturbation is probably due to a low-mass companion.

The long-period orbit cannot have a large amplitude. The 26 radial velocity measurements made during > 12 years with the CORAVEL spectrometer do not reveal any significant variation at a 200 m s^{-1} level. Intensive monitoring of 51 Peg is in progress to confirm this long-period orbit.

In Fig. 4 a short-period circular orbit is fitted to the data after correction of the variation in γ-velocity. Leaving the eccentricity as a free parameter would have given $e = 0.09 \pm 0.06$ with almost the same standard deviation for the r.m.s. residual (13 m s^{-1}). Therefore we consider that a circular orbit cannot be ruled out. At present the eccentricity range is between 0 and about 0.15. Table 1 lists the orbital parameters of the circular-orbit solution.

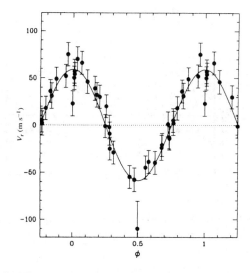

Fig. 4. Orbital motion of 51 Peg corrected from the long-term variation of the γ-velocity. The solid line represents the orbital motion computed from the parameters of Table 1.

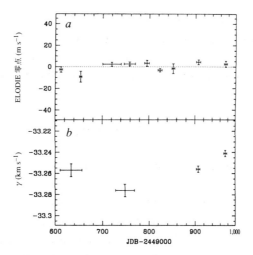

图 3. *a*, ELODIE 零点由样本中的 87 颗恒星计算所得，这些恒星都至少经过了两次测量，并且未显示出速度的变化。并没有发现存在仪器本身的零点漂移。*b*, 由图 2 中拟合出的飞马座 51 恒星轨道中的 γ 速度变化。考虑到 ELODIE 的长期稳定性，这个扰动很可能来自一个小质量的伴星。

长周期轨道运动不可能有大振幅。在长于 12 年的 26 个视向速度测量中，相关式视向速度仪（CORAVEL）并未显示出存在 $200 \ \mathrm{m \cdot s^{-1}}$ 量级上显著的变化。对飞马座 51 恒星已经开始了更高密集程度的监测用来确定该长周期轨道。

在图 4 中，对修正了 γ 速度变化的数据使用一个短周期的圆轨道进行拟合。保留轨道偏心率作为自由参数，可以得到 $e = 0.09 \pm 0.06$，拟合的标准均方根残差几乎不变（$13 \ \mathrm{m \cdot s^{-1}}$），因此我们认为圆形轨道不可能被排除。现在偏心率的范围在 0 到 0.15 之间。表 1 列出了在解圆轨道时所用的轨道参量。

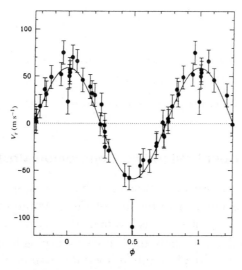

图 4. 飞马座 51 恒星经过 γ 速度长期变化项修正后的轨道运动。实线表示了由表 1 参量计算得到的轨道运动结果。

An orbital period of 4.23 days is rather short, but short-period binaries are not exceptional among solar-type stars. (Five spectroscopic binaries have been found with a period < 4 days in a volume-limited sample of 164 G-type dwarfs in the solar vicinity[6].) Although this orbital period is not surprising in binary stars, it is puzzling when we consider the mass obtained for the companion:

$$M_2 \sin i = 0.47 \pm 0.02 \; M_J$$

where i is the (unknown) inclination angle of the orbit.

51 Peg (HR8729, HD217014 or Gliese 882) is a 5.5 mag star, quite similar to the Sun (see Table 2), located 13.7 pc (45 light yr) away. Photometric and spectroscopic analyses indicate a star slightly older than the Sun, with a similar temperature and slight overabundance of heavy elements. The estimated age[11] derived from its luminosity and effective temperature is typical of an old galactic-disk star. The slight overabundance of heavy elements in such an old disk star is noteworthy. But this is certainly not a remarkable peculiarity in view of the observed scatter of stellar metallicities at a given age.

Table 2. Physical parameters of 51 Peg compared with those of the Sun

		51 Peg		
	Sun	Geneva photometry*	Spectroscopy†	Strömgren photometry and spectroscopy[11]
T_{eff} (K)	5,780	5,773	5,724	5,775
$\log g$	4.45	4.32	4.30	4.18
Fe/H	0		0.19	0.06‡
M/H	0	0.20		
M_v	4.79	4.60		
$R/R_\odot$	1	1.29		

M/H is the logarithmic ratio of the heavy element abundance compared to the Sun (in dex).
* M. Grenon (personal communication).
† J. Valenti (personal communication).
‡ But other elements such as Na I, Mg I, Al I are overabundant, in excess of 0.20.

Upper Limit for the Companion Mass

A priori, we could imagine that we are confronted with a normal spectroscopic binary with an orbital plane almost perpendicular to the line of sight. Assuming a random distribution of binary orbital planes, the probability is less than 1% that the companion mass is larger than 4 M_J, and 1/40,000 that it is above the hydrogen-burning limit of 0.08 $M_\odot$. Although these probability estimates already imply a low-mass companion for 51 Peg, an even stronger case can be made from considerations of rotational velocity. If we assume that the

一个轨道的周期只有 4.23 天实在有点短，但是在类太阳恒星中短周期双星也不是特殊的情况。(在太阳附近 [6] 164 颗 G 型矮星的限定体积样本中，已经发现了 5 组分光双星，其周期都小于 4 天。) 尽管在双星系统中这种轨道周期一点都不特殊，但是当我们考虑这些伴星的质量时还是会令人困惑：

$$M_2 \sin i = 0.47 \pm 0.02 \ M_J$$

其中 i 是轨道倾角(未知)。

飞马座 51 恒星(HR8729、HD217014 或者 Gliese 882)是一颗光度为 5.5 mag 的恒星，和太阳很相似(见表 2)，距离为 13.7 pc(45 光年)。测光分析和光谱分析显示该星比太阳要稍微年老一些，温度相近，重元素有点过丰。从光度和有效温度上得到的估计年龄 [11] 表明该星为典型的年老银盘恒星。对于这样一个老年盘族星，其重元素丰度轻微高于太阳丰度是值得注意的。但是对于这种给定年龄的恒星，被观测到金属元素丰度存在偏离也并不是很特别。

表 2. 飞马座 51 恒星和太阳各项物理参数的比较

	太阳	飞马座 51 恒星		
		日内瓦测光 *	光谱 †	斯特龙根测光和光谱 [11]
T_{eff} (K)	5,780	5,773	5,724	5,775
log g	4.45	4.32	4.30	4.18
Fe/H	0		0.19	0.06‡
M/H	0	0.20		
M_v	4.79	4.60		
$R/R_\odot$	1	1.29		

M/H 是飞马座 51 恒星和太阳的重元素丰度对数比比值(指数)。
* 格勒农(个人交流)。
† 瓦伦蒂(个人交流)。
‡ 但是其他元素，例如 Na I、Mg I、Al I 等元素都过丰，超过 0.20。

伴星的质量上限

先验地，假设我们面对着一个普通的分光双星系统，其轨道平面是和我们的视线方向垂直的。假设双星的轨道面随机分布，则伴星质量大于 4 M_J 的概率小于 1%，伴星质量超过氢燃烧极限，即 0.08 $M_\odot$ ($M_\odot$太阳质量)的概率是 1/40,000。尽管这些概率估计暗示了飞马座 51 恒星存在一个小质量的伴星，但是从自转速度考虑可以更有力地支持这个情况。如果我们假设飞马座 51 恒星的自转轴和轨道平面方向是平行的，

rotational axis of 51 Peg is aligned with the orbital plane, we can derive $\sin i$ by combining the observed projected rotational velocity ($v \sin i$) with the equatorial velocity $V_{equ} = 2\pi R / P$ ($v \sin i = V_{equ} \cdot \sin i$).

Three independent precise $v \sin i$ determinations of 51 Peg have been made: by line-profile analysis[12], $v \sin i = 1.7 \pm 0.8$ km s^{-1}; by using the cross-correlation function obtained with the CORAVEL spectrometer[13], $v \sin i = 2.1 \pm 0.6$ km s^{-1}; and by using the cross-correlation function obtained with ELODIE, $v \sin i = 2.8 \pm 0.5$ km s^{-1}. The unweighted mean $v \sin i$ is 2.2 ± 0.3 km s^{-1}. The standard error is probably not significant as the determination of very small $v \sin i$ is critically dependent on the supposed macroturbulence in the atmosphere. We accordingly prefer to admit a larger uncertainty: $v \sin i = 2.2 \pm 1$ km s^{-1}.

51 Peg has been actively monitored for variability in its chromospheric activity[14]. Such activity, measured by the re-emission in the core of the Ca II lines, is directly related to stellar rotation via its dynamo-generated magnetic field. A very low level of chromospheric activity is measured for this object. Incidentally, this provides an independent estimate of an age of 10 Gyr (ref. 14), consistent with the other estimates. No rotational modulation has been detected so far from chromospheric emission, but a 30-day period is deduced from the mean chromospheric activity level S-index. A V_{equ} value of 2.2 ± 0.8 km s^{-1} is then computed if a 25% uncertainty in the period determination is assumed

Using the mean $v \sin i$ and the rotational velocity computed from chromospheric activity, we finally deduce a lower limit of 0.4 for $\sin i$. This corresponds to an upper limit for the mass of the planet of 1.2 M_J. Even if we consider a misalignment as large as 10°, the mass of the companion must still be less than 2 M_J, well below the mass of brown dwarfs.

The 30-day rotation period of 51 Peg is clearly not synchronized with the 4.23-day orbital period of its low-mass companion, despite its very short period. (Spectroscopic binaries with similar periods are all synchronized.) The lack of synchronism on a timescale of 10^{10}yr is a consequence of the q^{-2} ($q = M_2/M_1$) dependence of the synchronization timescale[15]. In principle this can be used to derive an upper limit to the mass of the companion. It does at least rule out the possibility of the presence of a low-mass stellar companion.

Alternative Interpretations?

With such a small amplitude of velocity variation and such a short period, pulsation or spot rotation might explain the observations equally well[16,17]. We review these alternative interpretations below and show that they can probably be excluded.

Spot rotation can be dismissed on the basis of the lack of chromospheric activity and the large period derived from the S chromospheric index, which is clearly incompatible with the observed radial-velocity short period. A solar-type star rotating with a period of 4.2 days would have a much stronger chromospheric activity than the currently observed

288

根据可观测的自转速度投影 ($v \sin i$) 和赤道速度 $V_{equ} = 2\pi R / P$ ($v \sin i = V_{equ} \times \sin i$) 我们可以得到 $\sin i$。

三种独立的方法都分别给出了飞马座 51 恒星 $v \sin i$ 的精确值：根据谱线轮廓 [12] 分析，得到 $v \sin i = 1.7 \pm 0.8$ km·s⁻¹；用互相关函数对相关式视向速度仪 [13] 得到的结果进行处理，得到 $v \sin i = 2.1 \pm 0.6$ km·s⁻¹；用互相关函数对 ELODIE 得到的结果进行处理，得到 $v \sin i = 2.8 \pm 0.5$ km·s⁻¹。$v \sin i$ 的非加权平均值为 $v \sin i = 2.2 \pm 0.3$ km·s⁻¹。我们认为因为 $v \sin i$ 这么小的数值会和大气中假想的宏观湍流有着精确的依赖关系，所以标准差并没有那么明显。因此我们宁愿承认一个更大的不确定度：$v \sin i = 2.2 \pm 1$ km·s⁻¹。

飞马座 51 恒星的色球活动 [14] 变化一直在被密切地监测着。这些活动是通过测量恒星核心中 CaII 线的再发射确定的，这直接和恒星发电机机制下磁场对其自转的影响相关。根据这种方式，在其色球观测到了较微弱的活动迹象。另外，这也独立地提供了 10 Gyr(参考文献 14)的年龄估计，这和其他估计值是一致的。从色球辐射上并未观测到自转的调制作用，但从色球平均活动水平中的 S 指数却推导出了 30 天的周期。如果假设该时间段内的不确定度为 25%，则可得到 $V_{equ} = 2.2 \pm 0.8$ km·s⁻¹。

根据观测色球活动计算得到的平均 $v \sin i$ 和自转速度，我们推导出了 $\sin i$ 的下限为 0.4。这和质量上限为 1.2 M_J 的行星是相对应的。就算我们考虑角度偏差达到 10°，相对应的伴星的质量仍然要小于 2 M_J，这也远低于褐矮星的质量。

尽管飞马座 51 恒星 30 天的自转周期很短，但也很显然和 4.23 天的低质量伴星的轨道运动周期是不同步的。(有着相似周期的分光双星都是同步的。)在 10^{10} 年尺度上的不同步来源于与同步化时标 [15] 有关的 q^{-2} ($q = M_2 / M_1$)。原则上是可以根据这来推导伴星的质量上限的。它起码可以排除低质量恒星伴星存在的可能性。

另一种解释?

基于很小的速度变化波动和很短的周期，脉动或黑子自转理论似乎也可以将观测解释得同样好 [16,17]。接下来我们就回顾一下这些解释，并证明它们可能被排除在外。

黑子自转理论被排除是基于其缺乏色球活动和由色球 S 指数推得的长周期，这长周期明显与观测到的短周期视向速度不符合。一个自转周期为 4.2 天的类太阳恒

value[14]. Moreover, a period of rotation of 4.2 days for a solar-type star is typical of a very young object (younger than the Pleiades) and certainly not of an old disk star.

Pulsation could easily yield low-amplitude velocity variations similar to the one observed, but would be accompanied by luminosity and colour variations as well as phase-related absorption line asymmetries. The homogeneous photometric survey made by the Hipparcos satellite provides a comprehensive view of the intrinsic variability of stars of different temperatures and luminosities. The spectral type of 51 Peg corresponds to a region of the Hertzsprung–Russell diagram where the stars are the most stable[18].

Among solar-type stars no mechanisms have been identified for the excitation of pulsation modes with periods as long as 4 days. Only modes with very low amplitude ($\ll 1$ m s^{-1}) and periods from minutes to 1 h are detected for the Sun.

Radial velocity variations of a few days and < 100 m s^{-1} amplitude have been reported for a few giant stars[19]. Stars with a similar spectral type and luminosity class are known to be photometric variables[18]. Their observed periods are in agreement with predicted pulsation periods for giant stars with radii $> 20\ R_{\odot}$. 51 Peg, with its small radius, can definitely not be compared to these stars. These giant stars also pulsate simultaneously in many short-period modes, a feature certainly not present in the one-year span of 51 Peg observations. It is worth noticing that 51 Peg is too cold to be in the δ Scuti instability strip.

G. Burki *et al.* (personal communication) made 116 photometric measurements of 51 Peg and two comparison stars in the summer of 1995 at ESO (la Silla) during 17 almost-consecutive nights. The observed magnitude dispersions for the three stars are virtually identical, respectively $V = 0.0038$ for 51 Peg, and $V = 0.0036$ and 0.0039 for the comparison stars. The fit of a sine curve with a period of 4.2293 days to the photometric data limits the possible amplitude to 0.0019 for V magnitude and 0.0012 for the $[B_2-V_1]$ Geneva colour index. Despite the high precision of these photometric measurements we cannot completely rule out, with these photometric data alone, the possibility of a very low-amplitude pulsation. In the coming months, stronger constraints can be expected from the numerous Hipparcos photometric data of this star.

Pulsations are known to affect the symmetry of stellar absorption lines. To search for such features we use the cross-correlation technique, as this technique is a powerful tool for measuring mean spectral line characteristics[20]. The difference in radial velocity of the lower and upper parts of the cross-correlation function is an indicator of the line asymmetry. The amplitude of a 4.2-day sine curve adjusted to this index is less than 2 m s^{-1}. The bisector of the cross-correlation function does not show any significant phase variation.

From all the above arguments, we believe that the only convincing interpretation of the observed velocity variations is that they are due to the orbital motion of a very-low-mass companion.

星会产生比现在观测值[14]更强的色球活动。而且，一个自转周期为 4.2 天的类太阳恒星是典型的年轻天体（比昴星团还要年轻），肯定不是一个老年盘族星。

脉动理论很容易与观测到的小幅速度变化符合，但是同时也伴随有光度和色指数变化以及出现相位相关不对称的吸收线。依巴谷卫星曾经进行过光度一致性的研究，对恒星不同温度和不同光度的本质性变化进行了广泛的采样调查。飞马座 51 恒星所属的光谱型对应赫罗图中最稳定[18]的恒星区域。

在类太阳恒星中没有什么机制被证明可以激发出周期为 4 天的脉动模式。在太阳上仅仅发现了具有很小振幅（$\ll 1\,\mathrm{m\cdot s^{-1}}$），周期为几分钟到 1 小时范围的模式。

周期为几天、振幅小于 $100\,\mathrm{m\cdot s^{-1}}$ 的视向速度变化已经在一些巨星[19]中被发现。这样光谱类型相似且光度同等级别的恒星都已知为测光变星[18]。它们被观测到的周期和对半径大于 $20\,R_\odot$ 的巨星预估出的脉动周期是一致的。由于飞马座 51 恒星的半径较小，所以完全无法与这类恒星相比。这类巨星同时还会有伴随许多短周期的脉动模式，这在飞马座 51 恒星为期一年时间跨度的观测中是没有明显出现的。值得注意的是，飞马座 51 恒星的温度相对于盾牌 δ 型星不稳定带也是过低的。

在 1995 年夏季几乎连续的 17 个夜晚，伯基等人（个人交流）在欧南台（智利拉西亚）对飞马座 51 恒星及两颗比较星进行了 116 次测光观测。这三颗星观测到的星等弥散看起来是一样的，飞马座 51 恒星及另两颗比较星的星等弥散分别是 $V=0.0038$，$V=0.0036$ 和 $V=0.0039$。用周期为 4.2293 天的正弦函数拟合测光数据，将 V 星等的可能振幅限制在 0.0019，以及日内瓦色指数 $[B_2-V_1]$ 振幅限制在 0.0012。尽管这些测光观测精度很高，但是我们不能只靠这些测光数据完全将这种小振幅脉动存在的可能性完全排除。在以后几个月，我们会通过依巴谷卫星得到更多关于该星的测光数据，以此进一步对其限制。

现已知脉动会对恒星吸收线的对称性产生影响。互相关法是一种测量平均光谱线特征[20]很有力的工具，所以我们用它来寻找脉动的特征。视向速度的互相关函数中上限和下限的差值就表征了光谱线的不对称性。周期为 4.2 天的正弦函数的振幅转化为这种指数会小于 $2\,\mathrm{m\cdot s^{-1}}$。互相关函数的等分线并没有表现出任何明显的相位变化。

根据以上的讨论，我们可以相信对于观测到的速度变化，唯一使人信服的解释就是：速度变化是由一个非常小质量伴星的轨道运动引起的。

Jupiter or Stripped Brown Dwarf?

At the moment we certainly do not have an understanding of the formation mechanism of this very-low-mass companion. But we can make some preliminary comments about the importance of evaporation as well as the dynamic evolution of the orbit.

If we compare 51 Peg b with other planets or G-dwarf stellar companions (Fig. 5) it is clear that the mass and the low orbital eccentricity of this object are in the range of heavy planets, but this certainly does not imply that the formation mechanism of this planet was the same as for Jupiter.

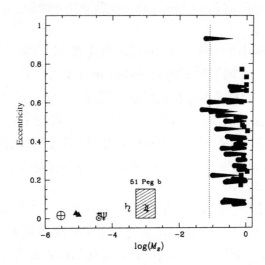

Fig. 5. Orbital eccentricities of planets as well as companion of G-dwarf binaries[8] in the solar vicinity as a function of their mass M_2. The planets of the Solar System are indicated with their usual symbols. The planets orbiting around the pulsar[24,25] PSR B 1257 + 12 are indicated by filled triangles. The uncertainties on the mass of SB1 (single-spectrum spectroscopic binaries), owing to their unknown orbital inclination, are indicated by an elongated line that thins to a sin i probability of 99%. SB2s are indicated by filled squares. (Only the stellar orbits not tidally circularized with periods larger than 11 days are indicated.) Note the discontinuity in the orbital eccentricities when planets are binary stars are compared, and the gap in masses between the giant planets and the lighter secondaries of solar-type stars. The dotted line at 0.08 $M_\odot$ indicates the location of the minimum mass for hydrogen burning. The position of 51 Peg b with its uncertainties is indicated by the hatched rectangle.

Present models for the formation of Jupiter-like planets do not allow the formation of objects with separations as small as 0.05 AU. If ice grains are involved in the formation of giant planets, the minimum semi-major axis for the orbits is about 5 AU (ref. 8), with a minimum period of the order of 10 yr. A Jupiter-type planet probably suffers some orbital decay during its formation by dynamic friction. But it is not clear that this could produce an orbital shrinking from 5 AU to 0.05 AU.

All of the planets in the Solar System heavier than 10^{-6} $M_\odot$ have almost circular orbits

292

木星或褐矮星剥离？

目前我们对这小质量伴星的形成机制还没有充分地理解。但是就蒸发机制的重要性和轨道的动力学演化而言，我们可以做出一些初步的解释。

如果我们将飞马座 51 的行星 b 和其他行星，甚至 G 型矮星的伴星进行比较（图 5），很显然这个天体的质量和较小的轨道偏心率属于重行星的范围，但是这并不表示它的形成机制和木星的一样。

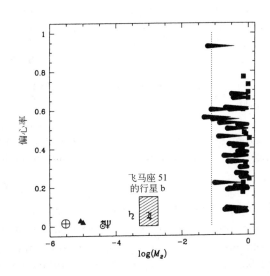

图 5. 行星及太阳附近 G 型矮星双星系统 [8] 中伴星的轨道偏心率与它们的质量 M_2 的函数关系。太阳系中的行星均用常用的符号进行了标注。围绕脉冲星 [24,25] PSR B 1257+12 做轨道运动的行星都用实心三角形进行了标注。由于不知道 SB1（单谱分光双星）的轨道倾角，所以 SB1 质量的不确定度用 $\sin i$ 置信度减小到 99% 的延长线进行了标注。SB2 用实心正方形标注出来了。（只有轨道未受潮汐圆化影响，且周期大于 11 天的恒星才被标注了。）值得注意的是，图中比较了双星系统中伴星轨道偏心率的不连续性和巨行星与类太阳恒星的较轻伴星之间的质量差距。在 $0.08\ M_\odot$ 处的点线表示了满足氢燃烧的最小质量所在位置。飞马座 51 恒星及其不确定度在图中用阴影线的长方形标注出来了。

现有关于类木行星的各种形成模式中，并不允许天体在 0.05 AU 距离内形成。如果在巨行星形成中包括冰颗粒，则轨道的最小半长轴约为 5 AU（参考文献 8），最小周期的量级为 10 年。一个类木行星在形成的过程中，可能会由于动力学摩擦而经历轨道衰减的过程。但是我们还不清楚这是否会使轨道从 5 AU 缩小到 0.05 AU。

太阳系内所有质量超过 $10^{-6} M_\odot$ 的行星，因为都形成于原行星气体盘而有一个接

as a result of their origin from a protoplanetary gaseous disk. Because of its close separation, however, the low eccentricity of 51 Peg b is not a proof of similar origin. Tidal dissipation acting on the convective envelope is known[15] to circularize the orbit and produce a secular shrinking of the semi-major axis of binary systems. The characteristic time is essentially proportional to $q^{-1}P^{16/3}$. For stars of the old open cluster M67, orbital circularization is observed for periods lower than 12.5 days (ref. 21). We derive for 51 Peg a circularization time of a few billion years, shorter than the age of the system. The low orbital eccentricity of 51 Peg b could result from the dynamic evolution of the system and not necessarily from its formation conditions.

A Jupiter-sized planet as close as 0.05 AU to 51 Peg should have a rather high temperature of about 1,300 K. To avoid a significant evaporation of a gaseous atmosphere, the escape velocity V_e has to be larger than the thermal velocity V_{th}: $V_e > \alpha V_{th}$. This imposes a minimum mass for a gaseous planet at a given separation:

$$\frac{M_P}{M_J} > \alpha^2 \left(\frac{kT_*}{m}\right)\left(\frac{GM_J}{R_P}\right)^{-1} (1-\gamma)^{1/4} \left(\frac{R_*}{2a}\right)^{1/2}$$

where γ denotes the albedo of the planet, R_p and M_p are its radius and mass, m is the mass of atoms in the planet atmosphere, and R_* and T_* are the radius and effective temperature of the star.

Our lack of knowledge of the detailed structure of the atmosphere of the planet prevents us from making an accurate estimate of α. A first-order estimate of $\alpha \approx 5-6$ is nevertheless made by analogy with planets of the Solar System[22]. We find that with a planetary radius probably increased by a factor of 2–3 owing to the high surface temperature (A. Burrows, personal communication), gaseous planets more massive than 0.6–1.0 M_J are at the borderline for suffering thermal evaporation. Moreover, for the Solar-System planets, non-thermal evaporative processes are known to be more efficient than thermal ones[9]. The atmosphere of 51 Peg b has thus probably been affected by evaporation.

Recent work[23] on the fragmentation of molecular clouds shows that binary stars can be formed essentially as close to each other as desired, especially if the effects of orbital decay are considered. We can thus speculate that 51 Peg b results from a strong evaporation of a very close and low-mass brown dwarf. In such a case 51 Peg b should mostly consist of heavy elements. This model is also not free of difficulties, as we expect that a brown dwarf suffers less evaporation owing to its larger escape velocity.

We are eager to confirm the presence of the long-period companion and to find its orbital elements. If its mass is in the range of a few times that of Jupiter and its orbit is also quasi-circular, 51 Peg could be the first example of an extrasolar planetary system associated with a solar-type star.

The search for extrasolar planets can be amazingly rich in surprises. From a complete planetary

近圆形的轨道。然而因为其过近的间距，飞马座 51 恒星的较小偏心率不足以成为类似形成过程的证据。已知作用于对流包层的潮汐摩擦作用 [15]，造成了轨道趋于圆形并使双星系统的半长轴在长期地缩小。实际上特征时间是和 $q^{-1}P^{16/3}$ 成比例的。年老的疏散星团 M67 中的恒星观测到的轨道圆化，其周期小于 12.5 天（参考文献 21）。我们推算飞马座 51 恒星的圆化时间为几十亿年，这要比整个系统的年龄短。飞马座 51 的行星 b 的较小轨道偏心率可能产生于系统的动力学演化，并不一定和形成条件有关。

和飞马座 51 恒星相距仅 0.05 AU 的一个木星大小的行星，其温度应该要高于 1,300 K。要避免明显的气态大气蒸发，逃逸速度 V_e 要比热力学速度 V_{th} 大：$V_e > \alpha V_{th}$。这就给出了一个气态行星在给定距离内的最小质量：

$$\frac{M_P}{M_J} > \alpha^2 \left(\frac{kT_*}{m}\right)\left(\frac{GM_J}{R_P}\right)^{-1}(1-\gamma)^{1/4}\left(\frac{R_*}{2a}\right)^{1/2}$$

其中 γ 表示行星的反照率，R_P 和 M_P 表示行星的半径和质量，m 是行星大气中的原子质量，R_* 和 T_* 是恒星的半径和有效温度。

我们对行星大气的细节结构了解还不够，这阻碍了我们对 α 做出精确的估计。最初级的估计是 $\alpha \approx 5\sim6$，这不过是我们根据太阳系内的行星进行的类推结果 [22]。我们发现由于表面温度较高，行星的半径可能增至 2~3 倍（伯罗斯，个人交流），质量超过 0.6~1.0 M_J 的气态行星处于发生热蒸发效应的边缘。而且，在太阳系行星中，已知的非热蒸发过程产生的作用远比热蒸发产生的大 [9]。因此，飞马座 51 的行星 b 的大气很可能会受到蒸发作用的影响。

最近关于分子云分裂的研究 [23] 表明，实际上双星系统中的双星在形成时是很接近的，这点当轨道衰减的作用被考虑后变得尤为明显。因此我们可以猜测飞马座 51 的行星 b 产生于一颗相距很近、小质量的褐矮星的强劲蒸发作用。在这种情况下，飞马座 51 的行星 b 应该由重元素组成。但是这个模型也不是没有问题，正如我们考虑到的褐矮星具有较大的逃逸速度，相应就只会有较微弱的蒸发效应。

我们迫切想确认是否存在长周期运动的伴星，并且得到其轨道组成。如果它的质量在木星质量的几倍范围之内，且其轨道是准圆形的，则飞马座 51 恒星将是第一个太阳系以外类太阳恒星的行星系统实例。

对于太阳系外行星的研究是充满惊喜的。从在脉冲星附近发现的完整的行星系

system detected around a pulsar[24,25], to the rather unexpected orbital parameters of 51 Peg b, searches begin to reveal the extraordinary diversity of possible planetary formation sites.

Note added in revision: After the announcement of this discovery at a meeting held in Florence, independent confirmations of the 4.2-day period radial-velocity variation were obtained in mid-October by a team at Lick Observatory, as well as by a joint team from the High Altitude Observatory and the Harvard–Smithsonian Center for Astrophysics. We are deeply grateful to G. Marcy, P. Butler, R. Noyes, T. Kennelly and T. Brown for having immediately communicated their results to us.

(**378**, 356-359; 1995)

Michel Mayor & Didier Queloz
Geneva Observatory, 51 Chemin des Maillettes, CH-1290 Sauverny, Switzerland

Received 29 August; accepted 31 October 1995.

References:

1. Walker, G. A. H., Walker, A. R. & Irwin, A. W., *Icarus* 116, 359-375 (1995).

2. Cochran, W. D. & Hatzes, A. P. *Astrophys. Space Sci.* 212, 281-291 (1994).

3. Marcy, G. W. & Butler, R. P. *Publ. Astron. Soc. Pacif.* 104, 270-277 (1992).

4. McMillan, R. S., Moore, T. L., Perry, M. L. & Smith, P. H. *Astrophys. Space Sci.* 212, 271- 280 (1994).

5. Marcy, G. W. & Butler, R. P. in *The Bottom of the Main Sequence and Beyond* (ESO Astrophys. Symp.) (ed. Tinney, C. G.) 98-108 (Springer, Berlin, 1995).

6. Duquennoy, A. & Mayor, M. *Astr. Astrophys.* 248, 485-524 (1991).

7. Mayor, M., Duquennoy, A., Halbwachs, J. L. & Mermilliod, J. C. in *Complementary Approaches to Double and Multiple Star Research* (eds McAlister, A. A. & Hartkopf, W. I.) (ASP Conf. Ser. 32, 73-81 (Astr. Soc. Pacific, California, 1992).

8. Boss, A. P. *Science* 267, 360-362 (1995).

9. Hunten, D. H., Donahue, T. M., Walker, J. C. G. & Kasting, J. F. in *Origin and Evolution of Planetary and Satellite Atmospheres* (eds Atreya, S. K., Pollack, J. B. & Matthews, M. S.) 386-422 (Univ. of Arizona Press, Tucson, 1989).

10. Baranne, A. *Astrophys. J. Suppl.* (submitted).

11. Edvardsson, B. *et al. Astr. Astrophys.* 275, 101-152 (1993).

12. Soderblom, D. R. *Astrophys. J. Suppl. Ser.* 53, 1-15 (1983).

13. Baranne, A., Mayor, M. & Poncet, J. L. *Vistas Astr.* 23, 279-316 (1979).

14. Noyes, R. W., Hartmann, L. W., Baliunas, S. L., Duncan, D. K. & Vaughan, A. H. *Astrophys. J.* 279, 763-777 (1984).

15. Zhan, J. P. *Astr. Astrophys.* 220, 112-116 (1989).

16. Walker, G. A. H. *et al. Astrophys. J.* 396, L91-L94 (1992).

17. Larson, A. M. *et al. Publ. Astron. Soc. Pac.* 105, 825-831 (1993).

18. Eyer, L., Grenon, M., Falin, J. L., Froeschlé, M. & Mignard, F. *Sol. Phys.* 152, 91-96 (1994).

19. Hatzes, A. P. & Cochran, W. D. *Proc. 9th Cambridge Workshop* (ed. Pallavicini, R.) (Astronomical Soc. of the Pacific) (in the press).

20. Queloz, D. in *New Developments in Array Technology and Applications* (eds Davis Philip, A. G. *et al.*) 221-229 (Int. Astr. Union, 1995).

21. Latham, D. W., Mathieu, R. D., Milone, A. A. E. & Davis, R. J. in *Binaries as Tracers of Stellar Formation* (eds Duquennoy, A. and Mayor, M.) 132-138 (Cambridge Univ. Press, 1992).

22. Lewis, J. S. & Prinn, R. G. *Planets and their Atmospheres—Origin and Evolution* (Academic, Orlando, 1984).

23. Bonnell, I. A. & Bate, M. R. *Mon. Not. R. astr. Soc.* 271, 999-1004 (1994).

24. Wolszczan, A. & Frail, D. A. *Nature* 355, 145-147 (1992).

25. Wolszczan, A. *Science* 264, 538-542 (1994).

Acknowledgements. We thank G. Burki for analysis of photometric data, W. Benz for stimulating discussions, A. Burrows for communicating preliminary estimates of the radius of Jupiter at different distances from the Sun, and F. Pont for his careful reading of the manuscript. We also thank all our colleagues of Marseille and Haute-Provence Observatories involved in the building and operation of the ELODIE spectrograph, namely G. Adrianzyk, A. Baranne, R. Cautain, G. Knispel, D. Kohler, D. Lacroix, J.-P. Meunier, G. Rimbaud and A. Vin.

统 [24,25]，到意外发现的飞马座 51 的行星 b 的轨道参数，研究正在揭开行星形成可能位置的非凡多样性。

修订中的说明：在我们于佛罗伦萨的会议上公布这一发现后，在 10 月中利克天文台的一个小组以及来自高山天文台及哈佛–史密森天体物理中心组成的联合团队分别对周期为 4.2 天的视向速度变化进行独立的确认。我们对马西、巴特勒、诺伊斯、肯内利和布朗及时与我们交流结果表示深深的谢意。

（冯翀 翻译；周济林 审稿）

Orbital Migration of the Planetary Companion of 51 Pegasi to Its Present Location

D. N. C. Lin *et al.*

Editor's Note

The recent discovery of a planet orbiting the Sun-like star 51 Pegasi raised serious issues, not least of which was why this Jupiter-like planet was so close to its parent star—0.04 astronomical units, or less than a tenth of the distance of Mercury from the Sun. Most of Jupiter's mass is in the form of gases in the atmosphere, but here Douglas Lin and colleagues demonstrate that, for the 51 Peg planet, such gases could not condense so close to the star. They find it a more likely explanation that the planet formed farther away from the star and then migrated inwards through an interaction with material in the disk out of which the planet formed.

The recent discovery[1] and confirmation[2] of a possible planetary companion orbiting the solar-type star 51 Pegasi represent a breakthrough in the search for extrasolar planetary systems. Analysis of systematic variations in the velocity of the star indicates that the mass of the companion is approximately that of Jupiter, and that it is travelling in a nearly circular orbit at a distance from the star of 0.05 AU (about seven stellar radii). Here we show that, if the companion is indeed a gas-giant planet, it is extremely unlikely to have formed at its present location. We suggest instead that the planet probably formed by gradual accretion of solids and capture of gas at a much larger distance from the star ($\sim$5 AU), and that it subsequently migrated inwards through interactions with the remnants of the circumstellar disk. The planet's migration may have stopped in its present orbit as a result of tidal interactions with the star, or through truncation of the inner circumstellar disk by the stellar magnetosphere.

THE first argument against the *in situ* formation of a planetary companion is based on models of the nebular disks[3] that are known to exist around young stars[4]. The standard picture of the formation of a giant planet involves the coagulation and accretion of small particles of ice and rock in the disk[5] until a core of about 15 Earth masses is built up; then gas, composed mainly of H and He, is accreted from the disk[6]. Standard disk models show that at 0.05 AU the temperature is about 2,000 K, too hot for the existence of any small solid particles. An alternative formation model[7] involves a massive disk, whose self-gravity is comparable to that of the central object, in which a gaseous subcondensation could form by contraction under its own gravity. But recent detailed calculations of such massive disks[8] indicate that they tend to form spiral arms and to transfer mass into the central star instead of fragmenting into subcondensations.

飞马座 51 的行星型伴星轨道迁移到现在位置的过程

林潮等

编者按

最近发现的绕类太阳恒星飞马座 51 公转的行星引发了很深刻的问题，不仅仅是为什么这颗类木行星与它的母星距离如此之近，只有 0.04 个天文单位，还不足水星到太阳距离的十分之一。木星大部分物质是以气体的形式存在于其大气层中，但在这篇文章中，林潮以及他的合作者证明，对于飞马座 51 的行星，这种气体无法在离恒星如此之近的距离处凝结。他们认为更为可能的解释是此行星在离恒星遥远的地方形成，之后通过与行星诞生的原恒星盘上物质的相互作用向内迁移。

近年来系外行星系统搜索的一个突破是，发现[1] 和证实[2] 可能存在行星型伴星环绕在太阳型恒星飞马座 51 的周围。对恒星视向速度的系统性变化的分析显示此伴星的质量与木星的质量相似，在距离恒星 0.05 AU（大约 7 倍恒星半径）处的近圆轨道上运行。这里我们证明，如果伴星的确是一颗气态巨行星，那么它极不可能在目前的位置形成。我们认为该行星可能是通过逐渐吸积固体颗粒和捕获距离恒星更远处（~5 AU）的气体形成，此后通过与星周盘遗迹作用向内迁移。由于恒星的潮汐相互作用，或者在由恒星磁层引起的星周盘内断层处，该行星可能停止迁移形成目前的轨道。

第一个不支持行星型伴星在原地形成的论点建立在星云盘[3] 模型上。我们知道，星云盘存在于年轻恒星周围[4]。巨行星形成的标准图景涉及盘内冰和石块小颗粒[5] 的凝结与吸积，直到形成一个大约 15 倍地球质量的核；吸积盘内的气体主要由氢和氦组成[6]。标准盘模型表明在 0.05 AU 处温度约为 2,000 K，太热而不可能存在任何固体小颗粒。另一个巨行星形成模型[7] 包含一个大质量盘，其自引力和中心天体的引力大小相当，在盘内可通过自身引力收缩形成气态的子凝聚物。但是最近对大质量盘的详细计算[8] 表明它们趋向于形成旋臂并将质量转移到中心恒星上而不是碎裂成子凝聚物。

A second problem with the formation of a planet at 0.05 AU is that although the present evaporation rate of the planet is negligible, this effect would have been of major importance in the past. At 0.05 AU, the companion's effective temperature, due to stellar irradiation, is ~1,300 K. In order to determine the planetary radius R_p in the presence of such heating, we calculated the evolution of objects in the mass range $M_p = 1$–$10\,M_J$ (jovian masses) using a standard stellar structure code[9,10] with a non-ideal interior equation of state[11]. The rotation of the planet is almost certainly tidally locked[12] so that the same hemisphere always faces 51 Peg. We assume that atmospheric motions and convection in the interior redistribute the heat so that the dark side and the bright side have nearly the same temperature.

In Fig. 1 we show the evolution of R_p for various M_p. At 8 Gyr, the estimated age of 51 Peg, $R_p = 8.3 \times 10^7$ m for $1\,M_J$, not much larger than the present radius of Jupiter $(7.0 \times 10^7$ m). For these values, the escape velocity of a hydrogen atom is 12 times larger than the mean thermal speed, and a simple calculation of the Jeans escape rate[13] shows that evaporation is negligible. A further process to be considered is hydrodynamic escape[14] in which ultraviolet and X-ray radiation from the star are absorbed by hydrogen atoms in the planetary atmosphere and drive a planetary wind. A rough estimate, based on the observed X-ray flux of young stars[14], shows that the effect of this process is also negligible. Thus the planet at present is quite safe against evaporation. The evaporation of a low-mass star or brown dwarf, another proposed explanation[15] for the existence of the companion, would be even more difficult because the object would have a much higher surface gravity. But during the early history of a planet[6] its radius is a factor of ten or more larger than the present radius, so the escape speed becomes much less and both evaporation mechanisms, along with ablation by the stellar wind, will prevent formation.

We propose that the companion was formed several AU away from the star through the standard process. Recent detailed calculations[16] for the accretion of Jupiter at 5 AU have shown that it is possible for that planet to build well before the nebula dissipates. The protoplanet interacts tidally with the disk during its growth[17]. Let v be the disk viscosity, M_* the stellar mass, ω the orbital frequency, and r_n the distance from the star at which the planet formed. If $M_p \gtrsim 40 v M_* / (\omega(r_n) r_n^2)$ when its tidal radius, $(M_p / 3 M_*)^{1/3} r_n$, exceeds H (the vertical scale height of the disk), the protoplanet induces the formation of a gap[18,19] in the disk near r_n so that growth of the planet stops. Standard disk models[3] give $H(r) \approx 0.1r$, where r is the distance from the star. The disk evolves viscously on a timescale $\tau_v \approx r_d^2 / v$ which is inferred to be ~5×10^6 yr from infrared observations[20]. The effective radius r_d which contains most of the disk mass observed in the infrared is[4] ~100 AU. Applying these estimates to the gap formation conditions, we find $M_p \approx M_J$.

在 0.05 AU 处形成行星的第二个问题是，尽管行星目前的蒸发率可忽略，但该效应在过去是一大重要因素。在 0.05 AU 处，由于恒星辐射，伴星的有效温度约为 1,300 K。为了确定在这样加热机制下的行星半径 R_P，我们使用非理想内部状态方程 [11] 的标准恒星结构程序 [9,10] 计算了质量在 $M_P = 1\ M_J \sim 10\ M_J$（$M_J$ 为木星质量）范围内天体的演化。行星的自转几乎已经被潮汐锁定 [12]，所以同一半球总是面对着飞马座 51。我们假设其内部的大气运动和对流使得热量重新分布，所以暗面和亮面几乎具有相同温度。

图 1 中我们展示了各种 M_P 下 R_P 的演化。在飞马座 51 的估计年龄为 8 Gyr 时，质量为 $1\ M_J$ 的天体，$R_P = 8.3 \times 10^7$ m，比木星目前的半径（7.0×10^7 m）大不了多少。从这些值得出，氢原子的逃逸速度是平均热运动速度的 12 倍，简单计算金斯逃逸率 [13] 结果显示蒸发是可忽略的。值得进一步考虑的机制是流体力学逃逸 [14]，恒星的紫外和 X 射线辐射被行星大气里的氢原子吸收并驱动行星风形成。根据观测的年轻恒星的 X 射线流量 [14] 的初步计算结果，这个过程的效应也可以忽略。因此，该行星目前非常安全，不会被蒸发。关于伴星存在的另一个解释——小质量恒星或褐矮星的蒸发 [15]——更难实现，因为这种天体需要具有更大的表面重力。但是在行星的早期历史中 [6]，其半径为目前半径的 10 倍甚至更多，所以逃逸速度变得很小，蒸发机制和星风的烧蚀会共同阻止行星的形成。

我们提出伴星通过标准机制在距离恒星几个天文单位处形成。最近对木星在 5 AU 处吸积的详细计算 [16] 结果表明，该行星有可能在星云消散之前就已经形成。原行星在成长过程中与盘发生潮汐作用 [17]。令 v 为盘的黏度，M_* 为恒星质量，ω 为轨道频率，r_n 为行星形成处与恒星的距离。假如 $M_P \geqslant 40 v M_* / (\omega(r_n) r_n^2)$，当潮汐半径 $(M_P / 3 M_*)^{1/3} r_n$ 大于 H（盘的垂直标高）时，原行星在盘内 r_n 附近处形成空隙 [18,19] 致使行星增长停止。标准盘模型 [3] 给出 $H(r) \approx 0.1r$，这里 r 是行星与恒星的距离。盘的黏性演化时标 $\tau_v \approx r_d^2 / v$，红外观测 [20] 得出这个值为 $\sim 5 \times 10^6$ yr。包含大部分红外观测到的盘质量 [4] 的有效半径 r_d 为 ~ 100 AU。将这些估计值应用到空隙形成条件，我们得到 $M_P \approx M_J$。

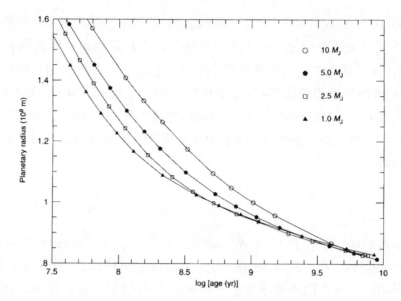

Fig. 1. Planetary radius as a function of log age for objects (bottom trace to top trace) of masses 1, 2.5, 5 and 10 M_J. The calculation assumes that the planet migrated to its present position during its first ten million years of existence. The plot shows the subsequent evolution, during which the orbit of the planet was stationary and it was heated by a central star whose luminosity was constant in time. For this evolutionary history, evaporation is not important.

After the gap formation, angular momentum transfer continues and the protoplanet undergoes orbital migration coupled to the viscous evolution of the disk[21,22]. The orbital radius of the planet (r_p) and that of the gap (both are still embedded in the disk) decrease on the timescale[22,23] of τ_v. The planet essentially follows the material of the inner disk as it evolves towards the star. We now propose two possible mechanisms which suggest that this migration can terminate at ~0.05 AU and that the planet will not plunge into 51 Peg.

Mechanism 1. As the planet approaches 51 Peg, tidal friction can induce angular momentum exchange between the planet's orbital motion and the spin of the star. If R_* is the stellar radius and P the orbital period of the planet, the timescale for tidal evolution is[12]

$$\tau_r = r_p \left(\frac{dr_p}{dt} \right)^{-1} = \frac{P}{9\pi} \left(\frac{r_p}{R_*} \right)^5 \left(\frac{M_*}{M_p} \right) Q_* \tag{1}$$

We estimate the dissipation parameter $Q_* = 1.5 \times 10^5$ for a main sequence star based on the observation[24] that the orbits of short-period pre-main-sequence binary stars and the main-sequence binary stars in the Pleiades cluster are circularized for $P \lesssim 5$ and 7 days, respectively. As young stars rotate more rapidly than their main-sequence counterparts[25], we assume that 51 Peg was rotating rapidly enough that the co-rotation point τ_{CR} (the distance from the star where an orbiting object has the same angular frequency as the stellar rotation) was inside 0.05 AU. The tidal effect then results in outward migration of the planet. Thus there may exist a radius r_c where the protoplanet's radial migration was

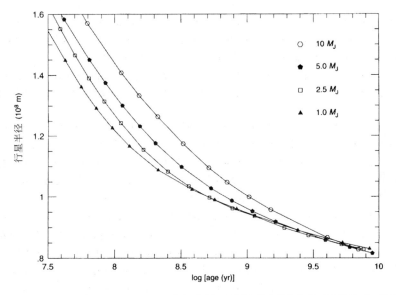

图 1. 质量分别为 $1\,M_J$、$2.5\,M_J$、$5\,M_J$ 和 $10\,M_J$ 的行星（由下到上）的半径与年龄对数的关系。该计算假设行星在形成后的一千万年内迁移到目前的位置。图中显示行星之后的演化，在此期间行星轨道稳定，且行星由亮度恒定的中心恒星加热。对这段演化历史，蒸发并不重要。

在空隙形成后，角动量继续转移，原行星开始进行轨道迁移[21,22]并与盘的黏性演化相耦合。行星的轨道半径（r_p）和空隙的半径（两者都位于盘内）随着时标 τ_v 减小[22,23]。行星随着盘内的物质流逐渐向恒星演化。我们现在提出两个可能的机制，表明此迁移会在 ~0.05 AU 处终止，行星不会落入飞马座 51。

机制 1。当行星趋近飞马座 51 时，潮汐摩擦导致角动量在行星轨道运动和恒星自转之间交换。假设 R_* 为恒星半径，P 为行星轨道周期，潮汐演化的时标为[12]：

$$\tau_r = r_p \left(\frac{dr_p}{dt} \right)^{-1} = \frac{P}{9\pi} \left(\frac{r_p}{R_*} \right)^5 \left(\frac{M_*}{M_p} \right) Q_* \tag{1}$$

根据观测，昴星团里短周期前主序双星和主序双星分别对于周期 $P \lesssim 5$ 天和 7 天的轨道是圆化的[24]，我们估计主序星的耗散参数为 $Q_* = 1.5 \times 10^5$。因为年轻恒星比其对应的主序星旋转得更快[25]，我们假设飞马座 51 旋转得足够快让共转点 τ_{CR}（天体公转角频率和恒星自转相同时与恒星的距离）位于 0.05 AU 内。潮汐效应将导致行星向外迁移。因此可能存在一个半径 r_c，在这点由盘引起的向内推力和来自飞马座 51 的向外推力达到平衡，原行星停止径向迁移。在这个半径处，整个盘内的角动量转移

halted by a balance between the inward push on it by the disk and the outward push from 51 Peg. At that point, the angular momentum transfer equilibrium throughout the disk[19] implies $\tau_v \approx \tau_r$ such that $r_c \equiv (9\pi\tau_v M_P / P_* Q_* M_*)^{2/13} R_*$, where P_* is the keplerian orbital period at R_*. Based on an estimate[26] of $R_* = 4R_\odot$ (where $R_\odot$ is the solar radius) during its early history, we find that this equilibrium can be established at 0.05 AU during the early epoch of 51 Peg.

But this equilibrium is only temporary. The disk material between the planet and the star will accrete onto the latter, leaving the planet with the remaining disk outside its orbit. The disk's surface density adjusts until a quasi-equilibrium state is attained in which the angular momentum flux is approximately constant with distance from the star. At this stage the planet's equilibrium radius is determined by the condition that the star's tidal torque on it, $M_p r_p^2 \omega_p / \tau_r$, is balanced by the angular momentum flux through the disk, $\sim M_d r_d^2 \omega_d / \tau_v$, where ω_d is a mean angular frequency of the disk and M_d is its mass. For $R_* = 4R_\odot$ we find that $r_c \approx 0.03 (M_p/M_d)^{1/6} (\tau_v / 5 \times 10^6 \text{ yr})^{1/6}$ AU. If the disk then dissipates sufficiently so that its mass $M_d \approx M_p$ and its evolution timescale lengthens so that $\tau_v \approx 10^8$ yr, then r_c could be close to the present orbital position of the planet. The dissipation must occur before the star contracts substantially ($< 10^7$ yr) or spins down ($> 10^8$ yr). In view of the rather precise timing and the relatively large R_* needed for this mechanism to work, we consider an alternative, as follows.

Mechanism 2. The spin periods of classical T Tauri stars are clustered[27] around 8 days, longer than those of the weak line T Tauri stars. One explanation for the 8-day periods is that the spin rate is controlled by coupling between the stellar magnetosphere and the disk[28]. The presence of the magnetosphere would also clear[29] the inner disk out to a point slightly less than r_{CR} (0.08 AU for an 8-day period). Once the planet has spiralled in to $r_p = 0.05$ AU, angular momentum exchange between it and the disk occurs only via the 2:1 resonance at a reduced (by $\sim M_p/M_*$) rate[17,30]. Because $r_p < r_{CR}$, the stellar tidal effect also continues to induce an inward migration. But as long as $R_* < 3R_\odot$, consistent with evolutionary tracks[26], τ_r is larger than the stellar contraction timescale, and the migration effectively stops near 0.05 AU.

After this time, in either case, τ_r and τ_v increase rapidly because the star contracts on a relatively short timescale, and the disk dissipates. During its contraction to the main sequence, 51 Peg may have spun up, if it conserved angular momentum, but once it reached the main sequence, the star would have spun down[31] because of angular momentum loss via stellar wind. Eventually in both cases r_p becomes less than r_{CR} and $R_* \approx R_\odot$, causing the companion to migrate inwards on the timescale $\tau_r \approx 14 \sin i_p$ Gyr, which is much longer than the age of the star for all reasonable values of the inclination angle i_p between the normal to the orbital plane and the line of sight. This is the configuration we observe today. The requirement that the tidal migration timescale (τ_r) be large compared with the life span of a typical solar-type star is a further piece of evidence that supports the interpretation that the companion is a planet with $M_p \approx M_J$ rather than a more massive object.

(**380**, 606-607; 1996)

平衡[19]，意味着 $\tau_v \approx \tau_r$，因此 $r_c \equiv (9\pi\tau_v M_p / P_* Q_* M_*)^{2/13} R_*$，这里 P_* 是在 R_* 处的开普勒轨道周期。根据早期历史过程中 $R_* = 4R_\odot$ 的估计[26]（这里 $R_\odot$ 是太阳半径），我们发现这个平衡在飞马座 51 的早期历元就能够在 0.05 AU 处建立。

但是这个平衡只是暂时的。行星和恒星之间盘内的物质将吸积到后者上，留下行星与其轨道外的剩余盘物质。盘的面密度不断调节直到实现准平衡态，该状态下距离恒星的角动量通量近似恒定。在这个阶段，行星的平衡半径由恒星对其潮汐力矩 $M_p r_p^2 \omega_p / \tau_r$ 和通过盘的角动量通量 $\sim M_d r_d^2 \omega_d / \tau_v$ 的平衡状况决定，这里 ω_d 是盘的平均角频率，M_d 是盘的质量。对于 $R_* = 4R_\odot$ 我们发现 $r_c \approx 0.03\,(M_p/M_d)^{1/6}\,(\tau_v/5\times10^6\,\mathrm{yr})^{1/6}\,\mathrm{AU}$。如果接下来盘完全消散，那么它的质量 $M_d \approx M_p$，其演化时标也增长，$\tau_v \approx 10^8\,\mathrm{yr}$，从而 r_c 接近于行星目前的轨道位置。耗散必然发生在恒星充分收缩（$<10^7\,\mathrm{yr}$）或者自转变慢（$>10^8\,\mathrm{yr}$）之前。这个机制需要相当精确的时间和相对较大的 R_*，因此我们考虑下面一种可能。

机制 2。经典金牛 T 型星的自转周期一般集中在 8 天左右[27]，比那些弱线金牛 T 型星的周期长。关于 8 天周期的一个解释是自转速度受到恒星磁层和盘耦合[28]的控制。磁层的存在也解释了在半径稍小于 r_{CR}（对于 8 天周期，为 0.08 AU）距离处的内盘[29]。一旦行星旋入 $r_p = 0.05\,\mathrm{AU}$ 内，行星和盘只能通过 2:1 共振以一个减少的速率（通过 $\sim M_p/M_*$）[17,30]发生角动量的交换。因为 $r_p < r_{CR}$，恒星潮汐效应也持续引起向内的迁移。但是只要 $R_* < 3R_\odot$ 和演化轨迹一致[26]，那么 τ_r 就会比恒星收缩时标大，且迁移在 0.05 AU 附近就会有效停止。

之后，在以上两种情形的任意一种中，由于恒星在相对较短的时标内收缩，τ_r 和 τ_v 迅速增大，并且盘开始消散。在恒星收缩到主序时，如果角动量守恒，飞马座 51 可能已经完成加速，但是一旦到达主序，恒星风带走的角动量损失将导致恒星自转变慢[31]。最后在两个情形下，r_p 比 r_{CR} 小，并且 $R_* \approx R_\odot$，导致伴星以时标 $\tau_r \approx 14\sin i_p\,\mathrm{Gyr}$ 向内迁移。这个时标远比任何轨道面法线和视线方向倾角 i_p 值对应的恒星年龄都长。这是我们现在观测到的状况。潮汐迁移时标（τ_r）比典型太阳型恒星的寿命要长的要求进一步支持了伴星是一颗 $M_p \approx M_J$ 的行星，而不是更大质量天体的观点。

（肖莉 翻译；周济林 审稿）

D. N. C. Lin[*], **P. Bodenheimer**[*] & **D. C. Richardson**[†]

[*] UCO / Lick Observatory, Board of Studies in Astronomy and Astrophysics, University of California, Santa Cruz, California 95064, USA

[†] Canadian Institute for Theoretical Astrophysics, McLennan Laboratories, University of Toronto, 60 St George Street, Toronto, Ontario, Canada M5S 1A7

Received 24 October 1995; accepted 5 March 1996.

References:

1. Mayor, F. & Queloz, D. *Nature* **378**, 355-359 (1995).

2. Marcy, G. & Butler, R. P. *IAU Circ. No.* 6251 (1995).

3. Lin, D. N. C. & Papaloizou, J. in *Protostars and Planets II* (eds Black, D. C. & Matthews, M. S.) 981-1072 (Univ. Arizona Press, Tucson, 1985).

4. Beckwith, S. V. W., Sargent, A. I., Chini, R. & Güsten, R. *Ast. J.* **99**, 924-945 (1990).

5. Wetherill, G. W. *A. Rev. Astr. Astrophys.* **18**, 77-113 (1980).

6. Bodenheimer, P. & Pollack, J. B. *Icarus* **67**, 391-408 (1986).

7. Cameron, A. G. W. *Moon Planets* **18**, 5-40 (1978).

8. Laughlin, G. & Bodenheimer, P. *Astrophys. J.* **436**, 335-354 (1994).

9. Laughlin, G. & Bodenheimer, P. *Astrophys. J.* **403**, 303-314 (1993).

10. Stringfellow, G., Black, D. C. & Bodenheimer, P. *Astrophys. J.* **349**, L59-L62 (1990).

11. Saumon, D., Chabrier, G. & Van Horn, H. M. *Astrophys. J. Suppl. Ser.* **99**, 713-741 (1995).

12. Goldreich, P. & Soter, S. *Icarus* **5**, 375-389 (1966).

13. Shu, F. H. *The Physical Universe* 441 (University Science Books, Mill Valley, CA, 1982).

14. Zahnle, K. in *Protostars and Planets III* (eds Levy, E. & Lunine, J.) 1305-1338 (Univ. Arizona Press, Tucson, 1993).

15. Burrows, A. & Lunine, J. *Nature* **378**, 333 (1995).

16. Pollack, J. B. *et al. Icarus* (submitted).

17. Lin, D. N. C. & Papaloizou, J. C. B. *Mon. Not. R. astr. Soc.* **186**, 799-812 (1979).

18. Papaloizou, J. C. B. & Lin, D. N. C. *Astrophys. J.* **285**, 818-834 (1984).

19. Lin, D. N. C. & Papaloizou, J. C. B. in *Protostars and Planets III* (eds Levy, E. & Lunine, J.) 749-836 (Univ. Arizona Press, Tucson, 1993).

20. Strom, S. E., Edwards, S. & Skrutskie, M. F. in *Protostars and Planets III* (eds Levy, E. & Lunine, J.) 837-866 (Univ. Arizona Press, Tucson, 1993).

21. Goldreich, P. & Tremaine, S. *Astrophys. J.* **241**, 425-441 (1980).

22. Lin, D. N. C. & Papaloizou, J. C. B. *Astrophys. J.* **309**, 846-857 (1986).

23. Takeuchi, T., Miyama, S. & Lin, D. N. C. *Astrophys. J.* (in the press).

24. Mathieu, R. D. *A. Rev. Astr. Astrophys.* **32**, 465-530 (1994).

25. Skumanich, A. *Astrophys. J.* **171**, 565-567 (1972).

26. D'Antona, F. & Mazzitelli, I. *Astrophys. J. Suppl. Ser.* **90**, 467-500 (1994).

27. Bouvier, J., Cabrit, S., Fernandez, M., Martin, E. L. & Matthews, J. M. *Ast. Astrophys.* **272**, 176-206.

28. Königl, A. *Astrophys. J.* **370**, L39-L43 (1991).

29. Shu, F. H. *et al. Astrophys. J.* **429**, 781-796 (1994).

30. Goldreich, P. & Tremaine, S. *Astrophys. J.* **233**, 857-871 (1979).

31. MacGregor, K. & Bremner, M. *Astrophys. J.* **376**, 204-213 (1991).

Acknowledgements. We thank G. Marcy for providing us with his data before publication, and P. Artymowicz, G. Basri, D. O. Gough, L. Hartmann, R. D. Mathieu, F. H. Shu, S. Sigurdsson and A. Title for conversations. D.C.R. was supported by a fellowship from the Natural Sciences and Engineering Research Council (Canada). This work was supported in part by the US NSF and in part by a NASA astrophysics theory programme which supports a joint Center for Star Formation Studies at NASA-Ames Research Center, UC Berkeley and UC Santa Cruz.

306

A 3.5-Gyr-old Galaxy at Redshift 1.55

J. Dunlop *et al.*

Editor's Note

The Hubble constant (quantifying the expansion rate of the universe) was accurately measured in the 1990s, enabling a good estimate of the age of the Universe. Here James Dunlop at the University of Edinburgh and colleagues identify a problem with that figure. They report a very red galaxy (containing old stars) at a redshift of 1.55, and determine that the galaxy is at least 3.5 billion years old. Yet, at that time, the age of the Universe was considered to be just 2 billion years. The researchers point out the inconsistency, but have no explanation. It became clear two years later that the answer lies in the fact that the expansion rate of the Universe was slower in the past.

One of the most direct methods of constraining the epoch at which the first galaxies formed—and thereby to constrain the age of the Universe—is to identify and date the oldest galaxies at high redshift. But most distant galaxies have been identified on the basis of their abnormal brightness in some spectral region[1-4]; such selection criteria are biased towards objects with pronounced nuclear activity or young star-forming systems, in which the spectral signature of older stellar populations will be concealed. Here we report the discovery of a weak and extremely red radio galaxy (53W091) at $z = 1.55$, and present spectroscopic evidence that its red colour results from a population of old stars. Comparing our spectral data with models of the evolution of stellar populations, we estimate that we are observing this galaxy at least 3.5 Gyr after star-formation activity ceased. This implies an extremely high formation redshift ($z > 4$) for 53W091 and, by inference, other elliptical galaxies. Moreover, the age of 53W091 is greater than the predicted age of the Universe at $z = 1.55$, under the assumption of a standard Einstein–de Sitter cosmology (for any Hubble constant greater than $50\,\text{km s}^{-1}\,\text{Mpc}^{-1}$), indicating that this cosmological model can be formally excluded.

IT is relatively easy for a short-lived burst of star-formation[5], or re-processed light from an active nucleus[6,7], to dominate the appearance of a galaxy in the rest-frame optical–ultraviolet, concealing the presence of an older underlying stellar population[8]. It is thus the reddest objects at a given redshift which are of greatest importance for constraining the first epoch of galaxy formation[9-11], and the correlation between the ultraviolet and radio properties of powerful radio galaxies[12,13] indicates that radio-based searches for passive elliptical galaxies at $z > 1$ should be confined to millijansky flux-density levels. We are therefore investigating the properties of radio galaxies with flux densities at 1.4 GHz $S_{1.4\text{GHz}} > 1$ mJy selected from the Leiden–Berkeley deep survey (LBDS) and

红移 1.55 处的一个年龄为 3.5 Gyr 的星系

邓洛普等

编者按

在二十世纪九十年代，哈勃常数（用以描述宇宙膨胀率）已被准确测量，这使得我们可以很好地估计出宇宙年龄。然而在本文中，爱丁堡大学的詹姆斯·邓洛普与他的同事们却在该图像下发现了一个问题。他们发现了红移 1.55 处的一个极红星系（包含老年恒星），其年龄至少可以达到 35 亿年，而当时人们估计出的宇宙年龄大约只有 20 亿年。虽然研究人员指出了两者的不一致性，但是却并未给出相应的解释。直到大约两年后，人们才开始意识到造成两者不一致的主要原因可能是宇宙膨胀率（过去宇宙拥有更慢的膨胀率）。

发现最古老的高红移星系并确定它们的年龄，是界定第一代星系形成时间并进而界定宇宙年龄最直接的方法之一。但是大多数遥远的星系都是依据它们某些频谱区 [1-4] 中的反常亮度来进行确定的，这些选择标准倾向于识别出具有明显核活动的天体或者那些年轻星形成的系统，在这些天体系统中较年老星族的光谱特征并不明显。本文中，我们报告了一个在红移 $z = 1.55$ 处发现的极红的弱射电星系（53W091），相应光谱分析结果证实其色指数偏红是由星系中的老年恒星星族造成的。通过比较光谱数据与恒星星族演化模型，我们估计被观测的这个星系在恒星形成活动停止后至少还经历了 3.5 Gyr 的演化。这意味着该星系是在极高红移 $(z > 4)$ 处形成的，由此推断其他的椭圆星系也是如此。另外，在标准爱因斯坦－德西特宇宙学的假设下（对于任何哈勃常数大于 $50 \ km \cdot s^{-1} \cdot Mpc^{-1}$ 的情况），53W091 的年龄会大于理论预言的红移 $z = 1.55$ 时的宇宙年龄，这就表明我们的观测结果正式排除了这个宇宙学模型。

在静止参考系的光学－紫外波段，短时标爆发的恒星形成过程 [5] 或活动核辐射的再发射 [6,7] 容易湮没年老星族 [8] 的贡献，从而主导星系的辐射全貌。因此在一个给定的红移处，最红的天体对于限制星系形成的最初时刻 [9-11] 具有重大意义。而强射电星系在紫外和射电波段特征的相关性 [12,13] 表明用射电方法寻找 $z > 1$ 的被动演化椭圆星系时，其流量密度应该被限制在毫央（mJy）的水平上。因此我们研究了在 1.4 GHz 处 $S_{1.4GHz} > 1$ mJy 的射电星系的性质，这些射电星系选自莱顿－伯克利深度巡天（LBDS）以及该巡天的后续扩展观测 [14-17]。现在我们拥有一个包含 77 个星系完

its extensions[14-17]. We now possess optical–infrared photometry down to $V \approx 26$ mag and $K \approx 20$ mag for a complete sample of 77 galaxies, enabling us to estimate redshifts both from spectral fitting[13] and from a modified version of the infrared Hubble diagram[18]. From this sample we have isolated a subset of 10 extremely red ($R - K > 5$) potentially high-redshift (estimated redshift $z_{est} > 1.5$) objects for intensive spectroscopic study.

The galaxy 53W091, one of the reddest in the sample ($R = 24.6 \pm 0.20$ mag and $K = 18.75 \pm 0.05$ mag within a 4-arcsec aperture) was observed for 1.5 hours on 25 July 1995 with the Low Resolution Imaging Spectrograph (LRIS) on the 10-m W. M. Keck Telescope on Mauna Kea, Hawaii. This observation yielded the detection of a faint and very red continuum, but no emission-line redshift. Therefore, in an attempt to constrain the galaxy redshift, we obtained deep J- and H-band images with IRCAM3 on the United Kingdom Infrared Telescope in August 1995. The ease with which the galaxy was detected at J and H ($J = 20.5 \pm 0.1$ mag, $H = 19.5 \pm 0.1$ mag; Fig. 1a) revealed that, if its red R–K colour was in part due to a redshifted 4,000 Å break (the most prominent spectral feature displayed by an old stellar population at optical wavelengths), this feature must lie at observed wavelength $\lambda_{obs} < 1.2$ µm, constraining the redshift to $z < 2$.

Finally, on 31 August and 1 September 1995 we re-observed 53W091 for a total of 4 hours, again with LRIS on the 10-m Keck Telescope (Fig. 1b). The spectrum produced by co-adding all 5.5 hours of integration is shown in Fig. 1c, plotted in the rest frame of the radio galaxy assuming $z = 1.552$. This unambiguous redshift was deduced from numerous absorption lines and two strong spectral breaks (at rest wavelength $\lambda_{rest} = 2,635$ and $2,897$ Å), features whose existence in the near-ultraviolet spectrum of the Sun and other low-mass dwarfs is long-established[19], and which are evident in the ultraviolet spectra of low-redshift ellipticals such as M32 (Fig. 1c). This is, to our knowledge, the first time that such a high galaxy redshift has been derived successfully from late-type absorption features; indeed, with $V \approx 26$ mag, this is probably the faintest galaxy for which an absorption-line redshift has ever been determined.

整样本的光学-红外测光数据，其 V 波段和 K 波段星等分别约为 26 mag（mag，星等）和 20 mag。通过该样本，我们可以同时利用光谱拟合[13]方法或基于修改后的红外哈勃图[18]来估计这些星系的红移。为了深入地进行光谱分光研究，我们从该样本中挑出了 10 个极红的（$R-K > 5$）、可能是高红移（估计红移 $z_{est} > 1.5$）的天体进行后续的分析。

星系 53W091 是该样本中最红的星系之一（在 4 arcsec 的孔径中，$R = 24.6 \pm 0.20$ mag，$K = 18.75 \pm 0.05$ mag）。我们于 1995 年 7 月 25 日利用夏威夷冒纳凯阿火山上的 10 m 凯克望远镜的低分辨率成像光谱仪（LRIS）对该星系进行了观测，曝光时间为 1.5 小时。虽然本次观测探测到该星系非常红的暗弱连续谱，但是我们却无法通过发射线测量出该星系的红移。为了试图限制星系的红移，我们在 1995 年 8 月利用英国红外望远镜配备的 IRCAM3 观测得到该星系在 J 波段和 H 波段上的深场图像。由于在 J 波段和 H 波段（$J = 20.5 \pm 0.1$ mag，$H = 19.5 \pm 0.1$ mag，图 1a）能轻易地探测到该星系，这表明如果该星系的 $R-K$ 色指数偏红是部分由红移后的 4,000 Å 不连续（年老星族在光学波段最明显的光谱特征）导致的，那么该特征一定位于观测波长 $\lambda_{obs} < 1.2$ μm 处，这将意味着该星系的红移 $z < 2$。

最后，在 1995 年 8 月 31 日和 9 月 1 日，我们再次利用 10 m 凯克望远镜的 LRIS 对 53W091 进行了总计 4 个小时的观测（图 1b）。对该星系总计 5.5 小时的曝光观测进行叠加并假设其红移为 $z = 1.552$，我们得到该射电星系在静止参考系中的光谱，如图 1c 所示。这里，该星系的准确红移是利用大量吸收线以及两个显著的光谱不连续特征（在静止波长 $\lambda_{rest} = 2,635$ Å 和 2,897 Å 处）确定下来的，很早之前我们就发现这两个近紫外波段显著的光谱不连续特征存在于太阳和其他小质量矮星中[19]，而在一些低红移的椭圆星系中，例如 M32（图 1c）的紫外光谱中也存在这一明显现象。据我们所知，利用晚型吸收特征成功确定如此高的星系红移尚属首例。事实上在 $V \approx 26$ mag 时，该星系可能是迄今为止用吸收线方法确定红移的最暗的星系。

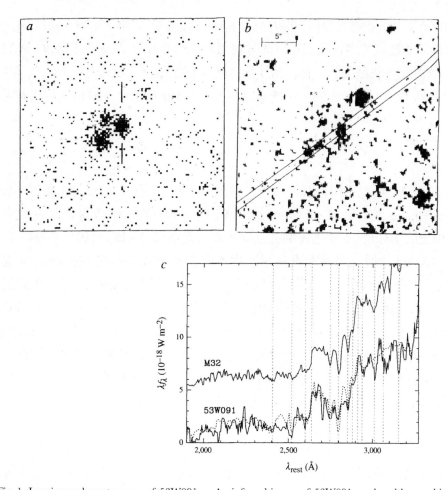

Fig. 1. Imaging and spectroscopy of 53W091. *a*, An infrared image of 53W091 produced by combining the IRCAM3 *J*- and *H*-band images. The field is 30-arcsec square and the radio galaxy identification is indicated by two bars. The position of this object is coincident with the centroid of the radio source (right ascension 17 h 21 min 17.81s, declination +50° 08′ 47.5″ (1950)) to within 0.5 arcsec. *b*, A 10-minute *R*-band image of the same field taken with the 10-m Keck Telescope, with lines superimposed to indicate the position and orientation of the LRIS slit which was used to obtain the optical spectrum. The position of the optical identification is coincident, to within the errors, with that of the near-infrared identification (in contrast to the highly wavelength-dependent morphologies often displayed by more powerful high-redshift galaxies such as 3C324[8]). In this 1-arcsec seeing optical image the radio galaxy is resolved, with a deconvolved full-width at half-maximum of 1.3 arcsec. The orientation of the LRIS slit (126°) was chosen to be close to the parallactic angle (which varied between 100° and 150° during the spectroscopic observations) and to enable a spectrum to also be obtained of a second red galaxy which lies a few arcsec southeast of 53W091. This galaxy appears to be at the same redshift as 53W091, ($z = 1.55$), and, interestingly, has an almost identical SED, providing further circumstantial evidence that the optical–infrared properties of 53W091 are essentially unaffected by its active nucleus. At radio wavelengths, 53W091 is a steep-spectrum source ($\alpha_{1.41\mathrm{GHz}}^{0.61\mathrm{GHz}} = 1.3 \pm 0.13$, where $f_v \propto v^{-\alpha}$) with $S_{1.41\mathrm{GHz}} = 22.5 \pm 1.0$ mJy (ref. 17), and is extended (by ~4 arcsec) along a position angle of 131° (ref. 16). *c*, The 5.5-h Keck spectrum of 53W091 (lower solid line) from 25 July, 31 August and 1 September, plotted in terms of rest wavelength (assuming $z = 1.552$) and compared both with the (scaled) ultraviolet spectrum of the nearby elliptical M32 (upper solid line) and with the best-fitting model spectrum (dotted line). (The model spectrum was produced by synthesizing the spectrum produced by a Scalo IMF with a main-sequence turn-off mass of $1.35M_\odot$ (equivalent to spectral type F2), corresponding to an age of 3.5 Gyr.) The data for M32 and

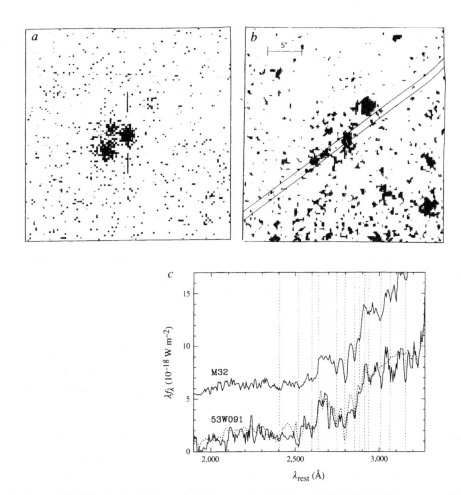

图 1. 53W091 的成像观测和光谱。*a*，结合 IRCAM3 的 *J* 波段和 *H* 波段观测所得到的 53W091 的红外图像。视场大小为 30 arcsec²。两条竖线之间的天体为识别出的射电星系。该天体的位置与射电源（赤经：17 h 21 min 17.81 s，赤纬：+50° 08′ 47.5″ (1950)）的中心重合，二者之间的位置偏离不超过 0.5 arcsec。*b*，用 10 m 凯克望远镜在 *R* 波段曝光 10 min，拍摄的相同视场所得到的图像。图上所画的两条直线表示的是拍摄光学光谱所用的 LRIS 狭缝的位置和方向。光学确认的位置与近红外确认的位置在误差范围内重合（与之不同的是，对于一些辐射更强的高红移星系进行观测，例如 3C324[8]，在不同波段进行观测其形态各异）。在视宁度为 1 arcsec 的观测条件下，经过半峰全宽为 1.3 arcsec 的退卷积之后，在光学图像中可以分辨出该射电星系。LRIS 狭缝（126°）方向的选取接近于星位角（在光谱分光观测期间，星位角在 100° 至 150° 之间变化），这样的选取还能同时获得位于 53W091 东南方，距离几个角秒的另一个红星系的光谱。该星系看起来与 53W091 具有相同的红移（*z* = 1.55），有趣的是它们几乎具有相同的光谱能量分布，这一点为证明 53W091 的光学－红外性质本质上不受活动核影响提供了进一步的旁证。在射电波段，53W091 是一个沿着 131° 的位置角延伸（~4 arcsec）[16]、$S_{1.41GHz} = 22.5 \pm 1.0$ mJy 的陡谱源（$\alpha_{1.41GHz}^{0.61GHz} = 1.3 \pm 0.13$，此处 $f_\nu \propto \nu^{-\alpha}$）（参考文献 17）。*c*，用凯克望远镜在 7 月 25 日，8 月 31 日和 9 月 1 日，观测 5.5 小时所得到的 53W091 的光谱（下方的实线）。图中所示为静止参考系的光谱（假设红移 *z* = 1.552），与之相比较的是近邻椭圆星系 M32（调整后）的紫外光谱（上方的实线）和最佳拟合给出的模型光谱（虚线）。（这里，通过假设主序星拐点质量为 $1.35M_\odot$（等效于 F2 光谱型，对应于 3.5 Gyr 年龄），并基于 Scalo 初始质量函数，通过光谱合成得到模型光谱。）为使 M32 和 53W091 的数据达到相

53W091 have been rebinned to the same (5 Å) resolution, with an additional median smooth being applied to the latter. For ease of comparison, the spectrum of M32 has been scaled to the same amplitude as that of 53W091 at $\lambda_{rest} = 2,897$ Å and then offset vertically by 5×10^{-18} W m^{-2}. The unambiguous nature of the redshift is demonstrated by the existence of at least 11 absorption features and two strong spectral breaks in the spectrum of 53W091 (indicated by vertical dashed lines) which are all reproduced in the rest-frame spectrum of M32 (in particular the "top-hat" feature between 2,640 and 2,750 Å is a feature unique to this spectral range, and rules out the (remote) possibility that the break observed at 7,400 Å could be the 4,000 Å break at lower redshift). These features (all except the reddest three of which are also reproduced in the model spectrum) are essentially those which are seen in the near-ultraviolet spectra of F and G stars, as can be judged by comparison with the near-ultraviolet spectrum of the Sun[19], vindicating our belief that the red colours of the millijansky radio galaxies are a result of their evolved stellar populations. Such features have not been detected before in the spectrum of a high-redshift galaxy, not even in the Keck spectrum of the reddest powerful radio galaxy at $z \sim 1$, 3C65, in which they are apparently swamped by broad Mg II emission[11].

The strength of such stellar features combined with the lack of detectable emission lines (Mg II is seen in absorption, in contrast to the situation in 3C65 (ref. 9)) indicates that the ultraviolet–optical light from radio galaxies selected at milijansky flux densities is essentially free from the contaminating effects (either direct or indirect) of their active nuclei. Moreover, the similarity of this near-ultraviolet spectrum to that of low-mass main-sequence stars suggest that the red optical–infrared colour of this galaxy results from an evolved stellar population rather than, for example, a significant contribution from a dust-obscured quasar[20-22] (a viewpoint supported by the very similar appearance of 53W091 at optical and infrared wavelengths; Fig. 1).

To investigate the extent to which our data can constrain the age of 53W091, we first calculated the best-fit age produced by an updated version of the evolutionary synthesis models of Guider-doni and Rocca-Volmerange[23]. Considering a model in which 53W091 is formed in a single burst of star-formation and evolves passively thereafter, we find that the observed optical spectrum and the infrared colours ($R - K = 5.8$, $J - K = 1.7$, $H - K = 0.8$) are all perfectly reproduced by the models only at a time of 4 Gyr after cessation of star-formation activity. Next, we considered the most recent versions of the models of Bruzual and Charlot[24]. It is known that red optical–infrared colours are produced more rapidly by these models, and indeed we found that at $z = 1.55$ the observed optical–infrared colour is reproduced after only 1.5 Gyr. But the dependence of $R - K$ colour on age is controversial (see below) and so it was with interest that we found these same models could not reproduce the shape of the Keck spectrum until an age > 3 Gyr, while to produce spectral breaks at $\lambda_{rest} = 2,635$ and 2,897 Å of the strength observed in 53W091 requires an age > 4 Gyr. Third, we considered the models of Bressan, Chiosi and Fagotto[25], which, assuming solar metallicity, yielded a best-fit age of 3 Gyr (again in good agreement with the above results).

We derived ages using these three alternative models of galaxy evolution because it is well known that different models can produce significantly different ages from a given set of data[26]. But much of the difference between these models lies in different treatments of post-main-sequence evolution, and although this is expected to have a significant effect

314

同的光谱分辨率 (5 Å)，我们进行了重新分区间统计，而对于后者我们还采用了额外的中值平滑。为了便于比较，我们还将 M32 的光谱进行缩放，使其振幅与 53W091 在 λ_{rest} = 2,897 Å 处的振幅相同，并将其垂直移动了 5×10^{-18} W·m^{-2}。53W091 存在至少 11 个吸收特征和 2 个很强的光谱不连续特征 (由垂直虚线表示)，因此可以证明红移的准确性，这些光谱特征全都可以在 M32 的静止参考系光谱中找到对应 (尤其是介于 2,640 Å 和 2,750 Å 之间该谱段独一无二的"高帽"特征，这一特征可以排除我们在 7,400 Å 观测到的不连续特征是由低红移处的 4,000 Å 不连续所导致的 (极小的) 可能性)。本质上，这些特征 (除了可以由模型光谱重现的最红的 3 个光谱特征以外) 可以在 F 型星和 G 型星的近紫外光谱中观测到，这一点可以通过与太阳的近紫外光谱[19] 比较得到验证，以上结果表明：毫央射电星系中偏红的色指数是由它们当中演化的星族所造成的。在此之前，在高红移星系的光谱中还没有探测到这些特征。即使在 z~1 处最红的强射电星系 3C65 的凯克光谱中，这些特征也明显被宽的 Mg II 发射所淹没，从而没有被探测到[11]。

考虑到该星系的这些恒星特征的强度以及在星系中没有探测到明显的发射线 (与 3C65 相反，这里 MgII 是吸收线[9])，这表明流量密度在毫央范围的射电星系的紫外–光学辐射实质上没有被活动星系核 (直接或间接地) 污染。此外，该星系的近紫外光谱与小质量主序星的光谱相似，这表明该星系的光学–红外色指数偏红是由演化的星族所导致，而并非源于尘埃遮蔽类星体[20-22] 的显著贡献 (53W091 在光学和红外波段的观测图像非常相似支持了这一观点；图 1)。

为了研究我们的数据对该星系年龄的限制能力，我们首先利用古德尔多尼和罗尔–沃尔默朗然[23] 的演化星族合成模型计算了该星系年龄的最佳拟合结果。假设 53W091 在一次单一的星暴中形成，随后被动演化，我们发现观测到的光学光谱和红外色指数 ($R-K$ = 5.8，$J-K$ = 1.7，$H-K$ = 0.8) 与恒星形成活动停止后 4 Gyr 的模型预言符合得相当好。接着我们考虑最近的布鲁苏阿尔和夏洛的模型[24]。众所周知，用这些模型可以更快地得到偏红的光学–红外色指数。事实上我们发现在红移 z = 1.55 处，观测到的光学–红外色指数可以由恒星形成活动停止后 1.5 Gyr 时的模型预言得到。但是这样得到的 $R-K$ 色指数对年龄的依赖是有争议的 (请见下文)，这是一个有趣的现象，因为我们发现只有当星系年龄大于 3 Gyr 时，这些相同的模型才能再现凯克望远镜观测到的谱形。不仅如此，为了拟合在 53W091 中观测到的在 λ_{rest} = 2,635 Å 和 2,897 Å 处的光谱不连续特征，这就要求星系年龄大于 4 Gyr。再者，我们还考虑了布雷森、基奥西和法戈托[25] 的模型，在假设太阳的金属丰度下给出的最佳拟合年龄是 3 Gyr (再次很好地符合了上面的结果)。

众所周知，对于一组给定的数据，因为不同的模型可以给出非常不同的年龄[26]，所以我们用三种不同的星系演化模型计算星系年龄。但是这些模型的主要差别在于对主序后演化采取了不同的处理方法，尽管我们预期这些不同会对理论预言的星系

on the predicted infrared–optical luminosity of the galaxy, its effect on the predicted near-ultraviolet spectral energy distribution (SED) should be minimal (an expectation apparently borne out by the fact that all three models are in good agreement over the age required to reproduce the ultraviolet SED). Indeed, because it is well documented that the near-ultraviolet spectrum ($\lambda < 3,000$ Å) of a stellar population with an age of a few Gyr should be determined simply by the turn-off point of the main sequence[27], and as the main-sequence lifetime of A $\rightarrow$ G stars is probably the best understood area of stellar evolution, it should be possible to date 53W091 in an appealingly model-independent manner by simply determining the spectral type of the main-sequence turn-off point. We have therefore used the latest stellar atmosphere models of Kurucz[28] to investigate the spectra produced by main-sequence stars at a variety of ages, and find that the single stellar spectral type which best describes the near-ultraviolet light from 53W091 between $\lambda \approx 2,000$ and 3,500 Å is F5 (with an effective temperature $T_{\mathrm{eff}} = 6,500$ K). Furthermore, an independent comparison with the International Ultraviolet Explorer satellite spectra of stars of various spectral types produces exactly the same result. However, to set a realistic limit on the age of this galaxy, one must integrate over the initial mass function (IMF) of stars from very low masses ($\sim 0.1 M_\odot$) up to the stellar mass at which the synthesized spectrum becomes unacceptably blue. Assuming a Scalo IMF[29], we find that the best-fitting main-sequence turn-off point occurs at spectral type F2 ($T_{\mathrm{eff}} = 6,900$ K), equivalent to a stellar mass of $1.35 M_\odot$. The main-sequence lifetime of a star of this mass is well established (to within 5%) to be 3.5 Gyr. Such an age is reassuringly consistent with the ages indicated by the different evolutionary synthesis codes considered above, and the fact that this simple model provides such an excellent description of the data confirms that we are justified in ignoring the contribution of post-main-sequence stars (Fig. 1c).

We have considered carefully the robustness of our result, paying particular attention to the ways in which we could possibly have over-estimated the age of 53W091. First, if the validity of the models at near-ultraviolet wavelengths is accepted, an age younger than 3 Gyr is strongly excluded by the overall shape of the ultra-violet SED of 53W091 (Fig. 2a). Of course, the inferred age may in principle be reduced by truncating the stellar IMF at almost exactly $1.35 M_\odot$, but this requires considerable fine tuning (any significant population of A stars will dominate the spectrum for ~ 3 Gyr) and, if true, it would be expected that the galaxy would display other signs of youth, such as strong emission lines which are not seen.

的红外-光学光度产生显著影响，但是它们对理论预言的近紫外光谱能量分布（SED）的影响应当非常小（为重现星系紫外光谱能量分布，三个模型均得到了相一致的星系年龄限制，因此这是一个很显然的推论）。事实上，有案可稽的是年龄为几十亿年的恒星星族的近紫外光谱（$\lambda < 3{,}000$ Å）是由主序的拐点确定的，同时 A 型星至 G 型星在主序阶段的寿命可能是在恒星演化领域中研究得最好的，因此我们有可能利用不依赖于模型的方法，通过简单地确定主序[27]拐点的光谱型来判断 53W091 的年龄。所以我们利用库鲁茨[28]最新的恒星大气模型来研究不同年龄主序星的光谱。结果发现能将 53W091 从波长 2,000 Å 至 3,500 Å 之间的近紫外辐射描述得最好的是单一恒星光谱型 F5 型星（有效温度 $T_{\text{eff}} = 6{,}500$ K）。此外，通过将 53W091 与国际紫外探测卫星（IUE）观测到的不同光谱型恒星光谱进行比较，我们得出了几乎完全相同的结果。然而，要对该星系的年龄给出一个真实限制，我们必须对初始质量函数（IMF）从非常小的质量（$\sim 0.1 \, M_{\odot}$）开始进行积分，直至某一质量上限，使得当时的合成光谱蓝到超出了可接受的范围。采用 Scalo 的初始质量函数[29]，我们发现最佳拟合的主星序拐点与 F2 光谱型相对应（$T_{\text{eff}} = 6{,}900$ K），等效恒星质量为 1.35 $M_{\odot}$。该质量范围的恒星其公认的主序星寿命为 3.5 Gyr（在 5% 的误差范围内）。这里得到的年龄结果再次被确认与上面提到的利用不同演化合成模型得到的年龄相一致。该简单模型能够如此好地描述观测数据，证实了我们忽略后主序星（图 1c）的贡献是合理的。

　　我们仔细地考虑了我们结果的稳定性和可靠性，特别关注了那些有可能高估 53W091 年龄的方法。首先，如果我们认为模型在近紫外波段是可靠的，那么 53W091 星系在紫外波段的光谱能量分布的整体形状（图 2a）可以强烈地排除该星系年龄小于 3 Gyr 的可能性。当然从原理上讲，虽然可以在近似等于 1.35 $M_{\odot}$ 处对恒星的初始质量函数进行截断来减小所推断的星系年龄，但是这就要求相当精确地调整参数（任何明显的 A 型星星族成分将会主导年龄 ～3 Gyr 的星系光谱特征）。而且如果以上想法确实是正确的，那么我们可以预期在该星系中探测到表征年轻的其他特征，例如强的发射线，然而我们并没有观测到。

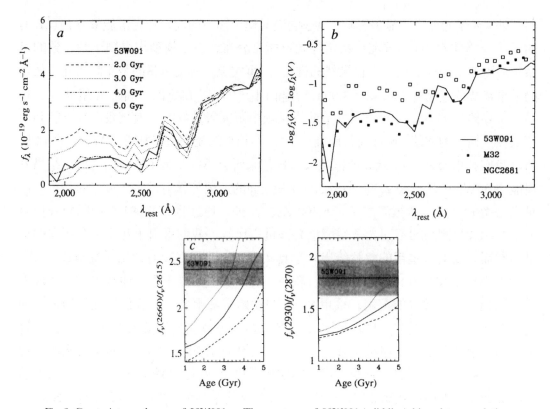

Fig. 2. Constraints on the age of 53W091. *a*, The spectrum of 53W091 (solid line), binned to a resolution of 50 Å, plotted in terms of f_λ, and compared with the SED produced by the passively evolving main-sequence model at ages of 2, 3, 4 and 5 Gyr after cessation of star-formation activity. The model spectra have been normalized to the observed flux density at 3,200 Å to illustrate why ages ≤ 3 Gyr can be formally rejected at a high level of significance on the basis of the overall shape of the ultraviolet SED. Models younger than 3 Gyr overpredict the flux density at $\lambda \approx 2,100$ Å by ~100% (the relative flux-calibration of the spectrum is accurate to ~10–15% across the full wavelength range). We stress that, in the regime of interest here, the connection between the spectral type of the main-sequence turn-off point and age is well-established (to within 5%), because stellar ages for main-sequence stars in the mass range 1–1.5 $M_\odot$ are essentially unaffected by uncertainties such as assumed mass loss, choice of convection theory or equation of state, and the opacities in the corresponding temperature range are well known. *b*, The spectrum of 53W091 (solid line), binned to a resolution of 50-Å, and this time plotted in terms of $\log f_\lambda\,(\lambda) - \log f_\lambda\,(V)$ to allow direct comparison with the normalized ultraviolet SEDs of two nearby early-type galaxies, M32 (filled squares) and NGC2681 (open squares)[30]. The rest-frame *V*-band flux density, $f_\lambda(V)$, of 53W091 was determined from a weighted average of the observed *J*- and *H*-band flux densities, measured through an aperture equivalent to that used in the flux calibration of the optical spectrum. Recent studies of M32 both at optical[31] and ultraviolet[27] wavelengths are in agreement that the most recent star-formation activity in this galaxy occurred 4–5 Gyr ago, whereas in NGC2681 star formation appears to have ceased 1–2 Gyr before the epoch of observation[32]. The level and shape of the rest-frame ultraviolet SED of 53W091 is more like that of M32, consistent with an age of 3–4 Gyr for this high-redshift galaxy. *c*, The observed strength of the spectral breaks at 2,635 Å (left-hand plot) and 2,897 Å (right-hand plot) compared with the break strengths predicted by the evolving main-sequence model as a function of age for three different choices of metallicity—solar (solid line), twice solar (dotted line) and 0.2 × solar (dashed line). In each plot the horizontal line indicates the break strength as measured from the spectrum shown in Fig. 1*c*, and the shaded area indicates the uncertainty in this estimate. The strength of the breaks in both the data and the models was determined from the ratio of the average value of f_v in 30-Å bins centred on the two wavelengths indicated on the vertical axis of each plot. Based on the (reddening-independent) strength of these features the inferred

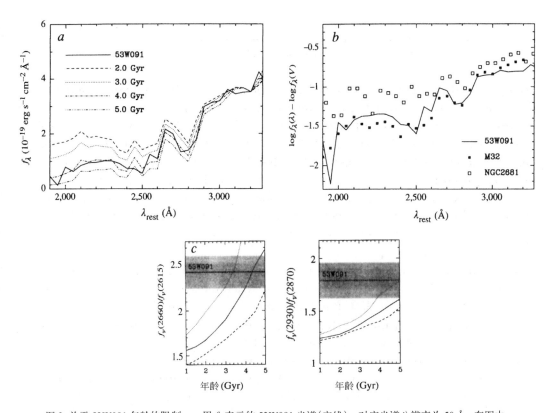

图 2. 关于 53W091 年龄的限制。a，用 f_λ 表示的 53W091 光谱（实线），对应光谱分辨率为 50 Å。在图中，我们将恒星形成活动停止后 2 Gyr、3 Gyr、4 Gyr、5 Gyr，由被动演化的主序星模型给出的光谱能量分布与 53W091 光谱进行了比较。我们对这些模型光谱进行归一化，使得 3,200 Å 处的模型光谱强度与观测流量密度一致，这样模型与观测所得的紫外光谱能量分布的整体形态的比较结果很自然地解释了可以以很高的置信度水平正式排除该星系年龄 ≤3 Gyr 的原因。年龄小于 3 Gyr 的模型所预言的流量密度在波长 $\lambda \approx 2{,}100$ Å 处高出观测结果 ~100%（在整个波长范围内，光谱相对流量定标精度约为 10%~15%）。我们强调，在本文所感兴趣的范围内，主序星拐点对应的光谱型与年龄的联系是公认的（在 5%内），这是因为对于质量范围在 1 $M_\odot$~1.5 $M_\odot$ 的主序星而言，恒星年龄本质上不受某些不确定因素影响，例如假设的质量损失，所选取的对流理论或物态方程；并且在对应温度范围内恒星的不透明度也是众所周知的。b，用 $\log f_\lambda(\lambda) - \log f_\lambda(V)$ 表示的 53W091 的光谱（实线），这样可以直接与两个近邻早型星系（M32，实心方块；NGC2681，空心方块）[30] 的归一化紫外光谱能量分布相比较，对应光谱分辨率为 50 Å。53W091 在静止参考系中 V 波段的流量密度 $f_\lambda(V)$ 是由观测到的 J 波段和 H 波段流量密度的加权平均确定的，测量孔径与光谱流量定标所采用的孔径一样。最近在光学 [31] 和紫光 [27] 波段对 M32 的研究结果支持该星系最近的恒星形成活动发生在 4~5 Gyr 之前，而 NGC2681 的恒星形成看起来停止于观测时期 [32] 之前的 1~2 Gyr。由于在静止参考系中，53W091 紫外光谱能量分布的强度和形状与 M32 更相像，因此这将符合该高红移星系的年龄为 3~4 Gyr 的推断。c，在 2,635 Å（左图）和 2,897 Å（右图）处观测得到的光谱不连续特征的强度。与之相比较的是演化主序星模型在三种不同金属丰度（太阳金属丰度，实线；2 倍太阳金属丰度，点线；0.2 倍太阳金属丰度，虚线）下所预言的光谱不连续的强度随年龄的变化。两幅图中的水平线表示利用图 1c 中光谱所测量出的光谱不连续特征的强度，阴影区域表示对应的误差范围。根据每幅图纵坐标，我们以两个波长为中心，在其 30 Å 范围内计算 f_ν 的平均值，再根据它们的比值分别得到了数据和模型光谱不连续的强度。根据这些特征的（与红化无关的）强度可以推断出 53W091 的年龄超过 4 Gyr，即使假设该天体拥有超太阳金属丰度，要用小于 3 Gyr 的年龄来解释我们

age of 53W091 is > 4 Gyr, and it is hard to reconcile the data with an age younger than 3 Gyr even if one assumes super-solar metallicity. The observed strengths of the corresponding breaks in the observed spectrum of M32 are 1.6 and 2.1 respectively, consistent with an age of ~4 Gyr.

Second, an independent check of our derived age is provided by comparing the ultraviolet properties of 53W091 with those of nearby early-type galaxies whose ages have been determined by other means. In Fig. 2b the normalized rest-frame ultraviolet SED of 53W091 is compared with the IUE spectra of M32 and NGC2681[30]. Recent studies of M32 both at optical[31] and ultraviolet[27] wavelengths agree that the most recent star-formation activity in this galaxy occurred 4–5 Gyr ago, whereas in NGC2681 star formation appears to have ceased 1–2 Gyr before the epoch of observation[32]. The level and shape of the rest-frame ultraviolet SED of 53W091 is undoubtedly more like that of M32, again consistent with an age of 3–4 Gyr for this high-redshift galaxy.

A third concern is that in comparing the ultraviolet SED of 53W091 both with models and with nearby ellipticals we may have over-estimated the age of 53W091 by ignoring the possible reddening effect of dust. But if this were the case, then it would be expected that reddening-independent features such as the strengths of the spectral breaks at $\lambda = 2,635$ Å, 2,897 Å would indicate a younger age. In fact these two features in the spectrum of 53W091 appear stronger than the corresponding breaks in the ultraviolet spectrum of M32, and when compared with the predictions of the main-sequence model, each break independently yields an age > 3.5 Gyr (Fig. 2c). This impressive agreement leads us to conclude that although a dusty torus may be present in the centre of this galaxy (obscuring the active nucleus) its effect on the integrated starlight of the galaxy is negligible.

The only remaining issue is the sensitivity of the derived age to the assumed value of metallicity. We have therefore investigated the effect of varying the metallicity both in our own main-sequence models, and in the full spectral synthesis code of Bressan. Both models yield very similar results; the derived age increases to 4.5 Gyr if 1/5 of solar metallicity is assumed, but is reduced by 0.5 Gyr (to 3 Gyr) if the metallicity is doubled to twice solar (Fig. 2c). Thus, although the age–metallicity degeneracy is hard to break (similar results are obtained by fitting either to the overall SED or to the strength of the two breaks), we conclude that very large values of metallicity are required to reduce the derived age of 53W091 to less than 3 Gyr.

Having considered the ways in which we might have overestimated the age of 53W091, we should stress that we regard it as more plausible that 3.5 Gyr may be a significant under-estimate of the true age of this galaxy. First, the spectral breaks indicate a larger age. Second, and more importantly, all of the model fits discussed above assume that this large elliptical galaxy was formed in an instantaneous burst after which star-formation completely ceased; if one assumes a burst of significant duration, or subsequent star-formation activity the derived age increases accordingly (for a 1-Gyr formation starburst,

的数据仍然十分困难。在 M32 的观测光谱中对应的光谱不连续的强度分别是 1.6 和 2.1，这与该星系的年龄约为 4 Gyr 是一致的。

其次，我们还通过将 53W091 与用其他方法确定年龄的近邻早型星系的紫外性质进行比较，来独立检验我们得到的星系年龄。图 2b 显示了 53W091 星系在静止参考系中归一化的紫外光谱能量分布与 M32 和 NGC2681[30] 的 IUE 光谱的比较结果。最近对 M32 在光学 [31] 和紫外 [27] 波段的研究证实在该星系中绝大多数恒星形成活动发生在 4~5 Gyr 之前。然而 NGC2681 中的恒星形成看上去停止于观测时刻前的 1~2 Gyr[32]。在静止参考系中 53W091 紫外光谱能量分布的幅度和形状无疑与 M32 更相似，这再次佐证了该高红移星系的年龄为 3~4 Gyr。

第三，通过比较 53W091 与模型和近邻椭圆星系的紫外光谱能量分布发现，我们也许由于忽略了可能的由尘埃引起的红化效应而高估了该星系的年龄。但如果确实如此，那么我们将期望那些不依赖红化的光谱特征，例如在 $\lambda = 2,635$ Å 和 2,879 Å 处光谱不连续的强度，能够表征更小的星系年龄。实际上，53W091 光谱中的这两个不连续特征看上去要强于 M32 光谱中的情况，因此当把 53W091 光谱与主序星模型预言的光谱相比较时，每一个不连续特征都能独立给出该星系年龄大于 3.5 Gyr 这一结果 (图 2c)。这个令人印象深刻的一致性使得我们得出结论：尽管尘埃环可能存在于该星系的中心 (遮挡住活动核)，但它对该星系的整体星光分布的影响可以忽略不计。

最后，剩下的唯一问题是我们得到的年龄是否敏感依赖于所假设的金属丰度。为此我们在主序模型和完整的布雷森光谱合成模型中研究了金属丰度变化所带来的影响。这两个模型均给出了非常类似的结果：如果假设金属丰度为太阳金属丰度的 1/5，那么所得到的年龄将增加到 4.5 Gyr；但是如果当金属丰度为太阳金属丰度的 2 倍时，所得年龄将减小 0.5 Gyr (即等于 3 Gyr) (图 2c)。因此，尽管很难打破年龄–金属丰度简并 (不管是通过拟合整个光谱能量分布，还是拟合那两个光谱不连续特征的强度时，都可以得到类似的简并结果)，我们仍然可以得出结论：非常高的金属丰度才能使得计算出的 53W091 年龄小于 3 Gyr。

在考虑了可能导致高估 53W091 年龄的不同因素之后，我们应当强调这里更合理的情形是 3.5 Gyr 的结论可能大大低估了该星系的真实年龄。首先，以上提到的两个光谱不连续特征暗示了比 3.5 Gyr 更大的年龄。其次，更为重要的是，以上所有讨论的模型拟合都假设这个大椭圆星系是由一个瞬时星暴形成，而在此之后恒星形成完全停止。如果假设星暴具有明显的持续时间或者随后仍有恒星形成活动存在，那么推导出的星系年龄也将会相应增加 (对于一个持续 1 Gyr 的星暴，即使采用太阳的

the best-fit age is 4.5 Gyr even if solar metallicity is assumed).

We now consider briefly the far-reaching implications of a 3.5-Gyr-old galaxy at $z = 1.55$. This is the first time that such an unambiguously old object has been discovered at such large look-back times, and its existence sets strong constraints both on the first epoch of galaxy formation and on cosmological models. The age of the Universe at this epoch is $1.6h^{-1}$ Gyr if $\Omega = 1$, increasing to $2.7h^{-1}$ Gyr for $\Omega = 0.2$, or at most $3.5h^{-1}$ Gyr for $\Omega = 0.2$, $\Lambda = 0.8$ ($h \equiv H_0 / 100$ km s^{-1} Mpc^{-1}). Obviously the Einstein-de Sitter model is in difficulty, and even in a low-density Universe an age > 3 Gyr at $z = 1.55$ requires a formation redshift $z_f > 4$ (for $h = 0.65$). It thus seems clear that at least some galaxies formed at redshifts greatly in excess of the recent formation era inferred from data on field galaxies[33], and the existence of similarly old galaxies at still higher redshifts would potentially allow one to infer a non-zero cosmological constant.

(**381**, 581-584; 1996)

James Dunlop[*], John Peacock[†], Hyron Spinrad[‡], Arjun Dey[§], Raul Jimenez[†], Daniel Stern[‡] & Rogier Windhorst[‖]

[*] Institute for Astronomy, Department of Physics and Astronomy, The University of Edinburgh, Edinburgh EH9 3HJ, UK
[†] Royal Observatory, Edinburgh EH9 3HJ, UK
[‡] Astronomy Department, University of California, Berkeley, California 94720, USA
[§] NOAO / KPNO, 950 North Cherry Avenue, Tucson, Arizona 85726, USA
[‖] Department of Physics and Astronomy, Arizona State University, Tempe, Arizona 85287-1504, USA

Received 17 January; accepted 23 April 1996.

References:

1. McCarthy, P. J. *A. Rev. Astr. Astrophys.* **31**, 639-688 (1993).

2. Steidel, C. C., Pettini, M. & Hamilton, D. *Astr. J.* **110**, 2519-2536 (1995).

3. Steidel, C. C., Giavalisco, M., Pettini, M., Dickinson, M. & Adelberger, K. L. *Astrophys. J.* **462**, L17-L20 (1996).

4. Petitjean, P., Pécontal, E., Valls-Gabaud, D. & Charlot, S. *Nature* **380**, 411-413 (1996).

5. Dunlop, J. S., Guiderdoni, B., Rocca-Volmerange, B., Peacock, J. A. & Longair, M. S. *Mon. Not. R. astr. Soc.* **240**, 257-284 (1989).

6. di Serego Alighieri, S., Fosbury, R. A. E., Quinn, P. J. & Tadhunter, C. N. *Nature* **341**, 307-309 (1989).

7. Tadhunter, C. N., Scarrott, S. M., Draper, P. & Rolph, C. *Mon. Not. R. astr. Soc.* **256**, 53P-58P (1992).

8. Rigler, M. A., Lilly, S. J., Stockton, A., Hammer, F., Le Fèvre, O. *Astrophys. J.* **385**, 61-82 (1991).

9. Hamilton, D. *Astrophys. J.* **297**, 371-389 (1985).

10. Lilly, S. J. *Astrophys. J.* **333**, 161-167 (1988).

11. Stockton, A., Kellogg, M. & Ridgway, S. *Astrophys. J.* **443**, L69-L72 (1995).

12. Rawlings, S. & Saunders, R. *Nature* **349**, 138-140 (1991).

13. Dunlop, J. S. & Peacock, J. A. *Mon. Not. R. astr. Soc.* **263**, 936-966 (1993).

14. Windhorst, R. A., van Heerde, G. M. & Katgert, P. *Astr. Astrophys. Suppl. Ser.* **58**, 1-37 (1984).

15. Oort, M. J. A. & van Langevelde, H. J. *Astr. Astrophys. Suppl. Ser.* **71**, 25-38 (1987).

16. Oort, M. J. A., Katgert, P., Steeman, F. W. M. & Windhorst, R. A. *Astr. Astrophys.* **179**, 41-59 (1987).

17. Neuschaefer, L. W. & Windhorst, R. A. *Astrophys. J. Suppl. Ser.* **96**, 371-399 (1995).

18. Dunlop, J. S., Peacock, J. A. & Windhorst, R. A. in *Galaxies in the Young Universe* 84-87 (Springer, Berlin, 1995).

19. Morton, D. C., Spinrad, H., Bruzual, G. A. & Kurucz, R. L. *Astrophys. J.* **212**, 438-445 (1977).

20. McCarthy, P. J. *Publs. astr. Soc. Pacif.* **105**, 1051-1057 (1993).

21. Lacy, M., Rawlings, S., Eales, S. & Dunlop, J. S. *Mon. Not. R. astr. Soc.* **273**, 821-826 (1995).

22. Webster, R. L., Francis, P. J., Peterson, B. A., Drinkwater, M. J. & Masci, F. J. *Nature* **375**, 469-471 (1995).

金属丰度，最佳拟合的年龄为 4.5 Gyr）。

现在我们简要地讨论在红移 $z = 1.55$ 处发现年龄为 3.5 Gyr 年老星系的深远意义。首先，这是首次在如此大的回溯时间上发现的一个确凿无疑的年老天体。该天体的存在对星系形成的初始时刻和宇宙学模型都给出了很强的限制。假设 $\Omega = 1$，那么在该时刻宇宙的年龄是 $1.6h^{-1}$ Gyr。对于 $\Omega = 0.2$，宇宙年龄增加到 $2.7h^{-1}$ Gyr，而在 $\Omega = 0.2$，$\Lambda = 0.8$（$h \equiv H_0/100$ km $\cdot$ s^{-1} $\cdot$ Mpc^{-1}）的情况下，宇宙年龄最多为 $3.5h^{-1}$ Gyr。很明显在爱因斯坦 – 德西特模型下解释 53W091 的年龄是有困难的，即使在一个低密度的宇宙中，在红移 $z = 1.55$ 处，星系年龄 > 3 Gyr 也将要求星系形成的红移 $z_f > 4$（当 $h = 0.65$ 时）。因此比较清楚的是，根据场星系的数据 [33]，我们发现在高红移处一些星系的形成时间超过了现在的形成纪元，加之在更高红移存在同样年老的星系，这可能使我们得到一个非零的宇宙学常数。

（李海宁 翻译；吴学兵 审稿）

23. Guiderdoni, B. & Rocca-Volmerange, B. *Astr. Astrophys.* **186**, 1-21 (1987).

24. Bruzual, G. B. & Charlot, S. *Astrophys. J.* **405**, 538-553 (1993).

25. Bressan, A., Chiosi, C. & Fagotto, F. *Astrophys. J. Suppl. Ser.* **94**, 63-115 (1994).

26. Charlot, S., Worthey, G. & Bressan, A. *Astrophys. J.* **457**, 626-644 (1996).

27. Magris, G. C. & Bruzual, G. A. *Astrophys. J.* **417**, 102-113 (1993).

28. Kurucz, R. *ATLAS9 Stellar Atmosphere Programs and 2km/s Grid CDROM* Vol. 13 (Smithsonian Astrophysical Observatory, Cambridge, MA, 1992).

29. Scalo, J. M. *Fund. Cosm. Phys.* **11**, 1-278 (1986).

30. Burstein, D., Bertola, F., Buson, L. M., Faber, S. M. & Lauer, T. R. *Astrophys. J.* **328**, 440-462 (1988).

31. Bressan, A., Chiosi, C. & Fagotto, F. *Astrophys. J. Suppl. Ser.* **94**, 63-115 (1994).

32. Windhorst, R. A. *et al. Astrophys. J.* **380**, 362-383 (1991).

33. Cowie, L. L., Hu, E. M. & Songaila, A. *Nature* **377**, 603-605 (1995).

Acknowledgements. We thank J. Davies for making the IRCAM3 service observations, A. Bressan for providing the age–metallicity models, and D. Burstein for supplying the IUE spectrum of M32. The United Kingdom Infrared Telescope is operated by the Royal Observatories on behalf of the UK Particle Physics and Astronomy Research Council. The W.M. Keck Observatory is a scientific partnership between the University of California and the California Institute of Technology, made possible by the generous gift of the W.M. Keck Foundation. This work was supported by the US National Science Foundation.

324

红移 1.55 处的一个年龄为 3.5 Gyr 的星系

Discovery of Ganymede's Magnetic Field by the Galileo Spacecraft

M. G. Kivelson *et al.*

Editor's Note

The Earth has a fairly strong magnetic field, and Mercury a weak one. Venus and Mars do not have magnetic fields. Although Jupiter's Galilean moon Io probably has one, it is a special case because of the extreme tidal forces on the moon created by Jupiter and the fact that it is immersed in Jupiter's massive field. It was quite surprising when, in this paper, Margaret Kivelson and her colleagues reported that Ganymede, the largest of the four Galilean moons of Jupiter, generates its own internal magnetic field, probably through convection in the molten iron core—a process similar to that which generates the Earth's magnetic field. The field is strong enough to carve out a small, separate magnetosphere within Jupiter's own magnetosphere.

The Galileo spacecraft has now passed close to Jupiter's largest moon—Ganymede—on two occasions, the first at an altitude of 838 km, and the second at an altitude of just 264 km. Here we report the discovery during these encounters of an internal magnetic field associated with Ganymede (the only other solid bodies in the Solar System known to have magnetic fields are Mercury, Earth and probably Io[1]). The data are consistent with a Ganymede-centred magnetic dipole tilted by ~10° relative to the spin axis, and an equatorial surface-field strength of ~750 nT. The magnetic field is strong enough to carve out a magnetosphere with clearly defined boundaries within Jupiter's magnetosphere. Although the observations require an internal field, they do not indicate its source. But the existence of an internal magnetic field should in itself help constrain models of Ganymede's interior.

ON Galileo's first inbound pass following orbital insertion, the magnetometer[2] measurements followed reasonably closely the predictions from a recent model of the magnetic field of Jupiter's magnetosphere[3] that we refer to as the KK96 model. (This model consists of the O6 model[4] of Jupiter's internal field plus the field of a warped and hinged current sheet parametrized to fit the magnetic field measured on the Pioneer 10 outbound pass near the dawn meridian.) The field increased in magnitude with approach to Jupiter and varied in orientation at Jupiter's rotation period. Data from 00:00 UT (universal time; h:min) to 12:00 UT on 27 June 1996 are plotted in Fig. 1; the Ganymede-associated perturbation is clearly apparent at 06:29 UT. The Ganymede encounter occurred well off the jovian magnetic equator (identifiable by the reversal of sign of B_r (the coordinate system is defined in Fig. 1 legend)) and thus in a region of relatively low

伽利略木星探测器发现木卫三的磁场

基维尔森等

编者按

地球有相当强的磁场，而水星磁场较弱。金星和火星没有磁场。尽管木星的伽利略卫星木卫一(Io)可能有磁场，但这是一个特殊的例子，因为这颗卫星受到了极强的潮汐力并且处于木星的强磁场中。在本文中，玛格丽特·基维尔森和她的同事们报道了相当令人吃惊的结果，木星的四颗伽利略卫星中最大的木卫三(Ganymede)有可能通过与熔融铁核作用产生了自己的内源磁场，该过程与地球磁场的产生相似。这个磁场强到足以在木星的磁层中开拓出一个小的独立磁层。

伽利略木星探测器已经两次掠过木星最大的卫星——木卫三。第一次距木卫三表面高度 838 km，第二次高度只有 264 km。在这里，我们报道在这两次与木卫三有关的内源磁场相遇期间的发现(在太阳系其他固态天体中已知具有磁场的只有水星、地球，可能还有木卫一 [1])。观测数据与以相对于木卫三自旋轴倾斜约 10° 的中心磁偶极子一致，且赤道处表面场强约为 750 nT(nT，纳特)。这个磁场的强度足以在木星磁层中开拓出一个具有明确边界的磁层。虽然观测结果需要用一个内源磁场来解释，但并不确定其来源。虽然如此，内源磁场的存在本身也可以帮助我们约束木卫三内部的模型。

在伽利略号入轨后的首次降轨过程中，磁强计 [2] 的测量结果与近期的木星磁层的磁场模型 [3]KK96 的预测结果符合得很好。(这个模型包括木星内源磁场的 O6 模型 [4] 加上一个翘曲铰连式的电流片，这个电流片的参数通过对先驱者 10 号在近黎明子午线附近升轨时的测量数据进行拟合得到。)在接近木星过程中场强增大，并且磁场方向变化的周期与木星自转周期相同。图 1 显示了从 1996 年 6 月 27 日 00:00 UT (世界时，小时:分钟)到 12:00 UT 的磁场数据，06:29 UT 出现了一个明显与木卫三有关的扰动。与木卫三的相遇发生在远离木星磁赤道的地方(从 B_r 的正负号反转可以看出来(坐标系的定义见图 1 注))，故在等离子体密度相对较低的区域内。最近点在木星磁层的等离子流中木卫三的下游。靠近木卫三，场强从 06:00 UT 的 107 nT

plasma density. Closest approach was downstream of Ganymede relative to the flowing plasma of Jupiter's magnetosphere. Near to Ganymede, the field increased from 107 nT at 06:00 UT to 480 nT at 06:29:07 UT (h:min:s) near closest approach, and then decreased to background, returning to 118 nT at ~07:00 UT. A considerable rotation accompanied the change in field magnitude.

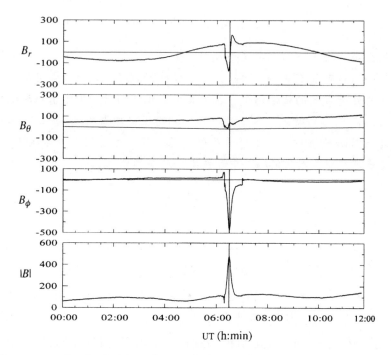

Fig. 1. Magnetometer data for the inbound pass to Jupiter on 27 June 1996 over a radial distance range from 17.44 to 13.14 jovian radii from Jupiter. The data are given in a Jupiter-centred right-handed spherical coordinate system (r, θ, ϕ) with r the radial distance, θ the angle from the spin axis, and ϕ the azimuthal angle. This is a variant of System III (1996)[14] with ϕ increasing westward. The ten-hour periodicity arises from Jupiter's rotation. Closest approach to Ganymede was at 06:29:07 UT. All times are given as spacecraft event time, which is universal time at the spacecraft. The magnetic field vector is $\mathbf{B} = (B_r, B_\theta, B_\phi)$.

The data near closest approach for this first pass are shown in Fig. 2. The field near closest approach is approximately that of a magnetic dipole centred at Ganymede with surface strength 750 nT at the equator added to the KK96 model. The dipole north pole is tilted by 10° from the spin axis towards 200° Ganymede east longitude. (Longitude of 180° is radially outward from Jupiter.) Currents flowing in the magnetospheric plasma in Ganymede's environment perturb the magnetic signature[5,6], so the actual magnetic moment of Ganymede may be slightly overestimated by the vacuum field estimate. In Fig. 3 we show projections of the trajectory and of the perturbation field vectors measured along it. (The perturbation field is the vector difference between the observed field and the KK96 model field.) The figure illustrates that the field perturbations converge towards Ganymede, as expected in the vicinity of an internal dipole.

增加到 06:29:07 UT(小时:分钟:秒)最近点附近的 480 nT,然后又回落到背景水平,在大约 07:00 UT 回到 118 nT。伴随磁场强度变化有可观的旋转。

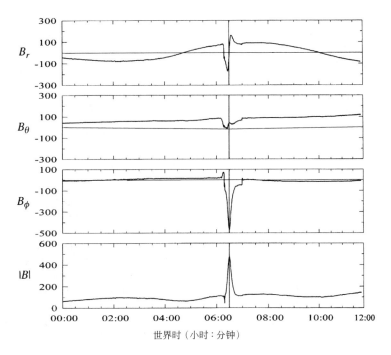

图 1. 1996 年 6 月 27 日向木星降轨过程中,径向距离从 17.44 个木星半径到 13.14 个木星半径的磁强计数据。数据基于以木星为中心的右手球坐标系 (r, θ, ϕ),其中 r 是径向距离,θ 是相对于自转轴的角度,ϕ 是方位角。这是 System III(1996)[14] 的一个方位角向西增大的变化形式。10 小时的周期来源于木星的自转。06:29:07 UT 到达距木卫三的最近点。所有时间都是飞船事件时间,也就是飞船上的世界时。磁场矢量 $\mathbf{B} = (B_r, B_\theta, B_\phi)$。

图 2 显示第一次经过距木卫三最近点附近的数据。最近点附近的场强大致相当于将一个以木卫三为中心且赤道表面场强为 750 nT 的磁偶极子叠加于 KK96 模型上的磁场强度。偶极子北极从自转轴向木卫三东经 200°(180°经度是从木星径向向外)方向倾斜 10°。由于木卫三周边磁层等离子体的电流扰动了磁场形态[5,6],所以按照真空场来估计,木卫三的实际磁矩可能被稍微高估。我们在图 3 中给出了伽利略号轨道的投影以及沿轨道测量的扰动场磁场矢量。(扰动场是指观测到的磁场与 KK96 模型场的矢量差。)此图显示扰动场汇聚于木卫三,符合内偶极子附近的预期。

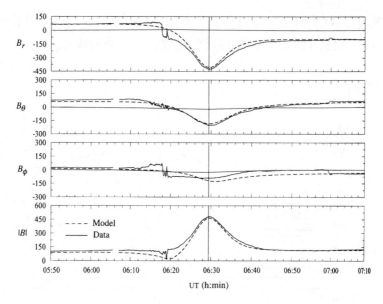

Fig. 2. Magnetic-field components and magnitude (solid lines) from 05:50 UT to 07:10 UT on 27 June 1996. The data are given in a Ganymede-centred coordinate system referenced to Ganymede's spin axis with r the radial distance, θ the angle from an axis parallel to Jupiter's spin axis and through Ganymede's centre, and ϕ the azimuthal angle, increasing eastward. Galileo's radial distance from the centre of Ganymede varied from $6.85\,R_G$ (radius of Ganymede, 2,634 km) at 05:50 UT to $1.32\,R_G$ at closest approach (marked by a vertical line) and out to $5.47\,R_G$ at 07:00 UT. Dashed lines show the superposition of a model of the jovian field[3] and the field of a Ganymede-centred magnetic moment described in the text. The data resolution is 0.333 s between 06:07 and 06:44 UT. Elsewhere the data resolution is 24 s.

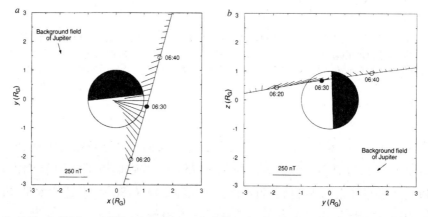

Fig. 3. Galileo's (27 June 1996) inbound towards Jupiter in the region near Ganymede (large circle, Ganymede; small circles, Galileo). The filled circle is near the closest approach to Ganymede. The plots use a coordinate system referenced to the direction of co-rotation (along $\hat{\mathbf{x}}$) and the spin axis of Jupiter, effectively the spin axis of Ganymede (along $\hat{\mathbf{z}}$). The unperturbed background field lies in the y–z plane roughly 50° outward from $-\hat{\mathbf{z}}$; $\hat{\mathbf{y}}$ is positive inward towards Jupiter. a, x–y projection, indicating the flow direction. The lines rooted along the trajectory are proportional to the projection of $\mathbf{B}-\mathbf{B}_{\text{model}}$ of Fig. 2 and the scale for the field perturbations is indicated. Key times are given. The projection direction of the background field is also shown. The night side of Ganymede is shaded. The terminator was crossed close to the centre of the wake, with the sunlit side corresponding to negative values of y. b, y–z projection of the trajectory and the perturbation field vectors. Note that the trajectory passes principally above Ganymede's equator.

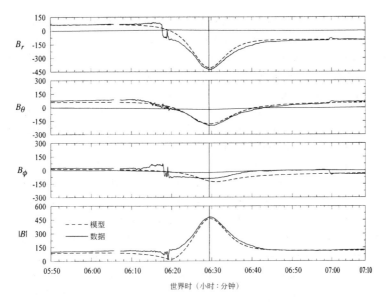

世界时（小时：分钟）

图 2. 1996 年 6 月 27 日 05：50 UT 到 07：10 UT 的磁场分量及其大小（实线）。数据基于以木卫三为中心、其自转轴为基准的坐标系。其中 r 是径向距离，θ 是相对一个平行于木星自转轴且通过木卫三中心的轴的角度，ϕ 是向东增大的方位角。伽利略号离开木卫三的径向距离从 05：50 UT 的 6.85 R_G（R_G 为木卫三半径，2,634 km）到最近点（垂直线）的 1.32 R_G，又在 07：00 UT 超过 5.47 R_G。虚线显示木星磁场模型 [3] 和正文中描述的木卫三为中心的磁矩的叠加。从 06：07 UT 到 06：44 UT，数据的分辨率为 0.333 s，其他地方数据的分辨率为 24 s。

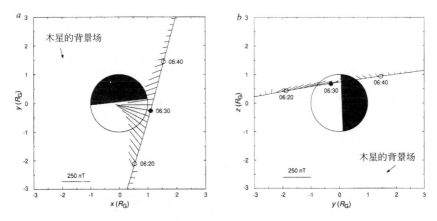

图 3. 1996 年 6 月 27 日伽利略号向木星降轨时经过木卫三附近区域（大圆圈，木卫三；小圆圈，伽利略号）。实心圆圈接近飞船距离木卫三的最近点。本图使用的坐标系是基于共转方向（沿 $\hat{x}$ 轴）以及木星自转轴，后者也基本上就是木卫三自转轴（沿 $\hat{z}$ 轴）。未受扰动的背景磁场在 y–z 平面上，从 $-\hat{z}$ 向外转大约 50°；$\hat{y}$ 轴正向向内指向木星。a，x–y 投影，显示流动方向。从轨迹发出的线段长度正比于图 2 中 $\mathbf{B} - \mathbf{B}_{model}$ 的投影，扰动场的比例尺和基本时刻点、背景场的投影方向也在图中标出。木卫三的夜半球用阴影表示。图中飞船轨迹的中点越过明暗界线，昼半球对应 y 的负值。b，轨迹的 y–z 投影以及扰动场矢量。注意轨迹主要经过木卫三赤道上方。

For the second pass, Ganymede was again located well above the jovian current sheet, probably in a region of low plasma density. The full data from the magnetometer and other particle and field instruments will be reported elsewhere, but measurements at ~1-minute resolution were acquired in advance of the full data set by operating the magnetometer[2] in a mode that averages and stores up to 200 field vectors for delayed transmission to the ground. Figure 4 presents these initial data and the vacuum field model previously described, which again provides a reasonable approximation to the data. Because of the low altitude of closest approach on this pass, the internal field dominates the signature (the field magnitude reached 1,146 nT at 18:59:45 on 6 September 1996). The small discrepancies between the data and the modelled field arise from various neglected effects. The vacuum superposition model does not allow for the fact that the external field is frozen into the flowing plasma of the jovian magnetosphere which modifies the interaction. In particular, currents that flow through the jovian plasma and close through Ganymede or a possible ionosphere[5,6] produce perturbations that tilt the field in the direction of the co-rotation flow. These neglected perturbations affect principally B_ϕ (see Fig. 2 legend) which is poorly fitted by the model. Possible contributions from higher-order multipoles of the internal field have not been considered.

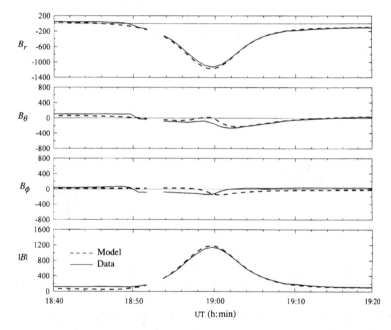

Fig. 4. Magnetic field components and magnitude (solid lines) from 18:40 to 19:20 UT on 6 September 1996 during Galileo's second fly-by of Ganymede. The coordinate system is defined in Fig. 2 legend. A short gap in the data removes artefacts associated with a programmed gain change. (The gap is covered in the recorded data.) The coordinate system is the same as in Fig. 2. Galileo's radial distance from the centre of Ganymede varied from $3.67R_G$ at 18:40 UT to $\sim1.10R_G$ at closest approach and out to $3.74R_G$ at 19:20 UT. Dashed lines show the superposition of the model fields discussed in the text. The data resolution is ~1 minute.

第二次经过时，木卫三同样处于木星电流片上方，可能位于低等离子体密度区域内。磁强计以及其他粒子和场测量仪器的完整数据将在另外的地方报告。但是在得到完整的数据之前，通过设定磁强计[2]模式为平均模式，并储存最多达200个场矢量以延迟输送到地面，先行得到了分辨率大约为1分钟的测量结果。图4展示了这些初始数据以及之前所述的真空场模型。模型与数据符合得也相当不错。由于这次经过木卫三时的高度很低，磁场特征主要由内源磁场决定（在1996年9月6日18:59:45，场强达到1,146 nT）。数据与模型的微小差别来源于各种被忽略的效应。真空叠加模型不能处理外部磁场被冻结在木星磁层的流动等离子体中的情况，这种情况对相互作用有修正。特别是穿过木星等离子体的电流和近距离流过木卫三或者可能的电离层[5,6]的电流会产生使磁场向共转流方向倾斜的扰动。这些被忽略的扰动主要影响B_ϕ（见图2注），在这个分量上模型和数据符合得很差。我们没有考虑可能的内源磁场的高阶多极分量的贡献。

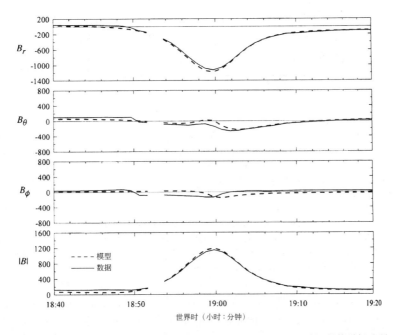

图4. 1996年9月6日18:40 UT到19:20 UT伽利略号第二次掠过木卫三时测得的磁场分量及其大小（实线）。坐标系定义见图2注。数据中间的小缺隙用来除去程序中设定的增益变化引起的假信号。（这个缺隙在记录的数据中被掩盖了。）所用坐标系与图2相同。伽利略号离开木卫三中心的径向距离从18:40 UT的3.67 R_G到最近点的大约1.10 R_G，并在19:20 UT超过3.74 R_G。虚线表示正文中讨论的叠加模型场。数据的分辨率为1分钟左右。

We have traced field lines near Ganymede in a simplified representation of the magnetic geometry (Fig. 5). The background (jovian) field is assumed uniform and tilted radially away from Jupiter. It has been added to the field of a Ganymede dipole pointing to 10° radially inward relative to the southward direction (which is the same as tilting the north pole 10° radially outward). This model neglects minor components of both the external field and the dipole moment out of the y–z plane (the coordinates are defined in Fig. 3 legend). A significant element of the schematic is the presence of field lines linked to Ganymede at both ends out to $\sim 2\,R_G$ (where R_G is the radius of Ganymede), implying that Ganymede carves out its own magnetosphere within the jovian magnetosphere. The interpretations of plasma wave emissions by Gurnett[7] are consistent with this interpretation.

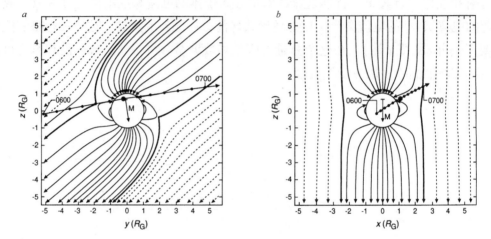

Fig. 5. Field lines in a vacuum superposition of a uniform external magnetic field of 120 nT lying in the y–z plane and tilted by 50° outward from $-z$ (a good approximation to the local value of the KK96 model field) and the field of a Ganymede-centred magnetic dipole with equatorial surface-field strength of 700 nT tilted 10° inward from $-z$. a, Field lines in the y–z plane. b, Field lines in the x–z plane. In both cases the projection of Galileo's trajectory is shown with dots every 5 minutes along the trajectory from 06:00 to 07:00 UT on 27 June 1996. Closest approach is marked with a large dot. Note that the projection in a appears to place closest approach at high latitude, but the projection in b makes clear that the actual latitude is much lower. The separatrices between field lines linked to Ganymede and field lines that do not intersect the moon, effectively the magnetopause, are shown as heavy lines. Dashed lines are jovian field lines not linked to Ganymede. Solid lines are field lines with at least one end on Ganymede. An arrow marked M is aligned with Ganymede's magnetic moment.

Figure 5 shows (heavy lines) the separatrix that encloses Ganymede's magnetosphere, identified as the region in which field lines link to Ganymede at least once. Galileo crossed into the Ganymede magnetosphere very near the boundary between field lines that intersect Ganymede's surface only once and field lines that intersect Ganymede's surface twice (see the trajectory in Fig. 5) and then passed through a region analogous to the cusp (or polar cusp) of a conventional magnetosphere[8]. The separatrix intersects the inbound trajectory at 06:20 UT, quite close to the time of the rather abrupt field rotation observed at ~06:17:45 UT, and intersects the outbound trajectory at 07:00 UT just at the time a small rotation was observed. The boundary is analogous to the magnetopause of a conventional magnetosphere, and its signature is also clear in the plasma wave data[7]. The trajectory passed

我们以简化的磁场几何表示追踪了木卫三附近的磁力线(图5)。假定背景(木星)磁场是均匀的，从木星沿径向向外倾斜。这个背景磁场被叠加到相对南方以 10°径向向内倾斜(相当于将北极沿径向向外倾斜 10°)的木卫三磁偶极子磁场上。这个模型忽略了外部磁场的次要分量和 y–z 平面之外的偶极矩次要分量(坐标系定义在图 3 注中)。本图的显著之处是在木卫三两边大约 2 R_G(R_G 是木卫三的半径)范围内存在与之相连的磁力线，暗示木卫三在木星磁层中开拓出了自己的磁层。这个说法符合古奈特的等离子波发射解释[7]。

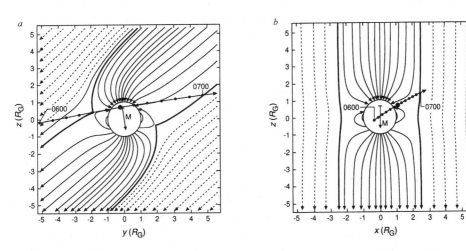

图 5. 真空叠加场的磁力线。 两个叠加磁场为：y–z 平面内从 –z 向外倾斜 50°(KK96 模型场在局部的不错的近似值)，强度为 120 nT 的均一外部磁场；木卫三为中心的赤道表面场强为 700 nT，从 –z 向内倾斜 10°的磁偶极子场。a, y–z 平面内的磁力线。b, x–z 平面内的磁力线。两图中伽利略号从 1996 年 6 月 27 日 06:00 UT 到 07:00 UT 的轨迹投影都以每 5 分钟一点标明。最近点用大一些尺寸的点表示。注意 a 图的投影中最近点看上去在高纬地区，但是 b 图的投影表明实际纬度要低得多。与木卫三相交的磁力线和不与这个卫星相交的磁力线的界面用粗实线表示，它实际上就相当于磁层顶。虚线是不与木卫三相交的木星磁场的磁力线。实线是与木卫三至少在一端相交的磁力线。以 M 标示的箭头显示木卫三的磁矩。

图 5 显示(粗实线)了木卫三磁层的界面，磁层定义为其磁力线至少与木卫三有一次相连的区域。伽利略号进入木卫三磁层的地点非常接近磁力线与木卫三相交一次和两次之间的边界区域(见图 5 中的轨迹)，然后穿越相当于一个普通磁层[8]的尖点(极尖)的区域。界面与飞船降轨轨迹在 06:20 UT 相交，这与在大约 06:17:45 UT 观测到的磁场较突然的旋转很接近。飞船升轨轨迹与界面在 07:00 UT 相交，同时观测到小的磁场旋转。界面相当于普通磁层的磁层顶，其特征在等离子波数据[7]中也很清楚。图中飞船轨迹穿过界面进入磁层的地点同时也恰好是它穿越磁中性线的位

through the magnetic neutral line in the schematic simultaneous with crossing of the inbound separatrix. In the data, the field magnitude dropped to 12.85 nT consistent with a near-neutral line encounter at 06:19:06 UT, more than one minute after the field rotation.

We have analysed the field rotation for the inbound magnetopause crossing. The normal to the surface lies almost in the y–z plane of Fig. 5a and is rotated radially outwards from the z-axis by 33°. This orientation suggests that in a more realistic model, the dipole centre might be shifted slightly towards positive z, the dimple in the separatrix would shift upward, and the trajectory would cross the separatrix just below the dimple. The small shift would produce a delay between the crossing of the separatrix and the encounter with the neutral line. It would account for the orientation of the magnetopause boundary during the inbound crossing, and it would hardly affect the outbound crossing. (Because the rotation at the outbound magnetopause is small, we are not able to evaluate the normal from the low-resolution data.)

It is not clear that we are justified in taking the details of a vacuum superposition model as far has been done in the above analysis, but, because the ambient jovian electron density was probably $\leqslant 1$ cm^{-3}, it is likely that neither the Alfvénic Mach number (flow speed/Alfvén speed) nor the plasma β (thermal pressure/magnetic pressure) was large enough for plasma effects to displace the boundary significantly. However, the Ganymede magnetosphere must divert the jovian plasma around the $\sim 4\,R_G$ (diameter) magnetic obstacle. This region of diverted flow is analogous to a traditional magnetosheath, although the diverted flow is not bounded by a shock. Just before the magnetopause crossing, low-frequency fluctuations were observed as is common in a magnetosheath. Elsewhere fluctuations are very small, although the time resolution of the data was sufficient to detect fluctuations on the scale of seconds between 06:07 UT and 06:44 UT. The absence of fluctuations suggests low plasma pressure within the magnetosphere, although a peak density of > 45 cm^{-3} has been reported[7].

The dipole moment of Ganymede (1.4×10^{13} T m^3 or 1.4×10^{20} A m^2) is close to that inferred for Io[1,9]. Thus Ganymede, whose angular momentum is also close to that of Io, follows the trend of the relation between the magnetic moment and the angular momentum for magnetized planets suggested by Blackett[10]. However, the discovery of an internal field at Ganymede does not bear directly on the interpretation of the signature at Io as the properties of the two bodies differ greatly. In some ways, the conditions at Io seem more favourable for field generation than those at Ganymede. Density is higher, a large core is known to be present[11], rotation is faster, and there is a source of heat that could drive convection, even though the heat is not deposited directly in the core. However, as details of the heating and transport processes are likely to be very different at the two jovian satellites, the discovery of a Ganymede field does little to support the arguments[1,9] for an internal magnetic field of Io.

Given that the dipole moments are comparable, it may seem puzzling that the evidence for an internal field at Ganymede is unambiguous whereas the evidence for an internal field at

336

置。在数据中，测到磁场旋转 1 分多钟后，在接近中性线的 06：19：06 UT，场强相应地降低到 12.85 nT。

我们分析了降轨穿越磁层顶时的磁场旋转。表面的法向矢量几乎处于图 5a 的 y–z 平面中，方向从 z 轴径向向外旋转 33°。这个定向表明在更接近实际的模型中，偶极子的中心应向 z 轴正方向稍微移动一些。界面凹窝中心应该略微上移，飞船轨迹应该在紧靠凹窝下方穿越界面。这个微小偏移会使轨迹穿越界面和中性线之间产生一个延迟。这还能解释降轨穿越时磁层顶边界的定向，但是对升轨穿越不会有什么影响。(因为穿出磁层顶的旋转很小，我们无法从低分辨率数据估计其法向。)

在以上分析中使用了很多真空叠加模型的细节，也许未必完全合理。但是因为周围的木星电子密度可能不大于 1 cm⁻³，不论是阿尔文–马赫数(流体速度／阿尔文速度)还是等离子体 β (热压／磁压)都不太可能大到使等离子体效应得以显著移动边界线位置。然而，木卫三的磁层必然使得木星等离子体流改道绕过自己这个 4 R_G (直径)的磁障碍物。改道后的流体区域相当于传统的磁鞘，只是流体没有激波边界。刚好在穿越磁层顶之前观测到的低频涨落是磁鞘中的常见现象。其他地方的涨落都十分微小，不过数据的时间分辨率还是能够探测出 06：07 UT 到 06：44 UT 之间几秒范围的涨落。虽然有等离子体最大密度大于 45 cm⁻³ 的报告 [7]，但缺少涨落表明磁层中等离子体压强很低。

木卫三的偶极矩(1.4×10^{13} T·m³ 或 1.4×10^{20} A·m²)和提到的木卫一的偶极矩 [1,9] 很接近，角动量也和木卫一的差不多，符合布莱克特提出的磁化行星的磁矩和角动量关系 [10]。但是由于两个天体的性质大为不同，木卫三内源磁场的发现不能直接用来解释木卫一的特征。从某些方面看，木卫一的条件比木卫三更适于产生磁场。它密度更高、内核更大 [11]、旋转更快，并且有一个能够推动对流的热源，尽管热量并不直接加载到内核上。但是由于木星的这两个卫星加热和运输的细节可能十分不同，所以木卫三磁场的发现对木卫一具有内源磁场的论证 [1,9] 不能提供什么支持。

由于二者磁矩差不多，故木卫三内源磁场的证据十分明确，而木卫一内源磁场的证据尽管很有力却不明确，这一点可能看起来有些让人感到迷惑。造成这种不明

Io, though compelling, is not unambiguous. The ambiguity arises both because the ratio of the magnitudes of the nominal surface equatorial magnetic field to the ambient jovian magnetic field is smaller by almost an order of magnitude at Io than at Ganymede and because the signature near Io is partially obscured by strong perturbations from ion pickup and charge exchange.

The source of Ganymede's magnetic field could be dynamo action in a molten iron core (or a salty-water internal ocean) or remanent magnetization in its interior. Ganymede's internal structure and thermal state determine which of these possibilities is the most likely. Schubert et al.[12] have considered how the field might be generated by dynamo action in an iron core or a watery mantle. The magnitude of Ganymede's magnetic field, almost an order of magnitude larger at the satellite's pole than the ambient jovian magnetic field, is in accord with theoretical expectations for a dynamo-generated field as is the direction of the dipole axis approximately aligned with the rotation axis[12]. The dominantly dipolar character of the field is consistent with generation deep within the body[13]. The large ratio between the polar field and the ambient field of the jovian magnetosphere at Ganymede cannot be explained by simple reactive processes such as paramagnetism or magneto-convection.

Remanent magnetization, though unlikely, cannot be ruled out completely. Whether in a core below the Curie temperature or in a magnetized shell in the interior of the moon, iron-rich material with a magnetic moment per unit volume $\mu = 40$–80 A m^{-1} would produce the observed dipole moment. The external field required to account for such strong magnetization is in excess of the field near Ganymede in its present environment, but cannot be ruled out during its evolution. The inferred values cannot be rejected as impossibly large because the natural remanent magnetization of iron-rich meteorites can be this large in some cases.

We note that the presence of a strong internal field carries significant implications for the form of plasma flow near Ganymede. There should be no direct access of torus plasma to the surface other than in regions near the poles. Thus, deposits of sulphur-rich material from the torus should be localized near the poles. Sputtering of neutral atoms or molecules from Ganymede or its atmosphere would occur only in the polar regions, and this might inhibit significant neutral cloud formation and would certainly impose constraints on the cloud shape near the source.

(**384**, 537-541; 1996)

M. G. Kivelson[*†], K. K. Khurana[*], C. T. Russell[*†], R. J. Walker[*], J. Warnecke[*], F. V. Coroniti[‡], C. Polanskey[§], D. J. Southwood[*‖] & G. Schubert[*†]

[*] Institute of Geophysics and Planetary Physics, [†]Department of Earth and Space Sciences, [‡]Department of Physics, University of California, Los Angeles, California 90095-1567, USA

[§] Jet Propulsion Laboratory, 4800 Oak Grove Drive, Pasadena, California 91109, USA

[‖] Department of Physics, Imperial College of Science, Technology, and Medicine, London SW7 2BZ, UK

Received 30 September; accepted 5 November 1996.

确的原因既是因为就表面赤道磁场强度与周围木星磁场强度之比来看，木卫一的数值比起木卫三几乎要小一个数量级，也是因为木卫一附近的磁场特征部分地被离子俘获和电荷交换带来的强烈扰动所掩蔽。

木卫三磁场的来源可能是熔融铁核（或者内部咸水体）中的发电机作用或者内部的剩余磁化强度。木卫三的内部结构和热学状态决定哪一个机制最为可能。舒伯特等 [12] 考虑了铁核或者水体幔层中发电机作用如何产生磁场的情形。木卫三的磁场在其极区几乎比周围木星磁场大一个数量级，偶极子轴与自转轴大致平行 [12]，这些都符合发电机产生磁场的理论预期。磁场的主要偶极特性与天体内部深处产生的场的性质相符 [13]。极区场强比起周围木星磁层在木卫三处的场强大很多，不能用诸如顺磁作用或磁对流这样的简单的电抗性过程来解释。

尽管发生剩余磁化强度的概率非常小，但是也不能完全排除。不论是居里温度之下的内核还是卫星内部磁化的壳层，富铁物质单位体积内磁矩达到 $\mu = 40 \sim 80\,A \cdot m^{-1}$ 就能产生观测到的偶极矩。要达到这样的磁化强度，需要的外部磁场大于现在环境下木卫三周围的磁场。但是不能排除在其演化历史中有过这样大的场强。由于一些富铁陨石的自然剩余磁化强度就有这么大，所以不能说这是不可能的。

我们注意到存在强的内源磁场对木卫三附近的等离子体流的形式有显著影响。除了极区之外，环形等离子流不能直接到达表面。因此环形流中的富硫物质只能沉积在极区。中性原子和分子从木卫三或者其大气层向外的溅射只能发生在极区，这会抑制显著中性云的产生，而且一定会限制源附近云的形状。

（何钧 翻译；杜爱民 审稿）

References:

1. Kivelson, M. G. *et al. Science* **273**, 337-340 (1996).

2. Kivelson, M. G., Khurana, K. K., Means, J. D., Russell, C. T. & Snare, R. C. *Space Sci. Rev.* **60**, 357-383 (1992).

3. Khurana, K. K. *J. Geophys. Res.* (submitted).

4. Connerney, J. E. P. in *Planetary Radio Emissions III* (eds Rucker, H. O., Bauer, S. J. & Kaiser, M. L.) 13-33 (Osterreichischen Akademie der Wissenschaftern, Vienna, 1992).

5. Neubauer, F. M. *J. Geophys. Res.* **85**, 1171-1178 (1980).

6. Southwood, D. J., Kivelson, M. G., Walker, R. J. & Slavin, J. A . *J. Geophys. Res.* **85**, 5959-5968 (1980).

7. Gurnett, D. A., Kurth, W. S., Roux, A., Bolton, S. J. & Kennel, C. F. *Nature* **384**, 535-537 (1996).

8. Hughes, W. in *Introduction to Space Physics* (eds Kivelson, M. G. & Russell, C. T.) 227-287 (Cambridge Univ. Press, New York, 1995).

9. Kivelson, M. G., Khurana, K. K., Walker, R. J., Warnecke, J. & Russell, C. T. *Science* **274**, 396-398 (1996).

10. Blackett, P. M. S. *Nature* **159**, 658 (1947).

11. Anderson, J. D., Shogren, W. L. & Schubert, G. *Science* **272**, 709-711 (1996).

12. Schubert, G., Zhang, K., Kivelson, M. G. & Anderson, J. D. *Nature* **384**, 544-545 (1996).

13. Elphic, R. C. & Russell, C. T. *Geophys. Res. Lett.* **5**, 211-214 (1978).

14. Dressler, A. J. in *Physics of the Jovian Magnetosphere* (ed. Dressler, A. J.) 498-504 (Cambridge Univ. Press, New York, 1983).

Acknowledgements. We thank S. Joy and A. Frederick for assistance in data preparation; R. L. Snare, J. Means, R. George, T. King and R. Silva for their varied contributions to the success of this effort; Y. Mei and D. Bindschadler for their support of magnetometer planning; V. Vasyliunas for discussions; and D. Stevenson for criticism. This work was supported by the Jet Propulsion Laboratory.

Correspondence should be addressed to M.G.K. (e-mail: mkivelson@igpp.ucla.edu).

Gravitational Scattering as a Possible Origin for Giant Planets at Small Stellar Distances

S. J. Weidenschilling and F. Marzari

Editor's Note

While inward migration of Jupiter-mass planets towards the parent star can explain the presence of such planets in tight, low-eccentricity (near-circular) orbits, it cannot explain those in high-eccentricity orbits. Here Stuart Weidenschilling and Francesco Marzari show that gravitational scattering—a kind of cosmic billiards—can explain these high-eccentricity systems. They say that a second distant planet in a high-eccentricity orbit, perhaps highly inclined to the plane of rotation of the star, would offer a telltale signature of such scattering. In most cases, one might expect there to be a third planet that is ejected completely. Such free-floating planets have since been reported, although with uncertain masses.

The recent discoveries[1-4] of massive planetary companions orbiting several solar-type stars pose a conundrum. Conventional models[5,6] for the formation of giant planets (such as Jupiter and Saturn) place such objects at distances of several astronomical units from the parent star, whereas all but one of the new objects are on orbits well inside 1 AU; these planets must therefore have originated at larger distances and subsequently migrated inwards. One suggested migration mechanism invokes tidal interactions between the planet and the evolving circumstellar disk[7]. Such a mechanism results in planets with small, essentially circular orbits, which appears to be the case for many of the new planets. But two of the objects have substantial orbital eccentricities, which are difficult to reconcile with a tidal-linkage model. Here we describe an alternative model for planetary migration that can account for these large orbital eccentricities. If a system of three or more giant planets form about a star, their orbits may become unstable as they gain mass by accreting gas from the circumstellar disk; subsequent gravitational encounters among these planets can eject one from the system while placing the others into highly eccentric orbits both closer and farther from the star.

THE most generally accepted model for the origin of the giant planets, Jupiter and Saturn, in our own Solar System assumes that solid planetesimals accumulated to form cores having masses of the order of ten times Earth's mass. Each core was then able to initiate accretion of gas from the surrounding solar nebula; the mass of the accreted gas became great enough for its own gravity to continue the process until that region of the nebula was exhausted, or tidal torques formed a gap at the planet's orbit[5]. Formation of such a core is believed to have required condensation of water (and perhaps other ices)

引力散射是近邻巨行星的可能起源

魏登席林，马尔扎里

编者按

尽管木星质量的行星向母星迁移可以解释此类行星存在于近邻的低偏心率（近圆）轨道中，但是这不能解释那些高偏心率轨道上的此类行星。在此，斯图尔特·魏登席林和弗朗切斯科·马尔扎里指出，引力散射——一种宇宙台球效应——可以解释这些高偏心率系统。他们指出，远处另外一颗处于高偏心率轨道（或许相对恒星自转平面有较大倾角）中的行星可能会显示出这种散射的迹象。在大部分情形中，有可能存在被完全抛射出去的第三颗行星。这种自由漂浮的行星已有报道，尽管质量不确定。

最近发现的几颗环绕类太阳恒星的大质量行星[1-4]给科学界提出了一个难题。巨行星（例如木星和土星）形成的传统模型[5,6]认为这类天体应该在距离母星几个天文单位处形成，不过这些新发现的行星除一颗外几乎都在1 AU内；故这些行星一定是在较远距离处形成，然后向内迁移的。一个可能的迁移机制是行星和演化中的星周盘之间的潮汐相互作用[7]。这样的机制导致行星具有小的圆形轨道，大部分新行星似乎是这样的情形。但是这些新行星中有两颗具有较大的轨道偏心率，这很难和潮汐相关模型一致。这里我们描述另外一个能解释这些大轨道偏心率的行星迁移模型。如果一个具有3个或者更多巨行星的系统在一颗恒星周围形成，由于它们通过吸积星周盘气体来获得质量，所以轨道可能变得不稳定；随后这些行星之间的引力交会可将一颗行星从系统内抛射出去，同时使其他行星进入距离恒星更近以及更远的高偏心率轨道。

关于太阳系巨行星（木星和土星）起源最为广泛接受的模型是固体星子聚集在一起形成10倍地球质量的核。接着每个核都能够从周围的太阳星云开始吸积气体；吸积气体的质量大到足以让自引力继续吸积气体直到那片区域的星云耗尽，或者潮汐力矩在行星轨道处形成一个空隙[5]。我们认为这种核的形成需要水的凝结（以及其他可能的冰物质）来提供足够的质量使得核能够在气体耗散之前增长。距太阳几个天文单位之内的高温将排除这种起源。广泛认为位于5.2 AU处的木星是在太阳星云的

to provide enough mass to allow growth of the core before dissipation of the gas. High temperatures within a few AU of the Sun would preclude such an origin. Jupiter, at 5.2 AU, is widely believed to have formed near the boundary of ice condensation in the solar nebula[8]. In contrast, at present six planets are known with masses comparable to Jupiter and orbits well inside 1 AU; these are companions of the stars 51 Pegasi, τ Boötis, v Andromedae, 55 Cancri, HD114762 and 70 Virginis. The first three all have semimajor axes $a \approx 0.05$ AU, the next has $a \approx 0.1$ AU, and the last two are at 0.4 and 0.47 AU. In addition, a planet of 47 Ursae Majoris is at a distance of 2.1 AU.

Lin *et al.*[7] suggested that giant planets migrated inwards while tidally linked to an evolving circumstellar disk that was accreting onto the star. This mechanism implies that planets must have formed simultaneously with the star itself, but it is difficult to account for the formation of planets on such short timescales[6]. Tidal linkage to the nebula also implies that any surviving planets should have essentially circular orbits. This appears to be the case for most of the newly-discovered planets, but 70 Vir B and HD114762 B have substantial orbital eccentricities ($e \approx 0.38$ and 0.25, respectively). This circumstance has led to the suggestion[9,10] that these objects are brown dwarfs that formed during the collapse of their pre-stellar clouds, rather than planets that formed within circumstellar disks.

We suggest an alternative origin for these bodies that is compatible with the more conventional model of formation of jovian planets. The accretion of their solid cores from planetesimals is stochastic, and should produce multiple embryonic cores with the potential to accrete gas if their masses become sufficiently large. There is no *a priori* reason why these cores should form with such wide separations that the orbits of the final planets would remain stable after accretion of gas increases their mass by an order of magnitude or more. The likely outcome in many cases would be that the first core to accrete gas outstrips its neighbours, and potential rivals are either accreted or ejected from the system by its perturbations. However, stochastic growth of the cores will yield some systems with multiple cores in orbits that are stable before they accrete gas, but too close to be permanently stable after they gain additional mass from the nebula.

In the classical three-body problem, two planets initially in circular orbits cannot make close approaches if their semimajor axes, a, differ by more than $2\sqrt{3} R_H$, where R_H is the so-called Hill radius, equal to $\bar{a}[(m_1 + m_2)/3M_*]^{1/3}$, where $\bar{a} = (a_1 + a_2)/2$ is their mean distance from the star of mass M_*, and m_1, m_2 are the planetary masses[11]. If their orbital separation is less than this critical value, close encounters can occur. Gravitational scattering can change their orbits, but cannot produce stability; the ultimate fate of such a system is collision between the two bodies. Conservation of angular momentum decrees that the orbit of the combined body is between those of the original pair, and so does not extend the range of allowed orbits.

Very different outcomes are possible for a system of more than two planets. Chambers *et al.*[12] found that for three or more planets of comparable mass, mutual perturbations

344

冰物质凝聚边界形成的 [8]。相反，目前知道的 6 颗质量和木星相当的行星轨道都在 1 AU 内；它们分别是恒星飞马座 51、牧夫座 τ、仙女座 ν、巨蟹座 55、HD114762 和室女座 70 的行星。前 3 颗行星的半长轴都为 $a \approx 0.05\,\text{AU}$，随后一颗的半长轴为 $a \approx 0.1\,\text{AU}$，最后两颗的半长轴分别为 0.4 AU 和 0.47 AU。此外，大熊座 47 的一颗行星位于 2.1 AU 处。

林潮等人 [7] 认为巨行星和正被吸积到恒星的演化中的星周盘发生潮汐相互作用时会向内迁移。这个机制表明行星一定是和恒星本身同时形成的，但是难以解释如何在这么短的时标内形成行星 [6]。与星云的潮汐联系也表明任何存在的行星本来应该有圆形轨道。大部分新发现的行星似乎都是这个情况，但室女座 70 B 和 HD114762 B 具有显著的轨道偏心率（分别为 $e \approx 0.38$ 和 $e \approx 0.25$）。这个情况表明这些天体是在其星前云坍缩时形成的褐矮星，而不是在星周盘内形成的行星 [9,10]。

我们提出这些天体的另外一种起源。这与更传统的类木行星形成模型一致。星子吸积形成固体核是随机的，并且应该形成多个胚胎核，当它们质量足够大时将能够吸积气体。没有先验的理由来解释为什么这些核应该在这么宽的间距内形成，以保证在吸积气体质量增加一个量级以上后，行星的轨道仍然保持稳定。很多情形中的可能结果是首个核吸积气体超过邻近核，潜在的竞争对手要么被吸积，要么因为受到这个核扰动而被抛射出系统。不过，核的随机增长将形成一些多核的系统，这些核的轨道在它们吸积气体之前稳定，但在它们从星云获得额外质量后由于距离太近而不能永久保持稳定。

在经典三体问题中，两颗开始处于圆轨道中的行星，如果它们的半长轴 a 相差大于 $2\sqrt{3}\,R_\text{H}$ 就不能靠得太近，这里 R_H 是所谓的希尔半径，等于 $\bar{a}[(m_1+m_2)/3M_*]^{1/3}$，此处 $\bar{a} = (a_1+a_2)/2$ 是它们距离质量为 M_* 的恒星的平均距离，m_1、m_2 是行星质量 [11]。如果它们的轨道间距比这个临界值小，则可能发生密近交会。引力散射能够改变它们的轨道，从而引起轨道不稳定；这样一个系统的最终命运是两个天体发生碰撞。角动量守恒使得合并后的天体的轨道位于它们初始轨道之间，并且不会超过允许的轨道范围。

超过两颗行星的系统的结果将非常不同。钱伯斯等人 [12] 发现对于三颗或者更多

could produce crossing orbits, even though each pair of bodies considered separately meets the Hill stability criterion. There appears to be no absolute stability in such systems; the timescale for orbits to become crossing merely increases with initial separation. With three (or more) planets, a greater variety of outcomes is possible as the result of close encounters. Angular momentum can be partitioned unequally among the planets, and the system need not remain completely bound to the star. The idea that systems of massive planets could be dynamically unstable was first mentioned by Farinella[13] in 1980. However, this suggestion was forgotten, as no such planets had been detected at that time, and there was no consensus as to how such systems could form.

We have carried out a series of numerical integrations of the orbits of three Jupiter-mass planets about a solar-mass star, using Everhart's[14] integrator. The initial orbits were circular, and separated by 4–5 Hill radii. Simulations were performed for coplanar orbits and for small (1–2°) initial inclinations, with similar outcomes. Mutual perturbations rapidly increased eccentricities until close approaches became possible. There followed a period of chaotic evolution with close encounters among the planets. The vast majority resulted in one planet being ejected on a hyperbolic trajectory. The remaining two planets usually had stable orbits. If the initial orbits were not strictly coplanar, the final orbits could also have significant mutual inclinations. There was no preference as to their fate; any of the three might be ejected, or end up in either the inner or outer orbit.

One example of such an outcome is shown in Fig. 1. Three Jupiter-mass planets had initially circular orbits at $a = 5.0$, 7.25 and 9.5 AU. After about 20,000 yr, the planet originally in the innermost orbit was ejected. The outermost original planet was left in a close orbit with $a = 2$ AU, $e = 0.78$, while the planet initially in the middle had a distant orbit with $a = 29$ AU, $e = 0.44$. These final orbits are stable against further encounters. The starting conditions were chosen to minimize computation time, and resulted in evolution on a timescale much shorter than the $\sim 10^6$ yr needed for the planets to reach their assumed masses by accretion of gas[6]. We have carried out other simulations with larger initial separations, in which orbits became crossing only after several million years.

Although ejection of one planet is by far the most common outcome, other end states are possible. A small fraction of cases end with collisions between planets, or sequential ejection of two planets, leaving a single planet orbiting the star. Impact of a planet onto the star itself cannot be ruled out, though we have no examples of such an outcome. It is beyond the scope of this Letter to explore the full range of possibilities. We note that such systems are chaotic, that is, the outcome is sensitive to small differences in initial conditions. Even for a fixed set of planetary masses and initial semimajor axes, different angular separations at the start of the simulation can lead to very different final configurations. Much additional work will be needed to map out the full range of outcomes and their statistical distributions as functions of the masses and initial orbits of the planets. The present work established that dynamical interactions can alter the cosmogonically imposed initial state of a planetary system.

颗质量相近行星，相互的扰动能够产生相交轨道，尽管分别对每对天体的考虑都符合希尔稳定性判据。在这样的系统中似乎没有绝对的稳定；轨道开始相交的时标仅随初始间距增加而增加。在三颗（或者更多颗）行星的系统中，由于密近交会而产生的结果可能会更为多种多样。角动量在这些行星间分配不均，系统不一定完全束缚于恒星。大质量行星系统可能动力学不稳定的想法首先是由法里内拉[13]于1980年提出的。不过，这个想法被人们遗忘，因为当时还没有探测到这样的行星，并且对如何能够形成这种系统没有共识。

我们使用埃弗哈特[14]的积分器对围绕太阳质量恒星的三颗具有木星质量的行星的轨道进行了一系列数值积分。初始轨道为圆形，间距4~5希尔半径。对共面轨道和初始小倾角轨道(1°~2°)的模拟结果类似。行星间的相互扰动使得偏心率迅速增大，直到行星间可能发生密近交会。接下来是行星发生密近交会的混沌演化时期。多数结果是一颗行星沿着双曲线轨道被抛射出去。剩下的两颗行星一般具有稳定的轨道。如果初始轨道不是严格共面，最终的轨道之间可能也具有明显的倾角。这对它们的命运没有太大影响；三颗行星中的任何一颗都有可能被抛射，或者最终处于靠内轨道或靠外轨道。

图1展示了这种结果的一个例子。三颗木星质量行星起初具有 $a = 5.0$ AU、7.25 AU 和 9.5 AU 的圆轨道。大约20,000年后，最初在最内轨道的行星被抛射出去。初始在最外的行星留存在 $a = 2$ AU，$e = 0.78$ 的近轨道上，开始处于中间的行星位于 $a = 29$ AU，$e = 0.44$ 的较远轨道上。这些最终轨道很稳定，不存在进一步的交会。初始条件的选择是为了使计算时间最小化，并且在比行星通过吸积气体[6]达到其假设质量所需要的一百万年短得多的时标上有演化。我们进行了更大初始间距的模拟，结果是轨道在仅仅几百万年后就开始相交。

尽管一颗行星的抛射是目前最常见的结果，但其他结局也是可能的。一小部分结果是行星之间发生碰撞或者相继抛射出两颗行星，只剩下一颗行星围绕恒星运转。一颗行星和恒星碰撞本身也不能够排除，尽管我们目前还没有这样的例子。探索所有可能性超过了本快报的范畴。我们注意到这样的系统是混沌的，即结果对初始条件的微小变化敏感。甚至对固定的行星质量和初始半长轴，模拟开始时不同的角间距也能导致非常不同的结果。想要得到全部可能的结果及其统计分布作为行星质量和初始轨道的函数，还需要很多额外的工作。目前的工作确定了动力学作用能够改变行星系统起源时的初始状态。

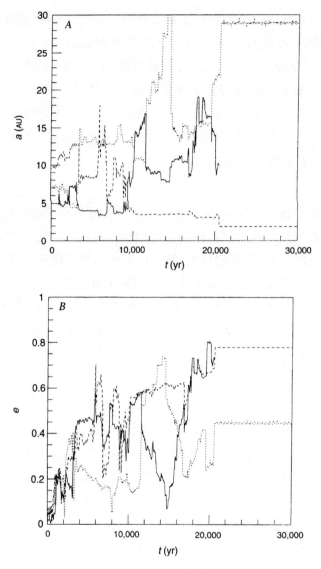

Fig. 1. Evolution of semimajor axes (panel *A*) and eccentricities (panel *B*) of a trio of Jupiter-mass planets orbiting a solar-mass star. Initial orbits are circular and coplanar, at distances of 5.0, 7.25 and 9.5 AU from the star. Their orbits become crossing, and after a series of close encounters the original inner planet is ejected hyperbolically at 21,000 yr. The planet originally in the outermost orbit is left in a close orbit ($a = 2$ AU, $e = 0.78$). The remaining planet has a distant orbit ($a = 29$ AU, $e = 0.44$).

Wetherill[15,16] has pointed out that there are problems with the timescale of formation of gas-giant planets by the core-accretion model (although these are less severe than for origin within a still-evolving disk). These considerations, and the lack of detection of such planets in earlier surveys, led him to suggest that they might be very rare[16]. If so, then it is surprising that our own system contains two such bodies, Jupiter and Saturn. It seems plausible that other planetary systems could produce three (or perhaps more) gas giants, particularly if their circumstellar nebulae were more massive than our own. If systems of multiple giant planets in unstable orbits are a common outcome of

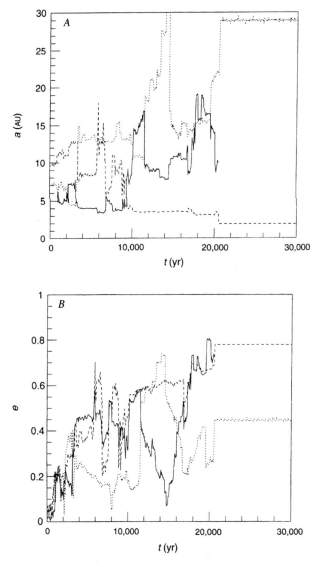

图 1. 一颗太阳质量恒星周围环绕的 3 颗木星质量行星的半长轴（A 图）和偏心率（B 图）的演化。初始轨道为圆形且共面，分别距离恒星 5.0 AU、7.25 AU 和 9.5 AU。它们的轨道开始相交，在经过一系列密近交会后，初始靠内的行星在 21,000 年通过双曲轨道被抛射。初始最靠外的行星则留在最近的轨道（$a = 2$ AU，$e = 0.78$）。另一颗行星的轨道很远（$a = 29$ AU，$e = 0.44$）。

    韦瑟里尔[15,16] 指出核吸积模型形成巨气体行星的时标有问题（尽管这些问题没有一个正在演化的盘的起源问题严重）。这些考虑，以及在较早的巡天中没有探测到这样的行星，导致他提出巨气体行星可能很少见[16]。如果是这样，那么我们自己的太阳系包含两个这样的天体——木星和土星——着实令人惊讶。其他行星系统能够形成三颗（或者可能更多）巨气体行星看起来是可能的，特别是如果它们的星周云比太阳系星云质量更大的话。如果不稳定轨道中的多颗巨行星系统是随机形成的核吸

stochastic core formation followed by gas accretion, then there are other implications for cosmogony. Many potential systems of terrestrial-type planets may be disrupted by a gas giant in their midst; we would owe our existence to the possibly fortuitous circumstance that only two cores in our own system, Jupiter and Saturn, grew large enough to accrete gas. It is not clear whether this fortunate outcome was purely stochastic, or influenced by the mass and density distribution of the solar nebula. There should also be a population of interstellar "Jupiters" that were ejected from forming systems.

Our model can account, at least qualitatively, for the large eccentricities observed for 70 Vir B and HD114762 B. It cannot explain the other planets unless some additional process acted to circularize their orbits after they were scattered inwards. Processes that could reduce eccentricities include tidal interactions with the circumstellar disk[17], drag of nebular gas enhanced by the planet's gravity[18], and accretion of a portion of the nebula by the planet after migration to a new orbit. Any or all of these might explain the low eccentricity observed for 47 UMa B, whereas the disks about 70 Vir and HD114762 may have had insufficient mass or lifetimes too short to accomplish such damping.

It is more difficult to account for very close orbits such as that of 51 Peg B. In principle, gravitational encounters can produce an eccentric orbit with very small periastron, which could then be circularized by the processes mentioned above, or by tidal dissipation within the star and the planet (Mayor and Queloz[1] derived a timescale of a few billion years for such tides to damp the eccentricity of 51 Peg B). The subsequent discoveries of other planets in similar orbits make this explanation unlikely, as such "star-grazing" orbits are rarely produced in our simulations. Observations suggest that a few per cent of solar-type stars may possess large planets at distances of 0.1 AU or less. Radial velocity surveys are most sensitive to close orbits, but it is unclear whether this bias can explain their apparent abundance. At present, there appear to be two classes of orbits: those that are circular and very near their stars, and more distant ones that may have significant eccentricities. If this trend is confirmed by additional discoveries, it would suggest that two mechanisms for planetary migration were effective. The close orbits may be the product of tidal locking to an evolving disk, while the eccentric ones are due to gravitational scattering.

One test of the two mechanisms is the existence of other planetary companions. Tidal locking does not require that other planets exist around a given star, but if they do, their orbits should be coplanar with the inner one. Gravitational scattering implies that a star with one close planet should have at least one other planet of comparable mass in a distant eccentric orbit; the two orbits may have high mutual inclination. Detection of such a distant companion will be difficult. Radial velocity perturbations vary in amplitude as $a^{-1/2}$, but their period increases as $a^{3/2}$. In the example of Fig. 1, the effect of the outer planet would be about 1/4 as large as that of the inner one, but its period more than 50 times longer. Detection of a planet by radial velocity variations requires a length of observation comparable to the orbital period. As high-precision radial velocity surveys have been conducted for about a decade, it is not surprising that the only planets discovered to date are close to their stars. More distant companions should be found

积气体的一般结果，那么这对天体演化还有其他意义。许多潜在的类地行星系统可能被巨气体行星瓦解；我们把我们的存在归功于可能的幸运环境——太阳系只有两个核（木星与土星）生长到足够大以吸积气体。目前还不清楚这样的幸运单纯是随机结果，还是受到太阳星云质量和密度分布的影响。应该存在一批从正在形成的系统抛射出去的星际"木星"。

我们的模型至少能够定性解释观测到的室女座 70 B 和 HD114762 B 的大偏心率。这不能解释其他行星，除非其他机制在它们被向内散射后发生作用以圆化其轨道。能够减小偏心率的机制包括与星周盘的潮汐相互作用[17]、被行星引力增强的星云气体的拖曳阻力[18] 和行星迁移到新轨道后对一部分星云气体的吸积。任何一个或者所有这些机制都可能解释观测到的大熊座 47 B 的低偏心率，而室女座 70 和 HD114762 的盘可能是由于没有足够的质量或是寿命太短，从而无法实现这样的偏心率减小。

对飞马座 51 B 这样非常近的轨道，解释起来更加困难。原则上，引力交会能够形成近星点非常小的偏心轨道，然后轨道可以通过上面提到的机制或者通过恒星与行星的潮汐耗散作用所圆化（梅厄和奎洛兹[1]推导出对这种潮汐阻尼衰减飞马座 51 B 偏心率的时标为几十亿年）。随后在类似轨道发现的其他行星排除了这个解释，因为这种"掠过恒星"轨道在我们的模拟中很少出现。观测表明百分之几的类太阳恒星可能在 0.1 AU 甚至更近的距离处拥有大行星。视向速度巡天对近轨道最灵敏，但是尚不清楚这种偏向性能否解释它们的视丰度。目前，似乎有两类轨道：圆形并且非常接近恒星的轨道，以及可能具有明显偏心率的较远的轨道。如果这个趋势被其他的发现证实，这将表明行星迁移的这两个机制都是有效的。近轨道可能是潮汐锁定到正在演化的盘的结果，而偏心轨道是引力散射的结果。

对这两个机制的一个检验是其他似行星伴天体的存在与否。潮汐锁定不需要在给定恒星周围存在其他行星，但是如果存在，它们的轨道应该跟靠内的轨道共面。引力散射表明具有较近行星的恒星应该至少另外有一颗在远距离偏心轨道中、质量相当的行星；这两个轨道的相对倾角可能很大。探测这样的远距离伴天体将是很困难的。视向速度扰动在幅度上以 $a^{-1/2}$ 变化，但是它们的周期以 $a^{3/2}$ 增加。在图 1 的例子中，外行星的作用是内行星作用的 1/4，其周期是后者的 50 多倍。通过视向速度变化来探测行星需要观测时间和轨道周期相当。因为高精度视向速度巡天已经进行了约十

eventually by long-term monitoring of stellar radial velocities. They may also be revealed by their perturbations on the orbits of the inner planets, by astrometry, or by direct imaging.

Note added in proof: Cochran *et al.*[19] have discovered another extra-solar planet with an eccentric orbit. This companion to the star 16 Cygni B has $a = 1.7$ AU, $e = 0.65$, and mass at least 1.6 times that of Jupiter. Rasio and Ford[20] have independently developed a model for planetary migration similar to ours.

(**384**, 619-621; 1996)

Stuart J. Weidenschilling* & Francesco Marzari†
* Planetary Science Institute / SJI, 620 North Sixth Avenue, Tucson, Arizona 85705, USA
† Dipartimento di Fisica, Universita di Padova, Via Marzolo 8, I-35131 Padova, Italy

Received 8 April; accepted 30 October 1996.

References:

1. Mayor, M. & Queloz, D. *Nature* **378**, 355-359 (1995).

2. Latham, D. W., Mazeh, T., Stefanik, R. P., Mayor, M. & Burki, G. *Nature* **339**, 38-40 (1989).

3. Marcy, G. W. & Butler, R. P. *Astrophys. J.* **464**, L147-L152 (1996).

4. Butler, R. P. & Marcy, G. W. *Astrophys. J.* **464**, L153-L156 (1996).

5. Lissauer, J. *Icarus* **69**, 249-265 (1987).

6. Pollack, J. *et al. Icarus* (in the press).

7. Lin, D. N. C., Bodenheimer, P. & Richardson, D. C. *Nature* **380**, 606-607 (1996).

8. Boss, A. *Science* **267**, 360-362 (1995).

9. Boss, A. *Nature* **379**, 397-398 (1995).

10. Beckwith, S. & Sargent, A. *Nature* **383**,139-144 (1996).

11. Gladman, B. *Icarus* **106**, 247-263 (1993).

12. Chambers, J., Wetherill, G. W. & Boss, A. *Icarus* **119**, 261-268 (1996).

13. Farinella, P. *Moon Planets* **22**, 25-29 (1980).

14. Everhart, E. in *Dynamics of Comets: Their Origin and Evolution* (eds Carusi, A. & Valsecchi, G.) 185-202 (Reidel, Dordrecht, 1985).

15. Wetherill, G. W. *Nature* **373**, 470 (1995).

16. Wetherill, G. W. *Astrophys. Space Sci.* **212**, 23-32 (1994).

17. Ward, W. R. *Icarus* **73**, 330-348 (1988).

18. Takeda, H., Matsuda, T., Sawada, K. & Hayashi, C. *Prog. Theor. Phys.* **74**, 272-287 (1985).

19. Cochran, W. D., Hatzes, A. P., Butler, R. P. & Marcy, G. W. *Bull. Am. Astron. Soc.* **28**, 1111 (1996).

20. Rasio, F. A. & Ford, E. B. *Science* (in the press).

Acknowledgements. We thank D. R. Davis and W. K. Hartmann for discussions and comments. This work was supported by NASA Planetary Geology and Geophysics Program and the Italian Space Agency.

Correspondence should be addressed to S.J.W. (e-mail: sjw@psi.edu).

年，目前仅发现距离恒星近的行星是不足为奇的。更多远距离伴天体通过长期恒星视向速度监测最终应该能被发现。它们也可能通过对内行星轨道的扰动、天体测量或者直接成像被发现。

　　附加说明：科克伦等[19]已经发现另外一颗具有偏心轨道的太阳系外行星——天鹅座 16 B，这颗行星的轨道为 $a = 1.7\,\text{AU}$，$e = 0.65$，质量至少是木星的 1.6 倍。拉西奥和福特[20]已经独立建立了一个和我们的模型类似的行星迁移模型。

（肖莉 翻译；周济林 审稿）

Transient Optical Emission from the Error Box of the γ-ray Burst of 28 February 1997

J. van Paradijs *et al.*

Editor's Note

Mysterious flashes of gamma rays in the sky have been reported since 1973, when it was thought that they originated within our solar system. Many theories were proposed to explain them, but the observation of a counterpart object at any other wavelength would constrain these considerably. Here Jan van Paradijs and colleagues describe such an object for the first time, using the Beppo-SAX satellite. They report a visible-light counterpart seen in less than 21 hours after the burst GRB 970228, which shows that the burst is associated with a faint galaxy, probably at high redshift. We know now that most "long" bursts like this are very distant, and seem to arise in peculiar types of supernovae (exploding stars).

For almost a quarter of a century[1], the origin of γ-ray bursts—brief, energetic bursts of high-energy photons—has remained unknown. The detection of a counterpart at another wavelength has long been thought to be a key to understanding the nature of these bursts (see, for example, ref. 2), but intensive searches have not revealed such a counterpart. The distribution and properties of the bursts[3] are explained naturally if they lie at cosmological distances (a few Gpc)[4], but there is a countervailing view that they are relatively local objects[5], perhaps distributed in a very large halo around our Galaxy. Here we report the detection of a transient and fading optical source in the error box associated with the burst GRB970228, less than 21 hours after the burst[6,7]. The optical transient appears to be associated with a faint galaxy[7,8], suggesting that the burst occurred in that galaxy and thus that γ-ray bursts in general lie at cosmological distance.

G RB970228 was detected[9] with the Gamma-ray Burst Monitor[10] on board the Italian–Dutch BeppoSAX satellite[11] on 1997 February 28, UT 02 h 58 min 01 s. The event lasted ~80 s and reached peak fluxes of ~4×10^{-6}, ~6×10^{-6} and ~10^{-7} erg cm^{-2} s^{-1} in the 40–600 keV, 40–1,000 keV and 1.5–7.8 keV ranges, respectively[9,12] (note that the peak flux of 0.23 Crab quoted in ref. 9 is in error). It occurred in the field of view of one of the BeppoSAX Wide Field Cameras (WFCs)[13]. The spectrum of the event is characteristic of classical γ-ray bursts (GRBs)[12]. Its position (about halfway between α Tauri and γ Orionis) was determined with an accuracy of 3′ (radius)[7] at right ascension (RA) 05 h 01 min 57 s, declination (dec.)+11° 46.4′. Application of the long-baseline timing technique[14] to the GRB data obtained with the Ulysses spacecraft, and with the BeppoSAX and the Wind satellites, respectively, constrained this location to be within each of two parallel annuli,

354

1997 年 2 月 28 日伽马射线暴误差框内的暂现光学辐射

帕拉基斯等

编者按

从 1973 年开始，神秘的伽马射线在天空中的闪光就有所报道。当时认为，闪光发生在我们的太阳系里。科学家们提出了很多理论去解释它们，但是在其他波段观测到的对应体会很大程度地限制这些理论。这里帕拉基斯及他的合作者们第一次用 BeppoSAX 卫星对它们进行描述。他们报道了在伽马射线暴 GRB970228 爆发后 21 小时内观测到的一个光学对应体，并表明该暴和一个可能处于高红移处的暗星系成协。我们现在知道，大多数类似的"长"暴都发生在很远的地方，并且似乎都产生于特殊类型的超新星（爆炸的恒星）之中。

将近 25 年的时间中 [1]，伽马射线暴——短暂的、巨大能量的高能光子爆发——的起源一直不为人所知。长期以来人们认为探测到其他波段的对应体将是理解这些暴本质的关键（例如，见参考文献 2），可是密集的搜寻并没有发现这样的对应体。如果这些暴位于宇宙学距离上（几十亿个秒差距）[4]，那么它们的分布和性质 [3] 可以很自然地得到解释。但是也有相反的观点，认为它们是相对邻近的天体 [5]，可能分布在银河系周围很大尺度的晕里。这里我们报道 GRB970228 爆发后 21 小时内在其误差圈里探测到一个暂现并衰减的光学源 [6,7]。这个光学暂现源似乎和一个暗星系成协 [7,8]，这表明伽马射线暴发生在那个星系里，因此伽马射线暴大体上是处于宇宙学距离上的。

GRB970228 是于 1997 年 2 月 28 日，世界时 (UT) 02 h 58 min 01 s，由安装在意大利 – 荷兰的 BeppoSAX 卫星 [11] 上的伽马射线暴监测器 [10] 探测到的 [9]。这个事件持续了 ~80 s，在 40~600 keV、40~1,000 keV 和 1.5~7.8 keV 能段的峰值流量分别为 ~4×10^{-6} erg·cm^{-2}·s^{-1}、~6×10^{-6} erg·cm^{-2}·s^{-1} 和 ~10^{-7} erg·cm^{-2}·s^{-1}[9,12]（注意：参考文献 9 引用的 0.23 倍蟹状星云的峰值流量是错误的）。它发生在 BeppoSAX 大视场照相机 (WFCs) [13] 的一个视场里。这个事件的光谱具有典型伽马射线暴 (GRBs) 的特征 [12]。它的位置（大约在金牛座 α 和猎户座 γ 的中间）确定为赤经 (RA) 05 h 01 min 57 s，赤纬 (dec.) +11°46.4′，精度为 3′（半径）[7]。分别对尤里西斯号太阳探测器 (Ulysses)、BeppoSAX 和 Wind 卫星观测到的伽马射线暴的数据应用长基线计时

with half-widths[15,16] of 31″ (3σ), and 30″ (3σ), respectively, which intersect the WFC error circle (Fig. 1).

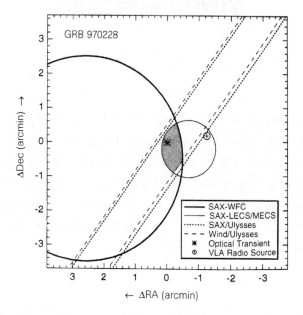

Fig. 1. The position of the optical transient, indicated with an asterisk, is shown with respect to the 3′ (radius) WFC location error circle, the 50″ (radius) error circles of the BeppoSAX X-ray transient, and the two annuli obtained from the differences between the times the GRB was detected with Ulysses, and with BeppoSAX and Wind, respectively. The area in common between these error regions in hatched. The coordinates are given in units of arcmin with respect to the position of the optical transient (RA 05 h 01 min 46.66 s, dec. +11° 46′ 53.9″, J2000). The position of an unrelated radio source[49] in the error circle of the X-ray transient is indicated with the square symbol.

Eight hours after the burst occurred, BeppoSAX was reoriented so that the GRB position could be observed with the LECS and MECS detectors[17,18]. A weak X-ray source was then found[19] at RA 05 h 01 min 44 s, dec. +11° 46.7′ (error radius 50″), near the edge of the WFC error circle[9] (Fig. 1). The 2–10 keV (MECS) flux of this source was 2.4×10^{-12} erg cm^{-2} s^{-1}. The LECS instrument measured a 0.1–10 keV source flux of $(2.6 \pm 0.6) \times 10^{-12}$ erg cm^{-2} s^{-1}. The source spectrum was consistent with a power-law model with photon index 2.7, reduced at low energy by a column density N_H of 5.6×10^{21} cm^{-2}. During an observation with the same instruments on March 3, UT 17 h 37 min this flux had decreased by a factor of 20 (ref. 19). With ASCA the X-ray source was detected[20] on 7 March at a 2–10 keV flux of $(0.8 \pm 0.2) \times 10^{-13}$ erg cm^{-2} s^{-1}.

On February 28, UT 23 h 48 min, 20.8 hours after the GRB occurred, before we had any knowledge of the X-ray transient, we obtained a V-band and an I-band image (exposure times 300 s each) of the WFC error box with the Prime Focus Camera of the 4.2-m William Herschel Telescope (WHT) on La Palma[21]. The 1,024 × 1,024 pixel CCD frames (pixel size 24 μm, corresponding to 0.421″) cover a 7.2′ × 7.2′ field, well matched to the size of the GRB error box. The limiting magnitudes of the images are $V = 23.7$, and

356

技术[14]，把位置限制在两个半宽[15,16]分别为 31″（3σ）和 30″（3σ）的平行环内，和 WFC 误差圈相交在一起（图 1）。

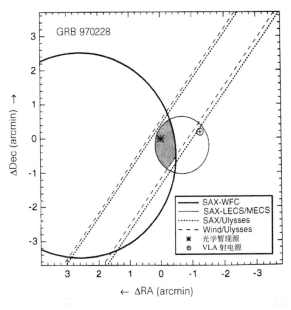

图 1. 光学暂现源的位置，用星号标示。相应的图中显示了 3′（半径）的 WFC 位置误差圈，50″（半径）的 BeppoSAX X 射线暂现源的误差圈，以及由 Ulysses 测量到的伽马射线暴时间分别与 BeppoSAX 和 Wind 测量到的伽马射线暴时间的差值得到的两个环。这些误差范围的重合区域画上了阴影。坐标表示光学暂现源的位置（RA 05 h 01 min 46.66 s, dec. +11°46′53.9″, J2000），以 arcmin 为单位。在 X 射线暂现源误差圈内一个不相干的射电源[49]的位置由 ⊕ 表示。

爆发 8 小时后，再次调整 BeppoSAX 的方向以使得伽马射线暴的位置能够被 LECS 和 MECS 探测器观测到[17,18]。在 WFC 误差圈[9]的边缘附近 RA 05 h 01 min 44 s, dec. +11°46.7′（误差半径为 50″）处发现了一个弱的 X 射线源[19]（图 1）。这个源在 2~10 keV（MECS）范围内的流量为 2.4×10^{-12} erg·cm^{-2}·s^{-1}。LECS 仪器测量源在 0.1~10 keV 范围内的流量为 $(2.6 \pm 0.6) \times 10^{-12}$ erg·cm^{-2}·s^{-1}。这个源的能谱和光子指数为 2.7 的幂律谱模型一致，在低能处流量由柱密度 N_H 为 5.6×10^{21} cm^{-2} 的吸收导致减少。在 3 月 3 日 UT 17 h 37 min，相同仪器观测到这个源的流量减少为原来的 1/20（参考文献 19）。3 月 7 日 ASCA 探测到的这个 X 射线源[20]，在 2~10 keV 的流量为 $(0.8 \pm 0.2) \times 10^{-13}$ erg·cm^{-2}·s^{-1}。

在 2 月 28 日 UT 23 h 48 min，伽马射线暴爆发 20.8 小时后，在不知道 X 射线暂现源之前，我们利用拉帕尔马的 4.2 m 威廉·赫歇尔望远镜（WHT）的主焦点照相机得到 WFC 误差圈的 V 波段和 I 波段图像（每幅图曝光时间 300 s）[21]。1,024 × 1,024 像素的 CCD 图像（像素大小为 24 μm，对应于 0.421″）覆盖了 7.2′ × 7.2′ 的天区，和伽马射线暴误差圈的大小符合得很好。成像的极限星等为 V = 23.7 和 I = 21.4。3 月

$I = 21.4$. We obtained a second I-band image on March 8, UT 21 h 12 min with the same instrument on the WHT (exposure time 900 s), and a second V-band image on March 8, UT 20 h 42 min with the Isaac Newton Telescope (INT) on La Palma (exposure 2,500 s). Photometric calibration was obtained from images of standard star number 336 and Landolt[22] field 104. The images were reduced using standard bias subtraction and flatfielding.

A comparison of the two image pairs immediately revealed one object with a large brightness variation[6]: it is clearly detected in both the V- and I-band images taken on 28 February, but not in the second pair of images taken on 8 March (Fig. 2). From a comparison with positions of nearby stars that were obtained using the Digitized Sky Survey we find for its location RA 05 h 01 min 46.66 s, dec. +11° 46' 53.9" (equinox J2000); this position has an estimated (internal) accuracy of 0.2". The object is located in the error box defined by the WFC position, the Ulysses/BeppoSAX/Wind annuli, and the transient X-ray source position (Fig. 1).

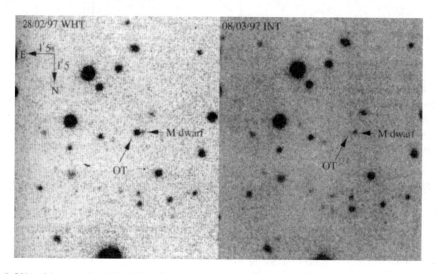

Fig. 2. V-band images of a 1.5' × 1.5' region of the sky containing the position of the optical transient. The left image was obtained with the WHT on 1997 28 February, UT 23 h 48 min, the right image with the INT on 8 March, UT 20 h 42 min. The optical transient is indicated by "OT". The M dwarf, separated from the optical transient by 2.9", is also indicated.

Using aperture photometry software we determined the magnitudes of the variable as follows[6]: $V = 21.3 \pm 0.1$, $I = 20.6 \pm 0.1$ on 28 February and $V > 23.6$, $I > 22.2$ on 8 March. The shape of the source in both the 28 February V- and I-band images is consistent with that of the point-spread function, as determined for 15 stars in the same images.

Close to the optical transient is a star, located 2.85" away at RA 05 h 01 min 46.47 s, dec. +11° 46' 54.0", with $V = 23.1$, $I = 20.5$. A spectrum of this star, taken on March 1, UT 0 h with the ESO 3.6-m telescope using the EFOSC1 spectrograph and the R1000 grating (resolution of 14 Å per pixel), covering the 5,600–11,000 Å region, reveals the

8 日 UT 21 h 12 min，我们使用 WHT 上相同的仪器得到第二幅 I 波段图像（曝光时间 900 s），3 月 8 日 UT 20 h 42 min 使用拉帕尔马的艾萨克·牛顿望远镜（INT）得到第二幅 V 波段图像（曝光时间 2,500 s）。测光定标通过 336 号标准恒星和朗多[22] 场星 104 的图像完成。图像通过标准偏差扣除与平场处理得到还原。

紧接着对两组图的比较显示出一个亮度大幅变化的源[6]：它在 2 月 28 日拍的 V 波段和 I 波段图像上都清楚地出现，但是在 3 月 8 日拍的第二对图像中（图 2）没有被探测到。通过比较数字化巡天得到的附近恒星的位置，我们得到其坐标为 RA 05 h 01 min 46.66 s，dec. +11° 46′ 53.9″（春分点 J2000）；这个位置估计（固有）精度为 0.2″。这个天体位于由 WFC 的位置、Ulysses/BeppoSAX/Wind 环和 X 射线暂现源位置决定的误差框区域内（图 1）。

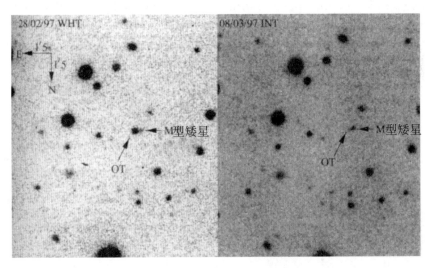

图 2. 包含光学暂现源位置 1.5′×1.5′ 天区的 V 波段图。左边图像是用 WHT 在 1997 年 2 月 28 日 UT 23 h 48 min 观测的，右边图像是用 INT 于 3 月 8 日 UT 20 h 42 min 观测的。光学暂现源用"OT"标示。距离光学暂现源 2.9″ 处的 M 型矮星也被标示出。

使用孔径测光软件我们得到变源的星等为[6]：2 月 28 日 $V=21.3\pm0.1$，$I=20.6\pm0.1$ 和 3 月 8 日 $V>23.6$，$I>22.2$。这个源在 2 月 28 日 V 波段和 I 波段的图像上的形状都和从相同图像上 15 颗恒星得到的点扩散函数一致。

靠近光学暂现源的是一颗恒星，位于 2.85″ 的距离，坐标为 RA 05 h 01 min 46.47 s，dec. +11° 46′ 54.0″，$V=23.1$，$I=20.5$。3 月 1 日 UT 0 h 利用 ESO（欧南台）3.6 m 望远镜获得了该恒星的一个光谱，其中利用了 EFOSC1 光谱仪和 R1000 光栅（分辨率为每像素 14 Å），覆盖了 5,600~11,000 Å 波段。光谱显示 TiO 带的存在，这表明它

presence of TiO bands, which indicate it is an M-type star. With foreground absorption $A_v = 0.4 \pm 0.3$ mag (ref. 23), (substantially smaller than the value inferred from the low-energy cut off in the LECS spectrum), its colour index, $V-I = 2.6$, corresponds to an M2 star[24]. It is most likely to be an M dwarf at a distance of ~3 kpc (an early M-type giant would be located at a distance of ~0.4 Mpc, which we consider much less likely).

Further images were obtained with the Nordic Optical Telescope (NOT, La Palma) on 4 March, with the INT on 9 March, and with the ESO New Technology Telescope (NTT) on 13 March (see Table 1 for a summary). The transient was not detected in these images, which puts a lower limit on its average decay rate (in 4 days) of 0.7 mag per day. The NTT image shows that at the location of the variable object there is an extended object, probably a galaxy[7,8] (Fig. 3); this object is also seen in the INT B- and R-band images. From differential astrometry relative to the nearby M star for both the V- and I-band images, we find that the centres of the optical transient and the galaxy have a relative distance $(0.22 \pm 0.12)''$ (1σ; quadratic addition of the errors in the two relative positions). The relative position of the optical transient did not change by more than 0.2″ between the 28 February V- and I-band images. From the NTT image and the 9 March INT image we measured[7] the galaxy's magnitude to be $R = 23.8 \pm 0.2$ and 24.0 ± 0.2, respectively, consistent with the value $R = 24.0$ reported by Metzger et al.[8], and $B = 25.4 \pm 0.4$ (Table 1).

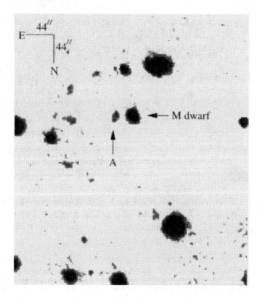

Fig. 3. R-band image of a 44″ × 44″ region of the sky containing the position of the optical transient, obtained with the NTT on 1997 13 March UT 0 h. The faint galaxy coincident with the optical transient (A) and the M-dwarf are indicated.

是一颗 M 型恒星。考虑前景吸收 $A_v = 0.4 \pm 0.3$ mag(参考文献 23,实质上的比 LECS 光谱低能截断得到的值小得多),它的色指数为 $V-I = 2.6$,对应于 M2 型恒星 [24]。这颗恒星很可能是颗距离为 ~3 kpc 的 M 型矮星(早期 M 型巨星位于距离 ~0.4 Mpc 处,我们认为这种可能性小很多)。

更多的图像于 3 月 4 日利用北欧光学望远镜(NOT,拉帕尔马),3 月 9 日利用 INT 以及 3 月 13 日利用 ESO 的新技术望远镜(NTT)得到(总结见表 1)。暂现源在这些图像中没有被探测到,从而得到其平均衰减率(4 天内)的下限为每天 0.7 mag。NTT 的图像显示在变源位置处存在一个延展天体,可能是一个星系 [7,8](图 3);这个天体也在 INT 的 B 波段和 R 波段图像上呈现。通过在 V 波段和 I 波段的图像上相对邻近的 M 型恒星的较差天体测量,我们发现光学暂现源的中心和星系的中心相对距离为 $(0.22 \pm 0.12)''$(1σ;两个相对位置误差的平方和)。光学暂现源在 2 月 28 日的 V 波段和 I 波段图像上的相对位置变化不超过 0.2″。从 NTT 的图像和 3 月 9 日 INT 的图像我们测量出 [7] 星系的星等分别为 $R = 23.8 \pm 0.2$ 和 24.0 ± 0.2,和梅茨格等人 [8] 报道的 $R = 24.0$ 和 $B = 25.4 \pm 0.4$ 一致(表 1)。

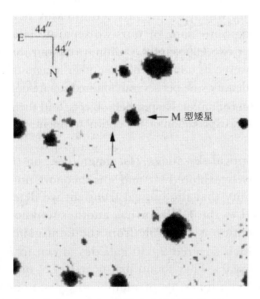

图 3. 1997 年 3 月 13 日 UT 0 h 利用 NTT 观测到的包含光学暂现源位置 44″×44″ 天区的 R 波段图。与光学暂现源重合的暗星系(A)和 M 型矮星都已被标出。

Table 1. Summary of optical observations

Date (1997)	Time (UT)	Telescope	Band	Integration time (s)	Magnitude	Remarks
28 Feb.	23 h 48 min	WHT	V	300	$V = 21.3$	Transient
28 Feb.	23 h 53 min	WHT	I	300	$I = 20.6$	Transient
04 Mar.	20 h 42 min	NOT	V	900	$V > 24.2$	–
08 Mar.	20 h 42 min	INT	V	2,500	$V > 23.6$	–
08 Mar.	21 h 12 min	WHT	I	300	$V > 22.2$	–
09 Mar.	21 h 30 min	INT	R	1,200	$R = 24.0$	Extended
09 Mar.	20 h 30 min	INT	B	2,500	$B = 25.4$	Extended
13 Mar.	0 h 0 min	NTT	R	3,600	$R = 23.8$	Extended

Known types of optical transient events (novae, supernovae, dwarf novae, flare stars) are unlikely to account for the optical transient for a variety of reasons, such as the amplitude and short timescale of its variability, its colour index, or its inferred distance.

The GRB source is located relatively close to the ecliptic, at latitude $-11°$, and this raises the possibility that the optical transient is an asteroid. However, on 1 March asteroids in the direction of the GRB have proper motions of at least $0.1°$ per day (T. Gehrels, personal communication), which would have led to easily detectable motion ($> 2.5''$) during the 600-s total exposure time of our two separate images. On the basis of its proper motion during our two exposures we cannot rule out that the optical transient is a Kuiper belt object. However, its non-detection on other images taken in the week following the GRB, the low surface density (one per several hundred square degrees at 21st magnitude; T. Gehrels, personal communication) of Kuiper belt objects, and their very red colours[25], make such objects a highly unlikely explanation of the optical transient.

The variability of the optical sky above 21st magnitude, on timescales of a few days, has not yet been extensively explored[26,27], and it is therefore not possible to make a firm estimate of the probability that the optical transient we detected is a faint, strongly variable AGN unrelated to the GRB, or has another (unknown) origin unrelated to the GRB. Some information is available from the faint-galaxy monitoring program of Kochanski et al.[28] which covered 2,830 galaxies (down to $B = 24.8$, $R = 23.3$) in a $16' \times 16'$ field during 10 years. They found that near $B = 24$ only ~0.5% of these varied by more than 0.03 mag on a timescale of months to years; none varied by more than a magnitude. Variability of blazars on timescales of minutes to hours does not exceed 0.3 mag; day-to-day variations of a factor of two or more have been observed[50].

Although we cannot firmly exclude that the optical transient in the error box of GRB970228 is caused by some unknown event unrelated to the GRB, the temporal coincidence between the optical and X-ray transients, and their spatial coincidence with the GRB lead us to believe that both the optical transient and the decaying BeppoSAX X-ray source are associated with GRB970228.

表 1. 光学观测总结

日期 (1997)	时间 (UT)	望远镜	波段	积分时间 (s)	星等	标注
2 月 28 日	23 h 48 min	WHT	V	300	$V = 21.3$	暂现源
2 月 28 日	23 h 53 min	WHT	I	300	$I = 20.6$	暂现源
3 月 4 日	20 h 42 min	NOT	V	900	$V > 24.2$	–
3 月 8 日	20 h 42 min	INT	V	2,500	$V > 23.6$	–
3 月 8 日	21 h 12 min	WHT	I	300	$V > 22.2$	–
3 月 9 日	21 h 30 min	INT	R	1,200	$R = 24.0$	延展
3 月 9 日	20 h 30 min	INT	B	2,500	$B = 25.4$	延展
3 月 13 日	0 h 0 min	NTT	R	3,600	$R = 23.8$	延展

光学暂现源的已知类型（新星、超新星、矮新星、耀星）因为各种原因，例如光变的幅度和短时标，色指数或者推断的距离等，都解释不了这个光学暂现源。

伽马射线暴源所在纬度为 $-11°$，相对接近于黄道，这提出了光学暂现源是小行星的可能性。不过，3 月 1 日在该星方向的小行星的自行至少为每天 0.1°（赫雷尔斯，个人交流），在我们两幅独立的、总曝光时间为 600 s 的图像上应该很容易探测到其运动（$> 2.5''$）。根据两次曝光测到的自行，我们不能排除这个光学暂现源为柯伊伯带天体。不过，在伽马射线暴之后的一个星期内拍摄的其他图像并没有探测到这个暂现源，以及柯伊伯带天体的低面密度（21 星等的天体每几百平方度中才有 1 个；赫雷尔斯，个人交流）和它们非常红的色指数 [25]，使得光学暂现源几乎不可能是柯伊伯带天体。

在 21 星等之上，时标为几天的光学变源目前还没有得到广泛的研究 [26,27]，因此不大可能准确估计所探测到的光学暂现源是暗的、和伽马射线暴无关的、光变幅度较大的 AGN 的可能性，或者是来自其他（未知）和伽马射线暴无关的起源的可能性。我们从科汉斯基等人 [28] 10 年内在 $16' \times 16'$ 的天区覆盖 2,830 个星系（最暗能探测到：$B = 24.8$，$R = 23.3$）的弱星系监测项目中得到一些信息。他们发现在 $B = 24$ 附近只有 ~0.5% 的星系在几个月到几年时标上变化超过 0.03 mag；没有星系变化超过一个星等。耀变体在几分钟到几小时时标上的变化不会超过 0.3 mag；以天为时标，幅度达到 2 倍或更多的变化已经被观测到 [50]。

尽管我们不能明确排除 GRB970228 误差圈内的光学暂现源是由一些和伽马射线暴无关的事件引起的可能性。但是光学暂现源和 X 射线暂现源在时间上的一致，以及它们和该暴在空间位置上的一致，让我们相信光学暂现源和衰减的 BeppoSAX X 射线源两者都与 GRB970228 成协。

Radio observations (at 6 cm wavelength) of the GRB error box[6] made with the Westerbork Radio Synthesis telescope on February 28, UT 23 h 17 min (for 1.2 h; 20.4 h after the GRB, simultaneous with the observations at the WHT), and on March 1, UT 18 h and March 2, UT 18 h (each lasting 12 hours) show that at the position of the optical transient there is then no radio point source with a flux exceeding 1.0, 0.33 and 0.33 mJy, respectively (2σ upper limit).

Some rough spectral information on the optical/X-ray transient can be obtained if we assume that between the two BeppoSAX X-ray observations, made 4 days apart, the X-ray flux of the transient decreased with time since the GRB as a power law. Approximating the spectrum by $F_v \propto v^\alpha$ we estimate from $F_X \simeq 0.04$ μJy (inter-polated) and $F_V \simeq 10$ μJy, that $\alpha = -0.7 \pm 0.1$. Extrapolation of this spectrum to the radio region would lead to an expected radio flux density of 10–100 mJy, far exceeding the observed upper limit. This indicates that the X-ray, optical and radio flux densities of the transient cannot be represented by a single power law.

The close positional coincidence between the optical transient and the galaxy suggests that the transient may be located in that galaxy. In an effort to quantify this we adopt a Bayesian approach. We consider three disjoint and exhaustive hypotheses for the source of the optical transient. The hypotheses are: H_c, the optical transient is in the centre of a galaxy, H_g, the optical transient is in a galaxy but not at its centre; and H_n, the optical transient is not in a galaxy.

We assume that there is a single optical transient detected in the field of view of angular area A and that n non-overlapping galaxies are detected in the field. The transient is at distance $r \pm \sigma_r$ from the nearest galaxy, where the error includes the uncertainty in the positions of the centroids of both the galaxy and the transient. The probability density at r under H_c is $P(r|H_c) = (2\pi\sigma_r^2 n)^{-1}\exp[-(r^2/2\sigma_r^2)]$. The probability density under H_g depends on the size, shape and inclination of the galaxy and the specifics of the model for the distribution of sources in the galaxy. For simplicity, we assume the probability density to be gaussian with width σ_g. Then $P(r|H_g) = (2\pi\sigma_g^2 n)^{-1}\exp[-(r^2/2\sigma_g^2)]$. The probability density under H_n is uniform over the field of view, so $P(r|H_n) = A^{-1}$. The posterior probability of each hypothesis H is $P(H|r) = kP(H)P(r|H)$, where $P(H)$ is the prior probability and the normalization constant k is obtained by the requirement that $P(H_c|r) + P(H_g|r) + P(H_n|r) = 1$.

For the NTT observation we find seven galaxies in $A = (44'')^2$ field, that is, $n = 13$ per arcmin², $r = 0.22''$, $\sigma_r = 0.12''$. A reasonable estimate for the galaxy width is $\sigma_g = 1''$. With these values the probability densities at r are $P(r|H_c) = 0.294$, $P(r|H_g) = 0.022$ and $P(r|H_n) = 0.0005$, all in units of arcsec⁻². Assuming equal priors $P(H_c) = P(H_g) = P(H_n) = 1/3$, the posterior probabilities are $P(H_c|r) = 0.928$, $P(H_g|r) = 0.070$ and $P(H_n|r) = 0.0016$. The posterior probability for H_g depends sensitively on the assumed σ_g. However, the posterior probability, $P(H_n|r)$, that the transient is not associated with a galaxy is in the range 0.09–0.18%, for any assumptions about the size of faint galaxies. Within the range of assumed values for σ_r between $0.08''$ and $0.2''$ the values of $P(H_n|r)$ increase by less than a factor of 3.

伽马射线暴误差圈的射电观测[6](在 6 cm 波长)于 2 月 28 日 UT 23 h 17 min(伽马射线暴爆发后 20.4 h，持续 1.2 h，和 WHT 观测同时进行)，3 月 1 日 UT 18 h 和 3 月 2 日 UT 18 h(每次持续 12 h)利用韦斯特博克综合孔径射电望远镜完成，结果显示在光学暂现源的位置上流量分别没有超过 1.0 mJy、0.33 mJy 和 0.33 mJy 的射电点源(2σ 上限)。

如果假设在两次间隔 4 天的 BeppoSAX X 射线观测之间，暂现源的 X 射线流量在伽马射线暴后随着时间呈幂律变化，我们能得到光学/X 射线暂现源一些大体的光谱信息。通过 $F_v \propto v^\alpha$ 近似拟合光谱，我们从 $F_X \simeq 0.04$ μJy(内插)和 $F_v \simeq 10$ μJy 估计得到 $\alpha = -0.7 \pm 0.1$。将这个谱外推到射电波段将得到预期的射电流量密度 10~100 mJy，远高于观测上限。这表明暂现源的 X 射线、光学和射电流量密度不能用单个幂律谱来表示。

光学暂现源和星系在位置上的一致性表明暂现源可能位于那个星系中。我们采用贝叶斯方法来定量计算这一点。关于光学暂现源的来源，我们提出三个不相干但完备的假设，分别是：H_c，光学暂现源位于星系中心；H_g，光学暂现源位于星系内但是不在中心；H_n，光学暂现源不在星系内。

我们假设在角面积 A 的视场内探测到一个光学暂现源和 n 个不重叠的星系。暂现源离最近的星系的距离为 $r \pm \sigma_r$，其中误差包括了星系中心位置和暂现源中心位置的误差。在 H_c 条件下暂现源在 r 处的概率密度为 $P(r|H_c) = (2\pi\sigma_r^2 n)^{-1}\exp[-(r^2/2\sigma_r^2)]$。在 H_g 条件下的概率密度依赖于星系的大小、形状和倾角以及星系内源分布的模型。为了简化，我们假设概率密度是宽度为 σ_g 的高斯分布。因而 $P(r|H_g) = (2\pi\sigma_g^2 n)^{-1}\exp[-(r^2/2\sigma_g^2)]$。在 H_n 条件下的概率密度是在视场内均匀分布的，所以 $P(r|H_n) = A^{-1}$。每个假设(H)的后验概率为 $P(H|r) = kP(H)P(r|H)$，这里 $P(H)$ 是先验概率，而归一化常数 k 是从 $P(H_c|r) + P(H_g|r) + P(H_n|r) = 1$ 的条件中获得的。

从 NTT 的观测结果中，我们在 $A = (44'')^2$ 的视场内发现 7 个星系，即 $n = 13$ arcmin^{-2}，$r = 0.22''$，$\sigma_r = 0.12''$。对星系宽度的合理估计为 $\sigma_g = 1''$。使用以上数值，在 r 处的概率密度为 $P(r|H_c) = 0.294$，$P(r|H_g) = 0.022$ 和 $P(r|H_n) = 0.0005$，全部以 arcsec^{-2} 为单位。假设先验概率相等 $P(H_c) = P(H_g) = P(H_n) = 1/3$，后验概率为 $P(H_c|r) = 0.928$，$P(H_g|r) = 0.070$ 和 $P(H_n|r) = 0.0016$。H_g 的后验概率非常灵敏地依赖于假设的 σ_g 值。不过对于任何暗星系大小的假设，后验概率 $P(H_n|r)$，即暂现源不和星系成协，都是在 0.09%~0.18% 的范围内。假设 σ_r 的值在 $0.08''$ 到 $0.2''$ 范围内，$P(H_n|r)$ 的值将增加为不到原来的 3 倍。

The above analysis suggests that the optical transient is related to the faint galaxy, which provides support for the cosmological distance scale for GRBs.

A rough estimate of the expected redshift, z, of the galaxy may be made by assuming that its absolute magnitude is in the range −21 to −16, which covers the bulk of normal galaxies[29]. For an assumed Hubble constant of 60 km s^{-1} Mpc^{-1}, this corresponds to z in the range 0.2–2.

The close proximity of the optical transient to the centre of the faint galaxy, and the presence of relatively bright quasars in the 8 arcmin^2 error box of GRB781119 ($V = 20$)[30-32], and in the 3′ (radius) error box[33] of GRB960720 ($R = 18.8$)[34-36] raise the possibility that GRBs occur preferentially, or exclusively, in or near galactic nuclei.

Searches for an optical counterpart to a GRB have been continually attempted for the past 20 years. Recent reviews and descriptions of serendipitous, rapid follow-up, and delayed searches for optical counterparts of GRBs[36-44] show that these previous searches were generally made a week or longer after the GRB, or they were not as deep ($V < 20$) as the images presented here. The most sensitive rapid follow-up observations so far had delay times (δt) and limiting magnitudes (m) as follows: $\delta t = 1.85$ d, $m < 23$ (ref. 45), and $\delta t = 4.0$ d, $m_B < 22$ (ref. 38).

It was not until the launch of BeppoSAX in 1996 that accurate (several arcmin) locations for GRBs became available within hours of detection, hence facilitating rapid follow-up observations at large ground-based optical telescopes for those bursts which happened to be in the field of the WFC. The continued operation of BeppoSAX and the approval of the High Energy Transient Explorer-2 (HETE-2) mission bode well for great progress in the rapid follow-up observations of GRBs. Also, near-real-time, fully automated optical systems linked to the BATSE–BACODINE[37] system are becoming operational, and their sensitivity is continually improving[46,47].

We expect that X-ray and optical transients associated with GRBs will again be seen (though perhaps not in all cases[48]) in the near future. This could be a turning point in GRB astronomy. Detailed studies (light curves and spectra) of such transients can be expected within a year, and we are optimistic that the distance scale as well as the mechanism behind the enigmatic GRBs are now within reach.

Note added in proof: After this paper was submitted, an HST observation was made of the optical transient (K. Sahu *et al.*, *IAU Circ. No.* 6606). This observation confirms that the transient is associated with an extended emission region, but seems to exclude that the transient is located at the centre of that region.

(**386**, 686-689; 1997)

以上的分析表明光学暂现源和暗星系相关，支持了伽马射线暴是处在宇宙学距离尺度上的说法。

对该星系红移 z 的粗略估计可以通过假设星系的绝对星等位于 $-21 \sim -16$ 的范围来获得，这个范围覆盖了大部分正常星系 [29]。假定哈勃常数为 $60\ \mathrm{km \cdot s^{-1} \cdot Mpc^{-1}}$，得到 z 的范围为 0.2～2。

光学暂现源接近暗星系的中心，以及在 GRB781119 $(V=20)$ [30-32] 的 8 $\mathrm{arcmin^2}$ 误差圈和 GRB960720 $(R=18.8)$ [34-36] 的 3' (半径) 误差圈 [33] 内存在相对较亮的类星体，这些都提出了伽马射线暴倾向于，或者说仅发生在星系核或附近的可能性。

在过去的 20 年内，人们一直致力于寻找到伽马射线暴的光学对应体。最近的综述和叙述文章表明，无论是偶然的、快速跟踪的还是延后的对伽马射线暴光学对应体的搜寻 [36-44]，之前的这些搜寻工作大部分都在伽马射线暴发生一周甚至更长时间之后进行，或者它们也没有像这里的图像这么深的曝光 $(V<20)$。目前最灵敏快速的跟踪观测得到的伽马射线暴的延后时间 (δt) 和极限星等 (m) 如下：$\delta t = 1.85$ 天，$m < 23$ (参考文献 45)，和 $\delta t = 4.0$ 天，$m_B < 22$ (参考文献 38)。

直到 1996 年 BeppoSAX 的发射才使得在探测到伽马射线暴几个小时之后就获得到伽马射线暴的精确定位 (几个角分) 成为可能，进而促使大型地面光学望远镜对这些发生在 WFC 视场内的伽马射线暴能进行快速跟踪观测。BeppoSAX 的继续运行和已经获得批准的高能暂现源探测器 2 号 (HETE-2) 项目预示着伽马射线暴的快速跟踪观测将取得很大的进步。另外，与 BATSE-BACODINE 系统 [37] 连接的，近实时、全自动的光学系统也正变得可操作，而且它们的灵敏度也在不断地提高 [46,47]。

我们预期和伽马射线暴成协的 X 射线和光学暂现源在不久的将来会再次被观测到 (尽管可能不是所有暴的对应体都能被发现 [48])。这可能是伽马射线暴天文学的一个转折点。对这些暂现源详细的研究 (光变曲线和光谱) 可望在一年内开展，我们乐观地认为，谜一样的伽马射线暴的距离尺度及其背后机制正在被揭开。

附加说明： 在这篇文章投稿后，HST 对这个光学暂现源进行了观测 (萨胡等，*IAU Circ. No.* 6606)。观测证实这个暂现源和一个延展的辐射区域成协，但是似乎排除了暂现源位于该区域中心的可能性。

（肖莉 翻译；黎卓 审稿）

J. van Paradijs[1,2], P. J. Groot[1], T. Galama[1], C. Kouveliotou[3,4], R. G. Strom[5,1], J. Telting[5,6], R. G. M. Rutten[5,6], G. J. Fishman[4], C. A. Meegan[4], M. Pettini[7], N. Tanvir[8], J. Bloom[8], H. Pedersen[9], H. U. Nørdgaard-Nielsen[10], M. Linden-Vørnle[10], J. Melnick[11], G. van der Steene[11], M. Bremer[12], R. Naber[13], J. Heise[14], J. in't Zand[14], E. Costa[15], M. Feroci[15], L. Piro[15], F. Frontera[16], G. Zavattini[16], L. Nicastro[17], E. Palazzi[17], K. Bennet[18], L. Hanlon[19] & A. Parmar[18]

[1] Astronomical Institute "Anton Pannekoek", University of Amsterdam, and Center for High Energy Astrophysics, Kruislaan 403, 1098 SJ Amsterdam, The Netherlands.

[2] Physics Department, University of Alabama in Huntsville, Huntsville, Alabama 35899, USA.

[3] Universities Space Research Association.

[4] NASA Marshall Space Flight Center, ES-84, Huntsville, Alabama 35812, USA.

[5] Netherlands Foundation for Research in Astronomy, Postbus 2, 7990 AA Dwingeloo, The Netherlands.

[6] Isaac Newton Group, Apartado de Correos 321, 38780 Santa Cruz de La Palma, Tenerife, Canary Islands.

[7] Royal Greenwich Observatory, Madingley Road, Cambridge CB3 0EZ, UK.

[8] Institute of Astronomy, Madingley Road, Cambridge CB3 0HA, UK.

[9] Copenhagen University Observatory, Juliane Maries Vej 30, 2100 Copenhagen, Denmark.

[10] Danish Space Research Institute, Juliane Maries Vej 30, 2100 Copenhagen, Denmark.

[11] European Southern Observatory, Casilla 19001, Santiago 19, Chile.

[12] Leiden Observatory, Postbus 9513, 2300 RA Leiden, The Netherlands.

[13] Kapteyn Astronomical Institute, Postbus 800, 9700 AV Groningen, The Netherlands.

[14] SRON Laboratory for Space Research, Sorbonnelaan 2, 3584 CA Utrecht, The Netherlands.

[15] Istituto di Astrofisica Spaziale CNR, Via Enrico Fermi 21/23, Frascati CP 67, Italy.

[16] Dipartimento di Fisica, Universita di Ferrara, Via Paradiso 12,44100 Ferrara, Italy.

[17] Istituto Tecnologie e Studio Radiazione Extraterrestrie CNR, Via Gobetti 101, 40129 Bologna, Italy.

[18] ESA/ESTEC, Space Science Department, Postbus 299, 2200 AG Noordwijk, The Netherlands.

[19] Physics Department, University College Dublin, Belfield, Stillorgan Road, Dublin, Ireland.

Received 25 March; accepted 29 March 1997.

References:

1. Klebesadel, R. W., Strong, I. B. & Olson, R. A. Observations of gamma-ray bursts of cosmic origin. *Astrophys. J.* **182**, L85-L88 (1973).

2. Fishman, G. J. & Meegan, C. A. Gamma-ray bursts. *Annu. Rev. Astron. Astrophys.* **33**,415-458 (1995).

3. Meegan, C. A. *et al.* Spatial distribution of gamma-ray bursts observed by BATSE. *Nature* **355**, 143-145 (1992).

4. Paczyński, B. Gamma-ray bursters at cosmological distances. *Astrophys. J.* **308**, L43-L46 (1986).

5. Podsiadlowski, Ph., Rees, M. J. & Ruderman, M. Gamma-ray bursts and the structure of the Galactic halo. *Mon. Not. R. Astron. Soc.* **273**, 755-771 (1995).

6. Groot, P. J. *et al. IAU Circ. No.* 6584 (1997).

7. Groot, P. J. *et al. IAU Circ. No.* 6588 (1997).

8. Metzger, M. R. *et al. IAU Circ. No.* 6588 (1997).

9. Costa, E. *et al. IAU Circ. No.* 6572 (1997).

10. Frontera, F. *et al.* The high-energy X-ray experiment PDS on board the SAX satellite. *Adv. Space Res.* **11**, 281-286 (1991).

11. Piro, L., Scarsi, L. & Butler, R. C. SAX: the wideband mission for X-ray Astronomy, *Proc. SPIE* **2517**, 169-181 (1995).

12. Palmer, D. *et al. IAU Circ. No.* 6577 (1997).

13. Jager, R., Heise, J., In't Zand, J. & Brinkman, A. C. Wide field cameras for SAX. *Adv. Space Res.* **13**, 315-318 (1995).

14. Hurley, K. *et al.* The Ulysses supplement to the BATSE 3B Catalog. *AIP Proc.* **384**, 422-426 (1996).

15. Cline, T. L. *et al. IAU Circ. No.* 6593 (1997).

16. Hurley, K. *et al. IAU Circ. No.* 6594 (1997).

17. Parmar, A. N. *et al.* The low-energy concentrator spectrometer on-board the SAX X-Ray astronomy satellite. *Astron. Astrophys.* (in the press).

18. Bonura, A. *et al.* Performance characteristics of the scientific model of the medium energy concentrator spectrometer on board the X-ray astronomy satellite SAX. *Proc. SPIE* **1743**, 510-522 (1992).

19. Costa, E. *et al. IAU Circ.No.* 6576 (1997).

20. Yoshida, A. *et al. IAU Circ.No.* 6593 (1997).

21. Groot, P. *et al. IAU Circ.No.* 6574 (1997).

22. Landolt, A.UBVRI photometric standard stars in the magnitude range 11.5-16.0 around the celestial equator. *Astron.J.* **104**, 340-376 (1992).

23. Hakkila, J., Myers, J.M., Stidham, B. J. & Hartmann, D. H. A computerized model of large-scale visual interstellar extinction. *Astron.J* (submitted).

24. Johnson, H.L. Astronomical measurements in the infrared. *Annu. Rev. Astron. Astrophys.* **4**, 191-206(1996).

25. Luu, J. & Jewitt, D. Color diversity among the Centaurs and Kuiper Belt Objects. *Astron. J.* **112**, 2310-2318 (1996).

26. Paczynski, B. *Variable Stars and the Astrophysical Returns of Microlensing Surveys* (ed. Ferlet, R. and Maillard, J.-P.)(Proc. 12th AIP Collop., in the press).

27. Trevese, D., Pittella, G., Kron, R. G., Koo, D. C. & Bershady, M. A survey for faint variable objects in SA 57. *Astron. J.* **98**, 108-116(1989).

28. Kochanski, G. P., Tyson, J. A. & Fischer, P. Flickering faint galaxies: few and far between. *Astron J.* (in the press).

29. Schechter, P. An analytic expression for the luminosity function for galaxies. *Astrophys. J.* **203**, 297- 306 (1976).

30. Pedersen, H. *et al.* Optical candidates for the 1978 November 19 gamma-ray burst source. *Astrophys. J.* **270**, L43-L47 (1983).

31. Pedersen, H. & Hansen, J. A quasar in the 1978 November 19 gamma-ray burst error box. *Astrophys. J.* (submitted).

32. Boer, M. *et al.* ROSAT detection and high precision localization of X-ray sources in the November 19, 1978 gamma-ray burst error box. *Astrophys. J.* (in the press).

33. In't Zand, J. *et al. IAU Circ. No.* 6969 (1997).

34. Greiner, J. & Heise, J. *IAU Circ. No.* 6570 (1997).

35. Piro, L. *et al. IAU Circ. No.* 6570 (1997).

36. Walsh, D. *et al.* Spectroscopy of 26 QSO candidates from the Jodrell Bank 966-MHz survey. *Mon. Not. R. Astron. Soc.* **211**, 105-109 (1984).

37. Barthelmy, S. *et al.* Progress with the real-time GRB coordinates distribution Network (BACODINE). *AIP Proc.* **384**, 580-584 (1996).

38. McNamara, B. *et al.* Ground-based gamma-ray burst follow-up efforts: results from the first two years of the BATSE/COMPTEL/NMSU Rapid Response Network. *Astrophys. J. Suppl. Ser.* **103**, 173-181 (1996).

39. Vrba, F., Hartmann, D. & Jennings, M. Deep optical counterpart searches of gamma-ray burst localizations. *Astrophys. J.* **446**, 115-149 (1995).

40. Vrba, F. Searches for gamma-ray burst counterparts: current status and future prospects. *AIP Proc.* **384**, 565-574 (1996).

41. Luginbuhl, C., Vrba, F., Hudec, R., Hartmann, D. & Hurley, K. Results from the USNO quiescent optical counterpart search of IPN[3] GRB and optical transient localizations. *AIP Proc.* **384**, 676-679 (1996).

42. McNamara, B. *et al.* Ground-based γ-ray burst follow-up efforts: the first three years of the BATSE/COMPTEL/NMSU γ-Ray Burst Rapid Response Network. *AIP Proc.* **384**, 680-684 (1996).

43. Castro-Tirado, A., Brandt, S., Lund, N. & Guziy, A. Optical follow-up of gamma-ray bursts observed by WATCH. *AIP Proc.* **307**, 404-407 (1994).

44. Klose, S. Search for an optical counterpart of the source of GRB911001. *Astrophys. J.* **446**, 357-360 (1995).

45. Schaefer, B. *et al.* Rapid searches for counterparts of GRB 930131. *Astrophys. J.* **422**, L71-L74 (1994).

46. Park, H. S. *et al.* Limits on Real-time Optical Emission from Gamma-Ray Bursts Measured by the GROCSE Experiment. *Astrophys. J.* (submitted).

47. Lee, B. *et al.* Results from GROCSE: a real-time search for gamma-ray burst optical counterparts. *Astrophys. J.* (submitted).

48. Castro-Tirado, A. *et al. IAU Circ. No.* 6598 (1997).

49. Frail, D. *et al. IAU Circ. No.* 6576 (1997).

50. Miller, H. R. & Noble, J. C. The microvariability of blazars and related AGN. *ASP Conf. Ser.* **110**, 17-29 (1996).

Acknowledgements. We thank T. Courvoisier, T. Gehrels, J. Hakkila, D. Hartmann, M. Kippen, S. Perlmutter, P. Sackett, T. Tyson and M. Urry for their helpful answers to our many questions. We also thank W. Lewin and M. van der Klis and the referee, F. Vrba, for their critical comments on this Letter.

Correspondence should be addressed to C.K. (e-mail: chryssa.kouveliotou@msfc.nasa.gov).

Spectral Constraints on the Redshift of the Optical Counterpart to the γ-ray Burst of 8 May 1997

M. R. Metzger *et al.*

Editor's Note

Although the first optical counterpart (an object seen at visible wavelengths) to a gamma-ray burst had been found a couple of months earlier than this work, there was still uncertainty about how far away the bursts are. Here Mark Metzger and colleagues report the spectrum of an optical object at the same position as the burst GRB 970508. There are numerous absorption lines in the spectrum, analogous to the ones seen in the spectra of distant quasars. The researchers interpret these lines as arising from gas inside the host galaxy of the burst, and assigned it a redshift of 0.835. This clearly demonstrated that such GRBs lie far beyond our galaxy.

Brief, intense bursts of γ-rays occur approximately daily from random directions in space, but their origin has remained unknown since their initial detection almost 25 years ago[1]. Arguments based on their observed isotropy and apparent brightness distribution[2] are not sufficient to constrain the location of the bursts to a local[3] or cosmological origin[4]. The recent detection of a counterpart to a γ-ray burst at other wavelengths[5,6] has therefore raised the hope that the sources of these energetic events might soon be revealed. Here we report spectroscopic observations of the possible optical counterpart[7,8] to the γ-ray burst GRB970508. The spectrum is mostly featureless, except for a few prominent absorption lines which we attribute to the presence of an absorption system along the line of sight at redshift $z = 0.835$. Coupled with the absence of Lyman-α forest features in the spectra, our results imply that the optical transient lies at $0.835 \leqslant z \leqslant 2.3$. If the optical transient is indeed the counterpart of GRB970508, our results provide the first direct limits on the distance to a γ-ray burst, confirming that at least some of these events lie at cosmological distances, and are thus highly energetic.

O N 8 May 1997 UT, a moderate-fluence classical γ-ray burst (GRB970508) was detected by instruments aboard the Italian–Dutch satellite BeppoSAX[9]; the burst was localized initially to an error region of 5-arcmin radius[10] and later to a region of 3-arcmin radius[11]. A potential optical counterpart was identified within two days[7,8], which we refer to as OT J065349+79163 (here OT stands for optical transient). Interest in this object was heightened when the appearance of a bright X-ray source was reported[12] that was not seen in a previous X-ray all-sky survey, and included OT J065349+79163 in the 45-arcsec-radius error circle. The presence of a new and bright X-ray source in a γ-ray burst (GRB) error circle has been seen in the recent three GRBs observed by BeppoSAX.

对 1997 年 5 月 8 日伽马射线暴光学对应体红移的光谱限制

梅茨格等

编者按

尽管第一个伽马射线暴的光学对应体(在光学波段观测到的天体)在这篇文章发表几个月前就被发现了，但是这些暴离我们到底有多远仍然不确定。这里马克·梅茨格以及他的合作者报道了和伽马射线暴 GRB 970508 相同位置的光学天体的光谱。在光谱中存在着许多的吸收线，这和遥远的类星体光谱中的吸收线类似。研究者认为这些吸收线是由该暴寄主星系内的气体造成的，并且确定其红移为 0.835。这很明确地证明这样的伽马射线暴发生在我们的星系之外很远的地方。

短暂、强烈的伽马射线暴几乎每天都在太空的不同方向随机发生，但是自从 25 年前首次被发现以来，它们的起源就一直处于未知的状态[1]。以观测到的伽马射线暴各向同性和视亮度分布为基础的论证[2]尚不足以限制这些暴的源头处于近邻区域[3]还是处在宇宙距离上[4]。最近在其他波段探测到伽马射线暴的对应体[5,6]唤起了可能很快揭示这些高能事件起源的希望。这里我们报道对伽马射线暴 GRB970508 可能的光学对应体[7,8]的光谱观测。这个光谱大体上是无特征的，除了一些明显的吸收线(这些吸收线我们认为是由于在视线方向上红移 $z = 0.835$ 处存在一个吸收系统)。结合光谱上缺少莱曼 α 森林特征的事实，我们的结果表明这个光学暂现源位于红移 $0.835 \leqslant z \leqslant 2.3$ 之内。假如这个光学暂现源真是 GRB970508 的对应体，我们的结果首次直接限制了伽马射线暴的距离，并证实至少有一些爆发事件处在宇宙学距离，因而具有很高的能量。

世界时(UT)1997 年 5 月 8 日，安装在意大利-荷兰卫星 BeppoSAX[9]上的仪器探测到一个中等能流的经典伽马射线暴(GRB970508)；这个暴开始定位于半径为 5 arcmin 的误差区域内[10]，随后定位于一个半径为 3 arcmin 的区域内[11]。在探测到这个伽马射线暴之后的 2 天内认证出一个可能的光学对应体[7,8]，记为 OT J065349+79163(这里 OT 代表光学暂现源)。当有报道发现了一个之前的 X 射线全天巡天中不存在的亮 X 射线源[12]，并且它的半径为 45 arcsec 的误差圆内包含了 OT J065349+79163 后，人们对这个光学暂现源的研究兴趣大为增大。在最近 BeppoSAX

In the first such case, a decaying optical source[5] was also seen and these authors suggested that such fading (X-ray and optical) sources are the afterglow of γ-ray bursts. It has been suggested[13] that because OT J065349+79163 exhibits unusual variability at optical and X-ray wavelengths, it represents a similar optical afterglow to GRB970508.

We obtained spectra of OT J065349+79163 with the Keck II 10-m telescope on 11 May 1997 UT, using the Low Resolution Imaging Spectrograph[14] (LRIS) with a 2,048 × 2,048 pixel CCD (charge-coupled device). Starting at 05:44 UTC, a sequence of three 10-minute spectra were obtained with a 1.0-arcsecond-wide slit oriented along the direction of the atmospheric dispersion. The spectrograph was configured with a 300 lines mm^{-1} grating blazed at 5,000 Å, covering the region 3,850–8,550 Å. We obtained two further spectra of 10 minutes each on 12 May 1997 UT. Calibration Hg–Kr–Ar and flat lamp spectra were taken at the end of each exposure sequence. Owing to the extremely low elevation angle of the telescope during the observations (~25°) and a problem with the Keck II lower shutter, approximately half of the telescope aperture was blocked during the exposures. Instrumental sensitivity was calibrated by an observation of the standard star[15] BD+284211, but owing to an approximate correction for the occulted aperture the absolute calibration is only rough.

The resulting combined spectra from 11 May is shown in Fig. 1a. It is customary to represent the spectral flux density (F_v) of a non-thermal source by a power law, $F_v \propto v^\alpha$; here v is the frequency. The optical index computed from the spectrum is $\alpha_O = -0.9 \pm 0.3$. The large uncertainty is due to the uncertain correction for atmospheric extinction in the blue region of the spectrum at the large zenith angles of our observations.

Several absorption features are evident; the strongest, near 7,600 and 6,870 Å, are due to telluric O_2. In the region between 4,300 and 5,300 Å (Fig. 1b), there are several significant absorption features[16] that we identify. The identifications were made based on Mg II doublet (5,129 and 5,143 Å) line ratios, and assigning further rough identification of other metal lines based on wavelength ratios between these and the Mg doublet. Table 1 shows the lines identified in the spectrum; independent redshifts are computed from each line. This reveals a relatively strong[17] metal line absorption system at $z = 0.8349 \pm 0.0002$, and a weaker Mg II system at $z = 0.768$. The eight lines present in the strong absorption system make the redshift assignment unambiguous. The continuum source is either more distant and absorbed by a gas cloud at this redshift, or perhaps is located physically within the cloud, but the absorption places a firm lower limit to the redshift of the source, $z \geqslant 0.835$.

探测到的三例伽马射线暴(GRB)中，我们都发现伽马射线暴的误差圆内存在新的 X 射线亮源。在第一个这种事例中同时发现有衰减的光源[5]，研究作者们认为这些衰减的(X 射线和光学)源是伽马射线暴的余晖。这是因为 OT J065349+79163 在光学和 X 射线波段都存在不寻常的变化，它的表现类似 GRB970508 的光学余晖[13]。

我们于 1997 年 5 月 11 日 UT 利用 Keck II 10 m 望远镜通过装载着 2,048 × 2,048 像素 CCD(电荷耦合器件)的低分辨率成像摄谱仪[14](LRIS) 得到 OT J065349+79163 的光谱。05：44 UTC(协调世界时)开始，通过沿着大气色散方向放置 1.0 arcsec 宽的狭缝获得了 3 次曝光 10 分钟的光谱。摄谱仪配置有每毫米 300 条狭缝的光栅，照射中心为 5,000 Å，覆盖波长范围 3,850~8,550 Å。我们于 1997 年 5 月 12 日 UT 得到另外两次曝光 10 分钟的光谱。在曝光之后对每段光谱都进行了 Hg–Kr–Ar 和平谱校准。由于观测时望远镜仰角极低($\sim25°$)以及 Keck II 低遮光板的问题，曝光时望远镜接近一半的孔径都被遮住。仪器灵敏度是通过对标准星[15]BD+284211 的观测来校准的，但是由于对孔径遮盖只是近似修正，这里绝对校准是较为粗糙的。

5 月 11 日得到的合并后光谱在图 1a 中展示。按惯例用幂律来描述非热源的光谱流量密度(F_v)，$F_v \propto v^\alpha$；这里 v 是频率。从光谱得到的光学指数为 $\alpha_0 = -0.9 \pm 0.3$。比较大的不确定度主要来自在大天顶角观测时对光谱蓝端大气消光修正的误差。

从光谱中可以明显看到一些吸收特征；在 7,600 Å 和 6,870 Å 附近的最强吸收线来自地球大气的 O_2。在 4,300~5,300 Å 区域(图 1b)我们认证出好几条吸收线[16]。这些认证是根据 Mg II 双重线(5,129 Å 和 5,143 Å)的谱线比来实现的，其他金属谱线的进一步粗略认证是根据这些金属谱线和 Mg 双重线的波长比来实现的。表 1 显示了光谱中认证出的谱线；对每条谱线都独立计算出红移。结果显示在 $z = 0.8349 \pm 0.0002$ 红移处是相对较强[17]的金属谱线吸收系统，在 $z = 0.768$ 处有相对而言较弱的 Mg II 吸收系统。强吸收系统的 8 条谱线很清楚地确定出红移。连续谱发射源可能位于更远处，谱线被处于这个红移处的一团气体云吸收，或者可能发射源本身就处于气体云里，但是吸收特征能够确定这个源红移的下限，即 $z \geq 0.835$。

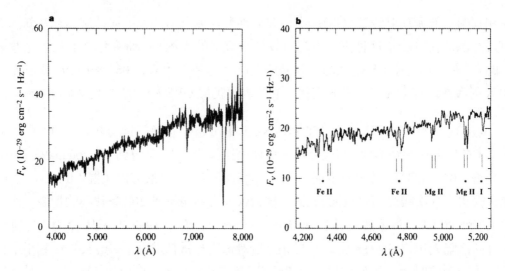

Fig. 1. The spectrum of the optical variable. **a**, Full spectrum; **b**, expansion of a limited region, with strong absorption lines and identifications indicated. The lines marked with an asterisk are identified with an absorption system at redshift $z = 0.835$, the others at $z = 0.767$. The spectrum has been smoothed with a three-pixel boxcar filter. A few additional weak features (not shown) have also been tentatively identified with the $z = 0.767$ system. F_v is the flux density, and d is the wavelength in Å.

Table 1. OT J065349+79163 absorption lines

λ_{vac} (Å)	Unc.	W_λ (Å)	Unc.	λ_{rest} (Å)	z	Assignment
4,302.5	1.8	1.3	0.3	2,344.2	0.8354 (8)	Fe II
4,359.7	1.4	1.3	0.3	2,374.5	0.8360 (6)	Fe II
4,372.2	1.5	1.4	0.3	2,382.8	0.8349 (6)	Fe II
4,746.7	1.7	1.0	0.4	2,586.7	0.8350 (7)	Fe II
4,769.7	1.3	2.3	0.2	2,600.2	0.8344 (5)	Fe II
4,941.1	1.5	1.3	0.3	2,796.4	0.7670 (5)	Mg II
4,953.9	1.5	1.0	0.4	2,803.5	0.7670 (5)	Mg II
5,130.4	1.1	2.7	0.2	2,796.4	0.8346 (4)	Mg II
5,144.0	1.1	3.0	0.2	2,803.5	0.8348 (4)	Mg II
5,232.6	1.3	1.8	0.2	2,853.0	0.8341 (5)	Mg I

Table gives measured parameters for identified absorption lines in OT J065349+79163 and the inferred redshift of each feature. λ_{vac} is the measured wavelength of each line, corrected to vacuum, and the following column is the uncertainty (in Å); W_λ is the observed (not rest frame) equivalent width of the line in Å, along with the corresponding uncertainty; the last three columns list the assigned physical absorption for each line, with rest vacuum wavelength (λ_{rest}), implied redshift, and element/ionization state.

Such absorption systems are commonly seen in the spectra of high-redshift quasi-stellar objects (QSOs)[17]. An imaging study of such systems[18] reveals that most are associated with normal galaxies close to the line of sight to the QSO. An analysis of these systems[19] at redshift similar to the system we identify in OT J065349+79163 indicates a correlation

374

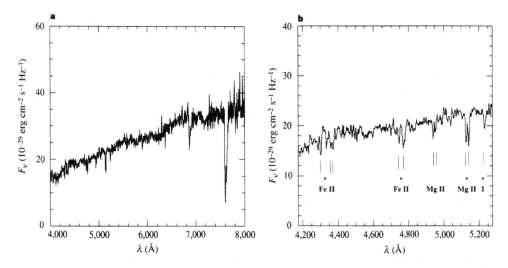

图 1. 光学变源的光谱。a，光谱整体；b，光谱一部分的放大图，强吸收线和认证被标示出。用星号标记的线被认为是红移 $z = 0.835$ 处的吸收系统，其他的则被认为是 $z = 0.767$ 处的吸收系统。光谱经过三个像素的 boxcar 滤波器的平滑处理。一些额外的弱特征（这里没有显示出）已经初步被认证为对应于 $z = 0.767$ 的系统。F_v 是流量密度，d 是以 Å 为单位的波长。

表 1. OT J065349+79163 吸收线

λ_{vac}(Å)	误差	W_λ(Å)	误差	λ_{rest}(Å)	z	认证
4,302.5	1.8	1.3	0.3	2,344.2	0.8354 (8)	Fe II
4,359.7	1.4	1.3	0.3	2,374.5	0.8360 (6)	Fe II
4,372.2	1.5	1.4	0.3	2,382.8	0.8349 (6)	Fe II
4,746.7	1.7	1.0	0.4	2,586.7	0.8350 (7)	Fe II
4,769.7	1.3	2.3	0.2	2,600.2	0.8344 (5)	Fe II
4,941.1	1.5	1.3	0.3	2,796.4	0.7670 (5)	Mg II
4,953.9	1.5	1.0	0.4	2,803.5	0.7670 (5)	Mg II
5,130.4	1.1	2.7	0.2	2,796.4	0.8346 (4)	Mg II
5,144.0	1.1	3.0	0.2	2,803.5	0.8348 (4)	Mg II
5,232.6	1.3	1.8	0.2	2,853.0	0.8341 (5)	Mg I

表格给出 OT J065349+79163 认证出的吸收线测量参数和对每个特征推测得到的红移。λ_{vac} 是每条线在真空里的测量波长，随后一列是波长的误差，以 Å 为单位；W_λ 是谱线的等值宽度的观测值（非静止系），以 Å 为单位，而对应的误差列在后一列；最后三列列出每条线所确定的物理吸收，包括静止真空波长（λ_{rest}）、确定的红移和元素（或电离态）。

这样的吸收系统经常出现在高红移类星体(QSOs)的光谱中 [17]。对这种系统 [18] 的成图研究显示大部分吸收系统和 QSO 视线方向附近的正常星系成协。对类似 OT J065349+79163 红移处这些系统 [19] 的分析显示谱线等值宽度和碰撞参数有一定的相关性（弥散较大）。正因为看到的吸收和 QSO 系统的类似，我们预期深度曝光的

(with significant scatter) between line equivalent width and impact parameter. As the absorption we see should be similar to QSO systems, we expect that deep images (perhaps taken after the transient fades) would reveal a galaxy responsible for this absorption system, though it is difficult to predict its brightness or separation from the transient. A hint of such an object has already been suggested[20]. Note that as the OT was far brighter than any other nearby object, any contamination of the spectrum is negligible and thus the OT features were a physical absorption.

At these redshifts, the number of Mg II absorption systems with rest equivalent widths $W_\lambda > 0.3$ Å per unit redshift is of the order of unity[17]. Detection of one or two such absorption systems in our spectrum is thus not unusual. However, the ratio of line strengths (Mg I / Mg II) seems unusually high, and combined with the high strength of the Mg II absorption system provides some evidence for a dense foreground interstellar medium. This implies either a small impact parameter[19], or, more likely, that the $z = 0.835$ system is due to the GRB host galaxy itself. We can also place an approximate upper limit to the source redshift from the absence of apparent Lyman-α absorption features in our spectra. The short-wavelength limit of our data corresponds to $z_{\text{Ly}\alpha} \approx 2.3$. In addition to the lack of individual lines, the mean observed continuum decrement at this redshift is[21,22] $D_A \approx 0.1$–0.2, and it increases with redshift. If present, such a continuum drop should be detected in our data for wavelength $\lambda > 4,000$ Å. We can thus place an approximate upper limit to the source redshift of $z \lesssim 2.3$.

One might ask whether from current observations we should expect to see a host galaxy for the burst, if such a galaxy were present. If we assume a minimum redshift of $z = 0.835$ in a standard Friedmann cosmology with $H_0 = 70 \text{ km s}^{-1} \text{ Mpc}^{-1}$ and $\Omega_0 = 0.2$, the luminosity distance is 1.49×10^{28} cm. The B band would be redshifted just slightly past the Gunn i band, and for observations[13] made on 10 May UT, the observed flux in the redshifted B band is ~ 39 μJy. For the assumed redshift and cosmology, this implies an absolute magnitude of $M_B \approx -22.6$ mag, or a lower limit for $L_B \approx 7 L_*$, where L_* is a characteristic galaxy luminosity[33]. Thus, OT J065349+79163 is still significantly outshining any host galaxy for the most probable host luminosity range. The properties of the Mg absorption lead us to expect that once the OT has faded, the galaxy could be identified optically.

Taken together, the source's compact optical appearance, a featureless continuum, X-ray emission and high redshift suggest a possible classification (independent of a burst event) as a BL Lac object[23]. One of the known characteristics of BL Lac objects is their variability from radio to γ-ray wavelengths. We now evaluate the a posteriori probability that we might be seeing a BL Lac object by random coincidence with the γ-ray error box. The surface density of BL Lac objects with Rosat X-ray flux $f_X \lesssim 10^{-12} \text{ erg cm}^{-2} \text{ s}^{-1}$ is not very well known, but there are indications that this distribution is quite flat at low flux densities. A simple extrapolation for the expected number of BL Lacs with $f_X > 6 \times 10^{-13} \text{ erg cm}^{-2} \text{ s}^{-1}$ is[24] 0.03 per square degree. Thus the probability of finding a BL Lac object within the 3-arcmin-radius localization region is $\sim 2 \times 10^{-4}$. The amplitude of the variability detected in the counterpart[13] over a few days is also larger than has been observed in studies of BL

图像(可能在暂现源衰退之后)将显示产生这个吸收系统的星系,尽管很难预计星系的亮度以及离开暂现源的距离。已经有人提供了存在这样一个星系的线索[20]。注意因为这个光学暂现源比附近任何天体都亮得多,任何光谱污染都可忽略,因此这个光学暂现源的光谱特征确实来自物理吸收。

在这些红移处,每单位红移静止等值宽度 $W_\lambda > 0.3$ Å 的 Mg II 吸收系统的数目是 1 的量级[17]。因此在我们的系统中探测到一个或两个这样的吸收系统是正常的。不过,谱线强度比(Mg I/Mg II)似乎过高,结合 Mg II 吸收系统的高强度特征绘出了一些存在前景稠密星际介质的证据。这表明或者碰撞参数很小[19],或者更可能的是这个红移 0.835 的系统是来自伽马射线暴寄主星系本身。我们也从光谱缺少明显的莱曼-α 吸收特征来估计源的红移上限。我们的数据在短波端的极限对应于 $z_{Ly\alpha} \approx 2.3$。除了缺少单独的谱线,这个红移处观测到的平均连续谱减幅为 $D_A \approx 0.1 \sim 0.2$,并且随着红移增大。这样的连续谱陡降如果存在,在波长 $\lambda > 4,000$ Å 处应该能在我们的数据中探测到。因此,我们能大概限制源的红移上限为 $z \lesssim 2.3$。

你可能会问如果存在这个伽马射线暴的寄主星系,那么我们是否能从目前的观测中观测到它。假设最小红移为 0.835,那么在 $H_0 = 70$ km · s^{-1} · Mpc^{-1} 和 $\Omega_0 = 0.2$ 的标准弗里德曼宇宙,光度距离为 1.49×10^{28} cm。B 波段将轻微红移过冈恩 i 波段,5 月 10 日 UT 在红移后的 B 波段观测[13]的流量为 ~39 mJy。在假设的红移和宇宙学模型下,得出绝对星等 $M_B \approx -22.6$ mag,或者下限 $L_B \approx 7L_*$,这里 L_* 是星系特征光度[33]。因此,在最可能的寄主星系光度范围内 OT J065349+79163 仍然比它的寄主星系都要闪耀。Mg 吸收的性质让我们预期一旦光学暂现源光度衰弱,我们就能在光学上探测到该星系。

综合考虑该源的致密光学外观、无特征的连续谱、X 射线辐射和高红移这些因素,可将其归类(与该爆发事件无关)为 BL Lac 天体[23]。目前知道的 BL Lac 天体的一个特征是它们从射电到 γ 射线波段都存在光变。我们现在估计 BL Lac 天体随机落在 γ 射线误差圆内的后验概率。虽然伦琴 X 射线流量为 $f_X \lesssim 10^{-12}$ erg · cm^{-2} · s^{-1} 的 BL Lac 天体的面密度目前不是很清楚,但是有迹象表明在低流量密度处面密度分布比较平。简单外推得到 $f_X > 6 \times 10^{-13}$ erg · cm^{-2} · s^{-1} 的 BL Lac 天体的预计个数[24]为每平方度 0.03 个。因此在半径为 3 arcmin 的范围内发现 BL Lac 天体的概率为 ~2×10^{-4}。所探测到的对应体在几天内的光变幅度[13]也比目前 BL Lac 天体光变研究中观测到

Lac object variability[25,26]. Although we cannot completely exclude the possibility that OT J065349+79163 is a chance coincidence of a BL Lac object with the GRB error circle, the probability of finding a random BL Lac object which also exhibits variability that is temporally correlated with a γ-ray burst is quite small. Thus we conclude that the OT is probably associated with GRB970508, regardless of classification, though the strongest constraints naturally come from higher-energy emission.

The high redshift of OT J065349+79163, its featureless spectrum and slowly decaying optical flux are consistent with the so-called fireball models for cosmological bursts[27-29], which are efficient at emitting γ-rays and produce power-law spectral energy distributions. The fluence[30] of GRB970508 in the energy range 20–1,000 keV was 3×10^{-6} erg cm^{-2}, and at the minimum redshift implied for OT J065349+79163, this burst would have a total γ-ray energy of 7×10^{51} erg (assuming isotropic emission). This falls in the general range of typical γ-ray burst energies from various cosmological models[31,32].

The remarkable progress in detecting X-ray and optical counterparts to GRBs has been made possible only by rapid localization of the burst by BeppoSAX and prompt dissemination of the coordinates by the BeppoSAX team. Further progress in understanding GRBs requires many more optical counterparts to be identified. It is clear from experience of the first two optical counterparts that, in order to obtain the critical data, the counterparts must be discovered and followed up spectroscopically within a few days. It now seems that an understanding of the physical mechanisms behind γ-ray bursts is within reach.

(**387**, 878-880; 1997)

M. R. Metzger[*], S. G. Djorgovski[*], S. R. Kulkarni[*], C. C. Steidel[*], K. L. Adelberger[*], D. A. Frail[†], E. Costa[‡] & F. Frontera[§]

[*] Palomar Observatory, 105-24, California Institute of Technology, Pasadena, California 91125, USA

[†] National Radio Astronomy Observatory, Socorro, New Mexico 87801, USA

[‡] Istituto di Astrofisica Spaziale CNR, 00044 Frascati, Italy

[§] Dipartimento di Fisica, Universita' di Ferrara and Istituto TESRE-CNR, 40129 Bologna, Italy

Received 21 May; accepted 3 June 1997.

References:

1. Klebesadel, R. W., Strong, I. B. & Olsen, R. A. Observations of gamma-ray bursts of cosmic origin. *Astrophys. J.* **182**, L85-L88 (1973).

2. Meegan, C. A. *et al.* Spatial distribution of gamma-ray bursts observed by BATSE. *Nature* **355**, 143- 145 (1992).

3. Lamb, D. Q. The distance scale to gamma-ray bursts. *Publ. Astron. Soc. Pacif.* **107**, 1152-1166 (1995).

4. Paczyński, B. How far away are gamma-ray bursters? *Publ. Astron. Soc. Pacif.* **107**,1167-1175 (1995).

5. van Paradijs, J. *et al.* Transient optical emission from the error box of the γ-ray burst of 28 February 1997. *Nature* **386**, 686-689 (1997).

6. Costa E. *et al. IAU Circ.* No. 6576 (1997).

7. Bond, H.E. *IAU Circ.* No. 6654 (1997).

8. Djorgovski, S. *et al. IAU Circ.* No. 6655 (1997).

9. Boella, G. *et al.* BeppoSAX, the wide band mission for X-ray astronomy. *Astron. Astrophys. Suppl.* **122**, 299 (1997).

10. Costa, E. *et al. IAU Circ.* No. 6649 (1997).

11. Heise, J. *et al. IAU Circ.* No. 6654 (1997).

12. Piro, L. *et al. IAU Circ.* No. 6656 (1997).

13. Djorgovski, S. G. *et al.* The optical counterpart to the γ-ray burst GRB970508. *Nature* **387**, 876-878 (1997).

378

的 [25,26] 大。尽管我们不能完全排除 OT J065349+79163 是偶然落入 GRB 误差圆的 BL Lac 天体的可能性，但是发现随机的一个 BL Lac 天体显示的光变在时间上与一个伽马射线暴相关的概率非常小。因此我们得出结论，不论是什么类别，这个光源暂现源可能和 GRB970508 有关，尽管最强的限制理应来自更高能的辐射。

OT J065349+79163 的高红移、无特征的光谱和缓慢衰减的光学流量都与宇宙学伽马射线暴的所谓火球模型 [27-29] 一致。火球模型能有效产生 γ 射线并产生幂律的能谱。GRB970508 在 20~1,000 keV 能量段的能量 [30] 为 3×10^{-6} erg·cm^{-2}，在 OT J065349+79163 可能的最小红移值处，这个暴产生的总 γ 射线能量为 7×10^{51} erg（假设各向同性辐射）。这个能量处于各种宇宙学模型下典型伽马射线暴的一般能量范围内 [31,32]。

伽马射线暴的 X 射线和光学对应体探测的显著进展是通过 BeppoSAX 快速定位伽马射线暴的位置和 BeppoSAX 团队即时发布坐标信息才变得可能。进一步理解伽马射线暴的性质需要发现更多的光学对应体。从最早的两例光学对应体的发现经验可知，为了得到关键的数据，必须在伽马射线暴发生的几天内利用光谱发现并跟踪对应体。目前看来，对伽马射线暴背后的物理机制的理解正逐步成为可能。

（肖莉 翻译；黎卓 审稿）

14. Oke, J. B. *et al.* The Keck low-resolution imaging spectrometer. *Publ. Astron. Soc. Pacif.* **107**, 375-385 (1995).

15. Massey, P., Strobel, K., Barnes, J. V. & Anderson, E. Spectrophotometric standards. *Astrophys. J.* **328**, 315-333 (1988).

16. Metzger, M. R. *et al. IAU Circ.* No. 6655 (1997).

17. Steidel, C. C. & Sargent, W. L. W. Mg II absorption in the spectra of 103 QSOs: implications for the evolution of gas in high-redshift galaxies. *Astrophys. J. Suppl. Ser.* **80**, 1-108 (1992).

18. Steidel, C. C., Dickinson, M. & Persson, S. E. Field galaxy evolution since z approximately 1 from a sample of QSO absorption-selected galaxies. *Astrophys. J.* **437**, L75-L78 (1994).

19. Steidel, C. C. in *QSO Absorption Lines* (ed. Meylan, G.) 139-152 (Springer, Berlin, 1995).

20. Djorgovski, S. *et al. IAU Circ.* No. 6660 (1997).

21. Oke, J. B. & Korycansky, D. G. Absolute spectrophotometry of very large redshift quasars. *Astrophys. J.* **255**, 11-19 (1982).

22. Zuo, L. & Lu, L. Measurements of D_t for a large QSO sample and determination of evolution of Lyman-alpha clouds. *Astrophys. J.* **418**, 601-616 (1993).

23. Stocke, J. T. *et al.* The Einstein Observatory extended medium-sensitivity survey. II – The optical identifications. *Astrophys. J. Suppl. Ser.* **76**, 813-874 (1991).

24. Nass, P. *et al.* BL Lacertae objects in the ROSAT All-Sky Survey: new objects and comparison of different search techniques. *Astron. Astrophys.* **309**, 419-430 (1996).

25. Miller, H. R. & Noble, J. C. The microvariability of blazars and related AGN. 17-29 (ASP Conf. Ser. 110, Astron. Soc. Pacif., 1996).

26. Heidt, J. & Wagner, S. J. Statistics of optical intraday variability in a complete sample of radio-selected BL Lacertae objects. *Astron. Astrophys.* **305**, 42-52 (1995).

27. Nészáros, P. & Rees, M. J. Gamma-ray bursts; multiwaveband spectral predictions for blast wave models. *Astrophys. J.* **418**, L59-L62 (1993).

28. Paczyński, B. & Rhoads, J. E. Radio transients from gamma-ray bursters. *Astrophys. J.* **418**, L5-L8 (1993).

29. Waxman, E. Gamma-ray burst after-glow: confirming the cosmological fireball model. *Astrophys. J.* (submitted).

30. Kouveliotou, C. *et al. IAU Circ.* No. 6660 (1997).

31. Fenimore, E. E. *et al.* The intrinsic luminosity of gamma-ray bursts and their host galaxies. *Nature* **366**, 40-42 (1993).

32. Fenimore, E. E. & Bloom, J. S. Determination of distance from time dilation of cosmological gamma-ray bursts. *Astrophys. J.* **453**, 25-36 (1995).

33. Schachter, P. An analytic expression for the luminosity function of galaxies. *Astrophys. J.* **203**, 297-306 (1976).

Acknowledgements. We thank the BeppoSAX team for their efforts in disseminating information to observers for rapid identification of an optical transient, making this work possible. We thank W. Sargent and M. Pahre for discussions, and the Keck Observatory staff for assistance at the telescope. This work is based on observations obtained at the W. M. Keck Observatory, which is jointly operated by the California Institute of Technology and the University of California. M.R.M. was supported by Caltech; S.R.K. was supported by NASA and NSF; S.G.D. was supported by NSF and the Bressler Foundation; C.C.S. was supported by NSF and the Sloan Foundation.

Correspondence should be addressed to M.R.M. (e-mail: mrm@astro.caltech.edu).

Discovery of a Supernova Explosion at Half the Age of the Universe

S. Perlmutter *et al.*

Editor's Note

In the early 1990s it became possible to use type Ia supernovae as "standard candles" to determine astronomical distances. Using this approach, Saul Perlmutter and coworkers here report a supernova at a redshift of 0.83 that is fainter than expected. Although they initially interpreted the faintness as evidence that the universe has a lower average density than was thought, it was soon realized that the best explanation is that the universe is expanding at an accelerating rate. This is now the accepted view, although the reason is unclear and represents one of the central puzzles in contemporary cosmology. One interpretation is that the universe is pervaded by "dark energy" that creates a repulsive force, counteracting gravitational attraction.

The ultimate fate of the Universe, infinite expansion or a big crunch, can be determined by using the redshifts and distances of very distant supernovae to monitor changes in the expansion rate. We can now find[1] large numbers of these distant supernovae, and measure their redshifts and apparent brightnesses; moreover, recent studies of nearby type Ia supernovae have shown how to determine their intrinsic luminosities[2-4]—and therefore with their apparent brightnesses obtain their distances. The > 50 distant supernovae discovered so far provide a record of changes in the expansion rate over the past several billion years[5-7]. However, it is necessary to extend this expansion history still farther away (hence further back in time) in order to begin to distinguish the causes of the expansion-rate changes—such as the slowing caused by the gravitational attraction of the Universe's mass density, and the possibly counteracting effect of the cosmological constant[8]. Here we report the most distant spectroscopically confirmed supernova. Spectra and photometry from the largest telescopes on the ground and in space show that this ancient supernova is strikingly similar to nearby, recent type Ia supernovae. When combined with previous measurements of nearer supernovae[2,5], these new measurements suggest that we may live in a low-mass-density universe.

SN1997ap was discovered by the Supernova Cosmology Project collaboration on 5 March 1997 UT, during a two-night search at the Cerro Tololo Interamerican Observatory (CTIO) 4-m telescope that yielded 16 new supernovae. The search technique finds such sets of high-redshift supernovae on the rising part of their light curves and guarantees the date of discovery, thus allowing follow-up photometry and spectroscopy of the transient supernovae to be scheduled[1]. The supernova light curves were followed

在宇宙年龄一半处发现的超新星爆发

珀尔马特等

编者按

在 20 世纪 90 年代的早期，使用 Ia 型超新星作为"标准烛光"测定天文学距离已成为可能。使用这种方法，索尔·珀尔马特以及他的合作者们在这里报道了在红移 0.83 处比预期的暗淡的一颗超新星。尽管他们最初把这颗超新星的暗淡解释为宇宙的平均密度比普遍认为的要低的证据，但是很快他们意识到最好的解释是宇宙的膨胀在加速。现在这个观点已经被普遍认可，尽管原因尚不知晓，这也是当代宇宙学最主要的难题之一。一种解释是宇宙中充斥着暗能量，产生排斥力，与引力相抗衡。

宇宙的命运最终是无限膨胀还是大挤压，可以通过测量遥远超新星的红移和距离，进而监测宇宙膨胀速率的变化来确定。我们现在发现了大量遥远的超新星[1]，并且测定了它们的红移和视亮度；而对较近 Ia 型超新星的研究已经找到了测定本征光度[2-4]的方法，通过它们的视亮度就能得到距离。至今已发现的 50 多个远距离 Ia 型超新星记录了过去几十亿年间宇宙膨胀速率的变化[5-7]。不过我们还需要进一步追溯更远（即时间上更早）的宇宙膨胀历史，从而找出宇宙膨胀速率变化的原因——是在宇宙质量密度的引力影响下变慢，或是由宇宙学常数的反作用而加速[8]。我们在此报道一颗已获光谱确认的最遥远的超新星。来自地面和空间中最大望远镜的光谱和测光数据表明这颗古老的超新星和较近的 Ia 型超新星非常相似。结合以前对较近距离的 Ia 型超新星的观测数据[2,5]，这些新测量表明我们可能生活在一个质量密度偏低的宇宙之中。

SN1997ap 于世界时（UT）1997 年 3 月 5 日被超新星宇宙学项目团队发现。位于智利的托洛洛山美洲天文台（CTIO）4 米望远镜在两个晚上的搜寻中总共发现了 16 颗新的超新星。搜寻技术能够在这些高红移超新星处于光变曲线上升阶段的时候发现它们，这就保证了发现的日期。因此我们可以安排对这些暂现的超新星进行进一步的测光和光谱观测[1]。利用 CTIO、WIYN、ESO 3.6 米和 INT 的望远镜，我们按

with scheduled R-, I- and some B-band photometry at the CTIO, WIYN, ESO 3.6-m, and INT telescopes, and with spectroscopy at the ESO 3.6-m and Keck II telescopes. (Here WIYN is the Wisconsin, Indiana, Yale, NOAO Telescope, ESO is the European Southern Observatory, and INT is the Isaac Newton Telescope.) In addition, SN1997ap was followed with scheduled photometry on the Hubble Space Telescope (HST).

Figure 1 shows the spectrum of SN1997ap, obtained on 14 March 1997 UT with a 1.5-h integration on the Keck II 10-m telescope. There is negligible (≤ 5%) host-galaxy light contaminating the supernova spectrum, as measured from the ground- and space-based images. When fitted to a time series of well-measured nearby type Ia supernova spectra[9], the spectrum of SN1997ap is most consistent with a "normal" type Ia supernova at redshift $z = 0.83$ observed 2 ± 2 supernova-restframe days (~4 observer's days) before the supernova's maximum light in the rest-frame B band. It is a poor match to the "abnormal" type Ia supernovae, such as the brighter SN1991T or the fainter SN1986G. For comparison, the spectra of low-redshift, "normal" type Ia supernovae are shown in Fig. 1 with wavelengths redshifted as they would appear at $z = 0.83$. These spectra show the time evolution from 7 days before, to 2 days after, maximum light.

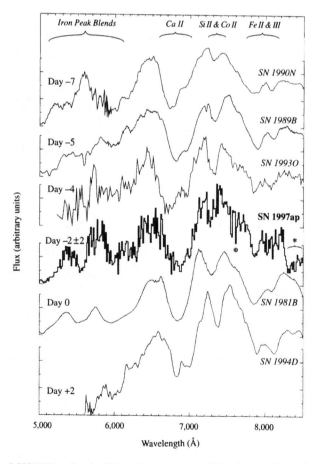

Fig. 1. Spectrum of SN1997ap placed within a time sequence of five "normal" type Ia supernovae. The

计划在 R、I 及 B 波段对超新星进行了测光从而得到光变曲线，同时还用 ESO 3.6 米和凯克 II 望远镜拍摄了光谱。(这里的 WIYN 是指威斯康星大学–印第安纳大学–耶鲁大学–美国国家光学天文台望远镜；ESO 是指欧洲南方天文台；INT 是指艾萨克·牛顿望远镜。)此外，我们还安排哈勃空间望远镜(HST)对 SN1997ap 进行测光。

图 1 是 SN1997ap 的光谱，它是由凯克 II 10 米天文望远镜于 1997 年 3 月 14 日 UT 持续 1.5 小时的观测数据积分而成。地面以及空间望远镜测量到的图像显示，寄主星系对超新星光谱的污染可以忽略不计 (≤ 5%)。我们把 SN1997ap 的光谱与一系列较近且已充分测量过的 Ia 型超新星光谱做时间序列拟合，发现它的光谱与红移 $z = 0.83$、在 B 波段极大亮度前 2±2 日(超新星静止系，约为 4 个观测者日)观测的"正常" Ia 型超新星最为一致。SN1997ap 的光谱与其他"非正常" Ia 型超新星(例如较亮的 SN1991T 或较暗的 SN1986G)的光谱并不匹配。图 1 中为了方便比较，低红移的"正常" Ia 型超新星的光谱被红移到 $z = 0.83$ 处。这些光谱显示了超新星从最大亮度的前 7 天到后 2 天的时间演化。

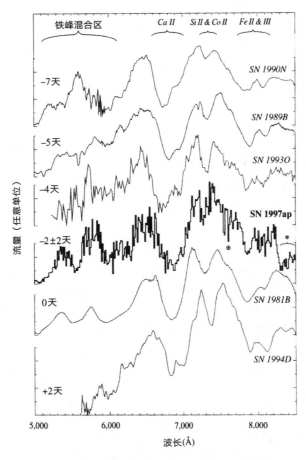

图 1. SN1997ap 的光谱与五个"正常" Ia 型超新星光谱放在一个时间序列中。SN1997ap 的光谱以 12.5 Å

data for SN1997ap have been binned by 12.5 Å ; the time series of spectra of the other supernovae[17-21] (the spectrum of SN1993O was provided courtesy of the Calán/Tololo Supernova Survey) are given as they would appear redshifted to $z = 0.83$. The spectra show the evolution of spectral features between 7 rest-frame days before, and 2 days after, rest-frame B-band maximum light. SN1997ap matches best at 2 ± 2 days before maximum light. The symbol $\oplus$ indicates an atmospheric absorption line and * indicates a region affected by night-sky line subtraction residuals. The redshift of $z = 0.83 \pm 0.005$ was determined from the supernova spectrum itself, as there were no host galaxy lines detected.

Figure 2 shows the photometry data for SN1997ap, with significantly smaller error bars for the HST observations (Fig. 2a) than for the ground-based observations (Fig. 2b and c). The width of the light curve of a type Ia supernova has been shown to be an excellent indicator of its intrinsic luminosity, both at low redshift[2-4] and at high redshift[5]: the broader and slower the light curve, the brighter the supernova is at maximum. We characterize this width by fitting the photometry data to a "normal" type Ia supernova template light curve that has its time axis stretched or compressed by a linear factor, called the "stretch factor"[1,5]; a "normal" supernova such as SN1989B, SN1993O or SN1981B in Fig. 1 thus has a stretch factor of $s \approx 1$. To fit the photometry data for SN1997ap, we use template U- and B-band light curves that have first been $1 + z$ time-dilated and wavelength-shifted ("K-corrected") to the R- and I-bands as they would appear at $z = 0.83$ (see ref. 5 and P.N. *et al.*, manuscript in preparation). The best-fit stretch factor for all the photometry of Fig. 2 indicates that SN1997ap is a "normal" type Ia supernova: $s = 1.03 \pm 0.05$ when fitted for a date of maximum at 16.3 March 1997 UT (the error-weighted average of the best-fit dates from the light curve, 15.3 ± 1.6 March 1997 UT, and from the spectrum, 18 ± 3 March 1997 UT).

It is interesting to note that we could alternatively fit the $1 + z$ time dilation of the event while holding the stretch factor constant at $s = 1.0^{+0.05}_{-0.14}$ (the best fit value from the spectral features obtained in ref. 10). We find that the event lasted $1 + z = 1.86^{+0.31}_{-0.09}$ times longer than a nearby $s = 1$ supernova, providing the strongest confirmation yet of the cosmological nature of redshift[9,11,12].

The best-fit peak magnitudes for SN1997ap are $I = 23.20 \pm 0.07$ and $R = 24.10 \pm 0.09$. (All magnitudes quoted or plotted here are transformed to the standard Cousins[13] R and I bands.) These peak magnitudes are relatively insensitive to the details of the fit: if the date of maximum is left unconstrained or set to the date indicated by the best-match spectrum, or if the ground- and space-based data are fitted alone, the peak magnitudes still agree well within errors.

为区间合并；其他超新星的光谱 [17-21](SN1993O 的光谱数据由 Calán/Tololo 超新星巡天提供）都被红移到 $z = 0.83$ 处。这些光谱反映了静止系 B 波段光强达到峰值的 7 天前至 2 天后之间的光谱特征变化。SN1997ap 的数据与最大亮度前 2 ± 2 天的光谱数据最吻合。符号 $\oplus$ 表示大气吸收线，* 表示受夜天光抵扣残余影响的区域。红移 $z = 0.83 \pm 0.005$ 得自超新星光谱，我们并没有检测到寄主星系的谱线。

图 2 显示的是超新星 SN1997ap 的测光数据。哈勃空间望远镜的观测（图 2a）误差显著小于地面望远镜的观测（图 2b 和 2c）。Ia 型超新星光变曲线的宽度被证明是"本征光度"的绝佳表征，无论是在低红移 [2-4] 还是在高红移 [5] 处：超新星的光变曲线越宽、变化越慢，那么最大亮度就越高。我们通过将测光数据与"正常"Ia 型超新星的光变曲线模板进行拟合来得到这个宽度，其中模板的时间轴由一个称为"伸展因子" [1,5] 的线性参数进行"拉伸"或"压缩"。图 1 中的"正常"超新星 SN1989B、SN1993O 及 SN1981B 的伸展因子 s 约等于 1。为了拟合 SN1997ap 的测光数据，我们使用 U 和 B 波段的光变曲线作为模板，将其经过 $1+z$ 倍的时间拉伸和波长平移（即"K 修正"）之后移动到 R 和 I 波段，就像它们在红移 0.83 处一样（见参考文献 5 和纽金特等人正在撰写的文章）。利用图 2 中所有测光数据对延展因子进行拟合得到 $s = 1.03 \pm 0.05$，这表明 SN1997ap 是一个"正常"的 Ia 型超新星：拟合的亮度极大值日期为 1997 年 3 月 16.3 日 UT（根据光变曲线得到的误差加权平均值为 1997 年 3 月 15.3 ± 1.6 日 UT，根据光谱得到的结果为 1997 年 3 月 18 ± 3 日 UT）。

值得一提的是，我们也可以保持伸展因子 $s = 1.0^{+0.05}_{-0.14}$ 不变（这个最佳拟合值来自文献 10 中的光谱数据），然后对事件的时间膨胀因子 $1+z$ 进行拟合。我们发现 SN1997ap 爆发事件的持续时间是一颗较近超新星（$s = 1$）的 $1+z = 1.86^{+0.31}_{-0.09}$ 倍，这一点为红移的宇宙学属性提供了迄今最强的确认 [9,11,12]。

超新星 SN1997ap 的峰值星等最佳拟合值为 $I = 23.20 \pm 0.07$ 以及 $R = 24.10 \pm 0.09$。（在本文中，所有提到和绘出的星等都已转换为标准的库森 [13]R 和 I 波段星等。）这些峰值星等对拟合的细节并不敏感：如果我们不限制最大值日期或是以最佳匹配光谱来设定最大值日期，或者是只用地面观测数据或空间观测数据来单独拟合，拟合的星等峰值仍然在误差以内。

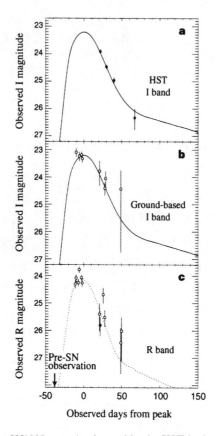

Fig. 2. Photometry points for SN1997ap. **a**, As observed by the HST in the F814W filter; **b**, as observed with ground-based telescopes in the Harris I filter; and **c**, as observed with the ground-based telescopes in the Harris R filter (open circles) and the HST in the F675W filter (filled circle); with all magnitudes corrected to the Cousins I or R systems[13]. The solid line shown in both **a** and **b** is the simultaneous best fit to the ground- and space-based data to the K-corrected, $(1+z)$ time-dilated Leibundgut B-band type Ia supernova template light curve[22], and the dotted line in **c** is the best fit to a K-corrected, time-dilated U-band type Ia supernova template light curve. The ground-based data was reduced and calibrated following the techniques of ref. 5, but with no host-galaxy light subtraction necessary. The HST data was calibrated and corrected for charge-transfer inefficiency following the prescriptions of refs 23, 24. K-corrections were calculated as in ref. 25, modified for the HST filter system. Correlated zero-point errors are accounted for in the simultaneous fit of the light curve. The errors in the calibration, charge-transfer inefficiency correction and K-corrections for the HST data are much smaller (~4% total) than the contributions from the photon noise. No corrections were applied to the HST data for a possible ~4% error in the zero points (P. Stetson, personal communication) or for nonlinearities in the WFPC2 response[26], which might bring the faintest of the HST points into tighter correspondence with the best-fit light curve in **a** and **c**. Note that the individual fits to the data in **a** and **b** agree within their error bars, providing a first-order cross-check of the HST calibration.

The ground-based data show no evidence of host-galaxy light, but the higher-resolution HST imaging shows a marginal detection (after co-adding all four dates of observation) of a possible $I = 25.2 \pm 0.3$ host galaxy 1 arcsec from the supernova. This light does not contaminate the supernova photometry from the HST and it contributes negligibly to the ground-based photometry. The projected separation is ~6 kpc (for $\Omega_M = 1$, $\Omega_\Lambda = 0$ and $h_0 = 0.65$, the dimensionless cosmological parameters describing the mass density,

388

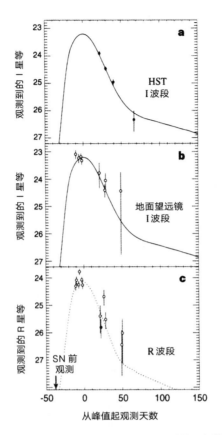

图 2. SN1997ap 的测光数据点。(**a**) 哈勃空间望远镜使用 F814W 滤镜观测的结果；(**b**) 地面望远镜使用哈里斯 I 波段滤镜的观测结果；(**c**) 地面望远镜使用哈里斯 R 波段滤镜所观测的结果 (空心点) 和哈勃空间望远镜使用 F675W 滤镜所观测的结果 (实心点)；所有星等已被转换到库森 I 或 R 波段[13]。图 **a** 和图 **b** 中的实线是同时将地面和空间的观测数据对经 K 修正和 $1+z$ 倍时间延展的 Ia 型超新星 Leibundgut B 波段光变曲线模板[22] 拟合的最佳结果；图 **c** 中的点线是对经 K 修正和 $1+z$ 倍时间延展的 Ia 型超新星 U 波段光变曲线模板拟合的最佳结果。地面观测的数据使用参考文献 5 中的方法进行处理和校准，但没有扣除寄主星系背景光的必要。哈勃空间望远镜的观测数据根据参考文献 23 和 24 中的方法进行了电荷迁移低效率的处理和修正。K 修正使用了参考文献 25 中的方法计算，并根据哈勃空间望远镜的滤镜系统做了调整。在对光变曲线的同步拟合中，我们考虑了零点误差。在哈勃空间望远镜数据的校正、电荷迁移低效率修正和 K 修正中的误差 (共约 4%) 远小于光子噪声的贡献。对于哈勃空间望远镜数据潜在的约 4% 的零点误差 (斯特森，个人交流) 和 WFPC2 响应的非线性[26] 没有做任何修正。这可能会使哈勃空间望远镜最暗的数据点与图 **a** 和图 **c** 中的最佳拟合曲线吻合得更好。注意图 **a** 和图 **b** 的每个数据点都在各自误差范围内与拟合结果一致，这提供了对哈勃空间望远镜校准的一阶交叉检验。

　　地面望远镜观测的数据没有显示寄主星系的光存在的证据，但是在分辨率更高的哈勃空间望远镜所拍摄的图像中有微弱的迹象 (在叠加全部四次观测之后) 表明，在距离该超新星 1 角秒的地方可能存在一个 $I = 25.2 \pm 0.3$ 的寄主星系。这些光并没有污染到哈勃空间望远镜的超新星测光结果，同时对地面望远镜测光的影响也微不足道。寄主星系到 SN1997ap 的投影距离约为 6 kpc (条件为 $\Omega_{M} = 1$、$\Omega_{\Lambda} = 0$ 和

vacuum energy density and Hubble constant, respectively) and the corresponding B-band rest-frame magnitude is $M_B \approx -17$ and its surface brightness is $\mu_B \approx 21$ mag arcsec^{-2}, consistent with properties of local spiral galaxies. We note that the analysis will need a final measurement of any host-galaxy light after the supernova has faded, in the unlikely event that there is a very small knot of host-galaxy light directly under the HST image of SN1997ap.

We compare the K-corrected $R-I$ observed difference of peak magnitudes (measured at the peak of each band, not the same day) to the $U-B$ colour found for "normal" low-redshift type Ia supernovae. We find that the rest-frame colour of SN1997ap [$(U-B)_{SN1997ap} = -0.28 \pm 0.11$] is consistent with an unreddened "normal" type Ia supernova colour, $(U-B)_{normal} = -0.32 \pm 0.12$ (see ref. 14 and also P.N. *et al.*, manuscript in preparation). In this region of the sky, there is also no evidence for Galactic reddening[15]. Given the considerable projected distance from the putative host galaxy, the supernova colour, and the lack of galaxy contamination in the supernova spectra, we proceed with an analysis under the hypothesis that the supernova suffers negligible host-galaxy extinction, but with the following caveat.

Although correcting for $E(U-B) \approx 0.04$ of reddening would shift the magnitude by only one standard deviation, $A_B = 4.8E(U-B) = 0.19 \pm 0.78$, the uncertainty in this correction would then be the most significant source of uncertainty for this one supernova. This is because of the large uncertainty in the $(U-B)_{SN1997ap}$ measurement, and the sparse low-redshift U-band reference data. HST J-band observations are currently planned for future $z > 0.8$ supernovae, to allow a comparison with the restframe $B-V$ colour, a much better indicator of reddening for type Ia supernovae. Such data will thus provide an important improvement in extinction correction uncertainties for future supernovae and eliminate the need for assumptions regarding host-galaxy extinction. In the following analysis, we also do not correct the lower-redshift supernovae for possible host-galaxy extinction, so any similar distribution of extinction would partly compensate for this possible bias in the cosmological measurements.

The significance of type Ia supernovae at $z = 0.83$ for measurements of the cosmological parameters is illustrated on the Hubble diagram of Fig. 3. To compare with low-redshift magnitudes, we plot SN1997ap at an effective rest-frame B-band magnitude of $B = 24.50 \pm 0.15$, derived, as in ref. 5, by adding a K-correction and increasing the error bar by the uncertainty due to the (small) width-luminosity correction and by the intrinsic dispersion remaining after this correction. By studying type Ia supernovae at twice the redshift of our first previous sample at $z \approx 0.4$, we can look for a correspondingly larger magnitude difference between the cosmologies considered. At the redshift of SN1997ap, a flat $\Omega_M = 1$ universe is separated from a flat $\Omega_M = 0.1$ universe by almost one magnitude, as opposed to half a magnitude at $z \approx 0.4$. For comparison, the uncertainty in the peak magnitude of SN1997ap is only 0.15 mag, while the intrinsic dispersion amongst stretch-calibrated type Ia supernovae is ~0.17 mag (ref. 5). Thus, at such redshifts even individual type Ia supernovae become powerful tools for discriminating amongst various world models, provided observations are obtained, such as those presented here, where the

$h_0 = 0.65$，这些无量纲宇宙学参数分别表示质量密度、真空能密度和哈勃常数），对应的 B 波段静止系星等为 $M_B \approx -17$，它的面亮度 $\mu_B \approx 21 \ \mathrm{mag \cdot arcsec^{-2}}$，这个属性与较近的旋涡星系的一致。需要指出的是，通常需要在超新星逐渐暗淡后，才能最终测量寄主星系的亮度，以免哈勃空间望远镜拍摄的 SN1997ap 图像正下方正好有个寄主星系的非常小的亮结，不过这种情况不太可能发生。

我们将 K 修正之后的 $R-I$ 峰值星等差（分别观测每个波段的亮度峰值，并非在同一天）和低红移"正常"Ia 型超新星的 $U-B$ 颜色结果做比较。发现 SN1997ap 的静止系颜色 $[(U-B)_{\mathrm{SN1997ap}} = -0.28 \pm 0.11]$ 与未红化的"正常"Ia 型超新星的颜色 $[(U-B)_{\mathrm{normal}} = -0.32 \pm 0.12]$（见参考文献 14 以及纽金特等人正在撰写的文章）相一致。在这片天区中，我们也没有发现"银河红化"[15]的证据。考虑到假定的寄主星系相当远的投影距离、超新星的颜色，以及光谱并没有受到寄主星系光的污染，我们后面的分析都假定寄主星系对该超新星的消光可以忽略。不过下面几点需要注意。

尽管 $E(U-B) \approx 0.04$ 的红化修正只会让星等偏移一个标准差，$A_B = 4.8E(U-B) = 0.19 \pm 0.78$，这个修正的不确定性是这颗超新星最主要的误差来源。这是因为 $(U-B)_{\mathrm{SN1997ap}}$ 的测量有很大的不确定性，而且还缺乏低红移处的 U 波段参考数据。哈勃空间望远镜计划对未来 $z > 0.8$ 的超新星进行 J 波段的观测，以便能够对比静止系 $B-V$ 颜色，它可以作为 Ia 型超新星更好的红化表征。这些数据可以帮助减小未来超新星消光修正的不确定性，并消除有关寄主星系消光的假定。在接下来的分析中，我们也没有对较低红移超新星进行可能的寄主星系的消光修正，所以任何相似的消光分布对宇宙学测量可能造成的偏差会部分地抵消。

红移 $z = 0.83$ 的 Ia 型超新星对宇宙参数测量的意义在图 3 的哈勃图中已经标示出。为了和低红移星等进行对比，我们将 SN1997ap 的数据画在有效静止系下 $B = 24.50 \pm 0.15$ 处，这个值的推导和文献 5 中一样，经过了 K 修正，并考虑了由（小的）宽度–光度修正的不确定性和修正之后遗留的内禀弥散度所引起的误差棒增加。通过研究我们此前 $z \approx 0.4$ 的第一个样本红移的两倍处的超新星，我们可以寻找不同宇宙学模型之间相应更大的星等差。在 SN1997ap 的红移处，$\Omega_M = 1$ 的平坦宇宙与 $\Omega_M = 0.1$ 的平坦宇宙模型相差将近一个星等，而在红移值 $z \approx 0.4$ 处，仅相差半个星等。相比之下，SN1997ap 的峰值星等的不确定性仅有 0.15 星等，然而 Ia 型超新星在延展修正后的内禀弥散度约为 0.17 星等（参考文献 5）。因此，在这样的红移处，只要观测能够提供本文这样的测光误差低于内禀弥散度的数据，即使是单个 Ia

photometric errors are below the intrinsic dispersion of type Ia supernova.

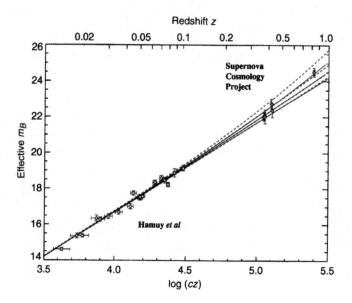

Fig. 3. SN1997ap at $z = 0.83$ plotted on the Hubble diagram from ref. 5. Also plotted are the 5 of the first 7 high-redshift supernovae that could be width–luminosity corrected, and the 18 of the lower-redshift supernovae from the Calán/Tololo Supernova Survey that were observed earlier then 5 d after maximum light[3]. Magnitudes have been K-corrected, and also corrected for the width-luminosity relation. The inner error bar on the SN1997ap point corresponds to the photometry error alone, while the outer error bar includes the intrinsic dispersion of type Ia supernovae after stretch correction (see text). The solid curves are theoretical m_B for $(\Omega_M, \Omega_\Lambda) = (0, 0)$ on top, $(1, 0)$ in middle and $(2, 0)$ on bottom. The dotted curves are for the flat-universe case, with $(\Omega_M, \Omega_\Lambda) = (0, 1)$ on top, $(0.5, 0.5)$, $(1, 0)$ and $(1.5, -0.5)$ on bottom.

By combining such data spanning a large range of redshift, it is also possible to distinguish between the effects of mass density Ω_M and cosmological constant Λ on the Hubble diagram[8]. The blue contours of Fig. 4 show the allowed confidence region on the Ω_Λ ($\equiv \Lambda/(3H_0^2)$) versus Ω_M plane for the $z \approx 0.4$ supernovae[5]. The yellow contours show the confidence region from SN1997ap by itself, demonstrating the change in slope of the confidence region at higher redshift. The red contours show the result of the combined fit, which yields a closed confidence region in the Ω_M–Ω_Λ plane. This fit corresponds to a value of $\Omega_M = 0.6 \pm 0.2$ if we constrain the result to a flat universe ($\Omega_\Lambda + \Omega_M = 1$), or $\Omega_M = 0.2 \pm 0.4$ if we constrain the result to a $\Lambda = 0$ universe. These results are preliminary evidence for a relatively low-mass-density universe. The addition of SN1997ap to the previous sample of lower-redshift supernovae decreases the best-fit Ω_M by approximately 1 standard deviation compared to the earlier results[5].

型超新星也能成为区分各种宇宙模型的有力工具。

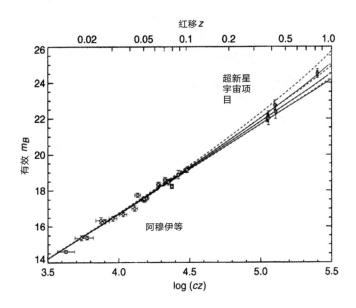

图 3. 红移 $z = 0.83$ 的 SN1997ap 标注在参考文献 5 中的哈勃图上。图上还画出首次能做宽度–光度修正的 7 颗高红移超新星中的 5 个，和 18 颗由 Calán/Tololo 超新星巡天观测到的较低红移超新星，它们是在亮度极大前和后五天的区间内被观测到的 [3]。星等已经经过 K 修正和宽度–光度关系修正。SN1997ap 数据点上的内侧误差棒只表示测光误差，而外侧误差棒则包含了经过伸展修正后的 Ia 型超新星内禀弥散度（见正文）。实线对应不同宇宙模型的理论 m_B：上，$(\Omega_M, \Omega_\Lambda) = (0, 0)$；中，$(\Omega_M, \Omega_\Lambda) = (1, 0)$；下，$(\Omega_M, \Omega_\Lambda) = (2, 0)$。虚线对应平坦宇宙模型：上，$(\Omega_M, \Omega_\Lambda) = (0, 1)$；下，$(\Omega_M, \Omega_\Lambda) = (0.5, 0.5)$、$(1, 0)$ 和 $(1.5, -0.5)$。

综合这些大范围红移的数据，我们还有可能在哈勃图中区分质量密度 Ω_M 和宇宙学常数 Λ 的影响 [8]。图 4 中的蓝色等值线显示了 $z \approx 0.4$ 的超新星所允许的 Ω_Λ（$\equiv \Lambda/(3H_0^2)$）与 Ω_M 平面的置信区域 [5]。黄色等值线显示的是 SN1997ap 独立限制的 Ω_Λ 和 Ω_M 置信区域，表明在较高红移处置信区域的斜率有所变化。红色等值线显示的是综合所有超新星进行拟合的结果，得到了一个 Ω_M-Ω_Λ 的封闭置信区域。如果我们限定宇宙是平坦的（$\Omega_\Lambda + \Omega_M = 1$），拟合可以得到 $\Omega_M = 0.6 \pm 0.2$；如果我们限定于 $\Lambda = 0$ 的宇宙，可以得到 $\Omega_M = 0.2 \pm 0.4$。这些结果是宇宙质量密度偏低的初步证据。将超新星 SN1997ap 的数据加入以往较低红移的超新星样本中，我们所得的最佳拟合值比以前的结果减少了大约一个标准差 [5]。

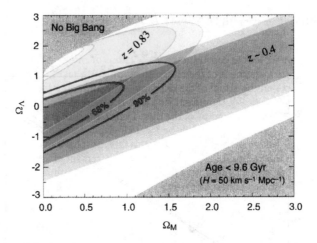

Fig. 4. Contour plot of the best fit confidence regions in the Ω_Λ versus Ω_M plane for SN1997ap and the five supernovae at $z \approx 0.4$ (see ref. 5). The 68% (1σ) and 90% confidence regions for (blue shading) the supernovae at $z \approx 0.4$, (yellow shading) SN1997ap at $z = 0.83$ by itself, and (red contours) all of these supernovae taken together. The two labelled corners of the plot are ruled out because they imply: (upper left corner) a "bouncing" universe with no Big Bang[27], or (lower right corner) a universe younger than the oldest heavy elements, $t_0 <9.6$ Gyr (ref. 28), for any value of $H_0 \geqslant 50$ km s^{-1} Mpc^{-1}.

Our data for SN1997ap demonstrate: (1) that type Ia supernovae at $z > 0.8$ exist; (2) that they can be compared spectroscopically with nearby supernovae to determine supernova ages and luminosities and check for indications of supernova evolution; and (3) that calibrated peak magnitudes with precision better than the intrinsic dispersion of type Ia supernovae can be obtained at these high redshifts. The width of the confidence regions in Fig. 4 and the size of the corresponding projected measurement uncertainties show that with additional type Ia supernovae having data of quality comparable to that of SN1997ap, a simultaneous measurement of Ω_Λ and Ω_M is now possible. It is important to note that this measurement is based on only one supernova at the highest ($z > 0.8$) redshifts, and that a larger sample size is required to find a statistical peak and identify any "outliers". In particular, SN1997ap was discovered near the search detection threshold and thus may be drawn from the brighter tail of a distribution ("Malmquist bias"). There is similar potential bias in the lower-redshift supernovae of the Calán/Tololo Survey[2], making it unclear which direction such a bias would change Ω_M.

Several more supernovae at comparably high redshift have already been discovered by the Supernova Cosmology Project collaboration, including SN1996cl, also at $z = 0.83$. SN1996cl can be identified as a very probable type Ia supernova, as a serendipitous HST observation (M. Donahue et al., personal communication) shows its host galaxy to be an elliptical or S0. Its magnitude and colour, although much more poorly constrained by photometry data, agree within uncertainty with those of SN1997ap. The next most distant spectroscopically confirmed type Ia supernovae are at $z = 0.75$ and $z = 0.73$ (ref. 16; these supernovae are awaiting final calibration data). In the redshift range $z = 0.3–0.7$, we have discovered over 30 additional spectroscopically confirmed type Ia supernovae, and followed them with two-filter photometry. (The first sample of supernovae with $z \approx 0.4$

394

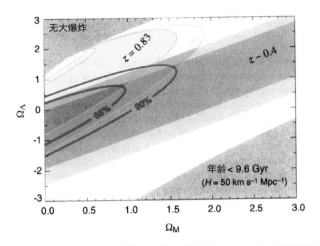

图 4. 由 SN1997ap 和 5 个 $z \approx 0.4$ 的超新星（参考文献 5）限制的 Ω_{M}-Ω_{Λ} 平面的最佳拟合置信区域等值
线图。图中画出了 68%（1 个标准差）和 90% 的置信区域：蓝色为五个 $z \approx 0.4$ 超新星的结果，黄色为
$z = 0.83$ 的 SN1997ap 单独限制的结果，红色等值线为所有超新星的结果。图中有标注的两角是被排除
的区域：左上角表示一个没有大爆炸的反弹宇宙 [27]，右下角表示了一个比最老的重元素年轻的宇宙，
在哈勃常数 $H_0 \geqslant 50$ km · s^{-1} · Mpc^{-1} 时，$t_0 < 96$ 亿年（参考文献 28）。

　　我们的 SN1997ap 观测数据说明：（1）存在红移值 $z > 0.8$ 的 Ia 型超新星；（2）
与较近超新星的光谱做对比可以确定超新星的年龄和光度，并检查超新星演化的迹
象；（3）在高红移处，可以获得比 Ia 型超新星内禀弥散度更加精确的校准后峰值星
等。图 4 中置信区域的宽度与其对应的投影测量的不确定性的大小表明，通过增加
与 SN1997ap 数据质量相当的 Ia 型超新星数据，现在是可能同时测量 Ω_{Λ} 和 Ω_{M} 的。
需要特别指出的是：这一测量只依赖于一颗最高红移超新星（$z > 0.8$），需要更大的
超新星数据样本来得到统计峰值并排除虚假的源。特别是，超新星 SN1997ap 的发
现接近搜寻探测极限，可能来自亮度分布的明亮尾段（所谓"马姆奎斯特偏差"）。在
Calán/Tololo 巡天观测的较低红移超新星中存在相似的潜在偏差 [2]，我们不清楚这些
偏差对 Ω_{M} 的测量的影响。

　　超新星宇宙学观测项目已经发现了其他几颗相对高红移的超新星。其中有
SN1996cl，红移值也是 $z = 0.83$，它很可能是一颗 Ia 型超新星。哈勃空间望远镜偶
然发现它的寄主星系可能是一个椭圆星系或者是 S0 星系（多纳休等，个人交流）。
SN1996cl 的星等和颜色，尽管测光数据限定得不太好，但与 SN1997ap 的数据在误
差范围内是一致的。距离上居次的已由光谱确定的两颗 Ia 型超新星的红移值分别为
$z = 0.75$ 和 $z = 0.73$（参考文献 16；正等待这两颗超新星的最后校准数据）。在红移值
$z = 0.3$ 到 0.7 区间，我们已经发现了 30 多颗额外的经光谱确认的 Ia 型超新星，并继
以双色测光。（第一组 $z \approx 0.4$ 的超新星样本不是全部经过光谱确认和双色测光的 [5]）

were not all spectroscopically confirmed and observed with two-filter photometry[5].) These new supernovae will improve both the statistical and systematic uncertainties in our measurement of Ω_M and Ω_Λ in combination. A matching sample of ≥ 6 type Ia supernovae at $z > 0.7$ is to be observed in two filters in Hubble Space Telescope observations due to start on 5 January 1998. SN1997ap demonstrates the efficacy of these complementary higher-redshift measurements in separating the contribution of Ω_M and Ω_Λ to the total mass-energy density of the Universe.

(**391**, 51-54; 1998)

S. Perlmutter[1,2], G. Aldering[1], M. Della Valle[3], S. Deustua[1,4], R. S. Ellis[5], S. Fabbro[1,6,7], A. Fruchter[8], G. Goldhaber[1,2], A. Goobar[9], D. E. Groom[1], I. M. Hook[1,10], A. G. Kim[1,11], M. Y. Kim[1], R. A. Knop[1], C. Lidman[12], R. G. McMahon[5], P. Nugent[1], R. Pain[1,6], N. Panagia[13], C. R. Pennypacker[1,4], P. Ruiz-Lapuente[14], B. Schaefer[15] & N. Walton[16]

[1] E. O. Lawrence Berkeley National Laboratory, 1 Cyclotron Road, MS 50-232, Berkeley, California 94720, USA
[2] Center for Particle Astrophysics, University of California, Berkeley, California 94720, USA
[3] Dipartimento di Astronomia, Universita' di Padova, Vicolo Osservatorio 5, 35122, Padova, Italy
[4] Space Sciences Laboratory, University of California, Berkeley, California 94720, USA
[5] Institute of Astronomy, Madingley Road, Cambridge CB3 0HA, UK
[6] LPNHE, Universites Paris VI & VII, T33 Rdc, 4, Place Jussieu, 75252 Paris Cedex 05, France
[7] Observatoire de Strasbourg, 11, Rue de l'Universite, 67000 Strasbourg, France
[8] Space Telescope Science Institute, 3700 San Martin Drive, Baltimore, Maryland 21218, USA
[9] Physics Department, Stockholm University, Box 6730, S-11385 Stockholm, Sweden
[10] European Souther Observatory, Karl-Schwarzschild-Strasse 2, D-85748 Garching bei Munchen, Germany
[11] Physique Corpusculaire et Cosmologie, Collège de France, 11, Place Marcelin-Berthelot, 75231 Paris, France
[12] European Southern Observatory, Alonso de Cordova, 3107, Vitacura, Casilla 19001, Santiago, 19, Chile
[13] Space Telescope Science Institute, 3700 San Martin Drive, Baltimore, Maryland 21218, USA; affiliated with the Astrophysics Division, Space Science Department of ESA
[14] Department of Astronomy, Faculty of Physics, University of Barcelona, Diagonal 647, E-08028 Barcelona, Spain
[15] Department of Physics, Yale University, 260 Whitney Avenue, JWG 463, New Haven, Connecticut 06520, USA
[16] Isaac Newton Group, Apartado 321, 38780 Santa Cruz de La Palma, The Canary Islands, Spain

Received 7 October; accepted 18 November 1997.

References:

1. Perlmutter, S. *et al.* in *Thermonuclear Supernovae* (eds Ruiz-Lapuente, P. *et al.*) 749-763 (Kluwer, Dordrecht, 1997).

2. Phillips, M. M. The absolute magnitudes of Type Ia supernovae. *Astrophys. J.* **413**, L105-L108 (1993).

3. Hamuy, M. *et al.* The absolute luminosities of the Calán/Tololo Type Ia supernovae. *Astron. J.* **112**, 2391-2397 (1996).

4. Riess, A. G., Press, W. H. & Kirshner, R. P. Using Type Ia supernova light curve shapes to measure the Hubble constant. *Astrophys. J.* **438**, L17-L20 (1995).

5. Perlmutter, S. *et al.* Measurements of the cosmological parameters Ω and Λ from the first seven supernovae at $z \geq 0.35$. *Astrophys. J.* **483**, 565-581 (1997).

6. Perlmutter, S. *et al. IAU Circ.* No. 6621 (1997).

7. Schmidt, B. *et al. IAU Circ.* No. 6646 (1997).

8. Goobar, A. & Perlmutter, S. Feasibility of measuring the cosmological constant Λ and mass density Ω using Type Ia supernovae. *Astrophys. J.* **450**, 14-18 (1995).

9. Riess, A. G. *et al.* Time dilation from spectral feature age measurements of Type Ia supernovae. *Astron. J.* **114**, 722-729 (1997).

10. Nugent, P. *et al.* Evidence for a spectroscopic sequence among Type Ia supernovae. *Astrophys. J.* **455**, L147-L150 (1993).

11. Goldhaber, G. *et al.* in *Thermonuclear Supernovae* (eds Ruiz-Lapuente, P. *et al.*) 777-784 (Kluwer, Dordrecht, 1997).

12. Leibundgut, B. *et al.* Time dilation in the light curve of the distant Type Ia supernova SN 1995K. *Astrophys. J.* **466**, L21-L44 (1996).

13. Bessell, M. S. UBVRI passbands. *Publ. Astron. Soc. Pacif.* **102**, 1181-1199 (1990).

14. Branch, D., Nugent, P. & Fisher, A. in *Thermonuclear Supernovae* (eds Ruiz-Lapuente, P. *et al.*) 715-734 (Kluwer, Dordrecht, 1997).

15. Burstein, D. & Heiles, C. Reddenings derived from H I and galaxy counts—accuracy and maps. *Astron. J.* **87**, 1165-1189 (1982).

16. Perlmutter, S. *et al. IAU Circ.* No. 6540 (1997).

这些新的超新星有利于在把 Ω_M 和 Ω_Λ 结合在一起测量时减小统计误差和系统误差。哈勃空间望远镜从 1998 年 1 月 5 日起将开始使用两片滤镜观测至少 6 颗 $z > 0.7$ 的 Ia 型超新星匹配样本。SN1997ap 表明，我们可以有效地使用额外的较高红移超新星的数据区分 Ω_M 和 Ω_Λ 在宇宙总质能密度中的贡献。

（余恒 翻译；陈阳 审稿）

17. Leibundgut, B. *et al.* Premaximum observations of the type Ia SN 1990N. *Astrophys. J.* **371,** L23-L26 (1991).

18. Wells, L. A. *et al.* The type Ia supernova 1989B in NGC 3627 (M66). *Astron. J.* **108,** 2233-2250 (1994).

19. Branch, D. *et al.* The type I supernova 1981b in NGC 4536: the first 100 days. *Astrophys. J.* **270,** 123-139 (1983).

20. Patat, F. *et al.* The Type Ia supernova 1994D in NGC 4526: the early phases. *Mon. Not. R. Astron. Soc.* **278,** 111-124 (1996).

21. Cappellaro, E., Turatto, M. & Fernley, J. in *IUE—ULDA Access Guide No. 6: Supernovae* (eds Cappellaro, E., Turatto, M. & Fernley, J.) 1-180 (ESA, Noordwijk, 1995).

22. Leibundgut, B., Tammann, G., Cadonau, R. & Cerrito, D. Supernova studies. VII. An atlas of light curves of supernovae type I. *Astron. Astrophys. Suppl. Ser.* **89,** 537-579 (1991).

23. Holtzman, J. *et al.* The photometric performance and calibration of WFPC2. *Publ. Astron. Soc. Pacif.* **107,** 1065-1093 (1995).

24. Whitmore, B. & Heyer, I. *New Results on Charge Transfer Efficiency and Constraints on Flat-Field Accuracy* (Instrument Sci. Rep. WFPC2 97-08, Space Telescope Science Institute, Baltimore, 1997).

25. Kim, A., Goobar, A. & Perlmutter, S. A generalized K-corrections for Type Ia supernovae: comparing R-band photometry beyond z = 0.2 with B, V, and R-band nearby photometry. *Publ. Astron. Soc. Pacif.* **108,** 190-201 (1996).

26. Stiavelli, M. & Mutchler, M. *WFPC2 Electronics Verification* (Instrument Sci. Rep. WFPC2 97-07, Space Telescope Science Institute, Baltimore, 1997).

27. Carrol, S., Press, W. & Turner, E. The cosmological constant. *Annu. Rev. Astron. Astrophys.* **30,** 499-542 (1992).

28. Schramm, D. in *Astrophysical Ages and Dating Methods* (eds Vangioni-Flam, E. *et al.*) 365-384 (Editions Frontières, Gif sur Yvette, 1990).

Acknowledgements. The authors are members of the Supernova Cosmology Project. We thank CTIO, Keck, HST, WIYN, ESO and the ORM–La Palma observatories for a generous allocation of time, and the support of dedicated staff in pursuit of this project; D. Harmer, P. Smith and D. Willmarth for their help as WIYN queue observers; and G. Bernstein and A. Tyson for developing and supporting the Big Throughput Camera which was instrumental in the discovery of this supernova.

Correspondence and requests for materials should be addressed to S.P. (e-mail: saul@lbl.gov).

Identification of a Host Galaxy at Redshift z=3.42 for the γ-ray Burst of 14 December 1997

S. R. Kulkarni *et al.*

Editor's Note

Gamma-ray bursts were first discovered by a satellite launched to look for detonations of nuclear weapons. The first counterparts at other wavelengths were found in 1997 by the BeppoSAX satellite, after which it became clear that the bursts were happening in distant and faint galaxies. But the energy release could not be calibrated without knowing just how far away they are. Here Shrinivas Kulkarni at the California Institute of Technology and coworkers determine that a gamma-ray burst of 14 December 1997 happened in a galaxy at a redshift of 3.42. This meant that if it was radiated uniformly, the energy released as gamma rays alone would be more than 10^{53} erg, or about 1,000 times greater than a supernova. Astronomers now believe that the burst is beamed through a small angle, reducing the total energy to about 10^{51} erg, or about ten times that of a normal supernova.

Knowledge of the properties of γ-ray bursts has increased substantially following recent detections of counterparts at X-ray, optical and radio wavelengths. But the nature of the underlying physical mechanism that powers these sources remains unclear. In this context, an important question is the total energy in the burst, for which an accurate estimate of the distance is required. Possible host galaxies have been identified for the first two optical counterparts discovered, and a lower limit obtained for the redshift of one of them, indicating that the bursts lie at cosmological distances. A host galaxy of the third optically detected burst has now been identified and its redshift determined to be $z = 3.42$. When combined with the measured flux of γ-rays from the burst, this large redshift implies an energy of 3×10^{53} erg in the γ-rays alone, if the emission is isotropic. This is much larger than the energies hitherto considered, and it poses a challenge for theoretical models of the bursts.

EVER since their discovery nearly three decades ago[1], it was understood that progress in solving the puzzle of γ-ray bursts (GRBs) depends on their identification at other—preferably optical—wavelengths, so that the distances could be measured using standard spectroscopic techniques. From distances and flux measurements one can then infer luminosities and other physical parameters, which can then be used to test theoretical models of the bursts and their origins.

A recent breakthrough in this field was the precise localization of bursts by the BeppoSAX

1997 年 12 月 14 日的伽马射线暴在红移 3.42 处的寄主星系的证认

库尔卡尼等

编者按

伽马射线暴最早是被一个探测核武器爆炸的卫星所发现的。1997 年，贝波 X 射线天文卫星首次探测到伽马射线暴在其他波段上的对应体。自那以后，人们逐渐意识到这些暴发生于遥远的暗弱星系当中。然而，如果我们不知道它们到底离我们有多远，我们就没办法确定它们释放的能量。这里，根据加州理工学院的什里尼沃斯·库尔卡尼与他的合作者的研究结果得知，1997 年 12 月 14 日发现的伽马射线暴产生于一个红移为 3.42 的星系当中。这意味着如果它各向均匀地向外辐射，那么单以 γ 射线形式释放的能量就将超过 10^{53} erg，大约是一个超新星能量的 1,000 倍。当前天文学家相信伽马射线暴是在一个小角度范围内定向爆发，这样使得该伽马射线暴所释放的能量降至大约 10^{51} erg，对应一个普通超新星能量的十倍左右。

随着近来对伽马射线暴在 X 射线、光学以及射电波段对应体的观测，我们对这种天体特征的认知大大地获得增加。然而我们对于提供暴能源的底层物理机制仍不清楚。这其中，一个重要的问题就是，要确定整个暴的总能量，需要精确地测量它相对我们的距离。到目前为止，人们已经辨认出了最先发现的两个光学对应体的可能寄主星系，并定出了其中一个星系的红移下限，这表明伽马射线暴发生在宇宙学距离上。在这里，我们也证认出了第三个从光学上探测到对应体的暴的寄主星系，并定出其红移为 3.42。根据实际探测到的 γ 射线流量，如果辐射是各向同性的，那么如此高的红移就意味着单 γ 射线上释放的能量就达到 3×10^{53} erg。这比先前考虑的能量都要大得多，因此对伽马射线暴的理论模型提出了挑战。

自三十年前发现伽马射线暴 [1] 起，人们就知道，要解决伽马射线暴（GRB）的谜团需要证认出其他波段——尤其是可见光波段的对应体，这样才能通过标准的光谱方法来确定它们的距离。由距离和流量我们就能推算出它的光度和其他物理量，用以验证伽马射线暴机制和起源的理论模型。

这个领域最新的突破是 BeppoSAX 卫星对暴的精确定位 [2]，它帮助我们首次在

satellite[2], which has led to the first identifications of GRBs at other wavelengths: X-rays[3], optical[4] and radio[5]. This has further led to the determination of the distance scale of GRBs, with the detection of intergalactic absorption lines[6] in the optical transient[7,8] (OT) of GRB970508 (refs 9, 10). Apparent host galaxies have been detected for the first two optical afterglows found[11-14].

Here we report follow-up studies of the OT[15] of a relatively bright burst, GRB971214 (refs 16–18). As the OT faded away, we found an extended object with a red-band magnitude $R = 25.6 \pm 0.15$ at the position of the OT. Based on the excellent positional coincidence, 0.06 ± 0.06 arcsec, we argue here that this is the host galaxy of GRB971214. Spectroscopic observations show that the host is a typical star-forming galaxy[19-21] at a redshift $z = 3.418$.

Given this high redshift, the γ-ray energy release of this burst is unexpectedly large, about 3×10^{53} erg, assuming isotropic emission, corresponding to about 16% of the rest-mass energy of our Sun. Energy released in other forms of radiation, for example, neutrinos or gravity waves, is not included in this energy budget. Nonetheless, the inferred energy release in γ-rays alone is so substantial that it may present difficulties for some of the currently popular theoretical models for the origin of the bursts (coalescence of neutron stars). We may be forced to consider even more energetic possibilities[22,23] or to find ways of extracting more electromagnetic energy in coalescence models.

The Optical Transient

Our follow-up at the Keck Observatory began on the night of 1997 December 15 UT, approximately 13 h after the burst. The only available imager was a "guide" camera with a small field-of-view and relatively low sensitivity; consequently, these data were not used in the analysis described below. For the subsequent nights we used the Low Resolution Imaging Spectrograph[24] (LRIS) mounted at the Cassegrain focus of the Keck II 10-m telescope. Our early observations were made in the I band. This choice was driven by the presence of a bright Moon in the sky. Later observations were done under darker sky conditions and with a R-band filter (OG570+KG3). In Table 1 we give a summary of our observations as well as other reported measurements[25-29].

Figure 1 shows an I-band image obtained on 1997 December 16.52 UT, when the OT[26] was still bright. I-band and R-band magnitudes of the OT are given in Table 1 and shown in Fig. 2. In order to establish consistency between our measurements and the earlier measurements reported by others, we measured instrumental magnitudes of a number of "secondary" stars in the general vicinity of the OT and then used them to define the zero-point of our instrumental magnitudes for all of our data sets. We calibrated the instrumental magnitudes by using observations of the standard star PG0231+051 (for which we assume I-band magnitude $I = 16.639$ mag; ref. 30) obtained on 1997 December 22 UT. We then derived the magnitudes for stars H1 and H2 (identified in Fig. 1) as 16.1 ± 0.1 mag and 18.6 ± 0.2 mag, respectively. For these two stars, Halpern et al.[15] obtained 15.93 mag and 18.46 mag.

其他波段（X 射线 [3]、光学波段 [4]、射电波段 [5]）证认出 GRB 的对应体。据此，我们可以进一步确定 GRB 的距离尺度，比如对 GRB970508 的光学暂现源 [7,8](OT) 的星系际吸收线 [6] 进行探测以确定 GRB970508(参考文献 9 和 10) 的距离。到目前为止，人们已经探测到最早发现的两个光学余辉所对应的寄主星系 [11-14]。

这里我们要报道的是对一个较亮的 GRB971214(参考文献 16 ~ 18) 的 OT[15] 的后续研究。在 OT 消失以后，我们在原来的位置上发现了一个 R 波段星等为 25.6±0.15 的延展天体。由于位置上惊人的一致——相差仅 0.06±0.06 角秒，我们推测这就是 GRB971214 的寄主星系。光谱分析表明这个星系是一个典型的恒星形成星系 [19-21]，红移为 3.418。

如此之高的红移意味着这个伽马射线暴所释放的 γ 射线的能量相当大，如果辐射是各向同性的，那么释放的能量约为 3×10^{53} erg，相当于 16% 的太阳静质能。这其中还不包括其他形式的辐射，比如中微子和引力波等。然而，单是这推算出的 γ 射线波段的能量如此之大就已经给目前流行的伽马射线暴爆发起源理论模型（中子星并合）带来了挑战。我们不得不考虑其他可以产生更高能量的理论模型 [22,23]，或者在并合模型中找到可以提取更多电磁能量的方式。

光学暂现源

我们于世界时 1997 年 12 月 15 日夜里（大约是伽马射线暴爆发后的 13 小时）在凯克观测站开始进行后续观测。唯一可用的成像设备是一个小视场、相对低灵敏度的导星相机，因此这批数据没有用于下面的分析。接下来的几个夜里我们使用装在 10 米凯克望远镜 II 卡塞格林焦点上的低分辨率成像摄谱仪 [24](LRIS)。前期的观测由于有月光干扰，因此只在 I 波段进行。在天空背景变暗后我们采用 R 波段滤光片（OG570+KG3）进行后续的观测。在表 1 中我们对我们的观测条件以及其他已有的测量结果 [25-29] 进行了简单的总结。

图 1 显示的是世界时 1997 年 12 月 16.52 日获得的 I 波段图像，图像中 OT[26] 还非常明亮，我们将 OT I 波段和 R 波段的星等列于表 1 中并标示于图 2 中。为了使所得数据格式同前人的测量结果相一致，我们测量了 OT 附近一系列二级标准星的仪器星等，用以确定我们所有数据集的仪器星等的零点。然后我们通过世界时 1997 年 12 月 22 日对标准星 PG0231+051(取其 I 波段星等为 16.639 星等；参考文献 30) 的观测对仪器星等进行定标。这样，我们得到恒星 H1 和 H2(在图 1 中标出) 的绝对星等分别为 16.1±0.1 星等和 18.6±0.2 星等。相应地，哈尔彭等人 [15] 给出的结果是 15.93 星等和 18.46 星等，两者在误差范围内一致，于是我们可以将我们 I 波段的测

Within errors, these two sets of measurements agree and thus allow us to link our I-band magnitudes to those of Halpern *et al.*[26]. The zero point for our R-band magnitudes was set by assuming $R = 20.1 \pm 0.1$ for object 2 of Henden *et al.*[29]; see also Fig. 1. Our R-band magnitude is in excellent agreement with that obtained by Diercks *et al.*[28] at about the same epoch.

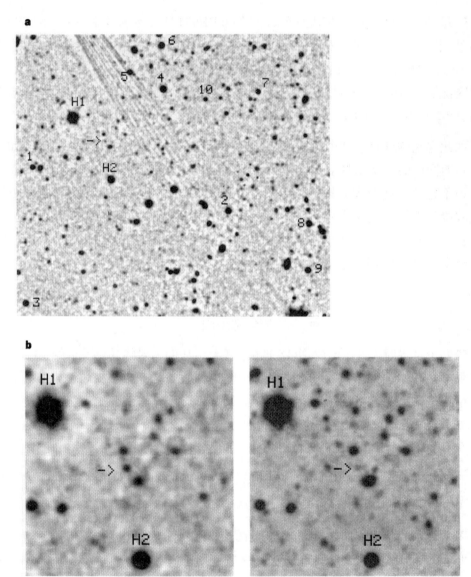

Fig. 1. Images of the field of the optical transient of GRB971214 and the associated host galaxy. **a**, I-band image of the field of the optical transient (OT) of GRB971214. The image is 3.5 arcmin by 3.0 arcmin and has been smoothed by a gaussian of full-width at half-maximum (FWHM) of 0.53 arcsec. The numbered stars are "secondary" stars used for achieving consistency within our various data sets. Stars H1 and H2 allow us to link our photometry with that of Halpern *et al.*[26] and Henden *et al.*[29]. The "rays" are light from the bright star (6.7 mag) SAO15663 diffracted by the telescope structure; this bright star is ~1 arcmin northeast of the OT. The rays rotate as the telescope tracks the source. In **b**, the left panel is an I-band image of the OT obtained from data taken on 1997 December 16 UT. The image is a square of side 1

量结果与他们的结果 [26] 联合起来使用。R 波段的零点是在假定亨登等人 [29] 的 2 号天体 R 波段星等为 20.1±0.1 的情况下而确定的；参见图 1。这样，我们获得的 R 波段数据与迪克斯等人 [28] 在同时期的测量结果相当吻合。

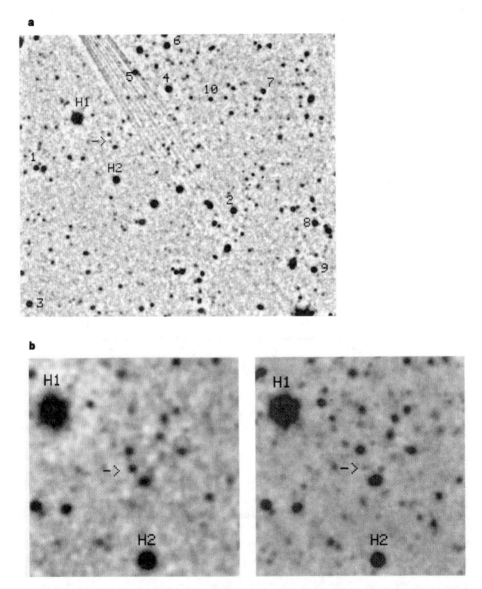

图 1. GRB971214 的 OT 及其寄主星系所在视场的图像。**a**. GRB971214 的 OT 所在视场的 I 波段图像。图像为 3.5 角分 ×3.0 角分，已做半峰全宽为 0.53 角秒的高斯平滑处理。已编号的恒星是用以同其他批次数据进行定标的二级标准星。恒星 H1 和 H2 使我们的测光数据得以同哈尔彭等人 [26] 和亨登等人 [29] 的数据进行比较。图中的"射线状特征"是被望远镜结构衍射的亮星 SAO15663(6.7 星等)的星光；这颗亮星位于 OT 东北约 1 角分处。随着望远镜跟踪目标时的视场旋转，射线也相应转动。**b**. 左图是世界时 1997 年 12 月 16 日时拍摄的 OT 的 I 波段图像。图像大小为 1 角分见方，经半峰全宽为 0.53 角秒的

arcmin, and has been smoothed by a guassian with FWHM of 0.53 arcsec. The OT is marked. The right panel is an R-band image of the field of OT obtained from data taken on 1998 January 10 UT. The image is a square of side 1 arcmin, and has been smoothed by a gaussian with FWHM of 0.53 arcsec. An extended object is seen at the position of the OT. We suggest that this is the host galaxy of GRB971214. For all three images, North is to the top and East to the left. In all three images the optical transient is marked by an arrow. We have also obtained the spectrum of the galaxy ~4.5 arcsec to the northeast of the OT. The spectrum shows a prominent [O II] 3,727 emission line and the usual absorption features at a redshift of 0.5023.

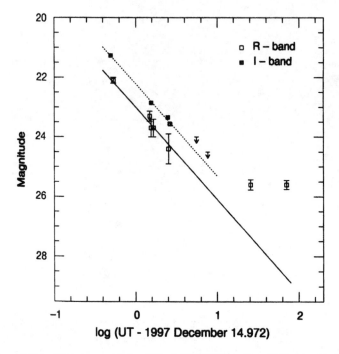

Fig. 2. The R- and I-band light curve of the OT of GRB971214. The OT is identified in Fig.1. The x-axis is $\log(t-t_0)$ where t is the UT date of the observation and t_0 is the UT date of the GRB. All times are measured in units of days. The GRB took place on 1997 December 14.9727 (ref.16). At I band the OT is well detected for the first three nights. Upper limits exist for the nights of UT 1997 December 20 and December 22. The dotted line is a linear least-squares fit to the I-band data but restricted to these first three nights. The assumed model is $I(t-t_0) = I_1 + s \log(t-t_0)$; here I is the magnitude at time t, s is the slope and I_1 the offset. The fits yield $I_1 = 22.22 \pm 0.18$ and $s = 3.09 \pm 0.52$. For reasons discussed in the text it is reasonable to assume that the R-band light curve has the same slope, s. The solid line is parallel to the dotted line but goes through our R-band magnitude of 1997 December 16.63 UT. This line describes adequately the R-band light curve for the first three nights. Clearly, the December 15.50 point of Henden *et al.*[29] is not well accounted for by this model. The R-band measurements of 1998 January 10 and February 24 lie well above the extrapolation of the dotted line. We attribute this excess over the decaying optical transient as arising from galaxy K shown in Fig. 1. In both the January 10 and February 24 images, the full width at half maximum (FWHM) of K is 1.07 arcsec which is larger than the same estimated from stars in the vicinity of K (Table 1). The rough size of K is thus ~0.65 arcsec.

The decay of previous OTs associated with GRBs appears to be well characterized by a power law: $S(t) \propto t^{-a}$, where $S(t)$ is the flux of the OT and t is the time since the γ-ray burst. Accepting this parametric form and considering only the time interval December 15–17,

高斯平滑处理。图中标出了 OT 的位置。右图是世界时 1998 年 1 月 10 日获得的 OT 所在视场的 R 波段图像，也是 1 角分见方，同样经半峰全宽为 0.53 角秒的高斯平滑处理。在 OT 原来的位置上可以看见一个延展天体，我们认为那就是 GRB971214 的寄主星系。所有的三幅图片均为上北左东，OT 由箭头标出。我们也获得了 OT 东北约 4.5 角秒处的星系的光谱，光谱中有明显的 [O ɪɪ] 3,727 发射线和红移 0.5023 处常见的吸收线。

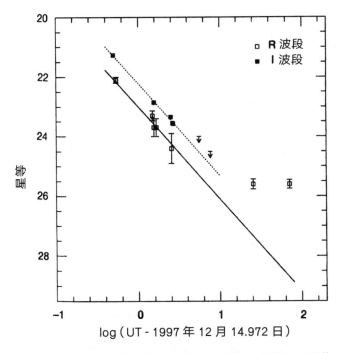

图 2. GRB971214 光学暂现源 R 和 I 波段的光变曲线。OT 已在图 1 中标出。x 轴是 $\log(t-t_0)$，其中 t 是观测的世界时，t_0 是 GRB 的世界时，均以天为单位。GRB 发生于 1997 年 12 月 14.9727 日（参考文献 16）。在 I 波段前三个晚上的探测都可得到很确切的结果，在 12 月 20 日和 22 日都只能得到上限。图中的虚线是 I 波段数据的线性最小二乘法拟合，不过仅限于对前三天的观测进行拟合。假定的模型是 $I(t-t_0) = I_1 + s \log(t-t_0)$，其中 I 是时间 t 时的星等，s 是斜率，I_1 是偏置。拟合的结果为 $I_1 = 22.22 \pm 0.18$，$s = 3.09 \pm 0.52$。由于正文中提到的原因，有理由相信 R 波段光变曲线有着同样的斜率 s。图中实线与虚线平行，且过世界时 1997 年 12 月 16.63 日测得的 R 波段星等，同前三天的 R 波段光变数据基本吻合。亨登等人 [29] 12 月 15.50 日的数据点显然不满足这个模型。1998 年 1 月 10 日和 2 月 24 日的 R 波段测量结果都在虚线的外推之上，我们认为这种对衰减的光学暂现源的亮度超出是由图 1 中显示的星系 K 引起的，在 1 月 10 日和 2 月 24 日的图像中 K 的半峰全宽为 1.07 角秒，这个值比它邻近的恒星的估值大得多（见表 1）。由此估计 K 的角直径约为 0.65 角秒。

　　之前提到的与 GRB 相关的 OT 的光度衰减可以按幂律形式进行描述，即 $S(t) \propto t^a$，其中 $S(t)$ 为 OT 的流量，t 是从伽马射线暴爆发算起的时间。基于这个参数表达式且仅考虑 12 月 15～17 日的数据，我们可以得到 $I(t) = (22.22 \pm 0.18) + (3.09 \pm 0.52)\log(t)$，

we obtain $I(t) = (22.22 \pm 0.18) + (3.09 \pm 0.52) \log(t)$: here t is in days. As magnitudes are defined to be $-2.5 \log(S) +$ constant, the value of a is 1.22 ± 0.2—remarkably similar to those reported for previous OTs[31-33].

From the data in Fig. 2 we see that initially the R-band flux appears to decrease in approximately the same fashion as the I-band flux. Such broad-band decay has been noted in previous OTs[32,33] and indeed is an expectation of simple fireball models[31,34]. Accepting that all optical bands decay with the same a, we obtain $R(t) = (23.0 \pm 0.22) + (3.09 \pm 0.52) \log(t)$. In arriving at this equation we have used our 16 December R-band measurement (Table 1) to set the zero point.

Identifying the Host Galaxy

Using the above decay formula we predict $R = 27.4 \pm 0.8$ mag on 1998 January 10 UT. In contrast, we find at the same epoch an object at the position of the OT with $R = 25.6 \pm 0.17$ mag (see Fig. 1). Furthermore, this object, which we will call "K", is extended in comparison to stars in the same field. We suggest that K is the host galaxy of the GRB and has become apparent now that the OT has dimmed.

It is important to assess how well K coincides with the OT. To this end, we determined the location of the OT relative to 15 "tertiary" stars in the vicinity of the OT in the December 16.52 I-band image, and likewise for K in the January 10 R-band image. The error in the measured angular difference between the OT and K is determined by two factors, as follows. (1) The error in determining the centroid of K, which primarily arises from the small number of signal photons and the sky background. We estimate this to be 0.20 pixel (r.m.s.); each LRIS pixel is a square of side 0.211 arcsec. The similar error for the OT is negligible because the OT is much brighter than K in the December 16 image that we used. (2) Errors due to coordinate transformation between the two images which are of different orientations. We determined a coordinate transformation between the frames using the measurements of the tertiary stars, and then transformed the centroid of the OT into the frame of January 10 image. This error turns out to be negligible compared to the error discussed in (1). The difference between the position of the OT in the transformed image and that of K is $\Delta X = 0.24 \pm 0.20$ pixel $= 0.05 \pm 0.04$ arcsec and $\Delta Y = 0.14 \pm 0.20$ pixel $= 0.03 \pm 0.04$ arcsec, corresponding to a radial separation of 0.06 ± 0.06 arcsec.

One may argue that the density of galaxies as faint as K is sufficiently high that the coincidence of the OT with a background galaxy is not improbable. The cumulative surface density[35] of galaxies with $R \leqslant 25.5$ mag is 3.9×10^5 per degree^2. Considering angular offsets up to the apparent radius of the galaxy, 0.35 arcsec (see Fig. 1), the chance coincidence probability is reasonably small ($\sim 10^{-3}$). We therefore suggest that galaxy K is the host of the OT and thus is the host of GRB971214.

其中 t 的单位是天。由于星等值的定义可以表示为 $-2.5\log(S)+$ 常数，因此 a 的值为 1.22 ± 0.2，这个结果同以前 OT 的研究结果 [31-33] 很接近。

从图 2 中的数据可以看出初始阶段 R 波段的流量衰减特征几乎与 I 波段的情况相同，这一宽波段的衰减现象也曾在以前对 OT 的观测 [32,33] 中被人们注意到，与此同时这个现象也符合简单火球模型的理论预言 [31,34]。如果所有光学波段的衰减的幂指数都对应相同的 a 值，那么 $R(t)=(23.0\pm0.22)+(3.09\pm0.52)\log(t)$。此处我们采用了 12 月 16 日 R 波段的测量结果（见表 1）来标定星等零点。

寄主星系证认

利用上面的衰减公式我们估计出世界时 1998 年 1 月 10 日的 OT R 波段星等为 27.4 ± 0.8，然而相比之下，我们在同一时期 OT 的位置上发现一个 R 波段星等为 25.6 ± 0.17 的天体（见图 1）。此外，该天体（称为"K"）相比于同视场的恒星而言是一个展源。因此我们推断 K 可能是这个 GRB 的寄主星系，正是在 OT 变得暗淡之后才得以显现出来。

明确天体 K 与 OT 的一一对应的关系是非常重要的。为此，我们在 12 月 16.52 日的 I 波段图像上确定了 OT 相对于它附近 15 个三级标准星的相对位置，并在 1 月 10 日的 R 波段图像上对天体 K 进行同样的操作。这里有两个因素决定 OT 和 K 之间角度差异的误差。(1) K 天体形心的定位误差，这个误差主要来源于过少数量的信号光子以及天空背景。我们估计其误差大小为 0.20 个像素（均方根），而 LRIS 望远镜的每个像素为长 0.211 角秒的方格。相比之下，OT 的此类误差则可以忽略，因为在选取的 12 月 16 日的图像中 OT 比 K 亮很多。(2) 将方位角不同的两幅图像进行坐标变换所引入的误差。我们通过测量三级标准星来对两个图像系统进行坐标变换，然后依此把 OT 的形心位置变换到 1 月 10 日的图像中去。相比于前面讨论的因素(1)，这个误差可以忽略不计。转换后的图像中 OT 与 K 的位置差别为 $\Delta X=0.24\pm0.20$ 像素 $=0.05\pm0.04$ 角秒，$\Delta Y=0.14\pm0.20$ 像素 $=0.03\pm0.04$ 角秒，这对应它们的径向距离为 0.06 ± 0.06 角秒。

有人也许会怀疑，像 K 这样暗淡的星系其分布密度相当高，OT 有可能刚好与一个背景星系位置相重合。但是所有 R 波段星等小于等于 25.5 的星系累积面密度 [35] 为每平方度 3.9×10^5 个。即使考虑到这个星系视半径的角度偏移（也只有 0.35 角秒，见图 1），这种意外重合的可能性也相当小（约为 10^{-3}）。因此我们认为星系 K 是 OT 的寄主星系，也是 GRB971214 的寄主星系。

Determining the Redshift

Spectroscopic observations of the OT and its host galaxy were obtained using LRIS[24]. Our initial attempt to obtain the spectrum of the OT on 1997 December 17 UT was unsuccessful, largely due to the bright moonlight. Subsequent observations were obtained under much better sky conditions, and after the OT had faded and object K dominated the light. Almost all spectra were obtained with the slit position angle close to the parallactic angle, and the wavelength-dependent slit losses are not important for the discussion below. The log and the technical details of the spectroscopic observations can be found in Table 2.

The spectrum, shown in Fig. 3, shows a prominent emission line at 5,382.1 Å with a clear continuum drop immediately on the blue (short-wavelength) side of the line. No other emission line is detected. The continuum rises slightly towards the red and is undetectable at about 4,000 Å.

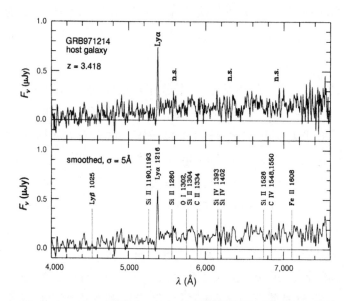

Fig. 3. The composite Keck spectrum of the host galaxy of GRB971214. The top panel shows the original data, with the locations of the prominent night sky (n.s.) emission lines indicated. The bottom panel shows the same spectrum smoothed with a gaussian with $\sigma = 5$ Å, which is approximately equal to the effective instrumental resolution. Locations of several absorption features commonly seen in the spectra of $z \approx 3$ galaxies are indicated. The redshift, $z = 3.418$, has been derived from the mean point of the Lyα emission and absorption features, as described in the text. The log of the observations and the details of the spectrograph settings can be found in Table 2. The "lower" resolution data were reduced completely independently by two of the authors (S.G.D. and K.L.A.), using independent reduction packages. The results are in an excellent mutual agreement. The "higher" resolution data are also fully consistent with them, showing essentially the same spectroscopic features. The useful wavelength range spanned by the lower-resolution data is ~4,000–7,600 Å, and the higher-resolution data spans ~4,900–7,300 Å. All of the spectra have been averaged with appropriate signal-to-noise weighting, after suitable resampling.

确 定 红 移

我们通过 LRIS[24] 获得了 OT 及其寄主星系的光谱观测结果。在世界时 1997 年 12 月 17 日，我们初次尝试拍摄 OT 光谱，但主要由于月光太亮而没能成功。后续观测的天气状况较好，且当时 OT 已减弱，亮度主要由天体 K 主导。在拍摄所有光谱时基本上狭缝位置角都接近于星位角，从而波长依赖的狭缝损失在下文中的讨论并不重要。我们在表 2 中给出了一些光谱观测日志和技术细节。

在图 3 显示的光谱中，我们在波长 5,382.1 Å 处发现了一条很强的发射线，并在紧挨该发射线的蓝（短波）端发现连续谱有明显的下降。同时并没有探测到其他的发射线。另外，连续谱朝着红端有轻微上升，并在大约 4,000 Å 处已不能被测到。

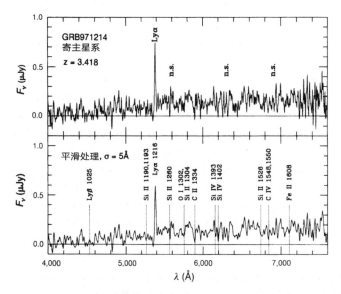

图 3. GRB971214 寄主星系的凯克望远镜合成光谱。上部分是原始数据，标出了明显的夜晚天光（n.s.）发射线的位置。下部分是经 σ = 5 Å 的高斯平滑处理之后所得到的光谱，与仪器的有效分辨率接近。图中标出了红移在 3 左右的星系中常见的吸收线，如正文中所述，通过 Lyα 发射线的中心和吸收线特征得出该星系的红移 z = 3.418。观测日志和光谱细节可以在表 2 中看到。"较低"分辨率数据由作者中的两位（乔尔戈夫斯基和阿德尔贝格尔）用完全不同的测光包各自独立处理得出，结果相当一致。"较高"分辨率的数据也与之吻合，显现出相同的光谱特征。低分辨率数据的有效波长范围约为 4,000 ～ 7,600 Å，高分辨率数据约为 4,900 ～ 7,300 Å。全部光谱都经适当重采样后对信噪比加权进行了平均操作。

There are only two plausible interpretations of this prominent emission line. All other choices would require a presence of other, stronger emission lines which are not seen in our spectrum. First, the line could be [O $_{II}$] 3,727 at z = 0.444. We would then expect to see comparably strong [O $_{III}$] 5,007 and 4,959 lines at 7,230 Å and 7,160 Å, Hβ 4,861 Å at 7,020 Å, and perhaps also other Balmer lines. Yet none of these lines are seen in our data, even though they fall in a reasonably clean part of the spectrum. It would be also hard to understand the abrupt drop in the continuum blueward of the emission line.

The alternative interpretation, and the interpretation which we favour, is that the emission line is Lyα 1,215.7 Å at a redshift z_{em} = 3.428. The overall appearance of the spectrum is typical of the known, star-forming galaxies at comparable redshifts[19-21]. In the absence of an active nucleus, no other strong emission lines would be expected within the wavelength range probed by our spectrum.

Additional arguments in favour of the large-redshift interpretation include:

(1) The drop across the Lyα line, due to the absorption by intervening hydrogen, is characteristic of objects at such high redshifts. The measured drop amplitude[36] D_A = 0.35 ± 0.05 is exactly the mean value seen in the spectra of quasars at this redshift[37].

(2) We expect no flux blueward of 4,030 Å, the redshifted Lyman continuum break. Additionally, we expect a relatively flat, star-formation-powered continuum redward of the Lyα line. Our spectrum is consistent with these expectations. The expected suppression of blue light is also supported by our imaging observations in the B band (Table 1); this band covers the range 3,900–4,800 Å. Our 3σ upper limit on the B magnitude of galaxy K is 26.8 mag, corresponding to a B-band flux of < 84 nJy. We find that the red portion of the spectrum can be approximated by $F_\nu = 174(\nu/\nu_R)^\alpha$ nJy with α = −0.7 ± 0.2; here F_ν is the spectral density at frequency ν and $\nu_R = 4.7 \times 10^{14}$ Hz, the centre frequency of the R band. The extrapolation of this spectrum to the B band ($\nu_B = 6.8 \times 10^{14}$ Hz) is 134 nJy, above our upper limit.

(3) We plot in Fig. 3 the expected locations of some of the commonly observed interstellar gas absorption lines seen in the spectra of star-forming galaxies at high redshifts[19-21]. We note several coincidences with dips in the observed spectrum.

Table 1. Summary of R- and I-band measurements

Epoch (UT)	Band	Magnitude	Notes and refs*
1997 Dec.15.47	I	21.27 ± 0.30	Ref. 26
1997 Dec.16.52	I	22.85 ± 0.33	9 × 120 s, 0.9″, R.G.
1997 Dec.17.45	I	23.34 ± 0.31	9 × 120 s, 1.1″, R.G. †
1997 Dec.17.60	I	23.55 ± 0.26	11 × 120 s, 0.8″, R.G.
1997 Dec.17.63	I	23.57 ± 0.40	11 × 120 s, 0.63″, R.G.‡

对于这条显著的发射线的解释只有两种可能，因为所有其他的解释都还要求存在其他更强的发射线，但它们未在我们光谱中出现。首先一种可能是这条谱线是红移为 0.444 的 [O II] 3,727。如果是这样，我们还将期待在 7,230 Å 和 7,160 Å 处看到差不多强的 [O III] 5,007 和 4,959 线，以及在 7,020 Å 处的 Hβ 4,861 Å，也许还有其他的巴耳末谱线。但是我们的数据中却并没有出现这些谱线，即使它们落在光谱中相对干净的一块区域。与此同时，这种解释也很难解释发射线蓝端连续谱的突降。

因此我们更倾向于另一种可能，即这条显著的发射线其实是红移 3.428 的 Lyα 1,215.7 Å。整个光谱的特征同具有差不多红移的已知恒星形成星系的典型光谱特征相一致 [19-21]。因为没有活跃的星系核，在我们所观测的光谱范围内不会看到其他更强的谱线。

支持高红移的其他理由还包括：

(1) 由传播过程中的氢吸收所导致的跨 Lyα 线的流量陡降是这类高红移天体的普遍特征。测量到的下降幅度 [36] $D_A = 0.35 \pm 0.05$，正是这个红移处观测类星体光谱所得到的平均值 [37]。

(2) 我们预期在 4,030 Å（即红移后的莱曼连续谱跳变）的更蓝端测不到流量。另外，我们期待在 Lyα 线的红端看到一个相对平坦的由恒星形成活动所贡献的连续谱。我们的光谱结果与这些预期相吻合。我们所预期的蓝光的压低也在 B 波段（3,900 ~ 4,800 Å，见表 1）的成像观测中得到了证实。天体 K 的 B 波段 3σ 置信度的星等上限为 26.8，对应的 B 波段流量小于 84 nJy。我们发现光谱红端部分的谱密度近似满足公式 $F_v = 174(v/v_R)^\alpha$ nJy，其中 F_v 为频率 v 处的谱密度，$\alpha = -0.7 \pm 0.2$，$v_R = 4.7 \times 10^{14}$ Hz 为 R 波段的中央频率；将此光谱外推到 B 波段（$v_B = 6.8 \times 10^{14}$ Hz），那么对应的流量值为 134 nJy，这将高于我们测得的流量上限。

(3) 我们在图 3 中标出了通常能在高红移的恒星形成星系中观测到的星际气体吸收线的位置 [19-21]，我们注意到它们与观测到的光谱中的几处凹陷区域相对应。

表 1. R 和 I 波段测量数据汇总

时期（世界时）	波段	星等	注释和参考文献 *
1997 年 12 月 15.47 日	I	21.27 ± 0.30	参考文献 26
1997 年 12 月 16.52 日	I	22.85 ± 0.33	9 × 120 s，0.9″，古德里奇
1997 年 12 月 17.45 日	I	23.34 ± 0.31	9 × 120 s，1.1″，古德里奇 †
1997 年 12 月 17.60 日	I	23.55 ± 0.26	11 × 120 s，0.8″，古德里奇
1997 年 12 月 17.63 日	I	23.57 ± 0.40	11 × 120 s，0.63″，古德里奇 ‡

Continued

Epoch (UT)	Band	Magnitude	Notes and refs*
1997 Dec. 20.53	I	> 24	10 × 360 s, A.V.F.1.4″ §
1997 Dec. 22.63	I	> 24.5	2 × 400 s, 0.9″, M.D.‖
1997 Dec.15.50	R	21.7 ± 0.10	Ref. 29
1997 Dec.15.51	R	22.1 ± 0.10	Ref. 28
1997 Dec.16.46	R	23.3 ± 0.17	6 × 180 s, 0.86″, R.G.
1997 Dec.16.63	R	23.7 ± 0.10	3 × 180 s, 0.56″, R.G. ‡
1997 Dec.16.52	R	23.7 ± 0.30	Ref. 28
1997 Dec.17.51	R	24.4 ± 0.50	Ref. 25
1998 Jan.10.62	R	25.6 ± 0.17	12 × 300 s, 0.86″, J. Aycock¶
1998 Feb. 24.52	R	25.6 ± 0.15	10 × 300 s, 0.87″, S.R.K.
1998 Feb. 24.58	B	> 26.8	6 × 300, 3 × 600,1.00″, S.R.K.#

* The entries in the last column are: (for Keck observations) the number of frames obtained, the integration time, the seeing specified as full-width at half-maximum (FWHM), the observer or the reference (for observations reported in the literature). Additional notes are indicated by a letter. The log of the December 15 "guide" camera observations is not included here as we did not use those data in any quantitative analysis. (The OT was detected in those data, though.) Successive frames are displaced by 5–15 arcsec and the final image is obtained by registering the frames and then adding. All the imaging analysis was carried out with IRAF, a software package supplied by the National Optical Astronomy Observatories.

† Rhoads[27] reports $I = 22.9 ± 0.4$ on 1997 Dec 17.37 UT. This is consistent, within errors, with our measurement of the same night. However, due to the large error bars, this is not included in our compilation (above).

‡ This observation was affected by light diffracted by the telescope structure contaminating the region in the vicinity of the OT (see Fig.1). The diffracted pattern was ~15 pixels wide and long. An image was formed by assigning each pixel a value equal to the median of a box 75×5 pixels, aligned along the diffraction spikes. All features smaller than this box disappear in the resultant image, leaving only the diffracted rays. This median image was then subtracted from the original image leaving an image free from the diffracted pattern. Photometry of the OT, reported in this table, was then performed on this subtracted image.

§ The data were taken in the polarimetric mode of LRIS and hence the light is split into two orthogonal polarization components—we call this the top and bottom channels. The images from each of the channels were reduced separately, averaged together and the two resultant images were co-added to give the final image. The seeing as estimated by observations of the standard star SA 104 440 (ref. 30) was exceptionally bad, ~1.4 arcsec. No object was found in the final image at the position of the OT. We do not have a good estimate of the true point-spread function in the final image as the useful field is only 25 arcsec across and there are no bright stars within this. Based on noise statistics within an aperture of radius 1.5 arcsec we set a 3σ detection limit of $I = 24$ mag for this data. Analysis of artificial stars with FWHM of 1.4 arcsec using the IRAF task DAOFIND led to a detection limit of 23 mag. However, visual inspection of the artificial stars knowing the position of the object results in a more stringent limit, closer to the $I = 24$ mag obtained above.

‖ 3σ upper limit estimated from the detection statistics of artificial stars which have the same point-spread function as that of nearby stars and embedded in the general vicinity of the OT. This observation suffered from high sky brightness (19.1 mag per square arcsec) and also lack of numerous exposures.

¶ The first four exposures suffered from brighter sky as the Moon was still up in the sky. The image shown in Fig. 1 is the sum of the next eight exposures.

3σ upper limit. See footnote ‖ for details of the procedure of deriving the upper limit.

If this interpretation is correct, a better estimate of the galaxy's redshift can be obtained as follows. The Lyα emission line tends to be shifted systematically to the red in such galaxies, due to resonant scattering and absorption in the ambient gas. In any given case,

时期（世界时）	波段	星等	注释和参考文献 *
1997 年 12 月 20.53 日	I	> 24	10×360 s，菲力片科 1.4″§
1997 年 12 月 22.63 日	I	> 24.5	2×400 s，0.9″，戴维斯 ‖
1997 年 12 月 15.50 日	R	21.7±0.10	参考文献 29
1997 年 12 月 15.51 日	R	22.1±0.10	参考文献 28
1997 年 12 月 16.46 日	R	23.3±0.17	6×180 s，0.86″，古德里奇
1997 年 12 月 16.63 日	R	23.7±0.10	3×180 s，0.56″，古德里奇 ‡
1997 年 12 月 16.52 日	R	23.7±0.30	参考文献 28
1997 年 12 月 17.51 日	R	24.4±0.50	参考文献 25
1998 年 1 月 10.62 日	R	25.6±0.17	12×300 s，0.86″，艾科克 ¶
1998 年 2 月 24.52 日	R	25.6±0.15	10×300 s，0.87″，库尔卡尼
1998 年 2 月 24.58 日	B	> 26.8	6×300，3×600，1.00″，库尔卡尼 #

* 最后一列的内容分别是：（对于凯克的观测数据来说）观测帧数、积分时间、以半峰全宽表示的视宁度、观测者或参考文献（对于文章报道的观测）。额外的注释由符号标出。12 月 15 日的导星照相机的观测数据没有用于任何定量分析，因此这里没有给出日志（尽管其中观测数据也已探测到 OT）。相邻各帧之间位移了 5～15 角秒，通过对齐和叠加来获得最终的图像。所有的图像分析工作都使用美国国家光学天文台开发的 IRAF 软件包完成。

† 罗兹 [27] 给出世界时 1997 年 12 月 17.37 日 $I = 22.9 \pm 0.4$。这与我们同天的结果在误差范围内一致。但是这个结果由于误差棒过大，所以没有列入上面的列表中。

‡ 望远镜结构造成的衍射光污染了 OT 邻近视场（见图 1），观测结果受到一定影响。衍射花纹长宽各大概 15 像素。我们给每一个像素赋值，该值取为 75×5 像素的方框区域内的中值，而该方框沿着衍射峰成直线。如此得到的图像中所有小于此方框的特征都会消失，只剩衍射线。将此图像从原始图像中减去，便可除去衍射花纹。表中所给出的 OT 的测光值，便是在相减后的图像上得到的。

§ 这些数据是由 LRIS 在偏振模式下获得的，因此光被分解为两个正交的极化分量，分别称为上、下通道。各通道的图像分开进行处理，求平均，然后将得到的两幅图像进行叠加得到最后的图像。视宁度是通过观测标准星 SA 104 440（参考文献 30）估计出来的，结果异常地差，约为 1.4 角秒。在最后的图像中 OT 的位置上并未出现任何天体。由于有效视场跨度只有 25 角秒，而且其中没有亮星，我们无法较好地估计最后的图像中的点扩展函数。根据孔径 1.5 角秒内的噪声统计，我们估计数据在 3σ 内的探测极限为 $I = 24$ 星等。用软件 IRAF 的 DAOFIND 函数包分析半峰全宽为 1.4 角秒的虚拟假星，所获得的探测极限可达 23 星等。不过若考虑假星的位置已知，按上述方法我们可以得到一个更加严格的限制，更接近上面得到的 24 星等。

‖ 我们对与邻近恒星具有同样点扩展函数且落在 OT 附近的假星进行探测统计所获得的 3σ 上限。此观测受到明亮天光的影响（每平方角秒 19.1 星等），且曝光次数也不够。

¶ 由于有月亮在天上，前四次曝光受到明亮天光的影响。图 1 中的图像是由随后八次曝光叠加而成。

\# 3σ 上限。处理细节参看脚注 ‖。

    如果这个解释正确，就可以用下面的方法更好地估计星系的红移值。在此类星系中 Lyα 发射线会因为环境气体的共振散射和吸收而系统地向红端移动。在任何特

the bias thus introduced in the centring of the peak of the Lyα emission depends on the exact geometry and kinematics of the gas and dust, which are not known; but on average the effect is to shift the peak of the Lyα emission to the red, relative to the systemic redshift of the galaxy[19-21]. We measure the wavelength of the apparent absorption dip immediately on the high-frequency side of the emission line as: $\lambda_{\text{obs,air}} = 5{,}356.1$ Å. Taking the mean of this wavelength and that of the emission line, and applying the standard air-to-vacuum correction, we obtain for the systemic redshift of this galaxy, $z_K = 3.418 \pm 0.010$.

As discussed above the continuum slope of the observed spectrum redward of the emission line is $F_\nu \propto \nu^a$ with $a = -0.7 \pm 0.2$. Unobscured star-forming galaxies have $a \approx 0.0$ to -0.5, depending on the initial mass function and the history of star formation. The slightly steeper slope of our spectrum suggests a modest amount of rest-frame extinction for the galaxy as a whole. This of course does not constrain the extinction along any particular line of sight, such as the direction to the OT itself. We note that similar extinctions are inferred for star-forming galaxies at comparable redshift[19].

Two corrections must be taken into account before a quantitative interpretation of the spectrum can be done. First, the Galactic reddening in this direction is estimated[33] to be $E_{B-V} \approx 0.016$ mag, implying a Galactic extinction correction to the observed fluxes of about 6% at the wavelength of the emission line, and about 4% in the R band. Second, by comparing the spectrum to our measured R-band photometry of the galaxy, we estimate the loss due to the finite size of the slit to be a factor of 1.48.

The Nature of the Host Galaxy

We now discuss the physical properties and nature of this galaxy. Assuming a standard Friedman model cosmology with $H_0 = 65$ km s^{-1} Mpc^{-1} and $\Omega_0 = 0.3$, we derive a luminosity distance (d_L) of 9.7×10^{28} cm (changing the cosmological parameters to $H_0 = 50$ km s^{-1} Mpc^{-1} and $\Omega_0 = 1$ yields $d_L = 8.6 \times 10^{28}$ cm). (H_0, the Hubble constant, is the expansion rate of the Universe at the present time. Ω_0 is the ratio of the mean density of the Universe to the closure density.) Assuming that the intrinsic spectrum F_ν is $\propto \nu^{-0.7}$ and scaling from the observed R-band magnitude corrected for the Galactic extinction, we find the restframe B-band flux to be ~ 0.45 µJy. The corresponding absolute B magnitude is $M_B \approx -20.9$, and is about equal to the absolute magnitude of a typical, L_* galaxy today. Given its high redshift, this galaxy probably has yet to produce most of its stars, and even with some evolutionary fading of its present-day luminosity it may evolve to an L_* galaxy.

The observed Lyα emission line flux is $F_{\text{Ly}\alpha} = (3.8 \pm 0.4) \times 10^{-18}$ erg cm^{-2} s^{-1}. Correcting for the flux calibration zero-point and the Galactic extinction, this becomes $F_{\text{Ly}\alpha} = (6.2 \pm 0.7) \times 10^{-18}$ erg cm^{-2} s^{-1}. For our assumed cosmology, the implied Lyα line luminosity is $L_{\text{Ly}\alpha} = (7.3 \pm 0.8) \times 10^{41}$ erg s^{-1}. Estimates of conversion of the Lyα line luminosity to the implied unobscured star-formation rate (SFR) are in the range $L_{\text{Ly}\alpha} = (7 \pm 4) \times 10^{41}$ erg s^{-1} for SFR $= 1$ $M_\odot$ yr^{-1} (where $M_\odot$ is the solar mass), depending on the stellar initial mass

定情况下，定位 Lyα 发射线的峰所引入的偏差都与具体的然而未知的气体和尘埃几何和动力学特征有关，但是平均来讲这些效应都会使 Lyα 发射线的峰相对星系的系统性红移更偏向红端[19-21]。我们测量了紧挨着发射线的高频端的明显吸收凹陷的波长为 5,356.1 Å。取此值和发射线波长的平均值，再应用标准大气–真空修正关系，我们可以得到此星系的系统性红移为 $z_K = 3.418 \pm 0.010$。

正如前文提到过的，观测上发射线红端的连续谱斜率满足 $F_\nu \propto \nu^\alpha$，其中 $\alpha = -0.7 \pm 0.2$。对于没有被遮挡的恒星形成星系，α 约为 $0.0 \sim -0.5$，这依赖于星系的初始质量函数和恒星形成历史。我们光谱数据中略陡的斜率表明该星系从总体来说受到了中等程度的静止参照系消光。这当然不能限制任何特定方向上的消光，比如 OT 的方向上。我们也注意到具有差不多红移的其他恒星形成星系也存在类似的消光[19]。

在对光谱进行定量解释之前还要做两个修正。首先，该方向上的银河系红化可以近似估计[33]为 $E_{B-V} \approx 0.016$ 星等，由此可推算出在发射线波长处应对流量进行 6% 的银河系消光修正，而在 R 波段进行 4% 的修正。其次，通过把光谱和我们得到的 R 波段的星系测光结果相比较，我们估计由狭缝尺度有限导致的光损失因子为 1.48。

寄主星系性质

下面来讨论这个星系的物理特征和性质。在 $H_0 = 65$ km·s^{-1}·Mpc^{-1}(H_0 是哈勃常数，指宇宙现在的膨胀速率)，$\Omega_0 = 0.3$(Ω_0 是宇宙平均密度与闭合密度的比值)的标准弗里德曼宇宙学模型下，我们可以得到该星系的光度距离为 9.7×10^{28} cm(如果将宇宙学参量改变为 $H_0 = 50$ km·s^{-1}·Mpc^{-1}，$\Omega_0 = 1$，则 $d_L = 8.6 \times 10^{28}$ cm)。假定本征光谱满足 $F_\nu \propto \nu^{-0.7}$，并由改正了银河系消光后的 R 波段星等进行推算，我们求得静止参照系中 B 波段的流量约为 0.45 μJy，对应的 B 波段绝对星等 $M_B \approx -20.9$，基本和目前典型的 L_* 型星系的绝对星等相同。考虑到它是一个高红移的星系，大部分恒星应该还没有产生，甚至它现在的光度受一些演化过程影响而变暗，它可能正向一个 L_* 型星系演化。

观测到的 Lyα 发射线的流量为 $F_{Ly\alpha} = (3.8 \pm 0.4) \times 10^{-18}$ erg·cm^{-2}·s^{-1}，经零点修正和银河系消光修正后变为 $F_{Ly\alpha} = (6.2 \pm 0.7) \times 10^{-18}$ erg·cm^{-2}·s^{-1}。在我们所采用的宇宙学模型框架下，我们可以推算出 Lyα 谱线的光度为 $L_{Ly\alpha} = (7.3 \pm 0.8) \times 10^{41}$ erg·s^{-1}。从谱线光度和暗示的未遮挡恒星形成率(SFR)之间的转换关系来看，对应 SFR = 1 $M_\odot$·yr^{-1}(其中 $M_\odot$ 为太阳质量)的区间是 $L_{Ly\alpha} = (7 \pm 4) \times 10^{41}$ erg·s^{-1}，具体依赖于恒星的初始质

function (IMF)[39,40]. We thus estimate the unobscured star-formation rate in this galaxy to be approximately $(1.0 \pm 0.5)\, M_\odot\, \mathrm{yr}^{-1}$.

Table 2. Summary of spectroscopic observations

Epoch (UT)	Grating (lines per mm)	Slit width (arcsec)	Int. time (s)	Notes
1997 Dec. 28	300	1.5	2×1800 s	T.K.
1998 Feb. 3	300	1.0	5×1800 s	S.R.K
1998 Feb. 22	600	1.0	2×1800 s	S.R.K.
1998 Feb. 23	600	1.0	6×1800 s	S.R.K.

Entries (from left to right) are: the UT date of the observation, the grating used, the slit width in arcsec, the number of exposures and the integration time per exposure, and the name of the principal observer. The 300 lines per mm grating is blazed to 5,000 Å and the blaze wavelength of the 600 lines per mm grating is 7,500 Å. The centre wavelength of the 300 lines per mm spectra (hereafter "lower" resolution data) is ~6,400 Å and the dispersion is ~2.45 Å per pixel. The centre wavelength of the 600 lines per mm spectra, the "higher resolution" data, is 6,100 Å and the dispersion is 1.25 Å per pixel. In both cases, the spectral resolution as parametrized by the FWHM is ~5 pixels. The spectrum of the standard star HZ 44 (ref. 45) obtained on February 3 was used to flux calibrate the lower-resolution spectra, and the spectrum of GD 153 (ref. 46), observed on February 22, was used to calibrate the higher-resolution spectra. The zero-point uncertainty of the flux calibration is estimated to be ~10%, judging by the internal agreement, but the slit losses are likely to be higher. Wavelength calibration was derived from arc lamp spectra taken immediately after the observations of the target. The random errors in the wavelength calibration are 0.3 Å and the estimated systematic errors due to instrument flexure are of the same order. Both these uncertainties are unimportant for our redshift determination discussed in the text.

However, the Lyα emission line is probably attenuated by resonant scattering in neutral hydrogen and by absorption by dust. An independent, and perhaps more robust, estimate of the star-formation rate can be obtained from the rest-frame continuum luminosity at 1,500 Å (ref. 41). We note that the observed flux at a wavelength of $1,500(1+z_K)$ Å is 0.22 μJy. This translates to a star-formation rate, under the usual assumption of a Salpeter IMF, of $5.2\, M_\odot\, \mathrm{yr}^{-1}$. These numbers should be regarded as lower limits given the unknown extinction in the rest-frame of galaxy K.

On the whole, the properties of galaxy K are typical of the known systems at comparable redshifts[19-21], thus giving us some confidence that our redshift interpretation is indeed correct.

The fluence of this GRB[17] above the observed photon energy of 20 keV is $F = 1.1 \times 10^{-5}\, \mathrm{erg\, cm}^{-2}$. The γ-ray fluence as observed by the Gamma-ray Burst Monitor (GRBM) on board BeppoSAX is $(0.9 \pm 0.9) \times 10^{-5}\, \mathrm{erg\, cm}^{-2}\, \mathrm{s}^{-1}$ (above 40 keV). The burst was also observed by the All Sky Monitor on board the X-ray satellite XTE (ref. 42). D. A. Smith (personal communication) estimates the fluence in the 2–12 keV band to be $1.8(\pm 0.03) \times 10^{-7}$ erg cm^{-2}. Thus the isotropic energy loss in γ-rays alone at the distance to the host is $4\pi d_L^2 F/(1+z) \approx 3 \times 10^{53}$ erg.

The currently favoured model[43] for GRBs is coalescence of neutron stars. The coalescence is expected to release most of the energy in neutrinos (as in type II supernovae) and about

量函数 (IMF)[39,40]。我们据此估计该星系的未遮挡恒星形成率约为 $(1.0\pm0.5)\ M_\odot\cdot\mathrm{yr}^{-1}$。

表 2. 光谱观测数据汇总

时期 (世界时)	光栅 (每毫米线数)	狭缝宽度 (角秒)	积分时间 (秒)	注释
1997 年 12 月 28 日	300	1.5	2×1800 秒	昆迪茨
1998 年 2 月 3 日	300	1.0	5×1800 秒	库尔卡尼
1998 年 2 月 22 日	600	1.0	2×1800 秒	库尔卡尼
1998 年 2 月 23 日	600	1.0	6×1800 秒	库尔卡尼

表头从左到右依次为：观测的标准世界时，所用光栅，缝宽（角秒为单位），曝光次数和每次曝光时间，主要观测者的名字。每毫米 300 线光栅的闪耀波长为 5,000 Å，每毫米 600 线的光栅闪耀波长则为 7,500 Å。300 线光栅光谱（以下称低分辨率数据）的中央波长约为 6,400 Å，色散约为每像素 2.45 Å。600 线光栅光谱（高分辨率数据）的中央波长为 6,100 Å，色散为每像素 1.25 Å。两种情况下由半峰全宽参数化表示的光谱分辨率约为 5 像素，2 月 3 日获得的标准星 HZ 44 的光谱（参考文献 45）用于低分辨率光谱流量校准，而 2 月 22 日获得的 GD 153 的光谱（参考文献 46）用于高分辨率光谱流量校准。流量校准时零点的不确定性，通过内部符合判断，约为 10%，但狭缝损失率可能会更高。波长修正是从观测完目标后立即拍摄的弧光灯光谱中得到的，随机误差为 0.3 Å，仪器弯沉也会带来同量级的系统误差。所有这些误差对文中所讨论的红移的确定并不重要。

但是，Lyα 发射线的强度可能被中性氢的共振散射和尘埃吸收所减弱。利用静止参考系下连续谱在 1,500 Å 处的光度可以对恒星形成率给出另一个独立的、更稳健的估计（参考文献 41）。我们注意到在波长 $1,500(1+z_K)$ Å 处的观测流量为 0.22 μJy，如果采用萨尔皮特初始质量函数的通常假定，得出的恒星形成率为 $5.2\ M_\odot\cdot\mathrm{yr}^{-1}$。鉴于星系 K 在静止参考系中的消光情况未知，这些结果应视为恒星形成率的下限。

总的来说，星系 K 的性质在已知的同等红移的星系[19-21]中非常典型，这使我们更加确信自己的红移推算是正确的。

这个 GRB[17] 在 20 keV 的光子能量以上的能流为 $F = 1.1\times10^{-5}\ \mathrm{erg}\cdot\mathrm{cm}^{-2}$。由装在 BeppoSAX 卫星上的伽马射线暴监测器 (GRBM) 观测到的 γ 射线流量 (40 keV 以上) 则为 $(0.9\pm0.9)\times10^{-5}\ \mathrm{erg}\cdot\mathrm{cm}^{-2}\cdot\mathrm{s}^{-1}$。X 射线时变探测器（参考文献 42）上的全天探测器同样观测到这次暴，史密斯（个人交流）估计该暴在 2~12 keV 能段的能流为 $1.8(\pm0.03)\times10^{-7}\ \mathrm{erg}\cdot\mathrm{cm}^{-2}$。因此在这个距离上各向同性释放的 γ 射线能量为 $4\pi d_L^2 F/(1+z)\approx3\times10^{53}\ \mathrm{erg}$。

目前倾向于用中子星并合模型[43]来解释 GRB。并合过程中大部分能量应以中微子的形式放出（与 II 型超新星相同），有大约 10^{51} erg 的能量以电磁能的形式释放。

10^{51} erg is supposed to be in the form of electromagnetic energy. The measured fluence of GRB971214 when combined with our proposed redshift for the GRB appears to be inconsistent with the expectations of the neutron-star merger model in its simplest form; however, it is possible that more elaborate versions of this model could be made to fit these observations. Non-spherical emission could reduce the strain on the energy budget but will not alter the fact that this fairly bright burst originated from such a large redshift.

The most significant implication of our hypothesis that K is the host galaxy of GRB971214 is the implied extreme energetics. The energy budget for GRBs goes up from the traditional 10^{51} erg to perhaps the entire energy available in the coalescence of neutron-star mergers[43]. Other energetic models, ranging from the death of an extremely massive star[23] to coalescence of black-hole neutron star binaries[22] may then become more attractive. In the latter models, GRBs are directly related to the formation rate of massive stars. If that is the case, then the typical redshift of GRBs is approximately 2 (ref. 44), consistent with the suggested high redshift for GRB971214. Moreover, this burst evidently did not originate in a highly obscured star-forming region in its host galaxy.

Regardless of the details of their genesis, GRBs appear to be the brightest known objects in the Universe, albeit over the limited duration of the burst. The term "hypernova"[23] can be justifiably used to describe these most extreme events, especially their after-glow. GRB971214 was not a particularly faint event, and thus statistically we expect many fainter events to arise from larger redshifts. The high brightness of GRBs and their optical transients offer us exciting and new opportunities to probe the Universe at high redshifts.

(**393**, 35-39; 1998)

S. R. Kulkarni[*], S. G. Djorgovski[*], A. N. Ramaprakash[*†], R. Goodrich[‡], J. S. Bloom[*], K. L. Adelberger[*], T. Kundic[*], L. Lubin,[*], D. A. Frail[§], F. Frontera[‖ #], M. Feroci[¶], L. Nicastro[☆], A. J. Barth[**], M. Davis[**], A. V. Filippenko[**] & J. Newman[**]

[*] Palomar Observatory 105-24, California Institute of Technology, Pasadena, California 91125, USA

[†] Inter-University Centre for Astronomy and Astrophysics, Ganeshkhind, Pune 411 007, India

[‡] W. M. Keck Observatory, 65-0120 Mamalahoa Highway, Kamuela, Hawaii 96743, USA

[§] National Radio Astronomy Observatory, Socorro, New Mexico 8801, USA

[‖] Istituto Tecnnologie Studio delle Radiazioni Extraterrestri, CNR, via Gobetti 101, Bologna I-40129, Italy

[#] Dipartimento di Fisica, Universita Ferrara, Via Paradiso 12, I-44100, Italy

[¶] Istituto di Astrofisica Spaziale, CNR, via Fosso del Cavaliere, Roma I-00133, Italy

[☆] Istituto Fisica Cosmica App. Info., CNR, via U. La Malfa 153, Palermo I-90146, Italy

[**] Department of Astronomy, University of California, Berkeley, California 94720, USA

Received 18 March; accepted 14 April 1998.

References:

1. Klebesadel, R. W., Strong, I. B. & Olson, R. A. Observations of gamma-ray bursts of cosmic origin. *Astrophys. J.* **182**, L85-L88 (1973).

2. Boella, G. *et al.* BeppoSAX, the wide band mission for x-ray astronomy. *Astron. Astrophys. Suppl. Ser.* **122**, 299-399 (1997).

3. Costa, E. *et al.* Discovery of an X-ray afterglow associated with the γ-ray burst of 28 February 1997. *Nature* **387**, 783-785 (1997).

4. van Paradijs, J. *et al.* Transient optical emission from the error box of the γ-ray burst of 28 February 1997. *Nature* **386**, 686-689 (1997).

5. Frail, D. A., Kulkarni, S. R., Nicastro, L., Feroci, M. & Taylor, G. B. The radio afterglow from the γ-ray burst of 8 May 1997. *Nature* **389**, 261-263 (1997).

而在 GRB971214 中，结合推算出的红移以及测量到的能量，我们发现结果与最简单的中子星并合模型预言不符，然而对模型作更细致的调整也许能使模型与观测数据相吻合。非球对称的辐射也会缓解释放能量不足的问题，但是仍很难解释在这样高的红移处有如此亮的爆发。

在 K 是 GRB971214 的寄主星系的前提下，最富意义的暗示就是所推算出的巨大能量。GRB 的能量需求从传统的 10^{51} erg 增加到了大概中子星并合事件所能释放的全部能量 [43]。诸如极大质量恒星死亡 [23]，黑洞-中子星并合事件 [22] 等其他模型因此变得更有吸引力。在后面的这两个模型中，GRB 将与大质量恒星的形成率直接相关，如果是这样，GRB 的典型红移应为 2 左右（参考文献 44），这与 GRB971214 给出的高红移的结果相一致。此外，这个暴很明显地并没有发生在该寄主星系被严重遮蔽的恒星形成区中。

不论它们的具体起源是什么，GRB 是宇宙中已知的最亮的天体，尽管只是在有限的持续时间之内。术语"极超新星" [23] 可以用来形容这类极端事件，特别是它们的余辉。GRB971214 并不是一个特别暗的事件，因此我们可以期望统计上在更高红移处会探测到很多更暗的事件。GRB 及其光学暂现源的巨大亮度为我们探索高红移处的宇宙提供了令人兴奋的新机会。

（余恒 翻译；黎卓 审稿）

6. Metzger, M. R. *et al.* Spectral constraints on the redshift of the optical counterpart to the γ-ray burst of 8 May 1997. *Nature* **387,** 878-880 (1997).

7. Bond, H. E. *et al. IAU Circ.* No. 6665 (1997).

8. Djorgovski, S. G. *et al.* The optical counterpart to theγ-ray burst 970508. *Nature* **387,** 876-878 (1997).

9. Costa, E. *et al. IAU Circ.* No. 6649 (1997).

10. Piro, L. *et al. IAU Circ.* No. 6656 (1997).

11. Bloom, J. S., Kulkarni, S. R., Djorgovski, S. G. & Frail, D. A. *GCN Note* No. 30 (1998).

12. Zharikov, S. V., Sokolov, V. V. & Baryshev, Y. V. *GCN Note* No. 31 (1998.).

13. Galama, T. J. *et al.* Optical followup of GRB 970508. *Astrophys. J.* (submitted).

14. Sahu, K. C. *et al.* Observations of GRB 970228 and GRB 970508 and the neutron star merger mode. *Astrophys. J.* **489,** L127-L131 (1997).

15. Halpern, J. P., Thorstensen, J. R., Helfand, D. J. & Costa, E. Optical afterglow of the γ-ray burst of 14 December 1997 *Nature* **393,** 41-43 (1998).

16. Heise, J. *et al. IAU Circ.* No. 6787 (1997).

17. Kippen, R. M. *et al. IAU Circ.* No. 6789 (1997).

18. Antonelli, L. A. *et al. IAU Circ.* No. 6792 (1997).

19. Steidel, C. C., Giavalisco, M., Pettini, M., Dickinson, M. & Adelberger, K. Spectroscopic confirmation of a population of normal star-forming galaxies at redshifts z > 3. *Astrophys. J.* **462,** L17-L21 (1996).

20. Steidel, C. C., Giavalisco, M., Dickinson, M. & Adelberger, K. L. Spectroscopy of Lyman break galaxies in the Hubble Deep Field. *Astron. J.* **112,** 352-358 (1996).

21. Steidel, C. C. *et al.* A large structure of galaxies at redshift z ~ 3 and its cosmological implications. *Astrophys. J.* **492,** 428-438 (1998).

22. Meszáros, P. & Rees, M. J. Poynting jets from black holes and cosmological gamma-ray bursts. *Astrophys. J.* **482,** L29-L31 (1997).

23. Paczyński, B. Are gamma-ray bursts in star-forming regions? *Astrophys. J.* **492,** L45-L48 (1998).

24. Oke, J. B. *et al.* The Keck low-resolution imaging spectrometer. *Publ. Astron. Soc. Pacif.* **107,** 375-385 (1995).

25. Castander, F. J. *et al. GCN Note* No. 11 (1997).

26. Halpern, J., Thorstensen, J., Helfand, D. & Costa, E. *IAU Circ.* No. 6788 (1997).

27. Rhoads, J. *IAU Circ.* No. 6793 (1997).

28. Diercks, A. *et al. IAU Circ.* No. 67921 (1997).

29. Henden, A. A., Luginbuhl, C. B. & Vrba, F. J. *GCN Note.* No. 16 (1997).

30. Landolt, A. U. UBVRI photometric standard stars in the magnitude range 11.5 < V < 16.0 around the celestial equator. *Astron. J.* **104,** 340-371 (1992).

31. Wijers, R. A. M. J., Rees, M. J. & Meszáros, P. Shocked by GRB 970228: the afterglow of a cosmological fireball. *Mon. Not. R. Astron Soc.* **288,** L51-L56 (1997).

32. Sokolov, V. V. *et al.* BVR$_c$I$_C$ photometry of GRB 970508 optical remnant: May–August, 1997. *Astron. Astrophys.* (in the press); preprint http://xxx.lanl.gov, astro-ph/0902341 (1998).

33. Galama, T. J. *et al.* Optical follow-up of GRB 970508. *Astrophys. J.* (in the press); preprint http:// xxx.lanl.gov, astro-ph/9802160 (1998).

34. Waxman, E. Gamma-ray-burst afterglow: supporting the cosmological fireball model, constraining parameters, and making predictions. *Astrophys. J.* **485,** L5-L8 (1997).

35. Hogg, D. W. *et al.* Counts and colors of faint galaxies in the U and R bands. *Mon. Not. R. Astron. Soc.* **288,** 404-410 (1997).

36. Oke, J. B. & Korycansky, D. Absolute spectrophotometry of very large redshift quasars. *Astrophys. J.* **255,** 11-19 (1996).

37. Kennefick, J. D., Djorgovski, S. G. & de Carvalho, R. R. The luminosity function of z > 4 quasars from the second Palomar sky survey. *Astron. J.* **110,** 2553-2565 (1995).

38. Schlegel, D. J., Finkbeiner, D. P. & Davis, M. Maps of dust IR emission for use in estimation of reddening and CMBR foregrounds. *Astrophys. J.* (in the press); preprint http://xxx.lanl.gov, astro-ph/0910327.

39. Thompson, D., Djorgovski, S. & Trauger, J. A narrow-band imaging survey for primeval galaxies. *Astron. J.* **110,** 963-981 (1995).

40. Charlot, S. & Fall, S. M. Lyman-alpha emission from galaxies. *Astrophys. J.* **415,** 580-588 (1993).

41. Leitherer, C., Robert, C. & Heckman, T. M. Atlas of synthetic ultraviolet-spectra of massive star populations. *Astrophys. J. Suppl.* **99,** 173-187 (1995).

42. Doty, J. P. The All Sky Monitor for the X-ray Timing Explorer. *Proc. SPIE* **982,** 164-172 (1988).

43. Narayan, R., Pacsyński, B. & Piran, T. Gamma-ray bursts as the death throes of massive binary stars. *Astrophys. J.* **395,** L83-L86 (1992).

44. Wijers, R. A. M. J., Bloom, J., Bagla, J. S. & Natarajan, P. Gamma-ray bursts from stellar remnants: probing the universe at high redshift. *Mon. Not. R. Astron. Soc.* **294,** L13-L17 (1998).

45. Massey, P., Strobel, K., Barnes, J. & Anderson, E. Spectrophotometric standards. *Astrophys. J.* **328,** 315-333 (1988).

46. Bohlin, R., Colina, L. & Finley, D. White dwarf standard stars: G191-B2B, GD 71, GD 153, HZ 43. *Astron. J.* **110,** 1316-1325 (1995).

Acknowledgements. The observations reported here were obtained at the W. M. Keck Observatory, which is operated by the California Association for Research in Astronomy, a scientific partnership among California Institute of Technology, the University of California and NASA. It was made possible by the financial support from W. M. Keck Foundation. We thank W. Sargent, Director of the Palomar Observatory, F. Chaffee, Director of the Keck Observatory and our colleagues for continued support of our GRB program. We thank J. C. Clemens and M. H. van Kerkwijk for help with observations and exchange of dark time. S.R.K.'s research is supported by the NSF and NASA. S.G.D. acknowledges partial support from the Bressler Foundation. A.N.R. is grateful to the International Astronomical Union for a travel grant.

Correspondence and requests for materials should be addressed to S.R.K. (e-mail: srk:surya.caltech.edu).

Observation of Contemporaneous Optical Radiation from a γ-ray Burst

C. Akerlof *et al.*

Editor's Note

Gamma-ray bursts are brief, extremely intense flashes of gamma rays emitted by astrophysical objects and associated with high-energy explosions. Since their discovery in the late 1960s, many have been observed, but their explanation is still debated. Although numerous visible-light objects coincident with the location of gamma-ray bursts had been seen since 1997, before this paper none had been observed while the burst was still active. Carl Akerlof and colleagues used a robotic, wide-field camera to find the optical counterpart of GRB990123 while the burst was still underway, with the observations starting just 22 seconds after the burst began. The visible emission decayed more slowly than the gamma rays. Such observations help to narrow down the identity of the emitting object.

The origin of γ-ray bursts (GRBs) has been enigmatic since their discovery[1]. The situation improved dramatically in 1997, when the rapid availability of precise coordinates[2,3] for the bursts allowed the detection of faint optical and radio afterglows—optical spectra thus obtained have demonstrated conclusively that the bursts occur at cosmological distances. But, despite efforts by several groups[4-7], optical detection has not hitherto been achieved during the brief duration of a burst. Here we report the detection of bright optical emission from GRB990123 while the burst was still in progress. Our observations begin 22 seconds after the onset of the burst and show an increase in brightness by a factor of 14 during the first 25 seconds; the brightness then declines by a factor of 100, at which point (700 seconds after the burst onset) it falls below our detection threshold. The redshift of this burst, $z \approx 1.6$ (refs 8, 9), implies a peak optical luminosity of 5×10^{49} erg s^{-1}. Optical emission from γ-ray bursts has been generally thought to take place at the shock fronts generated by interaction of the primary energy source with the surrounding medium, where the γ-rays might also be produced. The lack of a significant change in the γ-ray light curve when the optical emission develops suggests that the γ-rays are not produced at the shock front, but closer to the site of the original explosion[10].

THE Robotic Optical Transient Search Experiment (ROTSE) is a programme optimized to search for optical radiation contemporaneous with the high-energy photons of a γ-ray burst. The basis for such observations is the BATSE detector on board the Compton Gamma-Ray Observatory. Via rapid processing of the telemetry data stream, the GRB Coordinates Network[11] (GCN) can supply estimated coordinates to

伽马射线暴同时的光学辐射的观测

阿克洛夫等

editor's note

编者按

伽马射线暴是一类由遥远天体在短时间内发出的极强伽马射线耀发，是一种高能爆发现象。自 20 世纪 60 年代末以来，人们虽已发现并观测到多个伽马射线暴，但是对于它们起源问题的解释却是众说纷纭。尽管自 1997 年以来在很多伽马射线暴的位置处人们发现了相应的发光天体，然而在本工作之前还没有人在伽马射线暴发生的过程中找到其发光对应体。卡尔·阿克洛夫和他的同事利用一个自动的宽场相机在 GRB990123 暴的过程中，即暴开始仅 22 秒之后，找到了它的光学对应体。结果发现，可见光辐射强度的衰减明显慢于伽马射线辐射。这样的观测将有助于我们证认发出辐射的天体。

自 γ 射线暴（GRB）被发现以来，它的起源问题就一直是个谜[1]。这个状况在 1997 年有了很大的改善，因为当时 γ 射线暴快速有效的精确定位技术[2,3]的发展使得探测暗弱的光学和射电余辉成为可能——而由此探测到的光谱则证实 γ 射线暴是在宇宙学距离上发生的。尽管几个团队一直在不断地努力[4-7]，然而迄今为止，人们都没能在短暂的暴持续时间内探测到 γ 射线暴的光学辐射。这里我们报道在 GRB990123 暴期间所发出的明亮的光学辐射的探测。我们的观测在暴开始 22 秒后进行，结果显示，在最初的 25 秒内其亮度增加到 14 倍；随后其亮度下降了 99%，从此（在暴开始 700 秒之后）其亮度下降到我们的探测阈值之下。这个 γ 射线暴的红移为 $z \approx 1.6$（参考文献 8 和 9），这意味着它的光学光度的峰值为 5×10^{49} erg·s^{-1}。一般认为，γ 射线暴的光学辐射产生于主要能量源和周围介质相互作用所形成的激波波前，而 γ 射线辐射也可能在这产生。由于光学辐射发展时 γ 射线的光变曲线并没有发生明显的变化，这表明 γ 射线不是在激波波前产生，而是在更接近原始爆发位置处产生[10]。

自动光学暂现源搜寻实验（ROTSE）是个以搜寻和 γ 射线暴的高能光子同时产生的光学辐射为目标的项目。这个观测的基础来自安装在康普顿 γ 射线天文台上的 BATSE 探测仪。通过对遥测数据流的快速处理，GRB 坐标定位网[11]（GCN）能够在

distant observatories within a few seconds of the burst detection. The typical error of these coordinates is 5°. A successful imaging system must match this field of view to observe the true burst location with reasonable probability.

The detection reported here was performed with ROTSE-I, a two-by-two array of 35 mm camera telephoto lenses (Canon $f/1.8$, 200 mm focal length) coupled to large-format CCD (charge-coupled device) imagers (Thomson 14 μm $2,048 \times 2,048$ pixels). All four cameras are co-mounted on a single rapid-slewing platform capable of pointing to any part of the sky within 3 seconds. The cameras are angled with respect to each other, so that the composite field of view is $16° \times 16°$. This entire assembly is bolted to the roof of a communications enclosure that houses the computer control system. A motor-driven flip-away cover shields the detector from precipitation and direct sunlight. Weather sensors provide the vital information to shut down observations when storms appear, augmented by additional logic to protect the instrument in case of power loss or computer failure. The apparatus is installed at Los Alamos National Laboratory in northern New Mexico.

Since March 1998, ROTSE-I has been active for ~75% of the total available nights, with most of the outage due to poor weather. During this period, ROTSE-I has responded to a total of 53 triggers. Of these, 26 are associated with GRBs and 13 are associated with soft γ-ray repeaters (SGRs). The median response time from the burst onset to start of the first exposure is 10 seconds.

During most of the night, ROTSE-I records a sequence of sky patrol images, mapping the entire visible sky with two pairs of exposures which reach a 5σ V magnitude threshold sensitivity (m_v) of 15. These data, approximately 8 gigabytes, are archived each night for later analysis. A GCN-provided trigger message interrupts any sequence in progress and initiates the slew to the estimated GRB location. A series of exposures with graduated times of 5, 75 and 200 seconds is then begun. Early in this sequence, the platform is "jogged" by $\pm 8°$ on each axis to obtain coverage of a four times larger field of view.

At 1999 January 23 09:46:56.12 UTC, an energetic burst triggered the BATSE detector. This message reached Los Alamos 4 seconds later and the first exposure began 6 seconds after this. Unfortunately, a software error prevented the data from being written to disk. The first analysable image was taken 22 seconds after the onset of the burst. The γ-ray light curve for GRB990123 was marked by an initial slow rise, so the BATSE trigger was based on relatively limited statistics. Thus the original GCN position estimate was displaced by 8.9° from subsequent localization, but the large ROTSE-I field of view was sufficient to contain the transient image. At 3.8 hours after the burst, the BeppoSAX satellite provided an X-ray localization[12] in which an optical afterglow was discovered by Odewahn et al.[13] at Mt Palomar. This BeppoSAX position enabled rapid examination of a small region of the large ROTSE-I field. A bright and rapidly varying transient was found in the ROTSE images at right ascension (RA) 15 h 25 min 30.2 s, declination (dec.) 44° 46′ 0″, in excellent agreement with the afterglow found by Odewahn et al. (RA 15 h 25 min 30.53 s, dec. 44° 46′ 0.5″). Multiple absorption lines in the spectrum of the optical afterglow

探测到暴的几秒之内向远处的天文台提供估计的位置坐标。这些坐标的典型误差为 5°。一个成功的成像系统必须覆盖这么大的视场，才能在适当的概率上观测到暴的真正位置。

这里我们所报道的探测是由 ROTSE-I 完成的，它是一个由 35 mm 相机摄远镜头（佳能 f/1.8，焦距 200 mm）组成的 2×2 阵，并配备大幅面 CCD（电荷耦合装置）成像仪（汤姆逊 14 μm 2,048 × 2,048 像素）。所有的 4 个相机都安装在一个快速回转平台上，这使得该仪器能够在 3 秒内指向天区的任何位置。相机彼此之间具有一定的倾角，这使得其仪器的综合视场为 16° × 16°。整个装置固定在装有计算机控制系统的通讯围罩的顶上。有一个电机驱动的滑动盖用以保护探测仪不受降雨和日光直射的影响。当暴风雨来临前，气象传感仪将提供重要信息，关闭探测仪器，有额外的措施处理断电和计算机故障，以增强对设备的保护。整个设备安装在新墨西哥州北部的洛斯阿拉莫斯国家实验室处。

自 1998 年 3 月开始，ROTSE-I 在约 75% 可观测的夜晚都在工作，而其他大多时候设备都是因为糟糕的天气而不得不停止。这段时期内，ROTSE-I 总共对 53 次触发作出了反应，其中 26 次和 GRB 成协，13 次和软 γ 重复暴成协。从暴开始到首次曝光开始之间的中值反应时间为 10 秒。

夜晚大部分时间内，ROTSE-I 记录一系列巡天图像：对整个可视天区进行两次成双的曝光，5σ V 波段极限星等 (m_v) 达到 15 星等。这样，每个晚上记录下约 8 千兆的数据有待后续分析。若 GCN 提供触发信息，将打断正在进行的巡天观测，并启动望远镜使其回转到所估计的 GRB 位置处。然后望远镜将开始对那块天区进行一系列 5 秒、75 秒和 200 秒的曝光。在这个观测序列早期，观测平台的每个轴都"慢跑" $\pm 8°$ 以覆盖达到视场 4 倍大的天区。

在 1999 年 1 月 23 日 09:46:56.12 UTC，一个能量巨大的暴触发了 BATSE 探测仪。这个信息在 4 秒后抵达洛斯阿拉莫斯，再 6 秒后首次曝光开始。不幸的是，一个软件错误导致这次观测的数据没有被写到硬盘里。而首幅可用于分析的图像是在暴开始 22 秒后得到的。由于 GRB990123 的 γ 射线光变曲线特征为初始缓慢的增加，所以 BATSE 触发的准确度相对有限。这使得初始的 GCN 位置估计和随后的定位位置相差 8.9°，但是 ROTSE-I 的大视场已足够包含住该暂现源的像。在暴发生 3.8 小时后，BeppoSAX 卫星给出了一个 X 射线定位 [12]，而在该位置上奥德万等人 [13] 在帕洛马山发现了一个光学余辉。这个 BeppoSAX 给出的位置使得我们可以快速检验 ROTSE-I 的大视场中的一个小区域。我们在 ROTSE 图像赤经 15 h 25 min 30.2 s、赤纬 44° 46′ 0″ 处发现了一个明亮且快速变化的暂现源，其位置和奥德万等人发现的光学余辉的位置（赤经 15 h 25 min 30.53 s、赤纬 44° 46′ 0.5″）符合得很好。从光学余辉

indicate a redshift of $z > 1.6$. Dark-subtracted and flattened ROTSE-I images of the GRB field are shown in Fig. 1. Details of the light curve are shown in Table 1.

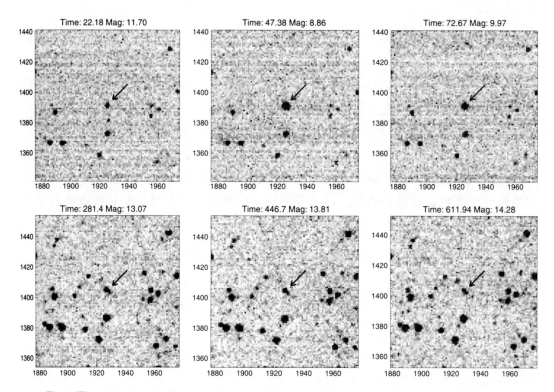

Fig. 1. Time series images of the optical burst. Each image is 24′ on a side, and represents 6×10^{-4} of the ROTSE-I field of view. The horizontal and vertical axes are the CCD pixel coordinates. The sensitivity variations are due to exposure time; the top three images are 5-s exposures, the bottom three are 75-s exposures. The optical transient (OT) is clearly detected in all images, and is indicated by the arrow. South is up, east is left. Thermal effects are removed from the images by subtracting an average dark exposure. Flat field images are generated by median averaging about 100 sky patrol (see text) images. Object catalogues are extracted from the images using SExtractor[18]. Astrometric and photometric calibrations are determined by comparison with the ~1,000 Tycho[19] stars available in each image. Residuals for stars of magnitude 8.5–9.5 are $< 1.2''$. These images are obtained with unfiltered CCDs. The optics and CCD quantum efficiency limit our sensitivity to a wavelength range between 400 and 1,100 nm. Because this wide band is non-standard, we estimate a "V equivalent" magnitude by the following calibration scheme. For each Tycho star, a "predicted ROTSE magnitude" is compared to the 2.5 pixel aperture fluxes measured for these objects to obtain a global zero point for each ROTSE-I image. For the Tycho stars, the agreement between our predicted magnitude and the measured magnitude is ± 0.15. These errors are dominated by colour variation. The zero points are determined to ± 0.02. With large pixels, we must understand the effects of crowding. (This is especially true as we follow the transient to ever fainter magnitudes.) To check the effect of such crowding, we have compared the burst location to the locations of known objects from the USNO A V2.0 catalogue[20]. The nearest object, 34″ away, is a star with R-band magnitude $R = 19.2$. More important is an $R = 14.4$ star, 42″ away. This object affects the measured magnitude of the OT only in our final detection. It can be seen in the final image to the lower right of the OT. A correction of $+0.15$ is applied to compensate for its presence. Magnitudes for the OT associated with GRB990123, measured as described, are listed in Table 1. Further information about the ROTSE-I observations is available at http://www.umich.edu/~rotse.

光谱的多条吸收线中，我们发现其红移为 $z > 1.6$。图 1 显示了扣除暗曝光并平场化的 ROTSE-I GRB 区域图像。光变曲线的细节则在表 1 中进行了展示。

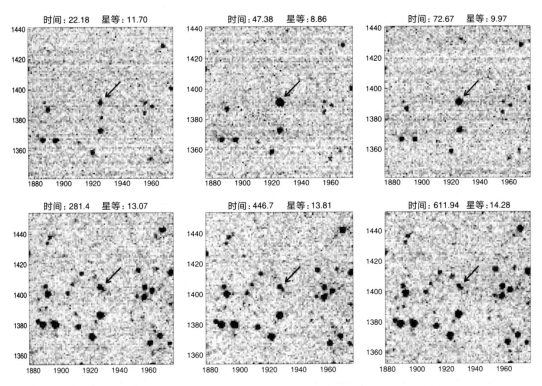

图 1. 光学暴的时间序列成像图。每幅图边长 24′，代表 6×10^{-4} 的 ROTSE-I 视场。水平轴和垂直轴为 CCD 像素坐标。灵敏度会随着曝光时间的变化而发生变化；上面三幅图曝光时间为 5 秒，下面三幅曝光时间为 75 秒。所有图上都能清楚地探测到光学暂现源 (OT)，以箭头标示。上边是南，左边是东。热效应已通过扣掉平均暗曝光从图像中去除。平场图像则是通过对大约 100 幅巡天图进行平均得到（见正文）。我们利用 SExtractor[18] 从图像中提取出目标源表，再通过与每幅图中存在的 1,000 颗左右的第谷恒星[19] 进行比较来进行天体测量与测光的定标。最终，星等为 8.5～9.5 的恒星的位置残差 < 1.2″。这些图像都是通过未经滤波的 CCD 得到的。光学和 CCD 量子效率限制了我们的灵敏度，使得我们的波长范围为 400～1,100 nm。因为这个宽波带是非标准的，我们通过下面的校准程序得到一个"等效 V"星等。对每颗第谷恒星而言，我们通过比较"ROTSE 预期星等"和测量到的 2.5 个像素孔径内的流量来确定每幅 ROTSE-I 图像的整体零点。这样，对第谷恒星而言，我们的预期星等和测量星等在 ±0.15 之内基本相符，这里误差主要来自色差。零点确定精确到 ±0.02 之内。对于大的像素，我们必须考虑成团效应（对更弱星等的暂现源这个效应尤其重要）。为了检查这种成团效应，我们比较了暴的位置和 USNO A V2.0 列表[20] 的已知天体的位置。其中距离最近的天体位于 34″ 开外，是一颗 R 波段星等为 19.2 的恒星。更为重要的是一颗 42″ 开外，$R = 14.4$ 的恒星，这个天体仅会在我们最后的观测中影响 OT 的测量星等。最后一张图中，可以在 OT 的右下看到该天体。为了弥补这个影响，我们对 OT 的星等进行了 +0.15 的修正。最终我们将如上所述与 GRB990123 成协的 OT 的观测星等列于表 1。更多关于 ROTSE-I 的观测信息请参见 http://www.umich.edu/~rotse。

Table 1. ROTSE-I observations

Exposure start	Exposure duration	Magnitude	Camera
−7,922.08	75	< 14.8	C
22.18	5	11.70 ± 0.07	A
47.38	5	8.86 ± 0.02	A
72.67	5	9.97 ± 0.03	A
157.12	5	11.86 ± 0.13	C
281.40	75	13.07 ± 0.04	A
446.67	75	13.81 ± 0.07	A
611.94	75	14.28 ± 0.12	A
2,409.86	200	< 15.6	A
5,133.98	800	< 16.1	A

Exposure start times are listed in seconds, relative to the nominal BATSE trigger time (1999 January 23.407594 UT). Exposure durations are in seconds. Magnitudes are in the "V equivalent" system described in Fig. 1 legend. Errors include both statistical errors and systematic errors arising from zero-point calibration. They do not include errors due to variations in the unknown spectral slope of the emission. Magnitude limits are 5σ. The final limit results from co-adding the last four 200-s exposures and is quoted at the mean time of those exposures. Camera entries record the camera in which each observation was made.

By the time of the first exposure, the optical brightness of the transient had risen to $m_v = 11.7$ mag. The flux rose by a factor of 13.7 in the following 25 seconds and then began a rapid, apparently smooth, decline. This decline began precipitously, with a power-law slope of ~−2.5 and gradually slowed to give a slope of ~−1.5. This decline, 10 minutes after the burst, agrees well with the power-law slope found hours later in early afterglow measurements[14]. These observations cover the transition from internal burst emission to external afterglow emission. The composite light curve is shown in Fig. 2.

A number of arguments establish the association of our optical transient with the burst and the afterglow seen later. First, the statistical significance of the transient image exceeds 160σ at the peak. Second, the temporal correlation of the light curve with the GRB flux and the spatial correlations to the X-ray and afterglow positions argue strongly for a common origin. Third, the most recent previous sky patrol image was taken 130 minutes before the burst and no object is visible brighter than $m_v = 14.8$ mag. This is the most stringent limit on an optical precursor obtained to date. Searches further back in time (55 images dating to 28 September 1998) also find no signal. Finally, the "axis jogging" protocol places the transient at different pixel locations within an image and even in different cameras throughout the exposure series, eliminating the possibility of a CCD defect or internal "ghost" masquerading as a signal.

表 1. ROTSE-I 的观测

曝光开始时间	曝光持续时间	星等	照相机
−7,922.08	75	< 14.8	C
22.18	5	11.70 ± 0.07	A
47.38	5	8.86 ± 0.02	A
72.67	5	9.97 ± 0.03	A
157.12	5	11.86 ± 0.13	C
281.40	75	13.07 ± 0.04	A
446.67	75	13.81 ± 0.07	A
611.94	75	14.28 ± 0.12	A
2,409.86	200	< 15.6	A
5,133.98	800	< 16.1	A

曝光开始时间是相对于 BATSE 触发时间 (1999 年 1 月 23.407594 日 UT) 而言的, 以秒为单位。曝光持续时间以秒为单位。星等系统为图 1 注中说明的"等效 V"系统。误差包括统计误差和零点校准引起的系统误差, 不包括由于未知的辐射谱斜率变化所引起的误差。星等极限为 5σ。最后的极限星等是通过叠加最后 4 个 200 秒的曝光结果得到的, 并按照这些曝光的平均时间给出。最后一列记录每次观测使用的照相机。

在首次曝光的时候, 暂现源的光学亮度已经增加到了 $m_v = 11.7$ 星等。流量在接下来的 25 秒内增加到 13.7 倍, 然后开始快速平滑的衰减。衰减开始很急剧, 其幂律斜率约为 −2.5, 然后逐渐变慢, 幂律斜率变为约 −1.5。同时, 暴 10 分钟后的衰减幂律和几个小时后的早期余辉测量获得的幂律斜率[14]一致。这些观测覆盖了从内部暴辐射到外部余辉辐射转换的整个过程。合成的光变曲线在图 2 中展示。

有多个论据表明我们观测到的光学暂现源和该暴及随后的余辉存在关联。首先, 暂现源图像的统计显著性在峰值处超过 160σ。其次, 其光变曲线和 GRB 流量在时间上的相关, 以及其位置和 X 射线及余辉位置在空间上的相关, 都强烈表明它们具有相同的起源。第三, 最近观测的巡天图拍摄于暴前 130 分钟, 而图中并没有亮度超过 $m_v = 14.8$ 星等的天体。这是迄今为止获得的对光学前兆最严格的限制。追溯以往的巡天图 (至 1998 年 9 月 28 日共 55 幅图像) 也没有发现信号。最后,"轴慢跑"方案使得暂现源出现在图像的不同像素位置, 甚至在一系列曝光中出现在不同的照相机中, 从而排除了因 CCD 缺损或内部"鬼影"产生假信号的可能性。

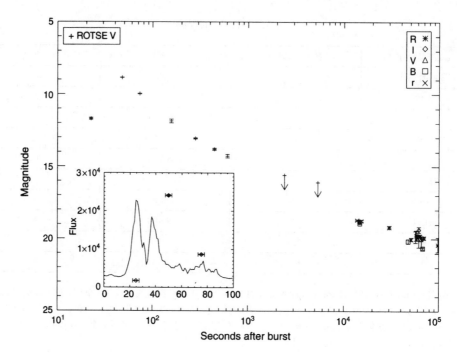

Fig. 2. A combined optical light curve. Afterglow data points are drawn from the GCN archive[21-33]. The early decay of the ROTSE-I light curve is not well fitted by a single power law. The final ROTSE limit is obtained by co-adding the final four 200-s images. The inset shows the first three ROTSE optical fluxes compared to the BATSE γ-ray light curve in the 100–300 keV energy band. The ROTSE-I fluxes are in arbitrary units. Horizontal error bars indicate periods of active observation. We note that there is no information about the optical light curve outside these intervals. Vertical error bars represent flux uncertainties. Further information about GCN is available at http://gcn.gsfc.nasa.gov/gcn.

The fluence of GRB990123 was exceptionally high (99.6 percentile of BATSE triggers; M. Briggs, personal communication), implying that such bright optical transients may be rare. Models of early optical emission suggest that optical intensity scales with γ-ray fluence[15-17]. If this is the case, ROTSE-I and similar instruments are sensitive to 50% of all GRBs. This translates to ~12 optically detected events per year. Our continuing analysis of less well-localized GRB data may therefore reveal similar transients. To date, this process has been hampered by the necessity of identifying and discarding typically 100,000 objects within the large field of view and optimizing a search strategy in the face of an unknown early time structure. The results we report here at least partially resolve the latter problem while increasing the incentive to complete a difficult analysis task. The ROTSE project is in the process of completing two 0.45-m telescopes capable of reaching 4 magnitudes deeper than ROTSE-I for the same duration exposures. If γ-ray emission in bursts is beamed but the optical emission is more isotropic, there may be many optical transients unassociated with detectable GRBs. These instruments will conduct sensitive searches for such events. We expect that ROTSE will be important in the exploration to come.

(**398**, 400-402; 1999)

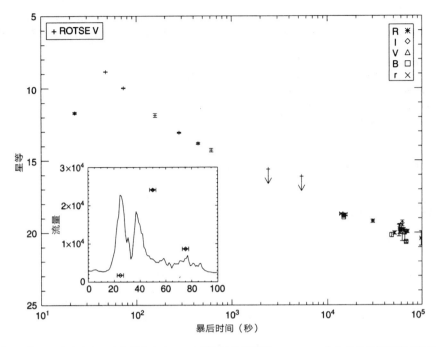

图 2. 联合的光学光变曲线。余辉数据来自 GCN 的归档数据 [21-33]。ROTSE-I 光变曲线的早期衰减无法用单个幂律来进行拟合。最后的 ROTSE 极限是通过叠加最后 4 个 200 秒曝光的图像得到的。内嵌插图显示前 3 个 ROTSE 光学流量与 100～300 keV 能段的 BATSE γ 射线光变曲线的比较。ROTSE-I 流量单位是任意选取的。水平误差棒表示有效的观测时段。注意这里没有在这些时段之外的光学光变曲线信息。垂直误差棒代表流量不确定度。更多关于 GCN 的信息请参见 http://gcn.gsfc.nasa.gov/gcn。

GRB990123 的能流非常大（占据 BATSE 触发事件的 99.6%；布里格斯，个人交流），这表明这样亮的光学暂现源可能很少见。早期光学辐射的模型显示光学强度和γ 射线能流成比例 [15-17]。事实如此的话，ROTSE-I 和类似设备将对 50% 的 GRB 都非常灵敏。这意味着每年我们将可以探测到约 12 例光学事件。因此我们接下来对那些定位不是很好的 GRB 数据进行分析可能可以发现类似的暂现源。 迄今为止，这个方法的困难之处在于必须证认并排除大视场内约 100,000 个天体，并且在未知早期时间结构的情况下必须最大限度地优化搜寻策略。我们这里报道的结果至少部分解决了第二个问题，并激励我们去完成这些困难的分析任务。ROTSE 项目目前正在搭建两个 0.45 米望远镜使得在相同曝光时间下的星等极限比 ROTSE-I 深 4 个星等。假如暴的 γ 射线辐射是束状的，而光学辐射更加各向同性，那么就可能有很多光学暂现源并没有成协的、能探测到的 GRB。这些仪器将对这种事件做出灵敏的搜寻。我们预期 ROTSE 将在未来的研究探索中发挥非常重要的作用。

（肖莉 翻译；黎卓 审稿）

C. Akerlof*, R. Balsano[†], S. Barthelmy[‡§], J. Bloch[†], P. Butterworth[‡‖], D. Casperson[†], T. Cline[‡], S. Fletcher[†], F. Frontera[¶], G. Gisler[†], J. Heise[#], J. Hills[†], R. Kehoe*, B. Lee*, S. Marshall[☆], T. McKay*, R. Miller[†], L. Piro[**], W. Priedhorsky[†], J. Szymanski[†] & J. Wren[†]

* University of Michigan, Ann Arbor, Michigan 48109, USA
[†] Los Alamos National Laboratory, Los Alamos, New Mexico 87545, USA
[‡] NASA/Goddard Space Flight Center, Greenbelt, Maryland 20771, USA
[§] Universities Space Research Association, Seabrook, Maryland 20706, USA
[‖] Raytheon Systems, Lanham, Maryland 20706, USA
[¶] Università degli Studi di Ferrara, Ferrara, Italy
[#] Space Research Organization, Utrecht, The Netherlands
[☆] Lawrence Livermore National Laboratory, Livermore, California 94550, USA
[**] Instituto Astrofisica Spaziale, Rome, Italy

Received 5 February; accepted 19 February 1999.

References:

1. Klebesadel, R. W., Strong, I. B. & Olson, R. A. Observations of gamma-ray bursts of cosmic origin. *Astrophys. J.* **182,** L85-L88 (1973).

2. Piro, L. *et al.* The first X-ray localization of a γ-ray burst by BeppoSAX and its fast spectral evolution. *Astron. Astrophys.* **329,** 906-910 (1998).

3. Costa, E. *et al.* Discovery of an X-ray afterglow associated with the γ-ray burst of 28 February 1997. *Nature* **387,** 783-785 (1997).

4. Krimm, H. A., Vanderspek, R. K. & Ricker, G. R. Searches for optical counterparts of BATSE gamma-ray bursts with the Explosive Transient Camera. *Astron. Astrophys. Suppl.* **120,** 251-254 (1996).

5. Hudec, R. & Soldán, J. Ground-based optical CCD experiments for GRB and optical transient detection. *Astrophys. Space Sci.* **231,** 311-314 (1995).

6. Lee, B. *et al.* Results from Gamma-Ray Optical Counterpart Search Experiment: a real time search for gamma-ray burst optical counterparts. *Astrophys. J.* **482,** L125-L129 (1997).

7. Park, H. S. *et al.* New constraints on simultaneous optical emission from gamma-ray bursts measured by the Livermore Optical Transient Imaging System experiment. *Astrophys. J.* **490,** L21-L24 (1997).

8. Kelson, D. D., Illingworth, G. D., Franx, M., Magee, D. & van Dokkum, P. G. *IAU Circ.* No. 7096 (1999).

9. Hjorth, J. *et al. GCN Circ.* No. 219 (1999).

10. Fenimore, E. E., Ramirez-Ruiz, E., Wu, B. GRB990123: Evidence that the γ-rays come from a central engine. Preprint astro-ph9902007 at ⟨http://xxx.lanl.gov⟩ (1999).

11. Barthelmy, S. *et al.* in *Gamma-Ray Bursts: 4th Huntsville Symp.* (eds Meegan, C. A, Koskut, T. M. & Preece, R. D.) 99-103 (AIP Conf. Proc. 428, Am. Inst. Phys., College Park, 1997).

12. Piro, L. *et al. GCN Circ.* No. 199 (1999).

13. Odewahn, S. C. *et al. GCN Circ.* No. 201 (1999).

14. Bloom, J. S. *et al. GCN Circ.* No. 208 (1999).

15. Katz, J. I. Low-frequency spectra of gamma-ray bursts. *Astrophys. J.* **432,** L107-L109 (1994).

16. Mészáros, P. & Rees, M. J. Optical and long-wavelength afterglow from gamma-ray bursts. *Astrophys. J.* **476,** 232-237 (1997).

17. Sari, R. & Piran, T. The early afterglow. Preprint astro-ph/9901105 at ⟨http://xxx.lanl.gov⟩ (1999).

18. Bertin, E. & Arnouts, S. SExtractor: Software for source extraction. *Astron. Astrophys. Suppl.* **117,** 393-404 (1996).

19. Høg, E. *et al.* The Tycho reference catalogue. *Astron. Astrophys.* **335,** L65-L68 (1998).

20. Monet, D. *et al. A Catalog of Astrometric Standards* (US Naval Observatory, Washington DC, 1998).

21. Zhu, J. & Zhang, H. T. *GCN Circ.* No. 204 (1999).

22. Bloom, J. S. *et al. GCN Circ.* No. 206 (1999).

23. Gal, R. R. *et al. GCN Circ.* No. 207 (1999).

24. Sokolov, V. *et al. GCN Circ.* No. 209 (1999).

25. Ofek, E. & Leibowitz, E. M. *GCN Circ.* No. 210 (1999).

26. Garnavich, P., Jha, S., Stanek, K. & Garcia, M. *GCN Circ.* No. 215 (1999).

27. Zhu, J. *et al. GCN Circ.* No. 217 (1999).

28. Bloom, J. S. *et al. GCN Circ.* No. 218 (1999).

29. Maury, A., Boer, M. & Chaty, S. *GCN Circ.* No. 220 (1999).

30. Zhu, J. *et al. GCN Circ.* No. 226 (1999).

31. Sagar, R., Pandey, A. K., Yadav, R. K. R., Nilakshi & Mohan, V. *GCN Circ.* No. 227 (1999).

32. Masetti, N. *et al. GCN Circ.* No. 233 (1999).

33. Bloom, J. S. *et al. GCN Circ.* No. 240 (1999).

Acknowledgements. The ROTSE Collaboration thanks J. Fishman and the BATSE team for providing the data that enable the GCN localizations which made this experiment possible; and we thank the BeppoSAX team for rapid distribution of coordinates. This work was supported by NASA and the US DOE. The Los Alamos National Laboratory is operated by the University of California for the US Department of Energy (DOE). The work was performed in part under the auspices of the US DOE by Lawrence Livermore National Laboratory. BeppoSAX is a programme of the Italian Space Agency (ASI) with participation of the Dutch Space Agency (NIVR).

Correspondence and requests for materials should be addressed to C.A. (e-mail: akerlof@mich.physics.lsa.umich.edu).

A Flat Universe from High-resolution Maps of the Cosmic Microwave Background Radiation

P. de Bernardis *et al.*

Editor's Note

In the early Universe, before subatomic particles united into atoms, photons, electrons and atomic nuclei were sloshing around in a kind of primordial soup. As with sloshing water in a bathtub, there were certain naturally resonant frequencies. Radiation from this time carries echoes of those resonances, preserved as characteristic ripples in the cosmic microwave background radiation. Paolo de Bernardis and his colleagues launched a high-altitude balloon in the Antarctic to measure these ripples. Here they report their findings: the ripples are best explained by a "flat" Universe, in which the sum of standard matter, dark matter and dark energy is equal to a "critical density", poised between positive (closed) and negative (open) curvature. Subsequently the WMAP satellite confirmed this conclusion.

The blackbody radiation left over from the Big Bang has been transformed by the expansion of the Universe into the nearly isotropic 2.73 K cosmic microwave background. Tiny inhomogeneities in the early Universe left their imprint on the microwave background in the form of small anisotropies in its temperature. These anisotropies contain information about basic cosmological parameters, particularly the total energy density and curvature of the Universe. Here we report the first images of resolved structure in the microwave background anisotropies over a significant part of the sky. Maps at four frequencies clearly distinguish the microwave background from foreground emission. We compute the angular power spectrum of the microwave background, and find a peak at Legendre multipole $l_{peak} = (197 \pm 6)$, with an amplitude $\Delta T_{200} = (69 \pm 8)$ µK. This is consistent with that expected for cold dark matter models in a flat (euclidean) Universe, as favoured by standard inflationary models.

PHOTONS in the early Universe were tightly coupled to ionized matter through Thomson scattering. This coupling ceased about 300,000 years after the Big Bang, when the Universe cooled sufficiently to form neutral hydrogen. Since then, the primordial photons have travelled freely through the Universe, redshifting to microwave frequencies as the Universe expanded. We observe those photons today as the cosmic microwave background (CMB). An image of the early Universe remains imprinted in the temperature anisotropy of the CMB. Anisotropies on angular scales larger than ~2° are dominated by the gravitational redshift the photons undergo as they leave the density fluctuations present at decoupling[1,2]. Anisotropies on smaller angular scales are enhanced by oscillations of the

高分辨率的宇宙微波背景辐射观测揭示我们的宇宙是平直的

在宇宙早期，亚原子粒子结合成原子之前，光子、电子以及原子核在某种原始汤中游荡。类似于浴缸中受到扰动的水，其中自然而然存在一些共振的频率。这个时期的辐射携带了这些共振的回音，以典型涟漪的形式保留在宇宙微波背景辐射上。保罗·德·贝尔纳迪斯和他的合作者们在南极洲发射了一个高海拔的气球来探测这些涟漪。这里报道了他们的结果：涟漪能够通过"平直"宇宙模型得到最好的解释。在这个模型中，标准物质、暗物质以及暗能量的总密度等于"临界密度"，介于正的（封闭的）以及负的（开的）曲率之间。随后，WMAP 卫星证实了这个结论。

随着宇宙的膨胀，大爆炸遗留下来的黑体辐射逐步演化成为几乎各向同性、温度为 2.73 K 的宇宙微波背景。早期宇宙的不均匀性在微波背景上留下了微小的各向异性的痕迹。这些各向异性蕴含着宇宙学基本参数的信息，尤其是宇宙整体能量密度与曲率。在本工作中，我们首次展示了一块足够大的天区上微波背景各向异性结构的高分辨率图像。四个波段下的图像可以清楚地将微波背景辐射与前景辐射区分开。我们计算了微波背景的角功率谱，并在勒让德阶数 $l_{peak} = (197 \pm 6)$ 处发现了一个峰，其幅度为 $\Delta T_{200} = (69 \pm 8)$ μK。该结果同平直（欧几里得）宇宙下冷暗物质模型的理论预期相吻合，也被标准的暴胀模型所支持。

在宇宙早期，光子通过汤姆孙散射与电离物质紧密耦合在一起。这种耦合状态一直持续到大爆炸之后 300,000 年，这个时候宇宙冷却到足以形成中性氢。自那时候开始，原初光子就在宇宙中自由地传播，随着宇宙膨胀而红移到微波波段。这就是我们今天所观测到的宇宙微波背景（CMB）。而 CMB 中的温度各向异性则保留了早期宇宙的基本图像。角尺度大于约 2° 的各向异性信息由退耦时离开密度扰动的光子经历的引力红移所主导[1,2]。而更小角尺度上各向异性的程度因退耦前光子–重子

photon–baryon fluid before decoupling[3]. These oscillations are driven by the primordial density fluctuations, and their nature depends on the matter content of the Universe.

In a spherical harmonic expansion of the CMB temperature field, the angular power spectrum specifies the contributions to the fluctuations on the sky coming from different multipoles l, each corresponding to the angular scale $\theta = \pi/l$. Density fluctuations over spatial scales comparable to the acoustic horizon at decoupling produce a peak in the angular power spectrum of the CMB, occurring at multipole l_{peak}. The exact value of l_{peak} depends on both the linear size of the acoustic horizon and on the angular diameter distance from the observer to decoupling. Both these quantities are sensitive to a number of cosmological parameters (see, for example, ref. 4), but l_{peak} primarily depends on the total density of the Universe, Ω_0. In models with a density Ω_0 near 1, $l_{peak} \approx 200/\Omega_0^{1/2}$. A precise measurement of l_{peak} can efficiently constrain the density and thus the curvature of the Universe.

Observations of CMB anisotropies require extremely sensitive and stable instruments. The DMR[5] instrument on the COBE satellite mapped the sky with an angular resolution of $\sim 7°$, yielding measurements of the angular power spectrum at multipoles $l < 20$. Since then, experiments with finer angular resolution[6-16] have detected CMB fluctuations on smaller scales and have produced evidence for the presence of a peak in the angular power spectrum at $l_{peak} \approx 200$.

Here we present high-resolution, high signal-to-noise maps of the CMB over a significant fraction of the sky, and derive the angular power spectrum of the CMB from $l = 50$ to 600. This power spectrum is dominated by a peak at multipole $l_{peak} = (197 \pm 6)$ (1σ error). The existence of this peak strongly supports inflationary models for the early Universe, and is consistent with a flat, euclidean Universe.

The Instrument

The Boomerang (balloon observations of millimetric extragalactic radiation and geomagnetics) experiment is a microwave telescope that is carried to an altitude of ~ 38 km by a balloon. Boomerang combines the high sensitivity and broad frequency coverage pioneered by an earlier generation of balloon-borne experiments with the long (~ 10 days) integration time available in a long-duration balloon flight over Antarctica. The data described here were obtained with a focal plane array of 16 bolometric detectors cooled to 0.3 K. Single-mode feedhorns provide two 18' full-width at half-maximum (FWHM) beams at 90 GHz and two 10' FWHM beams at 150 GHz. Four multi-band photometers each provide a 10.5', 14' and 13' FWHM beam at 150, 240 and 400 GHz respectively. The average in-flight sensitivity to CMB anisotropies was 140, 170, 210 and 2,700 μK s$^{1/2}$ at 90, 150, 240 and 400 GHz, respectively. The entire optical system is heavily baffled against terrestrial radiation. Large sunshields improve rejection of radiation from $> 60°$ in azimuth from the telescope boresight. The rejection has been measured to be greater than 80 dB at all angles occupied by the Sun during the CMB observations. Further details on

流体振荡而加剧[3]。这些振荡由原初密度扰动驱动，而它们的本质依赖于宇宙中的物质成分。

对 CMB 温度场进行球谐展开，其角功率谱表征了全天不同阶数 l（对应角尺度 $\theta = \pi/l$）处扰动的强度。与退耦时声学视界相比拟的空间尺度上的密度扰动在 CMB 的角功率谱上产生一个峰，对应阶数为 l_{peak}。l_{peak} 的精确取值依赖于声学视界的线性尺度以及观测者（现在）到退耦时期的角直径距离。虽然这些量与一系列宇宙学参数密切相关（比如参考文献 4），但是 l_{peak} 主要取决于宇宙整体密度 Ω_0。对于密度 Ω_0 接近 1 的宇宙学模型，$l_{peak} \approx 200/\Omega_0^{1/2}$。因此，对 l_{peak} 进行精确测量可有效地限制宇宙整体的密度，从而可以限制宇宙的曲率。

对 CMB 各向异性的观测需要极高分辨率和稳定性的观测仪器。COBE 卫星上的 DMR[5] 设备的巡天角分辨率约为 7°，这使得对 $l < 20$ 的角功率谱的测量成为可能。自此之后，具有更高角分辨率的实验[6-16]探测到更小尺度上的 CMB 扰动，并且得到的证据表明在角功率谱上 l_{peak} 约 200 处存在一个峰。

这里我们展示了一块足够大天区的 CMB 高分辨率高信噪比的图像，得到了 $l = 50$ 到 600 的 CMB 角功率谱。在这个范围内，功率谱在 $l_{peak} = 197 \pm 6$（1σ 误差范围）处存在一个峰。这个峰的存在极大地支持了早期宇宙的暴胀模型，与一个平直的欧几里得宇宙的理论预言相吻合。

观 测 设 备

Boomerang（毫米波段河外辐射和地磁场球载观测）实验利用球载微波望远镜在海拔约 38 千米的高空进行观测。Boomerang 在上一代球载实验开创的高分辨率和宽波段覆盖特性的基础上，更具在南极上空长时间的飞行中可用积分时间长（约 10 天）的能力。本工作所采用的数据来源于由 16 个冷却到 0.3 K 的热辐射探测器所组成的焦平面阵列。通过单模喇叭馈源，我们在 90 GHz 处得到了两个半峰全宽为 18 角分的波束，并在 150 GHz 处得到了两个半峰全宽为 10 角分的波束。而四个多波段光度计均在 150、240 以及 400 GHz 处得到了半峰全宽分别为 10.5、14 以及 13 角分的波束。CMB 各向异性在 90、150、240 以及 400 GHz 处运行中的平均灵敏度分别为 140、170、210 以及 2,700 $\mu K \cdot s^{1/2}$。整个光学系统可以很好地抵御地球辐射的影响。大型遮阳板有助于遮挡来自望远镜视轴方向且方位角大于 60° 的辐射。我们测

the instrument can be found in refs 17–21.

Observations

Boomerang was launched from McMurdo Station (Antarctica) on 29 December 1998, at 3:30 GMT. Observations began 3 hours later, and continued uninterrupted during the 259-hour flight. The payload approximately followed the 79° S parallel at an altitude that varied daily between 37 and 38.5 km, returning within 50 km of the launch site.

We concentrated our observations on a target region, centred at roughly right ascension (RA) 5 h, declination (dec.) −45°, that is uniquely free of contamination by thermal emission from interstellar dust[22] and that is approximately opposite the Sun during the austral summer. We mapped this region by repeatedly scanning the telescope through 60° at fixed elevation and at constant speed. Two scan speeds ($1° s^{-1}$ and $2° s^{-1}$ in azimuth) were used to facilitate tests for systematic effects. As the telescope scanned, degree-scale variations in the CMB generated sub-audio frequency signals in the output of the detector[23]. The stability of the detector system was sufficient to allow sensitive measurements on angular scales up to tens of degrees on the sky. The scan speed was sufficiently rapid with respect to sky rotation that identical structures were observed by detectors in the same row in each scan. Detectors in different rows observed the same structures delayed in time by a few minutes.

At intervals of several hours, the telescope elevation was interchanged between 40°, 45° and 50° in order to increase the sky coverage and to provide further systematic tests. Sky rotation caused the scan centre to move and the scan direction to rotate on the celestial sphere. A map from a single day at a single elevation covered roughly 22° in declination and contained scans rotated by ± 11° on the sky, providing a cross-linked scan pattern. Over most of the region mapped, each sky pixel was observed many times on different days, both at $1° s^{-1}$ and $2° s^{-1}$ scan speed, with different topography, solar elongation and atmospheric conditions, allowing strong tests for any contaminating signal not fixed on the celestial sphere.

The pointing of the telescope has been reconstructed with an accuracy of 2′ r.m.s. using data from a Sun sensor and rate gyros. This precision has been confirmed by analysing the observed positions of bright compact H II regions in the Galactic plane (RCW38[24], RCW57, IRAS08576 and IRAS1022) and of radio-bright point sources visible in the target region (the QSO 0483–436, the BL Lac object 0521–365 and the blazar 0537–441).

440

量发现，在整个 CMB 观测中各角度的太阳光均被遮挡了 80 dB 以上。更多关于设备的细节请参看参考文献 17 ~ 21。

观　测

格林尼治标准时间 1998 年 12 月 29 日 3 时 30 分，Boomerang 从南极洲麦克默多站发射。观测始于 3 小时后，并在 259 小时的飞行过程中保持不间断。探测器每天大致在离发射场 50 千米范围内沿着南纬 79° 飞行，海拔高度在 37 ~ 38.5 千米范围内。

我们将观测集中在以赤经 5 h、赤纬 −45° 为中心的目标区域内。这块区域非常独特，不仅可以免受星际尘埃热辐射的污染[22]，且在整个南半球夏季都会背对着太阳。我们利用望远镜以 60° 倾角和恒定的速度进行重复扫描来对该区域成图。为帮助测试其中的系统误差，我们使用了两种扫描速度（每秒方位角 1° 和 2°）。随着望远镜对天区进行扫描，CMB 的度尺度上的变化在探测器输出中产生了亚音频信号[23]。整个探测器系统的稳定性很好，使得对天区进行大到数十度角尺度的高灵敏度的测量成为可能。所采用的扫描速度相对于天空旋转而言是足够快的，这样在每次扫描过程中同一行的探测器观测到的是相同的结构。而不同行的探测器则会在延迟几分钟之后也观测到这些相同的结构。

为了增加天区覆盖并且进行更多系统误差的测试，望远镜的倾角每隔几小时会在 40°、45° 以及 50° 之间交替变化。而天空的旋转不仅会使扫描中心发生移动，同时还会使扫描方向在天球上发生旋转。同一天同一倾角的图像大致覆盖赤纬 22° 的范围，包含天空中旋转了 ±11° 的扫描模式，这就形成了交联的扫描模式。对于大部分成图的区域而言，每个天空格点都会在不同时间里以两种扫描速度（每秒 1° 和 2°）在不同地形、太阳距角以及大气条件下被观测多次，这就使得我们可以对天球上各种可能的污染信号进行强有力的测试。

利用太阳传感器和速率陀螺的数据，望远镜的指向已进行重构，其均方根精度为 2 角分。通过对银道面上明亮的致密 H II 区（RCW38[24]、RCW57、IRAS08576 以及 IRAS1022）以及目标天区可见的射电明亮点源（类星体 QSO 0483–436、BL Lac 天体 0521–365 以及耀变体 0537–441）观测位置的分析，我们进一步确认了望远镜的指向精度。

Calibrations

The beam pattern for each detector was mapped before flight using a thermal source. The main lobe at 90, 150 and 400 GHz is accurately modelled by a gaussian function. The 240 GHz beams are well modelled by a combination of two gaussians. The beams have small shoulders (less than 1% of the total solid angle), due to aberrations in the optical system. The beam-widths were confirmed in flight via observations of compact sources. By fitting radial profiles to these sources we determine the effective angular resolution, which includes the physical beamwidth and the effects of the 2′ r.m.s. pointing jitter. The effective FWHM angular resolution of the 150 GHz data that we use here to calculate the CMB power spectrum is $(10 \pm 1)'$, where the error is dominated by uncertainty in the pointing jitter.

We calibrated the 90, 150 and 240 GHz channels from their measured response to the CMB dipole. The dipole anisotropy has been accurately (0.7%) measured by COBE-DMR[25], fills the beam and has the same spectrum as the CMB anisotropies at smaller angular scales, making it the ideal calibrator for CMB experiments. The dipole signal is typically ~3 mK peak-to-peak in each 60° scan, much larger than the detector noise, and appears in the output of the detectors at $f = 0.008$ Hz and $f = 0.016$ Hz in the $1° \, s^{-1}$ and $2° \, s^{-1}$ scan speeds, respectively. The accuracy of the calibration is dominated by two systematic effects: uncertainties in the low-frequency transfer function of the electronics, and low-frequency, scan-synchronous signals. Each of these is significantly different at the two scan speeds. We found that the dipole-fitted amplitudes derived from separate analysis of the $1° \, s^{-1}$ and $2° \, s^{-1}$ data agree to within $\pm 10\%$ for every channel, and thus we assign a 10% uncertainty in the absolute calibration.

From Detector Signals to CMB Maps

The time-ordered data comprises 5.4×10^7 16-bit samples for each channel. These data are flagged for cosmic-ray events, elevation changes, focal-plane temperature instabilities, and electromagnetic interference events. In general, about 5% of the data for each channel are flagged and not used in the subsequent analysis. The gaps resulting from this editing are filled with a constrained realization of noise in order to minimize their effect in the subsequent filtering of the data. The data are deconvolved by the bolometer and electronics transfer functions to recover uniform gain at all frequencies.

The noise power spectrum of the data and the maximum-likelihood maps[26-28] were calculated using an iterative technique[29] that separates the sky signal from the noise in the time-ordered data. In this process, the statistical weights of frequencies corresponding to angular scales larger than 10° on the sky are set to zero to filter out the largest-scale modes of the map. The maps were pixelized according to the HEALPix pixelization scheme[30].

校　准

在进行飞行试验前我们利用一个热源对每一个探测器的波束模式进行了成像。在 90、150 以及 400 GHz 的主瓣可以精确地用高斯函数进行拟合，而在 240 GHz 的波束则可以用两个高斯函数进行叠加。因为光学系统存在光行差，波束存在小的肩状分布（小于整个立体角的 1%）。在整个飞行过程中通过对致密源进行观测进一步确认了波束宽度。通过对这些源的径向轮廓进行拟合，我们得到了有效的角分辨率，这其中不仅包含其本身波束宽度的贡献，还包含 2 角分均方根指向抖动带来的影响。本工作中我们计算 CMB 功率谱所采用的 150 GHz 数据的有效半峰全宽角分辨率为（10±1）角分，其误差主要来源于指向抖动的不确定性。

我们通过测量 90、150 以及 240 GHz 波段对 CMB 偶极子的响应对相应波段的观测进行校准。偶极子各向异性充满整个波束，并且在更小的角尺度上与 CMB 各向异性拥有相同的谱结构，而之前 COBE-DMR 已经对偶极子的各向异性进行了准确的测量（0.7%）[25]，因此偶极子各向异性是校准 CMB 实验观测的理想工具。每次 60° 扫描的典型偶极子信号总幅度大约为 3 mK，比探测器噪声大很多，且在每秒 1° 以及 2° 的扫描速度下分别在探测器输出的 $f = 0.008$ Hz 以及 $f = 0.016$ Hz 处出现。校准的准确性主要受到两个系统误差的影响：电子设备低频转移函数的不确定性以及低频扫描同步信号的不确定性。我们发现，对不同波段以及不同扫描速度（每秒 1° 以及每秒 2°）下的数据单独分析得到的偶极子拟合幅度在 ±10% 范围内相互吻合。因此我们绝对校准的不确定度定为 10%。

从探测器信号到 CMB 图像

每个通道的时间序列数据包含 5.4×10^7 个 16 位样本。我们将受到宇宙线事件、倾角变化、焦平面温度不稳定性和电磁干扰事件影响的数据进行了标记。整体来说，每个通道大约有 5% 的数据被标记，这些数据不用于后续分析。为了降低后续数据平滑过滤过程中上述操作所带来的影响，我们利用一个约束噪声来填补这个过程所造成的数据缺失。随后，通过测辐射热计和电子设备转移函数对数据进行解卷积，我们重新得到了在所有频段具有均匀增益的数据样本。

通过一个可以在时间序列数据中将天空信号和噪声分离的迭代技术[29]，我们计算得到了数据的噪声功率谱以及最大似然图像[26-28]。在这个过程中，我们将对应天空角尺度大于 10° 的频率的统计权重设为 0，以此来滤除图像中的最大尺度模式。整个图像根据 HEALPix 像素化方法进行像素化[30]。

Figure 1 shows the maps obtained in this way at each of the four frequencies. The 400 GHz map is dominated by emission from interstellar dust that is well correlated with that observed by the IRAS and COBE/DIRBE satellites. The 90, 150 and 240 GHz maps are dominated by degree-scale structures that are resolved with high signal-to-noise ratio. A qualitative but powerful test of the hypothesis that these structures are CMB anisotropy is provided by subtracting one map from another. The structures evident in all three maps disappear in both the $90-150$ GHz difference and in the $240-150$ GHz difference, as expected for emission that has the same spectrum as the CMB dipole anisotropy used to calibrate the maps.

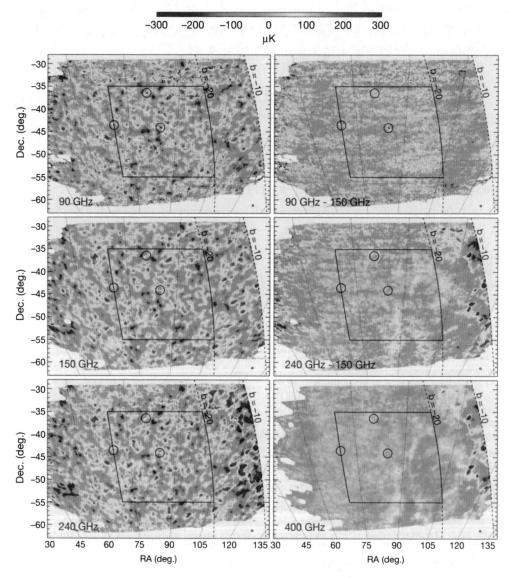

Fig. 1. Boomerang sky maps (equatorial coordinates). The sky maps at 90, 150 and 240 GHz (left panels) are shown with a common colour scale, using a thermodynamic temperature scale chosen such that CMB anisotropies will have the same amplitude in the three maps. Only the colour scale of the 400 GHz map

　　图 1 中显示了利用上述方法得到的四个频段的图像。400 GHz 的图像由星际尘埃的辐射主导，与 IRAS 和 COBE/DIRBE 卫星的观测结果存在很好的相关性。而 90、150 以及 240 GHz 的图像则由高信噪比的度尺度结构所主导。假设这些结构是 CMB 各向异性的，一个定性但有效检验这一假设的方法是将这些图像相互去减。那些在三个图像中均明显存在的结构在 90−150 GHz 的残差图以及 240−150 GHz 的残差图中都消失了，这从辐射中可以预见出，这些辐射与用来校准图像的 CMB 偶极子各向异性具有相同谱结构。

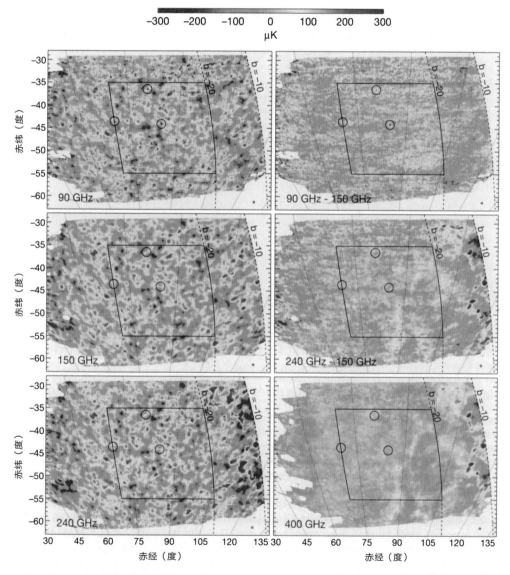

图 1. Boomerang 天图（赤道坐标）。左图中 90、150 以及 240 GHz 的天图具有相同的颜色显示标度，均采用相同的热力学温标，这样在三幅天图中 CMB 各向异性具有相同的幅度。只有右下角的 400 GHz 天

(bottom right) is 14 times larger than the others: this has been done to facilitate comparison of emission from interstellar dust (ISD), which dominates this map, with ISD emission present in the lower-frequency maps. The maps at 90 and 400 GHz are each from a single detector, while maps at 150 and 240 GHz have each been obtained by co-adding data from three detectors. For purposes of presentation, the maps have been smoothed with gaussian filters to obtain FWHM effective resolution of 22.5′ (small circle in the bottom right side of each panel). Structures along the scan direction larger than 10° are not present in the maps. Several features are immediately evident. Most strikingly, the maps at 90, 150 and 240 GHz are dominated by degree-scale structures that fill the map, have well-correlated morphology and are identical in amplitude in all three maps. These structures are not visible at 400 GHz. The 400 GHz map is dominated by diffuse emission which is correlated with the ISD emission mapped by IRAS/DIRBE[22]. This emission is strongly concentrated towards the right-hand edge of the maps, near the plane of the Galaxy. The same structures are evident in the 90, 150 and 240 GHz maps at Galactic latitude $b > -15°$, albeit with an amplitude that decreases steeply with decreasing frequency. The large-scale gradient evident especially near the right edge of the 240 GHz map is a result of high-pass-filtering the very large signals near the Galactic plane (not shown). This effect is negligible in the rest of the map. The two top right panels show maps constructed by differencing the 150 and 90 GHz maps and the 240 and 150 GHz maps. The difference maps contain none of the structures that dominate the maps at 90, 150 and 240 GHz, indicating that these structures do indeed have the ratios of brightness that are unique to the CMB. The morphology of the residual structures in the 240 − 150 GHz map is well-correlated with the 400 GHz map, as is expected if the residuals are due to the ISD emission. Three compact sources of emission are visible in the lower-frequency maps, as indicated by the circles. These are known radio-bright quasars from the SEST pointing catalogue at 230 GHz. The boxed area has been used for computing the angular power spectrum shown in Fig. 2.

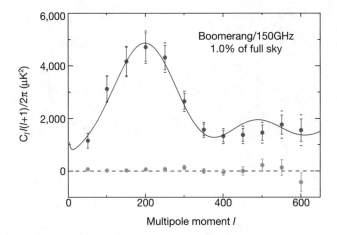

Fig. 2. Angular power spectrum measured by Boomerang at 150 GHz. Each point is the power averaged over $\Delta l = 50$ and has negligible correlations with the adjacent points. The error bars indicate the uncertainty due to noise and cosmic/sampling variance. The errors are dominated by cosmic/sampling variance at $l < 350$; they grow at large l due to the signal attenuation caused by the combined effects[39] of the 10′ beam and the 14′ pixelization (0.87 at $l = 200$ and 0.33 at $l = 600$). The current $\pm 10\%$ uncertainty in the calibration corresponds to an overall re-scaling of the y-axis by $\pm 20\%$, and is not shown. The current 1′ uncertainty in the angular resolution of the measurement creates an additional uncertainty—indicated by the distance between the ends of the red error bars and the blue horizontal lines—that is completely correlated and is largest (11%) at $l = 600$. The green points show the power spectrum of a difference map obtained as follows. We divided the data into two parts corresponding to the first and second halves of the timestream. We made two maps (A and B) from these halves, and the green points show the power spectrum computed from the difference map, $(A-B)/2$. Signals originating from the sky should disappear in this map, so this is a test for contamination in the data (see text). The solid curve has parameters $(\Omega_b, \Omega_m, \Omega_\Lambda, n_s, h) = (0.05, 0.31, 0.75, 0.95, 0.70)$. It is the best-fit model for the Boomerang test flight data[15,16], and is shown for comparison only. The model that best fits the new data reported here will be presented elsewhere.

图的颜色标度是其他图像的 14 倍，这么做的原因是为了方便将此天图与其他更低频段天图的星际尘埃辐射进行对比。对 400 GHz 天图而言，星际尘埃的辐射占主导。90 和 400 GHz 的天图的数据各由单独探测器获得，而 150 和 240 GHz 的天图则是将三个探测器的数据进行叠加之后所得。为了便于展示，所有天图均经过高斯滤波平滑处理，使得有效半峰全宽分辨率为 22.5 角分（每幅图右下角的小圆圈的大小）。沿扫描方向大于 10°的结构没有显示在天图中。图中有几个特征非常明显。其中最为显著的是，90、150 以及 240 GHz 的天图由充满天图的度尺度结构主导，在三幅天图中它们的形态存在很好的相关性，且幅度完全相同。然而在 400 GHz 的天图中却看不到这些结构。400 GHz 天图由弥散的辐射所主导，与 IRAS/DIRBE[22] 观测得到的星际尘埃辐射存在很好的相关性。该辐射主要集中在天图的右边缘，靠近银道面的区域。在 90、150 以及 240 GHz 的天图对应银纬 $b > -15°$ 的区域中，这些结构同样显著，尽管随着频率的下降其幅度急剧减小。在 240 GHz 的天图中，特别是右边缘附近存在较为明显的大尺度梯度，这是由对银道面（未显示）附近非常高的信号进行高通滤波所造成的结果。这个效应对天图其他的区域所造成的影响可以忽略不计。右上的两幅图分别显示了 150 GHz 天图同 90 GHz 天图相减以及 240 GHz 同 150 GHz 天图相减所得的残差图。这些残差图中并不存在那些在 90、150 以及 240 GHz 天图中占主导的结构，这意味着这些结构确实具有 CMB 所特有的亮度比例。而 240–150 GHz 残差图中剩余结构的形态分布同 400 GHz 天图存在很好的相关性，这与我们假设残差主要源自星际尘埃辐射所得到的理论预期相符。在低频段的天图中，三个致密的辐射源清晰可见，具体位置已用圆圈标记。它们均为 230 GHz 的 SEST 星表中已知的射电明亮类星体。方框内的区域被用来计算图 2 中显示的角功率谱。

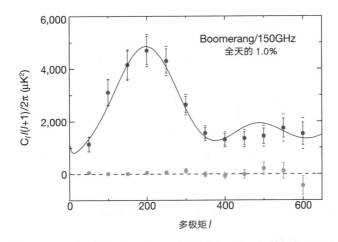

图 2. 150 GHz 处 Boomerang 测量得到的角功率谱。每个点为对应区间内功率的平均值，区间宽度 $\Delta l = 50$，相邻两点间的相关性可以忽略不计。误差棒反映了因噪声以及宇宙/样本方差所造成的不确定性。对于 $l < 350$ 的数据点而言，其误差主要来自宇宙/样本方差的不确定性；由于将 10 角分的波束同 14 角分的格点相结合导致的信号衰减[39]（$l = 200$ 时为 0.87，而 $l = 600$ 时为 0.33），误差会随着 l 的增加而变大。目前校准过程带来的 ±10% 的不确定性对应 y 轴整体改变 ±20%，这在本图中并未显示。而目前测量中角分辨率中存在的 1 角分的不确定性还会导致一个额外的不确定因素（表现在红色误差棒端与蓝色水平线之间的距离），这些不确定因素完全相关，且在 $l = 600$ 处最大（可达 11%）。绿色的点显示了用下面方法得到的残差图的功率谱。我们将数据按时间流分为两半，分别独立成图（A 和 B），而绿色的点显示了残差图（A－B）/2 的功率谱。在这幅残差图中所有来自天空的信号都将消失，因此这是检验数据中污染源的一个测试。实线对应的宇宙学参数（Ω_b，Ω_m，Ω_Λ，n_s，h）=（0.05，0.31，0.75，0.95，0.70）。这是与 Boomerang 测试飞行数据[15,16] 最相符的拟合参数，在此显示只作比较之用。与本工作新数据最相符的宇宙学模型将在别处讨论。

To quantify this conclusion, we performed a "colour index" analysis of our data. We selected the ~18,000 14′ pixels at Galactic latitude $b < -15°$, and made scatter plots of 90 GHz versus 150 GHz and 240 GHz versus 150 GHz. A linear fit to these scatter plots gives slopes of 1.00 ± 0.15 and 1.10 ± 0.16, respectively (including our present 10% calibration error), consistent with a CMB spectrum. For comparison, free–free emission with spectral index -2.35 would produce slopes of 2.3 and 0.85, and was therefore rejected with > 99% confidence; emission from interstellar dust with temperature $T_d = 15$ K and spectral index of emissivity $\alpha = 1$ would produce slopes of 0.40 and 2.9. For any combination of $T_d > 7$ K and $1 < \alpha < 2$, the dust hypothesis is rejected with > 99% confidence. We conclude that the dominant source of structure that we detect at 90, 150 and 240 GHz is CMB anisotropy.

We further argue that the 150 GHz map at $b < -15°$ is free of significant contamination by any known astrophysical foreground. Galactic synchrotron and free–free emission is negligible at this frequency[31]. Contamination from extragalactic point sources is also small[32]; extrapolation of fluxes from the PMN survey[33] limits the contribution by point sources (including the three above-mentioned radio-bright sources) to the angular power spectrum derived below to < 0.7% at $l = 200$ and < 20% at $l = 600$. The astrophysical foreground that is expected to dominate at 150 GHz is thermal emission from interstellar dust. We placed a quantitative limit on this source of contamination as follows. We assumed that dust properties are similar at high ($b < -20°$) and moderate ($-20° < b < -5°$) Galactic latitudes. We selected the pixels at moderate Galactic latitudes and correlated the structure observed in each of our four bands with the IRAS/DIRBE map, which is dominated by dust in cirrus clouds. The best-fit slope of each of the scatter plots measures the ratios of the dust signal in the Boomerang channels to the dust signal in the IRAS/DIRBE map. We found that the 400 GHz map is very well correlated to the IRAS/DIRBE map, and that dust at $b < -20°$ can account for at most 10% of the signal variance at 240 GHz, 3% at 150 GHz and 0.5% at 90 GHz.

Angular Power Spectra

We compared the angular power spectrum of the structures evident in Fig. 1 with theoretical predictions. In doing so, we separated and removed the power due to statistical noise and systematic artefacts from the power due to the CMB anisotropies in the maps. The maximum-likelihood angular power spectrum of the maps was computed using the MADCAP[34] software package, whose algorithms fully take into account receiver noise and filtering.

Full analysis of our entire data set is under way. Because of the computational intensity of this process, we report here the results of a complete analysis of a limited portion of the data chosen as follows. We analysed the most sensitive of the 150 GHz detectors. We restricted the sky coverage to an area with RA > 70°, $b < -20°$ and $-55° < $ dec. $ < -35°$, and we used only the ~50% of the data from this detector that was obtained at a scan speed of $1° \text{ s}^{-1}$. We used a relatively coarse pixelization of 8,000 14-arcmin pixels as a compromise

为了将这个结论定量化，我们对数据进行了"色指数"分析。我们挑选了银纬 $b < -15°$ 的大约 18,000 个 14 角分的格点，并对其作 90 GHz 相对于 150 GHz 以及 240 GHz 相对于 150 GHz 的散点图。对这些散点图进行线性拟合，其斜率分别为 1.00 ± 0.15 以及 1.10 ± 0.16（其中包含我们之前提及的 10% 的校准误差），与 CMB 谱结果相吻合。作为比较，谱指数为 −2.35 的自由−自由辐射产生的斜率分别为 2.3 和 0.85，因此以 > 99% 的置信度被排除；温度为 $T_d = 15$ K、发射率谱指数为 $\alpha = 1$ 的星际尘埃辐射产生的斜率分别为 0.40 和 2.9。满足 $T_d > 7$ K 以及 $1 < \alpha < 2$ 范围中的尘埃辐射都会以 > 99% 的置信度被排除。因此我们得出结论：我们在 90、150 以及 240 GHz 图像上探测到的结构，其主要来源为 CMB 各向异性。

进一步，我们认为银纬 $b < -15°$ 的 150 GHz 图像几乎免受任何已知前景天体的严重污染。银河系同步辐射以及自由−自由辐射在这个频段可以忽略不计[31]。河外点源的污染也很小[32]；对 PMN 巡天[33] 的光通量进行外推，可以限制点源（包括三个上面提及的射电明亮天体）对角功率谱的贡献：$l = 200$ 处低至 < 0.7%，而 $l = 600$ 处低至 < 20%。在 150 GHz 频段，我们预测最主要的前景天体应该是星际尘埃的热辐射。我们采用如下方法对这一污染源进行定量化限制。首先我们假设中 ($−20° < b < −5°$)、高 ($b < −20°$) 银纬的尘埃性质类似。我们选取中等银纬的天区图像格点，将我们四个波段所观测到的由卷云尘埃主导的结构同 IRAS/DIRBE 图像进行相关分析。这几个散点图的最佳拟合斜率表征了 Boomerang 各通道测量得到的尘埃信号与 IRAS/DIRBE 图像中的尘埃信号的比值。我们发现 400 GHz 图像同 IRAS/DIRBE 图像存在很好的相关性，同时银纬 $b < −20°$ 的尘埃分别至多可以贡献 240 GHz、150 GHz 以及 90 GHz 图像中 10%、3% 以及 0.5% 的信号变化。

角 功 率 谱

我们将图 1 中明显结构的角功率谱同理论预言进行比较。在这个过程中，我们从图像中将统计噪声以及系统误差的贡献和 CMB 各向异性的信号分离并去除。利用 MADCAP[34] 软件包计算了这些图像的最大似然角功率谱，该软件的算法充分考虑了接收器噪声以及滤波过程所带来的影响。

我们正在对整个数据进行详尽的分析。由于整个过程的计算量很大，这里我们只展示如下所选的有限的一部分数据的完整分析结果。我们选择分析最为敏感的 150 GHz 探测器所得到的数据。我们将天区范围限制在赤经 > 70°，$b < −20°$ 以及 $−55° < 赤纬 < −35°$ 的区域内，并只分析该探测器扫描速度为每秒 1° 所得到的近一半的数据。作为计算速度与高的多级覆盖的折中选择，我们使用一套相对粗糙且包含 8,000 个 14 角分的格点。最后，我们将分析限制在 $l \leqslant 600$ 的范围内，因为在该

between computation speed and coverage of high multipoles. Finally, we limited our analysis to $l \lesssim 600$ for which the effects of pixel shape and size and our present uncertainty in the beam size (1′) are small and can be accurately modelled.

The angular power spectrum determined in this way is shown in Fig. 2 and reported in Table 1. The power spectrum is dominated by a peak at $l_{peak} \approx 200$, as predicted by inflationary cold dark matter models. These models additionally predict the presence of secondary peaks. The data at high l limit the amplitude, but do not exclude the presence of a secondary peak. The errors in the angular power spectrum are dominated at low multipoles ($l \lesssim 350$) by the cosmic/sampling variance, and at higher multipoles by detector noise.

Table 1. Angular power spectrum of CMB anisotropy

l range	150 GHz ([1st half] + [2nd half])/2	150 GHz ([1st half] − [2nd half])/2
26–75	$1,140 \pm 280$	63 ± 32
76–125	$3,110 \pm 490$	16 ± 20
126–175	$4,160 \pm 540$	17 ± 28
176–225	$4,700 \pm 540$	59 ± 44
226–275	$4,300 \pm 460$	68 ± 59
276–325	$2,640 \pm 310$	130 ± 82
326–375	$1,550 \pm 220$	-7 ± 92
376–425	$1,310 \pm 220$	-60 ± 120
426–475	$1,360 \pm 250$	0 ± 160
476–525	$1,440 \pm 290$	220 ± 230
526–575	$1,750 \pm 370$	130 ± 300
576–625	$1,540 \pm 430$	-430 ± 360

Shown are measurements of the angular power spectrum of the cosmic microwave background at 150 GHz, and tests for systematic effects. The values listed are for $\Delta T_l^2 = l(l+1)c_l/2\pi$, in μK^2. Here $c_l = \langle a_{lm}^2 \rangle$, and a_{lm} are the coefficients of the spherical harmonic decomposition of the CMB temperature field: $\Delta T(\theta, \phi) = \Sigma a_{lm} Y_{lm}(\theta, \phi)$. The stated 1σ errors include statistical and cosmic/sample variance, and do not include a 10% calibration uncertainty.

The CMB angular power spectrum shown in Fig. 2 was derive from 4.1 days of observation. As a test of the stability of the result, we made independent maps from the first and second halves of these data. The payload travels several hundred kilometres, and the Sun moves 2° on the sky, between these maps. Comparing them provides a stringent test for contamination from sidelobe pickup and thermal effects. The angular power spectrum calculated for the difference map is shown in Fig. 2. The reduced χ^2 of this power spectrum with respect to zero signal is 1.11 (12 degrees of freedom), indicating that the difference map is consistent with zero contamination.

范围内格点形状、大小以及目前波束尺寸（1 角分）的不确定性所带来的影响很小而且能够利用模型准确进行描述。

图 2 及表 1 中展示了用这种方法得到的角功率谱结果。该功率谱在 $l_{peak} \approx 200$ 处存在一个主峰，这与冷暗物质暴胀模型的预言相符。这些模型还预言了次峰的存在。高 l 的数据的幅度受到了限制，但并不排除次峰存在的可能性。角功率谱中低的多级区域（$l \leqslant 350$）的误差主要来自宇宙/样本方差，而高的多级区域的误差则主要来自探测器噪声。

表 1. CMB 各向异性的角功率谱

l 范围	150 GHz （[前一半]+[后一半]）/2	150 GHz （[前一半]−[后一半]）/2
26~75	1,140±280	63±32
76~125	3,110±490	16±20
126~175	4,160±540	17±28
176~225	4,700±540	59±44
226~275	4,300±460	68±59
276~325	2,640±310	130±82
326~375	1,550±220	−7±92
376~425	1,310±220	−60±120
426~475	1,360±250	0±160
476~525	1,440±290	220±230
526~575	1,750±370	130±300
576~625	1,540±430	−430±360

这里显示的是 150 GHz 处测量得到的宇宙微波背景的角功率谱以及系统误差的测试结果。这里显示的值对应 $\Delta T_l^2 = l(l+1) c_l/2\pi$，单位为 μK^2。这里 $c_l = \langle a_{l_m}^2 \rangle$，$a_{l_m}$ 为 CMB 温度场的球谐分解系数：$\Delta T(\theta, \phi) = \Sigma a_{l_m} Y_{l_m}(\theta, \phi)$。所展示的 1σ 误差包含统计误差以及宇宙/样本方差，但没有包含 10% 的校准不确定性。

图 2 中显示的 CMB 角功率谱是基于 4.1 天的观测数据分析所得。为了测试结果的稳定性，我们将所有数据按时间分为两半，分别独立成图。在这两套图像之间，观测设备移动了数百千米，而太阳在天空中移动了 2°。通过比较这些图像可以对旁瓣污染以及热效应等污染因素进行严格的测试。图 2 还展示了两套图像残差结果的功率谱。该功率谱相对于零信号的约化 χ^2 为 1.11（12 个自由度），这个结果表明残差结果与零污染结果相吻合。

A Peak at *l*≈200 Implies a Flat Universe

The location of the first peak in the angular power spectrum of the CMB is well measured by the Boomerang data set. From a parabolic fit to the data at $l = 50$ to 300 in the angular power spectrum, we find $l_{peak} = (197 \pm 6)$ (1σ error). The parabolic fit does not bias the determination of the peak multipole: applying this method to Monte Carlo realizations of theoretical power spectra we recover the correct peak location for a variety of cosmological models. Finally, the peak location is independent of the details of the data calibration, which obviously affect only the height of the peak and not its location. The height of the peak is $\Delta T_{200} = (69 \pm 4) \pm 7$ μK (1σ statistical and calibration errors, respectively).

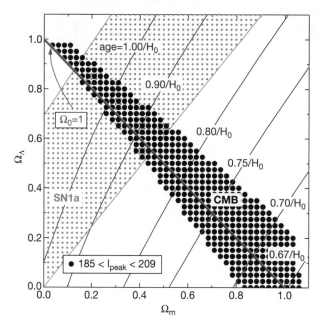

Fig. 3. Observational constraints on Ω_m and Ω_Λ. All the cosmological models (from our data base) consistent with the position of the peak in the angular power spectrum measured by Boomerang (95% confidence intervals) define an "allowed" region in the Ω_m–Ω_Λ plane (marked by large black dots). Such a region is elongated around the $\Omega_0 = 1$ line identifying a flat geometry, euclidean Universe. The blue lines define the age of the Universe for the considered models. The green-dotted region is consistent (95% confidence contour) with the recent results of the high-redshift supernovae surveys[40,41].

The data are inconsistent with current models based on topological defects (see, for example, ref. 35) but are consistent with a subset of cold dark matter models. We generated a database of cold dark matter models[36,37], varying six cosmological parameters (the range of variation is given in parentheses): the non-relativistic matter density, Ω_m (0.05–2); the cosmological constant, Ω_Λ (0–1); the Hubble constant, h (0.5–0.8); the baryon density, $h^2\Omega_b$ (0.013–0.025); the primordial scalar spectral index, n_s (0.8–1.3); and the overall normalization A (free parameter) of the primordial density fluctuation power spectrum. We compared these models with the power spectrum we report here to place constraints on allowed regions in

$l \approx 200$ 的峰意味着一个平直宇宙

通过 Boomerang 数据，我们很好地测量到 CMB 角功率谱第一个峰的位置。通过对角功率谱 $l = 50 \sim 300$ 范围内的数据进行抛物线拟合，我们发现 $l_{peak} = (197 \pm 6)$（1σ 误差范围）。这种抛物线拟合不会对多级峰值的测量带来偏差：将这种方法应用于理论功率谱的蒙特卡洛实现，我们恢复出一系列宇宙学模型正确的峰值位置。最后可知，峰值位置不受数据校准细节的影响，因为数据校准细节仅仅影响峰的高度而并不影响它的位置。峰的高度为 $\Delta T_{200} = (69 \pm 4) \pm 7\ \mu K$（分别对应 1σ 的统计误差和校准误差）。

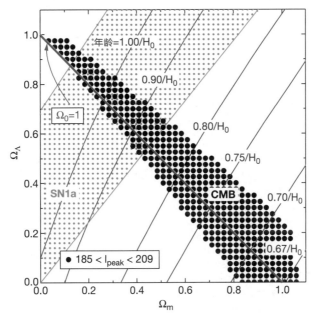

图 3. 对 Ω_m 和 Ω_Λ 的观测限制。所有与 Boomerang 测量得到的角功率谱的峰位置一致（95% 置信区间）的宇宙学模型（来自我们模型库）在 Ω_m-Ω_Λ 平面定义了一个"允许"区域（用大的黑色点进行标记）。这样的一个区域沿 $\Omega_0 = 1$ 的线进行延伸，这意味着宇宙是平直的，符合欧几里得几何。蓝色的线表征了所考虑的模型的宇宙年龄。绿色点的区域在 95% 置信度上符合最近高红移超新星巡天观测结果[40,41]。

这些数据与目前基于拓扑缺陷所建立的模型（比如参考文献 35）预言不吻合，但同冷暗物质模型的一个子集的理论预言相吻合。我们通过改变六个宇宙学参数构建了冷暗物质模型的模型库[36,37]（括号中显示参数的变化范围）：非相对论性的物质密度 Ω_m（0.05 ~ 2）、宇宙学常数 Ω_Λ（0 ~ 1）、哈勃常数 h（0.5 ~ 0.8）、重子密度 $h^2\Omega_b$（0.013 ~ 0.025）、原初标量谱指数 n_s（0.8 ~ 1.3）以及原初密度扰动功率谱的归一化因子 A（自由参数）。通过将这些模型同我们得到的功率谱进行比较，我们可以在这个 6 参数空间中对所允许的区域进行限制。图 3 中我们用大的黑色圆点标记了 Ω_m-Ω_Λ

this 6-parameter space. In Figure 3 we mark with large black dots the region of the Ω_m–Ω_Λ plane where some combination of the remaining four parameters within the ranges defined by our model space gives a power spectrum consistent with our 95% confidence interval for l_{peak}. This region is quite narrow, and elongated along the "flat Universe" line $\Omega_m + \Omega_\Lambda = 1$. The width of this region is determined by degeneracy in the models, which produce closely similar spectra for different values of the parameters[38]. We further evaluated the likelihood of the models given the Boomerang measurement and the same priors (constraints on the values of the cosmological parameters) as in ref. 16. Marginalizing over all the other parameters, we found the following 95% confidence interval for $\Omega_0 = \Omega_m + \Omega_\Lambda$: $0.88 < \Omega_0 < 1.12$. This provides evidence for a euclidean geometry of the Universe. Our data clearly show the presence of power beyond the peak at $l = 197$, corresponding to smaller-scale structures. The consequences of this fact will be fully analysed elsewhere.

(**404**, 955-959; 2000)

P. de Bernardis[1], P. A. R. Ade[2], J. J. Bock[3], J. R. Bond[4], J. Borrill[5,12], A. Boscaleri[6], K. Coble[7], B. P. Crill[8], G. De Gasperis[9], P. C. Farese[7], P. G. Ferreira[10], K. Ganga[8,11], M. Giacometti[1], E. Hivon[8], V. V. Hristov[8], A. Iacoangeli[1], A. H. Jaffe[12], A. E. Lange[8], L. Martinis[13], S. Masi[1], P. V. Mason[8], P. D. Mauskopf[14,15], A. Melchiorri[1], L. Miglio[16], T. Montroy[7], C. B. Netterfield[16], E. Pascale[6], F. Piacentini[1], D. Pogosyan[4], S. Prunet[4], S. Rao[17], G. Romeo[17], J. E. Ruhl[7], F. Scaramuzzi[13], D. Sforna[1] & N. Vittorio[9]

[1] Dipartimento di Fisica, Universita' di Roma "La Sapienza", P.le A. Moro 2, 00185 Roma, Italy

[2] Department of Physics, Queen Mary and Westfield College, Mile End Road, London E1 4NS, UK

[3] Jet Propulsion Laboratory, Pasadena, California 91109, USA

[4] CITA University of Toronto, Toronto M5S 3H8, Canada

[5] NERSC-LBNL, Berkeley, California 94720, USA

[6] IROE–CNR, Via Panciatichi 64, 50127 Firenze, Italy

[7] Department of Physics, University of California at Santa Barbara, Santa Barbara, California 93106, USA

[8] California Institute of Technology, Mail Code 59-33, Pasadena, California 91125, USA

[9] Dipartimento di Fisica, Universita' di Roma Tor Vergata, Via della Ricerca Scientifica 1, 00133 Roma, Italy

[10] Astrophysics, University of Oxford, Keble Road, OX1 3RH, UK

[11] PCC, College de France, 11 pl. Marcelin Berthelot, 75231 Paris Cedex 05, France

[12] Center for Particle Astrophysics, University of California at Berkeley, 301 Le Conte Hall, Berkeley, California 94720, USA

[13] ENEA Centro Ricerche di Frascati, Via E. Fermi 45, 00044 Frascati, Italy

[14] Physics and Astronomy Department, Cardiff University, Cardiff CF2 3YB, UK

[15] Department of Physics and Astronomy, University of Massachusetts, Amherst, Massachusetts 01003, USA

[16] Department of Physics and Astronomy, University of Toronto, Toronto M5S 3H8, Canada

[17] Istituto Nazionale di Geofisica, Via di Vigna Murata 605, 00143, Roma, Italy

Received 24 March; accepted 3 April 2000.

References:

1. Sachs, R. K. & Wolfe, A. M. Perturbations of a cosmological model and angular variations of the microwave background. *Astrophys. J.* **147**, 73-90 (1967).

2. Weinberg S., *Gravitation and Cosmology* (Wiley & Sons, New York, 1972).

3. Hu, W., Sugiyama, N. & Silk, J. The physics of cosmic microwave background anisotropies. *Nature* **386**, 37-43 (1997).

4. Bond, J. R., Efstathiou, G. & Tegmark, M. Forecasting cosmic parameter errors from microwave background anisotropy experiments. *Mon. Not. R. Astron. Soc.* **291**, L33-L41 (1997).

5. Hinshaw, G. *et al.* Band power spectra in the COBE-DMR four-year anisotropy map. *Astrophys. J.* **464**, L17-L20 (1996).

6. Scott, P. F. *et al.* Measurement of structure in the cosmic background radiation with the Cambridge cosmic anisotropy telescope. *Astrophys. J.* **461**, L1-L4 (1996).

平面上"允许"的区域；该区域中，在我们参数空间定义的范围内，通过对剩余四个参数进行组合，所得的功率谱在 95% 的置信区间内与得到的 l_{peak} 结果相吻合。这个区域相当窄，而且沿着"平直宇宙"线 $\Omega_m + \Omega_\Lambda = 1$ 进行延伸。这个区域的宽度由模型的简并性决定，当这些参数取不同的数值时可以得到非常相似的光谱[38]。我们进一步基于 Boomerang 的测量结果以及同参考文献 16 中相同的先验信息（对宇宙学参数值的限制）评估了不同模型的似然性。通过对所有其他参数进行边缘化分析，我们得到了对 $\Omega_0 = \Omega_m + \Omega_\Lambda$ 的欧几里得几何学在 95% 置信区间内的限制结果：$0.88 < \Omega_0 < 1.12$。这为宇宙模型的欧几里得几何学提供了依据。我们的结果也清楚地显示出在峰值 $l = 197$ 位置之外存在功率，这对应着更小尺度的结构。我们将在别处对这一现象展开详细的分析。

（刘项琨 翻译；李然 审稿）

7. Netterfield, C. B. *et al.* A measurement of the angular power spectrum of the anisotropy in the cosmic microwave background. *Astrophys. J.* **474**, 47-66 (1997).

8. Leitch, E. M. *et al.* A measurement of anisotropy in the cosmic microwave background on 7-22 arcminute scales. *Astrophys. J.* (submitted); also as preprint astro-ph/9807312 at ⟨http://xxx.lanl.gov⟩ (1998).

9. Wilson, G. W. *et al.* New CMB power spectrum constraints from MSAMI. *Astrophys. J.* (submitted); also as preprint astro-ph/9902047 at ⟨http://xxx.lanl.gov⟩ (1999).

10. Baker, J. C. *et al.* Detection of cosmic microwave background structure in a second field with the cosmic anisotropy telescope. *Mon. Not. R. Astron. Soc.* (submitted); also as preprint astro-ph/9904415 at ⟨http://xxx.lanl.gov⟩ (1999).

11. Peterson, J. B. *et al.* First results from Viper: detection of small-scale anisotropy at 40 GHZ. Preprint astro-ph/9910503 at ⟨http://xxx.lanl.gov⟩ (1999).

12. Coble, K. *et al.* Anisotropy in the cosmic microwave background at degree angular scales: Python V results. *Astrophys. J.* **519**, L5-L8 (1999).

13. Torbet, E. *et al.* A measurement of the angular power spectrum of the microwave background made from the high Chilean Andes. *Astrophys. J.* **521**, 79-82 (1999).

14. Miller, A. D. *et al.* A measurement of the angular power spectrum of the CMB from $l = 100$ to 400. *Astrophys. J.* (submitted); also as preprint astro-ph/9906421 at ⟨http://xxx.lanl.gov⟩ (1999).

15. Mauskopf, P. *et al.* Measurement of a peak in the CMB power spectrum from the test flight of BOOMERanG. *Astrophys. J.* (submitted); also as preprint astro-ph/9911444 at ⟨http://xxx.lanl. gov⟩ (1999).

16. Melchiorri, A. *et al.* A measurement of Ω from the North American test flight of BOOMERanG. *Astrophys. J.* (submitted); also as preprint astro-ph/9911445 at ⟨http://xxx.lanl.gov⟩ (1999).

17. de Bernardis, P. *et al.* Mapping the CMB sky: the BOOMERanG experiment. *New Astron. Rev.* **43**, 289-296 (1999).

18. Mauskopf, P. *et al.* Composite infrared bolometers with Si_3N_4 micromesh absorbers. *Appl. Opt.* **36**, 765-771 (1997).

19. Bock, J. *et al.* Silicon nitride micromesh bolometer arrays for SPIRE. *Proc. SPIE* **3357**; 297-304 (1998).

20. Masi, S. *et al.* A self contained ^{3}He refrigerator suitable for long duration balloon experiments. *Cryogenics* **38**, 319-324 (1998).

21. Masi, S. *et al.* A long duration cryostat suitable for balloon borne photometry. *Cryogenics* **39**, 217-224 (1999).

22. Schlegel, D. J., Finkbeiner, D. P. & Davis, M. Maps of dust IR emission for use in estimation of reddening and CMBR foregrounds. *Astrophys. J.* **500**, 525-553 (1998).

23. Delabrouille, J., Gorski, K. M. & Hivon, E. Circular scans for CMB anisotropy observation and analysis. *Mon. Not. R. Astron. Soc.* **298**, 445-450 (1998).

24. Cheung, L. H. *et al.* 1.0 millimeter maps and radial density distributions of southern HII/molecular cloud complexes. *Astrophys. J.* **240**, 74-83 (1980).

25. Kogut, A. *et al.* Dipole anisotropy in the COBE DMR first-year sky maps. *Astrophys. J.* **419**, 1-6 (1993).

26. Tegmark, M. CMB mapping experiments: a designer's guide. *Phys. Rev. D* **56**, 4514-4529 (1997).

27. Bond, J. R., Crittenden, R., Jaffe, A. H. & Knox, L. E. Computing challenges of the cosmic microwave background. *Comput. Sci. Eng.* **21**, 1-21 (1999).

28. Borrill, J. in *Proc. 3K Cosmology EC-TMR Conf.* (eds Langlois, D., Ansari, R. & Vittorio, N.) 277 (American Institute of Physics Conf. Proc. Vol. 476, Woodbury, New York, 1999).

29. Prunet, S. *et al.* in *Proc. Conf. Energy Density in the Universe* (eds Langlois, D., Ansari, R. & Bartlett, J.) (Editiones Frontieres, Paris, 2000).

30. Gorski, K. M., Hivon, E. & Wandelt, B. D. in *Proc. MPA/ESO Conf.* (eds Banday, A. J., Sheth, R. K. & Da Costa, L.) (European Southern Observatory, Garching); see also ⟨http://www.tac.dk/~healpix/⟩.

31. Kogut, A. in *Microwave Foregrounds* (eds de Oliveira Costa, A. & Tegmark, M.) 91-99 (Astron. Soc. Pacif. Conf. Series. Vol 181, San Francisco, 1999).

32. Toffolatti, L. *et al.* Extragalactic source counts and contributions to the anisotropies of the CMB. *Mon. Not. R. Astron. Soc.* **297**, 117-127 (1998).

33. Wright, A. E. *et al.* The Parkes-MIT-NRAO (PMN) surveys II. Source catalog for the southern survey. *Astrophys. J. Supp. Ser.* **91**, 111-308 (1994); see also ⟨http://astron.berkeley.edu/wombat/foregrounds/radio.html⟩.

34. Borrill, J. in *Proc. 5th European SGI/Cray MPP Workshop* (CINECA, Bologna, 1999); Preprint astro-ph/9911389 at ⟨xxx.lanl.gov⟩ (1999); see also ⟨http://cfpa.berkeley.edu/~borrill/cmb/madcap.html⟩.

35. Durrer, R., Kunz, M. & Melchiorri, A. *Phys. Rev. D* **59**, 1-26 (1999).

36. Seljak, U. & Zaldarriaga, M. A line of sight approach to cosmic microwave background anisotropies. *Astrophys. J.* **437**, 469-477 (1996).

37. Lewis, A., Challinor, A. & Lasenby, A. Efficient computation of CMB anisotropies in closed FRW models. Preprint astro-ph/9911177 at ⟨http://xxx.lanl.gov⟩ (1999).

38. Efstathiou, G. & Bond, R. Cosmic confusion: degeneracies among cosmological parameters derived from measurements of microwave background anisotropies. *Mon. Not. R. Astron. Soc.* **304**, 75-97 (1998).

39. Wright, E. *et al.* Comments on the statistical analysis of excess variance in the COBE-DMR maps. *Astrophys. J.* **420**, 1-8 (1994).

40. Perlmutter, S. *et al.* Measurements of and Λ from 42 high-redshift supernovae. *Astrophys. J.* **517**, 565-586 (1999).

41. Schmidt, B. P. *et al.* The high-Z supernova search: measuring cosmic deceleration and global curvature of the Universe using type Ia supernovae. *Astrophys. J.* **507**, 46-63 (1998).

Acknowledgements. The Boomerang experiment was supported by Programma Nazionale di Ricerche in Antartide, Universita' di Roma "La Sapienza", and Agenzia Spaziale Italiana in Italy, by the NSF and NASA in the USA, and by PPARC in the UK. We thank the staff of the National Scientific Ballooning Facility, and the United States Antarctic Program personnel in McMurdo for their preflight support and an effective LDB flight. DOE/NERSC provided the supercomputing facilities.

Correspondence and requests for materials should be addressed to P. d. B. (e-mail: debernardis@roma1.infn.it). Details of the experiment and numerical data sets are available at the web sites ⟨http://oberon.roma1.infn.it/boomerang⟩ and ⟨http://www.physics.ucsb.edu/~boomerang⟩.

456

The Accelerations of Stars Orbiting the Milky Way's Central Black Hole

A. M. Ghez *et al.*

Editor's Note

It had been widely believed for several years that black holes with masses of a million or more solar masses reside at the centres of most galaxies. But proving that notion was challenging. Here Andrea Ghez and colleagues track the motions of stars very close to the centre of the Milky Way, seeing them move in their orbits. Subsequent work has tracked these stars through their entire orbits; one travels around the Galactic Centre at the tremendous speed of over 5,000 km/s. These studies strongly suggest that the stars are gravitationally influenced by an object such as a supermassive black hole; the only (increasingly unlikely) alternative is a massive star made of some kind of exotic matter.

Recent measurements[1-4] of the velocities of stars near the centre of the Milky Way have provided the strongest evidence for the presence of a supermassive black hole in a galaxy[5], but the observational uncertainties poorly constrain many of the black hole's properties. Determining the accelerations of stars in their orbits around the centre provides much more precise information about the position and mass of the black hole. Here we report measurements of the accelerations of three stars located ~0.005 pc (projected on the sky) from the central radio source Sagittarius A* (Sgr A*); these accelerations are comparable to those experienced by the Earth as it orbits the Sun. These data increase the inferred minimum mass density in the central region of the Galaxy by an order of magnitude relative to previous results, and localize the dark mass to within 0.05 ± 0.04 arcsec of the nominal position of Sgr A*. In addition, the orbital period of one of the observed stars could be as short as 15 years, allowing us the opportunity in the near future to observe an entire period.

IN 1995, we initiated a programme of high-resolution 2.2-μm (K band) imaging of the inner 5 arcsec × 5 arcsec of the Galaxy's central stellar cluster with the W. M. Keck 10-m telescope on Mauna Kea, Hawaii (1 arcsec = 0.04 pc at the distance to the Galactic Centre, 8 kpc, ref. 6). From each observation, several thousand short exposure ($t_{exp} = 0.137$ s) frames were collected, using the facility near-infrared camera, NIRC[7,8], and combined to produce a final diffraction-limited image having an angular resolution of 0.05 arcsec. Between 1995 and 1997, images were obtained once a year with the aim of detecting the stars' velocities in the plane of the sky. The results from these measurements are detailed in ref. 4; in summary, two-dimensional velocities were measured for 90 stars with simple

458

环绕银河系中心黑洞运动的恒星的加速度

盖兹等

编者按

近年来，人们普遍相信在大部分的星系中心都存在质量超过百万太阳质量的黑洞。然而要想证明这一观点却是极具挑战性的。在本文中，安德烈娅·盖兹和她的同事们追踪了那些非常接近银河系中心的恒星的运动，以观测这些恒星的运动轨迹。其中有一颗围绕银河系中心运动的恒星，其速度竟可高达 5,000 km/s。这些研究结果表明银河系中心很有可能存在一个超大质量的黑洞，而这些恒星正是在这个超大质量黑洞的引力影响下围绕银河系中心运动。除此之外，唯一的可能性（越来越不可能）就是银河系中心存在一个由某种奇异物质组成的大质量恒星。

近年来对银河系中心附近恒星速度的测量[1-4]为星系中心存在超大质量黑洞提供了最有力的证据[5]，但是观测误差使得我们很难对黑洞的许多性质给出强有力的限制。确定恒星环绕中心轨道的加速度，将向我们提供更精确的黑洞位置和质量信息。在本文中，我们报道了距离中心射电源人马座 A*（简称 Sgr A*）约 0.005 秒差距（pc）（投影到天球上）处三颗恒星加速度的测量结果；这些加速度和地球环绕太阳经历的加速度相似。相对以前的结果，从这些数据推导出的银河系中心的最小质量密度增加了一个量级，并将暗质量区域限定在 Sgr A* 标定位置的 0.05 ± 0.04 角秒以内。此外，其中一颗恒星的轨道周期可能只有短短的 15 年，这让我们在未来有机会观测它的整个运动轨道周期。

我们在 1995 年启动了一个项目，利用夏威夷冒纳凯阿山的 10 米凯克望远镜在 2.2 μm 波段（K 波段）对银河系中心星团内部 5 角秒×5 角秒的区域做高分辨率的成图观测（距离银河系中心 8 kpc 处，1 角秒 = 0.04 pc，参考文献 6）。对于每次观测，我们使用近红外照相机设备 NIRC[7,8]收集了几千幅短曝光（$t_{exp} = 0.137$ s）图像，并将这些图像合在一起最终得到一幅角分辨率为 0.05 角秒的衍射极限图像。为了测量恒星在天球切面上的速度，我们从 1995 年到 1997 年每年成图一次。这些测量的具体结果罗列在参考文献 4 中；总的来说，我们对 90 颗恒星的位置与时间关系进行简单的线性拟合得到它们的二维速度。这些速度，有的可以高达 1,400 km · s⁻¹，

linear fits to the positions as a function of time. These velocities, which reach up to 1,400 km s^{-1}, implied the existence of a $2.6 \times 10^6 \, M_\odot$ black hole coincident (± 0.1 arcsec) with the nominal location of Sgr A* (ref. 9), the unusual radio source[10-12] long believed to be the counterpart of the putative black hole.

The new observations presented here were obtained several times a year from 1997 to 1999 to improve the sensitivity to accelerations. With nine independent measurements, we now fit the positions of stars as a function of time with second order polynomials (see Fig. 1). The resulting velocity uncertainties are reduced by a factor of 3 compared to our earlier work, primarily as a result of the increased time baseline, and, in the central square arcsecond, by a factor of 6 compared to that presented in ref. 3, owing to our higher angular resolution. Among the 90 stars in our original proper motion sample[4], we have now detected significant accelerations for three stars, S0-1, S0-2 and S0-4 (see Table 1); specifically, these are the sources for which the reduced chi-squared value for a quadratic fit is smaller than that for the linear model of their motions by more than 1. These three stars are independently distinguished in our sample, being among the fastest moving stars ($v = 570$ to $1,350$ km s^{-1}) and among the closest to the nominal position of Sgr A* ($r_{1995} = 0.004–0.013$ pc). With accelerations of 2–5 milliarcsec yr^{-2}, or equivalently $(3–6) \times 10^{-6}$ km s^{-2}, they are experiencing accelerations similar to the Earth in its orbit about the Sun.

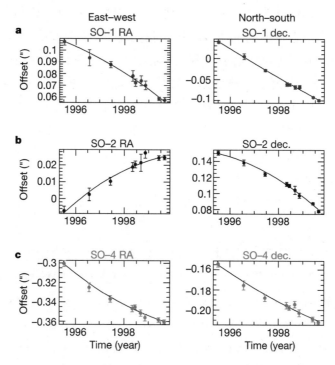

Fig. 1. East–west and north–south positional offsets from the nominal location of Sgr A* versus time for (**a**) S0-1, (**b**) S0-2, and (**c**) S0-4. The offset range shown is scaled to the points in each plot and therefore varies from ~−0.4″ to ~0.15″. Each of these stars is located with a precision of about 1–5 mas in the individual maps and the alignment of these positions, which is carried out by minimizing

意味着在 Sgr A* 的标定位置(参考文献 9)处(±0.1 角秒)很可能存在一个质量为 $2.6 \times 10^6 \, M_\odot$ 的黑洞。而 Sgr A* 这个不寻常的射电源[10-12]一直以来被认为是假定黑洞的对应体。

这里所采用的新数据是从 1997 年到 1999 年每年观测好几次得到的,为的是提高测量加速度的灵敏度。利用 9 次独立的测量结果,我们现在用二阶多项式来拟合恒星的位置–时间关系(见图 1)。主要因为时间基线变长,得到的速度误差减少为我们早期工作的 1/3;又因为角分辨率的提高,在图像中央的平方角秒区域内,速度误差减少为参考文献 3 的 1/6。在原初样本中有自行信息的 90 颗恒星中[4],我们现在已经探测到 S0-1、S0-2 和 S0-4 这三颗恒星的显著加速度(见表 1);特别是,用二次多项式拟合这些源的运动所得的约化 χ^2 值比线性拟合的约化 χ^2 值要少 1 以上。这三颗星作为运动最快($v = 570 \sim 1,350 \, km \cdot s^{-1}$)且最接近 Sgr A* 标定位置的恒星($r_{1995} = 0.004 \sim 0.013$ pc),在我们的样本中被单独区分出来。它们的加速度为 $2 \sim 5$ milliarcsec $\cdot$ yr^{-2},或等效于 $(3 \sim 6) \times 10^{-6} \, km \cdot s^{-2}$,正在经历与地球环绕太阳类似的加速运动过程。

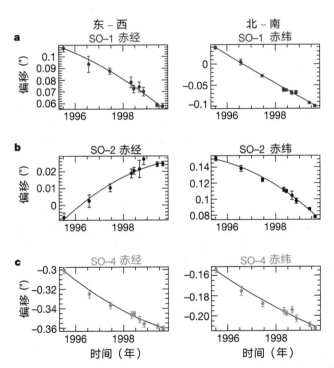

图 1. 恒星(**a**)S0-1、(**b**)S0-2 和(**c**)S0-4 相对于 Sgr A* 标定位置在东–西、北–南方向上的偏移随时间的变化。每幅图中偏移范围都拟合在数据点上,其变化范围约在 −0.4″ 到 0.15″ 之间。这三颗恒星在各自图中的位置精度为 1～5 毫角秒,按参考文献 4 描述的方法,我们通过将所有恒星的净位移最小化来将位置对准,半抽样自举重采样方法[22]带来大约 3 毫角秒的对准误差(对准误差在图中间最小,随着半径

the net displacement of all stars as described in ref. 4, has an uncertainty of about 3 mas based on the half sample bootstrap resampling method[22] (the alignment uncertainty is at a minimum at the centre of the map and grows linearly with radius). These two uncertainty terms are added in quadrature; the results are used in the fitting process and depicted as errorbars here. In each plot, the solid line shows the second order polynomial used to derive the acceleration term. These plots demonstrate that we are able to measure, for the first time, accelerations of 2 to 5 mas yr^{-2} (0.3–0.6 cm s^{-2}) for stars orbiting a supermassive black hole.

Table 1. Measurements for stars with significant accelerations

	S0-1 (S1)	S0-2 (S2)	S0-4 (S8)
Radius from Sgr A*-magnitude (milliparsecs)	4.42 ± 0.05	5.83 ± 0.04	13.15 ± 0.04
Radius from Sgr A*-position angle (degrees)	290.1 ± 0.7	3.1 ± 0.4	117.3 ± 0.2
Velocity-magnitude (km s^{-1})	$1,350 \pm 40$	570 ± 20	990 ± 30
Velocity-position angle (degrees)	168 ± 2	241 ± 2	129 ± 2
Acceleration-magnitude (milliarcsec yr^{-2})	2.4 ± 0.7	5.4 ± 0.3	3.2 ± 0.5
Acceleration-position angle (degrees)	80 ± 15	154 ± 4	294 ± 7

The primary nomenclature adopted here and in the text is from ref. 4; in parentheses star names from ref. 1 are also given. All quantities listed are derived from the second-order polynomial fit to the data and are given for epoch 1995.53. All position angles are measured east of north. All uncertainties are 1σ and are determined by the jackknife resampling method[22]. The radius uncertainties listed include only the relative positional uncertainties and not the uncertainty in the origin used (the nominal position of Sgr A*; see text for offset to measured dynamical centre).

Acceleration vectors, in principle, are more precise tools than the velocity vectors for studying the central mass distribution. Even projected onto the plane of the sky, each acceleration vector should be oriented in the direction of the central mass, assuming a spherically symmetric potential. Thus, the intersection of multiple acceleration vectors is the location of the dark mass. Figure 2 shows the acceleration vector's direction for the three stars. With in 1σ, these vectors do indeed overlap, and furthermore, the intersection point lies a mere 0.05 ± 0.03 arcsec east and 0.02 ± 0.03 arcsec south of the nominal position of Sgr A*, consistent with the identification of Sgr A* as the carrier of the mass. Previously, with statistical treatments of velocities only, the dynamical centre was located to within ± 0.1 arcsec (1σ) of Sgr A*'s position[4]. This velocity-based measurement is unaffected by the increased time baseline, as its uncertainty is dominated by the limited number of stars at a given radius. The accelerations thus improve the localization of our Galaxy's dynamical centre by a factor of 3, which is essential for reliably associating any near-infrared source with the black hole given the complexity of the region.

的增加而线性增大）。对这两个误差项取平方和开根号；得到拟合时所用的总误差，并在图中用误差棒描述。每幅图中，实线表示用二阶多项式拟合的最佳结果，用来推导恒星的加速度。这些结果显示，对环绕超大质量黑洞运动的恒星，我们首次能够测得它们量级在 $2 \sim 5$ mas·yr^{-2}($0.3 \sim 0.6$ cm·s^{-2}) 的运动加速度。

表 1. 有明显加速度的恒星的测量结果

	S0-1(S1)	S0-2(S2)	S0-4(S8)
相对 Sgr A* 的位置－大小 (milliparsec)	4.42 ± 0.05	5.83 ± 0.04	13.15 ± 0.04
相对 Sgr A* 的位置－位置角 (度)	290.1 ± 0.7	3.1 ± 0.4	117.3 ± 0.2
速度－大小 (km·s^{-1})	$1,350 \pm 40$	570 ± 20	990 ± 30
速度－位置角 (度)	168 ± 2	241 ± 2	129 ± 2
加速度－大小 (milliarcsec·yr^{-2})	2.4 ± 0.7	5.4 ± 0.3	3.2 ± 0.5
加速度－位置角 (度)	80 ± 15	154 ± 4	294 ± 7

这里和正文中采用的主要命名方式来自参考文献 4；参考文献 1 所给出的恒星名字也在圆括号里给出。所有列出量均来源于二阶多项式拟合所得数据，对应历元为 1995.53。所有方位角由北向东起量。误差都是 1σ，由刀切法确定[22]。列出的半径误差仅包括相对位置误差，没有包括中心的位置误差(Sgr A* 的标定位置；关于测量到的动力学中心与标定位置的偏移的具体细节请见正文)。

原则上来说，比起速度矢量，利用加速度矢量可以更精确地研究中心质量分布。假设一个球对称的引力势，就算投影到天球切面上，每个加速度矢量还将指向中心质量的方向。因此，多个加速度矢量的交叉点就是暗质量所在的位置。图 2 显示了这三颗恒星的加速度矢量方向。这些矢量在 1σ 误差范围内确实发生重叠，而且交叉点正好位于 Sgr A* 标定位置往东 0.05 ± 0.03 角秒、往南 0.02 ± 0.03 角秒处，与暗质量位于 Sgr A* 的结论一致。之前，在仅有对速度的相关统计分析的情况下，人们大致将动力学中心确定在位于 Sgr A* 位置 ± 0.1 角秒(1σ)的范围内[4]。这种基于速度的测量并不依赖时间基线的长短，因为它的误差主要来自给定半径内恒星数目的限制。因此，对恒星进行加速度矢量分析将银河系动力学中心位置的定位精度提高为原来的 3 倍。考虑到这个区域的复杂度，这点对于研究近红外源和黑洞的可靠成协关系来说很重要。

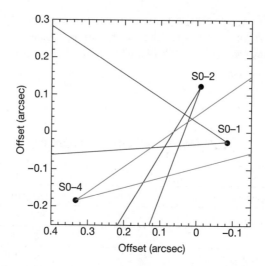

Fig. 2. The acceleration uncertainty cones and their intersections. The cones' edges represent the directions for which the accelerations deviate by 1σ from their best-fit values and their vertices are the time-averaged positions, measured relative to the nominal position of Sgr A*, rather than the positions listed in Table 1. If the accelerations are caused by a single supermassive black hole, or even a spherically symmetric mass distribution, these vectors should intersect at a common location, the centre of the mass. The 1σ intersection point lies 0.05 ± 0.03 arcsec east and 0.02 ± 0.03 arcsec south of the nominal position of Sgr A*. The existence of an intersection point suggests that there is indeed a common origin for the acceleration and pinpoints the location of the black hole to within 0.03 arcsec.

Like the directions, the magnitudes of the acceleration vectors also constrain the central black hole's properties. In three dimensions, the acceleration and radius vectors (a_{3D} and r_{3D}, respectively) provide a direct measure of the enclosed mass simply by $a_{3D} = G \times M/r_{3D}^2$. For a central potential, the acceleration and radius vectors are co-aligned and, with a projection angle to the plane of the sky θ, the two-dimensional projections place the following lower limit on the central mass: $M\cos^3(\theta) = a_{2D} \times r_{2D}^2/G$. This analysis is independent of the star's orbital parameters, although θ is in fact a lower limit for the orbital inclination angle. Figure 3 shows the minimum mass implied by each star's two-dimensional acceleration as a function of projected radius, with dashed curves displaying how the implied mass and radius grow with projection angle. For each point, the uncertainty in the position of the dynamical centre dominates the minimum mass uncertainties. Also plotted are the results from the statistical analysis of the velocity vectors measured in the plane of the sky, which imply a dark mass of $(2.3–3.3) \times 10^6\ M_\odot$ inside a radius of 0.015 pc[3,4]. If, as has been assumed, this dark mass is in the form of a single supermassive black hole, the enclosed mass should remain level at smaller radii. Projection angles of 51–56° and 25–37° for S0-1 and S0-2, respectively, would yield the mass inferred from velocities and place these stars at a mere ~0.008 pc (solid line portions of the limiting mass curves in Fig. 3), thus increasing the dark mass density implied by velocities by an order of magnitude to $8 \times 10^{12}\ M_\odot\ \mathrm{pc}^{-3}$. With smaller projection angles, these two stars also allow for the enclosed mass to decrease at smaller radii, as would occur in the presence of an extended distribution of dark matter surrounding a less massive black hole[13-17]. In contrast to the agreement between the mass distribution inferred from

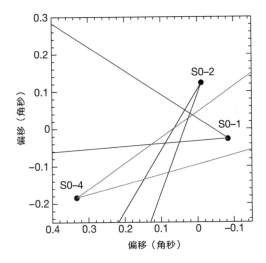

图 2. 加速度误差锥以及它们的交叠区域。锥的边缘代表加速度偏离它们的最佳拟合值 1σ 的方向，顶点则是它们相对 Sgr A* 标定位置的时间平均位置，而不是表 1 中所列举的位置。假如加速运动是由单个超大质量黑洞或者甚至是一个球形对称质量分布所引起的，这些矢量应该会在质量中心处相交。1σ 交叉点位于 Sgr A* 标定位置东边 0.05 ± 0.03 角秒和南边 0.02 ± 0.03 角秒处。交叉点的存在表明加速度确实来自一个共同的来源，并将黑洞位置精确定位在 0.03 角秒内。

　　和方向一样，加速度矢量的大小也能够用来限制中心黑洞的性质。在三维空间中，利用加速度和径向矢量（分别是 a_{3D} 和 r_{3D}），仅通过 $a_{3D} = G \times M/r_{3D}^2$，我们就可以直接测量所考虑半径范围内的质量。在中心引力势作用下，加速度和径向矢量的方向一致，若它们在天球切面上的投影角为 θ，那么二维投影将给出中心质量的下限为：$M\cos^3(\theta) = a_{2D} \times r_{2D}^2/G$。这个分析与恒星的轨道参数无关，尽管 θ 实际上就是轨道倾角的下限。图 3 显示由每颗恒星的二维加速度所推导的中心黑洞的最小质量随恒星与中心的投影距离的变化，其中虚线显示了导出的质量和距离随投影角的变化规律。对图中每个点，这个质量下限的误差主要来自动力学中心的位置误差。同时图中还画上了对投影在天球切面上的速度矢量的统计分析结果，表明在半径 0.015 pc 内存在 $(2.3 \sim 3.3) \times 10^6 \, M_\odot$ 的暗质量[3,4]。如果和假设的一样，暗质量是一个超大质量黑洞，那么这些暗质量应该被限制在一个更小半径之内。S0-1 和 S0-2 的投影角分别为 51° ~ 56° 和 25° ~ 37°，这些投影角使得我们可以从速度推导得质量并且将这些恒星的位置限定在距离中心仅仅约 0.008 pc 以内的区域（图 3 中质量下限曲线的实线部分），因此由速度推导得出的暗质量密度将增加一个量级，达到 $8 \times 10^{12} \, M_\odot \cdot pc^{-3}$。假如这两颗恒星具有更小的投影角，那么这两颗恒星的观测结果也同样允许出现所考虑半径范围内暗质量随着半径减小而减少的可能性，就和在一个质量较小的黑洞周围存在一个延展暗物质分布所造成的情形一样[13-17]。虽然从 S0-1 和 S0-2 的速度和加速度推断得到的质量分布相互吻合，可是从 S0-4 加速度

the velocities and the accelerations for S0-1 and S0-2, the minimum mass implied by S0-4's acceleration is inconsistent at least at the 1σ level. We note, in support of the validity of S0-4's acceleration vector, that its orientation is consistent with the intersection of the S0-1 and S0-2 acceleration vectors (see Fig. 2). Nonetheless, continued monitoring of S0-4 will be important for assessing this possible discrepancy. Overall, the individual magnitudes of the three acceleration vectors support the existence of a central black hole with a mass of approximately $3 \times 10^6 \, M_\odot$.

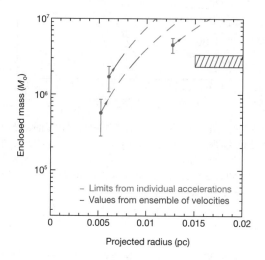

Fig. 3. The minimum enclosed mass implied by each star's acceleration measurement versus projected distance from the newly determined dynamical centre position. If the projection between the plane of the sky and both the radius and the acceleration vectors is θ, then the true mass increases as $1/\cos^3(\theta)$ and the true acceleration and radius vectors increase as $1/\cos(\theta)$. Also plotted is the mass range implied from a statistical analysis of the approximately 100 velocity vectors that have been measured in the plane of the sky[3,4].

Using acceleration measurements, it is now possible to constrain the individual orbits. We checked the conclusions of the previous studies[3,4] by assuming that the stars are bound to a central mass of $(2.2–3.3) \times 10^6 \, M_\odot$ located within 0.03 arcsec of the nominal position of Sgr A*. Excellent orbital fits were found for S0-1 and S0-2 for the entire range of masses and for true foci within 0.01 arcsec of the nominal position of Sgr A*, suggesting that we now have comparable accuracy in determining the infrared location of Sgr A* (ref. 9) and the dynamical centre of our Galaxy (see Fig. 4). The orbital solutions for these stars have eccentricities ranging from 0 (circular) to 0.9 for S0-1 and 0.5 to 0.9 for S0-2 and periods in the range 35–1,200 and 15–550 years, respectively, raising the possibility of seeing a star make a complete journey around the centre of the Galaxy within the foreseeable future. Although the fits are not yet unique, they impose a maximum orbital distance from the black hole, or apoapse, of 0.1 pc. This suggests that S0-1 and S0-2 might have formed locally. If these stars are indeed young[2], then their small apoapse distance presents a challenge to classical star formation theories in light of the strong tidal forces created by the central black hole and might require a collisional or compressional star formation scenario[18,19]. However, dynamical friction may be able to act on a short enough timescale to bring these stars in from a much larger distance[20]. More accurate orbital

466

推出的暗质量下限在 1σ 水平内，和其他两颗星得出的结果不一致。值得注意的是，S0-4 的加速度矢量方向在 S0-1 和 S0-2 的加速度矢量的交叠范围内（见图 2），这表明 S0-4 的加速度矢量是正确的。虽然如此，继续监测 S0-4 对判断这个差异是否存在非常重要。总体来看，三个加速度矢量的大小都支持银河系中心存在一个质量约为 $3 \times 10^6 \ M_\odot$ 的超大质量黑洞。

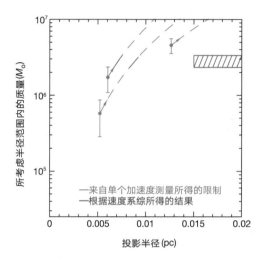

图 3. 根据每颗恒星的加速度测量值推导出的所考虑半径范围内的质量下限同新确定的动力学中心位置的投影距离的关系图。假如径向矢量和加速度矢量在天球切面的投影角均为 θ，那么真实的质量随着 $1/\cos^3(\theta)$ 增加，真实的加速度和径向矢量随着 $1/\cos(\theta)$ 增加。对在天球切面上测量到的接近 100 个速度矢量进行统计分析所得到的质量范围也画在图中[3,4]。

通过测量加速度，我们现在能够分别限制这三颗恒星各自的轨道。通过假设恒星束缚在位于 Sgr A* 标定位置 0.03 角秒内且质量为 $(2.3 \sim 3.3) \times 10^6 \ M_\odot$ 的中心质量周围，我们检查了前人的研究结论[3,4]。假定轨道实焦点在 Sgr A* 标定位置 0.01 角秒以内，对于整个可能的中心质量范围，我们都可以对 S0-1 和 S0-2 的轨道给出很好的拟合结果，这表明我们现在确定的 Sgr A* 红外位置（参考文献 9）和银河系动力学中心的精确度都相当好（见图 4）。在得到的轨道解当中，S0-1 解出的轨道偏心率分布在 0（圆）到 0.9 之间，S0-2 的轨道偏心率分布在 0.5 到 0.9 之间，周期分别分布在 35 ~ 1,200 年和 15 ~ 550 年的范围之内。这意味着在可预见的将来，也许我们能够完整地看到某颗恒星环绕银河系中心的运动轨迹。虽然所得到的拟合结果并不是唯一的，但是它们给出了离黑洞的最大轨道距离（或远质心点）为 0.1 pc。这表明 S0-1 和 S0-2 可能是在中心附近形成。如果这些恒星实际上很年轻[2]，那么这些恒星拥有这么小的远质心点距离将对经典恒星形成理论提出挑战。考虑到中心黑洞产生的强潮汐力，我们可能需要一个在碰撞或压缩条件下的恒星形成理论[18,19]。不过，

parameters are needed to fully address these problems and others such as the distance to the Galactic Centre[21]. The determinations of these parameters will be considerably improved when radial velocities are measured for these stars using adaptive optics techniques, as the current solutions based on the proper motion data alone predict radial velocities ranging from 200 and 2,000 km s^{-1}.

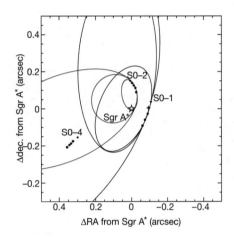

Fig. 4. The measured motion of S0-1, S0-2 and S0-4 and several allowed orbital solutions. Only the measurements obtained in June of each of the 5 years are shown. S0-1 and S0-2 are moving clockwise about Sgr A* and S0-4 is travelling radially outward. In the orbital analysis, two constraints are applied, a central mass of $2.6 \times 10^6 M_\odot$ and a true focus located at the position of Sgr A*. Displayed are orbital solutions with periods of 17, 80 and 505 years for S0-2 and 63, 200 and 966 years for S0-1.

(**407**, 349-351; 2000)

A. M. Ghez, M. Morris, E. E. Becklin, A. Tanner & T. Kremenek
Department of Physics and Astronomy, UCLA, Los Angeles, California 90095-1562, USA

Received 24 May; accepted 19 July 2000.

References:

1. Eckart, A. & Genzel, R. Stellar proper motions in the central 0.1 pc of the Galaxy. *Mon. Not. R. Astron. Soc.* **284**, 576-598 (1997).

2. Genzel, R., Eckart, A., Ott, T. & Eisenhauer, F. On the nature of the dark mass in the centre of the Milky Way. *Mon. Not. R. Astron. Soc.* **291-234**, 219 (1997).

3. Genzel, R., Pichon, C., Eckart, A., Gerhard, O. E. & Ott, T. Stellar dynamics in the galactic centre: Proper motions and anisotropy. *Mon. Not. R. Astron. Soc.* (in the press).

4. Ghez, A. M., Klein, B. L., Morris, M. & Becklin, E. E. High proper-motion stars in the vicinity of Sagittarius A*: Evidence for a supermassive black hole at the center of our galaxy. *Astrophys. J.* **509**, 678-686 (1998).

5. Maoz, E. Dynamical constraints on alternatives to supermassive black holes in galactic nuclei. *Astrophys. J.* **494**, L181-L184 (1998).

6. Reid, M. J. The distance to the center of the Galaxy. *Annu. Rev. Astron. Astrophys.* **31**, 345-372 (1993).

7. Matthews, K. & Soifer, B. T. in *Astronomy with Infrared Arrays: The Next Generation* (ed. McLean, I.) Vol. 190, 239-246 (Astrophysics and Space Sciences Library, Kluwer Academic, Dordrecht, 1994).

8. Matthews, K., Ghez, A. M., Weinberger, A. J. & Neugebauer, G. The diffraction-limited images from the W. M. Keck Telescope. *Proc. Astron. Soc. Pacif.* **108**, 615-619 (1996).

9. Menten, K. M., Reid, M. J., Eckart, A. & Genzel, R. The position of Sgr A*: Accurate alignment of the radio and infrared reference frames at the galactic center. *Astrophys. J.* **475**, L111-L114 (1997).

动力学摩擦也可能在足够短的时标内将这些恒星从更远距离处拉进来[20]。要解决这些问题，以及对其他诸如距银河系中心的距离[21]等问题有更加准确的理解，我们还需要更精确的轨道参数的测量。应用自适应光学技术测量这些恒星的视向速度，将极大地提高这些恒星轨道参数的测量精度，因为目前仅依赖自行数据得到的恒星视向速度只能限制在 200 到 2,000 km·s^{-1} 的广阔范围内。

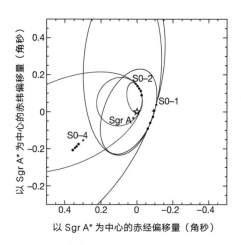

图 4. 测量到的 S0-1、S0-2 和 S0-4 的运动轨迹和几个可能的轨道解。图中只显示 5 年中每年 6 月份的观测。S0-1 和 S0-2 围绕 Sgr A* 顺时针运动，S0-4 沿径向向外运动。在轨道分析中我们应用了中心质量为 2.6×10^6 $M_{\odot}$ 和实焦点位于 Sgr A* 处这两个限制条件。图中显示了 S0-2 的周期分别为 17、80 和 505 年的轨道解，以及 S0-1 的周期分别为 63、200 和 966 年的轨道解。

（肖莉 翻译；周礼勇 审稿）

10. Lo, K. Y., Shen, Z.-Q., Zhao, J.-H. & Ho, P. T. Intrinsic size of SGR A*: 72 Schwarzschild radii. *Astrophys. J.* **508**, L61-L64 (1998).

11. Reid, M. J., Readhead, A. C. S., Vermeulen, R. C. & Treuhaft, R. N. The proper motion of Sagittarius A*. I. First VLBA results. *Astrophys. J.* **524**, 816-823 (1999).

12. Backer, D. C. & Sramek, R. A. Proper motion of the compact, nonthermal radio source in the galactic center, Sagittarius A*. *Astrophys. J.* **524**, 805-815 (1999).

13. Salati, P. & Silk, J. A stellar probe of dark matter annihilation in galactic nuclei. *Astrophys. J.* **338**, 24-31 (1989).

14. Tsiklauri, D. & Viollier, R. D. Dark matter concentration in the galactic center. *Astrophys. J.* **500**, 591-595 (1998).

15. Gondolo, P. & Silk, J. Dark matter annihilation at the galactic center. *Phys. Rev. Lett.* **83**, 1719-1722 (1999).

16. Torres, D. F., Capozziello, S. & Lambiase, G. A supermassive boson star at the galactic center? *Astrophys. J.* (in the press).

17. Romanowsky, A. & Kochanek, C. in *Proceedings of Dynamics of Star Clusters and the Milky Way* (ASP Conference Series, Astronomical Society of the Pacific, San Francisco, in the press).

18. Sanders, R. H. The case against a massive black hole at the Galactic Centre. *Nature* **359**, 131-132 (1992).

19. Morris, M., Ghez, A. M. & Becklin, E. E. The galactic center black hole: clues for the evolution of black holes in galactic nuclei. *Adv. Space Res.* **23**, 959-968 (1999).

20. Gerhard, O. The galactic center He I stars: remains of a dissolved young cluster? *Astrophys. J.* (submitted).

21. Salim, S. & Gould, A. Sagittarius A* "Visual Binaries": A direct measurement of the galactocentric distance. *Astrophys. J.* **523**, 633-641 (1999).

22. Babu, G. J. & Feigelson, E. D. *Astrostatistics* (Chapman & Hall, London, 1996).

Acknowledgements. This work was supported by the National Science Foundation and the Packard Foundation. We are grateful to J. Larkin for exchanging telescope time; O. Gerhard, M. Jura, and A. Weinberger for useful input; telescope observing assistants J. Aycock, T. Chelminiak, G. Puniwai, C. Sorenson, W. Wack, M. Whittle and software/instrument specialists A. Conrad and B. Goodrich for their help during the observations. The data presented here were obtained at the W. M. Keck Observatory, which is operated as a scientific partnership among the California Institute of Technology, the University of California and the National Aeronautics and Space Administration. The Observatory was made possible by the financial support of the W. M. Keck Foundation.

Correspondence and requests for materials should be addressed to A.G. (e-mail: ghez@astro.ucla.edu).

Rapid X-ray Flaring from the Direction of the Supermassive Black Hole at the Galactic Centre

F. K. Baganoff *et al.*

Editor's Note

Although it is generally accepted that there is a black hole with a mass of several million solar masses at the centre of our Milky Way galaxy, there is less light (especially X-rays) coming from the gas swirling into the black hole than standard theory predicts. This led to some uncertainty about what was powering the source of radio emission identified with the Galactic Centre, called Sagittarius A*. Here Frederick Baganoff, lead scientist for one of the instruments on board the Chandra space-based X-ray observatory, and his colleagues report rapid X-ray flares from the Galactic Centre, which imply that the X-rays must be generated within a tiny volume of space. This is consistent with its origin in gas that is falling into the supermassive black hole.

The nuclei of most galaxies are now believed to harbour supermassive black holes[1]. The motions of stars in the central few light years of our Milky Way Galaxy indicate the presence of a dark object with a mass of about 2.6×10^6 solar masses (refs 2, 3). This object is spatially coincident with the compact radio source Sagittarius A* (Sgr A*) at the dynamical centre of the Galaxy, and the radio emission is thought to be powered by the gravitational potential energy released by matter as it accretes onto a supermassive black hole[4,5]. Sgr A* is, however, much fainter than expected at all wavelengths, especially in X-rays, which has cast some doubt on this model. The first strong evidence for X-ray emission was found only recently[6]. Here we report the discovery of rapid X-ray flaring from the direction of Sgr A*, which, together with the previously reported steady X-ray emission, provides compelling evidence that the emission is coming from the accretion of gas onto a supermassive black hole at the Galactic Centre.

OUR view of Sgr A* in the optical and ultraviolet wavebands is blocked by the large visual extinction, $A_V \approx 30$ magnitudes[7], caused by dust and gas along the line of sight. Sgr A* has not been detected in the infrared owing to its faintness and to the bright infrared background from stars and clouds of dust[8]. We thus need to detect X-rays from Sgr A* in order to constrain the spectrum at energies above the radio-to-submillimetre band and to test whether gas is accreting onto a supermassive black hole (see above).

We first observed the Galactic Centre on 21 September 1999 with the imaging array of the Advanced CCD (charge-coupled device) Imaging Spectrometer (ACIS-I) aboard the

472

银河系中心超大质量黑洞方向上的
快速 X 射线耀发

巴加诺夫等

编者按

尽管人们普遍都接受了在我们的银河系中心存在着一个几百万太阳质量的黑洞这一观点，但是气体旋转着被黑洞吸积所释放的光（尤其是 X 射线）比标准理论所预测的要少。这使得被证认来自银河系中心人马座 A* 的射电辐射的来源究竟是什么有了一些不确定性。在这篇文章中，弗雷德里克·巴加诺夫（钱德拉空基 X 射线天文台上装载的一架仪器的首席科学家）和他的同事们报道了来自银河系中心的快速 X 射线耀发，这也意味着这些 X 射线肯定是来自于一个很小的空间体积内。这个结论与它是来自于正在被超大质量黑洞吸积的气体的理论相符合。

当前，人们普遍相信大部分星系的中心存在超大质量黑洞[1]。而我们银河系中非常靠近中心（距离约几光年以内）的恒星运动表明银河系中心存在一个质量大约为 2.6×10^6 太阳质量的暗天体（参考文献 2 和 3）。该天体在空间分布上正好与位于银河系动力学中心的致密射电源人马座 A*（Sgr A*）一致，其射电辐射则被认为来源于一个超大质量黑洞吸积物质所释放的引力势能[4,5]。然而，Sgr A* 在各个波段，特别是在 X 射线波段，要暗于预期，这使得人们对上述射电辐射来源产生了一些疑问。直到最近，人们才首次发现了该天体存在 X 射线辐射的强有力证据[6]。在本工作中，我们报告在 Sgr A* 方向发现了快速 X 射线耀发。连同之前发现的稳定 X 射线辐射，这一发现为 Sgr A* 致密射电源的射电辐射来源于银河系中心超大质量黑洞的气体吸积过程提供了极具说服力的证据。

我们在光学和紫外波段对 Sgr A* 的观测受限于视线方向上尘埃和气体产生的显著消光，A_V 约为 30 星等[7]。在红外波段，Sgr A* 没有被探测到，其原因一方面在于它本身暗弱，另一方面在于恒星以及尘埃云的红外背景明亮[8]。因此，为了限制 Sgr A* 的能量在射电到亚毫米波段以上的光谱，并验证气体是否吸积到超大质量黑洞上（见上），我们需要对 Sgr A* 的 X 射线进行探测。

1999 年 9 月 21 日，我们利用安装在钱德拉 X 射线天文台的高级 CCD（电荷耦合器件）成像光谱仪的成像阵列（ACIS-I）[9]对银河系中心进行了首次观测，发现了

473

Chandra X-ray Observatory[9] and discovered an X-ray source coincident within $0.35'' \pm 0.26''$ (1σ) of the radio source[6]. The luminosity in 1999 was very weak—$L_X \approx 2 \times 10^{33}$ erg s^{-1} in the 2–10 keV band—after correction for the inferred neutral hydrogen absorption column $\mathcal{N}_H \approx 1 \times 10^{23}$ cm^{-2}. This is far fainter than previous X-ray observatories could detect[6].

We observed the Galactic Centre a second time with Chandra ACIS-I from 26 October 2000 22:29 until 27 October 2000 08:19 (UT), during which time we saw a source at the position of Sgr A* brighten dramatically for a period of approximately 10,000 s (10 ks). Figure 1 shows surface plots for both epochs of the 2–8 keV counts integrated over time from a $20'' \times 20''$ region centred on the radio position of Sgr A*. The modest peak of the integrated counts at Sgr A* in 1999 increased by a factor of about 7 in 2000, despite the 12% shorter exposure. The peak integrated counts of the fainter features in the field show no evidence of strong variability, demonstrating that the flaring at Sgr A* is intrinsic to the source.

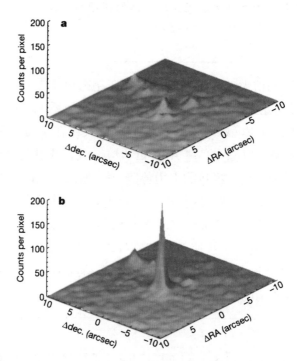

Fig. 1. Surface plots of the 2–8 keV counts within a $20'' \times 20''$ field centred on Sgr A* at two epochs. The data were taken with Chandra ACIS-I on (**a**) 21 September 1999 and (**b**) 26–27 October 2000. The effective exposure times were 40.3 ks and 35.4 ks, respectively. The spatial resolution is 0.5'' per pixel. An angle of 1'' on the sky subtends a projected distance of about 0.04 pc at the galactocentric distance of 8.0 kpc (ref. 30). The peak integrated counts per pixel at the position of Sgr A* increased by a factor of 7 from the first epoch to the second, despite the slightly smaller exposure time ($\sim$12%) in the second epoch. The low-level peak a few arcseconds to the southwest of Sgr A* is the infrared source IRS 13, and the ridge of emission to the northwest is from a string of unresolved point sources. The fainter features in the field are reasonably consistent between the two epochs, considering the limited Poisson statistics and the fact that these stellar sources may themselves be variable; this consistency shows that the strong variations at Sgr A* are intrinsic to the source.

与射电源位置在 $0.35'' \pm 0.26''$ (1σ) 以内相吻合的一个 X 射线源[6]。根据推断的柱密度为 $N_H \approx 1 \times 10^{23}$ cm^{-2} 的中性氢吸收进行修正之后,该 X 射线源的光度是非常低的(2 ~ 10 keV 波段内,光度 $L_X \approx 2 \times 10^{33}$ erg·s^{-1}),远比之前各 X 射线天文台的探测极限暗弱[6]。

2000 年 10 月 26 日 22:29 到 2000 年 10 月 27 日 08:19(UT) 这段时间里,我们利用钱德拉 ACIS-I 对银河系中心进行了第二次观测。这段时间中,我们看到在 Sgr A* 的位置上有一个在大约 10,000 s(10 ks) 的时段内剧烈变亮的源。图 1 显示了以 Sgr A* 射电源为中心的一个 $20'' \times 20''$ 区域内两次观测对时间积分的 2 ~ 8 keV 光子计数曲面分布图。尽管曝光时间少了 12%,但是 1999 年的计数曲面分布图上 Sgr A* 位置处的不太大的峰值在 2000 年增加到了大约 7 倍。视场中其他较暗的峰没有明显的光变,表明 Sgr A* 的这个耀发来自其自身的内禀光变。

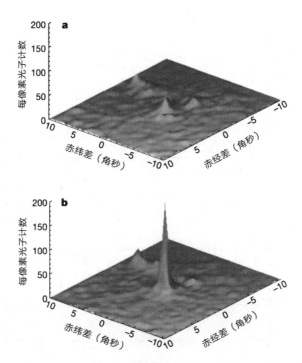

图 1. 两个不同时期以 Sgr A* 为中心的 $20'' \times 20''$ 的天区中 2 ~ 8 keV 波段光子计数的曲面分布图。图中所用数据来源于钱德拉 ACIS-I 于 (a) 1999 年 9 月 21 日和 (b) 2000 年 10 月 26 ~ 27 日的观测数据。有效曝光时间分别为 40.3 ks 和 35.4 ks。空间分辨率为每像素 0.5''。在距离银心 8.0 kpc 远的位置处(参考文献 30),天空中 1'' 的角度约对应 0.04 pc 的投影距离。Sgr A* 位置处每像素的峰值积分光子计数从第一个时期到第二个时期增加到了 7 倍,尽管第二个时期的曝光时间略短(约少 12%)。在 Sgr A* 西南方向外几个角秒处的较低水平的峰值对应的是红外源 IRS 13,而西北方向出现的脊状辐射来源于一串未分辨的点源。考虑到受限的泊松统计,以及这些恒星级天体本身的光度也可能发生变化,两个时期中视场内较暗的特征大致是一致的;这种一致性反映出 Sgr A* 位置处的强烈光变来自源的内禀原因。

Figure 2 shows light curves of the photon arrival times from the direction of Sgr A* during the observation in 2000. Figure 2a and b shows hard-band (4.5–8 keV) and soft-band (2–4.5 keV) light curves constructed from counts within an angular radius of 1.5″ of Sgr A*. Both bands exhibit roughly constant, low-level emission for the first 14 ks or so, followed by a 6-ks period of enhanced emission beginning with a 500-s event (4.4σ significance using 150-s bins). At 20 ks, flare(s) of large relative amplitude occur, lasting about 10 ks, and finally the emission drops back to the low state for the remaining 6 ks or so. About 26 ks into the observation, the hard-band light curve drops abruptly by a factor of 5 within a span of around 600 s and then partially recovers within a period of 1.2 ks. The soft-band light curve shows a similar feature, but it appears to lag the hard-band event by a few hundred seconds and is less sharply defined.

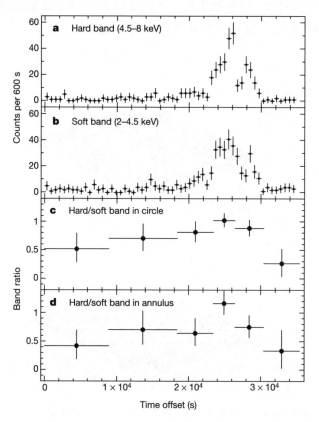

Fig. 2. Light curves of the photon arrival times and band ratios from the direction of Sgr A* on 26–27 October 2000. **a**, Hard-band (4.5–8 keV) counts. **b**, Soft-band (2–4.5 keV) counts. **c**, Hard/soft band ratio within a circle of radius 1.5″. **d**, Hard/soft band ratio within an annulus of inner and outer radii 0.5″ and 2.5″, respectively. The x axis shows the time offset from the start of the observation at 26 October 2000 22:29 (UT). The data are shown with 1σ error bars. The single 2.8-hour period of flaring activity which we have detected so far during a total of 21 hours of observations yields a (poorly determined) duty cycle of approximately 1/8. Our continuing observations of this source with Chandra will permit us to refine this value. During the quiescent intervals at the beginning and end of this observation, the mean count rate in the 2–8-keV band was $(6.4 \pm 0.6) \times 10^{-3}$ counts s^{-1}, consistent with the count rate we measured in

图 2 显示了 2000 年观测期间 Sgr A* 方向上关于光子到达时间的光变曲线。图 2a 和 2b 分别显示了根据 Sgr A* 的 1.5″ 角半径范围内的光子计数构建得到的硬 X 射线波段(4.5 ~ 8 keV)和软 X 射线波段(2 ~ 4.5 keV)的光变曲线。两个波段在最初的大概 14 ks 内都展现出大致稳定的低水平辐射,随后以一个 500 s 的事件作为开始呈现出一个时段为 6 ks 的辐射增强(以 150 s 分区间,其显著性为 4.4σ)。在 20 ks 的时候,发生相对大幅度的耀发,持续大约 10 ks。在剩余的约 6 ks 里,整体辐射回落到之前较低的水平。在观测到大约 26 ks 的时候,硬 X 射线波段的光变曲线在约 600 s 的时间间隔内突然下降为原来的五分之一,然后在 1.2 ks 的时段内得到了部分的恢复。软 X 射线波段显示了类似的特征,但是似乎要比硬 X 射线波段的事件滞后几百秒,且特征没有那么分明。

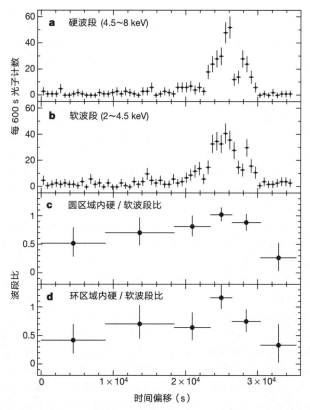

图 2. 2000 年 10 月 26 ~ 27 日 Sgr A* 方向上的光子到达时间的光变曲线以及波段比。**a**,硬 X 射线波段(4.5 ~ 8 keV)光子计数。**b**,软 X 射线波段(2 ~ 4.5 keV)光子计数。**c**,半径为 1.5″ 的圆区域内硬/软 X 射线波段比。**d**,环区域(内环半径 0.5″、外环半径 2.5″)内硬/软 X 射线波段比。*x* 轴表示从 2000 年 10 月 26 日 22:29(UT)起算的时间偏移。数据显示了 1σ 的误差棒。目前我们在整个 21 小时的观测中仅探测到一次 2.8 小时的耀发活动,占空比约为 1/8(结果仍不确定)。我们利用钱德拉对这个源进行后续观测将能够改进这一数值。在本次观测开始和结束的平静状态下,2 ~ 8 keV 波段的平均光子计数率为 $(6.4 \pm 0.6) \times 10^{-3}$ 个/秒,这个结果与我们在 1999 年测量的计数率 $(5.4 \pm 0.4) \times 10^{-3}$ 个/秒相一致。在提

1999, which was $(5.4 \pm 0.4) \times 10^{-3}$ counts s^{-1}. The detected count rate at the peak of the flare was 0.16 ± 0.01 counts s^{-1} within an extraction circle of radius 1.5″, but this is about 30% less than the true incident count rate owing to pile-up of X-rays in the detector during the 3.2-s integration time for each CCD (charge-coupled device) read out.

The band-ratio time series in Fig. 2c, defined as the ratio of hard-band counts to soft-band counts, suggests that the spectrum "hardened" (that is, became flatter and extended more strongly to higher energies) during the flare. The difference between the band ratio measured at the peak of the flare and the average of the band ratios during the quiescent periods at the beginning and end of the observation is 0.63 ± 0.21 (that is, 3σ). The peak-flare band ratio in Fig. 2c is affected to some extent by the effects of pile-up (see Fig. 2 legend), which would tend to harden the spectrum; however, the band ratios in Fig. 2d, which were computed using the non-piled-up data extracted from the wings of the point spread function, also show evidence for spectral hardening with 2.7σ significance. The sizes of dust-scattering haloes in the Galactic Centre are typically greater than 1′ (ref. 10), so dust-scattered X-rays from the source contribute a negligible fraction of the emission within the source extraction region that we used; hence dust-scattered X-rays cannot account for the spectral variations. We therefore conclude that the spectral hardening during the flare is likely to be real.

The quiescent-state spectra in 1999 and 2000 and the peak flaring-state spectrum in 2000 are shown in Fig. 3. We fitted each spectrum individually using a single power-law model with corrections for the effects of photoelectric absorption and dust scattering[10]. The best-fit values and 90% confidence limits for the parameters of each fit are presented in the first three lines of Table 1. The column densities for the three spectra are consistent, within the uncertainties, as are the photon indices of both quiescent-state spectra. Next, we fit a double power-law model to the three spectra simultaneously, using a single photon index for both quiescent spectra, a second photon index for the flaring spectrum, and a single column density for all three spectra. The best-fit models for each spectrum from the simultaneous fits are shown as solid lines in Fig. 3; the parameter values are given in the last line of Table 1. Using these values, we derive an absorption-corrected 2–10 keV luminosity of $L_X = (2.2^{+0.4}_{-0.3}) \times 10^{33}$ erg s^{-1} for the quiescent-state emission and $L_X = (1.0 \pm 0.1) \times 10^{35}$ erg s^{-1} for the peak of the flaring-state emission, or around 45 times the quiescent-state luminosity. We note that previous X-ray observatories did not have the sensitivity to detect such a short-duration, low-luminosity flare in the Galactic Centre[6]. The best-fit photon index $\Gamma_f = 1.3^{+0.5}_{-0.6}$ $(N(E) \propto E^{-\Gamma})$ for the flaring-state spectrum is slightly flatter than, but consistent with, systems thought to contain supermassive black holes[11]. Here E is the energy of a photon in keV, $N(E)$ is the number of photons with energies in the differential interval E to $E+dE$, and Γ is the power-law index of the photon number distribution.

取数据的半径为 1.5″ 的圆区域里，耀发峰值处探测的计数率为 0.16±0.01 个/秒。因为在每个 CCD(电荷耦合器件)读出时有 3.2 s 的积分时间，X 射线会在探测器中有堆积效应，因此这个数值比真实的入射光子计数率低大约 30%。

    图 2c 显示的波段比(其定义为硬 X 射线波段光子计数与软 X 射线波段光子计数之比)随时间的一连串变化表明在耀发的过程中光谱有"硬化"的现象(即光谱变得更平且更强地延伸到高能端)。耀发时峰值处与观测初期以及末期平静时期的波段比的差别为 0.63±0.21(即 3σ)。图 2c 中的耀发峰的波段比在一定程度上受堆积效应的影响(见图 2 图注)，该效应倾向于使光谱硬化；然而，图 2d 中的波段比利用从点扩散函数的两翼提取的无堆积效应的数据计算得到，它也在 2.7σ 的显著性上显示光谱硬化的证据。银河系中心尘埃散射晕的尺度通常认为大于 1′(参考文献 10)，因此，这个源被尘埃散射的 X 射线对我们用来提取数据的区域内辐射的贡献可以忽略；所以尘埃散射的 X 射线不足以解释所观测到的光谱变化。我们因而得到这样的结论：耀发过程中的光谱硬化很可能是真实的。

    图 3 中显示了 1999 年和 2000 年平静状态下的光谱以及 2000 年峰值耀发状态下的光谱。我们利用单一幂律模型对这些光谱逐个进行了拟合，其中对光电吸收和尘埃散射的效应做了修正[10]。表 1 的前三行分别列举了每个拟合的最佳拟合值及其 90% 置信限。三条光谱的柱密度在不确定度范围内是一致的，同时两条平静状态下光谱的光子幂指数也是一致的。随后，我们同时对三条光谱进行双幂律模型拟合，过程中对两条平静状态下的光谱采用一个统一的光子幂指数，对耀发光谱采用另一个光子幂指数，而对三条光谱采用统一的柱密度。图 3 中的实线显示了同时拟合情况下每条谱线的最佳拟合模型；表 1 的最后一行列举了对应的参数值。利用这些结果，我们得到了这个源在 2~10 keV 波段做了吸收修正后的光度：平静状态下 $L_X = (2.2^{+0.4}_{-0.3}) \times 10^{33}$ erg·s^{-1}，耀发状态下的峰值 $L_X = (1.0\pm0.1) \times 10^{35}$ erg·s^{-1}(大约是平静状态下的 45 倍)。我们注意到过去的 X 射线天文台的灵敏度不够高，无法探测到银河系中心这样持续时间短、光度低的耀发信号[6]。对于耀发状态下的光谱，其光子幂指数的最佳拟合结果为 $\Gamma_f = 1.3^{+0.5}_{-0.6}$ ($N(E) \propto E^{-\Gamma}$)，这个结果比被认为包含超大质量黑洞的系统的理论预言[11]稍平，但仍相互一致。这里，E 是以 keV 为单位的光子能量，$N(E)$ 是能量在 E 到 $E+dE$ 微分区间内的光子数目，Γ 是光子数分布的幂律指数。

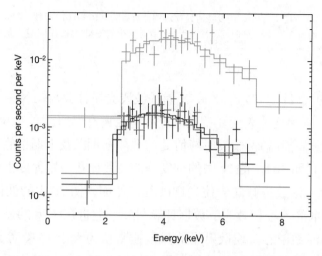

Fig. 3. X-ray spectra of the Chandra source at the position of Sgr A*. The data are shown as crosses with vertical bars indicating the 1σ errors in the count rate and horizontal bars the energy range of each bin. The events have been grouped to yield 10 counts per bin. The counts in the 1999 (black) and 2000 (red) quiescent-state spectra were extracted using a source radius of 1.5″. A non-piled-up, peak flaring-state spectrum (green) was extracted from the wings of the point spread function, using an annulus with inner and outer radii of 0.5″ and 2.5″, during the time interval 23.7–26.3 ks after the start of the 2000 observation. The solid lines are the best-fit models for each spectrum, obtained by fitting an absorbed, dust-scattered, double power-law model to the three spectra simultaneously. The best-fit values and 90% confidence intervals for the model parameters are given in the last line of Table 1. The spectra are well-fitted using a single photon index for both quiescent spectra, a second photon index for the flaring spectrum, and a single column density for all three spectra. The column density from the simultaneous fits corresponds to a visual extinction $A_V = 29.6^{+3.0}_{-8.1}$ magnitudes, which agrees well with infrared-derived estimates of $A_V \approx 30$ magnitudes[7]; we thus find no evidence in our X-ray data of excess gas and dust localized around the supermassive black hole. This places an important constraint on the maximum contribution to the infrared spectrum of Sgr A* produced by local dust reprocessing of higher energy photons from the accretion flow. The spectral models used in ref. 6 did not account for dust scattering; hence a higher column density was needed to reproduce the low-energy cut-off via photoelectric absorption alone. We note that there is no sign of an iron Kα emission line in the flaring-state spectrum.

Table 1. Spectral fits

Spectrum	$\mathcal{N}_H$	Γ_q	Γ_f	x^2/ν
1999 quiescent	$5.8^{+1.3}_{-1.4}$	$2.5^{+0.8}_{-0.7}$	...	19/22
2000 quiescent	$5.0^{+1.9}_{-2.3}$	$1.8^{+0.7}_{-0.9}$	...	7.6/12
2000 flaring	$4.6^{+2.0}_{-1.6}$	...	$1.0^{+0.8}_{-0.7}$	12/17
All	$5.3^{+0.9}_{-1.1}$	$2.2^{+0.5}_{-0.7}$	$1.3^{+0.5}_{-0.6}$	45/55

Best-fit parameter values and 90% confidence intervals for power-law models, corrected for photoelectric absorption and dust scattering[10]. $\mathcal{N}_H$ is the neutral hydrogen absorption column in units of 10^{22} H atoms cm^{-2}. Γ_q and Γ_f are the photon-number indices of the quiescent-state and peak flaring-state spectra ($\mathcal{N}(E) \propto E^{-\Gamma}$). x^2 is the value of the fit statistic for the best-fit model, and ν is the number of degrees of freedom in the fit. The parameter values for the spectrum marked "All" were derived by fitting an absorbed, dust-scattered, double power-law model to the three spectra simultaneously (see text and Fig. 3).

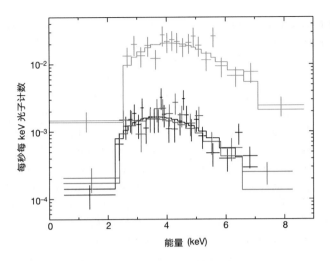

图 3. Sgr A* 位置处钱德拉源的 X 射线光谱。数据用交叉十字表示，其中纵向棒代表了光子计数率的 1σ 的误差，而横向棒代表了每个区间的能量范围。这些事件被分组以保证每个区间有 10 个计数。我们在源附近半径为 1.5″ 的范围内从 1999 年（黑色）和 2000 年（红色）的平静状态下的光谱中提取了光子计数，并于 2000 年观测开始的 23.7 ~ 26.3 ks 的时间间隔内使用一个环形范围（内环半径 0.5″、外环半径 2.5″）从点扩散函数的两翼中提取了耀发状态下无堆积效应的峰值光谱（绿色）。实线是每条光谱的模型最佳拟合结果，这是基于一个考虑吸收和尘埃散射的双幂律模型对三条光谱同时进行拟合得到的。模型参数的最佳拟合结果以及 90% 置信区间列在了表 1 的最后一行。在对两条平静状态下的光谱采用统一的光子幂指数，对耀发光谱采用另一个光子幂指数并对三条光谱采用统一的柱密度的情况下，我们可以很好地拟合这些光谱。同时拟合所得到的柱密度对应目视波段消光 $A_v = 29.6^{+5.0}_{-6.1}$ 星等，这个结果与从红外数据推断的 $A_v \approx 30$ 星等[7] 相吻合；因此我们在 X 射线数据中并没有发现超大质量黑洞附近存在额外气体和尘埃的证据。局域尘埃会吸收吸积流中更高能量的光子再发出红外辐射，这里得到的结果对这个过程为 Sgr A* 红外光谱所提供的最大贡献给出非常重要的限制。文献 6 中使用的光谱模型没有考虑尘埃散射；因此若只用光电吸收来再现低能端的截断，就需要更高的柱密度。我们注意到，在耀发状态下的光谱中没有铁的 Kα 发射线的迹象。

表 1. 谱线拟合

光谱	N_H	Γ_q	Γ_f	x^2/ν
1999 年的平静状态	$5.8^{+1.3}_{-1.4}$	$2.5^{+0.8}_{-0.7}$	…	19/22
2000 年的平静状态	$5.0^{+1.9}_{-2.3}$	$1.8^{+0.7}_{-0.9}$	…	7.6/12
2000 年的闪耀状态	$4.6^{+2.0}_{-1.6}$	…	$1.0^{+0.8}_{-0.7}$	12/17
总体	$5.3^{+0.9}_{-1.1}$	$2.2^{+0.5}_{-0.7}$	$1.3^{+0.5}_{-0.6}$	45/55

利用考虑了光电吸收和尘埃散射修正[10] 的幂律模型进行拟合的最佳参数值以及 90% 置信区间。N_H 为中性氢吸收柱密度，单位是 10^{22} 个氢原子/cm^2。Γ_q 和 Γ_f 分别为平静状态和峰值耀发状态下光谱的光子数指数（$N(E) \propto E^{-\Gamma}$）。x^2 是最佳拟合模型的拟合统计值，而 ν 是拟合中自由度的个数。标记为"总体"的光谱的参数值是基于一个考虑吸收和尘埃散射的双幂律模型对三条光谱同时进行拟合得到的（详情见正文和图 3）。

If we view the outburst as a single event, the rise/fall timescales of a few hundred seconds and the 10-ks duration are consistent with the light-crossing and dynamical timescales for the inner part (at less than 10 Schwarzschild radii; $R_S \equiv 2GM/c^2$) of the accretion flow around a black hole of 2.6×10^6 solar masses ; here R_S is the radius of the black-hole event horizon (that is, the boundary at which the escape velocity equals the speed of light, c), G is the gravitational constant and M is the mass of the black hole. Although we cannot strictly rule out an unrelated contaminating source as the origin of the flare (for example, an X-ray binary, for which little is known about such short-timescale, low-luminosity events as we have detected; W. Lewin, personal communication), this explanation seems unlikely, as the characteristic angular scales of the young and old stellar clusters around Sgr A* are 5–20″ (ref. 7), whereas the flaring source lies within 1/3″ of the radio position. These clusters contain up to a million solar masses of stars and stellar remnants[2]; hence it is rather improbable that there would be only one very unusual stellar X-ray source in the image and that it would be fortuitously superposed on Sgr A*. Furthermore, it is not clear that X-ray binaries can be easily formed or long endure near Sgr A*, given the high velocity dispersion and high spatial density of the stars in its deep gravitational potential well[6,12].

Strong, variable X-ray emission is a characteristic property of active galactic nuclei; factors of 2–3 variations on timescales ranging from minutes to years are typical for radio-quiet active galactic nuclei[11]. Moderate- to high-luminosity active galactic nuclei (that is, Seyfert galaxies and quasars) show a general trend of increasing variability with decreasing luminosity[13]. However, this trend does not extend to low-luminosity active galactic nuclei (LLAGN), which show little or no significant variability on timescales less than a day[14]. Assuming the X-ray flare is from Sgr A*, it is remarkable that this source— generally thought to be the nearest and least luminous example of accretion onto a central supermassive black hole—has shown a factor of 45 variation that is an order of magnitude more rapid than the fastest observed variation of similar relative amplitude by a radio-quiet active galactic nucleus of any luminosity class[15]. We note that flares of similar luminosity would be undetectable by Chandra in the nucleus of even the nearest spiral galaxy, M31. LLAGN emit $L_X \gtrsim 10^{38}$ erg s^{-1} (ref. 14), so it should be kept in mind that the astrophysics of accretion onto even the LLAGN may differ substantially from that of Sgr A*. This makes Sgr A* a useful source for testing the theory of accretion onto supermassive black holes in galactic nuclei.

The faintness of Sgr A* at all wavelengths requires that the supermassive black hole be in an extremely quiet phase, either because the accretion rate is very low, or because the accretion flow is radiatively inefficient, or both[5]. A variety of theoretical scenarios, usually based on advective accretion models[16-19], jet–disk models[20], or Bondi–Hoyle models[21,22], have developed this idea. An important prediction of the advective accretion models is that the X-ray spectrum in the Chandra energy band should be dominated by thermal bremsstrahlung emission from hot gas in the outer regions of the accretion flow ($R \gtrsim 10^3 R_S$), but a region this large could not produce the rapid, large-relative-amplitude variations we have seen. Thus, the properties of the X-ray flare are inconsistent with the advective accretion flow models. The low luminosity and short timescales of this event are also inconsistent with

如果我们将爆发看作是一次单一事件，那么一个几百秒的增强/减弱的时标以及 10 ks 的持续时间与一个 2.6×10^6 太阳质量的黑洞周围的吸积流内部区域（小于 10 倍施瓦西半径；$R_S \equiv 2GM/c^2$）的光线穿越以及动力学时标相符合；这里 R_S 是黑洞事件视界的半径（即逃逸速度等于光速 c 的边界），G 是引力常数，而 M 是黑洞质量。尽管我们不能严格地排除此次耀发起源于一个无关的污染源（比如一个 X 射线双星，对于探测到它的这类短时标、低光度的事件，我们知之甚少；与卢因的个人交流），但是这个解释的可能性很低，因为 Sgr A* 周围年轻和年老星团的特征角尺度大约为 5″ ~ 20″（参考文献 7），而耀发源位于射电源位置附近 1/3″ 以内。这些星团包含多达百万太阳质量的恒星以及恒星残迹[2]，因此，在图像中仅出现一个非常特殊的恒星级 X 射线源，且其很偶然地与 Sgr A* 的位置重合，这种概率是相当低的。此外，在 Sgr A* 深引力势阱中[6,12]，恒星拥有高速度弥散以及高空间密度，在这种情况下，X 射线双星能否在 Sgr A* 附近顺利形成或者长时间存活，目前并不清楚。

强且变化的 X 射线辐射是活动星系核的一个特征属性；而射电宁静活动星系核一般会在从分钟到年的时标上产生 2 ~ 3 倍的光变[11]。对于中到高光度活动星系核（即赛弗特星系和类星体）而言，它们大体的趋势是光度越低，光变越大[13]。然而，这种趋势并不能用于描述低光度活动星系核（LLAGN），因为它们在短于一天的时标下几乎没有显著的光变[14]。假设 X 射线耀发来源于 Sgr A*，那么值得注意的是，这个源（一般被认为是最近的而且是最暗的中心超大质量黑洞吸积的例子）展现了一个 45 倍的快速光变，这比拥有类似相对幅度的任何光度级的射电宁静活动星系核观测到的最快光变[15]还要快一个量级。我们注意到，即使是最近的漩涡星系 M31 的核存在类似光度的耀发，钱德拉望远镜也无法探测到。LLAGN 的 X 射线光度为 $L_X \gtrsim 10^{38}$ erg·s⁻¹（参考文献 14），即使是吸积到 LLAGN 上的天体物理过程也可能与吸积到 Sgr A* 上的大不相同，这一点我们也需要牢记在心。因此，Sgr A* 成为检验星系核超大质量黑洞吸积理论的有用的源。

在各个波段 Sgr A* 都非常的暗，这就需要超大质量黑洞处在一个极端宁静的状态，这或者是因为吸积率非常低，或者是因为吸积流的辐射效率低下，又或者两个原因都有[5]。基于径移吸积模型[16-19]、喷流–盘模型[20]或是邦迪–霍伊尔模型[21,22]等，大量理论图景都提出了这一想法。径移吸积模型的一个重要的预言是钱德拉能量波段的 X 射线光谱应当由吸积流外围区域（$R \gtrsim 10^3 R_S$）高温气体的热轫致辐射所主导，但是这样大的区域无法产生我们所看到的快速、相对大幅度的光变。因此，X 射线耀发的性质与径移吸积流模型不相符。这一事件的低光度、短时标特征，也与

tidal disruption of a star by a central supermassive black hole[23].

In all models, the radio-to-submillimetre spectrum of Sgr A* is cyclo-synchrotron emission from a combination of sub-relativistic and relativistic electrons (and perhaps positrons) spiralling around magnetic field lines either in a jet or in a static region within the inner $10R_S$ of the accretion flow. The electron Lorentz factor inferred from the radio spectrum of Sgr A* is $\gamma_e \approx 10^2$. If the X-ray flare were produced via direct synchrotron emission, then the emitting electrons would need $\gamma_e \gtrsim 10^5$. For the 10–100 G magnetic field strengths predicted by the models, the cooling time of the particles would be ~1–100 s. Thus, the approximately 10-ks duration of the flare would require repeated injection of energy to the electrons. On the other hand, if the X-rays were produced via up-scattering of the submillimetre photons off the relativistic electrons, a process called synchrotron self-Comptonization (SSC), then $\gamma_e \approx 10^2$–10^3 would be required, and the cooling time would be of the order of hours, which is consistent with the duration of the flare. The rapid turn-off of the X-ray emission might then be attributed to the dilution of both photon and electron densities in an expanding plasma.

The X-ray spectra of radio-quiet quasars and active galactic nuclei are thought to be produced by thermal Comptonization of infrared-to-ultraviolet seed photons from a cold, optically thick, geometrically thin accretion disk by hot electrons in a patchy corona above the disk[13,24]. The X-ray spectra of these sources generally "soften" (that is, become steeper and extend less strongly to higher energies) as they brighten[15]. In contrast, the extremely low luminosity of Sgr A* precludes the presence of a standard, optically thick accretion disk[5]; hence, the dominant source of seed photons would be the millimetre-to-submillimetre synchrotron photons.

The energy released by an instability in the mass accretion rate or by a magnetic reconnection event near the black hole would shock-accelerate the electrons, causing the synchrotron spectrum to intensify and to extend farther into the submillimetre band. Consequently, the Compton up-scattered X-ray emission would harden as the X-ray intensity increased, exactly as observed. We note that the millimetre-band spectrum of Sgr A* has been observed to harden during one three-week flare[25] and one three-day flare[26], as would be required by the current SSC models for Sgr A* (refs 20, 22).

To test the SSC models, we measured the flux density of Sgr A* at a wavelength of 3 mm with the Millimeter Array at the Owens Valley Radio Observatory, simultaneous with part of the 2000 Chandra measurement. Unfortunately, the available observing window (20:10–02:30 UT) preceded the X-ray flare (04:03–06:50 UT) by a few hours. The observed flux density of Sgr A* was 2.05 ± 0.3 Jy, consistent with previously reported measures[27,28]. Recently, a 106-day quasi-periodicity has been reported in the centimetre band from an analysis of 20 years of data taken with the Very Large Array (VLA)[29]. A weekly VLA monitoring program detected a 30% increase in the radio flux density of Sgr A* beginning around 24 October 2000 and peaking on 5 November 2000. This increase was seen at 2 cm, 1.3 cm, and 7 mm (R. McGary, J.-H. Zhao, W. M. Goss and G. C. Bower, personal communication).

中心超大质量黑洞潮汐瓦解恒星[23]不相符。

在所有模型中，Sgr A* 从射电到亚毫米波段的光谱来源于亚相对论性和相对论性电子（或许还有正电子）组合沿喷流中或者吸积流 $10R_s$ 以内稳定区域中的磁场线旋进所产生的回旋–同步加速辐射。从 Sgr A* 的射电谱推断，其电子的洛伦兹因子 γ_e 约为 10^2。如果 X 射线耀发是通过直接的同步加速辐射产生的，那么发出辐射的电子的洛伦兹因子需要达到 $\gamma_e \gtrsim 10^5$。对于模型所预言的 $10 \sim 100$ G 的磁场强度，粒子的冷却时间大约为 $1 \sim 100$ s。因此只有向电子重复注入能量，才能维持大约 10 ks 的耀发。另一方面，如果 X 射线是通过相对论性电子对亚毫米波光子的向上散射（此过程称为同步加速辐射的自康普顿效应（SSC））产生的，那么就需要 $\gamma_e \approx 10^2 \sim 10^3$，而冷却时间将是小时量级，这与耀发所持续的时间相符。X 射线辐射的快速减弱则可以归因于膨胀等离子体中光子和电子密度的稀释。

射电宁静类星体与活动星系核的 X 射线光谱被认为是由来自冷的、光学厚的、几何薄的吸积盘的从红外到紫外的种子光子被盘上方零散的冕中的热电子热康普顿化产生的[13,24]。这些源的 X 射线谱通常随着亮度的增强会出现"软化"的现象（即光谱变得更陡，更少能延伸到较高能量）[15]。相比之下，Sgr A* 的极低的光度排除了一个标准的、光学厚的吸积盘存在[5]的可能性；因此，种子光子的主要来源将是从毫米波到亚毫米波的同步加速辐射光子。

黑洞附近质量吸积率的不稳定性或者磁重联事件释放的能量将通过激波的方式加速电子，导致同步加速辐射谱增强并进一步延伸到亚毫米波段。因此，正如观测到的，随着 X 射线辐射强度的增加，康普顿向上散射造成的 X 射线辐射将会硬化。我们注意到已经观测到 Sgr A* 毫米波段的光谱在一次历时三周[25]和一次历时三天[26]的耀发过程中发生了硬化，而这正是目前 Sgr A* 的 SSC 模型（参考文献 20 和 22）所要求的。

为了检验 SSC 模型，我们利用欧文斯谷射电天文台的毫米波阵列以及部分钱德拉 2000 年的观测数据对 Sgr A* 在波长为 3 mm 处的流量密度进行了测量。不巧的是，可用的观测时段（20:10 至 02:30 UT）比 X 射线耀发事件（04:03 至 06:50 UT）早了几个小时。观测到 Sgr A* 的流量密度为 2.05 ± 0.3 Jy，这与以前公布的结果[27,28]是一致的。最近报道，甚大阵（VLA）20 年数据[29]分析得到了 Sgr A* 在厘米波段的一个 106 天的准周期性。一个 VLA 的每周监控项目探测到 Sgr A* 的射电流量密度大约从 2000 年 10 月 24 日起开始了一次 30% 的增强，在 2000 年 11 月 5 日达到峰值。这次增强在 2 cm、1.3 cm 以及 7 mm 的波长上看到了（与麦加里、赵军辉、戈斯和

The timing of the X-ray flare and the rise in the radio flux density of Sgr A* suggests that there is a connection between the two events, providing additional indirect support for the association of the X-ray flare with Sgr A* and further strengthening the case that it was produced via either the SSC or direct synchrotron processes. Definitive evidence for these ideas will require detection of correlated variations in the radio-to-submillimetre and X-ray wavebands through future coordinated monitoring projects.

(**413**, 45-48; 2001)

F. K. Baganoff[*], M. W. Bautz[*], W. N. Brandt[†], G. Chartas[†], E. D. Feigelson[†], G. P. Garmire[†], Y. Maeda[†‡], M. Morris[§], G. R. Ricker[*], L. K. Townsley[†] & F. Walter[‖]

[*] Center for Space Research, Massachusetts Institute of Technology, Cambridge, Massachusetts 02139-4307, USA

[†] Department of Astronomy and Astrophysics, Pennsylvania State University, University Park, Pennsylvania 16802-6305, USA

[‡] Institute of Space and Astronautical Science, 3-1-1 Yoshinodai, Sagamihara, 229-8501, Japan

[§] Department of Physics and Astronomy, University of California at Los Angeles, Los Angeles, California 90095-1562, USA

[‖] Department of Astronomy, California Institute of Technology, Pasadena, California 91125, USA

Received 27 April; accepted 2 August 2001.

References:

1. Richstone, D. *et al.* Supermassive black holes and the evolution of galaxies. *Nature* **395** (suppl. on optical astronomy) A14-A19 (1998).

2. Genzel, R., Pichon, C., Eckart, A., Gerhard, O. E. & Ott, T. Stellar dynamics in the Galactic Centre: proper motions and anisotropy. *Mon. Not. R. Astron. Soc.* **317**, 348-374 (2000).

3. Ghez, A. M., Morris, M., Becklin, E. E., Tanner, A. & Kremenek, T. The accelerations of stars orbiting the Milky Way's central black hole. *Nature* **407**, 349-351 (2000).

4. Lynden-Bell, D. & Rees, M. J. On quasars, dust and the Galactic Centre. *Mon. Not. R. Astron. Soc.* **152**, 461-475 (1971).

5. Melia, F. & Falcke, H. The supermassive black hole at the Galactic Center. *Annu. Rev. Astron. Astrophys.* **39**, 309-352 (2001).

6. Baganoff, F. K. *et al.* Chandra X-ray spectroscopic imaging of Sgr A* and the central parsec of the Galaxy. *Astrophys. J.* (submitted); also preprint astro-ph/0102151 at ⟨xxx.lanl.gov⟩ (2001).

7. Morris, M. & Serabyn, E. The galactic center environment. *Annu. Rev. Astron. Astrophys.* **34**, 645-702 (1996).

8. Menten, K. M., Reid, M. J., Eckart, A. & Genzel, R. The position of Sagittarius A*: accurate alignment of the radio and infrared reference frames at the Galactic Center. *Astrophys. J.* **475**, L111-L114 (1997).

9. Weisskopf, M. C., O'Dell, S. L. & van Speybroeck, L. P. Advanced X-Ray Astrophysics Facility (AXAF). *Proc. SPIE* **2805**, 2-7 (1996).

10. Predehl, P. & Schmitt, J. H. M. M. X-raying the interstellar medium: ROSAT observations of dust scattering halos. *Astron. Astrophys.* **293**, 889-905 (1995).

11. Mushotzky, R. F., Done, C. & Pounds, K. A. X-ray spectra and time variability of active galactic nuclei. *Annu. Rev. Astron. Astrophys.* **31**, 717-761 (1993).

12. Davies, M. B., Blackwell, R., Bailey, V. C. & Sigurdsson, S. The destructive effects of binary encounters on red giants in the Galactic Centre. *Mon. Not. R. Astron. Soc.* **301**, 745-753 (1998).

13. Nandra, K., George, I. M., Mushotzky, R. F., Turner, T. J. & Yaqoob, T. ASCA observations of Seyfert 1 galaxies. I. Data analysis, imaging, and timing. *Astrophys. J.* **476**, 70-82 (1997).

14. Ptak, A., Yaqoob, T., Mushotzky, R., Serlemitsos, P. & Griffiths, R. X-ray variability as a probe of advection-dominated accretion in low-luminosity active galactic nuclei. *Astrophys. J.* **501**, L37-L40 (1998).

15. Ulrich, M.-H., Maraschi, L. & Urry, C. M. Variability of active galactic nuclei. *Annu. Rev. Astron. Astrophys.* **35**, 445-502 (1997).

16. Narayan, R., Mahadevan, R., Grindlay, J. E., Popham, R. G. & Gammie, C. Advection-dominated accretion model of Sagittarius A*: evidence for a black hole at the Galactic center. *Astrophys. J.* **492**, 554-568 (1998).

17. Quataert, E. & Narayan, R. Spectral models of advection-dominated accretion flows with winds. *Astrophys. J.* **520**, 298-315 (1999).

18. Ball, G. H., Narayan, R. & Quataert, E. Spectral models of convection-dominated accretion flows. *Astrophys. J.* **552**, 221-226 (2001).

19. Blandford, R. D. & Begelman, M. C. On the fate of gas accreting at a low rate on to a black hole. *Mon. Not. R. Astron. Soc.* **303**, L1-L5 (1999).

20. Falcke, H. & Markoff, S. The jet model for Sgr A*: radio and X-ray spectrum. *Astron. Astrophys.* **362**, 113-118 (2000).

21. Melia, F. An accreting black hole model for Sagittarius A*. II: A detailed study. *Astrophys. J.* **426**, 577-585 (1994).

22. Melia, F., Liu, S. & Coker, R. Polarized millimeter and submillimeter emission from Sagittarius A* at the Galactic center. *Astrophys. J.* **545**, L117-L120 (2000).

23. Rees, M. J. Tidal disruption of stars by black holes of 10^6–10^8 solar masses in nearby galaxies. *Nature* **333**, 523-528 (1988).

鲍尔的个人交流）。X 射线耀发的时间与 Sgr A* 射电波段流量密度的增强的时间相一致，这意味着两个事件之间存在联系，这为本次 X 射线耀发与 Sgr A* 之间的关联提供了额外的间接支持，并且进一步强化了该事件起源于 SSC 或者直接的同步加速辐射过程的情形。要获得这些想法的决定性证据，我们还需要通过未来的协调监测项目探测射电到亚毫米波段与 X 射线波段之间的相关光变。

（刘项琨 翻译；陈阳 审稿）

24. Haardt, F., Maraschi, L. & Ghisellini, G. X-ray variability and correlations in the two-phase disk-corona model for Seyfert galaxies. *Astrophys. J.* **476**, 620-631 (1997).

25. Tsuboi, M., Miyazaki, A. & Tsutsumi, T. in *The Central Parsecs of the Galaxy* (ed. Falcke, H. *et al.*) Vol. 186, 105-112 (ASP Conf. Ser., Astronomical Society of the Pacific, San Francisco, 1999).

26. Wright, M. C. H. & Backer, D. C. Flux density of Sagittarius A at $\lambda = 3$ millimeters. *Astrophys. J.* **417**, 560-564 (1993).

27. Serabyn, E. *et al.* High frequency measurements of the spectrum of SGR A*. *Astrophys. J.* **490**, L77-L81 (1997).

28. Falcke, H. *et al.* The simultaneous spectrum of Sagittarius A* from 20 centimeters to 1 millimeter and the nature of the millimeter excess. *Astrophys. J.* **499**, 731-734 (1998).

29. Zhao, J.-H., Bower, G. C. & Goss, W. M. Radio variability of Sagittarius A*—a 106 day cycle. *Astrophys. J.* **547**, L29-L32 (2001).

30. Reid, M. J. The distance to the center of the Galaxy. *Annu. Rev. Astron. Astrophys.* **31**, 345-372 (1993).

Acknowledgements. We thank M. Begelman for useful comments. This work has been supported by a grant from NASA.

Correspondence and requests for materials should be addressed to F.K.B.(e-mail: fkb@space.mit.edu).

Detection of Polarization in the Cosmic Microwave Background Using DASI

J. M. Kovac *et al.*

Editor's Note

The characterization of the ripples in the cosmic microwave background radiation, the "afterglow" of the Big Bang, led to quite accurate determination of various cosmological parameters, such as the density of matter and energy, and the speed of expansion. The very success of the standard cosmological model that emerged prompted researchers to look for independent ways of testing it. The model makes very specific predictions about how the microwave background with be polarized (that is, how the oscillating electromagnetic fields of the microwave photons will be aligned). Here John Kovac and coworkers used a telescope at the South Pole to observe this polarization, and confirm that it is just as predicted.

The past several years have seen the emergence of a standard cosmological model, in which small temperature differences in the cosmic microwave background (CMB) radiation on angular scales of the order of a degree are understood to arise from acoustic oscillations in the hot plasma of the early Universe, arising from primordial density fluctuations. Within the context of this model, recent measurements of the temperature fluctuations have led to profound conclusions about the origin, evolution and composition of the Universe. Using the measured temperature fluctuations, the theoretical framework predicts the level of polarization of the CMB with essentially no free parameters. Therefore, a measurement of the polarization is a critical test of the theory and thus of the validity of the cosmological parameters derived from the CMB measurements. Here we report the detection of polarization of the CMB with the Degree Angular Scale Interferometer (DASI). The polarization is deteced with high confidence, and its level and spatial distribution are in excellent agreement with the predictions of the standard theory.

THE CMB radiation provides a pristine probe of the Universe roughly 14 billion years ago, when the seeds of the complex structures that characterize the Universe today existed only as small density fluctuations in the primordial plasma. As the physics of such a plasma is well understood, detailed measurements of the CMB can provide critical tests of cosmological models and can determine the values of cosmological parameters with high precision. The CMB has accordingly been the focus of intense experimental and theoretical investigations since its discovery nearly 40 years ago[1]. The frequency spectrum of the CMB was well determined by the COBE FIRAS instrument[2,3]. The initial detection of temperature anisotropy was made on large angular scales by the COBE DMR

利用 DASI 探测宇宙微波背景的偏振

科瓦奇等

编者按

宇宙微波背景辐射中的涟漪特征，又被称为宇宙大爆炸的"余辉"。我们可以非常准确地通过它来测定各种宇宙学参数，例如物质和能量的密度，以及膨胀速度等。标准宇宙学模型的成功进一步促使研究人员去寻找独立检测它的方法。该模型对微波背景如何被偏振（即微波光子产生的振荡电磁场如何排列）进行了非常具体的预测。约翰·科瓦奇及其同事在南极使用望远镜观测了这种偏振，并且确认观测结果与预测一致。

在过去几年已经出现了的标准宇宙学模型中，宇宙微波背景（CMB）辐射在一定的度角尺度上的小温差被理解为是由早期宇宙中热等离子体的声学振荡引起的，而后者又来源于原初密度扰动。最近在该模型下对温度扰动的测量导出了关于宇宙起源、演化和组成的深刻结论。仅仅利用温度扰动的测量，理论框架对预测 CMB 的偏振水平基本上没有自由参数可言。因此，偏振的测量是理论以及从 CMB 测量得到的宇宙学参数的有效性的重要检验。本文中，我们将报道用度角尺度干涉仪（DASI）检测 CMB 所得的探测结果。偏振的检测置信度很高，其水平和空间分布与标准理论的预测非常一致。

CMB 辐射可以作为约 140 亿年前宇宙状态的原始探测仪。现今宇宙复杂结构的种子在当时仅存在于原初等离子体的微小密度扰动中。由于人们已经很好地理解了这种等离子体的物理特性，所以对 CMB 更细节性地测量可以提供宇宙学模型的关键检测，并且可以高精度地确定宇宙学参数的值。自发现至今的将近 40 年时间里，CMB 一直是大量实验和理论研究关注的焦点[1]。CMB 的频谱由宇宙背景探测器（COBE）远红外绝对分光光度计（FIRAS）很好地确定[2,3]。通过 COBE 较差微波辐射计（DMR）[4]，研究人员在大角度尺度上进行了温度各向异性的初始检测，而最近

instrument[4] and recently there has been considerable progress in measuring the anisotropy on finer angular scales[5-10].

In the now standard cosmological model (see, for example, ref. 11), the shape of the CMB angular power spectrum directly traces acoustic oscillations of the photon-baryon fluid in the early Universe. As the Universe expanded, it cooled; after roughly 400,000 years, the intensity of the radiation field was no longer sufficient to keep the Universe ionized, and the CMB photons decoupled from the baryons as the first atoms formed. Acoustic oscillations passing through extrema at this epoch are observed in the CMB angular power spectrum as a harmonic series of peaks. Polarization is also a generic feature of these oscillations[12-18], and thus provides a model-independent test of the theoretical framework[19-21]. In addition, detection of polarization can in principle triple the number of observed CMB quantities, enhancing our ability to constrain cosmological parameters.

CMB polarization arises from Thompson scattering by electrons of a radiation field with a local quadrupole moment[22]. In the primordial plasma, the local quadrupole moment is suppressed until decoupling, when the photon mean free path begins to grow. At this time, the largest contribution to the local quadrupole is due to Doppler shifts induced by the velocity field of the plasma[23]. CMB polarization thus directly probes the dynamics at the epoch of decoupling. For a spatial Fourier mode of the acoustic oscillations, the velocities are perpendicular to the wavefronts, leading only to perpendicular or parallel alignment of the resulting polarization, which we define as positive and negative respectively. We refer to these polarization modes as scalar E-modes in analogy with electric fields; they have no curl component. Because the level of the polarization depends on velocity, we expect that the peaks in the scalar E-mode power spectrum correspond to density modes that are at their highest velocities at decoupling and are thus at minimum amplitude. The location of the harmonic peaks in the scalar E-mode power spectrum is therefore expected to be out of phase with the peaks in the temperature (T) spectrum[15-17].

In light of the above discussion, it is clear that the CMB temperature and polarization anisotropy should be correlated at some level[24]. For a given multipole, the sign of the TE correlation should depend on whether the amplitude of the density mode was increasing or decreasing at the time of decoupling. We therefore expect a change in the sign of the TE correlation at maxima in the T and the E power spectra, which correspond to modes at maximum and minimum amplitude, respectively. The TE correlation therefore offers a unique and powerful test of the underlying theoretical framework.

Primordial gravity waves will lead to polarization in the CMB[14,25] with an E-mode pattern as for the scalar density perturbations, but will also lead to a curl component, referred to as B-mode polarization[26-28]. The B-mode component is due to the intrinsic polarization of the gravity waves. In inflationary models, the level of the B-mode polarization from gravity waves is set by the fourth power of the inflationary energy scale. While the detection of B-mode polarization would provide a critical test of inflation, the signal is likely to be very weak and may have an amplitude that is effectively unobservable[29]. Furthermore, distortions

在更精细的角度尺度上测量各向异性的工作也已经取得了相当大的进展 [5-10]。

在现代标准的宇宙学模型中（如参考文献 11），CMB 角功率谱的形状可以直接追踪到早期宇宙中光子–重子流体的声学振荡。随着宇宙的扩张，它冷却了下来；大约在此 40 万年后，辐射场的强度已经不足以使宇宙保持电离状态，从而在第一个原子形成的同时，CMB 光子与重子退耦。声学振荡在这个时期达到了极值，并在 CMB 角功率谱中的一系列简谐波峰被观察到。偏振也是这些振荡 [12-18] 的普遍特征，因而为理论框架 [19-21] 提供了与模型无关的检测。此外，偏振的探测原则上可以使观察到的 CMB 数量增加到原来的三倍，从而也增强了我们约束宇宙学参数的能力。

CMB 偏振是由具有局部四极矩的辐射场中电子的汤普森散射产生的 [22]。在原始等离子体中，当光子平均自由程开始增加时，局部四极矩被抑制直到退耦。此时，对局部四极矩贡献最大的是由等离子体的速度场引起的多普勒频移 [23]。因此，CMB 偏振将直接探测到退耦时期的动力学过程。对于声学振荡的空间傅里叶模式，速度垂直于波阵面，导致所得偏振按垂直或平行方向排列，我们将其分别定义为正向和负向。这些偏振模式称为标量 E 模式，类似于电场；它们是无旋的。因为偏振水平取决于速度，我们期望标量 E 模式功率谱中的峰值对应于退耦时处于其最高速度的密度模式，也就是处于最小的幅度之时。因此可以预期标量 E 模式功率谱中的谐波峰值的位置与温度 (T) 频谱中的峰值异相 [15-17]。

鉴于上述讨论，很明显可以看出 CMB 温度和偏振的各向异性应该在某种程度上相关联 [24]。对于给定的多极矩，TE 关联的迹象应当取决于退耦时的密度幅度是增加还是减小。所以我们期望 TE 关联中 T 和 E 功率谱中最大值处的变化，分别对应于最大和最小幅度的模式。因此，TE 相关性为基础理论框架提供了独特而强大的检测手段。

原初引力波将导致 CMB 中具有与标量密度扰动有关的 E 模式的偏振 [14,25]，但也将产生一个旋度分量，称为 B 模式偏振 [26-28]。B 模式分量源于引力波的内禀偏振。在暴胀模型中，来自引力波的 B 模式偏振的水平由暴胀能量标度的四次幂决定。虽然 B 模式偏振的探测将为暴胀提供关键的检测，但是其信号可能非常微弱，而且事实上其幅度可能无法有效观测 [29]。此外，宇宙大尺度结构引力透镜的干预对标量 E

to the scalar E-mode polarization by the gravitational lensing of the intervening large scale structure in the Universe will lead to a contaminating B-mode polarization signal which will severely complicate the extraction of the polarization signature from gravity waves[30-33]. The possibility, however, of directly probing the Universe at energy scales of order 10^{16} GeV by measuring the gravity-wave induced polarization (see, for example, ref. 34) is a compelling goal for CMB polarization observations.

Prior to the results presented in this paper, only upper limits have been placed on the level of CMB polarization. This is due to the low level of the expected signal, demanding sensitive instruments and careful attention to sources of systematic uncertainty (see ref. 35 for a review of CMB polarization measurements).

The first limit to the degree of polarization of the CMB was set in 1965 by Penzias and Wilson, who stated that the new radiation that they had discovered was isotropic and unpolarized within the limits of their observations[1]. Over the next 20 years, dedicated polarimeters were used to set much more stringent upper limits on angular scales of order several degrees and larger[36-41]. The current best upper limits for the E-mode and B-mode polarizations on large angular scales are 10 µK at 95% confidence for the multipole range $2 \leqslant l \leqslant 20$, set by the POLAR experiment[42].

On angular scales of the order of one degree, an analysis of data from the Saskatoon experiment[43] set the first upper limit (25 µK at 95% confidence for $l \approx 75$); this limit is also noteworthy in that it was the first limit that was lower than the level of the CMB temperature anisotropy. The current best limit on similar angular scales was set by the PIQUE experiment[44]—a 95% confidence upper limit of 8.4 µK to the E-mode signal, assuming no B-mode polarization. A preliminary analysis of cosmic background imager (CBI) data[45] indicates an upper limit similar to the PIQUE result, but on somewhat smaller scales. An attempt was also made to search for the TE correlation using the PIQUE polarization and Saskatoon temperature data[46].

Polarization measurements have also been pursued on much finer angular scales (of the order of an arcminute), resulting in several upper limits (for example, refs 47 and 48). However, at these angular scales, corresponding to multipoles of about 5,000, the level of the primary CMB anisotropy is strongly damped and secondary effects due to interactions with large-scale structure in the Universe are expected to dominate[11].

In this paper, we report the detection of polarized anisotropy in the CMB radiation with the Degree Angular Scale Interferometer (DASI) located at the National Science Foundation (NSF) Amundsen–Scott South Pole research station. The polarization data were obtained during the 2001 and 2002 austral winter seasons. DASI was previously used to measure the temperature anisotropy from $140 < l < 900$, during the 2000 season. We presented details of the instrument, the measured power spectrum and the resulting cosmological constraints in a series of three papers: refs 49, 6 and 50. Prior to the start of the 2001 season, DASI was modified to allow polarization measurements in all four Stokes parameters over the same

模式偏振的扭曲将导致 B 模式偏振信号的污染，这将严重地使引力波 [30-33] 的偏振特征的提取复杂化。然而，探索能否通过测量引力波引起的偏振（如参考文献 34），进而在 10^{16} GeV 的能量尺度上直接探测宇宙，仍然是 CMB 偏振观测的一个引人入胜的目标。

在本文得出的结果之前，只有关于 CMB 偏振水平上限的报道。这是由于预期信号的水平较低，需要敏锐的仪器并仔细地留意系统不确定性的来源（参考文献 35 给出了有关 CMB 偏振测量的文献综述）。

CMB 偏振程度的第一个极限是由彭齐亚斯和威耳孙在 1965 年确定的，他们阐明其发现的新辐射在观察极限内是各向同性和非偏振的 [1]。在接下来的 20 年中，人们借助专用的起偏器以确定更严格的上限，其角度的量级为几度或者更大 [36-41]。对于多极矩而言，在 $2 \leqslant l \leqslant 20$ 的范围中，当前大角度范围内 E 模式和 B 模式偏振的最佳上限为 10 μK，由 POLAR 实验 [42] 给出。

在 1 度量级的角度尺度上，来自萨斯卡通实验的数据分析 [43] 确定了第一个上限（25 μK，置信度为 95%，$l \approx 75$）；这个极限也是值得注意的，因为它是第一个低于 CMB 温度各向异性水平的限制。假设没有 B 模式偏振，当前最佳的情况是由设置了相类似角度标度的 PIQUE 实验 [44] 给出的——E 模式信号上限为 8.4 μK，置信度为 95%。宇宙背景成像仪（CBI）数据 [45] 的初步分析显示了一个类似于 PIQUE 结果的上限，但是在稍微小的尺度上给出的。另外一个尝试是使用 PIQUE 偏振和萨斯卡通温度数据 [46] 研究 TE 相关性。

偏振测量也在更精细的角度范围（1 角分量级）内进行，研究人员已经得到了几个上限（如参考文献 47 和 48）。然而，在这些角度尺度上，对应于大约 5,000 的多极矩，主要的 CMB 各向异性效果被强烈阻尼掉，并且预计与宇宙中的大尺度结构相互作用而产生的次级效应会占主导地位 [11]。

在本文中，我们报告了使用位于美国国家科学基金会（NSF）阿蒙森–斯科特南极研究站的度角尺度干涉仪（DASI）检测到的 CMB 辐射中的偏振各向异性。偏振数据是在 2001 年和 2002 年南半球冬季获得的。DASI 前期用于测量 2000 年观测季中 140 < l < 900 的温度各向异性。我们在三篇系列文章（参考文献 49、6 和 50）中提供了仪器的详细信息、测量的功率谱和由此产生的宇宙学约束。在 2001 观测季开始之前，DASI 被调整为可以在所有四个斯托克斯参数下进行偏振观测，其中 l 的取值范

l range as the previous measurements. The modifications to the instrument, observational strategy, data calibration and data reduction are discussed in detail in a companion paper in this issue[51].

The measurements reported here were obtained within two 3.4° full-width at half-maximum (FWHM) fields separated by 1 h in right ascension. The fields were selected from the subset of fields observed with DASI in 2000 in which no point sources were detected, and are located in regions of low Galactic synchrotron and dust emission. The temperature angular power spectrum is found to be consistent with previous measurements and its measured frequency spectral index is −0.01 (−0.16 to 0.14 at 68% confidence), where 0 corresponds to a 2.73 K Planck spectrum. Polarization of the CMB is detected at high confidence ($\geq 4.9\sigma$) and its power spectrum is consistent with theoretical predictions, based on the interpretation of CMB anisotropy as arising from primordial scalar adiabatic fluctuations. Specifically, assuming a shape for the power spectrum consistent with previous temperature measurements, the level found for the E-mode polarization is 0.80 (0.56 to 1.10 at 68% confidence interval), where the predicted level given previous temperature data is 0.9 to 1.1. At 95% confidence, an upper limit of 0.59 is set to the level of B-mode polarization parameterized with the same shape and normalization as the E-mode spectrum. The TE correlation of the temperature and E-mode polarization is detected at 95% confidence and is also found to be consistent with predictions.

With these results contemporary cosmology has passed a long anticipated and crucial test. If the test had not been passed, the underpinnings of much of what we think we know about the origin and early history of the Universe would have been cast into doubt.

Measuring Polarization with DASI

DASI is a compact interferometric array optimized for the measurement of CMB temperature and polarization anisotropy[49,51]. Because they directly sample Fourier components of the sky, interferometers are well suited to measurements of the CMB angular power spectrum. In addition, an interferometer gathers instantaneous two-dimensional information while inherently rejecting large-scale gradients in atmospheric emission. For observations of CMB polarization, interferometers offer several additional features. They can be constructed with small and stable instrumental polarization. Furthermore, linear combinations of the data can be used to construct quantities with essentially pure E- and B-mode polarization response patterns on a variety of scales. This property of the data greatly facilitates the analysis and interpretation of the observed polarization in the context of cosmological models.

DASI is designed to exploit these advantages in the course of extremely long integrations on selected fields of sky. The 13 horn/lens antennas that comprise the DASI array are compact, axially symmetric, and sealed from the environment, yielding small and extremely stable instrumental polarization. Additional systematic control comes from multiple levels of phase

围与之前的研究相同。另一篇与本文一并发表的文章详细讨论了我们在仪器、观测策略、数据校准和数据归算方面的修改 [51]。

本文报告的测量结果是在两个视场获得的，二者具有 3.4° 半高宽（FWHM），赤经上间隔 1 小时。这些视场是从 2000 年利用 DASI 观测到的视场的子集中选出的。在这之中没有检测到点源，且位于银河系同步辐射和尘埃辐射较低的区域。温度角功率谱与先前的测量值一致，并且其测量的频率谱指数为 −0.01（置信度为 68% 的参数区间为 −0.16 至 0.14），其中 0 对应 2.73 K 的普朗克谱。基于对原始标量绝热扰动引起的 CMB 各向异性的解释，探测到的 CMB 的偏振具有很高的可信度（≥ 4.9σ），并且其功率谱与理论预测一致。具体来讲，假设功率谱的形状与先前的温度测量一致，针对 E 模式的偏振水平为 0.80（置信度为 68% 的参数区间为 0.56 至 1.10），若温度数据为先前给定的数据，预测水平将为 0.9 至 1.1。在 95% 置信度下，与 E 模式谱有着相同的形状和归一化的 B 模式偏振参数化水平的上限被限定为 0.59。温度与 E 模式偏振的 TE 相关性以 95% 置信度被探测到，并且发现与预测一致。

通过这些结果，当代宇宙学已经通过了期待已久的关键的检验。如果没有通过这些检验，那么我们对于宇宙的起源及其早期历史的认知基础恐怕就要动摇了。

用 DASI 测量偏振

DASI 是一款紧凑型干涉测量阵列，是专为测量 CMB 温度和偏振各向异性而优化的 [49,51]。因为它们直接采样天空的傅里叶分量，所以干涉仪非常适合测量 CMB 角功率谱。而且干涉仪会收集瞬时二维信息，同时本身还具有排除大气辐射中大尺度梯度的特性。对于 CMB 偏振的观察，干涉仪提供了几个附加功能。观测数据可以用微小且稳定的仪器偏振构造出来。此外，数据的线性组合可用于在各种尺度上构建具有基本上纯的 E 和 B 模式的偏振响应模式。数据的这种性质极大地方便了在宇宙学模型下分析和解释观察到的偏振。

DASI 旨在利用这些优势，在选定的天空视场进行极长的累积观测。构成 DASI 阵列的 13 个喇叭/透镜天线结构紧凑、轴对称、与环境隔离且密封，产生的仪器偏振非常小而极其稳定。额外的系统化的控制来自多级相位切换和视场作差，目的是

switching and field differencing designed to remove instrumental offsets. The DASI mount is fully steerable in elevation and azimuth with the additional ability to rotate the entire horn array about the faceplate axis. The flexibility of this mount allows us to track any field visible from the South Pole continuously at constant elevation angle, and to observe it in redundant faceplate orientations which allow sensitive tests for residual systematic effects.

Instrumental response and calibration

Each of DASI's 13 receivers may be set to admit either left or right circular polarization. An interferometer measures the correlations between the signals from pairs of receivers, called visibilities; as indicated by equation (4) in the "Theory covariance matrix" subsection recovery of the full complement of Stokes parameters requires the correlation of all four pairwise combinations of left and right circular polarization states (RR, LL, RL and LR), which we refer to as Stokes states. The co-polar states (RR, LL) are sensitive to the total intensity, while the cross-polar states (RL, LR) measure linear combinations of the Stokes parameters Q and U.

Each of DASI's analogue correlator channels can accommodate only a single Stokes state, so measurement of the four combinations is achieved via time-multiplexing. The polarizer for each receiver is switched on a 1-h Walsh cycle, with the result that over the full period of the cycle, every pair of receivers spends an equal amount of time in all four Stokes states.

In ref. 51, we detail the calibration of the polarized visibilities for an interferometer. In order to produce the calibrated visibilities as defined in equation (4) below, a complex gain factor which depends on the Stokes state must be applied to each raw visibility. Although the cross-polar gain factors could easily be determined with observations of a bright polarized source, no suitable sources are available, and we therefore derive the full calibration through observations of an unpolarized source. The gains for a given pair of receivers (a baseline) can be decomposed into antenna-based factors, allowing us to construct the cross-polar gains from the antenna-based gain factors derived from the co-polar visibilities. DASI's calibration is based on daily observations of the bright H_{II} region RCW38, which we described at length in ref. 49. We can determine the individual baseline gains for all Stokes states with statistical uncertainties < 2% for each daily observation. Systematic gain uncertainties for the complete data set are discussed in the "Systematic uncertainties" section.

The procedure for deriving the baseline gains from antenna-based terms leaves the cross-polar visibilities multiplied by an undetermined overall phase offset (independent of baseline). This phase offset effectively mixes Q and U, and must be measured to obtain a clean separation of CMB power into E- and B-modes. Calibration of the phase offset requires a source whose polarization angle is known, and we create one by observing RCW38 through polarizing wire grids attached to DASI's 13 receivers. From the wire-grid observations, we can derive the phase offset in each frequency band with an uncertainty of $\leqslant 0.4°$.

498

消除仪器错位。DASI 承载底座在高度和方位角上具有完全可操纵性，并且能够围绕面板对整个喇叭阵列进行轴旋转。这种承载架的灵活性使我们能够以恒定的仰角连续追踪来自南极的任何可见的视场，并以其他冗余的面板方向观察它，从而可以对剩余的系统效应进行灵敏性检测。

仪器响应和校准

DASI 的 13 个接收器中的每一个都允许被设置为左或右圆偏振。干涉仪测量的是来自成对的接收器信号之间的相关，称为可视性；如本文"理论协方差矩阵"部分中的等式(4)所示，斯托克斯参数的完整获得要求左右圆偏振状态(RR、LL、RL 和 LR)的所有四个成对组合相关，我们称之为斯托克斯状态。共极态(RR、LL)对总强度敏感，而交叉极态(RL、LR)可以测量斯托克斯参数 Q 和 U 的线性组合。

每个 DASI 的模拟相关器通道只能容纳单个斯托克斯状态，因此通过时分多路复用实现四种状态组合的测量。每个接收器的起偏器接通 1 小时沃尔什周期，最终使得在整个周期内，每对接收器在所有的四个斯托克斯状态下花费相等的时间。

在参考文献 51 中，我们详细说明了干涉仪偏振可视性的校准。为了产生如下文的等式(4)中定义的校准可视性，必须将依赖于斯托克斯状态的复增益因子应用在每个初始可视性上。尽管通过观察明亮的偏振源可以很容易地确定交叉偏振增益因子，但是没有合适的源可用，因此我们通过观察非偏振源得到完全的校准。一对给定的接收器(即一条基线)的增益可以分解为多个天线增益因子，这允许我们从来自共极可视性的天线增益因子来导出交叉偏振增益。DASI 的校准基于每天对 RCW38 的 H_{II} 区域的观测，我们在参考文献 [49] 中对其进行了详细的描述。我们可以确定所有斯托克斯状态的单个基线增益，每日观察的系统不确定性 <2%。"系统不确定性"部分讨论了完整数据集的系统增益不确定性。

从基于天线的项获得基线增益的过程中剩下了交叉偏振可视性乘以未确定的总相位偏移(与基线无关)。该相位偏移有效地混合 Q 和 U，并且必须进行测量才能获得从 CMB 功率到 E 模式和 B 模式的无污染的分离。相位偏移的校准需要一个偏振角度已知的光源，我们通过观测 DASI 的 13 个接收器附属的偏振丝栅 RCW38 来创建一个这样的光源。根据丝栅观测，我们可以推导出每个频带的相位偏移，不确定度 ≤0.4°。

As an independent check of this phase offset calibration, the Moon was observed at several epochs during 2001–02. Although the expected amplitude of the polarized signal from the Moon is not well known at these frequencies, the polarization pattern is expected to be radial to high accuracy, and this can be used to determine the cross-polar phase offset independently of the wire grid observations. We show in ref. 51 that these two determinations of the phase offset are in excellent agreement.

On-axis leakage

For ideal polarizers, the cross-polar visibilities are strictly proportional to linear combinations of the Stokes parameters Q and U. For realistic polarizers, however, imperfect rejection of the unwanted polarization state leads to additional terms in the cross-polar visibilities proportional to the total intensity I. These leakage terms are the sum of the complex leakage of the two antennas which form each baseline. During the 2000–01 austral summer, DASI's 13 receivers were retrofitted with multi-element broadband circular polarizers designed to reject the unwanted polarization state to high precision across DASI's 26–36 GHz frequency band. Before installation on the telescope, the polarizers were tuned to minimize these leakages.

At several epochs during 2001–02, the antenna-based leakages were determined with a fractional error of 0.3% from deep observations of the calibrator source RCW38. We show in ref. 51 that antenna-based leakages are $\lesssim 1\%$ (of I) at all frequency bands except the highest, for which they approach 2%; this performance is close to the theoretical minimum for our polarizer design. Comparison of the measurements from three epochs separated by many months indicates that the leakages are stable with time.

Given the low level of DASI's leakages, the mixing of power from temperature into polarization in the uncorrected visibilities is expected to be a minor effect at most (see the "Systematic uncertainties" section). Nonetheless, in the analysis presented in this paper, the cross-polar data have in all cases been corrected to remove this effect using the leakages determined from RCW38.

Off-axis leakage

In addition to on-axis leakage from the polarizers, the feeds will contribute an instrumental polarization that varies across the primary beam. Offset measurements of RCW38 and the Moon indicate that the off-axis instrumental polarization pattern is radial, rising from zero at the beam centre to a maximum of about 0.7% near 3°, and then tapering to zero (see also ref. 51).

With the on-axis polarizer leakage subtracted to $\lesssim 0.3\%$ (see above), this residual leakage, while still quite small compared to the expected level of polarized CMB signal (again,

500

作为这种相位偏移校准的独立检查，在 2001 年 ~ 2002 年期间的几个时期将月球作为观测对象。虽然在这些频率下，来自月球的偏振信号的预期幅度并不十分清楚，但在高精度下预计的偏振模式是发散状的，这可以用来独立地确定丝栅观测的交叉偏振相位偏移。我们在参考文献 51 中表明这两个相位偏移的测定非常一致。

共轴渗漏

对于理想的起偏器，交叉偏振可视性严格正比于斯托克斯参数 Q 和 U 的线性组合。然而，对于真实的起偏器，不需要的偏振态的不完全抑制导致交叉偏振可视性中的附加项与总强度 I 成正比。这些渗漏项是从每个基线形成的两个天线的复合渗漏的总和。在 2000 年 ~ 2001 年南半球夏季期间，研究人员采用多元宽带圆形起偏器更新改造了 DASI 的 13 个接收器，这是为了使通过 DASI 的 26 GHz ~ 36 GHz 频带在高精度时抑制不必要的偏振态。在安装到望远镜上之前，起偏器被调整到最大限度地减少这些渗漏的状态。

在 2001 年 ~ 2002 年间的几个时期内，通过对校准器源 RCW38 的深度观察，天线基线的渗漏被确定会导致 0.3% 的相对误差。参考文献 51 显示天线基线的渗漏在除最高频段之外的所有频段都 $\leqslant 1\%(I)$，而最高频段接近 2% ；这种性能接近我们起偏器设计的理论最小值。通过三个时期的测量值(时期之间相隔数个月)的比较，表明渗漏随时间推移是稳定的。

鉴于 DASI 渗漏水平较低，预计未校正的可视性中，从温度到极化的功率混合最多也只是产生次要的影响(见"系统不确定性"部分)。尽管如此，在本文提出的分析方案中，所有情况下都使用 RCW38 确定的渗漏来纠正交叉极化数据，从而消除上述影响。

离轴渗漏

除了来自起偏器的共轴渗漏之外，反馈将导致仪器偏振，这个偏振在横穿初级光束的过程中不断变化。RCW38 和月球的偏移测量表明，离轴仪器偏振模式是径向的，光束中心为 0，在接近 3° 时上升到最大值 0.7%，然后逐渐减小到 0(参考文献 51)。

随着共轴起偏器的渗漏减去 $\leqslant 0.3\%$(见上文)的值，这种残余的渗漏虽然与偏振 CMB 信号的预期水平相比仍然非常小(见"系统不确定性"部分)，但却是仪器贡献

see the "Systematic uncertainties" section), is the dominant instrumental contribution. Although the visibilities cannot be individually corrected to remove this effect (as for the on-axis leakage), it may be incorporated in the analysis of the CMB data. Using fits to the offset data (see ref. 51 for details), we account for this effect by modelling the contribution of the off-axis leakage to the signal covariance matrix as described in the "Theory covariance matrix" subsection.

CMB Observations and Data Reduction

Observations

For the observations presented here, two fields separated by 1 h of right ascension, at RA = 23 h 30 min and RA = 00 h 30 min with declination $-55°$, were tracked continuously. The telescope alternated between the fields every hour, tracking them over precisely the same azimuth range so that any terrestrial signal can be removed by differencing. Each 24-h period included 20 h of CMB observations and 2.3 h of bracketing calibrator observations, with the remaining time spent on skydips and miscellaneous calibration tasks.

The fields were selected from the 32 fields previously observed by DASI for the absence of any detectable point sources (see ref. 49). The locations of the 32 fields were originally selected to lie at high elevation angle and to coincide with low emission in the IRAS 100 μm and 408 MHz maps of the sky[52].

The data presented in this paper were acquired from 10 April to 27 October 2001, and again from 14 February to 11 July 2002. In all, we obtained 162 days of data in 2001, and 109 days in 2002, for a total of 271 days before the cuts described in the next section.

Data cuts

Observations are excluded from the analysis, or cut, if they are considered suspect owing to hardware problems, inadequate calibration, contamination from Moon or Sun signal, poor weather or similar effects. In the "Data consistency and χ^2 tests" section, we describe consistency statistics that are much more sensitive to these effects than are the actual results of our likelihood analysis, allowing us to be certain that the final results are unaffected by contamination. Here we briefly summarize the principal categories of data cuts; we describe each cut in detail in ref. 51.

In the first category of cuts, we reject visibilities for which monitoring data from the telescope indicate obvious hardware malfunction, or simply non-ideal conditions. These include cryogenics failure, loss of tuning for a receiver, large offsets between real/imaginary multipliers in the correlators, and mechanical glitches in the polarizer stepper motors. All data are rejected for a correlator when it shows evidence for large offsets, or excessive noise.

502

的主要成分。虽然可视性不能通过独立地校正以消除这种影响（对于共轴渗漏），但它却可能包含在 CMB 数据的分析中。使用拟合偏移数据（参考文献 51），我们通过模拟离轴渗漏对信号协方差矩阵的贡献来解释这种影响，如"理论协方差矩阵"部分所述。

CMB 观测和数据归算

观测

这里呈现的观测是在赤经为 23 h 30 min 和 00 h 30 min，赤纬为 −55°的位置上，对赤经相隔一个小时的两个视场连续追踪所获得的。望远镜每小时在各个视场之间交替观测，在精确相同的方位角上追踪它们，以便通过作差去除所有地面信号。每个 24 小时周期包括 20 小时的 CMB 观察和 2.3 小时的托架校准器观察，剩余的时间用于天空倾角校准和其他校准任务。

这些视场是从 DASI 先前观察到的 32 个视场中选择的，因为缺乏任何可检测的点源（参考文献 49）。这 32 个视场的位置最初被选择为处于高仰角并且与 IRAS 100 μm 和 408 MHz 天图中的低辐射一致的地方[52]。

本文提供的数据是在 2001 年 4 月 10 日至 10 月 27 日以及 2002 年 2 月 14 日至 7 月 11 日分两次获得的。我们在 2001 年总共获得了 162 天的数据，在 2002 年获得了 109 天的数据，在下一节描述的数据舍弃之前，共计 271 天。

数据舍弃

如果观测结果被认为具有硬件问题、校准不充分、月亮或太阳信号污染、恶劣天气或其他类似的可疑影响，将被排除在分析之外，或者说被舍弃。在"数据一致性和 χ^2 检验"部分中我们指出，比较而言，一致性统计对于这些影响的敏感度要远高于获得这一统计量的似然分析的实际结果，这要求我们要能够确定最终结果不受污染的影响。在这里，我们简要总结一下数据舍弃的主要类别；在参考文献 51 中我们详细描述了每种数据的舍弃。

第一类舍弃的可视性是在具有明显的硬件故障的或仅仅是非理想条件下的望远镜的监测数据。其中包括低温失效、接收机失调、相关器中实/虚乘法器之间的大偏移，以及起偏器步进电机中的机械假信号。当显示有过大偏移或过大噪声的证据时，

An additional cut, and the only one based on the individual data values, is a $> 30\sigma$ outlier cut to reject rare ($\ll 0.1\%$ of the data) hardware glitches. Collectively, these cuts reject about 26% of the data.

In the next category, data are cut on the phase and amplitude stability of the calibrator observations. Naturally, we reject data for which bracketing calibrator observations have been lost due to previous cuts. These cuts reject about 5% of the data.

Cuts are also based on the elevation of the Sun and Moon. Co-polar data are cut whenever the Sun was above the horizon, and cross-polar data whenever the solar elevation exceeded $5°$. These cuts reject 8% of the data.

An additional cut, which is demonstrably sensitive to poor weather, is based on the significance of data correlations as discussed in the "Noise model" subsection. An entire day is cut if the maximum off-diagonal correlation coefficient in the data correlation matrix exceeds 8σ significance, referred to gaussian uncorrelated noise. A total of 22 days are cut by this test in addition to those rejected by the solar and lunar cuts.

Reduction

Data reduction consists of a series of steps to calibrate and reduce the data set to a manageable size for the likelihood analysis. Phase and amplitude calibrations are applied to the data on the basis of the bracketing observations of our primary celestial calibrator, RCW38. The raw 8.4-s integrations are combined over each 1-h observation for each of 6,240 visibilities (78 complex baselines $\times$ 10 frequency bands $\times$ 4 Stokes states). In all cases, on-axis leakage corrections are applied to the data, and sequential 1-h observations of the two fields in the same $15°$ azimuth range are differenced to remove any common ground signal, using a normalization $(\text{field}1-\text{field}2)/\sqrt{2}$ that preserves the variance of the sky signal. Except in the case where the data set is split for use in the χ^2 consistency tests in the "χ^2 tests" subsection, observations from different faceplate rotation angles, epochs and azimuth ranges are all combined, as well as the two co-polar Stokes states, LL and RR. The resulting data set has $N \leqslant 4,680$ elements ($6,240 \times 3/4 = 4,680$, where the 3/4 results from the combination of LL and RR). We call this the uncompressed data set, and it contains all of the information in our observations of the differenced fields for Stokes parameters I, Q and U.

Data Consistency and χ^2 Tests

We begin our analysis by arranging the data into a vector, considered to be the sum of actual sky signal and instrumental noise: $\mathbf{\Delta} = \mathbf{s} + \mathbf{n}$. The noise vector $\mathbf{n}$ is hypothesized to be gaussian and random, with zero mean, so that the noise model is completely specified by a known covariance matrix $C_N \equiv \langle \mathbf{nn}^t \rangle$. Any significant excess variance observed in the

该相关器所有数据都被拒绝。额外的数据舍弃，以及唯一一种基于数据值个体本身的舍弃是 $>30\sigma$ 的异常值，以拒绝罕见的（$<<0.1\%$ 的数据）硬件假信号。总的来说，这些舍弃拒绝了大约 26% 的数据。

下一类数据舍弃的依据是校准器观察的相位和幅度稳定性。于是我们拒绝因先前数据舍弃导致托架校准器观察丢失的数据。这些削减拒绝了大约 5% 的数据。

数据也会因为太阳和月亮的升高带来的影响而被舍弃。每当太阳高于地平线时，同极数据都会被舍弃，而当太阳升高超过 5° 时，就会出现交叉偏振数据。这些舍弃拒绝了 8% 的数据。

另一个明显对恶劣天气敏感的舍弃，依据的是"噪声模型"部分对数据相关的显著性的讨论。如果数据相关矩阵中的最大非对角线相关系数的显著性超过 8σ，则视为高斯不相关噪声，全天的数据都将被舍弃。除了被太阳和月球影响而拒绝那些数据之外，该检测总共减少了 22 天的数据量。

归算

数据归算包括一系列步骤，用于校准数据集并将数据集压缩到可管理的大小以进行似然分析。根据我们的初级天体校准器 RCW38 的基础观测，相位和幅度校准将应用在数据上。原始的 8.4 s 集成结合了每小时观测中所有 6,240 种可视性（78 个复基线 × 10 个频带 × 4 个斯托克斯状态）。共轴渗漏校正被正应用于所有情况下的数据，两个视场在相同的 15° 方位角范围内的连续 1 h 观测值相减以消除任何常见的地面信号，并且使用（视场 1 − 视场 2）$/\sqrt{2}$ 来归一化，从而保留天空信号的方差。除了在"χ^2 检验"部分对数据集进行分划以实现 χ^2 一致性检验的情况之外，来自不同的面盘旋转角度、时期和方位角范围的观测结果都被组合起来，同时被组合起来的还包括两个共极斯托克斯状态，LL 和 RR。得到的数据集具有 $N \leqslant 4,680$ 个元素（$6,240 \times 3/4 = 4,680$，其中 3/4 由 LL 和 RR 的组合产生）。我们将其称为未压缩数据集，它包含经过作差处理之后的视场斯托克斯参数 I、Q 和 U 的所有观测信息。

数据一致性和 χ^2 检验

我们通过将数据排列成一个矢量来开始分析，该矢量 Δ 被认为是实际天空信号和仪器噪声的总和：$\Delta = s + n$。噪声矢量 n 被假设为高斯的且随机的，具有零均值，因此噪声模型完全由已知的协方差矩阵 $C_N \equiv \langle nn^t \rangle$ 确定。在数据矢量 Δ 中观察到的

data vector Δ will be interpreted as signal. In the likelihood analysis of the next section, we characterize the total covariance of the data set $C = C_T(\kappa) + C_N$ in terms of parameters κ that specify the covariance C_T of this sky signal. This is the conventional approach to CMB data analysis, and it is clear that for it to succeed, the assumptions about the noise model and the accuracy of the noise covariance matrix must be thoroughly tested. This is especially true for our data set, for which long integrations have been used to reach unprecedented levels of sensitivity in an attempt to measure the very small signal covariances expected from the polarization of the CMB.

Noise model

The DASI instrument and observing strategy are designed to remove systematic errors through multiple levels of differencing. Slow and fast phase switching as well as field differencing is used to minimize potentially variable systematic offsets that could otherwise contribute a non-thermal component to the noise. The observing strategy also includes Walsh sequencing of the Stokes states, observations over multiple azimuth ranges and faceplate rotation angles, and repeated observations of the same visibilities on the sky throughout the observing run to allow checks for systematic offsets and verification that the sky signal is repeatable. We hypothesize that after the cuts described in the previous section, the noise in the remaining data is gaussian and white, with no noise correlations between different baselines, frequency bands, real/imaginary pairs, or Stokes states. We have carefully tested the noise properties of the data to validate the use of this model.

Noise variance in the combined data vector is estimated by calculating the variance in the 8.4-s integrations over the period of 1 h, before field differencing. To test that this noise estimate is accurate, we compare three different short-timescale noise estimates: calculated from the 8.4-s integrations over the 1-h observations before and after field differencing and from sequential pairs of 8.4-s integrations. We find that all three agree within 0.06% for co-polar data and 0.03% for cross-polar data, averaged over all visibilities after data cuts.

We also compare the noise estimates based on the short-timescale noise to the variance of the 1-h binned visibilities over the entire data set (up to 2,700 1-hour observations, over a period spanning 457 days). The ratio of long-timescale to short-timescale noise variance, averaged over all combined visibilities after data cuts, is 1.003 for the co-polar data and 1.005 for the cross-polar data, remarkably close to unity. Together with the results of the χ^2 consistency tests described in the "χ^2 tests" subsection, these results demonstrate that the noise is white and integrates down from timescales of a few seconds to thousands of hours. We find that scaling the diagonal noise by 1% makes a negligible difference in the reported likelihood results (see the "Systematic uncertainties" section).

To test for potential off-diagonal correlations in the noise, we calculate a $6{,}240 \times 6{,}240$ correlation coefficient matrix from the 8.4-s integrations for each day of observations. To increase our sensitivity to correlated noise, we use only data obtained simultaneously for

任何显著的方差超出将被解释为信号。在下一部分的似然分析中，我们依据确定该天空信号协方差 C_T 的参数 κ 来表征数据集的总协方差 $C = C_T(\kappa) + C_N$。这是 CMB 数据分析的传统方法，为了能成功实现，很显然必须彻底检测噪声模型的假设和噪声协方差矩阵的准确性。这对于我们的数据集尤甚，因而为了测量由 CMB 偏振产生的非常小的信号的协方差，我们使用长期积分来达到前所未有的灵敏度水平。

噪声模型

DASI 仪器和观测策略被设计为通过多级作差消除系统误差。慢速和快速相位切换以及视场作差被用于最小化潜在可变的系统偏移，否则可能增加噪声的非热分量。观测策略还包括斯托克斯状态的沃尔什测序，多个方位角范围和面盘旋转角度的观测，以及通过允许检查系统与对偏移与天空信号的检查是可重复的观测，进而实现在整个观测过程中对天空的相同可视性进行重复观测。我们假设在如前一节所描述的数据舍弃之后，剩余数据中的噪声是高斯的白噪声，不同基线、频带、实/虚对或斯托克斯状态之间没有噪声相关。我们仔细检测了数据的噪声属性，以验证该模型的实用性。

通过在视场作差之前 1 小时的时段内计算 8.4 s 积分的方差，组合而成的数据矢量中的噪声方差可以被估算出来。为了检测这种噪声估计是否准确，我们比较了三段不同的短时间尺度内的噪声估计：计算视场差前后 1 h 观测值的 8.4 s 积分以及 8.4 s 积分的连续对。我们发现三者在共极数据中均为 0.06%，在交叉偏振数据中为 0.03%，在数据舍弃后平均了所有可视性。

我们还将基于短时间尺度的噪声估计值与整个数据集中 1 小时分格可视性的方差进行比较（在时间跨度为 457 天的情况下，最多可得到 2,700 个 1 小时观测值）。数据舍弃后，所有组合可视性的长时间尺度与短时间尺度噪声方差的比率可以给出，其中共极数据为 1.003，而交叉偏振数据为 1.005，非常接近于 1。这与"χ^2 检验"部分描述的 χ^2 一致性检验的结果一起表明噪声是白色的，并且随着时间尺度从几秒合并为几千小时，噪声也同时下降。我们发现将对角线噪声缩放 1% 使似然结果产生的差异可以忽略不计（参见"系统不确定性"部分）。

为了检测噪声中潜在的非对角线相关性，我们计算了每天观测的 8.4 s 积分的 6,240 × 6,240 相关系数矩阵。为了提高数据对相关噪声的灵敏度，我们仅使用特定的一对数据矢量元素同时获得的数据。由于 8.4 s 积分的变量数 M 曾用于计算每个

a given pair of data vector elements. Owing to the variable number of 8.4-s integrations M used to calculate each off-diagonal element, we assess the significance of the correlation coefficient in units of $\sigma = 1/\sqrt{M-1}$. Our weather cut statistic is the daily maximum off-diagonal correlation coefficient significance (see "Data cuts" subsection).

We use the mean data correlation coefficient matrix over all days, after weather cuts, to test for significant correlations over the entire data set. We find that 1,864 (0.016%) of the off-diagonal elements exceed a significance of 5.5σ, when about one such event is expected for uncorrelated gaussian noise. The outliers are dominated by correlations between real/imaginary pairs of the same baseline, frequency band, and Stokes state, and between different frequency bands of the same baseline and Stokes state. For the real/imaginary pairs, the maximum correlation coefficient amplitude is 0.14, with an estimated mean amplitude of 0.02; for interband correlations the maximum amplitude and estimated mean are 0.04 and 0.003, respectively. We have tested the inclusion of these correlations in the likelihood analysis and find that they have a negligible impact on the results, see the "Systematic uncertainties" section.

χ^2 tests

As a simple and sensitive test of data consistency, we construct a χ^2 statistic from various splits and subsets of the visibility data. Splitting the data into two sets of observations that should measure the same sky signal, we form the statistic for both the sum and difference data vectors, $\chi^2 = \Delta' C_N^{-1} \Delta$, where $\Delta = (\Delta_1 \pm \Delta_2)/2$ is the sum or difference data vector, and $C_N = (C_{N1} + C_{N2})/4$ is the corresponding noise covariance matrix. We use the difference data vector, with the common sky signal component subtracted, to test for systematic offsets and mis-estimates of the noise. The sum data vector is used to test for the presence of a sky signal in a straightforward way that is independent of the likelihood analyses that will be used to parameterize and constrain that signal.

We split the data for the difference and sum data vectors in five different ways:

(1) Year—2001 data versus 2002 data;

(2) Epoch—the first half of observations of a given visibility versus the second half;

(3) Azimuth range—east five versus west five observation azimuth ranges;

(4) Faceplate position—observations at a faceplate rotation angle of $0°$ versus a rotation angle of $60°$; and

(5) Stokes state—co-polar observations in which both polarizers are observing left circularly polarized light (LL Stokes state) versus those in which both are observing right circularly polarized light (RR Stokes state).

508

非对角线元素，故我们以 $\sigma = 1/\sqrt{M-1}$ 为单位评估相关系数的显著性。我们基于天气因素的数据舍弃统计量是每日非对角线相关系数的显著性的最大值（参见"数据舍弃"部分）。

在天气因素的舍弃数据后，所有观测日均使用平均数据相关系数矩阵来检测整个数据集的显著相关性。我们发现 1,864 个（0.016%）非对角线元素具有超过 5.5σ 的显著性，而原本这个相关系数矩阵应该只存在不相关的高斯噪声。异常值主要受相同基线、频带和斯托克斯状态的实/虚对之间的相关性，以及相同基线和斯托克斯状态的不同频带之间的相关性影响。对于实/虚对，最大相关系数幅度为 0.14，估计平均幅度为 0.02；对于带间相关性，最大幅度和估计平均值分别为 0.04 和 0.003。我们已经在似然分析中检测了这些相关性，发现它们对结果的影响可以忽略不计，请参阅"系统不确定性"部分。

χ^2 检验

作为对数据一致性的简单且灵敏的检验，我们从可视性数据的各种划分和子集中构建 χ^2 统计量。将数据分成两组观测值，它们应该测量相同的天空信号，我们构造了和与差数据矢量的统计量，$\chi^2 = \Delta' C_N^{-1} \Delta$，其中 $\Delta = (\Delta_1 \pm \Delta_2)/2$，是和或差数据矢量，而 $C_N = (C_{N1} + C_{N2})/4$ 是相应的噪声协方差矩阵。我们使用差数据矢量，减去公共天空信号分量，来检验系统偏移和噪声的误差估计。数据矢量之和是一种直接用于检验天空信号存在的方式，该方式独立于即将应用于参数化和信号约束的似然分析。

我们以五种不同的方式划分数据矢量的差与和：

（1）年份——2001 年的数据与 2002 年的数据；

（2）时期——对于给定可视性的前半次观测和后半次观测；

（3）方位角范围——东方的和西方的五个观测方位角；

（4）面盘位置——观察面板旋转角度为 0° 和旋转角度为 60°；

（5）斯托克斯状态——共极观测，其中两个起偏器都观察到左旋圆偏振光（LL 斯托克斯状态），而另两个起偏器都观察到右旋圆偏振光（RR 斯托克斯状态）。

These splits were done on the combined 2001–02 data set and (except for the first split type) on 2001 and 2002 data sets separately, to test for persistent trends or obvious differences between the years. The faceplate position split is particularly powerful, since the six-fold symmetry of the (u, v) plane coverage allows us to measure a sky signal for a given baseline with a different pair of receivers, different backend hardware, and at a different position on the faceplate with respect to the ground shields, and is therefore sensitive to calibration and other offsets that may depend on these factors. The co-polar split tests the amplitude and phase calibration between polarizer states, and tests for the presence of circularly polarized signal.

For each of these splits, different subsets can be examined: co-polar data only, cross-polar data only (for all except the Stokes state split), various l-ranges (as determined by baseline length in units of wavelength), and subsets formed from any of these which isolate modes with the highest expected signal to noise (s/n). In constructing the high s/n subsets, we must assume a particular theoretical signal template in order to define the s/n eigenmode basis[53] appropriate for that subset. For this we use the concordance model defined in the "Likelihood parameters" subsection, although we find the results are not strongly dependent on the choice of model. Note that the definitions of which modes are included in the high s/n subsets are made in terms of average theoretical signal, without any reference to the actual data. In Table 1, we present the difference and sum χ^2 values for a representative selection of splits and subsets. In each case we give the degrees of freedom, χ^2 value, and probability to exceed (PTE) this value in the χ^2 cumulative distribution function. For the 296 different split/subset combinations that were examined, the χ^2 values for the difference data appear consistent with noise; among these 296 difference data χ^2 values, there are two with a PTE < 0.01 (the lowest is 0.003), one with a PTE > 0.99, and the rest appear uniformly distributed between this range. There are no apparent trends or outliers among the various subsets or splits.

Table 1. χ^2 Consistency tests for a selection of data splits and subsets

Temperature data						
Split type	Subset	Difference			Sum	
		d.f.	χ^2	PTE	χ^2	PTE
Year	Full	1,448	1,474.2	0.31	23,188.7	$< 1 \times 10^{-16}$
	$s/n > 1$	320	337.1	0.24	21,932.2	$< 1 \times 10^{-16}$
	l range 0–245	184	202.6	0.17	10,566.3	$< 1 \times 10^{-16}$
	l range 0–245 high s/n	36	38.2	0.37	10,355.1	$< 1 \times 10^{-16}$
	l range 245–420	398	389.7	0.61	7,676.0	$< 1 \times 10^{-16}$
	l range 245–420 high s/n	79	88.9	0.21	7,294.4	$< 1 \times 10^{-16}$
	l range 420–596	422	410.5	0.65	3,122.5	$< 1 \times 10^{-16}$
	l range 420–596 high s/n	84	73.5	0.79	2,727.8	$< 1 \times 10^{-16}$
	l range 596–772	336	367.8	0.11	1,379.5	$< 1 \times 10^{-16}$

这些划分整合了 2001 年 ~ 2002 年的数据集(除第一种划分类型之外),是 2001 年和 2002 年分开进行的,以检测这两年持续的趋势或者明显的差异。面盘位置的划分方法特别有效,因为 (u, v) 平面覆盖的六重对称允许我们在给定基线上使用不同的接收器对、不同的后端硬件和相对于接地屏蔽的面盘的不同位置来测量天空信号,因此校准和其他偏移可能敏感地取决于这些因素。共极划分检测了起偏器状态之间的幅度和相位校准,并检测圆偏振信号是否存在。

对于上述每一个划分方法,可以检查不同的子集:单一共极数据、单一交叉偏振数据(除了斯托克斯状态划分之外的所有数据)、各种多级矩 l 范围(由波长为单位的基线长度确定),以及由其中具有任意最高的预估信噪比(s/n)的孤立模式形成的子集。在构造高 s/n 子集时,我们必须假设特定的理论信号拟合模板,以便定义适合于该子集的 s/n 基本本征模[53]。为此,我们使用"似然参数"部分定义的一致性模型,尽管发现结果并不强烈依赖于模型的选择。需要注意的是,高 s/n 子集中包括哪些模式的定义是根据平均理论信号进行的,而不参考实际数据。在表 1 中,我们给出了划分和子集的代表选择的差值与和值的 χ^2。在每种情况下,我们都给出自由度、χ^2 值以及这个值在 χ^2 累积分布函数的超出概率(PTE)。对于测试的 296 种不同的划分/子集组合,差值数据的 χ^2 值似乎与噪声一致;在这 296 个差值数据的 χ^2 值中,有两个 PTE < 0.01(最低为 0.003),一个 PTE > 0.99,其余均匀分布在该范围之间。各种子集或划分中没有明显的趋势或异常值。

表 1. 用于选择数据划分与数据子集的 χ^2 一致性检验

		温度数据				
划分方式	子集	差值			和值	
		d.f.	χ^2	PTE	χ^2	PTE
年份	完整	1,448	1,474.2	0.31	23,188.7	$< 1 \times 10^{-16}$
	$s/n > 1$	320	337.1	0.24	21,932.2	$< 1 \times 10^{-16}$
	l 范围 0 ~ 245	184	202.6	0.17	10,566.3	$< 1 \times 10^{-16}$
	l 范围 0 ~ 245 高 s/n	36	38.2	0.37	10,355.1	$< 1 \times 10^{-16}$
	l 范围 245 ~ 420	398	389.7	0.61	7,676.0	$< 1 \times 10^{-16}$
	l 范围 245 ~ 420 高 s/n	79	88.9	0.21	7,294.4	$< 1 \times 10^{-16}$
	l 范围 420 ~ 596	422	410.5	0.65	3,122.5	$< 1 \times 10^{-16}$
	l 范围 420 ~ 596 高 s/n	84	73.5	0.79	2,727.8	$< 1 \times 10^{-16}$
	l 范围 596 ~ 772	336	367.8	0.11	1,379.5	$< 1 \times 10^{-16}$

Continued

		Temperature data				
		Difference			Sum	
Split type	Subset	d.f.	χ^2	PTE	χ^2	PTE
Year	l range 596–772 high s/n	67	82.3	0.10	991.8	$< 1 \times 10^{-16}$
	l range 772–1,100	108	103.7	0.60	444.4	$< 1 \times 10^{-16}$
	l range 772–1,100 high s/n	21	22.2	0.39	307.7	$< 1 \times 10^{-16}$
Epoch	Full	1,520	1,546.3	0.31	32,767.2	$< 1 \times 10^{-16}$
	$s/n > 1$	348	366.5	0.24	31,430.0	$< 1 \times 10^{-16}$
Azimuth range	Full	1,520	1,542.6	0.34	32,763.8	$< 1 \times 10^{-16}$
	$s/n > 1$	348	355.2	0.38	31,426.9	$< 1 \times 10^{-16}$
Faceplate position	Full	1,318	1,415.2	0.03	27,446.5	$< 1 \times 10^{-16}$
	$s/n > 1$	331	365.3	0.09	26,270.1	$< 1 \times 10^{-16}$
Stokes state	Full	1,524	1,556.6	0.27	33,050.6	$< 1 \times 10^{-16}$
	$s/n > 1$	350	358.2	0.37	31,722.5	$< 1 \times 10^{-16}$

		Polarization data				
		Difference			Sum	
Split type	Subset	d.f.	χ^2	PTE	χ^2	PTE
Year	Full	2,896	2,949.4	0.24	2,925.2	0.35
	$s/n > 1$	30	34.4	0.27	82.4	8.7×10^{-7}
	l range 0–245	368	385.9	0.25	315.0	0.98
	l range 0–245 high s/n	73	61.0	0.84	64.5	0.75
	l range 245–420	796	862.2	0.05	829.4	0.20
	l range 245–420 high s/n	159	176.0	0.17	223.8	5.4×10^{-7}
	l range 420–596	844	861.0	0.33	837.3	0.56
	l range 420–596 high s/n	168	181.3	0.23	189.7	0.12
	l range 596–772	672	648.1	0.74	704.4	0.19
	l range 596–772 high s/n	134	139.5	0.35	160.0	0.06
	l range 772–1,100	216	192.3	0.88	239.1	0.13
	l range 772–1,100 high s/n	43	32.3	0.88	47.6	0.29
Epoch	Full	3,040	2,907.1	0.96	3,112.2	0.18
	$s/n > 1$	34	29.2	0.70	98.6	3.3×10^{-8}
Azimuth range	Full	3,040	3,071.1	0.34	3,112.9	0.17
	$s/n > 1$	34	38.7	0.27	98.7	3.3×10^{-8}
Faceplate position	Full	2,636	2,710.4	0.15	2,722.2	0.12
	$s/n > 1$	32	43.6	0.08	97.5	1.6×10^{-8}

Results of χ^2 consistency tests for a representative selection of splits and subsets of the combined 2001–02 data set. Visibility data containing the same sky signal is split to form two data vectors; using the instrument noise model, the χ^2 statistic is then calculated on both the difference and sum data vectors. Also tabulated are the number of degrees

	温度数据					
划分方式	子集	差值			和值	
		d.f.	χ^2	PTE	χ^2	PTE
年份	l 范围 596～772 高 s/n	67	82.3	0.10	991.8	$<1\times10^{-16}$
	l 范围 772～1,100	108	103.7	0.60	444.4	$<1\times10^{-16}$
	l 范围 772～1,100 高 s/n	21	22.2	0.39	307.7	$<1\times10^{-16}$
时期	完整	1,520	1,546.3	0.31	32,767.2	$<1\times10^{-16}$
	$s/n>1$	348	366.5	0.24	31,430.0	$<1\times10^{-16}$
方位角范围	完整	1,520	1,542.6	0.34	32,763.8	$<1\times10^{-16}$
	$s/n>1$	348	355.2	0.38	31,426.9	$<1\times10^{-16}$
面盘位置	完整	1,318	1,415.2	0.03	27,446.5	$<1\times10^{-16}$
	$s/n>1$	331	365.3	0.09	26,270.1	$<1\times10^{-16}$
斯托克斯状态	完整	1,524	1,556.6	0.27	33,050.6	$<1\times10^{-16}$
	$s/n>1$	350	358.2	0.37	31,722.5	$<1\times10^{-16}$
	偏振数据					
划分方式	子集	差值			和值	
		d.f.	χ^2	PTE	χ^2	PTE
年份	完整	2,896	2,949.4	0.24	2,925.2	0.35
	$s/n>1$	30	34.4	0.27	82.4	8.7×10^{-7}
	l 范围 0～245	368	385.9	0.25	315.0	0.98
	l 范围 0～245 高 s/n	73	61.0	0.84	64.5	0.75
	l 范围 245～420	796	862.2	0.05	829.4	0.20
	l 范围 245～420 高 s/n	159	176.0	0.17	223.8	5.4×10^{-7}
	l 范围 420～596	844	861.0	0.33	837.3	0.56
	l 范围 420～596 高 s/n	168	181.3	0.23	189.7	0.12
	l 范围 596～772	672	648.1	0.74	704.4	0.19
	l 范围 596～772 高 s/n	134	139.5	0.35	160.0	0.06
	l 范围 772～1,100	216	192.3	0.88	239.1	0.13
	l 范围 772～1,100 高 s/n	43	32.3	0.88	47.6	0.29
时期	完整	3,040	2,907.1	0.96	3,112.2	0.18
	$s/n>1$	34	29.2	0.70	98.6	3.3×10^{-8}
方位角范围	完整	3,040	3,071.1	0.34	3,112.9	0.17
	$s/n>1$	34	38.7	0.27	98.7	3.3×10^{-8}
面盘位置	完整	2,636	2,710.4	0.15	2,722.2	0.12
	$s/n>1$	32	43.6	0.08	97.5	1.6×10^{-8}

结合了 2001 年～2002 年数据集划分和子集的代表性选择的 χ^2 一致性检验结果。包含相同天空信号的可视性数据被划分成两个数据矢量；利用仪器噪声模型，我们计算了数据矢量差值与和值的 χ^2 统计量。这里还列出了自由度（d.f.）的大小与 χ^2 累积分布函数中的超出概率（PTE），来显示结果的显著性（在我们计算的 χ^2 累积分布函数的精度

of freedom (d.f.), and probability to exceed (PTE) the value in the χ^2 cumulative distribution function, to show the significance of the result (PTE values indicated as $< 1 \times 10^{-16}$ are zero to the precision with which we calculate the χ^2 cumulative distribution function). Difference data χ^2 values test for systematic effects in the data, while comparisons with sum data values test for the presence of a repeatable sky signal. Temperature (co-polar) data are visibility data in which the polarizers from both receivers are in the left (LL Stokes state) or right (RR Stokes state) circularly polarized state; polarization (cross-polar) data are those in which the polarizers are in opposite states (LR or RL Stokes state). The "$s/n > 1$" subset is the subset of s/n eigenmodes > 1 and the l range high s/n subsets are the 20% highest s/n modes within the given l range. See "χ^2 tests" section for further description of the data split types and subsets. We have calculated 296 χ^2 values for various split types and subsets, with no obvious trends that would indicate systematic contamination of the data.

The high s/n mode subsets are more sensitive to certain classes of systematic effects in the difference data vector and more sensitive to the expected sky signal in the sum data vector, that otherwise may be masked by noise. Also, the number of modes with expected $s/n > 1$ gives an indication of the power of the experiment to constrain the sky signal. The co-polar data, which are sensitive to the temperature signal, have many more high s/n modes than the cross-polar data, which measure polarized radiation. Within the context of the concordance model used to generate the s/n eigenmode basis, we have sensitivity with an expected $s/n > 1$ to ~ 340 temperature (co-polar) modes versus ~ 34 polarization (cross-polar) modes.

Detection of signal

Given that the data show remarkable consistency in χ^2 tests of the difference data vectors, the χ^2 values of the sum data vectors can be used to test for the presence of sky signal, independently of the likelihood analysis methods described below. In the co-polar data, all splits and subsets show highly significant χ^2 values (PTE $< 1 \times 10^{-16}$, the precision to which we calculate the cumulative distribution function).

For the cross-polar data, the sum data vector χ^2 values for the high s/n subsets show high significance, with the PTE $< 1 \times 10^{-6}$ for all $s/n > 1$ subsets in Table 1. This simple and powerful test indicates that we have detected, with high significance, the presence of a polarized signal in the data, and that this signal is repeatable in all of the data splits. The polarization map shown in Fig. 1 gives a visual representation of this repeatable polarization signal. Shown are the epoch split sum and difference polarization maps, constructed using the same using the same 34 modes with $s/n > 1$ of the polarization data in the concordance model s/n eigenmode basis that appear in Table 1. The sum map shows a repeatable polarized signal, while the difference map is consistent with instrument noise.

下，PTE 的值 $< 1 \times 10^{-16}$ 即为零）。差值数据 χ^2 值检验的是数据中的系统效应，同时与数据值的和值对照检验是否存在可重复的天空信号。温度（共极）数据是可视性数据，来自两个接收器的起偏器处于左（LL 斯托克斯状态）或右（RR 斯托克斯状态）圆偏振状态；偏振（交叉偏振）数据是起偏器处于相反状态（LR 或 RL 斯托克斯状态）的数据。"$s/n > 1$" 子集是 s/n 本征模 > 1 的子集，而 l 范围内的高 s/n 子集是给定 l 范围内最高的 20% 的 s/n 模式。有关数据划分类型和子集的进一步说明，请参阅 "χ^2 检验" 部分。我们计算了各种划分类型和子集的 296 个 χ^2 值，没有明显的趋势表明数据存在系统污染。

高 s/n 模式的子集对差值数据矢量中的某些类别的系统效应更敏感，并且对和值数据矢量中预期的天空信号更敏感，否则它就有可能被噪声掩盖。此外，$s/n > 1$ 预期的模式数量表示实验约束天空信号的能力。对温度信号敏感的共极数据比测量偏振辐射的交叉偏振数据的 s/n 模式更高。根据生成 s/n 基础本征模的一致性模型，我们预期约 340 种温度（共极）模式与约 34 种偏振（交叉偏振）模式具有 $s/n > 1$ 的灵敏度。

信号检测

假设数据在差值数据矢量的 χ^2 检验中出现了显著的一致性，则可以使用和值数据矢量的 χ^2 值来检测天空信号的存在，而与下面描述的似然分析方法无关。在共极数据中，所有划分和子集显示高度显著的 χ^2 值（PTE $< 1 \times 10^{-16}$，这个精度是我们计算累积分布函数的精度）。

对于交叉偏振数据，高 s/n 子集的和值数据向量 χ^2 值显示出高显著性，表 1 中所有 $s/n > 1$ 子集的 PTE $< 1 \times 10^{-6}$。这个简单而强大的检测表示我们已经高度显著地检测到数据中存在偏振信号，并且该信号在所有数据拆分方法中都是可重复的。图 1 中所示的偏振图给出了该可重复偏振信号的视觉呈现，展示出了根据时期划分的和值与差值偏振图，使用相同的 34 种模式构建，其中 $s/n > 1$ 的偏振数据在一致性模型 s/n 基础本征模中，如表 1 所示。和值图显示了可重复的偏振信号，而差值图与仪器噪声一致。

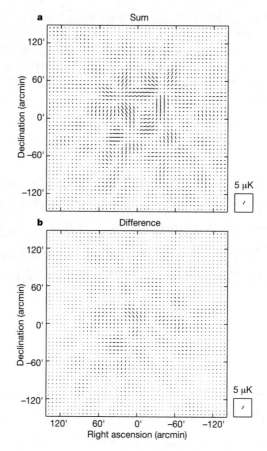

Fig. 1. Polarization maps formed from high signal/noise eigenmodes. Shown are maps constructed from polarized data sets that have been split by epoch, and formed into sum (**a**) or difference (**b**) data vectors, as reported in section "χ^2 tests". In order to isolate the most significant signal in our data, we have used only the subset of 34 eigenmodes which, under the concordance model, are expected to have average signal/noise (s/n) > 1. Because the maps have only 34 independent modes, they exhibit a limited range of morphologies, and unlike conventional interferometer maps, these s/n selected eigenmodes reflect the taper of the primary beam, even when no signal is present. This is visually apparent in the difference map (**b**), which is statistically consistent with noise. Comparison of the difference map to the sum map (**a**) illustrates a result also given numerically for this split/subset in Table 1: that these individual modes in the polarized data set show a significant signal.

It is possible to calculate a similar χ^2 statistic for the data vector formed from the complete, unsplit data set. Combining all the data without the requirement of forming two equally weighted subsets should yield minimal noise, albeit without an exactly corresponding null (that is, difference) test. Recalculating the s/n eigenmodes for this complete cross-polar data vector gives 36 modes with expected s/n > 1, for which $\chi^2 = 98.0$ with a PTE = 1.2×10^{-7}. This significance is similar to those from the sum data vectors under the various splits, which actually divide the data fairly equally and so are nearly optimal. It should be noted that our focus so far has been to test the instrumental noise model, and we have not dealt with the small cross-polar signal expected as a result of the off-axis leakage. As noted in the "Off-axis leakage" subsection, it is not possible to correct the data elements directly for this effect,

516

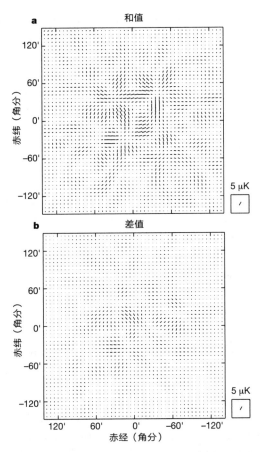

图 1. 根据高信号/噪声本征模绘制的偏振图。图中显示了由已经按时期划分的偏振数据集所构建的映射，并且生成了和值（**a**）或差值（**b**）数据矢量，如"χ^2 检验"部分中所报告的。为了分离数据中最重要的信号，我们只使用了 34 个本征模的子集，在协调模型下，预计平均信号/噪声（s/n）>1。因为这个图只有 34 个独立模式，它们表现的形态范围是有限的，并且与传统的干涉仪图不同，即使没有信号存在，这些 s/n 选择的本征模也反映了初级光束的锥度。这在差值图（**b**）中表现得很明显，其在统计上与噪声一致。差值图与和值图（**a**）的比较阐释了表 1 对于该划分/子集数值给出的结果：偏振数据集中的这些独立模式表现为显著的信号。

从一个完整的未划分数据集形成的数据矢量计算出相似的 χ^2 统计量是可能的。没有形成两个同等加权的子集的情况下，组合所有数据应该产生最小的噪声，尽管没有完全对应的零（即差值）检测。重新计算该完整交叉偏振数据矢量的 s/n 本征模，得到 36 个预期 $s/n>1$ 的模式，其中 $\chi^2=98.0$，PTE $=1.2\times10^{-7}$。这种显著性类似于那些来自各种划分下的和值数据矢量，其实际上相当均等地划分了数据，因此这几乎是最佳的方式。应该指出的是，到目前为止我们的重点是已经检测了的仪器噪声模型，而我们还没有处理能预期到的由于离轴渗漏而产生的小的交叉偏振信号。如"离轴渗漏"部分中所述，不可能直接针对此效应校正数据元素，但我们可以通过计

but we can account for it in calculating these χ^2 results by including the expected covariance of this leakage signal (see "Theory covariance matrix" subsection) in the noise matrix C_N. Again recalculating the s/n eigenmodes, we find 34 cross-polar modes with $s/n > 1$ which give a $\chi^2 = 97.0$ and a PTE $= 5.7 \times 10^{-8}$, a significance similar to before. The off-axis leakage is also included in the likelihood analyses, where it is again found to have an insignificant impact on the results.

The likelihood analysis described in the following sections makes use of all of the information in our data set. Such an analysis, in principle, may yield statistically significant evidence of a signal even in cases of data sets for which it is not possible to isolate any individual modes which have average $s/n > 1$. However, the existence of such modes in our data set, which has resulted from our strategy of integrating deeply on a limited patch of sky, allows us to determine the presence of the signal with the very simple analysis described above. It also reduces sensitivity to the noise model estimation in the likelihood results that we report next. Finally, it gives our data set greater power to exclude the possibility of no signal than it might have had if we had observed more modes but with less s/n in each.

Likelihood Analysis Formalism

The preceding section gives strong evidence for the presence of a signal in our polarization data. We now examine that signal using the standard tool of likelihood analysis. In such an analysis, the covariance of the signal, $C_T(\kappa)$, is modelled in terms of parameters κ appropriate for describing the temperature and polarization anisotropy of the CMB. The covariance of the data vector is modelled $C(\kappa) \equiv C_T(\kappa) + C_N$; where C_N is the noise covariance matrix. Given our data vector $\mathbf{\Delta}$, the likelihood of the model specified by the parameter vector κ is the probability of our data vector given that model:

$$L(\kappa) = P(\mathbf{\Delta}|\kappa) \propto \det(C(\kappa))^{-1/2}\exp(-\frac{1}{2}\,\mathbf{\Delta}^t C(\kappa)^{-1}\mathbf{\Delta}) \tag{1}$$

Although the full likelihood function itself is the most basic result of the likelihood analysis, it is useful to identify and report the values of the parameters that maximize the likelihood (so-called maximum likelihood (ML) estimators). Uncertainties in the parameter values can be estimated by characterizing the shape of the likelihood surface, as discussed further in the "Reporting of likelihood results" subsection.

The CMB power spectra

The temperature and polarization anisotropy of the CMB can be characterized statistically by six angular power spectra: three that give the amplitudes of temperature, E-mode and B-mode polarization anisotropy as a function of angular scale, and three that describe correlations between them. These spectra are written C_l^X, with $X = \{T, E, B, TE, TB, EB\}$. In our likelihood analyses, we choose various parameterizations of these spectra to constrain.

算这些噪声矩阵 C_N 中渗漏信号的预期协方差来计算 χ^2 结果（见"理论协方差矩阵"部分）。接着重新计算 s/n 本征模，我们发现 34 个交叉偏振模式，其中 $s/n > 1$，其给出 $\chi^2 = 97.0$ 及 PTE $= 5.7 \times 10^{-8}$，具有与之前类似的显著性。离轴渗漏也包括在似然分析中，而且再次发现它对结果的影响不大。

以下部分中描述的似然分析使用了我们数据集中的所有信息。即使对于整个数据可能无法分离出任何平均信噪比 $s/n > 1$ 的独立模式，但是原则上这样的分析还是可能得到一个信号具有统计学显著性的证据。但是，在我们数据集当中这种模式的存在，是我们在天空的有限区域中做深度的积分策略造成的，这使我们能够通过上述非常简单的分析来确定信号的存在。它还降低了我们接下来报告的结果中对噪声模型估计敏感的可能性。最后，它使我们的数据集可以更好地排除那些没有信号的可能性，而不是排除那些如果我们观测更多的模且在每个模中具有更小的 s/n 就可以观测到的情况。

似然分析形式化

上一节为我们的偏振数据中存在信号提供了有力证据。我们现在使用似然分析这一标准方法来检查信号。在这样的分析中，信号的协方差 $C_T(\kappa)$ 根据适合于描述 CMB 的温度和各向异性偏振的参数 κ 来建模。建模数据矢量的协方差为 $C(\kappa) \equiv C_T(\kappa) + C_N$；其中 C_N 是噪声协方差矩阵。考虑到我们的数据矢量为 Δ，由参数矢量 κ 给出的模型的似然函数是该模型的数据矢量的概率：

$$L(\kappa) = P(\Delta|\kappa) \;\propto\; \det(C(\kappa))^{-1/2}\exp(-\frac{1}{2}\Delta^t C(\kappa)^{-1}\Delta) \tag{1}$$

虽然完整的似然函数本身是似然分析的最基本结果，但确定和报告最大似然函数的参数值（所谓的最大似然（ML）估计量）是很有用的。参数值的不确定性可以通过似然函数的边缘概率函数来估计，"似然函数结果报告"部分中会有进一步的讨论。

CMB 功率谱

CMB 的温度和偏振各向异性可以通过六个角功率谱来统计表征：其中三个是温度幅度、E 模式和 B 模式各向异性偏振作为角度尺度的函数，另外三个是它们之间的相关性的描述。这些谱写成 C_l^X, $X = \{T, E, B, TE, TB, EB\}$。在似然分析中，我们选择这些功率谱的不同参数化来进行约束。

For a given cosmological model, these spectra can be readily calculated using efficient, publicly available Boltzmann codes[54]. Details of how to define these spectra in terms of all-sky multipole expansions of the temperature and linear polarization of the CMB radiation field are available in the literature (see refs 15 and 16). For DASI's 3.4° field of view, a flat sky approximation is appropriate[55], so that the spectra may be defined somewhat more simply[26]. In this approximation the temperature angular power spectrum is defined:

$$C_l^T \simeq C^T(u) \equiv \left\langle \frac{\widetilde{T}^*(\mathbf{u})\widetilde{T}(\mathbf{u})}{T_{\mathrm{CMB}}^2} \right\rangle \tag{2}$$

where $\widetilde{T}(\mathbf{u})$ is the Fourier transform of $T(\mathbf{x})$, T_{CMB} is the mean temperature of the CMB, and $l/2\pi = u$ gives the correspondence between multipole l and Fourier radius $u = |\mathbf{u}|$. The order spectra in the flat sky approximation are similarly defined, for example, $C^{TE}(u) \equiv \langle \widetilde{T}^*(\mathbf{u})\widetilde{E}(\mathbf{u})/T_{\mathrm{CNB}}^2 \rangle$. The relationship between $\widetilde{E}$, $\widetilde{B}$ and the linear polarization Stokes parameters Q and U is:

$$\widetilde{Q}(\mathbf{u}) = \cos(2\chi)\,\widetilde{E}(\mathbf{u}) - \sin(2\chi)\,\widetilde{B}(\mathbf{u})$$
$$\widetilde{U}(\mathbf{u}) = \sin(2\chi)\,\widetilde{E}(\mathbf{u}) - \cos(2\chi)\,\widetilde{B}(\mathbf{u}) \tag{3}$$

where $\chi = \arg(\mathbf{u})$ and the polarization orientation angle defining Q and U are both measured on the sky from north through east.

Theory covariance matrix

The theory covariance matrix is the expected covariance of the signal component of the data vector, $\mathbf{C}_T \equiv \langle \mathbf{s}\mathbf{s}' \rangle$. The signals measured by the visibilities in our data vector for a given baseline $\mathbf{u}_i$ (after calibration and leakage correction) are:

$$V^{RR}(\mathbf{u}_i) = \alpha_i \int d\mathbf{x}\, A(\mathbf{x}, \nu_i)[T(\mathbf{x}) + V(\mathbf{x})]e^{-2\pi i \mathbf{u}_i \cdot \mathbf{x}}$$
$$V^{LL}(\mathbf{u}_i) = \alpha_i \int d\mathbf{x}\, A(\mathbf{x}, \nu_i)[T(\mathbf{x}) - V(\mathbf{x})]e^{-2\pi i \mathbf{u}_i \cdot \mathbf{x}}$$
$$V^{RL}(\mathbf{u}_i) = \alpha_i \int d\mathbf{x}\, A(\mathbf{x}, \nu_i)[Q(\mathbf{x}) + iU(\mathbf{x})]e^{-2\pi i \mathbf{u}_i \cdot \mathbf{x}}$$
$$V^{LR}(\mathbf{u}_i) = \alpha_i \int d\mathbf{x}\, A(\mathbf{x}, \nu_i)[Q(\mathbf{x}) - iU(\mathbf{x})]e^{-2\pi i \mathbf{u}_i \cdot \mathbf{x}} \tag{4}$$

where $A(\mathbf{x}, \nu_i)$ specifies the beam power pattern at frequency ν_i, $T(\mathbf{x})$, $Q(\mathbf{x})$, $U(\mathbf{x})$, and $V(\mathbf{x})$ are the four Stokes parameters in units of CMB temperature (μK), and $\alpha_i = \partial B_{\mathrm{Plank}}(\nu_i, T_{\mathrm{CMB}})/\partial T$ is the appropriate factor for converting from these units to flux density (Jy). The co-polar visibilities V^{RR} and V^{LL} are sensitive to the Fourier transform of the temperature signal $T(\mathbf{x})$ and circular polarization component $V(\mathbf{x})$ (expected to be zero). The cross-polar visibilities V^{RL} and V^{LR} are sensitive to the Fourier transform of the linear polarization components

对于给定的宇宙学模型，使用有效的公开可用的玻尔兹曼代码[54]可以很容易地计算这些功率谱。有关如何根据 CMB 辐射场的温度和线性偏振的全天多极展开来定义这些功率谱的细节可在文献中获得（参考文献 15 和 16）。对于 DASI 的 3.4° 视场，平坦天空近似是合适的[55]，因此可以更简单地定义功率谱[26]。在该近似中，温度角功率谱定义为：

$$C_l^T \simeq C^T(u) \equiv \left\langle \frac{\widetilde{T}^*(\mathbf{u})\,\widetilde{T}(\mathbf{u})}{T_{\mathrm{CMB}}^2} \right\rangle \tag{2}$$

其中 $\widetilde{T}(\mathbf{u})$ 是 $T(\mathbf{x})$ 的傅里叶变换，T_{CMB} 是 CMB 的平均温度，$l/\pi = u$ 给出了多极矩 l 和傅里叶半径 $u = |\mathbf{u}|$ 之间的对应关系。平坦天空的有序谱定义也是类似的，例如，$C^{TE}(u) \equiv \langle \widetilde{T}^*(\mathbf{u})\widetilde{E}(\mathbf{u})/T_{\mathrm{CNB}}^2 \rangle$。$\widetilde{E}$、$\widetilde{B}$，以及线性偏振斯托克斯参数 Q 和 U 之间的关系是：

$$\widetilde{Q}(\mathbf{u}) = \cos(2\chi)\,\widetilde{E}(\mathbf{u}) - \sin(2\chi)\,\widetilde{B}(\mathbf{u})$$
$$\widetilde{U}(\mathbf{u}) = \sin(2\chi)\,\widetilde{E}(\mathbf{u}) - \cos(2\chi)\,\widetilde{B}(\mathbf{u}) \tag{3}$$

其中用来定义 Q 和 U 的 $\chi = \arg(\mathbf{u})$ 与偏振方向角都是从天空上由北到东测量的。

理论协方差矩阵

理论协方差矩阵是数据矢量的信号分量的预期协方差，$\mathbf{C}_T \equiv \langle \mathbf{ss}' \rangle$。对于给定的基线 $\mathbf{u}_i$，我们的数据矢量中可视性测量的信号（在校准和渗漏校正之后）是：

$$V^{RR}(\mathbf{u}_i) = \alpha_i \int d\mathbf{x}\, A(\mathbf{x}, v_i)[T(\mathbf{x}) + V(\mathbf{x})]e^{-2\pi i \mathbf{u}_i \cdot \mathbf{x}}$$
$$V^{LL}(\mathbf{u}_i) = \alpha_i \int d\mathbf{x}\, A(\mathbf{x}, v_i)[T(\mathbf{x}) - V(\mathbf{x})]e^{-2\pi i \mathbf{u}_i \cdot \mathbf{x}}$$
$$V^{RL}(\mathbf{u}_i) = \alpha_i \int d\mathbf{x}\, A(\mathbf{x}, v_i)[Q(\mathbf{x}) + iU(\mathbf{x})]e^{-2\pi i \mathbf{u}_i \cdot \mathbf{x}} \tag{4}$$
$$V^{LR}(\mathbf{u}_i) = \alpha_i \int d\mathbf{x}\, A(\mathbf{x}, v_i)[Q(\mathbf{x}) - iU(\mathbf{x})]e^{-2\pi i \mathbf{u}_i \cdot \mathbf{x}}$$

其中 $A(\mathbf{x}, v_i)$ 是指定频率 v_i 下的光束功率模式，$T(\mathbf{x})$、$Q(\mathbf{x})$、$U(\mathbf{x})$ 和 $V(\mathbf{x})$ 是以 CMB 温度（μK）为单位的四个斯托克斯参数，且 $\alpha_i = \partial B_{\mathrm{Plank}}(v_i, T_{\mathrm{CMB}})/\partial T$ 是从这些单位转换为通量密度（Jy）的适当因子。共极性 V^{RR} 和 V^{LL} 对温度信号 $T(\mathbf{x})$ 和圆偏振分量 $V(\mathbf{x})$（预期为零）的傅里叶变换敏感。交叉极性可视性 V^{RL} 和 V^{LR} 对线性偏振分量 Q 和 U 的傅里叶变换敏感。由等式（3）可以看出，可视性的成对组合是天空上几乎纯

Q and U. Using equation (3), it can be seen that pairwise combinations of the visibilities are direct measures of nearly pure T, E and B Fourier modes on the sky, so that the data set easily lends itself to placing independent constraints on these power spectra.

We construct the theory covariance matrix as the sum of components for each parameter in the analysis:

$$C_{\mathrm{T}}(\kappa) = \sum_p \kappa_p B_{\mathrm{T}}^p \qquad (5)$$

From equations (2)–(4), it is possible to derive a general expression for the matrix elements of a theory matrix component:

$$B_{\mathrm{T}ij}^p = \frac{1}{2}\alpha_i\alpha_j T_{\mathrm{CMB}}^2 \int d\mathbf{u}\, C^X(u)\widetilde{A}(\mathbf{u}_i-\mathbf{u},\nu_i) \times [\zeta_1\widetilde{A}(\mathbf{u}_j-\mathbf{u},\nu_j)+\zeta_2\widetilde{A}(\mathbf{u}_j+\mathbf{u},\nu_j)] \qquad (6)$$

The coefficients ζ_1 and ζ_2 can take values $\{0,\pm1,\pm2\}\times\{\cos\{2\chi,4\chi\},\sin\{2\chi,4\chi\}\}$ depending on the Stokes states (RR, LL, RL, LR) of each of the two baselines i and j and on which of the six spectra (T, E, B, TE, TB, EB) is specified by X. The integration may be limited to annular regions which correspond to l-ranges over which the power spectrum C^X is hypothesized to be relatively flat, or else some shape of the spectrum may be postulated.

Potentially contaminated modes in the data vector may be effectively projected out using a constraint matrix formalism[53]. This formalism can be used to remove the effect of point sources of known position without knowledge of their flux densities, as we described in ref. 6. This procedure can be generalized to include the case of polarized point sources. Although we have tested for the presence of point sources in the polarization power spectra using this method, in the final analysis we use constraint matrices to project point sources out of the temperature data only, and not the polarization data (see "Point sources" subsection).

The off-axis leakage, discussed in the "Off-axis leakage" subsection and in detail in ref. 51, has the effect of mixing some power from the temperature signal T into the cross-polar visibilities. Our moolel of the off-axis leakage allow us to write an expression for it analogous to equation (4), and to construct a corresponding theory covariance matrix component to account for it. In practice, this is a small effect, as discussed in the "Systematic uncertainties" section.

Likelihood parameters

In the "Likelihood results" section we present the results from nine separate likelihood analyses involving the polarization data, the temperature data, or both. Our choice of parameters with which to characterize the six CMB power spectra is a compromise between maximizing sensitivity to the signal and constraining the shape of the power spectra. In

的 T、E 和 B 的傅里叶模的直接测量，这使得数据集很容易对这些功率谱设置独立的约束。

在分析中，我们将理论协方差矩阵构造为每个参数部分的总和：

$$C_T(\kappa) = \sum_p \kappa_p\, B_T^P \tag{5}$$

从等式（2）~（4），可以推导出理论矩阵分量的矩阵元素的一般表达式：

$$B_{Tij}^p = \frac{1}{2}\alpha_i\alpha_j T_{CMB}^2 \int du\, C^X(u)\, \widetilde{A}\,(\mathbf{u}_i - \mathbf{u}, v_i) \times [\zeta_1\, \widetilde{A}\,(\mathbf{u}_j - \mathbf{u}, v_j) + \zeta_2\, \widetilde{A}\,(\mathbf{u}_j + \mathbf{u}, v_j)] \tag{6}$$

系数 ζ_1 和 ζ_2 可取 $\{0, \pm 1, \pm 2\} \times \{\cos\{2\chi, 4\chi\},\ \sin\{2\chi, 4\chi\}\}$，取决于两个基线 i 和 j 各自的斯托克斯状态（RR，LL，RL，LR），同时取决于 X 的谱（T, E, B, TE, TB, EB）。积分可以限于环形区域，其对应于范围为 l 的功率谱 C^X；C^X 被假设为相对平坦的，或者可以假设为某种形状的功率谱。

可以使用约束矩阵形式 [53] 有效地投射出数据矢量中的潜在污染模式。这种形式可以用来消除已知位置，但不知道它们的通量密度的点源的影响，正如我们在参考文献 6 中所描述的那样。这个程序可以运用推广到包括偏振点源的情况中。虽然我们已经使用这种方法检测了偏振功率谱中点源的存在，但在最后的分析中，我们仅从温度数据中用约束矩阵投射点源，而不是用偏振数据（参见"点源"部分）。

正如在"离轴渗漏"部分中讨论过的，详见参考文献 51，离轴渗漏具有来自温度信号 T 的一些功率混合到交叉偏振可视性中的效果。我们的离轴渗漏的模式允许我们为它写出类似于等式（4）的表达式，并构造相应的理论协方差矩阵分量来解释它。在实践中，这是一个很小的影响因素，如"系统不确定性"部分所述。

似然参数

在"似然结果"部分中，我们呈现了九组包括对偏振数据、温度数据或两者联合数据的不同的似然分析结果。对用于描述六张 CMB 功率谱的参数的选择，我们针对信号灵敏度最大化和约束功率谱形状这两方面进行折中。针对不同的分析，我们

the different analyses we either characterize various power spectra with a single amplitude parameter covering all angular scales, or split the l-range into five bands over which spectra are approximated as piecewise-flat, in units of $l(l+1)C_l/(2\pi)$. Five bands were chosen as a compromise between too many for the data to bear and too few to capture the shape of the underlying power spectra. The l-ranges of these five bands are based on those of the nine-band analysis we used in ref. 6; we have simply combined the first four pairs of these bands, and kept the ninth as before. In some analyses we also constrain the frequency spectral indices of the temperature and polarization power spectra as a test for foreground contamination.

The l-range to which DASI has non-zero sensitivity is $28 < l < 1,047$. That range includes the first three peaks of the temperature power spectrum, and within it the amplitude of that spectrum, which we express in units $l(l+1)C_l/(2\pi)$, varies by a factor of about 4. Over this same range, the E-mode polarization spectrum is predicted to have four peaks while rising roughly as l^2 (in the same units), varying in amplitude by nearly two orders of magnitude[17]. The TE correlation is predicted to exhibit a complex spectrum that in fact crosses zero five times in this range.

For the single bandpower analyses which maximize our sensitivity to a potential signal, the shape of the model power spectrum assumed will have an effect on the sensitivity of the result. In particular, if the assumed shape is a poor fit to the true spectrum preferred by the data, the results will be both less powerful and difficult to interpret. For temperature spectrum measurements, the most common choice in recent years has been the so-called flat bandpower, $l(l+1)C_l \propto$ constant, which matches the gross large-angle "scale-invariant" power-law shape of that spectrum. Because of extreme variations predicted in the E and TE spectra over DASI's l-range, we do not expect a single flat bandpower parameterization to offer a good description of the entire data set (although in the "E/B analysis" subsection we describe results of applying such an analysis to limited l-range subsets of data). A more appropriate definition of "flat bandpower" for polarization measurements sensitive to large ranges of $l < 1,000$ might be $C_l \propto$ constant (or $l(l+1)C_l \propto l^2$). Other shapes have been tried, notably the gaussian autocorrelation function (by the PIQUE group[56]) which reduces to $C_l \propto$ constant at large scales and perhaps offers a better fit to the gross shape of the predicted E spectrum.

In our single band analyses, we have chosen a shape for our single bandpower parameters based on the predicted spectra for a cosmological model currently favoured by observations. The specific model that we choose—which we will call the concordance model—is a ΛCDM model with flat spatial curvature, 5% baryonic matter, 35% dark matter, 60% dark energy, and a Hubble constant of 65 km s^{-1}Mpc^{-1}, ($\Omega_b = 0.05$, $\Omega_{cdm} = 0.35$, $\Omega_\Lambda = 0.60$, $h = 0.65$) and the exact normalization $C_{10} = 700 \ \mu K^2$. This concordance model was defined in ref. 50 as a good fit to the previous DASI temperature power spectrum and other observations. The concordance model spectra for T, E, and TE are shown in Fig. 4. The five flat bandpower likelihood results shown in Fig. 4, and discussed in the next section, suggest that the concordance shaped spectra do indeed characterize the data better than any

或假设功率谱在全部的角尺度范围上具有相同的幅度参数，或将功率谱的多极矩 l 分为 5 个频段并假设每一段的功率谱（以 $l(l+1)C_l/(2\pi)$ 为单位）基本平坦。之所以选择 5 个频段，是考虑到若分段太多，则需加载的数据太多；若分段太少，则无法捕捉到功率谱的基本形状。这五个频段的 l 的范围是基于我们在参考文献 6 用过的九频段分析。这里，我们只是将这九个频段的前八个频段按四对合并在一起，并保持第九个频段不变。在某些分析中，我们会约束温度和偏振功率谱的谱指数以检验前景污染。

DASI 实验的多极矩探测范围为 $28 < l < 1{,}047$。这个范围包含了温度功率谱的前三个震荡峰，其中谱的幅度（以 $l(l+1)C_l/(2\pi)$ 为单位）约为四。在同样的多极矩范围内，E 模式偏振功率谱预计会包含四个峰，大致随 l^2（和温度谱相同的单位）上升而上升，其幅度的变化程度几乎达到两个量级 [17]。TE 相关功率谱预计会比较复杂，实际上，它在这段 l 的范围内会有五次等于零。

对于单频段功率分析实现对一个潜在信号的灵敏度的最大化，其分析结果的灵敏度会受到假定的模型功率谱形状的影响，尤其当假设的形状和数据支持的真实功率谱的形状相差甚远时，结果就不太可靠且难以被解释。近几年来，人们普遍选择所谓的"平坦频段功率"方法，即 $l(l+1)\,C_l \propto$ 常数，来测量温度谱。"平坦频段功率"和温度谱的大角度"标度不变"幂律分布的形状相匹配。根据预测，在 DASI 的 l 范围内 E 和 TE 功率谱变化剧烈，因此我们并不指望一个单平坦频段功率参数化就能很好地描述整个数据集（尽管在"E/B 分析"部分中，我们所描述的结果就是将这种分析方法用于有限 l 范围的数据子集给出的）。对偏振测量敏感的范围（$l < 1{,}000$），更准确的"平坦频段功率"的定义可以是 $C_l \propto$ 常数（或者 $l(l+1)\,C_l \propto l^2$）。其他已经被测试过的形状，特别是高斯自相关函数（被 PIQUE[56] 组使用过），在大尺度上近似为 $C_l \propto$ 常数，或许会更符合 E 谱的总体形状。

作单频段分析的时候，我们会根据目前观测所支持的宇宙学模型预测的功率谱，为单频带功率参数选定一个形状。被选择的特定模型——我们称之为协调模型——是一个 ΛCDM 模型，该模型具有平坦的空间曲率、5% 重子物质、35% 暗物质、60% 暗能量、65 km $\cdot$ s^{-1} $\cdot$ Mpc^{-1} 的哈勃常数（即 $\Omega_b = 0.05$，$\Omega_{cdm} = 0.35$，$\Omega_\Lambda = 0.60$，$h = 0.65$），以及确定的归一化 $C_{10} = 700$ μK^2。这个协调模型定义在参考文献 50 中，与先前 DASI 温度功率谱和其他观测匹配得很好。图 4 展示了协调模型的 T、E 和 TE 功率谱。五个平坦频段功率似然结果（参见图 4）表明协调模型给出的形状的确比任何幂律分布近似都更符合数据，我们会在下一节对该结果进行讨论。在"E/B 分

power-law approximation. In the "E/B analysis" subsection, we explicitly test the likelihood of the concordance model parameterization against that of the two power laws mentioned above, and find that the concordance model shape is strongly preferred by the data.

It should be noted that the likelihood analysis is always model dependent, regardless of whether a flat or shaped model is chosen for parameterization. To evaluate the expectation value of the results for a hypothesized theoretical power spectrum, we must use window functions appropriate for the parameters of the particular analysis. The calculation of such parameter window functions has previously been described, both generally[57,58], and with specific reference to polarization spectra[59]. In general, the parameter window function has a non-trivial shape (even for a flat bandpower analysis) which is dependent on the shape of the true spectra as well as the intrinsic sensitivity of the instrument as a function of angular scale. Parameter window functions for the E/B and $E5/B5$ polarization analysis are shown in Fig. 2, and will also be made available on our website (http://astro.uchicago.edu/dasi).

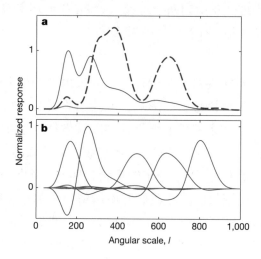

Fig. 2. Parameter window functions, which indicate the angular scales over which the parameters in our analyses constrain the power spectra. **a**, The functions for the E parameter of our E/B analysis, with the solid blue curve indicating response to the E power spectrum and the solid red (much lower) curve indicating response of the same E parameter to the B spectrum. The blue dashed curve shows the result of multiplying the E window function by the concordance E spectrum, illustrating that for this CMB spectrum, most of the response of our experiment's E parameter comes from the region of the second peak ($300 \lesssim l \lesssim 450$), with a substantial contribution also from the third peak and a smaller contribution from the first. **b**, The $E1$ to $E5$ parameter window functions for the E power spectrum (blue) and B power spectrum (red, again much lower) from our $E5/B5$ analysis. All of these window functions are calculated with respect to the concordance model discussed in the text. DASI's response to E and B is very symmetric, so that E and B parameter window functions which are calculated with respect to a model for which $E = B$ are nearly identical, with the E and B spectral response reversed.

Likelihood evaluation

Prior to likelihood analysis, the data vector and the covariance matrices can be compressed

析"部分，我们会有明确地检测以比较协调模型参数化和前面提到过的两种幂律参数化的似然。结果表明，数据强烈支持协调模型给出的形状。

需要注意的是，不管采用平坦模型或者给定成形模型来参数化，似然分析总是依赖于模型的。对于一个假设的理论功率谱，为了估计其分析结果的期望值，我们必须采用适用于这个特定分析的参数的窗口函数。这种参数窗口函数的计算方法曾经被概括地描述过[57,58]，也有特别针对偏振功率谱的文献[59]。一般而言，参数窗口函数具有非平凡的形状（即使对平坦频段功率分析也是如此），其形状依赖于真实功率谱的形状和仪器的内秉灵敏度，而后者是一个关于角度的函数。E/B、$E5/B5$ 偏振分析的参数窗口函数如图 2 所示，也可以在我们的网站（http://astro.uchicago.edu/dasi）上找到。

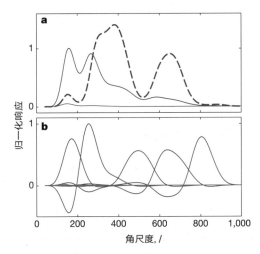

图 2. 参数窗口函数，它标示我们分析中的参数约束的是哪一段角度范围上的功率谱。**a**，E/B 分析的 E 参数窗口函数。其中蓝色实线表示对 E 功率谱的响应，红色实线（位置低得多）表示相同的 E 参数对 B 功率谱的响应。蓝色虚线表示的是将 E 窗口函数和协调模型 E 谱相乘的结果，这表明对于此 CMB 功率谱，我们实验的 E 参数的响应大部分来自第二峰的对应的 l 区域（$300 \lesssim l \lesssim 450$），也有相当一部分来自第三峰，第一峰的贡献更小些。**b**，E 功率谱（蓝色）、B 功率谱（红色，同样位置但低很多）的 $E1$ 到 $E5$ 参数窗口函数。所有这些窗口函数都是根据文中提到的协调模型计算的。DASI 对 E 和 B 模式的响应非常对称，因此根据 $E = B$ 的模型计算给出的 E 和 B 参数窗口函数基本相同，而 E 和 B 谱响应是相反的。

似然评估

做似然分析之前，通过合并 (u, v) 平面上邻近的、信号高度相关的数据点，我们可以压缩数据矢量和协方差矩阵，实现在不显著丢失信号信息的同时，减少分析

by combining visibility data from nearby points in the (u, v) plane, where the signal is highly correlated. This reduces the computational time required for the analyses without a significant loss of information about the signal. All analyses were run on standard desktop computers.

For each analysis, we use an iterated quadratic estimator technique to find the ML values of our parameters[53]. To further characterize the shape of the likelihood function, in ref. 6 we used an offset log-normal approximation. Here, for improved accuracy in calculating confidence intervals and likelihood ratios, we explicitly map out the likelihood function by evaluating equation (1) over a uniform parameter grid large enough to enclose all regions of substantial likelihood. A single likelihood evaluation typically takes several seconds, so this explicit grid evaluation is impractical for the analyses which include five or more parameters. For each analysis we also implement a Markov chain evaluation of the likelihood function[60]. We find this to be a useful and efficient tool for mapping the likelihoods of these high-dimensional parameter spaces in the region of substantial likelihood. We have compared the Markov technique to the grid evaluation for the lower-dimensional analyses and found the results to be in excellent agreement. In all cases, the peak of the full likelihood evaluated with either technique is confirmed to coincide with the ML values returned by the iterated quadratic estimator.

Simulations and parameter recovery tests

The likelihood analysis software was extensively tested through analysis of simulated data. The analysis software and data simulation software were independently authored, as a check for potential coding errors.

Simulated sky maps were generated from realizations of a variety of smooth CMB power spectra, including both the concordance spectrum and various non-concordance models, both with and without E and B polarization and TE correlations. Independent realizations of the sky were "observed" to construct simulated visibilities with Fourier-plane sampling identical to the real data. The simulations were designed to replicate the actual data as realistically as possible and include artefacts of the instrumental polarization response and calibration, such as the on-axis and off-axis leakages and the cross-polar phase offset, described in the "Measuring polarization with DASI" section, allowing us to test the calibration and treatment of these effects implemented in the analysis software.

Each of the analyses described in the "Likelihood results" section was performed on hundreds of these simulated data sets with independent realizations of sky and instrument noise, both with noise variances that matched the real data, and with noise a factor of ten lower. In all cases, we found that the means of the ML estimators recovered the expectation values $\langle \kappa_p \rangle$ of each parameter without evidence of bias, and that the variance of the ML estimators was found to be consistent with the estimated uncertainty given by F^{-1} evaluated at $\langle \kappa \rangle$, where F is the Fisher matrix.

所需要的计算时间。所有分析计算均在标准台式计算机上运行。

对每一次的分析，我们使用一个迭代二次估计量寻找参数的最大似然(ML)值[53]。为进一步表征似然函数的形状，在参考文献 6 中，我们采用了一个偏移对数正态近似。这里，为提高置信区间和似然比的计算精度，我们在足够大以至覆盖所有实际可能区域的均匀参数网格上求解方程(1)以明确地画出似然函数。一次似然估计一般需要花费几秒钟，因此这种显式网格评估对于包含五个或更多参数的分析是不切实际的。对每次分析我们也用马尔可夫链求似然函数[60]，我们发现它是将这些高维度参数空间的似然函数在实际可能的区域内映射出来的一个高效且实用的工具。对于低维度分析，我们发现马尔可夫法和格点估计法的结果完全吻合。在所有情况下，这两种方法给出的似然函数的峰值和迭代二次估计量给出的最大似然值都相符。

模拟和参数恢复检测

我们用模拟数据对似然分析软件进行了大量检测，分析软件和数据模拟软件被独立编写，以检测潜在的编程错误。

模拟的天图是通过各种光滑的 CMB 功率谱的实现生成的，功率谱包括协调模型的，各种非协调模型的，包括或者不包括 E、B 模式偏振和 TE 相关性的。我们通过和真实数据同样的傅里叶平面采样"观测"独立的天图的实现来构造模拟的可视度。模拟旨在尽可能真实地重复实际数据，并加入仪器偏振响应和校准的人造信号，比如"用 DASI 测量偏振"部分中讲到过的共轴渗漏、离轴渗漏和交叉偏振相位偏移效应，这样做使我们可以检测分析软件执行过程中对这些效应的校准和处理。

"似然结果"部分中描述的每个分析都是在数百个模拟数据集上进行的，模拟数据都由独立的天图实现和仪器误差给出，两者的噪声方差都和实际数据相匹配，其中噪声低于信号的十分之一。我们发现，在任何情况下，最大似然估计量给出的均值和每个参数的均值 $\langle \kappa_p \rangle$ 相比均无明显偏差，并且最大似然估计量的方差和用 F^{-1} 方法在 $\langle \kappa \rangle$ 处估算给出的不确定度一致，这里，F 是费希尔矩阵。

Reporting of likelihood results

Maximum likelihood (ML) parameter estimates reported in this paper are the global maxima of the multidimensional likelihood function. Confidence intervals for each parameter are determined by integrating (marginalizing) the likelihood function over the other parameters; the reported intervals are the equal-likelihood bounds which enclose 68% of this marginal likelihood distribution. This prescription corresponds to what is generally referred to as the highest posterior density (HPD) interval. When calculating these intervals we consider all parameter values, including non-physical ones, because our aim is simply to summarize the shape of the likelihood function. Values are quoted in the text using the convention "ML (HPD-low to HPD-high)" to make clear that the confidence range is not directly related to the maximum likelihood value.

In the tabulated results, we also report marginalized uncertainties obtained by evaluating the Fisher matrix at the maximum likelihood model, that is, $(F^{-1})_{ii}^{1/2}$ for parameter i. Although in most cases, the two confidence intervals are quite similar, we regard the HPD interval as the primary result.

For parameters which are intrinsically positive we consider placing (physical) upper limits by marginalizing the likelihood distribution as before, but excluding the unphysical negative values. We then test whether the 95% integral point has a likelihood smaller than that at zero; if it does we report an upper limit rather than a confidence interval.

We also report the parameter correlation matrices for our various likelihood analyses to allow the reader to gauge the degree to which each parameter has been determined independently. The covariance matrix is the inverse of the Fisher matrix and the correlation matrix, R, is defined as the covariance matrix normalized to unity on the diagonal, that is, $C = F^{-1}$ and $R_{ij} = C_{ij}/\sqrt{C_{ii}C_{jj}}$.

Goodness-of-fit tests

Using the likelihood function, we wish to determine if our results are consistent with a given model. For example, we would like to examine the significance of any detections by testing for the level of consistency with zero signal models, and we would like to determine if the polarization data are consistent with predictions of the standard cosmological model. We define as a goodness-of-fit statistic the logarithmic ratio of the maximum of the likelihood to its value for some model $\mathcal{H}_0$ described by parameters κ_0:

$$\Lambda(\mathcal{H}_0) \equiv -\log\left(\frac{L(\kappa_{\mathrm{ML}})}{L(\kappa_0)}\right)$$

The statistic Λ simply indicates how much the likelihood has fallen from its peak value down

530

似然结果汇报

本文中报告的最大似然(ML)参数估计是多维似然函数的全局最大值。每个参数的置信区间是将似然函数对其他参数求积分(边缘化)确定的。本文报告的区间包含 68% 边缘似然分布的等概率边界,这个区间一般被认为是最高后验概率密度(HPD)区间。我们的目标仅仅是给出似然函数的形状,因此在计算这个区间的时候,我们会考虑所有的参数值,包括非物理的参数。本文约定用"ML(HPD 低到 HPD 高)"的格式引述参数值,以此来说明置信区间和最大似然值之间并没有直接关系。

通过计算最大似然模型处的费希尔矩阵,即对参数 i 求 $(F^{-1})_{ii}^{1/2}$,我们给出边缘化后的不确定度,也将其报告在结果列表中。虽然在大多数情况下,这两个置信区间非常相近,但我们采用 HPD 区间作为基本结果。

对于那些固有取正的参数,我们考虑通过如前所述的似然函数边缘化的方法来设置(物理)上界,消去非物理的负值。然后我们检查 95% 积分点的似然是否比参数值为零处的小。如果是,则报告一个上限而不是置信区间。

我们还报告了各个似然分析的参数相关矩阵,方便读者能够评估每个参数被独立测定的程度。协方差矩阵是费希尔矩阵的逆,把它的对角元素归一化后的矩阵被定义为相关矩阵 R,即 $C = F^{-1}$,$R_{ij} = C_{ij}/\sqrt{C_{ii}C_{jj}}$。

拟合优度检验

我们想使用似然函数来确定我们的结果是否支持一个给定的模型。比如说,我们想检查所有探测信号的显著性,则需要检测与零信号模型一致的程度。此外,我们想确定偏振数据是否和标准宇宙学模型预言的一致。我们定义了一个拟合优度统计量,它是似然函数的最大值和某些模型 $\mathscr{H}_0$ 的似然的对数比,记模型 $\mathscr{H}_0$ 的参数为 κ_0,则有:

$$\Lambda(\mathscr{H}_0) \equiv -\log\left(\frac{L(\kappa_{\mathrm{ML}})}{L(\kappa_0)}\right)$$

简单来说,统计量 Λ 表明当参数从最大似然处的值变为 κ_0 时,似然函数下降的程度。

to its value at κ_0. Large values indicate inconsistency of the likelihood result with the model $\mathcal{H}_0$. To assess significance, we perform Monte Carlo (MC) simulations of this statistic under the hypothesis that $\mathcal{H}_0$ is true. From this, we can determine the probability, given $\mathcal{H}_0$ true, to obtain a value of Λ that exceeds the observed value, which we hereafter refer to as PTE.

When considering models which the data indicate to be very unlikely, sufficient sampling of the likelihood statistic becomes computationally prohibitive; our typical MC simulations are limited to only 1,000 realizations. In the limit that the parameter errors are normally distributed, our chosen statistic reduces to $\Lambda = \Delta\chi^2/2$. The integral over the χ^2 distribution is described by an incomplete gamma function;

$$\text{PTE} = \frac{1}{\Gamma(N/2)} \int_{\Lambda}^{\infty} e^{-x} x^{\frac{N}{2}-1} dx$$

where $\Gamma(x)$ is the complete gamma function, and N is the number of parameters. Neither the likelihood function nor the distribution of the ML estimators is, in general, normally distributed, and therefore this approximation must be tested. In all cases where we can compute a meaningful PTE with MC simulations, we have done so and found the results to be in excellent agreement with the analytic approximation. Therefore, we are confident that adopting this approximation is justified. All results for PTE in this paper are calculated using this analytic expression unless otherwise stated.

Likelihood Results

We have performed nine separate likelihood analyses to constrain various aspects of the signal in our polarization data, in our temperature data, or in both analysed together. The choice of parameters for these analyses and the conventions used for reporting likelihood results have been discussed in "Likelihood parameters" and "Reporting of likelihood results" subsections. Numerical results for the analyses described in this section are given in Tables 2, 3 and 4.

Polarization data analyses and E and B results

E/B analysis. The E/B analysis uses two single-bandpower parameters to characterize the amplitudes of the E and B polarization spectra. As discussed in the "Likelihood parameters" subsection, this analysis requires a choice of shape for the spectra to be parameterized. DASI has instrumental sensitivity to E and B that is symmetrical and nearly independent. Although the B spectrum is not expected to have the same shape as the E spectrum, we choose the same shape for both spectra in order to make the analysis also symmetrical.

We first compute the likelihood using a $l(l+1)C_l/2\pi = $ constant bandpower (commonly referred to as "flat" bandpower) including data only from a limited l range in which DASI

532

Λ 值大，则暗示似然结果不支持模型 $\mathscr{H}_0$。为了评估显著性程度，我们在 $\mathscr{H}_0$ 为真的假设下对该统计量进行蒙特卡罗（MC）模拟。这样，我们可以求出当 $\mathscr{H}_0$ 为真的情况下得到一个比观测到的 Λ 值更大的值的概率大小。我们后面称这个概率为 PTE。

考虑在数据十分不支持的模型的情况下，似然统计的充分抽样在计算上会变得难以承受，我们标准的 MC 模拟次数限制在仅 1,000 次实现。在参数误差是正态分布的情况下，我们选择的统计量 Λ 简化为 $\Lambda = \Delta\chi^2/2$。对 χ^2 分布的积分可以描述为一个不完全伽马函数：

$$\text{PTE} = \frac{1}{\Gamma(N/2)} \int_{\Lambda}^{\infty} e^{-x} x^{\frac{N}{2}-1} dx$$

其中 $\Gamma(x)$ 是完全伽马函数，N 是参数的数目。由于似然函数和最大似然估计量的分布一般都不是正态分布，我们还需要验算这个近似是否成立。任何情况下，我们都可以用 MC 模拟计算出一个有意义的 PTE，这样处理后，我们发现 MC 给出的结果和解析近似的结果非常一致。因此我们认为这个近似是可信的。若无特殊说明，本文中 PTE 的结果都是用这个解析表达式计算得到的。

似然结果

我们已经进行了九次不同的似然分析，对偏振数据、温度数据或两者的联合数据的各个方面作出约束。我们已经在"似然参数"和"似然结果报告"部分中对这些分析的参数的选择和报告似然结果所使用的约定进行了讨论。本节讨论似然分析的数值结果并在表 2、3 和 4 中给出。

偏振数据分析和 E、B 结果

E/B 分析　E/B 分析用到两个单频带功率参数描述 E 和 B 偏振谱的幅度。如同"似然参数"部分所讨论过的，该分析需选择要被参数化的功率谱的形状。DASI 对 E 和 B 的仪器灵敏度是对称的且几乎独立的。虽然 B 功率谱的形状预计和 E 功率谱的不同，但为使该分析是对称的，我们还是为这两种功率谱选择了相同的形状。

我们首先用 $l(l+1)C_l/2\pi = $ 常数频段功率（一般称为"平坦频段功率"）来计算似然函数，数据仅包括来自 DASI 且具有高灵敏度的有限的 l 范围上的数据，见图 2。利

has high sensitivity; see Fig. 2. Using the range $300 < l < 450$ which includes 24% of the complete data set, we find the ML at flat bandpower values $E = 26.5\ \mu K^2$ and $B = 0.8\ \mu K^2$. The likelihood falls dramatically for the zero polarization "nopol" model $E = 0$, $B = 0$. Marginalizing over B, we find $\Lambda(E=0) = 16.9$ which, assuming the uncertainties are normally distributed, corresponds to a PTE of 5.9×10^{-9} or a significance of E detection of 5.8σ. As expected, changing the l range affects the maximum likelihood values and the confidence of detection; for example, shifting the centre of the above l range by ± 25 reduces the confidence of detection to 5.6σ and 4.8σ, respectively. Clearly it is desirable to perform the analysis over the entire l range sampled by DASI using a well motivated bandpower shape for the parameterization.

We considered three *a priori* shapes to check which is most appropriate for our data: the concordance E spectrum shape (as defined in "Likelihood parameters"), and two power law alternatives, $l(l+1)C_l \propto$ constant (flat) and $l(l+1)C_l \propto l^2$. For each of these three cases, the point at $E = 0$, $B = 0$ corresponds to the same zero-polarization "nopol" model, so that the likelihood ratios Λ(nopol) may be compared directly to assess the relative likelihoods of the best-fit models. For the $l(l+1)C_l \propto$ constant case, the ML flat bandpower values are $E = 6.8\ \mu K^2$ and $B = -0.4\ \mu K^2$, with Λ(nopol) $= 4.34$. For the $l(l+1)C_l \propto l^2$ case, the ML values are $E = 5.1\ \mu K^2$ and $B = 1.2\ \mu K^2$ (for $l(l+1)C_l/2\pi$) at $l = 300$, with Λ(nopol) $= 8.48$. For the concordance shape, the ML values are $E = 0.80$ and $B = 0.21$ in units of the concordance E spectrum amplitude, with Λ(nopol) $= 13.76$. The likelihood of the best-fit model in the concordance case is a factor of 200 and 12,000 higher than those of the $l(l+1)C_l \propto l^2$ and $l(l+1)C_l \propto$ constant cases, respectively, and so compared to the concordance shape either of these is a very poor model for the data. The data clearly prefer the concordance shape, which we therefore use for the E/B and other single bandpower analyses presented in our results tables.

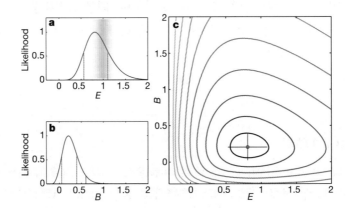

Fig. 3. Results from the two-parameter shaped bandpower E/B polarization analysis. An E-mode power spectrum shape as predicted for the concordance model is assumed, and the units of amplitude are relative to that model. The same shape is assumed for the B-mode spectrum. **c**, The point shows the maximum likelihood value with the cross indicating Fisher matrix errors. Likelihood contours are placed at levels $\exp(-n^2/2)$, $n = 1,2...$, relative to the maximum, that is, for a normal distribution, the extrema of these contours along either dimension would give the marginalized n-sigma interval. **a**, **b**, The corresponding single parameter likelihood distributions marginalized over the other parameter. Note

用 $300 < l < 450$(含完整数据集的 24%)的数据，我们发现平坦频段功率的最大似然值为 $E = 26.5\ \mu K^2$，$B = 0.8\ \mu K^2$。零偏振($E = 0$，$B = 0$)的"无偏振"模型的似然值急剧下降。把 B 边缘化后，我们得到 $\Lambda(E = 0) = 16.9$。若假设误差呈正态分布，则该值对应的 PTE 为 5.9×10^{-9}，或者说探测到 E 的显著程度为 5.8σ。正如我们预计的，改变 l 范围会改变最大似然值和探测的显著水平。举个例子，把上面的 l 范围中心值移动 ± 25，则探测的显著水平分别下降至 5.6σ 和 4.8σ。采用一个合理的频段功率形状进行参数化，并在 DASI 采样的整个 l 范围内做这样的分析，这显然是我们所希望实现的。

我们考虑了三个先验形状来检查哪个最适合我们的数据，它们分别是协调模型的 E 谱(定义在"似然参数"部分中)、$l(l+1)C_l \propto$ 常数(平坦)和 $l(l+1)C_l \propto l^2$。后两者为幂律分布。对这三种情况，$E = 0$，$B = 0$ 的点都对应偏振为零的"无偏振"模型，因此直接比较似然比 Λ(无偏振)就可以估计最佳拟合模型的相对似然值。对 $l(l+1)C_l \propto$ 常数的情况，其在 $l = 300$ 处的最大似然值为 $E = 6.8\ \mu K^2$，$B = -0.4\ \mu K^2$，Λ(无偏振)= 4.34；对 $l(l+1)C_l \propto l^2$ 模型，同样在 $l = 300$ 处的最大似然值是 $E = 5.1\ \mu K^2$，$B = 1.2\ \mu K^2$，Λ(无偏振)= 8.48。对于协调模型形状，取协调模型 E 谱幅度为单位幅度，ML 值为 $E = 0.8$，$B = 0.21$，Λ(无偏振)= 13.76。协调模型最佳拟合下的似然比模型 $(l+1)C_l \propto l^2$ 和 $l(l+1)C_l \propto$ 常数分别增大为原来的 200 倍和 12,000 倍，因此比起协调形状，其他两个幂律模型都很糟糕。数据明显支持协调模型形状，故我们将此模型用于那些展示在结果列表中的 E/B 及其他单频段功率的分析。

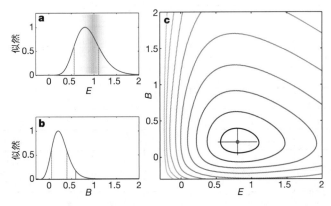

图 3. 双参数成形频段功率 E/B 偏振分析结果。假设 E 模式功率谱形状由协调模型给出，其幅度的单位是相对于该模型的幅度，假设 B 模式功率谱也具有相同的形状。c，图中的点表示最大似然值，十字表示费希尔矩阵误差。似然等高线设置为最大似然值的 $\exp(-n^2/2)$ 水平上，其中 $n = 1$，$2\cdots$。对于一个正态分布，沿着任何一个方向，这些等高线的极值就是边缘化的 $n\text{-}\sigma$ 区间。a 与 b，单参数似然分布(将其余参数边缘化)图。注意似然函数朝着低功率值快速下降，这种似然函数形状(和 χ^2 分布类似)是那

the steep fall in likelihood toward low power values; this likelihood shape (similar to a χ^2 distribution) is typical for positive-definite parameters for which a limited number of high s/n modes are relevant. The grey lines enclose 68% of the total likelihood. The red line indicates the 95% confidence upper limit on B-mode power. The green band shows the distribution of E expectation values for a large grid of cosmological models weighted by the likelihood of those models given our previous temperature result (see ref. 50).

Figure 3 illustrates the result of the E/B concordance shape polarization analysis. The maximum likelihood value of E is 0.80 (0.56 to 1.10 at 68% confidence). For B, the result should clearly be regarded as an upper limit; 95% of the $B > 0$ likelihood (marginalized over E) lies below 0.59.

Figure 2a shows the parameter window functions relevant for this analysis. Note that the E parameter has very little sensitivity to B and vice versa—the purity with which DASI can separate these is remarkable. This is also demonstrated by the low correlation (-0.046) between the E and B parameters (see Table 2).

Table 2. Results of likelihood analyses from polarization data

Analysis	Parameter	l_{low}–l_{high}	ML est.	68% interval				Units
				$(F^{-1})_{ii}^{1/2}$ error	Lower	Upper	UL(95%)	
E/B	E	–	0.80	± 0.28	0.56	1.10	–	Fraction of concordance E
	B	–	0.21	± 0.18	0.05	0.40	0.59	Fraction of concordance E
E5/B5	$E1$	28–245	-0.50	± 0.8	-1.20	1.45	2.38	μK^2
	$E2$	246–420	17.1	± 6.3	11.3	31.2	–	μK^2
	$E3$	421–596	-2.7	± 5.2	-10.0	4.3	24.9	μK^2
	$E4$	597–772	17.5	± 16.0	3.8	40.3	47.2	μK^2
	$E5$	773–1,050	11.4	± 49.0	-32.5	92.3	213.2	μK^2
	$B1$	28–245	-0.65	± 0.65	-1.35	0.52	1.63	μK^2
	$B2$	246–420	1.3	± 2.4	-0.7	5.0	10.0	μK^2
	$B3$	421–596	4.8	± 6.5	-0.6	13.5	17.2	μK^2
E5/B5	$B4$	597–772	13.0	± 14.9	1.6	31.0	49.1	μK^2
	$B5$	773–1,050	-54.0	± 28.9	-77.7	-4.4	147.4	μK^2
E/β_E	E	–	0.84	± 0.28	0.55	1.08	–	Fraction of concordance E
	β_E	–	0.17	± 1.96	-1.63	1.92	–	Temperature spectral index
Scalar/ Tensor	S	–	0.87	± 0.29	0.62	1.18	–	Fraction of concordance S
	T	–	-14.3	± 7.5	-20.4	-3.9	25.4	$T/(S=1)$

些只和有限个高信噪比模式相关的、定义为正的参数所特有的。灰色线是 68% 置信区间边界，红色线表示 B 模式功率的 95% 置信上限。绿色的带是大量的宇宙学模型给出的 E 期望值的分布，其权重是用以前的温度结果计算得到的这些模型的似然值（参考文献 50）。

    图 3 示意了 E/B 协调形状偏振分析的结果。E 的最大似然值为 0.8（置信度为 68% 的参数区间为 0.56 至 1.10）。对于 B，结果显然应该视为上限；B > 0 的概率为 95%（将 E 边缘化）时，B 小于 0.59。

    图 2a 展示了和本次分析相关的参数窗口函数。注意 E 参数几乎和 B 参数无关，反之亦然——DASI 区分两者的能力是非凡的，这一点也反映在 E 和 B 参数之间相关关系（相关系数为 0.046）弱上（见表 2）。

表 2. 偏振数据似然分析结果

分析	参数	$l_{low} \sim l_{high}$	ML est.	68% 区间				
				$(F^{-1})_{ii}^{1/2}$ 误差	下界	上界	UL(95%)	单位
E/B	E	–	0.80	± 0.28	0.56	1.10	–	一致的 E 的比例
	B	–	0.21	± 0.18	0.05	0.40	0.59	一致的 E 的比例
E5/B5	$E1$	28 ~ 245	−0.50	± 0.8	−1.20	1.45	2.38	μK^2
	$E2$	246 ~ 420	17.1	± 6.3	11.3	31.2	–	μK^2
	$E3$	421 ~ 596	−2.7	± 5.2	−10.0	4.3	24.9	μK^2
	$E4$	597 ~ 772	17.5	± 16.0	3.8	40.3	47.2	μK^2
	$E5$	773 ~ 1,050	11.4	± 49.0	−32.5	92.3	213.2	μK^2
	$B1$	28 ~ 245	−0.65	± 0.65	−1.35	0.52	1.63	μK^2
	$B2$	246 ~ 420	1.3	± 2.4	−0.7	5.0	10.0	μK^2
	$B3$	421 ~ 596	4.8	± 6.5	−0.6	13.5	17.2	μK^2
E5/B5	$B4$	597 ~ 772	13.0	± 14.9	1.6	31.0	49.1	μK^2
	$B5$	773 ~ 1,050	−54.0	± 28.9	−77.7	−4.4	147.4	μK^2
E/β_E	E	–	0.84	± 0.28	0.55	1.08	–	一致的 E 的比例
	β_E	–	0.17	± 1.96	−1.63	1.92	–	温度谱指数
标量/张量	S	–	0.87	± 0.29	0.62	1.18	–	一致的 S 的比例
	T	–	−14.3	± 7.5	−20.4	−3.9	25.4	$T/(S=1)$

Continued

					The four corresponding parameter correlation matrices						
E1	E2	E3	E4	E5	B1	B2	B3	B4	B5	E	B
1	−0.137	0.016	−0.002	0.000	−0.255	0.047	−0.004	0.000	0.000	1	−0.046
	1	−0.117	0.014	−0.002	0.024	−0.078	0.004	0.000	0.000		1
		1	−0.122	0.015	−0.003	0.010	−0.027	0.003	−0.001		
			1	−0.119	0.000	−0.001	0.002	−0.016	0.003	E	β_E
				1	0.000	0.000	0.000	0.002	−0.014	1	−0.046
					1	−0.226	0.022	−0.002	0.000		1
						1	−0.097	0.011	−0.002		
							1	−0.111	0.018	S	T
								1	−0.164	1	−0.339
									1		1

ML est., maximum likelihood estimate. $(F^{-1})_{ii}^{1/2}$, Fisher matrix uncertainty for parameter i is evaluated at ML. UL, upper limit.

Assuming that the uncertainties in E and B are normally distributed ("Goodness-of-fit tests" section), the likelihood ratio $\Lambda(\text{nopol}) = 13.76$ implies a probability that our data are consistent with the zero-polarization hypothesis of $PTE = 1.05 \times 10^{-6}$. Our data are highly incompatible with the no-polarization hypothesis. Marginalizing over B, we find $\Lambda(E = 0) = 12.1$ corresponding to detection of E-mode polarization at a PTE of 8.46×10^{-7} (or a significance of 4.92σ).

The likelihood ratio for the concordance model gives $\Lambda(E = 1, B = 0) = 1.23$, for which the Monte Carlo and analytic PTE are both 0.28. We conclude that our data are consistent with the concordance model.

However, given the precision to which the temperature power spectrum of the CMB is currently known, even within the ~7-parameter class of cosmological models often considered, the shape and amplitude of the predicted E-mode spectrum are still somewhat uncertain. To quantify this, we have taken the model grid generated for ref. 50 and calculated the expectation value of the shaped band E parameter for each model using the window function shown in Fig. 2. We then take the distribution of these predicted E amplitudes, weighted by the likelihood of the corresponding model given our previous temperature results (using a common calibration uncertainty for the DASI temperature and polarization measurements). This yields a 68% credible interval for the predicted value of the E parameter of 0.90 to 1.11. As illustrated in Fig. 3a, our data are compatible with the expectation for E on the basis of existing knowledge of the temperature spectrum.

E5/B5. Figure 4a and b show the results of a ten-parameter analysis characterizing the E and B-mode spectra using five flat bandpowers for each. Figure 2b shows the corresponding parameter window functions. Note the extremely small uncertainty in the measurements of the first bands $E1$ and $B1$.

续表

					四个对应参数的相关矩阵						
$E1$	$E2$	$E3$	$E4$	$E5$	$B1$	$B2$	$B3$	$B4$	$B5$	E	B
1	−0.137	0.016	−0.002	0.000	−0.255	0.047	−0.004	0.000	0.000	1	−0.046
	1	−0.117	0.014	−0.002	0.024	−0.078	0.004	0.000	0.000		1
		1	−0.122	0.015	−0.003	0.010	−0.027	0.003	−0.001		
			1	−0.119	0.000	−0.001	0.002	−0.016	0.003	E	β_E
				1	0.000	0.000	0.000	0.002	−0.014	1	−0.046
					1	−0.226	0.022	−0.002	0.000		1
						1	−0.097	0.011	−0.002		
							1	−0.111	0.018	S	T
								1	−0.164	1	−0.339
									1	1	1

ML est.，最大似然估计。$(\mathrm{F}^{-1})_{ii}^{1/2}$，在 ML 处参数 i 的费希尔矩阵不确定度。UL，上限。

假定 E 和 B 的不确定度呈正态分布（见"拟合优度检验"部分），似然比 Λ（无偏振）$= 13.76$ 意味着我们的数据和零偏振假设一致的概率 PTE $= 1.05 \times 10^{-6}$，我们的数据和无偏振的假设非常不相容。将 B 边缘化，我们发现 $\Lambda(E = 0) = 12.1$，意味着探测到 E 模式偏振的 PTE 为 8.46×10^{-7}（或者说显著程度为 4.92σ）。

协调模型给出的似然比为 $\Lambda(E = 1, B = 0) = 1.23$，蒙特卡洛和解析近似给出的 PTE 都是 0.28，由此推断我们的数据支持协调模型。

然而，考虑到目前 CMB 的温度功率谱的精度，即使是经常被考虑的具有 ~ 7 个参数的这类宇宙学模型，预计给出 E 模式谱的形状和幅度仍然有些不确定。为了量化这种不确定性，我们采用了为参考文献 50 构造的模型网格，并使用图 2 所示的窗口函数计算每个模型的成形的频带 E 参数的期望值。然后我们取这些 E 模式幅度的预测值，并根据原先给出温度结果的对应模型下的似然值作加权（对 DASI 的温度和偏振测量使用相同的校准不确定度）。这样做给出一个 E 参数的 68% 置信区间为 0.90 到 1.11。如图 3a 所示，基于目前我们对温度谱的知识给出的 E 的期望值和我们的数据是相匹配的。

E5/B5 图 4a 和 b 是 E 和 B 模式在分别使用五个平坦频段功率时的十参数分析结果。表 2b 是对应的参数窗口函数。注意第一频带 $E1$ 和 $B1$ 的测量不确定度非常小。

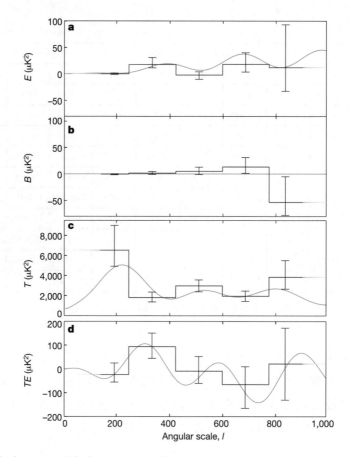

Fig. 4. Results from several likelihood analyses. The ten-parameter $E5/B5$ polarization analysis is shown in **a** and **b**. The $T5$ temperature analysis is shown in **c** and the five TE bands from the $T/E/TE5$ joint analysis are shown in **d**. All the results are shown in flat bandpower units of $l(l+1)C_l/(2\pi)$. The blue line shows the piecewise flat bandpower model for the maximum likelihood parameter values, with the error bars indicating the 68% central region of the likelihood of each parameter, marginalizing over the other parameter values (analogous to the grey lines in Fig. 3a and b). In each case the green line is the concordance model.

Table 3. Results of likelihood analyses from temperature data

Analysis	Parameter	$l_{low}-l_{high}$	ML est.	68% interval			
				$(F^{-1})_{ii}^{1/2}$ error	Lower	Upper	Units
T/β_T	T	–	1.19	±0.11	1.09	1.30	Fraction of concordance T
	β_T	–	−0.01	±0.12	−0.16	0.14	Temperature spectral index
$T5$	$T1$	28–245	6,510	±1,610	5,440	9,630	μK^2
	$T2$	246–420	1,780	±420	1,480	2,490	μK^2
	$T3$	421–596	2,950	±540	2,500	3,730	μK^2
	$T4$	597–772	1,910	±450	1,530	2,590	μK^2
	$T5$	773–1,050	3,810	±1,210	3,020	6,070	μK^2

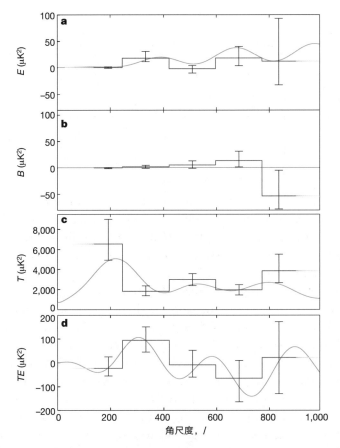

图 4. 数个似然分析的结果。**a** 和 **b** 展示了十参数 $E5/B5$ 偏振分析。$T5$ 温度分析结果在 **c** 中，$T/E/TE5$ 联合分析的五个 TE 频段结果在 **d** 中。图示所有结果都以 $l(l+1)C_l/2\pi$ 为单位。蓝色线表示最大似然参数值给出的分段平坦模型下的形状，误差棒是各个参数 68% 置信区间范围的区域大小（将其他参数边缘化，和图 3a 和 b 中的灰色线类似）。每个图中的绿色线，都对应协调模型给出的形状。

表 3. 温度数据似然分析结果

分析	参数	$l_{low} \sim l_{high}$	ML est.	68% 区间			单位
				$(F^{-1})_{ii}^{1/2}$ 误差	下界	上界	
T/β_T	T	–	1.19	±0.11	1.09	1.30	一致的 T 的比例
	β_T	–	−0.01	±0.12	−0.16	0.14	温度谱指数
$T5$	$T1$	$28 \sim 245$	6,510	±1,610	5,440	9,630	μK^2
	$T2$	$246 \sim 420$	1,780	±420	1,480	2,490	μK^2
	$T3$	$421 \sim 596$	2,950	±540	2,500	3,730	μK^2
	$T4$	$597 \sim 772$	1,910	±450	1,530	2,590	μK^2
	$T5$	$773 \sim 1,050$	3,810	±1,210	3,020	6,070	μK^2

Continued

The two corresponding parameter correlation matrices						
$T1$	$T2$	$T3$	$T4$	$T5$	T	β_T
1	−0.101	0.004	−0.004	−0.001	1	0.023
	1	−0.092	−0.013	−0.011		1
		1	−0.115	−0.010		
			1	−0.147		
				1		

To test whether these results are consistent with the concordance model, we calculate the expectation value for the nominal concordance model in each of the five bands, yielding $E = (0.8, 14, 13, 37, 16)$ and $B = (0, 0, 0, 0, 0)\ \mu K^2$. The likelihood ratio comparing this point in the ten-dimensional parameter space to the maximum gives $\Lambda = 5.1$, which for ten degrees of freedom results in a PTE of 0.42, indicating that our data are consistent with the expected polarization parameterized in this way. The $E5/B5$ results are highly inconsistent with the zero-polarization "nopol" hypothesis, for which $\Lambda = 15.2$ with a PTE = 0.00073. This statistic is considerably weaker than the equivalent one obtained for the single band analysis in the "E/B analysis" section, as expected from the higher number of degrees of freedom in this analysis. In this ten-dimensional space, all possible random deviations from the "nopol" expectation values $E = (0, 0, 0, 0, 0)$, $B = (0, 0, 0, 0, 0)$ are treated equally in constructing the PTE for our Λ statistic. Imagining the "nopol" hypothesis to be true, it would be far less likely to obtain a result in this large parameter space that is both inconsistent with "nopol" at this level and at the same time is consistent with the concordance model, than it would be to obtain a result that is merely inconsistent with "nopol" in some way at this level. It is the latter probability that is measured by the PTE for our Λ(nopol), explaining why this approach to goodness-of-fit weakens upon considering increasing numbers of parameters.

E/β_E. We have performed a two-parameter analysis to determine the amplitude of the E-mode polarization signal as above and the frequency spectral index β_E of this signal relative to CMB (Fig. 5). As expected, the results for the E-mode amplitude are very similar to those for the E/B analysis described in the previous section. The spectral index constraint is not strong; the maximum likelihood value is $\beta_E = 0.17$ (−1.63 to 1.92). The result is nevertheless interesting in the context of ruling out possible foregrounds (see the "Diffuse foregrounds" subsection below).

542

续表

两个对应参数的相关矩阵						
$T1$	$T2$	$T3$	$T4$	$T5$	T	β_T
1	−0.101	0.004	−0.004	−0.001	1	0.023
	1	−0.092	−0.013	−0.011		1
		1	−0.115	−0.010		
			1	−0.147		
				1		

为了检测这些结果是否与协调模型一致，我们计算了五个频段中每个频段的（名义上的）协调模型的期望值，得到 $E = (0.8, 14, 13, 37, 16)$，$B = (0, 0, 0, 0, 0)\ \mu K^2$。该点（在十维参数空间上）和最大处相比得到似然比 $\Lambda = 5.1$，对于十个自由度下对应的 PTE 为 0.42，表明我们的数据和这种参数化方式给出的预测是相符的。$E5/B5$ 结果和零偏振的"无偏振"模型高度不一致，其中 $\Lambda = 15.2$，PTE = 0.00073。该统计量远低于"E/B 分析"部分中单频段分析给出的对应的值，正如从该分析中较高的自由度数目所能预期的那样。在这个十维空间中，为我们的 Λ 统计量构造 PTE 时候，所有来自"无偏振"期望值 $E = (0, 0, 0, 0, 0)$，$B = (0, 0, 0, 0, 0)$ 的可能的随机偏离都有相同的贡献。假设"无偏振"是正确的，在这样一个大参数空间中获得一个既在上述的水平上与"无偏振"模型相悖，同时又和协调模型匹配的结果，比按照同样的方法得到一个仅仅和"无偏振"模型相悖的结果的可能性要小得多，我们的 Λ（无偏振）对应的 PTE 是用后者的概率给出的，这解释了为什么增加参数数量情况下该拟合优度方法的结论会减弱。

E/β_E　我们已经进行了双参数分析以确定如上所述的 E 模式极化信号的幅度以及该信号相对于 CMB 的谱指数 β_E（图 5）。正如我们预测的，E 模式幅度的结果和上一部分"E/B 分析"给出的差不多，谱指数的约束不是很强，最大似然估计值为 $\beta_E = 0.17$（置信度为 68% 的参数区间为 −1.63 到 1.92）。不管怎样，若不考虑前景（参见"弥散前景"部分）的可能性，这个结果还是值得关注的。

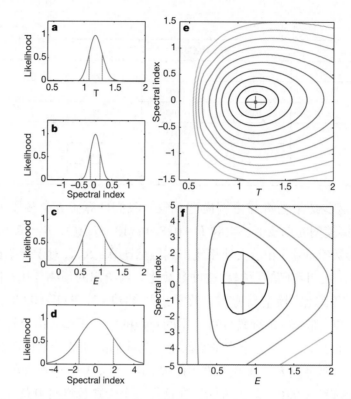

Fig. 5. Results of shaped bandpower amplitude/spectral-index analyses. **a**, **b**, **e**, The T/β_T temperature analysis assuming the T power spectrum shape as predicted for the concordance model, and in units relative to that model. The layout of the plot is analogous to Fig. 3. Spectral index is relative to the CMB blackbody—in these units, synchrotron emission would be expected to have an index of approximately -3. **c**, **d**, **f**, Results of the similar E/β_E analysis performed on the polarization data.

Scalar/tensor. Predictions exist for the shape of the E and B-mode spectra which would result from primordial gravity waves, also known as tensor perturbations, although their amplitude is not well constrained by theory. In a concordance-type model such tensor polarization spectra are expected to peak at $l \approx 100$. Assuming reasonable priors, current measurements of the temperature spectrum (in which tensor and scalar contributions will be mixed) suggest $T/S < 0.2$ (ref. 61), where this amplitude ratio is defined in terms of the tensor and scalar contributions to the temperature quadrupole C_2^T. We use the distinct polarization angular power spectra for the scalars (our usual concordance E shape, with $B = 0$) and the tensors (E_T and B_T) as two components of a likelihood analysis to constrain the amplitude parameters of these components. In principle, because the scalar B-mode spectrum is zero, this approach avoids the fundamental sample variance limitations arising from using the temperature spectrum alone. However, the $E5/B5$ analysis (see subsection "$E5/B5$") indicates that we have only upper limits to the E- and B-mode polarization at the angular scales most relevant ($l \lesssim 200$) for the tensor spectra. It is therefore not surprising that our limits on T/S derived from the polarization spectra as reported in Table 2 are quite weak.

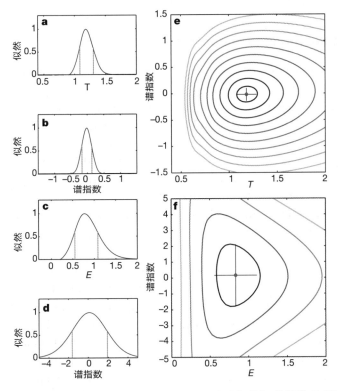

图 5. 成形带功率幅度/谱指数分析结果。图 a、b、e，假设温度功率谱满足协调模型下的 T/β_T 温度分析结果，与模型相对单位一致。此图的布局和图 3 相似，谱指数对应 CMB 黑体谱。在这些单位下，预计同步加速辐射的谱指数约为 -3。图 c、d、f，对偏振数据做类似 E/β_E 分析的结果。

标量/张量 对于 E 和 B 模式谱的形状存在多个预测，预测认为这些谱可能由原初引力波（即张量扰动）产生，尽管理论并不能很好地就它们的幅度给出约束。在协调型模型中，这种张量偏振谱预计在 $l = 100$ 处有峰值。假定一个合理的先验，目前的温度谱（张量和标量的贡献混合在一起）测量暗示 $T/S < 0.2$（参考文献 61），这个幅度比的定义是依据张量和标量对温度四极矩 C_2^T 的贡献给出的：我们把标量（我们惯用的协调模型 E 的形状，$B = 0$）和张量（E_T 和 B_T）各自的偏振角功率谱作为似然分析的两个部分来约束这两个组分的幅度参数。原则上，由于标量 B 模式功率谱是零，这个方法能够避免单独使用温度谱引入基本样本方差的局限性。但是 $E5/B5$ 分析（参见 "$E5/B5$" 部分）暗示我们在张量功率谱最相关的 $l \leqslant 200$ 的角尺度上，只能得到 E 和 B 模式的上限。这样，在表 2 中报告中，我们从偏振谱得到的 T/S 的限制相当弱也就不令人费解了。

Temperature data analyses and results for *T* spectrum

***T*/*β~T~*.**   Results are shown in Fig. 5 for a two-parameter analysis to determine the amplitude and frequency spectral index of the temperature signal. The bandpower shape used is that of the concordance T spectrum, and the amplitude parameter is expressed in units relative to that spectrum. The spectral index is relative to the CMB, so that 0 corresponds to a 2.73-K Planck spectrum. The amplitude of T has a maximum likelihood value of 1.19 (1.09 to 1.30), and the spectral index $\beta_T = -0.01$ (−0.16 to 0.14). Although the uncertainty in the temperature amplitude is dominated by sample variance, the spectral index is limited only by the sensitivity and fractional bandwidth of DASI. Owing to the extremely high s/n of the temperature data, the constraints on the spectral index are superior to those from previous DASI observations (ref. 6).

***T5*.**   Fig. 4c shows the results of an analysis using five flat bands to characterize the temperature spectrum. These results are completely dominated by the sample variance in the differenced field. They are consistent with, although less precise than our previous temperature power spectra described in ref. 6; we include them here primarily to emphasize that DASI makes measurements simultaneously in all four Stokes parameters and is therefore able to measure temperature as well as polarization anisotropy. Note that these results and those for T/β_T have not been corrected for residual point sources.

Joint analyses and cross spectra results

***T*/*E*/*TE*.**   Figure 6 shows the results of a three-parameter single bandpower analysis of the amplitudes of the T and E spectra, and the TE cross-correlation spectrum. As before, bandpower shapes based on the concordance model are used. The T and E constraints are, as expected, very similar to those from the E/B, E/β_E and T/β_T analyses described above. The new result here is TE which has a maximum likelihood value of 0.91 (0.45 to 1.37). Note that in contrast to the two-dimensional likelihoods shown in other figures, here the contours show apparent evidence of correlation between the two parameters; the parameter correlation coefficients from Table 4 are 0.28 for E/TE and 0.21 for T/TE.

Table 4. Results of likelihood analyses from joint temperature-polarization data set

Analysis	Parameter	l_{low}–l_{high}	ML est.	68% interval			
				$(F^{-1})_{ij}^{1/2}$ error	Lower	Upper	Units
T/*E*/*TE*	T	–	1.13	±0.10	1.05	1.29	Fraction of concordance T
	E	–	0.77	±0.27	0.57	1.10	Fraction of concordance E
	TE	–	0.91	±0.38	0.45	1.37	Fraction of concordance TE

温度数据分析和 T 谱结果

T/β_T　限制温度信号的幅度和谱指数的双参数分析，结果如图 5 所示，选用协调模型 T 谱的频带功率形状，幅度参数以该谱相对应的单位来表示。该谱的谱指数是相对于 CMB 的，因此 0 对应 2.73 K 的普朗克谱。T 幅度的最大似然估计值为 1.19（置信度为 68% 的参数区间为 1.09 到 1.30），谱指数 $\beta_T = -0.01$（置信度为 68% 的参数区间为 -0.16 到 0.14）。尽管温度幅度的误差主要来源于样本误差，但是谱指数的误差仅受限于 DASI 的灵敏度和带宽比。得益于温度数据极高的信噪比，温度谱指数的限制优于从先前 DASI 观测给出的（参考文献 6）。

$T5$　图 4c 展示了一个用五平坦频带描述温度谱的分析结果。这些结果完全由离散区域内的样本方差主导。尽管不如我们先前在参考文献 6 中所描述的温度功率谱那么精确，但它们是吻合的。我们把它们列在此处，主要是为了强调 DASI 同时测量四个斯托克斯参数，因此能够在测量温度的同时测量偏振的各向异性。注意，这些结果以及 T/β_T 的结果还未做残留点光源校准。

联合分析和交叉谱结果

$T/E/TE$　图 6 为功率 T 谱、E 谱和 TE 相关谱的幅度的三参数单频带分析结果。和前面一样，频带功率选用协调模型的形状。和预想的一样，T 和 E 的约束与前面描述过的 E/B、E/β_E 和 T/β_T 分析非常相似。这部分的新结果是 TE 的最大似然估计值，为 0.91（置信度为 68% 的参数区间为 0.45 到 1.37）。注意和其他图中的二维似然分布相反，这里两个参数的二维（似然）等高线图呈现明显的相关关系，从表 4 可知，参数的相关系数为 0.28（E/TE）和 0.21（T/TE）。

表 4. 温度－偏振数据集联合似然分析结果

分析	参数	$l_{low} \sim l_{high}$	ML est.	68% 区间			
				$(F^{-1})_{ii}^{1/2}$ 误差	下界	上界	单位
$T/E/TE$	T	–	1.13	± 0.10	1.05	1.29	一致的 T 的比例
	E	–	0.77	± 0.27	0.57	1.10	一致的 E 的比例
	TE	–	0.91	± 0.38	0.45	1.37	一致的 TE 的比例

Continued

Analysis	Parameter	l_{low}–l_{high}	ML est.	68% interval			
				$(F^{-1})_{ij}^{1/2}$ error	Lower	Upper	Units
T/E/TE5	T	–	1.12	±0.10	1.09	1.31	Fraction of concordance T
	E	–	0.81	±0.28	0.71	1.36	Fraction of concordance E
	TE1	28–245	−24.8	±32.2	−55.3	24.7	μK²
	TE2	246–420	92.3	±38.4	44.9	151.1	μK²
	TE3	421–596	−10.5	±48.2	−60.1	52.0	μK²
	TE4	597–772	−66.7	±74.3	−164.6	9.5	μK²
	TE5	773–1,050	20.0	±167.9	−130.3	172.3	μK²
T/E/B/ TE/TB/EB	T	–	1.13	±0.10	1.03	1.27	Fraction of concordance T
	E	–	0.75	±0.26	0.59	1.19	Fraction of concordance E
	B	–	0.20	±0.18	0.11	0.52	Fraction of concordance E
	TE	–	1.02	±0.37	0.53	1.49	Fraction of concordance TE
	TB	–	0.53	±0.32	0.08	0.82	Fraction of concordance TE
	EB	–	−0.16	±0.16	−0.38	0.01	Fraction of concordance E

The three parameter correlation matrices

T	E	TE1	TE2	TE3	TE4	TE5	T	E	B	TE	TB	EB
1	0.026	−0.071	0.202	−0.018	−0.075	0.008	1	0.026	0.004	0.230	0.136	0.033
	1	−0.067	0.339	−0.023	−0.090	0.008		1	−0.027	0.320	−0.040	−0.182
		1	−0.076	0.006	0.011	−0.001			1	−0.027	0.219	−0.190
			1	−0.078	−0.039	0.004				1	−0.150	0.109
				1	−0.056	0.004					1	0.213
					1	−0.066						1
						1						

T	E	TE
1	0.017	0.207
	1	0.282
		1

Marginalizing over T and E, we find that the likelihood of TE peaks very near 1, so that $\Lambda(TE = 1) = 0.02$ with a PTE of 0.857. For the "no cross-correlation" hypothesis, $\Lambda(TE = 0) = 1.85$ with an analytic PTE of 0.054 (the PTE calculated from Monte Carlo simulations is 0.047). This result represents a detection of the expected TE correlation at 95% confidence and is particularly interesting in that it suggests a common origin for the observed temperature and polarization anisotropy.

It has been suggested that an estimator of TE cross-correlation constructed using a $TE = 0$ prior may offer greater immunity to systematic errors[59]. We have confirmed that applying

分析	参数	$l_{low} \sim l_{high}$	ML est.	68% 区间			
				$(F^{-1})_{ij}^{1/2}$ 误差	下界	上界	单位
T/E/TE5	T	–	1.12	±0.10	1.09	1.31	一致的 T 的比例
	E	–	0.81	±0.28	0.71	1.36	一致的 E 的比例
	$TE1$	$28 \sim 245$	−24.8	±32.2	−55.3	24.7	μK²
	$TE2$	$246 \sim 420$	92.3	±38.4	44.9	151.1	μK²
	$TE3$	$421 \sim 596$	−10.5	±48.2	−60.1	52.0	μK²
	$TE4$	$597 \sim 772$	−66.7	±74.3	−164.6	9.5	μK²
	$TE5$	$773 \sim 1,050$	20.0	±167.9	−130.3	172.3	μK²
T/E/B/TE/TB/EB	T	–	1.13	±0.10	1.03	1.27	一致的 T 的比例
	E	–	0.75	±0.26	0.59	1.19	一致的 E 的比例
	B	–	0.20	±0.18	0.11	0.52	一致的 E 的比例
	TE	–	1.02	±0.37	0.53	1.49	一致的 TE 的比例
	TB	–	0.53	±0.32	0.08	0.82	一致的 TE 的比例
	EB	–	−0.16	±0.16	−0.38	0.01	一致的 E 的比例

三参数的相关矩阵

T	E	$TE1$	$TE2$	$TE3$	$TE4$	$TE5$	T	E	B	TE	TB	EB
1	0.026	−0.071	0.202	−0.018	−0.075	0.008	1	0.026	0.004	0.230	0.136	0.033
	1	−0.067	0.339	−0.023	−0.090	0.008		1	−0.027	0.320	−0.040	−0.182
		1	−0.076	0.006	0.011	−0.001			1	−0.027	0.219	−0.190
			1	−0.078	−0.039	0.004				1	−0.150	0.109
				1	−0.056	0.004					1	0.213
					1	−0.066						1
						1						

T	E	TE
1	0.017	0.207
	1	0.282
		1

将 T 和 E 边缘化，我们发现 TE 峰处的似然非常接近 1，因而 $\Lambda(TE=1)=0.02$，PTE 为 0.857。在 T 和 E"无交叉相关关系"的假设下，$\Lambda(TE=1)=1.85$，其解析 PTE 为 0.054(用蒙特卡洛模拟计算出的 PTE 为 0.047)。这个结果表明在 95% 置信度下，TE 具有相关性。更为有趣的是，这暗示着观测到的温度和偏振各向异性具有相同的起源。

也有人提出，使用 $TE=0$ 的先验构造的一个 TE 交叉相关估计量或许对系统误差更不敏感[59]。我们已经证明对我们的数据使用这样的技巧会产生和上面的似然分

such a technique to our data yields similar results to the above likelihood analysis, with errors slightly increased as expected.

$T/E/TE$5. We have performed a seven-parameter analysis using single shaped band powers for T and E, and five flat bandpowers for the TE cross-correlation; the TE results from this are shown in Fig. 4d. In this analysis the B-mode polarization has been explicitly set to zero. Again, the T and E constraints are similar to the values for the other analyses where these parameters appear. The TE bandpowers are consistent with the predictions of the concordance model.

$T/E/B/TE/TB/EB$. Finally, we describe the results of a six shaped bandpower analysis for the three individual spectra T, E and B, together with the three possible cross-correlation spectra TE, TB and EB. We include the B cross-spectra for completeness, though there is little evidence for any B-mode signal. Because there are no predictions for the shapes of the TB or EB spectra (they are expected to be zero), we preserve the symmetry of the analysis between E and B by simply parameterizing them in terms of the TE and E spectral shapes. The results for T, E, B and TE are similar to those obtained before, with no detection of EB or TB.

Systematic Uncertainties

Noise, calibration, offsets and pointing

To assess the effect of systematic uncertainties on the likelihood results, we have repeated each of the nine analyses with alternative assumptions about the various effects that we have identified which reflect the range of uncertainty on each.

Much of the effort of the data analysis presented in this paper has gone into investigating the consistency of the data with the noise model as discussed in the "Noise model" subsection. We find no discrepancies between complementary noise estimates on different timescales, to a level $\ll 1\%$. As discussed in the "χ^2 tests" subsection, numerous consistency tests on subsets of the co-polar and cross-polar visibility data show no evidence for an error in the noise scaling to a similar level. When we re-evaluate each of the analyses described in the "Likelihood results" section with the noise scaled by 1%, the shift in the maximum likelihood values for all parameters is entirely negligible.

In the "Noise model" subsection, we reported evidence of some detectable noise correlations between real/imaginary visibilities and between visibilities from different bands of the same baseline. When either or both of these noise correlations are added to the covariance matrix at the measured level, the effects are again negligible: the most significant shift is in the highest-l band of the E spectrum from the $E5/B5$ analysis (see the "$E5/B5$" subsection), where the power shifts by about 2 μK^2.

析类似的结果，不过和预测的一样，误差会稍稍增大。

T/*E*/*TE*5　我们也进行了一个七参数分析，其中对 *T* 和 *E* 采用了单形状频带功率，对 *TE* 交叉相关谱采用了五平坦频段功率的形状，*TE* 结果见图 4d。此分析中，*B* 模式偏振被明确地设置为零。同样，对 *T* 和 *E* 的约束与其他出现这两个参数的分析类似，*TE* 频段功率和调和模型给出的理论预言一致。

T/*E*/*B*/*TE*/*TB*/*EB*　最后，我们描述了对 *T*、*E* 和 *B* 三个功率谱以及三个可能的交叉相关谱 *TE*, *TB* 和 *EB* 的六个成型频带功率分析。虽然几乎没有迹象表明存在 *B* 模式信号，但是为了完备性，我们还是把 *B* 交叉谱列入了分析。由于没有对 *TB* 谱和 *EB* 谱的形状的预测（它们被认为是零），我们简单地根据 *TE* 和 *E* 谱形状对它们进行参数化，以保持 *E* 和 *B* 之间分析的对称性。*T*、*E*、*B* 和 *TE* 的结果与之前得到结果的类似，*EB* 和 *TB* 没有被探测到。

系统不确定度

噪声、校准、补偿和指向

为了评估系统不确定度对似然结果的影响，我们重复了这九组的每一个分析，但是对那些我们已经知道的反映各自不同的不确定范围的效应作了替代的假设。

本文展示的大部分数据分析都尝试研究数据和噪声模型（在"噪声模型"部分讨论）的一致性。我们发现不同时间尺度的互补噪声估计量之间并无差异，均为 $\ll 1\%$ 的水平。正如"χ^2 检验"部分讨论过的，我们对自相关偏振和交叉相关偏振的可视性数据的子集进行了大量的一致性检验，没有迹象表明有一个噪声误差能够达到类似的水平。我们将噪声按比例缩放 1%，重复"似然结果"部分的每一个分析，发现所有参数的最大似然估计值的偏移完全可以忽略不计。

在"噪声模型"部分，我们报告了存在于实部/虚部可视度之间，以及相同基线上不同频带之间可视性的某些可探测噪声相关性的证据。把这些噪声相关的其中一个或两个全部添加到协方差矩阵（在实际测量到的水平上），它们的影响再次可以忽略不计：最显著的偏移出现在 *E*5/*B*5 分析中 *E* 谱的最高 *l* 频段上（参见"*E*5/*B*5"部分），其中功率偏移了约 2 μK^2。

Errors in the determination of the absolute cross-polar phase offsets will mix power between E and B; these phase offsets have been independently determined from wire-grid calibrations and observations of the Moon, and found to agree to within the measurement uncertainties of about 0.4° (ref. 51). Reanalysis of the data with the measured phase offsets shifted by 2° demonstrates that the likelihood results are immune to errors of this magnitude: the most significant effect occurs in the highest-l band of the TE spectrum from the T, E, $TE5$ analysis (see the "$T/E/TE5$" subsection), where the power shifts by about 30 μK^2.

The on-axis leakages described in the "On-axis leakage" subsection will mix power from T into E and B, and the data are corrected for this effect in the course of reduction, before input to any analyses. When the likelihood analyses are performed without the leakage correction, the largest effects appear in the shaped TE amplitude analysis (see "$T/E/TE$" subsection), and the lowest-l band of $TE5$ from the T, E, $TE5$ analysis (see the "$T/E/TE5$" subsection); all shifts are tiny compared to the 68% confidence intervals. As the leakage correction itself has little impact on the results, the uncertainties in the correction, which are at the < 1% level, will have no noticeable effect.

As described in the "Off-axis leakage" subsection, the off-axis leakage from the feeds is a more significant effect, and is accounted for in the likelihood analysis by modelling its contribution to the covariance matrix. When this correction is not applied, the E, B results (see "E/B analysis" subsection) shift by about 4% and 2%, respectively, as expected from simulations of this effect. Although this bias is already small, the simulations show that the correction removes it completely to the degree to which we understand the off-axis leakage. Uncertainties in the leakage profiles of the order of the fit residuals (see ref. 51) lead to shifts of less than 1%.

The pointing accuracy of the telescope is measured to be better than 2 arcmin and the root-mean-square tracking errors are < 20 arcsec; as we discussed in refs 49 and 6, this is more than sufficient for the characterization of CMB anisotropy at the much larger angular scales measured by DASI.

Absolute calibration of the telescope was achieved through measurements of external thermal loads, transferred to the calibrator RCW38. The dominant uncertainty in the calibration is due to temperature and coupling of the thermal loads. As reported in ref. 6, we estimate an overall calibration uncertainty of 8% (1σ), expressed as a fractional uncertainty on the C_l bandpowers (4% in $\Delta T/T$). This applies equally to the temperature and polarization data presented here.

Foregrounds

Point sources. The highest-sensitivity point-source catalogue in our observing region is the 5-GHz PMN survey[62]. For our first-season temperature analysis described in refs 49

确定绝对交叉−偏振相位补偿时的误差会混合在 E 和 B 的功率之间，这些相位补偿是通过导线网格校准和对月球的观测独立确定的，且发现其在约 $0.4°$ 的测量误差值（参考文献 51）之内。用测量到的偏移量为 $2°$ 的相位补偿对数据进行重新分析表明，似然结果不受这个程度的误差影响：最显著的影响发生在 T、E、$TE5$ 分析（参见"$T/E/TE$"部分）中 TE 谱的最高 l 频段上，其中功率偏移了约 $30\ \mu K^2$。

"共轴渗漏"部分描述过的共轴渗漏会将 T 功率混入到 E 和 B 功率中，在做任何分析之前，数据会在还原过程中校正掉该效应。若似然分析之前并未做渗漏校准，受影响最显著的是在成形的 TE 幅度分析（参见"$T/E/TE$"部分）和 T、E、$TE5$ 分析（参见"$T/E/TE5$"部分）中 $TE5$ 的最低 l 频段上。相比较于 68% 置信区间，所有的偏移都是微小的。由于渗漏校准本身对结果几乎没有影响，校准时候的误差——小于 1% 的程度——将不会有明显的影响。

如同在"离轴渗漏"部分中所描述的，来自馈源的离轴渗漏是一个更加显著的效应。我们根据其在协方差矩阵的贡献进行建模以解释该效应对似然分析的影响。若不使用该校准，如同模拟该效应时所预料的那样，E、B 结果（参见"E/B 分析"部分）分别偏移约 4% 和 2%。虽然这种偏差已经很小，但模拟表明，就我们所理解的离轴渗漏的程度，校正能将其完全消除。达到拟合残差（参考文献 51）量级的渗漏带来的不确定性会导致的偏移小于 1%。

我们测得望远镜的指向精度优于 2 角分，均方根跟踪误差 <20 角秒。正如我们在参考文献 49 和 6 中所讨论的那样，这个精度对刻画在 DASI 测量的更大角尺度上的 CMB 各向异性已经足够了。

望远镜的绝对校准是通过测量外部热负荷给出的，并传输到 RCW38 定标器上。校准的最重要的不确定性来源于温度和热负荷的耦合。如参考文献 6 中报告的，我们估计存在 8%(1σ) 的总体校准不确定度，C_l 频段功率的不确定度表示为分数（在 $\Delta T/T$ 中占 4%）。这同样适用于温度和偏振数据。

前景

点源 我们观测区域中灵敏度最高的点源星表是 5 GHz PMN 巡天[62]。对于参考文献 49 和 6 所描述的第一季度温度分析，我们就是利用此目录来剔除已知点源

and 6 we projected out known sources using this catalogue. We have kept this procedure for the temperature data presented here, projecting the same set of sources as before.

Unfortunately the PMN survey is not polarization sensitive. We note that the distribution of point-source polarization fractions is approximately exponential (see below). Total intensity is thus a poor indicator of polarized intensity and it is therefore not sensible to project out the PMN sources in our polarization analysis.

Our polarization fields were selected for the absence of any significant point-source detections in the first-season data. No significant detections are found in the 2001–02 data, either in the temperature data, which are dominated by CMB anisotropy, or in the polarization data.

To calculate the expected contribution of undetected point sources to our polarization results we would like to know the distribution of polarized flux densities, but unfortunately no such information exists in our frequency range. However, to make an estimate, we use the distribution of total intensities, and then assume a distribution of polarization fractions. We know the former distribution quite well from our own first-season 32-field data where we detect 31 point sources and determine that $dN/dS_{31} = (32 \pm 7)S^{(-2.15\pm0.20)}$ $Jy^{-1}sr^{-1}$ in the range 0.1 to 10 Jy. This is consistent, when extrapolated to lower flux densities, with a result from the CBI experiment valid in the range 5–50 mJy (ref. 63). The distribution of point source polarization fractions at 5 GHz can be characterized by an exponential with a mean of 3.8% (ref. 64); data of somewhat lower quality at 15 GHz are consistent with the same distribution[65]. Qualitatively, we expect the polarization fraction of synchrotron-dominated sources initially to rise with frequency, and then reach a plateau or fall, with the break point at frequencies $\ll$5 GHz (see ref. 66 for an example). In the absence of better data we have conservatively assumed that the exponential distribution mentioned above continues to hold at 30 GHz.

We proceed to estimate the effect of point sources by Monte Carlo simulation, generating realizations using the total intensity and polarization fraction distributions mentioned above. For each realization, we generate simulated DASI data by adding a realization of CMB anisotropy and appropriate instrument noise. The simulated data are tested for evidence of point sources and those realizations that show statistics similar to the real data are kept. The effect of off-axis leakage, which we describe and quantify in ref. 51, is included in these calculations.

When the simulated data are passed through the E/B analysis described in the "E/B analysis" subsection, the mean bias of the E parameter is 0.04 with a standard deviation of 0.05; in 95% of cases the shift distance in the E/B plane is less than 0.13. We conclude that the presence of point sources consistent with our observed data has a relatively small effect on our polarization results.

Diffuse foregrounds. In ref. 51, we find no evidence for contamination of the

的。我们对这里所呈现的温度数据保留了同样的处理，和以前一样剔除掉了相同的点源集。

很遗憾的是，PMN 巡天对偏振并不敏感，我们注意到，点源偏振部分的分布大致呈指数分布（见下文），故而总强度不能代表偏振强度，因此做偏振分析时，剔除 PMN 源是不明智的。

在第一季度数据中没有发现任何显著的点源信号，据此我们选择偏振视场。在 2001 年～2002 年数据中，无论是由 CMB 各向异性主导的温度数据还是偏振数据，我们未发现显著的信号。

为了预先计算未被检测到的点源对我们的偏振结果的贡献，我们希望知道被偏振的通量密度的分布，但遗憾的是在我们的频率范围内并不存在这样的信息。但是，为了对其进行估算，我们使用总强度的分布，然后假设偏振部分的分布函数。从第一季度 32 视场数据中，我们对前者的分布有了很好的了解，在那里我们探测到 31 个点源并确定在 0.1 到 10 Jy 范围内有 $dN/dS_{31} = (32 \pm 7)S^{(-2.15 \pm 0.20)} Jy^{-1} \cdot sr^{-1}$。当外推到较低的通量密度时，它的结果和从 CBI 实验（参考文献 63）在 5～50 mJy 范围的结果是一致的。5 GHz 处的点源偏振部分的分布可以用平均值为 3.8%（参考文献 64）的指数分布表征，15 GHz 处的那些质量稍低的数据具有相同的分布 [65]。定性地说，我们预计同步加速机制主导的源的偏振部分首先随着频率上升而增大，然后达到平稳水平或达到频率 <<5 GHz 的转折点后下降（例子请参见参考文献 66）。在没有更好的数据的情况下，我们保守地假设上面提到的指数分布直到 30 GHz 仍适用。

我们继续使用蒙特卡罗模拟估计点源的影响，使用上面提到的总强度和偏振部分的分布生成实现。对于每个实现，我们通过加入一个 CMB 各向异性信号和适当的仪器噪声来生成模拟的 DASI 数据。我们就点源的痕迹对模拟数据进行检测，然后保留那些具有同实际数据相近的统计量的实现。离轴渗漏——我们在文献 51 中对其进行描述和量化——的影响已经包含在这些计算中了。

当模拟数据通过"E/B 分析"部分描述的方法进行 E/B 分析时，E 参数的平均偏差为 0.04，其中标准差为 0.05；在占 95% 的案例中，E/B 平面中的偏移距离小于 0.13。我们得出结论，若点源与我们观察到的数据一致，则它们的存在对我们的偏振结果的影响相对较小。

弥散前景   正如"T/β_T"部分对温度谱指数的限制所证实的，在参考文献 51 中，

temperature data by synchrotron, dust or free–free emission, as confirmed by the limits on the temperature spectral index presented in the "T/β_T" subsection. The expected fractional polarization of the CMB is of order 10%, while the corresponding number for free–free emission is less than 1%. Diffuse thermal dust emission may be polarized by several per cent (see for example ref. 67), although we note that polarization of the admixture of dust and free–free emission observed with DASI in NGC 6334 is $\ll 1\%$ (see ref. 51). Likewise, emission from spinning dust is not expected to be polarized at detectable levels[68]. Therefore if free–free and dust emission did not contribute significantly to our temperature anisotropy results they are not expected to contribute to the polarization. Synchrotron emission on the other hand can in principle be up to 70% polarized, and is by far the greatest concern; what was a negligible contribution in the temperature case could be a significant one in polarization.

There are no published polarization maps in the region of our fields. Previous attempts to estimate the angular power spectrum of polarized synchrotron emission have been guided by surveys of the Galactic plane at frequencies of 1–3 GHz (refs 69 and 70). These maps show much more small-scale structure in polarization than in temperature, but this is mostly induced by Faraday rotation[71], an effect which is negligible at 30 GHz. In addition, because synchrotron emission is highly concentrated in the disk of the Galaxy it is not valid to assume that the angular power spectrum at low Galactic latitudes has much to tell us about that at high latitudes[72].

Our fields lie at Galactic latitude $-58.4°$ and $-61.9°$. The brightness of the IRAS 100 μm and Haslam 408 MHz (ref. 52) maps within our fields lie at the 6% and 25% points, respectively, of the integral distributions taken over the whole sky. There are several strong pieces of evidence from the DASI data set itself that the polarization results described in this paper are free of significant synchrotron contamination. The significant TE correlation shown in Fig. 6 indicates that the temperature and E-mode signal have a common origin. The tight constraints on the temperature anisotropy spectral index require that this common origin has a spectrum consistent with CMB. Galactic synchrotron emission is known to have a temperature spectral index of -2.8 (ref. 73), with evidence for steepening to -3.0 at frequencies above 1–2 GHz (ref. 74). At frequencies where Faraday depolarization is negligible (> 10 GHz), the same index will also apply for polarization. The dramatically tight constraint on the temperature spectral index of $\beta_T = 0.01$ (-0.16 to 0.14) indicates that any component of the temperature signal coming from synchrotron emission is negligibly small in comparison to the CMB. More directly, the constraint on the E-mode spectral index $\beta_E = 0.17$ (-1.63 to 1.92) disfavours synchrotron polarization at nearly 2σ. A third, albeit weaker, line of argument is that a complex synchrotron emitting structure is not expected to produce a projected brightness distribution which prefers E-mode polarization over B-mode[75]. Therefore, the result in Fig. 3 could be taken as further evidence that the signal we are seeing is not due to synchrotron emission.

我们没有发现同步加速、尘埃或者自由－自由辐射污染温度数据的迹象。CMB 的偏振部分预计达到 10%的程度，而自由－自由辐射对应的偏振部分小于 1%。尽管我们注意到用 DASI 观察到的 NGC 6334 尘埃和自由－自由辐射的混合偏振 <<1%(参考文献 51)，但是弥散的热尘埃辐射可能会被偏振几个百分点(如参考文献 67)。同样地，自旋尘埃辐射的偏振程度预计不会达到可探测水平[68]。因此，如果自由－自由和尘埃热辐射对我们的温度各向异性结果没有显著贡献，则可以预料它们不会对偏振产生影响。另一方面，同步加速辐射的偏振率原则上可以高达 70%，这也是目前最被关注的问题。它对温度的贡献可以忽略不计，但可能对偏振有显著贡献。

我们视场区域不存在已经公布过的偏振天图。之前尝试估计计划的同步辐射角功率谱是根据频率为 1–3 GHz 的银道面巡天(参考文献 69 和 70)来做的。这些天图显示的偏振结构尺度比温度的更小，但这主要是由法拉第旋转[71]引起的，该效应在 30 GHz 处可以忽略不计。此外，由于同步辐射高度集中在银河星系盘中，因此通过假设的低银河系纬度的角功率谱是无法知道高纬度地区的角功率谱的[72]。

我们的视场位于银河纬度 −58.4°和 −61.9°之间。在我们的视场内，IRAS 100 μm 和哈斯拉姆 408 MHz(参考文献 52)天图的亮度分别占整个天空 6%和 25%点的分布。几个来自 DASI 自身数据集的有力证据表明本文所述的偏振结果没有受到显著的同步辐射污染。图 6 所示的 TE 强相关性表明温度和 E 模信号具有共同的起源。对温度各向异性谱指数的严格约束要求这个共同起源具有与 CMB 相一致的频谱。已知银河同步辐射的温度谱指数为 −2.8(参考文献 73)，有证据表明它在频率高于 1～2 GHz 时变陡峭，谱指数为 −3.0(参考文献 74)。在对法拉第消偏振可忽略不计 (>10 GHz)的频率处，相同的谱指数同样适用于偏振。在温度谱指数 $\beta_T = 0.01$ (置信度为 68%的参数区间为 −0.16 至 0.14)的严格约束下，与 CMB 相比，任何来自同步辐射的温度信号都可忽略不计。更直接地，E 模式谱指数 $\beta_E = 0.17$(置信度为 68%的参数区间为 −1.63 至 1.92)的限制在近 2σ 水平下和同步辐射偏振不符。第三个(虽然较弱的)论证思路是，复杂的同步辐射结构预计不会产生一个 E 模式优于 B 模式的投影亮度分布[75]。因此，图 3 中的结果可以进一步证明我们看到的信号不是来自同步辐射。

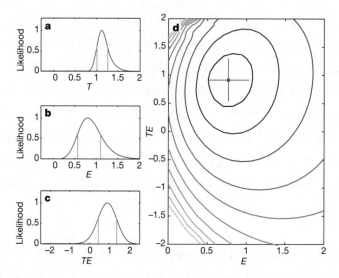

Fig. 6. Results from the three-parameter shaped bandpower $T/E/TE$ joint analysis. Spectral shapes as predicted for the concordance model are assumed (**a–c**) and the units are relative to that model. The layout of the plot is analogous to Fig. 3. The two-dimensional distribution in **d** is marginalized over the T dimension.

Discussion

This paper presents several measures of the confidence with which CMB polarization has been detected with DASI. Which measure is preferred depends on the desired level of statistical rigour and independence from *a priori* models of the polarization. The χ^2 analyses in the "χ^2 tests" subsection offer the most model-independent results, although the linear combinations of the data used to form the s/n eigenmodes are selected by consideration of theory and the noise model. For the high s/n eigenmodes of the polarization data, the probability to exceed (PTE) the measured χ^2 for the sum of the various data splits ranges from 1.6×10^{-8} to 8.7×10^{-7}, while the χ^2 for the differences are found to be consistent with noise. The PTE for the χ^2 found for the total (that is, not split) high s/n polarization eigenmodes, corrected for the beam offset leakage, is 5.7×10^{-8}.

Likelihood analyses are in principle more model dependent, and the analyses reported make different assumptions for the shape of the polarization power spectrum. Using theory to select the angular scales on which DASI should be most sensitive, we calculate the likelihood for a flat bandpower in E and B over the multipole range $300 < l < 450$ and find that data are consistent with $B = 0$ over this range, but that $E = 0$ can be rejected with a PTE of 5.9×10^{-9}.

The choice of the model bandpower is more important when the full l range of the DASI data is analysed. In this case, the likelihood analyses indicate that a $l(l+1)C_l \propto l^2$ model for the E-mode spectrum is 60 times more likely than a $l(l+1)C_l = $ constant model. Further, a bandpass shape based on the concordance model is found to be 12,000 times more

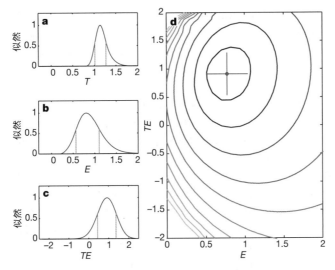

图 6. 三参数型频带功率 $T/E/TE$ 联合分析的结果。假设谱形状满足协调模型($\mathbf{a} \sim \mathbf{c}$),而单位则是相对于该模型的。该图的布局类似于图 3。$\mathbf{d}$ 中的二维分布是边缘化 T 维度得到的。

讨　论

本文提出了用 DASI 检测 CMB 偏振的几种置信度测量方法。首选哪种测量方法取决于所需的统计精度和是否独立于偏振的先验模型。尽管用于构成信噪比 (s/n) 本征模的数据的线性组合是基于理论和噪声模型来选择的,"χ^2 检验"部分中的 χ^2 分析还是给出了一个最不依赖于模型的结果。对于偏振数据的高信噪比本征模,用不同数据分割的求和计算给出的超过测量得到的 χ^2 的概率(PTE)范围从 1.6×10^{-8} 到 8.7×10^{-7} 不等。另外,对分割数据的差值计算给出的 χ^2 被发现和噪声是一致的。将光束偏移渗漏校准后,全部的(即不是分割的)高信噪比偏振本征模给出的 χ^2 的 PTE 是 5.7×10^{-8}。

似然分析原则上更加依赖于模型,且就本文报告的分析,我们对偏振功率谱的形状做出了不同的假设。根据理论来选择 DASI 理应最为敏感的角度范围,我们计算了多极矩范围为 $300 < l < 450$ 的一个平坦频段模型的 E 和 B 的似然函数。我们发现,在此范围内数据与 $B = 0$ 一致,但是 $E = 0$ 的 PTE 为 5.9×10^{-9},故该假设可以被拒绝。

分析完整 l 范围的 DASI 数据时,模型频段功率的选择更为重要。在这种情况下,似然分析表明对于 E 模式频谱,$l(l+1)C_l \propto l^2$ 的模型是 $l(l+1)C_l =$ 常数的模型的可能性的 60 倍。此外我们发现,基于协调模型的带通形状是平坦频段功率的可能性

likely than the flat bandpower. The likelihood ratio test leads to a PTE of 8.5×10^{-7} for the concordance shaped bandpower, corresponding to a confidence of detection of 4.9σ. In all three of these tests, B is found to be consistent with zero, as expected in the concordance model.

The concordance model is also supported by the results of the five-band piecewise-flat analyses, $E5/B5$. The upper limit for the first E band at $28 \leq l \leq 245$ is only 2.38 μK^2, a factor of 30 lower in power than the previous upper limit. The next band at $246 \leq l \leq 420$ is detected with a maximum likelihood value of 17.1 μK^2. Such a sharp rise in power with increasing l is expected in the E-mode spectrum (see Fig. 4) owing to the length scale introduced by the mean free path to photon scattering. The polarization of the larger modes is suppressed because the velocity differences are not as large on the scale of the mean free path. In fact, the E spectrum is expected to increase as $l(l+1)C_l \propto l^2$ at small l. This dependence is not expected to continue to the higher l values to which DASI is sensitive, owing to diffusion damping, which suppresses power on scales smaller than the mean free path. The maximum likelihood values of the higher l bands of the DASI five-band E analysis are consistent with the damped concordance model, but lie below a simple extrapolation of the l^2 power law. Again, the five-band B-mode spectrum is consistent with the concordance prediction of zero.

The TE analysis provides further confidence in our detection of CMB polarization and for the concordance model. From the $T/E/TE$ likelihood analysis, the $TE = 0$ hypothesis is rejected with a PTE of 0.054. Note that TE could be negative as well as positive and therefore the $TE \leq 0$ hypothesis is rejected with higher confidence.

Lastly, the measured T frequency spectral index, 0.01 (-0.16 to 0.14), is remarkably well constrained to be thermal and is inconsistent with any known foregrounds. The E frequency spectral index, 0.17 (-1.63 to 1.92), is also consistent with the CMB, and although less well constrained than the T index, is inconsistent with diffuse foreground synchrotron emission at nearly 2σ.

In summary, the analyses reported in this paper all indicate a robust detection of E-mode CMB polarization with a confidence level $\geq 4.9\sigma$. The measured properties of the polarization are in good agreement with predictions of the concordance model and, as discussed in the "Foregrounds" subsection, are inconsistent with expectations from known sources of foreground emission. These results therefore provide strong support for the underlying theoretical framework for the generation of CMB anisotropy. They lend confidence to the values of the cosmological parameters and to the extraordinary picture of the origin and structure of the Universe derived from CMB measurements. The prospect of further refining our understanding of the Universe using precision polarization measurements is the goal of many ongoing and planned CMB experiments. The detection of polarization at the predicted level reported in this paper points to a promising future for the field.

(**420**, 772-787; 2002)

的 12,000 倍。似然比检验给出一致形状频段功率的 PTE 为 8.5×10^{-7}，对应探测置信度为 4.9σ。在所有这三个检验中，我们发现，正如在协调模型中所预期的那样，B 与零一致。

五频段分段–平坦分析（$E5/B5$）的结果同样支持协调模型。在 $28 \le l \le 245$ 处的第一个 E 频段的上限仅为 $2.38~\mu K^2$，功率是前一个上限的三十分之一。在 $246 \le l \le 420$ 处的第二个频段的最大似然值为 $17.1~\mu K^2$。由于光子散射的平均自由程引入的长度标度，出现在 E 模式功率谱（参见图 4）的这种随着 l 的增加功率急剧上升的现象是在预料之中的。由于较大模式上的速度差异没有平均自由程尺度上的那么大，因此这些模式的偏振会被抑制。实际上，l 较小时，E 谱预计会以 $l(l+1)C_l \propto l^2$ 形式上升，但由于扩散阻尼抑制了小于平均自由程尺度上的功率，故这种依赖性不会保持到 DASI 敏感的较高的 l 值。在 DASI 五频段 E 分析中，较高 l 频段的最大似然值的和带阻尼的协调模型一致，但位于 l^2 幂律的简单外推模型的下方。同样，正如协调模型所预测的，五频段 B 模式功率谱与零一致。

TE 分析进一步使我们增强了对于 CMB 偏振信号的存在和协调模型的信心。根据 $T/E/TE$ 似然分析，TE 为零的假设以 0.054 的 PTE 被拒绝。注意，TE 可正可负，因此 $TE \le 0$ 假设以更高的置信度被拒绝。

最后，T 的谱指数的测量值为 0.01（置信度为 68% 的参数区间为 -0.16 至 0.14），极好地被约束在热的且不属于任何已知的前景。E 的谱指数为 0.17（置信度为 68% 的参数区间为 -1.63 至 1.92）也与 CMB 一致，虽然比起 T、E 的谱指数约束得不那么好，但与弥散前景同步辐射存在近 2σ 的不一致性。

总之，本文报道的分析均表明存在一个显著的 CMB 的 E 偏振模式信号，其中置信水平 $\ge 4.9\sigma$。测量到的偏振的特性和调和模型给出的预测非常一致，并且如"前景"部分所讨论的那样，与目前已知的前景辐射都不匹配。因此，这些结果为 CMB 各向异性的起源的基础理论框架提供了强而有力的支持，从而使得从 CMB 测量推导得出的宇宙学参数的值以及宇宙起源和结构的庞大图景更为可信。使用精确偏振测量进一步完善我们对宇宙的理解是许多正在进行的和计划中的 CMB 实验的目标。本文报道的探测到预计水平的偏振信号为该领域的光明未来指明了方向。

（谭秀慧 吕孟珍 翻译；夏俊卿 审稿）

J. M. Kovac[*†‡], **E. M. Leitch**[§†‡], **C. Pryke**[§†‡‖], **J. E. Carlstrom**[§*†‡‖], **N. W. Halverson**[¶†] & **W. L. Holzapfel**[¶†]

[*] Department of Physics; [†] Center for Astrophysical Research in Antarctica; [‡] Center for Cosmological Physics; [§] Department of Astronomy & Astrophysics and [‖] Enrico Fermi Institute, University of Chicago, 5640 South Ellis Avenue, Chicago, Illinois 60637, USA

[¶] Department of Physics, University of California at Berkeley, Le Conte Hall, California 94720, USA

Received 7 October; accepted 5 November 2002; doi:10.1038/nature 01269.

References:

1. Penzias, A. A. & Wilson, R. W. A measurement of excess antenna temperature at 4080 Mc/s. *Astrophys. J.* **142**, 419-421 (1965).

2. Mather, J. C. *et al.* Measurement of the cosmic microwave background spectrum by the COBE FIRAS instrument. *Astrophys. J.* **420**, 439-444 (1994).

3. Fixsen, D. J. *et al.* The cosmic microwave background spectrum from the full COBE FIRAS data set. *Astrophys. J.* **473**, 576-587 (1996).

4. Smoot, G. F. *et al.* Structure in the COBE differential microwave radiometer first-year maps. *Astrophys. J.* **396**, L1-L5 (1992).

5. Miller, A. D. *et al.* A measurement of the angular power spectrum of the cosmic microwave background form l = 100 to 400. *Astrophys. J.* **524**, L1-L4 (1999).

6. Halverson, N. W. *et al.* Degree angular scale interferometer first results: A measurement of the cosmic microwave background angular power spectrum. *Astrophys. J.* **568**, 38-45 (2002).

7. Netterfield, C. B. *et al.* A measurement by BOOMERANG of multiple peaks in the angular power spectrum of the cosmic microwave background. *Astrophys. J.* **571**, 604-614 (2002).

8. Lee, A. T. *et al.* A high spatial resolution analysis of the MAXIMA-1 cosmic microwave background anisotropy data. *Astrophys. J.* **561**, L1-L5 (2001).

9. Pearson, T. J. *et al.* The anisotropy of the microwave background to l = 3500: Mosaic observations with the cosmic background imager. *Astrophys. J.* (submitted); preprint astro-ph/0205388 at ⟨http://xxx.lanl.gov⟩ (2002).

10. Scott, P. F. *et al.* First results from the Very Small Array - III. The CMB power spectrum. *Mon. Not. R. Astron. Soc.* (submitted); preprint astro-ph/0205380 at ⟨http://xxx.lanl.gov⟩ (2002).

11. Hu, W. & Dodelson, S. Cosmic microwave background anisotropies. *Annu. Rev. Astron. Astrophys.* **40**, 171-216 (2002).

12. Kaiser, N. Small-angle anisotropy of the microwave background radiation in the adiabatic theory. *Mon. Not. R. Astron. Soc.* **202**, 1169-1180 (1983).

13. Bond, J. R. & Efstathiou, G. Cosmic background radiation anisotropies in universes dominated by nonbaryonic dark matter. *Astrophys. J.* **285**, L45-L48 (1984).

14. Polnarev, A. G. Polarization and anisotropy induced in the microwave background by cosmological gravitational waves. *Sov. Astron.* **29**, 607-613 (1985).

15. Kamionkowski, M., Kosowsky, A. & Stebbins, A. Statistics of cosmic microwave background polarization. *Phys. Rev. D.* **55**, 7368-7388 (1997).

16. Zaldarriaga, M. & Seljak, U. All-sky analysis of polarization in the microwave background. *Phys. Rev.D.* **55**, 1830-1840 (1997).

17. Hu, W. & White, M. A CMB polarization primer. *New Astron.* **2**, 323-344 (1997).

18. Kosowsky, A. Introduction to microwave background polarization. *New Astron. Rev.* **43**, 157-168 (1999).

19. Hu, W., Spergel, D. N. & White, M. Distinguishing causal seeds from inflation. *Phys. Rev. D* **55**, 3288-3302 (1997).

20. Kinney, W. H. How to fool cosmic microwave background parameter estimation. *Phys. Rev. D* **63**, 43001 (2001).

21. Bucher, M., Moodley, K. & Turok, N. Constraining isocurvature perturbations with cosmic microwave background polarization. *Phys. Rev. Lett.* **87**, 191301 (2001).

22. Rees, M. J. Polarization and spectrum of the primeval radiation in an anisotropic universe. *Astrophys. J.* **153**, L1-L5 (1968).

23. Zaldarriaga, M. & Harari, D. D. Analytic approach to the polarization of the cosmic microwave background in flat and open universes. *Phys. Rev. D* **52**, 3276-3287 (1995).

24. Coulson, D., Crittenden, R. G. & Turok, N. G. Polarization and anisotropy of the microwave sky. *Phys. Rev. Lett.* **73**, 2390-2393 (1994).

25. Crittenden, R., Davis, R. L. & Steinhardt, P. J. Polarization of the microwave background due to primordial gravitational waves. *Astrophys. J.* **417**, L13-L16 (1993).

26. Seljak, U. Measuring polarization in the cosmic microwave background. *Astrophys. J.* **482**, 6-16 (1997).

27. Kamionkowski, M., Kosowsky, A. & Stebbins, A. A probe of primordial gravity waves and vorticity. *Phys. Rev. Lett.* **78**, 2058-2061 (1997).

28. Seljak, U. & Zaldarriaga, M. Signature of gravity waves in the polarization of the microwave background. *Phys. Rev. Lett.* **78**, 2054-2057 (1997).

29. Lyth, D. H. What would we learn by detecting a gravitational wave signal in the Cosmic Microwave Background anisotropy? *Phys. Rev. Lett.* **78**, 1861-1863 (1997).

30. Zaldarriaga, M. & Seljak, U. Gravitational lensing effect on cosmic microwave background polarization. *Phys. Rev. D* **58**, 23003 (1998).

31. Hu, W. & Okamoto, T. Mass reconstruction with cosmic microwave background polarization. *Astrophys. J.* **574**, 566-574 (2002).

32. Knox, L. & Song, Y. A limit on the detectability of the energy scale of inflation. *Phys. Rev. Lett.* **89**, 011303 (2002).

33. Kesden, M., Cooray, A. & Kamionkowski, M. Separation of gravitational-wave and cosmic-shear contributions to cosmic microwave background polarization. *Phys. Rev. Lett.* **89**, 011304 (2002).

34. Kamionkowski, M. & Kosowsky, A. The cosmic microwave background and particle physics. *Annu. Rev. Nucl. Part. Sci.* **49**, 77-123 (1999).

35. Staggs, S. T., Gunderson, J. O. & Church, S. E. *ASP Conf. Ser. 181, Microwave Foregrounds* (eds de Oliveira-Costa, A. & Tegmark, M.) 299-309 (Astronomical Society of the Pacific, San Francisco, 1999).

36. Caderni, N., Fabbri, R., Melchiorri, B., Melchiorri, F. & Natale, V. Polarization of the microwave background radiation. II. An infrared survey of the sky. *Phys. Rev. D* **17**, 1908-1918 (1978).

37. Nanos, G. P. Polarization of the blackbody radiation at 3.2 centimeters. *Astrophys. J.* **232**, 341-347 (1979).

38. Lubin, P. M. & Smoot, G. F. Search for linear polarization of the cosmic background radiation. *Phys. Rev. Lett.* **42**, 129-132 (1979).

39. Lubin, P. M. & Smoot, G. F. Polarization of the cosmic background radiation. *Astrophys. J.* **245**, 1-17 (1981).

40. Lubin, P., Melese, P. & Smoot, G. Linear and circular polarization of the cosmic background radiation. *Astrophys. J.* **273**, L51-L54 (1983).

41. Sironi, G. *et al.* A 33 GHZ polarimeter for observations of the cosmic microwave background. *New Astron.* **3**, 1-13 (1997).

42. Keating, B. G. *et al.* A limit on the large angular scale polarization of the cosmic microwave background. *Astrophys. J.* **560**, L1-L4 (2001).

43. Wollack, E. J., Jarosik, N. C., Netterfield, C. B., Page, L. A. & Wilkinson, D. A measurement of the anisotropy in the cosmic microwave background radiation at degree angular scales. *Astrophys. J.* **419**, L49-L52 (1993).

44. Hedman, M. M. *et al.* New limits on the polarized anisotropy of the cosmic microwave background at subdegree angular scales. *Astrophys. J.* **573**, L73-L76 (2002).

45. Cartwright, J. K. *et al.* Polarization observations with the cosmic background imager. in *Moriond Workshop 37, The Cosmological Model* (in the press).

46. de Oliveira-Costa, A. *et al.* First attempt at measuring the CMB cross-polarization. *Phys. Rev. D* (submitted); preprint astro-ph/0204021 at ⟨http://xxx.lanl.gov⟩ (2002).

47. Partridge, R. B., Richards, E. A., Fomalont, E. B., Kellermann, K. I. & Windhorst, R. A. Small-scale cosmic microwave background observations at 8.4 GHz. *Astrophys. J.* **483**, 38-50 (1997).

48. Subrahmanyan, R., Kesteven, M. J., Ekers, R. D., Sinclair, M. & Silk, J. An Australia telescope survey for CMB anisotropies. *Mon. Not. R. Astron. Soc.* **315**, 808-822 (2000).

49. Leitch, E. M. *et al.* Experiment design and first season observations with the degree angular scale interferometer. *Astrophys. J.* **568**, 28-37 (2002).

50. Pryke, C. *et al.* Cosmological parameter extraction from the first season of observations with the degree angular scale interferometer. *Astrophys. J.* **568**, 46-51 (2002).

51. Leitch, E. M. *et al.* Measurement of polarization with the Degree Angular Scale Interferometer. *Nature* **420**, 763-771 (2002).

52. Haslam, C. G. T. *et al.* A 408 MHz all-sky continuum survey. I—Observations at southern declinations and for the north polar region. *Astron. Astrophys.* **100**, 209-219 (1981).

53. Bond, J. R., Jaffe, A. H. & Knox, L. Estimating the power spectrum of the cosmic microwave background. *Phys. Rev. D* **57**, 2117-2137 (1998).

54. Seljak, U. & Zaldarriaga, M. A line-of-sight integration approach to cosmic microwave background anisotropies. *Astrophys. J.* **469**, 437-444 (1996).

55. White, M., Carlstrom, J. E., Dragovan, M. & Holzapfel, W. H. Interferometric observation of cosmic microwave background anisotropies. *Astrophys. J.* **514**, 12-24 (1999).

56. Hedman, M. M., Barkats, D., Gundersen, J. O., Staggs, S. T. & Winstein, B. A limit on the polarized anisotropy of the cosmic microwave background at subdegree angular scales. *Astrophys. J.* **548**, L111-L114 (2001).

57. Knox, L. Cosmic microwave background anisotropy window functions revisited. *Phys. Rev. D* **60**, 103516 (1999).

58. Halverson, N. W. *A Measurement of the Cosmic Microwave Background Angular Power Spectrum with DASI.* PhD thesis, Caltech (2002).

59. Tegmark, M. & de Oliveira-Costa, A. How to measure CMB polarization power spectra without losing information. *Phys. Rev. D* **64**, 063001 (2001).

60. Christensen, N., Meyer, R., Knox, L. & Luey, B. Bayesian methods for cosmological parameter estimation from cosmic microwave background measurements. *Class. Quant. Gravity* **18**, 2677-2688 (2001).

61. Wang, X., Tegmark, M. & Zaldarriaga, M. Is cosmology consistent? *Phys. Rev. D* **65**, 123001 (2002).

62. Wright, A. E., Griffith, M. R., Burke, B. F. & Ekers, R. D. The Parkes-MIT-NRAO (PMN) surveys. 2: Source catalog for the southern survey ($-87.5° < \delta < -37°$). *Astrophys. J. Suppl.* **91**, 111-308 (1994).

63. Mason, B. S., Pearson, T. J., Readhead, A. C. S., Shepherd, M. C. & Sievers, J. L. The anisotropy of the microwave background to l = 3500: Deep field observations with the cosmic background imager. *Astrophys. J.* (submitted); preprint astro-ph/0205384 at ⟨http://xxx.lanl.gov⟩ (2002).

64. Zukowski, E. L. H., Kronberg, P. P., Forkert, T. & Wielebinski, R. Linear polarization measurements of extragalactic radio sources at λ6.3 cm. *Astron. Astrophys. Suppl.* **135**, 571-577 (1999).

65. Simard-Normandin, M., Kronberg, P. P. & Neidhoefer, J. Linear polarization observations of extragalactic radio sources at 2 cm and at 17-19 cm. *Astron. Astrophys. Suppl.* **43**, 19-22 (1981).

66. Simard-Normandin, M., Kronberg, P. P. & Button, S. The Faraday rotation measures of extragalactic radio sources. *Astrophys. J. Suppl.* **45**, 97-111 (1981).

67. Hildebrand, R. H. *et al.* A primer on far-infrared polarimetry. *Publ. Astron. Soc. Pacif.* **112**, 1215-1235 (2000).

68. Lazarian, A. & Prunet, S. *AIP Conf. Proc. 609, Astrophysical Polarized Backgrounds* (eds Cecchini, S., Cortiglioni, S., Sault, R. & Sbarra, C.) 32-43 (AIP, Melville, New York, 2002).

69. Tegmark, M., Eisenstein, D. J., Hu, W. & de Oliveria-Costa, A. Foregrounds and forecasts for the cosmic microwave background. *Astrophys. J.* **530**, 133-165 (2000).

70. Giardino, G. *et al.* Towards a model of full-sky Galactic synchrotron intensity and linear polarisation: A re-analysis of the Parkes data. *Astron. Astrophys.* **387**, 82-97 (2002).

71. Gaensler, B. M. *et al.* Radio polarization from the inner galaxy at arcminute resolution. *Astrophys. J.* **549**, 959-978 (2001).

72. Gray, A. D. *et al.* Radio polarimetric imaging of the interstellar medium: Magnetic field and diffuse ionized gas structure near the W3/W4/W5/HB 3 complex. *Astrophys. J.* **514**, 221-231 (1999).

73. Platania, P. *et al.* A determination of the spectral index of galactic synchrotron emission in the 1–10 GHz range. *Astrophys. J.* **505**, 473-483 (1998).

74. Banday, A. J. & Wolfendale, A. W. Fluctuations in the galactic synchrotron radiation—I. Implications for searches for fluctuations of cosmological origin. *Mon. Not. R. Astron. Soc.* **248**, 705-714 (1991).

563

75. Zaldarriaga, M. The nature of the E-B decomposition of CMB polarization. *Phys. Rev. D* (submitted); preprint astro-ph/0106174 at ⟨http://xxx.lanl.gov⟩ (2001).

Acknowledgements. We are grateful for the efforts of B. Reddall and E. Sandberg, who wintered over at the National Science Foundation (NSF) Amundsen–Scott South Pole research station to keep DASI running smoothly. We are indebted to M. Dragovan for his role in making DASI a reality, and to the Caltech CBI team led by T. Readhead, in particular, to S. Padin, J. Cartwright, M. Shepherd and J. Yamasaki for the development of key hardware and software. We are indebted to the Center for Astrophysical Research in Antarctica (CARA), in particular to the CARA polar operations staff. We are grateful for contributions from K. Coble, A. Day, G. Drag, J. Kooi, E. LaRue, M. Loh, R. Lowenstein, S. Meyer, N. Odalen, R. Pernic, D. Pernic and E. Pernic, R. Spotz and M. Whitehead. We thank Raytheon Polar Services and the US Antarctic Program for their support of the DASI project. We have benefitted from many interactions with the Center for Cosmological Physics members and visitors. In particular, we gratefully acknowledge many conversations with W. Hu on CMB polarization and suggestions from S. Meyer, L. Page, M. Turner and B. Winstein on the presentation of these results. We thank L. Knox and A. Kosowsky for bringing the Markov technique to our attention. We thank the observatory staff of the Australia Telescope Compact Array, in particular B. Sault and R. Subrahmanyan, for providing point source observations of the DASI fields. This research was initially supported by the NSF under a cooperative agreement with CARA, a NSF Science and Technology Center. It is currently supported by an NSF-OPP grant. J.E.C. gratefully acknowledges support from the James S. McDonnell Foundation and the David and Lucile Packard Foundation. J.E.C. and C.P. gratefully acknowledge support from the Center for Cosmological Physics.

Competing interests statement. The authors declare that they have no competing financial interests.

Correspondence and requests for materials should be addressed to J.M.K. (e-mail: john@hyde.uchicago.edu).

An Extended Upper Atmosphere Around the Extrasolar Planet HD209458b

A. Vidal-Madjar *et al.*

Editor's Note

The extrasolar planet HD 209458b crosses (transits) its parent star as seen from our perspective on Earth, which enables astronomers to determine its mass and radius. Here Alfred Vidal-Madjar and colleagues show that the planet has an extended atmosphere—the first such observation for an extrasolar planet. They obtained a spectrum of the parent star during three transits, and saw that some of the light was absorbed by the planet's atmosphere. They propose that the extended atmosphere probably consists of hydrogen atoms escaping from the planet's gravitational field, though it was later proposed that the atoms were ions from the stellar wind that pick up electrons in the outer atmosphere of the planet. Which explanation is correct is still debated.

The planet in the system HD209458 is the first one for which repeated transits across the stellar disk have been observed[1,2]. Together with radial velocity measurements[3], this has led to a determination of the planet's radius and mass, confirming it to be a gas giant. But despite numerous searches for an atmospheric signature[4-6], only the dense lower atmosphere of HD209458b has been observed, through the detection of neutral sodium absorption[7]. Here we report the detection of atomic hydrogen absorption in the stellar Lyman α line during three transits of HD209458b. An absorption of $15 \pm 4\%$ (1σ) is observed. Comparison with models shows that this absorption should take place beyond the Roche limit and therefore can be understood in terms of escaping hydrogen atoms.

FAR more abundant than any other species, hydrogen is well-suited for searching weak atmospheric absorptions during the transit of an extrasolar giant planet in front of its parent star, in particular over the strong resonant stellar ultraviolet Lyman α emission line at 1,215.67 Å. Depending upon the characteristics of the planet's upper atmosphere, an H_I signature much larger than that for Na_I at 0.02% (ref. 7) is foreseeable. Three transits of HD209458b (named A, B and C hereafter) were sampled in 2001 (on 7–8 September, 14–15 September and 20 October, respectively) with the Space Telescope Imaging Spectrograph (STIS) onboard the Hubble Space Telescope (HST); the data set is now public in the HST archive. To partially overcome contamination from the Earth's Lyman α geocoronal emission, we used the G140M grating with the $52'' \times 0.1''$ slit (medium spectral resolution: $\sim$20 km s^{-1}). For each transit, three consecutive HST orbits (named 1, 2 and 3 hereafter) were scheduled such that the first orbit (1,780 s exposure) ended before the first contact to serve as a reference, and the two following ones (2,100 s exposures each) were

太阳系外行星 HD209458b 扩展的
高层大气研究

维达尔–马贾尔等

编者按

在地球上观测到太阳系外行星 HD209458b 穿越其母恒星圆面的现象（凌星）可以帮助天文学家确定行星的质量与半径。本文中，阿尔弗雷德·维达尔–马贾尔与他的同事报道了一颗具有扩展大气层的行星，这也是第一次在太阳系外观测到这样的星体。他们从三次凌星过程中获得了母恒星的光谱，发现某些光线被行星的大气层吸收了。他们提出行星扩展大气层中可能包含有从行星重力场中逃逸出来的氢原子。尽管不久之后有研究指出这些原子是源于行星外层大气中携带电子的恒星风的离子，然而两种假说的正确性尚无定论。

位于 HD209458 系统中的行星是第一颗被观测到多次横越恒星圆面的行星[1,2]。结合凌星观测与视向速度法[3]，可以确定这颗行星的半径和质量，从而确认该行星是一颗气态巨行星。尽管人们对 HD209458 系统的大气特征进行了大量研究[4-6]，但这些研究仅是通过中性钠原子的吸收探测到 HD209458b 存在稠密的低层大气[7]。本文报道的是在 HD209458b 三次凌星过程中，探测到恒星莱曼 α 线被氢原子吸收，观测到的吸收强度为 $15\pm4\%$ (1σ)。通过与模型的模拟结果对比，发现这种吸收应该在超过洛希极限时才会发生，因此可以认为是逃逸的氢原子的吸收。

在宇宙中，氢的丰度远高于其他元素。因此，在太阳系外巨行星的凌星过程中，氢非常适合于探测较弱的大气吸收，尤其是在强共振恒星紫外莱曼 α 发射线（1,215.67Å）波段。根据该行星的上层大气特征，可以预测 H_1 的特征谱线比 Na_1 的特征谱线要强 0.02%[7]。通过哈勃空间望远镜（HST）上装载的空间望远镜成像光谱仪（STIS），分别在 2001 年 9 月 7 日～8 日、9 月 14 日～15 日和 10 月 20 日对 HD209458b 的三次凌星（分别命名为 A、B、C）进行了采样，这些数据现在已经在 HST 档案库中公开。为了部分地避免地冕莱曼 α 发射线的干扰，我们采用了 G140M 光栅，其缝隙大小为 $52''\times0.1''$（光谱分辨率约为 20 km·s^{-1}）。在每次凌星过程中，安排哈勃望远镜进行连续三个轨道周期（下文以 1、2、3 编号）的观测，其中第一周期（曝光 1,780 s）在发现凌星前就结束以便提供参考，而随后两个周期的观测（分别

partly or entirely within the transit.

The observed Lyman α spectrum of HD209458 is typical for a solar-type star, with a double-peaked emission originating from the stellar chromosphere (Fig. 1). It also shows a wide central absorption feature due to neutral hydrogen in the interstellar medium. The geocoronal emission filled the aperture of the spectrograph, resulting in an extended emission line perpendicular to the dispersion direction. The extent of this emission along the slit allowed us to remove it at the position of the target star. We evaluated its variation both along the slit and from one exposure to another, and excluded the wavelength domain where the corresponding standard deviation per pixel is larger than 20% of the final spectrum. We concluded that the geocoronal contamination can be removed with high enough accuracy outside the central region 1,215.5 Å < λ < 1,215.8 Å, labelled "Geo" in Figs 2, 4).

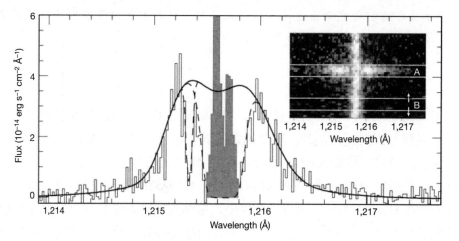

Fig. 1. The HD209458 Lyman α emission line. This high-resolution spectrum (histogram) was obtained with the E140M echelle grating and the $0.2'' \times 0.2''$ wide slit with a spectral resolution of 5 km s^{-1}; it was not used in the present analysis, but it allows the different components of the line profile to be seen. The continuum is a double-peaked emission line originating from the stellar chromosphere: the temperature increase in the lower chromosphere causes an emission line with a central dip due to the high opacity of the abundant hydrogen atoms (solid line). The observed spectrum also has a narrow absorption line (1,215.3 Å, barely seen at lower resolution) and a central wide absorption line (1,215.6 Å) due to the interstellar deuterium and hydrogen, respectively (dashed lines). The grey zone represents the fraction of the spectrum contaminated by the geocoronal emission, which is double-peaked in that case because the plotted spectrum is the average of four exposures obtained at two different epochs. The inset shows a small portion of the 2D image of a G140M first-order spectrum containing the stellar Lyman α profile and a sample of the geocoronal signal. This spectrum is one of the nine spectra used for this analysis. The G140M spectra have lower spectral resolution but higher signal-to-noise ratio. The stellar spectrum is seen as a horizontal line where the two peaks are resolved from the geocoronal emission (vertical line along the slit). The one-dimensional spectra are obtained by vertically adding around ten pixels within the A band. Measurements along the slit direction (~800 pixels available), for example at the position of the B band, allow us to estimate the geocoronal contamination and the background subtraction as well as the corresponding uncertainties.

曝光 2,100 s）则部分或全部在凌星期间内。

如图 1 所示，观测到的 HD209458 莱曼 α 谱线是典型的类太阳恒星光谱，具有恒星色球层产生的特征双峰发射。同时我们发现在中心位置附近还存在一个较宽的吸收带，这是因为在星际介质中存在中性氢。由于地冕发射线进入摄谱仪孔径，垂直色散方向的发射线发生扩展。沿光栅缝隙的发射线的这种扩展使得我们能够将其从观测目标恒星的位置移除。我们计算出发射线沿缝隙以及两次不同曝光之间的变化情况，同时去除波长域中对应单位像素标准差比最终光谱大 20% 的部分。我们从而得出以下结论：对于波长中心区域（1,215.5 Å < λ < 1,215.8 Å，在图 2、4 中用"Geo"标出）以外的波段，地冕引起的干扰可以很大程度地被消除掉。

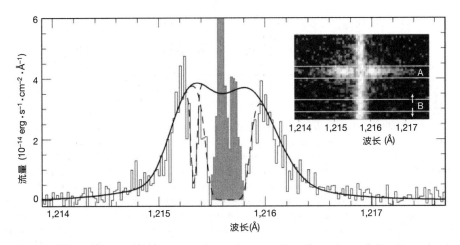

图 1. HD209458 的莱曼 α 发射线。这幅高分辨率光谱图（直方图）由 E140M 阶梯光栅得到的，该光栅的缝隙尺寸为 0.2″×0.2″，光谱分辨率为 5 km·s⁻¹，可以分离谱线轮廓的不同分量（尽管本次分析中并未使用）。图中连续发射谱具有双峰结构，这是由恒星色球层产生的，即由于大量氢原子的不透明度较高，底层色球层温度升高，这将使得发射线具有中心下陷结构（实线）。同时，观察到光谱还具有一个非常窄的吸收线（位于 1,215.3 Å，较低分辨率下很难观察到）和一个较宽的中心吸收线（1,215.6 Å，虚线），二者分别是由于星际间的氘和氢所引起的。图中灰色区域表示混杂着地冕干扰的光谱部分，由于光谱是在两个不同的纪元得到的四次曝光的平均，因此显示出双峰结构。右上角小框图给出了 G140M 一级光谱二维图像的一小部分，包含了恒星莱曼 α 廓线和部分地冕干扰信号。该光谱是我们用于分析的九个光谱之一，尽管 G140M 光谱具有较低的谱分辨率，但是具有较高的信噪比。该恒星光谱可以看作一条水平线，其中可以在地冕发射线中分辨出两个光谱峰（沿缝隙的垂直方向）。通过将 A 条带区域内的大约 10 个像素相加可以得到一维光谱图。通过沿缝隙方向测量（大约 800 个像素），例如在 B 条带位置，可以使我们估算出地冕干扰和背景影响，以及相应的不确定度。

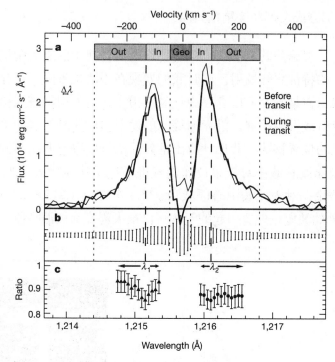

Fig. 2. The HD209458 Lyman α profile observed with the G140M grating. The geocoronal emission has been subtracted; the propagated errors are consequently larger in the central part of the profile, particularly in the Geo domain (see text). $\Delta\lambda$ represents the spectral resolution. **a,** The thin line shows the average of the three observations performed before the transits (exposures A1, B1 and C1); the thick line shows the average of the three observations recorded entirely within the transits (exposures A2, B3 and C3). Variations are seen in the In domain as absorption over the blue peak of the line and partially over the red peak (between -130 km s^{-1} and 100 km s^{-1}). Quoted velocities are in the stellar reference frame, centred on -13 km s^{-1} in the heliocentric reference frame. **b,** $\pm 1\sigma$ error bars. **c,** The ratio of the two spectra in the In domain, the spectra being normalized such that the ratio is 1 in the Out domain. This ratio is plotted as a function of λ_1 using $\lambda_2 = 1{,}216.10$ Å (triangles), and as a function of λ_2 using $\lambda_1 = 1{,}215.15$ Å (circles). The ratio is always significantly below 1, with a minimum at $\lambda_1 = 1{,}215.15$ Å (-130 km s^{-1}) and $\lambda_2 = 1{,}216.10$ Å (100 km s^{-1}). In the domain defined by these values, the Lyman α intensity decreases during the transits by $15 \pm 4\%$. The detection does not strongly depend on a particular selection of the domain. While the decrease of the Lyman α intensity is not sensitive to the position of λ_2, it is more sensitive to the position of λ_1, showing that most of the absorption occurs in the blue part of the line. Using the whole domain where the absorption is detected, the exoplanetary atmospheric hydrogen is detected at more than 3σ.

From the two-dimensional (2D) images of the far-ultraviolet (FUV) multi-anode microchannel array (MAMA) detector, the STIS standard pipeline extracts one-dimensional spectra in which the dark background has not been removed. The background level was systematically increasing from one exposure to the next within each of the three visits, but still remained below 2% of the peak intensity of the stellar signal. We therefore reprocessed the 2D images by using two independent approaches. The first one uses the 2D images provided by the standard pipeline and interpolates the background (including the Earth's geocoronal emission) with a polynomial fitted per column above and below the spectrum region. The second method starts from the 2D raw images to which is applied a dark

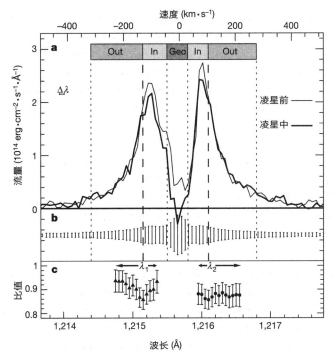

图 2. 利用 G140M 光栅观测到的 HD209458 的莱曼 α 轮廓。其中，地冕发射的干扰已经被剔除，因此传递误差在廓线中心附近变大，特别是在"Geo"区域，Δλ 表示光谱分辨率。**a.** 图中细线给出了凌星前进行的三次观测（即 A1、B1、C1 三次曝光）的平均；粗线表示凌星中的三次观测（即 A2、B3、C3 三次曝光）的平均。很明显，两条曲线在"In"域存在较大变化，特别是在谱线蓝峰以及部分红峰附近（速度在 $-130\ \mathrm{km \cdot s^{-1}}$ 到 $100\ \mathrm{km \cdot s^{-1}}$ 之间）。括号里的速度是相对恒星参考系的速度，在以太阳为中心的参考系中心速度为 $-13\ \mathrm{km \cdot s^{-1}}$。**b.** 该图给出了 $\pm 1\sigma$ 的误差棒。**c.** 该图给出了两条光谱曲线在"In"域的比值，归一化满足"Out"域比值为 1。其中，带三角标志的曲线表示 $\lambda_2 = 1{,}216.10\ \text{Å}$ 条件下比值与 λ_1 的关系，而带实心圆标志的曲线给出了 $\lambda_1 = 1{,}215.15\ \text{Å}$ 条件下比值与 λ_2 的关系，两种情况下比值的最小值分别于 $\lambda_1 = 1{,}215.15\ \text{Å}$（对应 $-130\ \mathrm{km \cdot s^{-1}}$）和 $\lambda_2 = 1{,}216.10\ \text{Å}$（对应 $100\ \mathrm{km \cdot s^{-1}}$）处获得。在这两个最小值定义域范围内，莱曼 α 谱线强度在凌星过程中减小了 $15 \pm 4\%$。值得注意的是，探测结果本身与特定的光谱域并没有很强的关联性。莱曼 α 谱线强度的减小对 λ_2 的位置并不敏感，而对于 λ_1 的位置则相对敏感，这说明更多地吸收发生于谱线蓝端部分。对整个探测到吸收的区域的分析表明，系外行星大气氢在大于 3σ 的情况下可以被测度。

根据远紫外（FUV）多阳极微通道阵（MAMA）探测器捕获到的二维图像，STIS 标准通道从二维图像中提取出一维光谱，其中的暗背景并未去除。尽管三次凌星的背景水平从某次曝光到下一次曝光会系统性地增加，但是仍然低于星体信号峰值强度的 2%。因此，我们采用了两种独立的方法对二维图像进行了重新处理。第一种方法采用标准通道提供的二维图像，利用多项式拟合方法对背景光谱区上下各区间（包括地冕发射）进行插值。第二种方法是对二维原始图像进行暗背景校正，这种校正参照一种超暗图像（通过周期观测建立），同时减去沿光栅缝隙测量得到的地冕发

background correction from a super-dark image (created for the period of the observations) as well as a subtraction of the geocoronal emission measured along the slit, away from the stellar spectrum. The differences between the results of both approaches are negligible, showing that systematic errors generated through the background corrections are small compared to the statistical errors. Those errors are dominated by photon counting noise to which we added quadratically the error evaluated for both the dark background and geocoronal subtractions.

The Lyman α line profiles observed before and during the transits are plotted in Fig. 2. The three exposures outside the transits (exposures A1, B1 and C1) and the three entirely within the transits (A2, B3 and C3) were co-added to improve the signal-to-noise ratio. An obvious signature in absorption is detected during the transits, mainly over the blue side of the line, and possibly at the top of the red peak.

To characterize this signature better, we have defined two spectral domains: "In" and "Out" of the absorption. The In domain is a wavelength interval limited by two variables λ_1 and λ_2 (excluding the Geo geocoronal region). The Out domain is the remaining wavelength coverage within the interval 1,214.4–1,216.8 Å, for which the Lyman α line intensity can be accurately measured at the time of the observation. The corresponding In/Out flux ratio derived for each exposure is shown in Fig. 3, revealing the absorption occurring during the transits. To evaluate whether this detection is sensitive to a particular choice of λ_1 and λ_2, we averaged the three ratios before the transits and the three ratios entirely within the transits. We calculated the averaged pre-transits over mid-transits ratio as a function of λ_1 and λ_2 and propagated the errors through boot-strap estimations of the ratio calculated with 10,000 randomly generated spectra according to the evaluated errors over each individual pixel (Fig. 2c). The averaged ratio is always significantly below 1, with the minimum at $\lambda_1 = 1,215.15$ Å and $\lambda_2 = 1,216.1$ Å. In the interval defined by these two wavelengths, the Lyman α line is reduced by $15 \pm 4\%$ (1σ) during the transit. This is a larger-than-3σ detection of an absorption in the hydrogen line profile during the planetary transits.

HD209458 (G0V) is close to solar type, for which time variations are known to occur in the chromospheric Lyman α line[8]. We thus evaluated the In/Out ratio in the solar Lyman α line profile as measured over the whole solar disk by the Solar Ultraviolet Measurements of Emitted Radiation (SUMER) instrument onboard the Solar and Heliospheric Observatory (SOHO)[9] during the last solar cycle from 1996 to 2001, that is, from quiet to active Sun. During this time, the total solar Lyman α flux varies by about a factor of two, while its In/Out ratio varies by less than $\pm 6\%$. Within a few months, a time comparable to our HD209458 observations, the solar In/Out ratio varies by less than $\pm 4\%$. This is an indication that the absorption detected is not of stellar origin but is due to a transient absorption occurring during the planetary transits.

A bright hot spot on the stellar surface hidden during the planetary transit is also excluded. Such a hot spot would have to contribute about 15% of the Lyman α flux over 1.5% of

射线的干扰。以上两种方法所得到的结果差异小到可以忽略不计，表明背景校正产生的系统误差小于统计误差。这些误差受光子计数噪声控制，我们在这些噪声中加入了暗背景和地冕删除两次误差评估。

图 2 给出了凌星前和凌星中观测到的莱曼 α 线廓线。为改善信噪比，我们把三次凌星前的曝光（分别表示为 A1、B1、C1）和三次凌星过程之中的曝光（分别表示为 A2、B3、C3）叠加在了一起。在凌星过程中，一个清晰的吸收信号被探测到，主要位于光谱的蓝端，也可能位于红端峰的顶部。

为了更好地描述该吸收信号特征，我们定义了光谱吸收域"In"和"Out"。其中"In"区域位于波长 λ_1 和 λ_2 之间（不包括"Geo"区域），"Out"区域位于波长 λ_1 和 λ_2 外侧，两区域均位于 1,214.4 ~ 1,216.8 Å 之间，该波段的莱曼 α 线强度在观测期间可以精确测量。图 3 给出了每次曝光相对应的 In/Out 通量比，结果表明在凌星过程中发生了吸收现象。为了评估这次探测对于我们选择的 λ_1、λ_2 值是否敏感，我们将凌星前的三个比值和凌星中的三个比值分别进行平均。然后计算出凌星前与凌星中比值的平均值，并将其作为关于 λ_1 和 λ_2 的函数。根据每个独立像素的估计误差，利用比值的自举估算来传递误差，其中比值是由 10,000 个随机产生的光谱计算产生（图 2c）。平均比值始终显著地小于 1，其中最小值位于位置 λ_1 = 1,215.15 Å 和 λ_2 = 1,216.1 Å。在 λ_1 和 λ_2 之间的波段，凌星过程中莱曼 α 线强度减弱了 15±4% (1σ)。这是一个在行星凌星中对氢吸收廓线所做的大于 3σ 的探测。

HD209458(G0V) 是类太阳恒星，其色球莱曼 α 线强度随时间发生变化[8]。在最近一个太阳活动周期（1996 ~ 2001 年），太阳活动从宁静变得逐渐活跃。在这段时间，太阳和日球层探测器（SOHO）上搭载的太阳紫外辐射光谱仪对整个日面进行了测量，从而可以求出太阳莱曼 α 线轮廓中的 In/Out 比率。结果表明，在该活动周期内，太阳莱曼 α 总通量变化因子大约为 2，其 In/Out 比值变化小于 ±6%。在和 HD209458 观测时间差不多的几个月内，太阳的 In/Out 比值起伏变化小于 ±4%。这说明，探测到的吸收光谱并不是由恒星本身产生，而是在凌星过程中产生的短暂吸收。

我们同时也排除了在凌星过程中恒星表面可能存在明亮热斑的可能性。因为

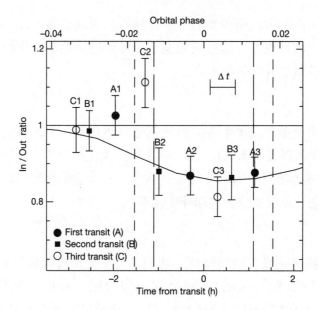

Fig. 3. Relative flux of Lyman α as a function of the HD209458's system phase. The averaged ratio of the flux is measured in the In (1,215.15–1,215.50 Å and 1,215.80–1,216.10 Å) and the Out (1,214.40–1,215.15 Å and 1,216.10–1,216.80 Å) domains in individual exposures of the three observed transits of HD209458b. The central time of each exposure is plotted relative to the transit time. The vertical dashed lines indicate the first and the second contact at the beginning and the end of the transit; the exposures A1, B1 and C1 were performed before the transits, and the exposures A2, B3 and C3 were entirely within the transits. The ratio is normalized to the average value of the three observations completed before the beginning of the transits. The $\pm 1\sigma$ error bars are statistical; they are computed through boot-strap estimations (see text). The In/Out ratio smoothly decreases by around 15% during the transit. The thick line represents the absorption ratio modelled through a particle simulation which includes hydrogen atoms escaping from the planet. In this simulation, hydrogen atoms are sensitive to the radiation pressure above an altitude of 0.5 times the Roche radius, where the density is assumed to be 2×10^5 cm^{-3}; these two parameters correspond to an escape flux of $\sim 10^{10}$ g s^{-1}. The stellar radiation pressure is taken to be 0.7 times the stellar gravitation. The mean lifetime of escaping hydrogen atoms is taken to be 4 h. The model yields an atom population in a curved comet-like tail, explaining why the computed absorption lasts well after the end of the transit.

the stellar surface occulted by the planet, in contradiction with Lyman α inhomogeneities observed on the Sun[10]. Furthermore, this spot would have to be perfectly aligned with the planet throughout the transit, at the same latitude as the Earth's direction, and with a peculiar narrow single-peaked profile confined over the In spectral region. It seems unlikely that a stellar spot could satisfy all these conditions.

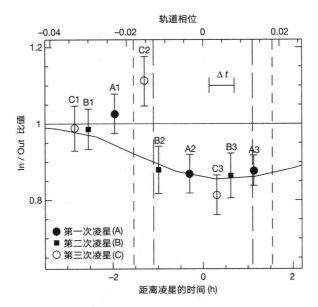

图 3. 莱曼 α 谱线的相对通量与 HD209458 系统相位的函数关系。HD209458b 每次凌星过程中，在 In (1,215.15 ~ 1,215.50 Å 和 1,215.80 ~ 1,216.10 Å) 和 Out(1,214.40 ~ 1,215.15 Å 和 1,216.10 ~ 1,216.80 Å) 域上测量每次曝光，可以得到通量平均比值。图中每次曝光过程的中心时间点是相对整个凌星的时间值，垂直虚线表示凌星开始和结束时的第一次和第二次接触。图中 A1、B1、C1 三次曝光是在凌星前进行，而 A2、B3、C3 是完全在凌星中进行。其中，图中比值相对凌星前完成的三次观测的平均值进行了归一化。$\pm 1\sigma$ 误差棒所示为统计误差，它们通过自举估算计算得到。可以看出，在整个凌星过程中，In/Out 比值平滑地降低了大约 15%。图中粗线表示通过粒子仿真模型计算得到的吸收比值，其中该模型考虑了从行星逃逸的氢原子。在该仿真中，氢原子对于海拔高度大于 0.5 倍洛希半径处的辐射压较为敏感，其密度假定为 $2 \times 10^5 \text{ cm}^{-3}$，对应的逃逸通量约为 $10^{10} \text{ g} \cdot \text{s}^{-1}$。恒星辐射压设为 0.7 倍于恒星引力。逃逸氢原子的平均寿命取 4 h。该模型计算表明，大量逃逸氢原子构成了一个类似弯曲彗尾的结构，这就解释了为什么凌星完成后还观测到吸收现象。

如果恒星表面被行星遮挡了 1.5% 的话，那么这样一个热斑将会贡献莱曼 α 通量的 15%，然而这与观测到的太阳的莱曼 α 非均匀性相矛盾 [10]。而且，如果有热斑存在的话，那么在整个凌星过程中，位于地球方向上的相同纬度，热斑都需要与该行星严格对准，此外在 In 光谱区域还会有一个宽度很窄的特征单峰。这些条件过于苛刻，恒星上的热斑不太可能满足所有的这些条件。

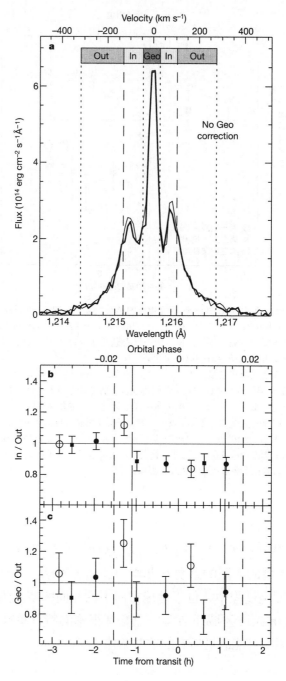

Fig. 4. Spectra and ratios with no geocoronal correction. To investigate the possibility that a bad estimate of the geocoronal correction may cause the detected signal, the spectra and ratios are plotted without this correction. First, note that the geocoronal emission is of the order of the stellar flux, and its peak value is never larger than three times the stellar one. Consequently, the correction is very small outside the Geo domain. **b** shows the same evaluation of the In/Out ratios as made in Fig. 3, with no geocoronal and no background correction. It appears that the impact of the geocoronal correction on these ratios is negligible. To further show that the "wings" of the geocoronal emission do not carry the detected transit signal, **c** shows the Geo/Out ratios, where Geo is the total flux due to the geocorona. If the flux in the In

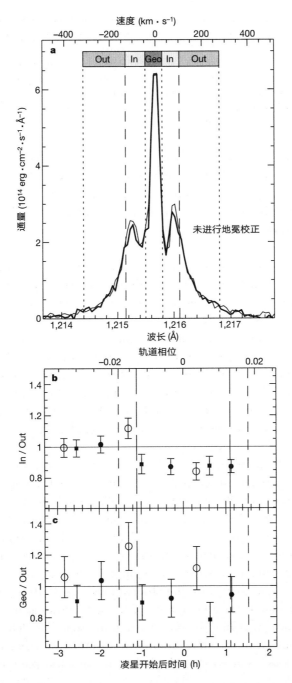

图 4. 未进行地冕校正的光谱和比值。为了研究不良的地冕校正对于探测信号的影响，图中给出了未经校正的光谱和比值。首先，值得注意的是地冕发射通量与恒星通量量级相同，而且其峰值不会高于恒星峰值的三倍。因此，在 Geo 域外的校正很小。图 b 给出了 In/Out 比值（与图 3 计算相同），但没有进行地冕和背景校正。从图中可见，地冕校正对于比值的影响可以忽略。为了进一步说明地冕发射的"两翼"并不具有探测到的凌星信号，图 c 给出了 Geo/Out 比值，其中 Geo 代表地冕发射的总通量。如果 In 区域内的通量受地冕发射的严重影响，则会发现 Geo/Out 比值与图 3 中比值相似。从图中可知，Geo/Out 比值在 A1、B1 和 C1 平均值附近随机起伏，并没有随地冕变化出现始终一致的凌星特征。类

577

domain was significantly affected by the geocorona, the Geo/Out ratios should present a signal similar to the In/Out ratios plotted in Fig. 3. The Geo/Out ratios are found to be random fluctuations around the average of the A1, B1 and C1 values. There is no consistent signature of the transit in the geocoronal variations. As in Fig. 2a, **a** shows the average of the three spectra obtained before and during the transits (thin and thick lines, respectively). Quoted velocities are in the stellar reference frame. Here the spectra are without any geocoronal correction. A simple constant has been subtracted to compensate for the mean background levels in order to have matching spectra in the Out domain. This clearly shows that the absorption signature is present even without the geocoronal correction.

Finally, we confirmed with various tests that there are no correlations between the geocoronal variations and the detected signature in absorption. One method is presented in Fig. 4, showing that a contamination of the In domain by the geocorona is excluded. We thus conclude that the detected profile variation can only be related to an absorption produced by the planetary environment.

The observed 15% intensity drop is larger than expected *a priori* for an atmosphere of a planet occulting only 1.5% of the star. Although the small distance (8.5 stellar radii) between the planet and the star results in an extended Roche lobe[11] with a limit at about 2.7 planetary radii (that is, 3.6 Jupiter radii), the filling up of this lobe gives a maximum absorption of about 10% during the planetary transit. Because a more important absorption is detected, hydrogen atoms must cover a larger area: a drop of 15% corresponds to an occultation by an object of 4.3 Jupiter radii. This is clearly beyond the Roche limit as theoretically predicted[6]. Thus some hydrogen atoms should escape from the planet. The spectral absorption width shows independently that the atoms have large velocities relative to the planet. Thus hydrogen atoms must be escaping the planetary atmosphere.

We have built a particle simulation in which we assumed that hydrogen atoms are sensitive to the stellar radiation pressure inside and outside the Roche lobe. Their motion is evaluated by taking into account both the planetary and stellar gravities. The Lyman α radiation pressure is known to be 0.7 times the stellar gravity, the escape flux and the neutral hydrogen lifetime being free parameters. The lifetime of hydrogen is limited to a few hours owing to stellar extreme ultraviolet ionization. Escaping hydrogen atoms expand in an asymmetric comet-like tail and progressively disappear when moving away from the planet. This simple scenario is consistent with the observations (Fig. 3). In this model, atoms in the evaporating coma and tail cover a large area of the star, and most are blueshifted because of the radiation pressure repelling them away from the star. The detection of most of the absorption in the blue part of the line is consistent with these escaping atoms. On the other hand, more observations are needed to clarify whether an absorption is also present in the red part of the line.

To account for the observed absorption depth, the particle simulation implies a minimum escape flux of around 10^{10} g s^{-1}. However, owing to saturation effects in the absorption line, a flux larger by several orders of magnitude would produce a similar absorption signature. So, to evaluate the actual escape flux, we need to estimate the vertical distribution of hydrogen atoms up to the Roche limit, in an atmosphere extended by the stellar tidal forces

似图 2a，图 **a** 给出了凌星前和凌星中获得的三个光谱的平均值，分别用细线和粗线画出。图中速度相对恒星参考系。这里的光谱没有进行任何地冕校正。为了在 Out 域获得匹配光谱，我们在光谱中减去了一个常数以补偿背景平均水平。这清楚地表明，即使没有经过地冕校正也存在特征吸收。

最终，我们通过各种检验确认探测到的特征吸收与地冕的变化没有相关性。图 4 给出了其中的一种方法，表明可以消除 In 区域内地冕造成的干扰。因此，我们得出以下结论：探测到的光谱廓线变化只可能与行星环境产生的吸收有关。

鉴于这颗恒星表面被行星大气遮挡住 1.5% 的情况，观测到的光谱强度减小了 15%，这比预期的值要大。尽管行星与恒星之间的短距离（8.5 恒星半径）造成了洛希瓣扩展 [11]，扩展的上限为 2.7 倍行星的半径（即木星半径的 3.6 倍），而对洛希瓣的填充会引起行星凌星过程的最大吸收，约为 10%。探测到的强吸收意味着氢原子一定覆盖着更大的面积，15% 的减小对应着一个 4.3 倍于木星半径的天体被遮挡，这显然超出了洛希极限的理论值 [6]。所以，一部分氢原子应该从这颗行星发生了逃逸。另外，光谱的吸收宽度独立地表明这些原子相对这颗行星具有较大的速度。因此氢原子必定从这颗行星的大气层逃逸了。

为此，我们建立了一个粒子仿真，在该仿真中我们假设氢原子对洛希瓣内外的恒星辐射压较为敏感。通过考虑行星和恒星的引力作用，可以计算出这些氢原子的运动情况。已知莱曼 α 辐射压是恒星引力的 0.7 倍，氢原子的逃逸通量和中性氢原子的寿命都是自由参量。由于恒星极端紫外电离，氢原子的寿命一般为数小时。逃逸的氢原子扩展形成一个非对称的彗状尾，并随着行星的远离逐渐消失。这一图景与图 3 中的观测相符。在这一模型中，位于蒸发中的彗发和彗尾中的逃逸氢原子覆盖了恒星很大的表面积，并且这些原子被辐射压从恒星表面排斥开，因此大多发生蓝移，蓝移可以解释为辐射压将这些原子从恒星上排斥出来。对蓝端部分谱线吸收的检测结果与这些逃逸中原子的检测结果相吻合。另一方面，对于大多数观测需要弄清是否在谱线红端部分也存在吸收现象。

考虑观测到的吸收深度，粒子仿真表明需要存在最小逃逸通量，其值大约为 10^{10} g · s^{-1}。由于吸收光谱存在饱和效应，即使比最小通量大若干个数量级的通量也会产生相似的特征吸收。所以，为了计算实际的逃逸通量，我们需要估计氢原子在洛希极限以内的大气层垂直分布。大气层由于受到恒星潮汐力作用而扩展，同时由于许多可能的机制而被加热，这些效应都可能引起更大的逃逸通量。对于通量的详

and heated by many possible mechanisms. These effects may lead to a much larger escape flux. A detailed calculation is beyond the scope of this Letter. This raises the question of the lifetime of evaporating extrasolar planets which may be comparable to the star's lifetime itself. If so, the so-called "hot Jupiters" could evolve faster than their parent star, eventually becoming smaller objects, which could look like "hot hydrogen-poor Neptune-mass" planets. This evaporation process, more efficient for planets close to their star, might explain the very few detections[12,13] of "hot Jupiters" with orbiting periods shorter than three days.

(**422**, 143-146; 2003)

A. Vidal-Madjar*, A. Lecavelier des Etangs*, J.-M. Désert*, G. E. Ballester†, R. Ferlet*, G. Hébrard* & M. Mayor‡

* Institut d'Astrophysique de Paris, CNRS/UPMC, 98bis boulevard Arago, F-75014 Paris, France
† Lunar and Planetary Laboratory, University of Arizona, 1040 E. 4th St., Rm 901, Tucson, Arizona 85721-0077, USA
‡ Observatoire de Genève, CH-1290 Sauverny, Switzerland

Received 13 September 2002; accepted 27 January 2003; doi:10.1038/nature01448.

References:

1. Henry, G. W., Marcy, G. W., Butler, R. P. & Vogt, S. S. A transiting "51 Peg-like" planet. *Astrophys. J.* **529**, L41-L44 (2000).

2. Charbonneau, D., Brown, T. M., Latham, D. W. & Mayor, M. Detection of planetary transits across a Sun-like star. *Astrophys. J.* **529**, L45-L48 (2000).

3. Mazeh, T. *et al.* The spectroscopic orbit of the planetary companion transiting HD 209458. *Astrophys. J.* **532**, L55-L58 (2000).

4. Bundy, K. A. & Marcy, G. W. A search for transit effects in spectra of 51 Pegasi and HD 209458. *Proc. Astron. Soc. Pacif.* **112**, 1421-1425 (2000).

5. Rauer, H., Bockelée-Morvan, D., Coustenis, A., Guillot, T. & Schneider, J. Search for an exosphere around 51 Pegasi B with ISO. *Astron. Astrophys.* **355**, 573-580 (2000).

6. Moutou, C. *et al.* Search for spectroscopical signatures of transiting HD 209458b's exosphere. *Astron. Astrophys.* **371**, 260-266 (2001).

7. Charbonneau, D., Brown, T. M., Noyes, R. W. & Gilliland, R. L. Detection of an extrasolar planet atmosphere. *Astrophys. J.* **568**, 377-384 (2002).

8. Vidal-Madjar, A. Evolution of the solar lyman alpha flux during four consecutive years. *Sol. Phys.* **40**, 69-86 (1975).

9. Lemaire, P. *et al.* in *Proc. Symp. SOHO 11, From Solar Min to Max: Half a Solar Cycle with SOHO* (ed. Wilson, A.) 219-222 (ESA SP-508, ESA Publications Division, Noordwijk, 2002).

10. Prinz, D. K. The spatial distribution of Lyman alpha on the Sun. *Astrophys. J.* **187**, 369-375 (1974).

11. Paczynski, B. Evolutionary processes in close binary systems. *Annu. Rev. Astron. Astrophys.* **9**, 183-208 (1971).

12. Cumming, A., Marcy, G. W., Butler, R. P. & Vogt, S. S. The statistics of extrasolar planets: Results from the Keck survey. Preprint astro-ph/0209199 available at ⟨http://xxx.lanl.gov⟩ (2002).

13. Konacki, M., Torres, G., Jha, S. & Sasselov, D. D. An extrasolar planet that transits the disk of its parent star. *Nature* **421**, 507-509 (2003).

Acknowledgements. This work is based on observations with the NASA/ESA Hubble Space Telescope, obtained at the Space Telescope Science Institute, which is operated by AURA, Inc. We thank M. Lemoine, L. Ben Jaffel, C. Emerich, P. D. Feldman and J. McConnell for comments, J. Herbert and W. Landsman for conversations on STIS data reduction, and J. Valenti for help in preparing the observations.

Competing interests statement. The authors declare that they have no competing financial interests.

Correspondence and requests for materials should be addressed to A.V.-M. (e-mail: alfred@iap.fr).

细计算已经超出了本文的研究范围。这也引出了关于太阳系外蒸发中的行星的寿命问题，其寿命可能与母恒星的寿命相当。如果确实如此，这些所谓的"热木星"就会比其母恒星演化更快，最终演变为更小的天体，而这些"热而缺氢"行星和海王星的质量差不多。如果行星距离恒星更近的话，以上蒸发过程将会更快。这也可能解释为什么轨道周期短于三天的"热木星"很少被探测到[12,13]。

（金世超 翻译；胡永云 审稿）

A Correlation between the Cosmic Microwave Background and Large-scale Structure in the Universe

S. Boughn and R. Crittenden

Editor's Note

The ripples in the cosmic microwave background radiation are relics of the seeds from which the large-scale structure of the Universe grew. If this is so, hot spots in this radiation should correspond to concentrations of galaxies, and cold spots to voids. This paper by Stephen Boughn and Robert Crittenden reports the first observation of such a correlation. X-rays from hot gas inside clusters of galaxies turn out to be correlated with fluctuations in the backgrounds. The distribution of galaxies that are bright radio sources is also correlated with the fluctuations. The authors point out that the correlation supports the idea that dark energy, which drives the acceleration of cosmic expansion, has left an imprint on large-scale structure.

Observations of distant supernovae and the fluctuations in the cosmic microwave background (CMB) indicate that the expansion of the Universe may be accelerating[1] under the action of a "cosmological constant" or some other form of "dark energy". This dark energy now appears to dominate the Universe and not only alters its expansion rate, but also affects the evolution of fluctuations in the density of matter, slowing down the gravitational collapse of material (into, for example, clusters of galaxies) in recent times. Additional fluctuations in the temperature of CMB photons are induced as they pass through large-scale structures[2] and these fluctuations are necessarily correlated with the distribution of relatively nearby matter[3]. Here we report the detection of correlations between recent CMB data[4] and two probes of large-scale structure: the X-ray background[5] and the distribution of radio galaxies[6]. These correlations are consistent with those predicted by dark energy, indicating that we are seeing the imprint of dark energy on the growth of structure in the Universe.

IN the standard model of the origin of structure, most of the fluctuations in the CMB were imprinted as the photons last scattered off free electrons, when the universe was just 400,000 years old (at a redshift of $z \approx 1,100$). However, once the dark energy (or the spatial curvature[7]) becomes important dynamically ($z \approx 1$), additional fluctuations are induced in the CMB photons by what is known as the integrated Sachs–Wolfe (ISW) effect[2]. The gravitational potentials of large, diffuse concentrations and rarefactions of matter begin to evolve, causing the energy of photons passing through them to change by an amount that depends on the depth of the potentials. The amplitude of these ISW fluctuations tends to

宇宙微波背景和大尺度结构的相关

博夫恩，克里滕登

编者按

宇宙微波背景辐射中的波纹是"种子"的遗迹，宇宙的大尺度结构便是从这些"种子"发育而成。如果真如上所言，这种辐射中的热斑应该对应于星系的聚集，而冷斑则对应巨洞。本文作者史蒂夫·博夫恩和罗伯特·克里滕登首次报道了这种相关性。来自星系团内的热气体的 X 射线与背景中的扰动相关。作为亮射电源的星系的分布也与扰动相关。作者指出，这种相关性支持以下观点：驱动宇宙膨胀加速的暗能量在大尺度结构上留下了印记。

对遥远的超新星和宇宙微波背景辐射（CMB）涨落的观测表明，宇宙很可能在"宇宙学常数"或者某种其他形式的暗能量作用下加速膨胀[1]。暗能量在当前宇宙中占主导地位，它不仅改变了宇宙的膨胀速率，也影响了物质密度扰动的演化，减慢了物质的引力塌缩（例如星系团的生成过程）。宇宙微波背景辐射温度的额外涨落来自 CMB 光子穿越大尺度结构时所受到的扰动[2]，因此这些涨落必然与所对应区域邻近物质的分布相关[3]。在这里我们报道最近的 CMB 数据[4] 和两种大尺度结构的观测结果（X 射线背景[5] 和射电星系分布[6]）之间的相关性。这些相关性与暗能量的预言相符，表明我们在宇宙结构演化的过程中观测到了暗能量的印记。

在结构起源的标准模型中，大部分 CMB 涨落都带有光子被自由电子最后一次散射时的特征，此刻宇宙只有 400,000 年的历史（红移大约为 1,100）。但是一旦暗能量（或者空间曲率[7]）逐渐变得重要（红移大约为 1），累计萨克斯-沃尔夫效应（ISW）将会在 CMB 中引入额外的扰动[2]。由于大量聚集和稀薄的弥散物质的引力势随时间开始演化，穿过引力势阱的光子能量发生变化，这个变化量取决于势阱的深度。除了在很大的尺度上以外，ISW 的扰动幅度要比源自最后散射的扰动小得多。然而

be small compared to the fluctuations originating at the epoch of last scattering except on very large scales. However, as ISW fluctuations were created more locally, it is expected that the CMB fluctuations should be partially correlated with galaxies, which serve as tracers of the large-scale matter distribution.

Detecting the relatively weak ISW effect requires correlating the CMB with the distribution of galaxies spread over a large fraction of the observable Universe. This necessitates a survey of galaxies covering much of the sky, out to distances of many billions of light years ($z \approx 1$). Focus thus has been on luminous active galaxies, which are believed to trace the mass distribution on large scales. Although active galaxies emit at a wide range of frequencies, the most useful maps are in the hard X-rays (2–10 keV), where they dominate the X-ray sky, and in the radio, where the number counts are dominated by sources at $z \lesssim 1$. The full sky map of the intensity of the hard X-ray background made by the HEAO-1 satellite[5] and a map of the number density of radio sources provided by the NRAO VLA sky survey (NVSS)[6] are two of the best maps in which to look for this effect. To predict the expected level of the ISW effect in these two surveys, it is essential to know both the inherent clustering of the sources and their distribution in redshift. The former can be determined from previous studies of the X-ray and radio auto-correlation functions[8,9] and the latter can be estimated from deep, pointed surveys[10,11].

Previous attempts at correlating the CMB with both the HEAO-1 and the NVSS maps have yielded only upper limits[8,9,12]. Here we repeat these analyses using the much improved CMB maps recently provided by the WMAP satellite mission[4]. We have compared two different CMB maps generated from this data (the "internal linear combination" map[13], and the "cleaned" map of ref. 14) with the HEAO and NVSS maps. The dominant "noise" in the maps is due to the fluctuations in the CMB itself and is well characterized, while the instrument noise is negligible on the angular scales relevant to the present analysis. To reduce possible contamination by emission from our Milky Way Galaxy as well as from other nearby radio sources, these maps were masked with the most aggressive foreground template provided by the WMAP team[13]. The X-ray map was similarly masked and was corrected for several large-scale systematics including a linear drift of the detectors[9]. Finally, the NVSS catalogue was corrected for a systematic variation of source counts with declination[8]. The sky coverage was 68% for the CMB maps, 56% for the NVSS map, and 33% for the X-ray map. The reduced coverage of the X-ray map was the result of its low resolution (3°) and a larger number of foreground sources. Significantly, however, our results do not depend on the corrections or the level of masking of the maps.

A standard measure of the correspondence of two data sets is the cross-correlation function, $\mathrm{CCF}(\theta)$. It represents the extent to which the two measures of the sky separated by an angle θ are correlated. Figures 1 and 2 show the CCFs of the WMAP data with the X-ray and radio surveys. The results using the "internal linear combination" and "cleaned" CMB maps were entirely consistent with each other, and we plot the averages of the results for the two maps. Four hundred Monte Carlo simulations of the CMB sky were also cross-correlated with the actual X-ray and radio maps. The r.m.s. values of these trials are

ISW 扰动是在局部空间产生的，CMB 涨落应与星系分布部分相关，故可以以此探测大尺度上的物质分布。

探测相对较弱的 ISW 效应要求研究 CMB 与占可观测宇宙中较大部分的星系分布的相关性。这需要星系巡天具有大比例的天空覆盖，且巡天深度延伸至几十亿光年的距离（红移为 1 左右）。因此巡天目标集中在明亮的活动星系，这被认为可以很好地示踪大尺度的质量分布。虽然活动星系在很宽的波段内都有辐射，但是最有效的成图在硬 X 射线波段（2 ~ 10 keV），它主导了 X 射线天空，同时在射电波段，射电源计数由红移 ⩽1 的源所主导。HEAO–1 卫星所作的亮硬 X 射线背景强度全天图 [5] 和 NRAO VLA 巡天（NVSS）所作的射电源数密度巡天图为寻找 ISW 效应提供了最好的图像。为了预测在这两个巡天中 ISW 效应所期望达到的水平，清楚这些源内在的成团性和它们随红移的分布是至关重要的。成团性可由过去已有的对 X 射线和射电源的自相关函数 [8,9] 的研究中得到，分布则可由深度指向巡天 [10,11] 来估计。

过去将 CMB 与 HEAO–1、NVSS 天图做交叉相关的尝试仅得到了相关的上限 [8,9,12]。在这里我们用 WMAP 卫星计划 [4] 最近提供的有显著改进的 CMB 图重复这种相关分析。我们将从 WMAP 得到的两种不同的 CMB 图（"内在线性组合"图 [13] 和"净化"图 [14]）与 HEAO 和 NVSS 的天图相比较。图中"噪声"的主要成分来自 CMB 自身的涨落，其特征能够被很好地描述，而在要分析的角尺度上仪器噪声则可以忽略。为了减少来自银河系和其他附近射电源的辐射污染，这些图均按 WMAP 团队所提供的最苛刻的前景模板进行了掩模 [13]。X 射线图也进行了类似的掩模，同时还修正了几个大尺度上类似探测器线性漂移的系统误差 [9]。最后，NVSS 星表也修正了源计数偏差引起的系统误差 [8]。CMB 图覆盖了整个天空的 68%，NVSS 图覆盖了 56%，X 射线图只覆盖了 33%。X 射线图的覆盖率偏低是因为较低的分辨率（3度）和大量的前景源。尽管如此，我们的结果与修正或对图像的掩模程度明显地关系不大。

度量两个数据集合相关性的标准方法是求交叉相关函数 CCF(θ)。它表征了角间距为 θ 的两个方向之间的空间相关性。图 1 与图 2 表示 WMAP 分别与 X 射线巡天图和射电巡天图之间的交叉相关函数。用 CMB 的"内在线性组合"图和"洁化"图得到的结果是完全符合的；我们还绘出了这两种图结果的平均。通过四百次蒙特卡洛方法模拟得到的 CMB 图被用于与实际的 X 射线巡天图和射电巡天图做交叉相关。

taken as the errors for each bin and these are highly correlated owing to the large-scale structure in the maps. For comparison, the uncertainties were also estimated from the data themselves by rotating the maps with respect to each other, and the errors determined in this way were very similar to those found using the simulated maps. The significance of the detection of the ISW effect is estimated by fitting the amplitude of the theoretical profile to the data. For the X-ray/CMB CCF, the correlations are detected at a confidence level (CL) of 99.9%, that is, a 3.0σ detection. The radio/CMB CCF is detected at a 99.4% CL, or a 2.5σ detection. These confidence limits are consistent with the number of times the CCF(θ) of the Monte Carlo trials exceed the values of the measured CCF(θ).

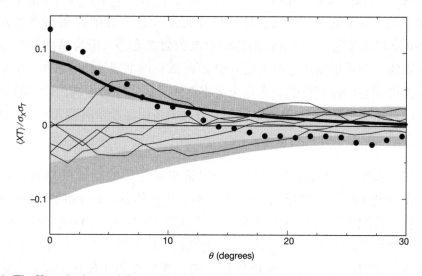

Fig. 1. The X-ray background fluctuations are correlated with the microwave sky at a higher level than would be expected by chance. Here we plot the cross correlation (filled circles) between the X-ray intensity (X) measured by HEAO-A1 and the CMB temperature fluctuations (T) measured by WMAP. We normalize by the r.m.s. levels of the two maps (σ_x and σ_T, respectively), which makes it independent of any linear biasing of the survey. To give an idea of the level of accidental correlations, the shaded areas show the 1σ and 2σ regions derived from simulated, uncorrelated CMB maps with the same power spectrum as the WMAP data. The thin solid lines show the results for five such Monte Carlo simulations, and we see that the signals in neighbouring bins are highly correlated for a given realization. The signal-to-noise ratio is greatest at smaller angular separations, even though the error is amplified because fewer pairs of pixels contribute to the correlation. For $\theta = 0°$, $1.3°$ and $2.6°$, the Monte Carlo trials exceed the amplitude of the actual X-ray/CMB correlation only 0.3%, 0.8% and 0.3% of the time, respectively. The bold line shows theoretical predictions for the ISW effect in a cosmological-constant-dominated Universe, using the best-fit WMAP model for scale-invariant fluctuations[19] (dark energy fraction $\Omega_\Lambda = 0.73$, Hubble constant $H_0 = 100\, h$ km s^{-1} Mpc^{-1} with $h = 0.72$, matter density $\Omega_m h^2 = 0.14$ and baryon density $\Omega_b h^2 = 0.024$). At larger angular separations, the observed correlations appear to fall faster than predicted by theory, but the low signal-to-noise ratio makes it difficult to say whether this is a real effect.

这些试验的均方根被认为是每一个数据间隔的误差，由于天图存在的大尺度结构，它们是高度相关的。作为比较，不确定性可以通过天图本身的相对旋转由对原始数据的分析获得，它所给出的误差与通过使用模拟图得到的误差非常接近。ISW 效应探测的显著性可以通过理论轮廓的大小和实际数据的拟合来估计。对于 CMB 与 X 射线巡天之间的交叉相关函数，探测的置信水平为 99.9%（即 3σ 探测）；对于 CMB 与射电巡天之间的交叉相关函数，探测的置信水平为 99.4%（即 2.5σ 探测）。这些置信限与大量蒙特卡洛试验得出的交叉相关函数是一致的。

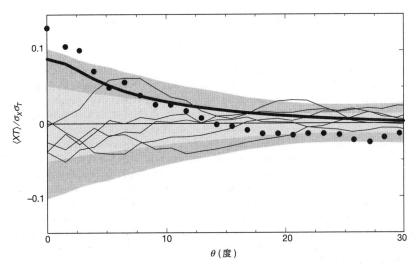

图 1. X 射线背景扰动与微波背景的相关性高于偶然事件的概率，这里我们给出了 HEAO-A1 测得的 X 射线强度 (X) 和 WMAP 观测的微波背景辐射温度扰动 (T) 的交叉相关（实心圆）。我们用两个图的均方根水平 (σ_X 和 σ_T) 做归一化，使得它与巡天的线性偏袒无关。为了了解偶然事件导致相关的可能性，阴影区域显示与 WMAP 具有相同功率谱且统计无关的 CMB 模拟图给出的 1σ 和 2σ 区域。细实线显示了 5 个这样的蒙特卡罗模拟，我们注意到，临近数据区间的信号与给定的模拟实现高度相关。尽管贡献相关性的像素较少而使误差被放大，这个信噪比在小角间距下还是最大的。对 $\theta = 1$ 度、1.3 度和 2.6 度，蒙特卡罗的结果只超过真实 X 射线/CMB 相关的 0.3%、0.8% 和 0.3%。图中粗线显示在宇宙学常数主导的宇宙中 ISW 效应的理论预测，它采用了尺度不变扰动的 WMAP 最佳拟合模型 [19]（暗物质分数 $\Omega_\Lambda = 0.73$，哈勃常数 $H_0 = 100h$ km·s^{-1}·Mpc^{-1}，$h = 0.72$，物质密度 $\Omega_m h^2 = 0.14$，重子密度 $\Omega_b h^2 = 0.024$）。对更大的角间距，观测到的相关性看上去比理论预言的相对性下降得更快，但是信噪比很低，因此很难说这是否是真实的结果。

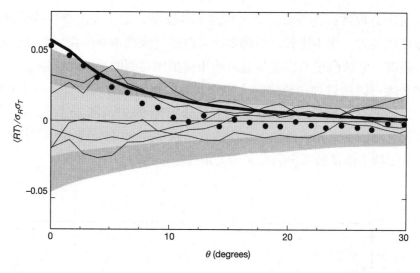

Fig. 2. The distribution of radio galaxies is correlated with the microwave sky. Here we plot the correlation (filled circles) between the NVSS radio galaxy number counts (R) and the WMAP temperature maps (T). The other curves are as in Fig. 1. The Monte Carlo trials exceed the amplitude of the actual radio/CMB correlation in the lowest three bins 1.2%, 1.9% and 3.4% of the time, respectively. Again, there is good agreement with the theoretical predictions, with the signal again falling off somewhat faster than predicted at larger angles. The lower amplitude compared to the X-ray result is due to the higher resolution of the radio map, which increases the radio r.m.s. (σ_R) without significantly increasing the cross-correlation. The consistency of the NVSS and X-ray CCFs suggests that the signal is not the result of unknown systematics in either the X-ray or the NVSS map. It is possible that microwave emission from the radio/X-ray sources themselves could result in the positive correlation of the maps. However, extrapolations of the frequency spectra of the radio galaxies indicate that the microwave emission is much smaller than the observed signal[8]. In addition, the clustering of radio sources is on a much smaller angular scale than the apparent signals, so any such contamination would appear only at $\theta \approx 0°$ and not at larger angles.

The theoretical curves for the ISW effect also are shown in Figs 1 and 2, and fall off with angular separation similarly to the observed correlations. For these calculations, we have assumed that the fluctuations in matter density, $\delta\rho$, are traced directly by the fluctuations in the number density of the sources. That is, $\delta N/N = b\delta\rho/\rho$ where the bias factor, b, is assumed to be constant and its value is derived from the auto-correlation function of the X-ray or radio sources. The predictions assume a scale-invariant primordial spectrum and the best-fit cosmology determined from WMAP. Although the derived bias factors depend on precisely how the sources are assumed to be distributed in redshift, the expected cross-correlation is only weakly dependent on this or on the possible evolution of the bias factor with redshift. Detailed considerations for calculating the amplitude of the ISW effect can be found elsewhere[3,12].

It is important to consider what else might be contributing to the observed correlation. Unmasked foreground sources are not a major contaminant, because the observed correlations do not change when the Galactic cuts are varied nor when the aggressiveness of the masking is changed. In fact, the amplitudes of the CCFs are essentially unchanged when no masking at all is applied to the CMB maps. Also, because the results from analysing

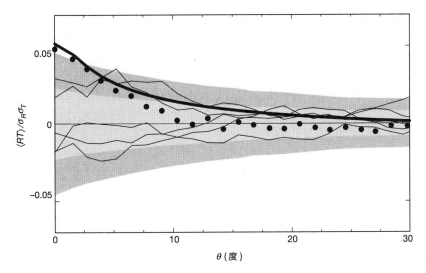

图 2. 射电星系的分布与微波背景的相关。这里我们绘出 NVSS 射电星系计数(R)和 WMAP 温度(T)的相关图(用实心圆表示)。其他曲线的含义与图 1 相同。蒙特卡罗试验中超过实际射电/CMB 相关的幅度在最低的三个区间分别为 1.2%、1.9%、3.4%。这与理论预言再一次保持一致,而在大角度处信号衰减速度再次快于预言。其幅度比 X 射线低是由于射电图更高的分辨率增加了射电的均方根(σ_R),而没有显著增加交叉相关。NVSS 和 X 射线交叉相关函数的一致暗示了信号不是来自 X 射线或者 NVSS 图像中未知的系统误差,而来自射电或者 X 射线源的微波辐射可能导致图像的正相关。尽管如此,射电星系的频谱外推指出这种微波辐射远小于观测结果 [8]。而且,射电源的成团性发生在比视信号小得多的角度上,因此所有这种污染应该出现在 $\theta \approx 0°$ 而非更大的角度。

ISW 效应的理论曲线同时在图 1 和图 2 中给出,它与观测所得的相关类似,随着角间隔增大而衰减。在这些计算中,我们已经假设物质密度扰动($\delta\rho$)可直接由辐射源的数密度扰动所示踪。即 $\delta N/N = b\delta\rho/\rho$,假设偏袒因子 b 为常数且它的值可通过 X 射线或者射电源的自相关函数得出。预测过程中假设了尺度不变的原初功率谱以及由 WMAP 数据的最优拟合得到的宇宙学。虽然由此得到的偏袒因子 b 严格依赖于对源的红移分布的假设,但对期望的交叉相关函数而言,其对红移分布的依赖性比较弱,对偏袒因子 b 可能的红移演化的依赖性也不强。对 ISW 效应幅度计算的详细讨论容易在其他文献中找到 [3,12]。

还有一点非常重要,即要考虑是否还有其他过程对观测到的相关有贡献。未掩模的前景源不是主要的污染,因为当改变银河系扣除方式,或者改变前述的严格掩模处理,观察到的相关性都不会改变。实际上,即使完全不对 CMB 做任何掩模处理,交叉相关函数的幅度都不会有本质的改变。同时,因为对不同半球的巡天图的

the different hemispheres of the maps separately are consistent, it is unlikely that a small number of diffuse foreground sources is responsible for the signal. Fluctuations caused by the Sunyaev–Zeldovich effect—the scattering of CMB photons as they travel through the ionized intergalactic medium of rich clusters of galaxies—are expected to contribute at smaller angles and be anti-correlated with the matter distribution[15]. Finally, microwave emission from the radio/X-ray sources could also produce correlations, but on much smaller angular scales than the observed signal.

The correlated signal evident in the figures is large enough to have been seen in our previous analyses using a COBE satellite combination map[8,9], albeit with a smaller signal-to-noise ratio because of that map's significant instrument noise and low angular resolution. However, we found no evidence for a correlation in the COBE analysis. When the differences in angular resolution are taken into account, the WMAP and COBE maps are consistent given COBE's large instrumental noise ($\sim$70 μK per pixel)[4]. The difference between the observed cross-correlations implies that a small part (r.m.s. $\approx$ 1–2 μK) of the (COBE-WMAP) difference map is anti-correlated with the X-ray and radio maps. This is an unlikely occurrence (1 in 100) if the COBE instrument noise model is correct. There are known systematics in the COBE data[16] above this level, but the large instrument noise makes it difficult to ascertain whether this is the origin of the problem. In any case, the source of the discrepancy almost certainly lies with the COBE data, which contain much larger noise and systematic effects than the WMAP data. Thus, we believe the WMAP correlations presented here are robust. Further discussion of this and other issues are available as Supplementary Information.

The confidence levels that we quote (99.9% and 99.4%) are not primarily limited by experimental error, but by the fact that we can only observe a finite region of the Universe. We find it significant that both signals are consistent with the theoretical predictions with no free parameters. However, it should be emphasized that these two measurements are by no means independent. Both of them use the same CMB maps and, furthermore, the X-ray background is highly correlated with the NVSS radio sources[17]. If the distribution of sources of one of the maps by chance coincided with the fluctuations of the CMB, one would expect the other map also to be correlated with the CMB to some degree. The coincidence of the expected amplitudes of the two signals is encouraging but by no means definitive. On the other hand, the detection of the ISW-type signal in both CCFs gives a strong indication that they are not due to unknown systematic effects in the maps. The radio and X-ray data were gathered by quite distinct methods, and it would be surprising if unknown systematics in the two maps were correlated in any way. We will present more detailed analysis of these issues elsewhere.

We conclude that we have observed the ISW effect. If so, these observations offer the first direct glimpse into the production of CMB fluctuations, and provide important, independent confirmation of the new standard cosmological model: an accelerating universe, dominated by dark energy. It should be pointed out that measurements of the power spectrum of CMB fluctuations do not show evidence of increased power on large

分析得到的结果是一致的，因此很难想象这些相关的信号是由少量的弥散前景源所导致的。苏尼阿耶夫–泽尔多维奇效应（CMB 光子在穿越富星系团中的电离星系际介质时发生的散射）引起的扰动应该贡献在小角度上，并且与物质分布反相关[15]。最后，射电源和 X 射线源发射的微波辐射也可能产生相关，但其发生的角尺度比观测信号要小得多。

图中给出的相关信号的证据已经足够大到即使使用宇宙背景探测器（COBE）卫星组合图也应该在我们前述的分析中被观察到[8,9]，尽管由于图像显著的仪器噪声和低角分辨率，信号的信噪比会比较低。但是我们并没有在对 COBE 的分析中发现存在相关性的证据。当考虑 WMAP 和 COBE 的角分辨率不同，以及在 COBE 具有较大仪器噪声（约 70 μK 每像素）[4]的情况下，COBE 和 WMAP 的图像是一致的。观测到交叉相关的差别意味着，在 COBE 与 WMAP 的图像差中的一小部分（均方根约为 1～2 μK）存在与 X 射线图和射电图的反相关。如果 COBE 的仪器噪声的模型是正确的，上述情况不大可能发生（发生概率为 1/100）。虽然已知在 COBE 的数据中系统误差高于这个水平，但仪器噪声过大让我们很难确定这是否是问题的根源。无论如何，这个不一致的来源都被认为来自 COBE 数据自身[16]，因为它包含的噪声和系统性效应比 WMAP 数据要大得多。因此，我们相信 WMAP 的相关性是稳健的。进一步的讨论和更多的问题详见补充材料。

我们报道的置信度（99.9% 和 99.4%）并非主要受限于试验误差，而是由于我们只能观测到宇宙的有限区域。重要的是，理论预言并不需要加入自由参数就能和两种信号很好地吻合。但应该强调，这两种方法并不完全独立。因为两者都使用了相同的 CMB 图，并且 X 射线背景与 NVSS 射电源[17]高度相关。如果某一种图的源分布偶然地与 CMB 的扰动相符，那么另一种图也会在某种程度上与 CMB 相关。虽然两个信号的期望幅度的一致性令人振奋，但还不是确定性的结果。另一方面，在两个交叉相关函数中都观测到 ISW 型的信号，强烈地表明这并非由图中未知的系统误差所引起的。射电数据和 X 射线数据是用非常不同的方法所采集的，如果两种巡天图中某种未知的系统误差存在关联，那将是令人惊奇的。我们会在其他论文中给出对这一问题的细致分析。

我们的结论是我们已观测到 ISW 效应。如果确实如此，这是第一次直接观测到 CMB 涨落的产生，也为新标准宇宙学模型——一个加速膨胀的、暗能量占主导的宇宙提供了重要的、独立的支持。应该指出的是，CMB 扰动功率谱的测量并没有如 ISW 效应所预言的那样在大的角尺度上（$\theta > 20°$）上升，而是可能存在扰动功

angular scales ($\theta > 20°$) as predicted by the ISW effect, but rather indicate that there may be power missing[4]. Although this deficit is only at the 2σ level, it is intriguing, and may be telling us something about the formation of the very largest structures in the Universe. The consequences of the ISW effect reported here are primarily on intermediate angular scales, and are not in direct conflict with the power deficit on larger angular scales. Finally, we note that the WMAP-NVSS results have recently been independently analysed by the WMAP team, who find a similar level of correlation[18].

(**427**, 45-47; 2004)

Stephen Boughn[1] & Robert Crittenden[2]

[1] Department of Astronomy, Haverford College, Haverford, Pennsylvania 19041, USA

[2] Institute of Cosmology and Gravitation, University of Portsmouth, Portsmouth PO1 2EG, UK

Received 15 April; accepted 8 October 2003; doi:10.1038/nature02139.

References:

1. Bahcall, N., Ostriker, J. P., Perlmutter, S. & Steinhardt, P. The cosmic triangle: Revealing the state of the universe. *Science* **284**, 1481-1488 (1999).

2. Sachs, R. K. & Wolfe, A. M. Perturbations of a cosmological model and angular variations of the microwave background. *Astrophys. J.* **147**, 73-90 (1967).

3. Crittenden, R. & Turok, N. Looking for a cosmological constant with the Rees-Sciama effect. *Phys. Rev. Lett.* **76**, 575-578 (1996).

4. Bennett, C. L. *et al.* First year *Wilkinson Microwave Anisotropy Probe* observations: Preliminary maps and basic results. *Astrophys. J. Suppl.* **148**, 1-27 (2003).

5. Boldt, E. The cosmic X-ray background. *Phys. Rep.* **146**, 215-257 (1987).

6. Condon, J. *et al.* The NRAO VLA sky survey. *Astron. J.* **115**, 1693-1716 (1998).

7. Kinkhabwala, A. & Kamionkowski, M. New constraint on open cold-dark-matter models. *Phys. Rev. Lett.* **82**, 4172-4175 (1999).

8. Boughn, S. & Crittenden, R. Cross correlation of the CMB with radio sources: Constraints on an accelerating universe. *Phys. Rev. Lett.* **88**, 021302 (2002).

9. Boughn, S., Crittenden, R. & Koehrsen, G. The large scale structure of the X-ray background and its cosmological implications. *Astrophys. J.* **580**, 672-684 (2002).

10. Cowie, L. L., Barger, A. J., Bautz, M. W., Brandt, W. N. & Garmire, G. P. The redshift evolution of the 2-8 keV X-ray luminosity function. *Astrophys. J.* **584**, 57-60 (2003).

11. Dunlop, J. S. & Peacock, J. A. The redshift cut-off in the luminosity function of radio galaxies and quasars. *Mon. Not. R. Astron. Soc.* **247**, 19-42 (1990).

12. Boughn, S., Crittenden, R. & Turok, N. Correlations between the cosmic X-ray and microwave backgrounds: Constraints on a cosmological constant. *New Astron.* **3**, 275-291 (1998).

13. Bennett, C. L. *et al.* First year *Wilkinson Microwave Anisotropy Probe* observations: Foreground emission. *Astrophys. J. Suppl.* **148**, 97-117 (2003).

14. Tegmark, M., de Oliveira-Costa, A. & Hamilton, A. A high resolution foreground cleaned CMB map from WMAP. Preprint at ⟨http://xxx.lanl.gov/astro-ph/0302496⟩ (2003).

15. Peiris, H. V. & Spergel, D. N. Cross-correlating the Sloan Digital Sky Survey with the microwave sky. *Astrophys. J.* **540**, 605-613 (2000).

16. Kogut, A. *et al.* Calibration and systematic error analysis for the COBE DMR 4 year sky maps. *Astrophys. J.* **470**, 653-673 (1996).

17. Boughn, S. Cross-correlation of the 2-10 keV X-ray background with radio sources: Constraining the large-scale structure of the X-ray background. *Astrophys. J.* **499**, 533-541 (1998).

18. Nolta, M.R. *et al.* First year *Wilkinson Microwave Anisotropy Probe* observations: Dark energy induced correlation with radio sources. Preprint at ⟨http://xxx.lanl.gov/astro-ph/0305907⟩ (2003).

19. Spergel, D. N. *et al.* First year *Wilkinson Microwave Anisotropy Probe* observations: Determination of cosmological parameters. *Astrophys. J. Suppl.* **148**, 175-194 (2003).

Supplementary Information accompanies the paper on www.nature.com/nature.

Acknowledgements. We are grateful to M. Nolta, L. Page and the rest of the WMAP team, as well as to N. Turok, B. Partridge and B. Bassett, for useful conversations. R.C. acknowledges financial support from a PPARC fellowship.

Competing interests statement. The authors declare that they have no competing financial interests.

Correspondence and requests for materials should be addressed to R.C. (Robert.Crittenden@port.ac.uk).

率丢失 [4]。虽然亏损的置信水平只有 2σ，但这一现象却十分神秘，这也许告诉了我们一些宇宙中非常大的结构形成的信息。这里对 ISW 效应所下的结论主要在中等角尺度上，并不与更大角尺度上的扰动功率丢失相矛盾。最后，WMAP 团队最近也对 WMAP–NVSS 结果做了独立的分析，他们也发现了类似水平的相关性 [18]。

（周杰 翻译；冯珑珑 审稿）

Infrared Radiation from an Extrasolar Planet

D. Deming *et al.*

Editors's Note

Many of the planets that have been found outside our solar system are "hot Jupiters", with masses similar to that of Jupiter and orbiting very close to their parent stars. Some of these planets' orbits are aligned such that the planet periodically crosses the face of the star (from our perspective) and is then eclipsed by it. Here Drake Deming and colleagues report infrared observations of the planet HD 209458b while it was neither crossing nor eclipsed, so that light from the planet contributed to the total light from the system. By subtracting the light when the planet was in eclipse, they were able for the first time to detect light coming directly from the atmosphere of an extrasolar planet.

A class of extrasolar giant planets—the so-called "hot Jupiters" (ref. 1)—orbit within 0.05 AU of their primary stars (1 AU is the Sun–Earth distance). These planets should be hot and so emit detectable infrared radiation[2]. The planet HD 209458b (refs 3, 4) is an ideal candidate for the detection and characterization of this infrared light because it is eclipsed by the star. This planet has an anomalously large radius (1.35 times that of Jupiter[5]), which may be the result of ongoing tidal dissipation[6], but this explanation requires a non-zero orbital eccentricity (~0.03; refs 6, 7), maintained by interaction with a hypothetical second planet. Here we report detection of infrared (24 μm) radiation from HD 209458b, by observing the decrement in flux during secondary eclipse, when the planet passes behind the star. The planet's 24-μm flux is 55 ± 10 μJy (1σ), with a brightness temperature of $1{,}130 \pm 150$ K, confirming the predicted heating by stellar irradiation[2,8]. The secondary eclipse occurs at the midpoint between transits of the planet in front of the star (to within ± 7 min, 1σ), which means that a dynamically significant orbital eccentricity is unlikely.

OPERATING cryogenically in a thermally stable space environment, the Spitzer Space Telescope[9] has sufficient sensitivity to detect hot Jupiters at their predicted infrared flux levels[8]. We observed the secondary eclipse (hereafter referred to as "the eclipse") of HD 209458b with the 24-μm channel of the Multiband Imaging Photometer for Spitzer (MIPS)[10]. Our photometric time series observations began on 6 December 2004 at 21:29 UTC (Coordinated Universal Time), and ended at approximately 03:23 UTC on 7 December 2004 (5 h 54 min duration). We analyse 1,696 of the 1,728 10-s exposures so acquired, rejecting 32 images having obvious flaws. The Supplementary Information contains a sample image, together with information on the noise properties of the data.

来自一颗系外行星的红外辐射

戴明等

编者按

太阳系外已经发现的行星中，大部分都是"热木星"。它们的质量和木星相近，并以非常近的距离围绕它们的主星公转。从我们的视角来看，这些行星当中的一部分有着"对齐"的公转轨道，这种"对齐"使得它们会周期性地先从主星面前经过，接着被主星所遮掩。本文中，德雷克·戴明和他的同事们报告了他们对行星 HD 209458b 在既没有经过恒星表面，也没有被恒星所遮掩时进行的红外观测。这时，来自该系统的所有光线中有一部分是行星所贡献的。通过扣除行星在被主星遮掩时的光线，他们第一次得以探测到直接来自一颗系外行星大气层的光线。

所谓的"热木星"[1]，是一类太阳系外巨行星。它们在距离主星 0.05 个天文单位的范围内（1 个天文单位指的是太阳到地球的平均距离）围绕主星公转。由于靠近主星，这些行星的温度较高，所以可以发出可探测的红外辐射 [2]。行星 HD 209458b[3,4] 因为会被其主星所遮掩，所以是探测并刻画这种红外辐射的理想候选体。这颗行星的半径较大，略显反常，是木星的 1.35 倍 [5]。这可能是正在发生的潮汐耗散作用的结果 [6]，但是这种解释需要假设存在第二颗行星，以使 HD 209458b 始终保持非零的轨道偏心率（约 0.03[6,7]）。这里我们报告通过观测行星经过恒星背面（即次食）时流量降低的程度而探测到的来自 HD 209458b 的红外辐射（24 微米）。这颗行星的 24 微米流量为 55 ± 10 μJy (1σ)，表面亮温度为 $1,130 \pm 150$ K。这确认了预测存在的由恒星辐射所产生的加热 [2,8]。行星在经过恒星表面的时候会产生凌星现象。由于次食恰好发生在两次凌星的中间时刻（正负偏差在 7 分钟以内，1σ），这就意味着该行星不太可能有一个在动力学上显著的轨道偏心率。

在热稳定的太空环境中低温运行的斯皮策空间望远镜 [9]，它的精度足够在理论预测的红外流量水平对热木星进行探测 [8]。我们利用斯皮策多波段成像光度计（MIPS）[10] 的 24 微米通道，对 HD 209458b 的次食进行了观测。我们的测光时序观测开始于 2004 年 12 月 6 日 UTC 时间（协调世界时）21:29，大致结束于 2004 年 12 月 7 日 UTC 时间 03:23（持续时间 5 小时 54 分钟）。在观测获得的 1,728 张 10 秒曝光图片中，我们舍弃 32 张有明显缺陷的图片，并对剩下的 1,696 张进行研究。补充信息中包含了一张示例图片，并附上了与数据噪声属性相关的信息。

595

We first verify that circumstellar dust does not contribute significantly to the stellar flux. Summing each stellar image over a 13×13 pixel synthetic aperture (33×33 arcsec), we multiply the average sum by 1.15 to account for the far wings of the point spread function (PSF)[11], deriving a flux of 21.17 ± 0.11 mJy. The temperature of the star is close to 6,000 K (ref. 12). At a distance of 47 pc (ref. 13), a model atmosphere[14] predicts a flux of 22 mJy, agreeing with our observed flux to within an estimated ~2 mJy error in absolute calibration. We conclude that the observed flux is dominated by photospheric emission, in agreement with a large Spitzer study of planet-bearing stars at this wavelength[11].

Our time series analysis is optimized for high relative precision. We extract the intensity of the star from each image using optimal photometry with a spatial weighting function[15]. Selecting the Tiny Tim[16] synthetic MIPS PSF for a 5,000-K source at 24 μm, we spline-interpolate it to 0.01 pixel spacing, rebin it to the data resolution, and centre it on the stellar image. The best centring is judged by a least-squares fit to the star, fitting to within the noise level. The best-centred PSF becomes the weighting function in deriving the stellar photometric intensity. We subtract the average background over each image before applying the weights. MIPS data includes per-pixel error estimates[17], which we use in the optimal photometry and to compute errors for each photometric point. The optimal algorithm[15] predicts the signal-to-noise ratio (SNR) for each photometric point, and these average to 119. Our data are divided into 14 blocks, defined by pre-determined raster positions of the star on the detector. To check our SNR, we compute the internal scatter within each block. This gives SNR in the range from 95 to 120 (averaging 111), in excellent agreement with the optimal algorithm. For each point we use the most conservative possible error: either the scatter within that block or the algorithm estimate, whichever is greater. We search for correlations between the photometric intensities and small fluctuations in stellar position, but find none. We also perform simple aperture photometry on the images, and this independent procedure confirms our results, but with 60% greater errors.

The performance of MIPS at 24 μm is known to be excellent[18]. Only one instrument quirk affects our photometry. The MIPS observing sequence obtains periodic bias images, which reset the detector. Images following resets have lower overall intensities (by ~0.1–1%), which recover in later images. The change is common to all pixels in the detector, and we remove it by dividing the stellar intensities by the average zodiacal background in each image. We thereby remove variations in instrument/detector response, both known and unknown. The best available zodiacal model[19] predicts a background increase of 0.18% during the ~6 h

我们首先确认了星周尘对恒星流量没有明显贡献。利用一个大小为 13×13 像素的合成孔径(33×33 角秒),我们在每一张恒星图像上进行求和。考虑到点扩散函数(PSF)线翼远端的贡献,我们又对和的平均值乘以系数 1.15[11],最终得到其流量为 21.17±0.11 mJy。这颗恒星的温度接近 6,000 K[12],在 47 pc[13] 的距离上,利用大气模型[14] 估计的流量为 22 mJy。这个值与我们观测到的流量相吻合,差别在绝对定标时对误差的估计(约为 2 mJy)以内。由此我们得出结论,观测到的流量中光球层发射占主导。在此波长上,该结论与一个利用斯皮策望远镜对带有行星的恒星所进行的大型研究相吻合[11]。

我们的时序分析对高相对精度进行了优化。在从每幅图像中抽取恒星强度的过程中,我们利用了带有空间权重函数的优化测光算法[15]。我们选择了利用 Tiny Tim 软件包[16] 合成 MIPS 点扩散函数,并将源的温度设定为 5,000 K,波长设定为 24 微米。接着,我们在将该点扩散函数样条插值至 0.01 像素的间距,并在重新合并至数据分辨率精度后,将其中心与恒星图像对齐。中心对齐的最好结果是由对恒星的最小二乘拟合来判定的,拟合程度为噪声范围以内。中心对齐最好的点扩散函数成为在计算恒星测光强度过程中的权重函数。在利用权重以前,我们从每幅图像中都扣除了背景平均值。MIPS 数据中包含了每个像素的误差估计[17],我们在优化测光算法和计算每个测光数据点的误差中都使用了这些误差估计。我们使用的优化算法[15]可以估计每个测光数据点的信噪比(SNR),它们的平均值为 119。我们的数据被分为 14 个区块,它们是利用恒星在传感器上预定的光栅位置来确定的。为了检验我们的信噪比,我们在每个区块的内部计算了内部的弥散。计算结果给出的信噪比范围为 95 到 120(平均值为 111),与优化算法给出的结果很好地吻合。在每一个测光点的误差选择上,我们采用了最为保守的方法:选择该点所在的区块内的弥散值和算法给出的误差估计中偏大的结果。在测光强度和细微的恒星位置波动之间,我们也没有找到相关性。我们在恒星图像上还使用了简单的孔径测光法来提取恒星强度。这种方法与前述方法之间是相互独立的,它也确认了我们得出的结果,但是得出的误差却高出 60%。

众所周知,MIPS 在 24 微米处的性能非常优秀[18],只有一处与仪器相关的问题影响测光精度。MIPS 在观测过程中,会周期性地拍摄偏置图像。该过程会重置传感器。传感器在重置过程后所拍摄的图像在总体强度上会有所偏小(约 0.1~1%),但会随着拍摄过程而逐渐恢复。该变化过程在传感器所有像素上都有体现,因此我们对其改正的方法为在每一幅图像上用得到的恒星强度除以平均黄道背景。由此,我们移除了仪器/设备上已知和未知的变化。在可用的黄道模型中,最好的模型[19] 在

of our photometry. Because the star will not share this increase, we remove a 0.18% linear baseline from the stellar photometry. Note that the eclipse involves both a decrease and increase in flux, and its detection is insensitive to monotonic linear baseline effects.

To detect weak signals reliably requires investigating the nature of the errors. We find that shot noise in the zodiacal background is the dominant source of error; systematic effects are undetectable after normalizing any individual pixel to the total zodiacal background. All of our results are based on analysis of the 1,696 individual photometric measurements versus heliocentric phase from a recent ephemeris[20] (Fig. 1a). We propagate the individual errors (not shown on Fig. 1a) through a transit curve fit to calculate the error on the eclipse depth. Because about half of the 1,696 points are out of eclipse, and half are in eclipse, and the SNR ≈ 111 per point, the error on the eclipse depth should be $\sim 0.009 \times 2^{0.5}/848^{0.5} = 0.044\%$ of the stellar continuum. Model atmospheres for hot Jupiters[2,8,21-24] predict eclipse depths in the range from 0.2–0.4% of the stellar continuum, so we anticipate a detection of 4–9σ significance. The eclipse is difficult to discern by eye on Fig. 1a, because the observed depth (0.26%) is a factor of 4 below the scatter of individual points. We use the known period (3.524 days) and radii[5] to fit an eclipse curve to the Fig. 1a data, varying only the eclipse depth, and constraining the central phase to 0.5. This fit detects the eclipse at a depth of 0.26% $\pm$ 0.046%, with a reduced χ^2 of 0.963, denoting a good fit. Note that the 5.6σ significance applies to the aggregate result, not to individual points. The eclipse is more readily seen by eye on Fig. 1b, which presents binned data and the best-fit eclipse curve. The data are divided into many bins, so the aggregate 5.6σ significance is much less for a single bin (SNR ≈ 1 per point). Nevertheless, the dip in flux due to the eclipse is apparent, and the observed duration is approximately as expected. As a check, we use a control photometric sequence (Fig. 1b) to eliminate false positive detection of the eclipse due to instrument effects. We also plot the distribution of points in intensity for both the in-eclipse and out-of-eclipse phase intervals (Fig. 1c). This shows that the entire distribution shifts as expected with the eclipse, providing additional discrimination against a false positive detection.

我们约 6 小时的测光观测中所预测的背景强度升高 0.18%。由于恒星本身并不受这种增加的影响，所以我们在恒星测光中移除了一条 0.18% 的线性基线。需要指出的是，次食过程既包含了流量下降的过程也包含了流量上升的过程，而探测次食现象并不受单调的线性基线影响。

探究误差的本质在可靠地探测较弱信号的过程中是必需的。我们发现黄道背景中的散粒噪声是误差的最主要来源；把所有像素以黄道背景总和为标准进行归一化后，来源于系统误差的影响均无法探测到。我们所有的结果都是从 1,696 张独立测光观测中获得的。这些测光观测对应于该系统最近的一个星历测定工作所得出的日心相位[20]（图 1a）。为了计算次食时掩食深度的误差，我们通过拟合凌星曲线来传递每个测光数据点的误差（误差并没有在图 1a 中标注）。因为在 1,696 个数据点中，次食内和次食外的点各占一半左右，并且每个点的信噪比约为 111，所以次食时的掩食深度误差应为恒星连续谱的 $\sim 0.009 \times 2^{0.5} / 848^{0.5} = 0.044\%$。热木星的大气模型[2,8,21-24]对次食时的掩食深度的预计范围为恒星连续谱的 0.2 ~ 0.4%，所以我们估计本次探测的显著程度应在 4 ~ 9σ 之间。由于观测到的次食深度（0.26%）仅为单独数据点的弥散程度的 1/4 左右，在图 1a 上很难用肉眼识别出次食的形状。我们使用已知的周期（3.524 天）和半径[5]在图 1a 的数据上进行次食曲线的拟合。拟合过程中，我们将相位中心固定至 0.5，只调整次食的掩食深度。该拟合结果较好，探测到的次食深度为 0.26% ± 0.046%，约化 χ^2 为 0.963。需要指出的是，5.6σ 的显著程度适用于累计的结果，并不适用于单独的数据点。图 1b 展示了合并以后的数据和最佳拟合的次食曲线，在图 1b 上更容易用肉眼观测到次食。因为数据点被分隔至很多个区间中，所以累计的 5.6σ 显著程度对每一个单独的区间要小很多（每个点信噪比约为 1）。尽管如此，由次食所造成的流量下降是明显的，而且观测到的次食时长也与预期基本相符。作为检查手段，我们使用了一个对照测光序列（图 1b）来消除由于可能的仪器效应所造成的次食假阳性探测。我们还按照强度区间，画出了次食相位内和次食相位外的数据点的分布（图 1c）。该图展示了整个分布会根据次食相位的不同而移动，这是与预期相吻合的，也提供了探测到的次食并非假阳性结果的额外证据。

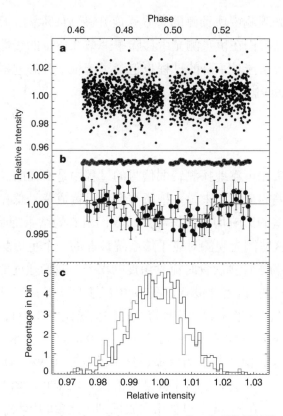

Fig. 1. Observations showing our detection of the secondary eclipse in HD 209458. **a**, Relative intensities versus heliocentric phase (scale at top) for all 1,696 data points. The phase is corrected for light travel time at the orbital position of the telescope. Error bars are suppressed for clarity. The gap in the data near phase 0.497 is due to a pause for telescope overhead activity. The secondary eclipse is present, but is a factor of ~4 below the ~1% noise level of a single measurement. **b**, Intensities from **a**, averaged in bins of phase width 0.001 (scale at top), with 1σ error bars computed by statistical combination from the errors of individual points. The red line is the best-fit secondary eclipse curve (depth = 0.26%), constrained to a central phase of 0.5. The points in blue are a control sequence, summing intensities over a 10×10-pixel region of the detector, to beat down the random errors and reveal any possible systematic effects. The control sequence uses the same detector pixels, on average, as those where the star resides, but is sampled out of phase with the variations in the star's raster motion during the MIPS photometry cycle. **c**, Histograms of intensity (lower abscissa scale) for the points in **a**, with bin width 0.1%, shown separately for the out-of-eclipse (black) and in-eclipse intervals (red).

We further illustrate the reality of the eclipse on Fig. 2. Now shifting the eclipse curve in phase, we find the best-fitting amplitude and χ^2 at each shift. This determines the best-fit central phase for the eclipse, and also further illustrates the statistical significance of the result. The thick line in Fig. 2a shows that the maximum amplitude (0.26%) is obtained at exactly phase 0.5 (which is also the minimum of χ^2). Further, we plot the eclipse "amplitude" versus central phase using 100 sets of synthetic data, consisting of gaussian noise with dispersion matching the real data, but without an eclipse. The amplitude (0.26%) of the eclipse in the real data stands well above the statistical fluctuations in the synthetic data.

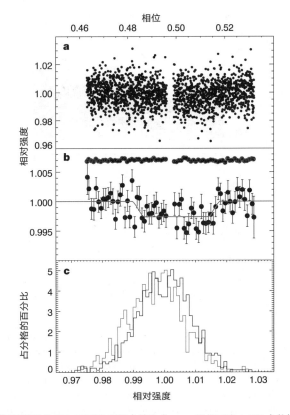

图 1. 观测展示了我们探测到的 HD 209458 系统中的次食。**a**，我们画出 1,696 个数据点的相对强度与该系统日心坐标系相位的关系图（横坐标在图的上方）。我们根据望远镜轨道位置的光传播时间对相位进行了修正。为将图像展示得更明了，我们没有画误差棒。相位 0.497 附近的数据间断是由于望远镜的非观测性活动造成的暂停产生的。系统的次食是存在的，但是仅为单独数据点噪声水平（约 1%）的 1/4 左右。**b**，将图 **a** 中的强度以 0.001 相位分格平均于每个小格中（横坐标在图的上方），同时也画出了 1σ 误差棒。误差棒是用统计的方法将单独点的误差合并计算得到的。红色线为最佳拟合的次食曲线（掩食深度为 0.26%），其中心相位被限制于 0.5。通过对传感器上 10×10 像素区域的强度求和，我们得到了一个对照序列，在图中是以蓝色显示的。我们使用该对照序列来减小随机误差的影响，同时揭示可能存在的系统误差影响。对照序列在传感器上使用的像素与恒星星像所处的像素基本相同，但在 MIPS 测光观测周期对它们采样时，恒星在传感器上的光栅位置是恰好错开的。**c**，用图 **a** 中的强度数据画出的直方图（横坐标位于图的下方），分格大小为 0.1%。次食区间内和次食区间外分别用红色和黑色标出。

我们利用图 2 来更深入地阐明次食的实际情况。我们移动次食曲线至不同的相位位置，并在每个移动的位置上找到拟合最好的幅度和 χ^2。该操作可以确定最佳拟合的次食中心相位，并可以进一步阐释结果的统计显著性。图 2a 中的粗线显示最大的幅度（0.26%）是准确地在 0.5 相位处获得的（该位置也是 χ^2 最小值获得处）。更进一步，我们用 100 组合成数据画出次食的"幅度"相对于中心相位的图像。这些数据包含了与真实数据相当的高斯噪声弥散，但并不含有次食事件。真实数据中次食的掩食深度（0.26%）显著高于合成数据的统计波动。

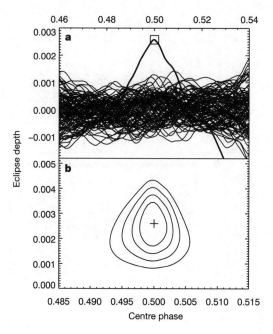

Fig. 2. Amplitude of the secondary eclipse versus assumed central phase, with confidence intervals for both. **a**, The darkest line shows the amplitude of the best-fit eclipse curve versus the assumed central phase (scale at top). The overplotted point marks the fit having smallest χ^2, which also has the greatest eclipse amplitude. The numerous thinner black lines show the effect of fitting to 100 synthetic data sets containing no eclipse, but having the same per-point errors as the real data. Their fluctuations in retrieved amplitude versus phase are indicative of the error in eclipse amplitude, and are consistent with $\sigma = 0.046\%$. Note that the eclipse amplitude found in the real data (0.26%) stands well above the error envelope at phase 0.5. **b**, Confidence intervals at the 1, 2, 3 and 4σ levels for the eclipse amplitude and central phase (note expansion of phase scale, at bottom). The plotted point marks the best fit (minimum χ^2) with eclipse depth of 0.26%, and central phase indistinguishable from 0.5. The centre of the eclipse occurs in our data at Julian day 2453346.5278.

Figure 2b shows confidence intervals on the amplitude and central phase, based on the χ^2 values. The phase shift of the eclipse is quite sensitive to eccentricity (e) and is given[25] as $\Delta t = 2Pe\cos(\omega)/\pi$, where P is the orbital period, and ω is the longitude of periastron. The Doppler data alone give $e = 0.027 \pm 0.015$ (Laughlin, G., personal communication), and allow a phase shift as large as ± 0.017 (87 min). We find the eclipse centred at phase 0.5, and we checked the precision using a bootstrap Monte Carlo procedure[26]. The 1σ phase error from this method is 0.0015 (~7 min), consistent with Fig. 2b. A dynamically significant eccentricity, $e \approx 0.03$ (refs 6, 7), constrained by our 3σ limit of $\Delta t < 21$ min, requires $|(\omega - \pi/2)| < 12$ degrees and is therefore only possible in the unlikely case that our viewing angle is closely parallel to the major axis of the orbit. A circular orbit rules out a promising explanation for the planet's anomalously large radius: tidal dissipation as an interior energy source to slow down planetary evolution and contraction[7]. Because the dynamical time for tidal decay to a circular orbit is short, this scenario posited the presence of a perturbing second planet in the system to continually force the eccentricity— a planet that is no longer necessary with a circular orbit for HD 209458b.

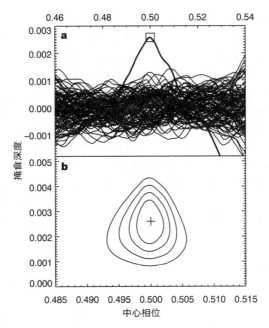

图 2. 次食深度与推定的中心相位关系图和二者各自的置信区间图。**a**，颜色最深的线展示了最佳拟合的次食曲线的幅度和推定的中心相位之间的关系（横坐标位于图的上方）。图中覆盖在粗线上方的点代表 χ^2 最小的拟合，该拟合结果的掩食深度也最大。旁边多条稍细的黑线展示了利用 100 组人造数据获得的结果。这些数据的单点误差和真实数据相同，但并不包含次食事件。人造数据拟合得到的掩食深度和相位之间的波动与 $\sigma = 0.046\%$ 相符，且可以大概标明从真实数据中得到的掩食深度的误差范围。可以看到，真实数据中得到的次食深度（0.26%）明显高于 0.5 相位处的误差范围。**b**，次食深度和中心相位分别在 1、2、3、4σ 处的置信区间（横坐标位于下方，请注意相位尺度的放大）。图中画出的点代表最佳拟合结果（最小的 χ^2），此时的次食深度为 0.26%，中心相位基本位于 0.5。在我们的数据中，次食中心发生在儒略日 2453346.5278。

　　图 2b 展示了幅度和中心相位基于 χ^2 的置信区间。次食的相移对于偏心率（e）较为敏感[25]，其表达式为 $\Delta t = 2Pe\cos(\omega)/\pi$，其中 P 为轨道周期，ω 为近星点经度。单独的多普勒数据给出的偏心率是 $e = 0.027 \pm 0.015$（劳克林，个人交流），该值所允许的最大相移为 ± 0.017（87 分钟）。我们发现掩食的中心在相位 0.5 处，然后利用自举蒙特卡罗方法来检测该结果的精度[26]。该方法所给出的 1σ 相位误差为 0.0015（约为 7 分钟），与图 2b 一致。利用前述结果可知 3σ 限制所对应的 $\Delta t < 21$ 分钟。在该约束条件下，若该系统有一个动力学显著的偏心率 $e \approx 0.03$[6,7]，则要求有 $|\omega - \pi/2| < 12°$。该条件成立的可能性不高，因为它只有在我们的观测方向和行星公转轨道的半长径方向较为平行的特殊情况下才可以满足。由于潮汐耗散作为内部能量源可以减慢行星演化和收缩，它是该行星拥有异常大半径的有效解释，但是该行星的圆轨道排除了这种可能性[7]。因为潮汐对轨道圆化的动力学时标较短，因此如果该行星有动力学显著的偏心率，则该偏心率需假定系统中还有第二个持续对 HD 209458b 摄动的行星来维持。但对于圆轨道的 HD 209458b 来说，第二颗行星便不再必要。

The infrared flux from the planet follows directly from our measured stellar flux (21.2 mJy) and the eclipse depth (0.26%), giving 55 ± 10 μJy. The error is dominated by uncertainty in the eclipse depth. Using the planet's known radius[5] and distance[13], we obtain a brightness temperature $T_{24} = 1,130 \pm 150$ K, confirming heating by stellar irradiation[2]. Nevertheless, T_{24} could differ significantly from the temperature of the equivalent blackbody (T_{eq}), that is, one whose bolometric flux is the same as the planet. Without measurements at shorter wavelengths, a model atmosphere must be used to estimate T_{eq} from the 24-μm flux. One such model is shown in Fig. 3, having $T_{eq} = 1,700$ K. This temperature is much higher than T_{24} (1,130 K) due to strong, continuous H_2O vapour absorption at 24 μm. The bulk of the planetary thermal emission derives ultimately from re-radiated stellar irradiation, and is emitted at 1–4 μm, between H_2O bands. However, our 24-μm flux error admits a range of models, including some with a significantly lower T_{eq} (for example, but not limited to, models with reflective clouds or less H_2O vapour).

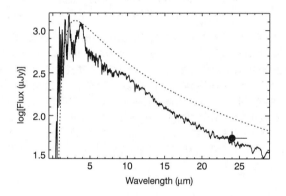

Fig. 3. Flux from a model atmosphere shown in comparison to our measured infrared flux at 24 μm. A theoretical spectrum (solid line) shows that planetary emission (dominated by absorbed and re-radiated stellar radiation) should be very different from a blackbody. Hence, models are required to interpret the 24-μm flux measurement in terms of the planetary temperature. The model shown has $T_{eq} = 1,700$ K and was computed from a one-dimensional plane-parallel radiative transfer model, considering a solar system abundance of gases, no clouds, and the absorbed stellar radiation re-emitted on the day side only. Note the marked difference from a 1,700-K blackbody (dashed line), although the total flux integrated over the blackbody spectrum is equal to the total flux integrated over the model spectrum. (The peaks at short wavelength dominate the flux integral in the atmosphere model, note log scale in the ordinate.) The suppressed flux at 24 μm is due to water vapour opacity. This model lies at the hot end of the range of plausible models consistent with our measurement, but the error bars admit models with cooler T_{eq}.

Shortly after submission of this Letter, we became aware of a similar detection for the TrES-1 transiting planet system[27] using Spitzer's Infrared Array Camera[28]. Together, these Spitzer results represent the first measurement of radiation from extrasolar planets. Additional Spitzer observations should rapidly narrow the range of acceptable models, and reveal the atmospheric structure, composition, and other characteristics of close-in extrasolar giant planets.

(**434**, 740-743; 2005)

在测量出该行星主星的红外流量（21.2 mJy）和掩食深度（0.26%）后，来自行星的红外流量便可以直接得出，为 55 ± 10 μJy。误差主要来源于掩食深度的不确定性。利用该行星已知的半径[5]和距离[13]，我们可以得出其亮温度为 $T_{24} = 1{,}130 \pm 150$ K。这个结果证实了恒星辐射的加热作用[2]。尽管如此，T_{24} 也有可能与等效的黑体温度（T_{eq}），即用行星的热流量计算出的温度，有着显著的差别。在缺乏短波长区域测量的情况下，我们必须使用行星大气模型结合 24 微米处的流量来估计 T_{eq}。图 3 中展示了我们使用的一种模型，它得出的 $T_{eq} = 1{,}700$ K。该温度明显高出 T_{24}（1,130 K），主要原因是在 24 微米处有着水蒸气连续且较强的吸收。行星热辐射的主要部分最终的源头是对恒星辐射的再辐射，主要集中在 1~4 微米之间，介于水的波段之间。但是，我们所计算出的 24 微米处流量的误差使得很多种模型都有可能，其中就包括一些可以产生明显更低的 T_{eq} 的模型（比如有着反射性云层或者更少水蒸气的模型，但并不局限于此）。

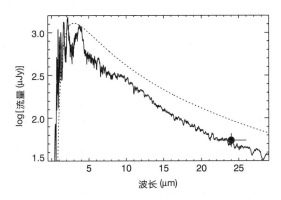

图 3. 利用大气模型得到的流量与我们在 24 微米处测量得到流量的比较。该理论光谱（实线）表明行星辐射（由吸收和再辐射恒星辐射占主导）应与黑体非常不同。因此，合理的模型必须能够用行星温度来解释 24 微米处测量到的流量。图中所展示的是一个一维平面平行辐射转移模型的结果。在计算中，我们设定的条件为已知太阳系的气体丰度，无云，且只有行星的向昼侧吸收恒星辐射并再辐射。此时得到的 $T_{eq} = 1{,}700$ K。注意一个 1,700 K 的黑体所发出的辐射（虚线）与模型的不同，尽管二者积分得到的总流量是相等的。（短波长附近的尖峰在大气模型的积分总流量中占据主导地位，请注意纵坐标为对数坐标）。24 微米处流量偏低是由于水蒸气的不透明度造成的。该模型在一系列与我们的测量相符合的可能模型中是偏热的，但是测量得到的误差棒是允许有其他 T_{eq} 较低的模型存在的。

在我们刚提交这篇文章后不久，我们了解到其他研究组使用了斯皮策望远镜[28]的红外阵列相机对 TrES-1 凌星行星系有过相似的观测[27]。与该工作一起，我们使用斯皮策望远镜得出的结果代表着首次对系外行星辐射的测量。更多使用斯皮策望远镜的观测应该会迅速地将可信模型的范围缩窄，同时也会揭示出密近太阳系外巨行星的大气结构、成分和其他的特性等信息。

（梁恩思 翻译；周济林 审稿）

Drake Deming[1], Sara Seager[3], L. Jeremy Richardson[2] & Joseph Harrington[4]

[1] Planetary Systems Laboratory and Goddard Center for Astrobiology, Code 693; [2] Exoplanet and Stellar Astrophysics Laboratory, Code 667, NASA's Goddard Space Flight Center, Greenbelt, Maryland 20771, USA

[3] Department of Terrestrial Magnetism, Carnegie Institution of Washington, 5241 Broad Branch Road NW, Washington DC 20015, USA

[4] Center for Radiophysics and Space Research, Cornell University, 326 Space Sciences Bldg, Ithaca, New York 14853-6801, USA

Received 3 February; accepted 28 February 2005; doi:10.1038/nature03507.
Published online 23 March 2005.

References:

1. Collier-Cameron, A. Extrasolar planets: what are hot Jupiters made of? *Astron. Geophys.* **43**, 421-425 (2002).

2. Seager, S. & Sasselov, D. D. Extrasolar giant planets under strong stellar irradiation. *Astrophys. J.* **502**, L157-L161 (1998).

3. Charbonneau, D., Brown, T. M., Latham, D. W. & Mayor, M. Detection of planetary transits across a sun-like star. *Astrophys. J.* **529**, L45-L48 (2000).

4. Henry, G. W., Marcy, G. W., Butler, R. P. & Vogt, S. S. A transiting "51 Peg-like" planet. *Astrophys. J.* **529**, L41-L44 (2000).

5. Brown, T. M., Charbonneau, D., Gilliland, R. L., Noyes, R. W. & Burrows, A. Hubble Space Telescope time-series photometry of the transiting planet of HD 209458. *Astrophys. J.* **552**, 699-709 (2001).

6. Bodenheimer, P., Lin, D. N. C. & Mardling, R. A. On the tidal inflation of short-period extrasolar planets. *Astrophys. J.* **548**, 466-472 (2001).

7. Laughlin, G. *et al.* A comparison of observationally determined radii with theoretical radius predictions for short-period transiting extrasolar planets. *Astrophys. J.* (in the press).

8. Burrows, A., Sudarsky, D. & Hubeny, I. in *The Search for Other Worlds: Proc. 14th Annu. Astrophys. Conf. in Maryland* (eds Holt, S. & Deming, D.) Vol. 713, 143-150 (American Institute of Physics, Melville, New York, 2003).

9. Werner, M. W. *et al.* The Spitzer Space Telescope mission. *Astrophys. J. Suppl.* **154**, 1-9 (2004).

10. Rieke, G. H. *et al.* The Multiband Imaging Photometer for Spitzer (MIPS). *Astrophys. J. Suppl.* **154**, 25-29 (2004).

11. Beichman, C. A. *et al.* Planets and IR excesses: preliminary results from a *Spitzer* MIPS survey of solar-type stars. *Astrophys. J.* (in the press).

12. Ribas, A. H., Solano, E., Masana, E. & Gimenez, A. Effective temperatures and radii of planet-hosting stars from IR photometry. *Astron. Astrophys.* **411**, L501-L504 (2003).

13. Perryman, M. A. C. (ed.) *The Hipparcos and Tycho Catalogues* (ESA SP-1200, European Space Agency, Noordwijk, 1997).

14. Kurucz, R. *Solar Abundance Model Atmospheres for 0, 1, 2, 4, and 8 km/s* CD-ROM 19 (Smithsonian Astrophysical Observatory, Cambridge, Massachusetts, 1994).

15. Horne, K. An optimal extraction algorithm for CCD spectrososcopy. *Publ. Astron. Soc. Pacif.* **98**, 609-617 (1986).

16. Krist, J. in *Astronomical Data Analysis Software and Systems IV* (eds Shaw, R. A., Payne, H. E. & Hayes, J. J. E.) Vol. 77, 349-352 (Astronomical Society of the Pacific, San Francisco, 1995).

17. Gordon, K. D. Reduction algorithms for the Multiband Imaging Photometer for SIRTF. *Publ. Astron. Soc. Pacif.* (in the press).

18. Rieke, G. H. *et al.* in *Proc. SPIE: Optical, Infrared, and Millimeter Space Telescopes* (ed. Mather, J. C.) Vol. 5487, 50-61 (SPIE, Bellingham, Washington, 2004).

19. Kelsall, T. *et al.* The COBE Diffuse Infrared Background Experiment (DIRBE) search for the cosmic infrared background. II. Model of the interplanetary dust cloud. *Astrophys. J.* **508**, 44-73 (1998).

20. Wittenmyer, R. A. *The Orbital Ephemeris of HD 209458b*. Master's thesis, San Diego State Univ. (2003).

21. Goukenleuque, C., Bezard, B., Joguet, B., Lellouch, E. & Freedman, R. A radiative equilibrium model of 51 Peg b. *Icarus* **143**, 308-323 (2000).

22. Seager, S., Whitney, B. A. & Sasselov, D. D. Photometric light curves and polarization of close-in extrasolar giant planets. *Astrophys. J.* **540**, 504-520 (2000).

23. Barman, T. S., Hauschildt, P. H. & Allard, F. Irradiated planets. *Astrophys. J.* **556**, 885-895 (2001).

24. Sudarsky, D., Burrows, A. & Hubeny, I. Theoretical spectra and atmospheres of extrasolar giant planets. *Astrophys. J.* **588**, 1121-1148 (2003).

25. Charbonneau, D. in *Scientific Frontiers in Research on Extrasolar Planets* (eds Deming, D. & Seager, S.) Vol. 294, 449-456 (Astronomical Society of the Pacific, San Francisco, 2003).

26. Press, W. H., Teukolsky, S. A., Vettering, W. T. & Flannery, B. P. *Numerical recipes in C* 2nd edn (Cambridge Univ. Press, Cambridge, 1992).

27. Alonso, R. *et al.* TrES-1, the transiting planet of a bright K0V star. *Astrophys. J.* **613**, L153-L156 (2004).

28. Charbonneau, D. *et al.* Detection of thermal emission from an extrasolar planet. *Astrophys. J.* (in the press).

Supplementary Information accompanies the paper on www.nature.com/nature.

Acknowledgements. We thank G. Laughlin for communicating the latest orbital eccentricity solutions from the Doppler data and for his evaluation of their status. We acknowledge informative conversations with D. Charbonneau, G. Marcy, B. Hansen, K. Menou and J. Cho. This work is based on observations made with the

Spitzer Space Telescope, which is operated by the Jet Propulsion Laboratory, California Institute of Technology, under contract to NASA. Support for this work was provided directly by NASA, and by its Origins of Solar Systems programme and Astrobiology Institute. We thank all the personnel of the Spitzer telescope and the MIPS instrument, who ultimately made these measurements possible. L.J.R. is a National Research Council Associate at NASA's Goddard Space Flight Center.

Competing interests statement. The authors declare that they have no competing financial interests.

Correspondence and requests for materials should be addressed to D.D. (Leo.D.Deming@nasa.gov).

Origin of the Cataclysmic Late Heavy Bombardment Period of the Terrestrial Planets

R. Gomes *et al.*

Editor's Note

The Moon seems to have experienced a large increase in strikes by comets and asteroids about 700 million years after the formation of the solar system. It is hard to understand why this is so; but here Alessandro Morbidelli and colleagues propose an explanation. They show that if the giant planets migrated slowly (Jupiter inward, Saturn, Uranus and Neptune outward), they could hit a "resonance" between the orbits of Jupiter and Saturn that would lead to very rapid migration. This gravitational churning of the solar system would cause objects from the disk of debris outside the orbit of Neptune to be pulled into the inner solar system, increasing the rate of impacts on planets and moons.

The petrology record on the Moon suggests that a cataclysmic spike in the cratering rate occurred ~700 million years after the planets formed[1]; this event is known as the Late Heavy Bombardment (LHB). Planetary formation theories cannot naturally account for an intense period of planetesimal bombardment so late in Solar System history[2]. Several models have been proposed to explain a late impact spike[3-6], but none of them has been set within a self-consistent framework of Solar System evolution. Here we propose that the LHB was triggered by the rapid migration of the giant planets, which occurred after a long quiescent period. During this burst of migration, the planetesimal disk outside the orbits of the planets was destabilized, causing a sudden massive delivery of planetesimals to the inner Solar System. The asteroid belt was also strongly perturbed, with these objects supplying a significant fraction of the LHB impactors in accordance with recent geochemical evidence[7,8]. Our model not only naturally explains the LHB, but also reproduces the observational constraints of the outer Solar System[9].

PREVIOUS work[9] explains the current orbital architecture of the planetary system by invoking an initially compact configuration in which Saturn's orbital period was less than twice that of Jupiter. After the dissipation of the gaseous circumsolar nebula, Jupiter's and Saturn's orbits diverged as a result of their interaction with a massive disk of planetesimals, and thus the ratio of their orbital periods, P_S/P_J, increased. When the two planets crossed their mutual 1:2 mean motion resonance (1:2 MMR, that is, $P_S/P_J = 2$) their orbits became eccentric. This abrupt transition temporarily destabilized the giant planets, leading to a short phase of close encounters among Saturn, Uranus and Neptune. As a result of these encounters, and of the interactions of the ice giants with the disk, Uranus

608

类地行星激变晚期重轰击期的起源

戈梅斯等

编者按

月球似乎在太阳系形成之后大约 7 亿年时，遭受到的彗星和小行星的撞击大大增加。这一现象的原因很难解释。但本文中亚历山德罗·莫尔比代利和他的同事就这一现象给出了一个解释。他们的研究显示，如果巨行星缓慢迁移（木星向内迁移，土星、天王星和海王星向外迁移），它们将使得木星和土星的轨道之间形成一种"共鸣"，从而导致非常迅速的迁移。这种太阳系重力的扰动会造成海王星轨道外碎屑盘的天体被吸引到内太阳系，从而增加了其对行星和月球造成影响的概率。

针对月球岩石学记录的研究表明，在行星形成后的大约 7 亿年，月球陨击率激增 [1]，这一事件被称为晚期重轰击（LHB）。行星形成理论尚不能合理解释为何在太阳系演化历史如此晚的阶段，会突然出现这样强的星子轰击期 [2]。为此，研究人员提出了若干理论解释这一晚期的撞击突增 [3-6]，然而目前还尚无一种理论建立在太阳系演化的自洽框架之内。本文中，我们提出 LHB 是由于巨行星的快速迁移触发的，该时期出现于长时间的宁静期之后。在这次迁移爆发的过程中，位于行星轨道外侧的星子盘变得不稳定，从而导致星子突然向太阳系内侧运动。同时，小行星带也受到强烈的扰动，其中的天体构成了 LHB 撞击体的绝大部分，这与最近发现的地球化学证据相符 [7-8]。我们的模型不仅自然解释了 LHB，而且重现了外太阳系的观测约束 [9]。

已有的研究 [9] 通过引入一个最初致密的构型，在这一构型中土星的轨道周期小于木星轨道周期的两倍，从而可以解释当前行星系统的轨道体系。当气态环日星云耗散后，木星、土星与大质量星子盘发生相互作用，导致二者的轨道发生背离，所以二者的轨道周期之比 P_S/P_J 也随之增加。当木星和土星越过其 1:2 平运动共振（MMR）时（即 $P_S/P_J = 2$），这两颗行星轨道变得越来越椭。这一突然转变暂时使得巨行星变得不稳定，从而导致土星、天王星以及海王星在短时间轨道发生近距离交会。由于上述交会以及带有盘的冰巨行星的相互作用，天王星和海王星到达目前的日心

609

and Neptune reached their current heliocentric distances and Jupiter and Saturn evolved to their current orbital eccentricities[9]. The main idea of this Letter is that the same planetary evolution could explain the LHB, provided that Jupiter and Saturn crossed the 1:2 MMR roughly 700 Myr after they formed. Thus, our goal is to determine if there is a generic mechanism that could delay the migration process.

In previous studies[9-12], planet migration started immediately because planetesimals were placed close enough to the planets to be violently unstable. Although this type of initial condition was reasonable for the goals of those studies, it is unlikely. Planetesimal-driven migration is probably not important for planet dynamics as long as the gaseous massive solar nebula exists. The initial conditions for the migration simulations should represent the system that existed at the time the nebula dissipated. Thus, the planetesimal disk should contain only those particles that had dynamical lifetimes longer than the lifetime of the solar nebula. In planetary systems like those we adopt from ref. 9, we find that they had to be beyond ~15.3 AU (Fig. 1), leading to the initial conditions illustrated in Fig. 2a.

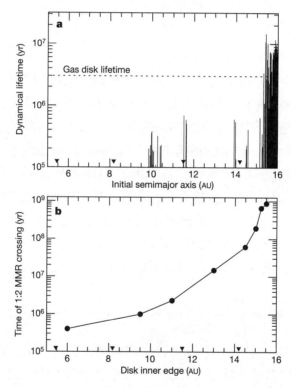

Fig. 1. Disk location and LHB timing. **a**, The histogram reports the average dynamical lifetime of massless test particles placed in a planetary system (shown as triangles) with Jupiter, Saturn and the ice giants on nearly circular, co-planar orbits at 5.45, 8.18, 11.5 and 14.2 AU, respectively. Initially, we placed 10 particles with $e = i = 0$ (where e is eccentricity and i is inclination) and random mean anomaly at each semimajor axis. Stable Trojans of the planets have been removed from this computation. Each vertical bar in the plot represents the average lifetime for those 10 particles. We define "dynamical lifetime" as the time required for a particle to encounter a planet within a Hill radius. A comparison between the histogram and the putative lifetime of the gaseous nebula[20] argues that, when the latter dissipated, the inner edge of

距，而木星和土星轨道则经过演化形成目前的偏心率[9]。本文的主要观点是，以上行星演化理论同样也可以用于解释 LHB，不过需要满足以下条件：木星和土星需要在形成后大约 7 亿年越过其 1:2 平运动共振。因此，我们的研究目的是确定是否存在可以推迟迁移进程的一般机制。

在以往的研究中[9-12]，星子与行星的距离被认为足够近，从而使行星变得非常不稳定，进而导致行星瞬间开始迁移。尽管这种初始条件对那些研究而言也是合理的，但是实际上却不可能发生。只要大量的气态大质量太阳星云还存在，那么星子驱动的迁移对于行星动力学而言可能就不会很重要。用于仿真模拟迁移过程的初始条件应该表征星云耗散时刻的系统状态。因此，星子盘应该只含有动力学寿命长于太阳星云的那些粒子。在类似被采纳的行星系统中[9]，我们发现这些星子的距离超过 ~15.3 AU（图 1），图 2a 对相应的初始条件进行了说明。

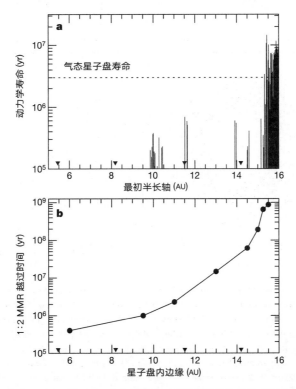

图 1. 星子盘的位置和 LHB 发生时间。**a**，直方图给出了无质量检验粒子的平均动力学寿命；这些粒子被置于与木星、土星和冷巨行星几乎共面的圆形轨道上，它们的轨道距离分别为 5.45 AU、8.18 AU、11.5 AU、14.2 AU，如图中三角所示。最初，我们放置了 10 颗粒子，其偏心率 e 和倾角 i 均为零，每个半长轴处的平近点角随机取值。在该计算中，我们没有考虑稳定的特洛伊型小行星。图中的每条垂直线代表着 10 颗粒子的平均寿命。我们将"动力学寿命"定义为粒子在希尔半径内交会行星所需的时间。通过对比直方图和推定的气体星云的寿命可知[20]，当星云耗散后，星子盘内边缘将超过最外层冰

the planetesimal disk had to be about 1–1.5 AU beyond the outermost ice giant. **b**, Time at which Jupiter and Saturn crossed the 1:2 MMR, as a function of the location of the planetesimal disk's inner edge, as determined from our first set of migration simulations. In all cases, the disk had a surface density equivalent to 1.9 M_E per 1 AU annulus. The outer edge of the disk was varied so that the total mass of the disk was 35 M_E. The disk was initially very dynamically cold, with $e = 0$ and $i < 0.5°$. A comparison between **a** and **b** shows that a disk that naturally should exist when the nebula dissipated would produce a 1:2 MMR crossing at a time comparable to that of the LHB event.

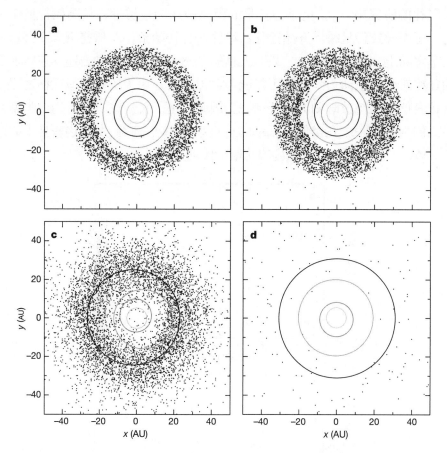

Fig. 2. The planetary orbits and the positions of the disk particles, projected on the initial mean orbital plane. The four panels correspond to four different snapshots taken from our reference simulation. In this run, the four giant planets were initially on nearly circular, co-planar orbits with semimajor axes of 5.45, 8.18, 11.5 and 14.2 AU. The dynamically cold planetesimal disk was 35 M_E, with an inner edge at 15.5 AU and an outer edge at 34 AU. Each panel represents the state of the planetary system at four different epochs: **a**, the beginning of planetary migration (100 Myr); **b**, just before the beginning of LHB (879 Myr); **c**, just after the LHB has started (882 Myr); and **d**, 200 Myr later, when only 3% of the initial mass of the disk is left and the planets have achieved their final orbits.

In this configuration, the initial speed of migration would be dependent on the rate at which disk particles evolve onto planet-crossing orbits. The time at which Jupiter and Saturn cross their 1:2 MMR depends on: (1) their initial distance from the location of the resonance, (2) the surface density of the disk near its inner edge, and (3) the relative location of the inner edge of the disk and the outer ice giant. On the basis of the above arguments, we initially

巨行星大约 1~1.5 AU。**b**，图中给出了木星和土星越过 1∶2 MMR 的时间与星子盘内边缘位置的函数关系，这一关系由我们第一组迁移的仿真模拟所确认。在所有的情况下，星子盘的密度等价于每 1 AU 环带 1.9 M_E。星子盘在最初动力学上是冷的，其偏心率 $e = 0$，倾角 $i < 0.5°$。比较图 **a** 和图 **b** 可知，当星云耗散后，自然存在的星子盘就会在与 LHB 事件相当的时间里越过 1∶2 MMR。

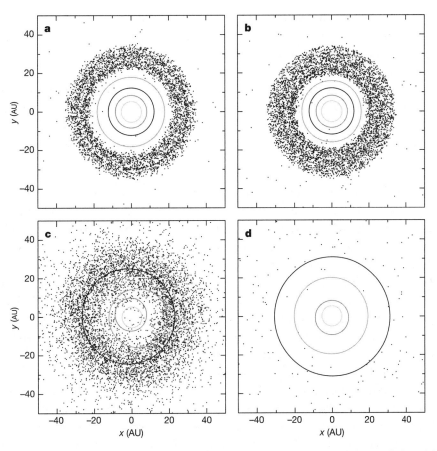

图 2. 在初始平均轨道平面上投影得到的行星轨道和星子盘粒子的位置。上面的四幅图片分别对应我们模拟过程中的四张快照。在我们的模拟中，四颗巨行星初始位置位于近圆形共面轨道上，其半长轴分别为 5.45 AU、8.18 AU、11.5 AU 和 14.2 AU。动力学上的冷星子盘的质量为 35 M_E，内边缘位于 15.5 AU，外边缘位于 34 AU。四幅图分别给出了行星系统在四个不同时期的状态：**a**，行星迁移的开始（1 亿年）；**b**，LHB 开始前不久（8.79 亿年）；**c**，LHB 开始后不久（8.82 亿年）；**d**，LHB 开始后的 2 亿年，这时星子盘只剩下 3% 的初始质量，各行星已经进入它们最终的轨道。

　　在这一构型中，行星迁移的初始速度将取决于星子盘粒子逸出至行星交会轨道的速率。木星和土星轨道越过 1∶2 MMR 的时间取决于：(1) 它们与共振位置的初始距离；(2) 靠近星子盘内边缘的表面密度；(3) 星子盘内边缘与外侧冰巨行星的相对位置。基于以上参数，我们进行了八个仿真模拟，其中星子盘内边缘的位置设为唯

performed a series of eight simulations where the location of the inner edge of the disk was set as the unique free parameter (Fig. 1). As expected, we found a strong correlation between the location of the inner edge and the time of the 1:2 MMR crossing. For disks with inner edges near 15.3 AU (see above), we find crossing times between 192 Myr and 880 Myr (since the beginning of the simulation).

We also performed eight simulations where we varied the initial location of the ice giants by ~1 AU, Saturn's location by ~0.1 AU, the total mass of the disk by 5 Earth masses ($5\ M_E$), and its initial dynamical state by pushing the particles' eccentricities up to 0.1 and inclinations up to $3.5°$. We found that we can delay the resonant crossing to 1.1 Gyr since the beginning of the simulation, although longer times are clearly possible for more extreme initial conditions. Therefore, we can conclude that the global instability caused by the 1:2 MMR crossing of Jupiter and Saturn could be responsible for the LHB, because the estimated date of the LHB falls in the range of the times that we found.

Figures 2 and 3 show the evolution of one of our runs from the first series of eight. Initially, the giant planets migrated slowly owing to leakage of particles from the disk (Fig. 3a). This phase lasted 880 Myr, at which point Jupiter and Saturn crossed the 1:2 MMR. After the resonance crossing event, the orbits of the ice giants became unstable and they were scattered into the disk by Saturn. They disrupted the disk and scattered objects all over the Solar System, including the inner regions. The solid curve in Fig. 3b shows the amount of material that struck the Moon as a function of time. A total of 9×10^{21} g struck the Moon after resonance crossing—roughly 50% of this material arrived in the first 3.7 Myr and 90% arrived before 29 Myr. The total mass is consistent with the estimate[4] of 6×10^{21} g, which was determined from the number and size distribution of lunar basins that formed around the time of the LHB epoch[1]. Such an influx spike happened in all our runs. The amount of cometary material delivered to the Earth is $\sim 1.8 \times 10^{23}$ g, which is about 6% of the current ocean mass. This is consistent with upper bounds on the cometary contribution to the Earth's water budget, based on D/H ratio measurement[13]. The average amount of material accreted by the Moon during this spike was $(8.4 \pm 0.3) \times 10^{21}$ g.

The above mass delivery estimate corresponds only to the cometary contribution to the LHB, as the projectiles originated from the external massive, presumably icy, disk. However, our scheme probably also produced an in flux of material from the asteroid belt. As Jupiter and Saturn moved from 1:2 MMR towards their current positions, secular resonances (which occur when the orbit of an asteroid processes at the same rate as a planet) swept across the entire belt[14]. These resonances can drive asteroids onto orbit with eccentricities and inclinations large enough to allow them to evolve into the inner Solar System and hit the Moon[4].

614

一的自由参量(图 1)。正如预期的那样，我们发现越过 1:2 MMR 的时间与星子盘内边缘的位置密切相关。当内边缘为 15.3 AU 时(见前述)，相应的越过时间介于 1.92 到 8.80 亿年之间(从模拟启动开始计)。

同时，我们也进行了另外八个仿真模拟。在这些模拟中，我们设定冰巨行星的起始位置变化约 1 AU，土星的位置变化约 0.1 AU，星子盘的总质量变化 5 倍地球质量($5\,M_E$)，并通过推动粒子的偏心率至 0.1 且推动倾角至 3.5°来改变星子盘的起始动力学状态。通过这些模拟，我们发现可以将共振穿越的时间推迟到 11 亿年(从启动模拟开始计算)。当然，如果采用更为极端的初始条件，也可能得到更长的时间。因此，我们得出以下结论：越过木星和土星的 1:2 MMR 引起的全局不稳定性将会导致 LHB，因为推算出的 LHB 发生时间恰好落在我们模拟的时间范围内。

图 2 和图 3 展示了第一批八个模拟中的一次演化过程。最初，由于星子盘中粒子的渗漏，巨行星迁移得很缓慢，参见图 3a。这个阶段持续了 8.8 亿年，同时木星和土星在该时间越过了 1:2 MMR。共振穿越事件发生后，冰巨行星的轨道变得不稳定，并且通过土星散射入星子盘；这导致了星子盘的瓦解，并将其中的天体散射到太阳系各处，包括靠近太阳的区域。图 3b 中的实线给出了撞击月球的物质数量随时间的变化。共振穿越之后，总共有质量为 9×10^{21} g 的物质撞击了月球——其中，在前 370 万年有大约 50% 的物质达到月球，而在前 2,900 万年有 90% 的物质达到月球。通过分析 LHB 时期形成的月球盆地的数量和尺寸分布[1]，计算出的撞击物质总质量与预测结果一致，为 6×10^{21}[4]。在我们所有的模拟中，均出现了这样的内向通量的峰值。该过程大约向地球释放了质量为 1.8×10^{23} g 的彗星物质，这大体相当于目前地球上海水总质量的 6%。该结果也与地球水量平衡中彗星贡献的上界相符，后者是基于 D/H 比值测量得到的[13]。在整个轰击峰值时期，月球增加的物质质量平均为 $(8.4\pm0.3)\times10^{21}$ g。

以上输运质量的估计仅仅是对应在 LHB 时期的彗星物质贡献，即这些彗星抛射物质源自外部大质量的，推测可能是冰质的盘。然而，我们的体系中也存在源自小行星带的物质内向通量。随着木星和土星从 1:2 MMR 移动到目前的位置，当小行星带中的一颗小行星的轨道进动速率(这里指近星点进动速率)与一颗行星的进动速率相同时，就会发生扫过整个小行星带的长期共振[14]。这些长期共振驱使部分小行星进入偏心率和倾角较大的轨道，从而使得这些小行星足以进入太阳系内部区域并撞击月球[4]。

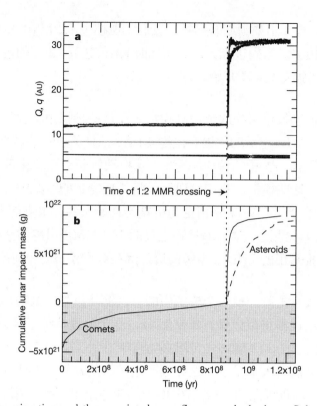

Fig. 3. Planetary migration and the associated mass flux towards the inner Solar System from a representative simulation. **a**, The evolution of the four giant planets. Each planet is represented by a pair of curves—the top and bottom curves are the aphelion and perihelion distances, Q and q, respectively. Jupiter and Saturn cross the 1:2 MMR at 880 Myr. The subsequent interaction between the planets and the disk led to the current planetary configuration as shown in ref. 9. **b**, The cumulative mass of comets (solid curve) and asteroids (dashed curve) accreted by the Moon. We have offset the comet curve so that the value is zero at the time of 1:2 MMR crossing. Thus, $\sim 5 \times 10^{21}$ g of comets was accreted before resonant crossing and 9×10^{21} g of cometary material would have struck the Moon during the LHB. Although the terrestrial planets were not included in our cometary simulations, we estimated the amount of material accreted by the Moon directly from the mass of the planetesimal disk by combining the particles' dynamical evolution with the analytic expressions in ref. 21. The impact velocity of these objects ranged from 10 to 36 km s^{-1} with an average of 21 km s^{-1}. Estimating the asteroidal flux first requires a determination of the mass of the asteroid belt before resonant crossing. This value was determined by first combining the percentage of asteroids remaining in the belt at the end of a simulation ($\sim 10\%$, very sensitive to planet migration rate and initial asteroid distribution) with estimates of the current mass of the belt to determine the initial asteroid belt mass ($\sim 5 \times 10^{-3}\ M_E$). The flux was then again determined by combining the particles' dynamical evolution with the analytic expressions in ref. 21. The dashed curve shows a simulation where class 2 particles dominate. The average asteroidal impact velocity is 25 km s^{-1}.

We investigated the role of asteroid impactors in our LHB model by the following numerical integrations. The orbits of an asteroid belt, composed of 1,000 massless particles with semimajor axes between 2.0 and 3.5 AU, were integrated under the gravitational influence of the Sun, Venus, Earth, Mars, Jupiter and Saturn. Because formation models[15,16] predict that the asteroid belt was partially depleted and dynamically excited well before the LHB, we set the particles' eccentricities between 0 and 0.3 and inclinations between 0° and 30°, but kept the perihelion distances, q, > 1.8 AU and aphelion distances, Q, < 4 AU. Jupiter

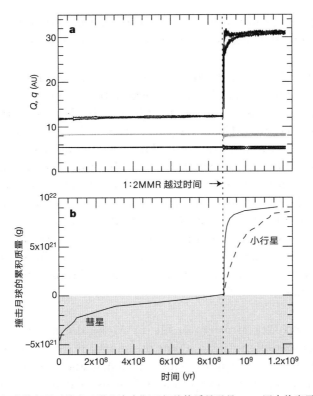

图 3. 一次典型模拟中的行星迁移以及进入内太阳系相关的质量通量。**a**，图中给出了四颗巨行星的演化。图中每颗行星由两条曲线表示，上下两条曲线分别代表远日点（Q）和近日点（q）距离。木星和土星在 8.80 亿年越过 1:2 MMR。随后，行星和星子盘之间的相互作用形成了现在的行星系统构型[9]。**b**，图中给出了彗星和小行星撞击月球的累积质量，分别用实线和虚线表示。我们对彗星曲线进行了偏移，使得 1:2 MMR 穿越发生时刻的累积质量为零。因此，在共振穿越发生前大约质量为 5×10^{21} g 的彗星物质落在月球上，而在 LHB 期间总共有质量为 9×10^{21} g 的彗星物质撞击了月球。尽管我们在彗星模拟过程中并没有将类地行星也包含进来，但我们可以将粒子的动力学演化和参考文献 21 中的解析表达式结合起来，进而估算出星子盘物质撞击月球的质量。这些小型天体的撞击速度介于 10 至 36 km/s，平均速度为 21 km/s。为了估计小行星流量，需要首先确定共振穿越前的小行星带的质量。我们可以将模拟结束后的小行星在带中的剩余比例（约 10%，对小行星迁移率和小行星初始分布十分敏感），与小行星带当前的估算质量综合起来，确定小行星带的初始质量（约 $5 \times 10^{-3} M_E$）。为了确定流量，我们可以将粒子动力学演化和参考文献 21 的解析表达式结合起来。图中虚线表明第二类粒子起主导作用的模拟结果。小行星的平均撞击速度为 25 km/s。

下面，我们借助数值积分的方法研究了 LHB 模型中小行星撞击体的作用。在太阳、金星、地球、火星、木星和土星的引力场的影响下，对以下小行星带进行积分：这是一条由 1,000 颗无质量粒子组成的小行星带，其中粒子的轨道半长轴介于 2.0 与 3.5 AU 之间。由于形成模型预测在 LHB 发生之前，这条小行星带就已经部分被清空并处于动力学上充分激发态[15-16]，因此我们将粒子的偏心率设在 0 到 0.3 之间，倾角

and Saturn were forced to migrate at rates that varied from run to run (adopted from ref. 9) by adding a suitably chosen drag-force term to their equations of motion.

We find that objects that reach Earth-crossing orbits follow one of two general paths. Some, referred to as class 1 particles, get trapped in the periapse secular resonance with Saturn (which affects eccentricities) and are driven directly onto Earth-crossing orbits. Other particles, referred to as class 2, stay in the asteroid belt, but are dynamically excited by resonant sweeping onto unstable orbits. These objects slowly leak out of the asteroid belt and can evolve into the inner Solar System. The two classes produce impact spikes with different temporal behaviours. Roughly 50% of class 1 particles arrive in the first 10 Myr, while 90% arrive within ~30 Myr. Conversely, the median arrival time for class 2 particles is ~50 Myr and 90% arrive within ~150 Myr. Class 2 particles dominated in our runs (Fig. 3). However, a preliminary investigation into this issue shows that this result is probably sensitive to the exact evolution of the giant planets and the dynamical state of the asteroid belt. Thus, the best we can conclude is that the impact spike due to asteroids is between these two extremes.

We find that $(3\text{--}8) \times 10^{21}$ g of asteroids hit the Moon during our simulations (Fig. 3). This amount is comparable to the amount of comets. So, our model predicts that the LHB impactors should have been a mixture of comets and asteroids. Unfortunately, we cannot say with any certainty the exact ratio of comets to asteroids in our model because, although the amount of cometary material is fairly well constrained (probably better than a factor of 2), the amount of asteroidal material is not well known (and could be outside the range reported above), because we do not have good estimates of the mass distribution in the asteroid belt before the LHB. It should also be noted that this ratio is probably a function of impactor size, because comets and asteroids probably have different size distributions. This ratio probably also varied with time. Within the first ~30 Myr comets dominated according to these simulations, but the last impactors were asteroidal. This is consistent with recent cosmochemical findings suggesting that some of the Moon's basins were formed by asteroids[7,8].

Our results support a cataclysmic model for the lunar LHB. Although many aspects of the LHB are not well known[1], our simulations reproduce two of the main characteristics attributed to this episode: (1) the 700 Myr delay between the LHB and terrestrial planet formation, and (2) the overall intensity of lunar impacts. Our model predicts a sharp increase in the impact rate at the beginning of the LHB. Unfortunately, the available lunar data are not yet capable of addressing this prediction.

Our model also has the advantage of supplying impactors that are a mixture of comets and asteroids. Our model predicts that the asteroid belt was depleted by a factor of ~10 during the LHB. This depletion does not contradict collisional evolution models[17,18]. On the contrary, the late secular resonance sweeping could explain why we do not see a large number of asteroid families that were produced during the LHB[18]. Our model predicts that the LHB lasted from between ~10 Myr and ~150 Myr. Correspondingly, the drop-off in

则介于 0 到 30°，近日点距离 q 满足 $q > 1.8\,\mathrm{AU}$，远日点距离 Q 满足 $Q < 4\,\mathrm{AU}$。通过在运动方程中加入合适的阻力项，木星和土星就会被迫以不断变化的速率迁移[9]。

我们发现，进入越地轨道的天体大致上遵循两条路径中的一条。一些天体被称为第一类粒子，它们会由于土星对偏心率的影响而被近星点长期共振俘获，而被直接拽入越地轨道。另一些粒子被称为第二类粒子，这类粒子位于小行星带中，但是会由共振清除作用被动力地激发到非稳定轨道上。这些小天体会逐渐脱离小行星带，然后进入内太阳系。以上两类粒子产生的撞击在时间分布上也不同。在第一类粒子中会有大约 50% 在最初的 1 千万年内发生撞击，大约 90% 会在约 3 千万年内发生撞击。对于第二类粒子，半数到达时间约为 5 千万年，90% 会在约 1.5 亿年内到达。如图 3 所示，在我们的计算机模拟中，第二类粒子起着主导作用。然而，我们的初步研究表明，以上问题的结果可能对巨行星的具体演化和小行星带的动力学状态十分敏感。因此，我们得出的最可信的结论是，小行星的撞击高峰期位于以上两个极值之间。

在模拟中，我们发现总质量为 $(3 \sim 8) \times 10^{21}\,\mathrm{g}$ 的小行星撞击了月球，如图 3 所示。这一质量可以与彗星的质量相比较。因此，我们的模型预测 LHB 撞击体应该包含彗星和小行星。但是，我们并不能确定模型中彗星和小行星的确切比例。尽管可以较为准确地限定彗星物质的质量（可能优于因子 2），但是尚不清楚小行星物质的质量（可能在上述范围之外），这是因为我们尚不能很好地估计出 LHB 之前小行星带的质量分布。需要指出的是，彗星和小行星二者质量的比值可能也是撞击体尺寸的函数，因为彗星和小行星可能具有不同的尺寸分布。同时，这一比值也可能随时间变化。根据我们的模拟可知，在最初的约 3 千万年，彗星起主导作用；但是最后一批撞击体主要由小行星构成。这与最近的宇宙化学发现相符，其认为部分月球盆地是由于小行星撞击形成的[7,8]。

我们的结果支持月球 LHB 的灾变模型。尽管对于 LHB，有许多问题亟待解决，但是我们的模拟重现了这一事件的两个主要特征：(1) LHB 要比地球形成晚 7 亿年；(2) 月球撞击的总强度。我们的模型也预测在 LHB 刚开始的时候，撞击率迅速增加。然而，目前的月球数据尚不能证实这一预测。

我们的模型另一个优点在于指出撞击体是彗星和小行星共同构成的。该模型预测，在 LHB 时期，整个小行星带经过耗损只剩下原先的约 1/10。这种耗损与碰撞演化模型并不矛盾[17,18]。相对地，晚期的长期共振清除机制可以解释为什么我们现在没有观察到大量产生于 LHB 时期的小行星族[18]。我们的模型预测，LHB 持续的时间介于 1 千万 ～ 1.5 亿年之间。相应的，这段时间内撞击率的衰减可以相当迅速

impact rates could be quite fast (with 50% of the impacts occurring in the first 3.7 Myr and 90% in 29 Myr) or moderately slow (with 50% of the impacts occurring in the first 50 Myr and 90% in 150 Myr). We are unable to pinpoint more exact values because the duration and the drop-off of the LHB depends on the relative contributions of class 1 asteroids, class 2 asteroids, and comets, which in turn are very sensitive to the pre-LHB orbital structure of the asteroid belt.

Most importantly, our scheme for the LHB is the result of a generic migration-delaying mechanism, followed by an instability, which is itself induced by a deterministic mechanism of orbital excitation of the planets[9]. This revised planetary migration scheme naturally accounts for the currently observed planetary orbits[9], the LHB, the present orbital distribution of the main-belt asteroids and the origin of Jupiter's Trojans[19].

(**435**, 466-469; 2005)

R. Gomes[1,2], H. F. Levison[2,3], K. Tsiganis[2] & A. Morbidelli[2]

[1] ON/MCT and GEA/OV/UFRJ, Ladeira do Pedro Antonio, 43 Centro 20.080-090, Rio de Janeiro, RJ, Brazil

[2] Observatoire de la Côte d' Azur, CNRS, BP 4229, 06304 Nice Cedex 4, France

[3] Department of Space Studies, Southwest Research Institute, 1050 Walnut Street, Suite 400, Boulder, Colorado, USA

Received 6 December 2004; accepted 18 April 2005.

References:

1. Hartmann, W. K., Ryder, G., Dones, L. & Grinspoon, D. in *Origin of the Earth and Moon* (eds Canup, R. & Righter, K.) 493-512 (Univ. Arizona Press, Tucson, 2000).

2. Morbidelli, A., Petit, J.-M., Gladman, B. & Chambers, J. A plausible cause of the Late Heavy Bombardment. *Meteorit. Planet. Sci.* **36**, 371-380 (2001).

3. Zappala, V., Cellino, A., Gladman, B. J., Manley, S. & Migliorini, F. Asteroid showers on Earth after family break-up events. *Icarus* **134**, 176-179 (1998).

4. Levison, H. F. *et al.* Could the lunar "Late Heavy Bombardment" have been triggered by the formation of Uranus and Neptune? *Icarus* **151**, 286-306 (2001).

5. Chambers, J. E. & Lissauer, J. J. A new dynamical model for the lunar Late Heavy Bombardment. *Lunar Planet. Sci. Conf.* **XXXIII**, abstr. 1093 (2002).

6. Levison, H. F., Thommes, E. W., Duncan, M. J., Dones, L. A. in *Debris Disks and the Formation of Planets: A Symposium in Memory of Fred Gillett (11-13 April 2002, Tucson, Arizona)* (eds Caroff, L., Moon, L. J., Backman, D. & Praton, E.) 152-167 (ASP Conf. Ser. 324, Astronomical Society of the Pacific, San Francisco, 2005).

7. Kring, D. A. & Cohen, B. A. Cataclysmic bombardment throughout the inner Solar System 3.9-4.0 Ga. *J. Geophys. Res. Planets* **107**(E2), 4-10 (2002).

8. Tagle, R. LL-ordinary chondrite impact on the Moon: Results from the 3.9 Ga impact melt at the landing site of Apollo 17. *Lunar Planet. Sci. Conf.* **XXXVI**, abstr. 2008 (2005).

9. Tsiganis, K., Gomes, R., Morbidelli, A. & Levison, H. F. Origin of the orbital architecture of the giant planets of the Solar System. *Nature* doi:10.1038/nature03539 (this issue).

10. Fernandez, J. A. & Ip, W.-H. Some dynamical aspects of the accretion of Uranus and Neptune—The exchange of orbital angular momentum with planetesimals. *Icarus* **58**, 109-120 (1984).

11. Hahn, J. M. & Malhotra, R. Orbital evolution of planets embedded in a planetesimal disk. *Astron. J.* **117**, 3041-3053 (1999).

12. Gomes, R. S., Morbidelli, A. & Levison, H. F. Planetary migration in a planetesimal disk: Why did Neptune stop at 30 AU? *Icarus* **170**, 492-507 (2004).

13. Morbidelli, A. *et al.* Source regions and timescales for the delivery of water to Earth. *Meteorit. Planet. Sci.* **35**, 1309-1320 (2000).

14. Gomes, R. S. Dynamical effects of planetary migration on the primordial asteroid belt. *Astron. J.* **114**, 396-401 (1997).

15. Wetherill, G. W. An alternative model for the formation of the asteroids. *Icarus* **100**, 307-325 (1992).

16. Petit, J., Morbidelli, A. & Chambers, J. The primordial excitation and clearing of the asteroid belt. *Icarus* **153**, 338-347 (2001).

17. Davis, D. R., Ryan, E. V. & Farinella, P. Asteroid collisional evolution: results from current scaling algorithms. *Planet. Space Sci.* **42**, 599-610 (1994).

18. Bottke, W. *et al.* The fossilized size distribution of the main asteroid belt. *Icarus* **175**(1), 111-140 (2005).

19. Morbidelli, A., Levison, H. F., Tsiganis, K. & Gomes, R. Chaotic capture of Jupiter's Trojan asteroids in the early Solar System. *Nature* doi:10.1038/nature03540 (this issue).

620

（50% 的撞击发生于开始的 370 万年，90% 的撞击发生于 2,900 万年内），或者适当减缓（50% 的撞击发生于开始的 5 千万年，90% 的撞击发生于 1.5 亿年内）。我们并不能更为精确的确定以上时间，因为 LHB 的持续和减弱依赖于第一类小行星、第二类小行星以及彗星的相对贡献，而这三者对 LHB 发生前的小行星带轨道结构颇为敏感。

最为重要的是，我们的 LHB 体系是由一般的迁移延迟机制造成的，随后出现的不稳定性本身又是由行星轨道的激发的决定性机制所诱发的 [9]。这一修正的行星迁移体系可以自然地将目前我们观测到的行星轨道 [9]、LHB、目前的主带小行星轨道分布以及木星–特洛伊族小行星的起源 [19] 纳入其中。

（金世超 翻译；季江徽 审稿）

20. Haisch, K. E., Lada, E. A. & Lada, C. J. Disk frequencies and lifetimes in young clusters. *Astrophys. J.* **553**, L153-L156 (2001).

21. Wetherill, G. W. Collisions in the asteroid belt. *J. Geophys. Res.* **72**, 2429-2444 (1967).

Acknowledgements. R.G. thanks Conselho Nacional de Desenvolvimento Científico e Tecnológico for support for his sabbatical year in the OCA observatory in Nice. K.T. was supported by an EC Marie Curie Individual Fellowship. A.M. and H.F.L. thank the CNRS and the NSF for funding collaboration between the OCA and the SWRI groups. H.F.L. was supported by NASA's Origins and PG&G programmes.

Author Information. Reprints and permissions information is available at npg.nature.com/reprintsandpermissions. The authors declare no competing financial interests. Correspondence and requests for materials should be addressed to A.M. (morby@obs-nice.fr).

Rain, Winds and Haze during the Huygens Probe's Descent to Titan's Surface

M. G. Tomasko *et al.*

Editor's Note

The Cassini spacecraft carried the Huygens mission, whose goal was to land on the surface of Saturn's largest moon, Titan. During Huygens' parachute descent, an array of instruments studied the moon's atmosphere, winds and surface. The lander survived on the surface for about three hours. This paper is one of two presenting the key findings of the mission. It reports that the moon's surface contains features resembling those made by flowing liquid on Earth, presumably due to liquid hydrocarbons. The landing site looks like a dry river or lake bed, scattered with small rounded "rocks", which are probably hydrocarbon-coated water ice. The atmospheric haze that obscures the surface from telescopes is made of small particles, most probably of complex and as yet poorly characterized organic matter.

The irreversible conversion of methane into higher hydrocarbons in Titan's stratosphere implies a surface or subsurface methane reservoir. Recent measurements from the cameras aboard the Cassini orbiter fail to see a global reservoir, but the methane and smog in Titan's atmosphere impedes the search for hydrocarbons on the surface. Here we report spectra and high-resolution images obtained by the Huygens Probe Descent Imager/Spectral Radiometer instrument in Titan's atmosphere. Although these images do not show liquid hydrocarbon pools on the surface, they do reveal the traces of once flowing liquid. Surprisingly like Earth, the brighter highland regions show complex systems draining into flat, dark lowlands. Images taken after landing are of a dry riverbed. The infrared reflectance spectrum measured for the surface is unlike any other in the Solar System; there is a red slope in the optical range that is consistent with an organic material such as tholins, and absorption from water ice is seen. However, a blue slope in the near-infrared suggests another, unknown constituent. The number density of haze particles increases by a factor of just a few from an altitude of 150 km to the surface, with no clear space below the tropopause. The methane relative humidity near the surface is 50 per cent.

THE surface of Titan has long been studied with various instruments, including those on the Hubble Space Telescope (HST) and ground-based adaptive optics systems[1]. More recently, Cassini investigations using the charge-coupled device (CCD) camera[2], the

惠更斯空间探测器下落至土卫六表面过程中的雨、风和雾霾

托马斯科等

编者按

卡西尼号土星探测器承担着惠更斯计划，惠更斯号探测器的目标是在土星最大的卫星土卫六的表面着陆。在惠更斯号打开降落伞的降落阶段，一系列探测仪器研究了土卫六的大气、风以及地表。着陆器在卫星表面工作了约3个小时。本文是报道本次探测任务主要发现的两篇文章之一，文章指出卫星的表面存在一些地貌特征，类似地球表面上液体流动留下的痕迹，推测可能和液态的烃有关。着陆点看起来像干涸的河床或者湖床，遍布圆球形的小"岩石"，这些很有可能是烃包被的水冰。由于大气中雾霾的存在，使得从望远镜只能观察到模糊的地表影像。雾霾是由细小粒子构成的，很可能成分复杂且存在迄今为止不为人知的有机物质。

在土卫六平流层探测到甲烷转变为高级烃的不可逆过程，这意味着土卫六的表面或地下存在甲烷储备。最近卡西尼轨道飞行器上搭载的照相机进行的观测并没能发现全球性的甲烷储备，但是土卫六大气中的甲烷和浓雾阻碍了在其表面寻找烃的工作。本文中，我们将报道惠更斯探测器在土卫六大气降落过程中，搭载的惠更斯降落成像仪/光谱辐射计测得的光谱和高分辨率图像。尽管这些图像并没有直接展现土卫六表面的液态烃湖，但是确实存在液态物质在土卫六表面流过留下的痕迹。土卫六上的地貌与地球表面有惊人的相似性，比较明亮的高地区域存在复杂交错的沟渠系统，它们汇入平坦而阴暗的低地。惠更斯号着陆后拍摄的照片显示的环境是一处干涸的河床。土卫六表面测得的红外反射光谱，不同于太阳系内任何其他已知的天体，在其光学波段的红端（较大波长）存在倾斜谱结构（红坡），这与一类有机物质（比如索林斯）的光谱特征相一致，水冰的吸收峰是可见的。尽管如此，近红外区域的蓝端倾斜结构（蓝坡）则预示另外一种未知的成分的存在。雾霾颗粒的数量密度从150 km的海拔到表面增大了数倍，并且在对流顶层以下没有明显的净空区域。甲烷在地表附近的相对湿度为50%。

借助包括哈勃空间望远镜（HST）搭载的仪器和地基自适应光学系统等在内的各种仪器，人们已经对土卫六的表面进行了长期的研究[1]。最近，卡西尼号飞船使用

Visible and Infrared Mapping Spectrometer (VIMS) instrument[3], and the Radio Detection and Ranging (RADAR) imaging system[4] have provided more detailed views of Titan's surface in the hope of revealing how the methane in Titan's atmosphere is replenished from the surface or interior of Titan. Of the Cassini imagers, the Imaging Science Subsystem (ISS) camera is potentially capable of the greatest spatial resolution, but Titan's obscuring haze limits its resolution on the surface to about 1 km, a value roughly similar to that available from VIMS and the radar imaging system. At this resolution, the bright and dark regions observed on the surface of Titan have proved difficult to interpret. Owing to its proximity to the surface, the Descent Imager/Spectral Radiometer (DISR) camera on the Huygens probe was capable of a linear resolution of some metres from a height of 10 km. In addition, the lower the probe descended, the less haze lay between the camera and the ground. The DISR was capable of linear resolution orders of magnitude better than has been available from orbit, although of a much smaller portion of Titan's surface. Also, a lamp was used at low altitude to measure the continuous reflectance spectrum of the surface without the complications introduced by observations through large amounts of methane and aerosol haze[5].

In addition to studying the surface of Titan, the DISR took measurements of solar radiation in the atmosphere. Spectrometers looking upward at continuum wavelengths (between the major methane absorptions) as well as downward measured the vertical distribution and wavelength dependence of the aerosol haze opacity. Measurements of the polarization of light at a scattering angle of 90° constrained the small dimension of the haze particles. Measurements of the brightness in the solar aureole around the Sun determined the projected area of the haze particles. Observations in the methane bands determined the methane mole fraction profile.

Data collection during the descent proceeded mostly, although not exactly, as planned. Turbulence during the first half of the descent tipped the probe more rapidly than expected, causing the Sun sensor to remain locked on the azimuth of the Sun for only a few successive rotations at a time. Below about 35 km, the signal from the direct solar beam was lost by the Sun sensor owing to the unexpectedly low temperature of this detector. These effects caused data from each of the DISR sub-instruments to be collected at mostly random, instead of specific, azimuths. Additionally, the probe rotated in the intended direction for only the first ten minutes before rotating in the opposite sense for the remainder of the descent. This resulted in ineffective baffling of the direct solar beam for the upward-looking visible spectrometer and the solar aureole camera. Consequently, some measurements made by the solar aureole camera are saturated, and the separation of the direct and diffuse solar beams in the visible spectral measurements must be postponed until a good model of the probe attitude versus time is available. Finally, the loss of one of the two radio communication channels in the probe receiver aboard the orbiter resulted in the loss of half the images as well as several other low-altitude spectrometer measurements.

Despite these misfortunes, the DISR instrument collected a unique and very useful data set. Images of the surface with unprecedented resolution were collected over the boundary

626

自身搭载的电荷耦合器件 (CCD) 照相机 [2]、可见光和红外测绘光谱仪 (VIMS) [3] 以及雷达成像系统 [4]，对土卫六表面进行了更加仔细的观测，并期望能够揭示土卫六上的甲烷是如何从地表或地下重新回到大气中的。在卡西尼号所有的成像仪中，成像科学子系统 (ISS) 有能力可以获得最高的空间分辨率，但是在土卫六表面雾霾的影响下，其分辨率被限制在与 VIMS 以及雷达成像系统大致相当的水平，仅为 1 km 左右。在这一分辨率下，土卫六表面上被观测到的明亮和阴暗的区域被证明是难以解译的。由于惠更斯空间探测器接近土卫六表面，它搭载的降落成像仪 / 光谱辐射计 (DISR) 相机在 10 km 的高度可以获得米量级的线性分辨率。此外，探测器高度越低，则相机与地面之间的雾霾的厚度越小。这样，DISR 可以获得比在轨道上高数个量级的线性分辨率，但观测到的土卫六地表范围减小很多。同时，惠更斯号在较低的高度借助照明灯测量了表面的连续反射光谱，这样避免了透过大量的甲烷和气溶胶薄雾进行观测所引起的光谱复杂性 [5]。

除了研究土卫六表面，DISR 系统还测量了大气中的太阳辐射。采用连续波段（甲烷主吸收峰之间的波段），光谱仪分别朝上和朝下测量了薄雾在垂直方向上的分布及其不同波长的不透明度。在 90° 散射角处测量光的偏振可以约束雾霾颗粒是小颗粒的，而对太阳周围的日晕亮度的测量可以确定雾霾颗粒的投影面积。甲烷波段的观测可以确定甲烷的摩尔分数。

在惠更斯号降落过程中，尽管数据的采集工作基本按预定计划进行，但也出现了一些意外情况。在降落过程的前半段，湍流造成的探测器的倾斜比预期要快很多，这导致探测器每次保持太阳传感器锁定在太阳所在的方向角的姿态只能持续几次探测器旋转周期。降至约 35 km 以下，由于探测器的温度出乎意料的低，太阳传感器无法感应来自太阳光直射的信号。以上意外导致 DISR 子设备得到的数据并没能像预计的一样从特定的方位角方向获取，大部分都有随机的方位角。另一件意料之外的事情是，探测器仅仅在前 10 分钟内按照预定的方向旋转，然后在剩余的下降时间就一直以相反的方向转动。这导致与朝上观测的可见光谱分析仪和太阳光晕照相机有关的直射太阳光没有被有效地遮挡。因此，太阳光晕照相机进行的某些测量过度曝光，可见光谱测量中的直射太阳光束与弥散太阳光束暂时也难以有效分离，除非找到关于探测器姿态与时间关系的合理模型。最后，轨道器搭载的探测器接收天线损失掉了两个无线电通信频道中的一个，造成一半图像丢失，同时也造成光谱仪部分低空测量数据的丢失。

尽管遭遇了这些不幸，DISR 中的设备还是采集到了一个独一无二且非常有用的数据集。轨道器拍摄到了明亮区域和暗区域之间的边界地带史无前例的高分辨率图

between bright and dark terrain seen from the orbiter. Owing to redundant transmission over both communication channels during most of the descent, almost all of the spectral and solar aureole observations were received. A very large set of high-quality spectra were obtained with good altitude resolution and with good coverage in azimuth both away from and towards the Sun. The images, the spectra, the Sun sensor pulses, the recording of the gain in the Cassini radio receiver, and information from Very Long Baseline Interferometry (VLBI) observations from Earth together will permit reconstruction of the probe attitude relative to the Sun as a function of time during the descent, enabling a full analysis of the spectral data. The large number of solar aureole measurements included several acquired near the Sun and many polarization measurements opposite to the Sun. The surface science lamp worked éxactly as planned, permitting surface reflection measurements even in strong methane absorption bands. Operations after landing included the collection of successive images as well as spectral reflectance measurements of the surface illuminated by the lamp from an assumed height of roughly 30 cm.

Taken together, the new observations shed substantial light on the role played by methane in forming the surface of Titan and how it is recycled into the atmosphere. The substantial relative humidity of methane at the surface and the obvious evidence of fluid flow on the surface provide evidence for precipitation of methane onto the surface and subsequent evaporation. Some indications of cryovolcanic flows are also seen. The vertical distribution and optical properties of Titan's haze have been characterized to aid the interpretation of remote measurements of the spectral reflection of the surface. The speed and direction of Titan's winds has also been measured for comparison with future dynamical models that include the radiative heating and cooling rates implied by the haze.

Physical Processes That Form the Surface

The imagers provided views of Titan's previously unseen surface, thus allowing a deeper understanding of the moon's geology. The three DISR cameras were designed to provide overlapping coverage for an unbroken 360°-wide swath stretching from nadir angles between 6° and 96°. Some 20 sets of such images were planned during the descent. Because of the opacity of the haze in the passband of our imager, surface features could be discerned in the images only below about 50 km, limiting the number of independent panoramic mosaics that can be made of the surface. The loss of half of the images meant that Titan's surface was not covered by systematic overlapping triplets, as expected. Three different views of Titan's surface are shown in Figs 1 to 3. A view of Titan's surface after the Huygens landing is shown in Fig. 4.

The highest view (Fig. 1), projected from an altitude of 34 km, displays an albedo variation very similar to the highest-resolution images provided by the ISS or VIMS cameras on the Cassini orbiter. It shows Titan's surface to consist of brighter regions separated by lanes or lineaments of darker material. No obvious impact features are visible, although craters less than roughly 10 km should not be abundant as a result of atmospheric shielding[6]. In the

像。在下降过程的大部分时间，两个通信频道都进行了冗余传输，这使得我们可以接收到大部分光谱以及太阳光晕的观测结果。惠更斯探测器获得了大量高质量的光谱，并且无论是朝向太阳还是背向太阳都具有极佳的高度分辨率和方位覆盖。通过获得的图像、光谱、太阳传感器脉冲、卡西尼无线电接收机的增益记录以及地球上甚长基线干涉测量(VLBI)信息，我们可以重建出下降阶段中探测器相对太阳的姿态随时间的变化函数，从而使我们能够全面地分析光谱数据。所获得的大量的太阳光晕测量包括在指向太阳方向附近进行的一些测量和大量的背向太阳方向的偏振测量。用于土卫六表面测量的科学照明灯完全按预期进行工作，使得即使是在强甲烷吸收波段中，也可以进行表面反射光谱的相关测量。探测器着陆后的操作包括采集连续图像以及测量地表的光谱反射率，其中反射率测量中照明灯在高度约为 30 cm 提供照明。

综上，惠更斯探测器进行的最新观测在相当程度上揭示了甲烷在土卫六地表形成过程中的作用，以及甲烷是如何通过循环进入大气的。地表附近甲烷有较高的相对湿度，而且地表明显存在液体流动的痕迹，这些证据表明甲烷经历了凝结降落以及再蒸发的循环过程。同时，也发现了冰火山作用相关的流动构造的迹象。土卫六上雾霾的垂直分布和光学属性也被确定，有助于解释地表谱反射率的遥测结果。另外，探测器对土卫六上的风速和风向也进行了测量，可以用于和将来的动力学模型的对比，模型将包含由雾霾得出的辐射加热速率与冷却速率。

形成土卫六表面的物理过程

惠更斯探测器上的成像仪提供了大量前所未有的关于土卫六地表的信息，使得更加深入地了解这颗卫星的地质情况成为可能。DISR 系统的三台照相机可以提供 360° 水平无缝覆盖全景图像，其天底角介于 6° 到 96° 之间。探测器原本预计在下降过程中拍摄大约 20 组这样的照片。由于雾霾对于成像仪的滤波通带的不透明性，只有在约 50 km 以下高度拍摄的照片中才可以分辨出地表特征，从而也限制了不同的表面全景拼接图像的数量。同时，前面提到的图像丢失一半意味着土卫六的表面不能够按预期那样被三台照相机所拍摄的照片系统的交叠全部覆盖。图 1 到图 3 给出了土卫六表面的三幅图片。图 4 给出了惠更斯探测器着陆后拍摄的一幅地表图片。

视角最高的视图(图 1)是以 34 km 高度作为基点进行的投影，表明星体反照率的变化，这种变化与卡西尼轨道器上的 ISS 或 VIMS 照相机提供的最高分辨率图像非常相似。从图像中可知，土卫六表面由明亮区域以及分割这些区域的较暗的水道或线性构造构成。尽管由于大气的遮挡可以观察到的直径小于 10 km 的陨石坑应该

rightmost (eastern) part of the mosaic the images become sharper as the lower altitude of the camera causes the scale to decrease and the contrast to increase. More than a dozen brighter areas in that region seem to be elongated along a direction parallel with the main bright/dark boundary of that region. At the limit of resolution, narrow dark channels cut the bright terrain.

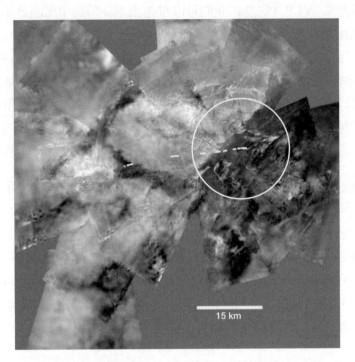

15 km

Fig. 1. View of Titan from 34 km above its surface. High-altitude (49 to 20 km) panoramic mosaic constructed from the DISR High and Medium Resolution Imagers (HRI and MRI) as projected from 34 km. The preliminary ground-track solution (indicated as small white points on gnomonic ground projection) represents the location of the probe when data were collected; north points up; scale indicated (although subsequent analysis indicates that north lies some 5–10° to the left of straight up in this and the two subsequent figures). Starting from the first surface image at 49 km, the probe moves in an east-northeastwardly direction at an initial speed of 20 m s^{-1}. Brighter regions separated by lanes or lineaments of darker material are seen. No obvious crater-like features are visible. The circle indicates the outline of the next-lowest pan, in Fig. 2. The method used for construction of panoramic mosaics incorporates knowledge of the probe's spatial location (longitude, latitude and altitude) and attitude (roll, pitch and yaw) at each image. With the exception of altitude, provided by the Huygens Atmospheric Structure Instrument[12] pressure sensor, none of these variables was directly measured. They are found through an iterative process in which a panorama is created, providing an improved ground-track and azimuth model, which results in an upgraded trajectory, which can improve the panorama, and so on. The current lack of pitch and roll knowledge constitutes the main source of error in the current composition and quality of the panoramas as well as the ground-track and wind speed determination reported below. Vigorous contrast-stretching in the images is required to reveal details washed out by the haze particle density at all altitudes in Titan's atmosphere. This contrast-stretching also displays the occasional ringing of the Discrete Cosine Transform data compressor, which appear as regular lines of bright and dark patterns, particularly in the MRI images.

The next view, Fig. 2, projected from an altitude of 8 km, reveals a large number of these channels (detailed in Fig. 5). The channel networks have two distinct patterns: short stubby

不会很多，图像中没有看到明显的陨石坑地貌[6]。在拼接图像的最右侧（东侧），图像变得锐度增加，这是由于照相机拍摄时的高度降低，从而使拍摄范围变小，对比度增加。图中十多块明亮区域看似沿主要的明暗边界方向被拉伸了。在现有的最高分辨率下，图中的较为阴暗的狭谷截切了明亮区域。

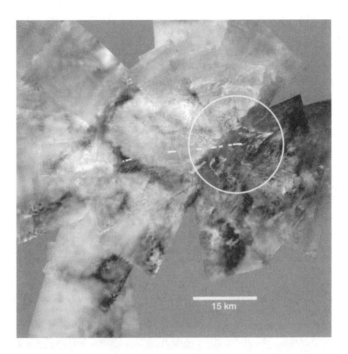

图1. 土卫六上空34 km高度拍摄的图像。通过DISR高分辨率成像仪和中分辨率成像仪（即HRI和MRI）构成的高空（49～20 km）全景拼接图，以34 km高度处作为基点投影。图中用小白点标出了探测器飞行轨迹在地面上的球心投影，这一初步的探测器飞行轨迹地表投影给出了探测器采集数据时的位置。图中正上方为北，比例尺标于图中（后续分析表明，图中实际的北方应该相对竖直向上方向向左偏移约5°～10°，后面的两幅图也是如此）。探测器从49 km的高度开始拍摄第一幅照片，然后以20 m/s的初始速度向东北偏东方向飞行。由图可知，较亮的区域被较暗的水道或线性构造分开，但没有发现明显的类似环形山的地貌。图中的圆圈表示下一个全景镜头的轮廓，如图2所示。全景图像拼接的构建方法将探测器在拍摄每张照片时的空间定位（经度、纬度和高度）与姿态（滚动、俯仰和偏航）两方面的数据纳入考量。对于以上变量，除了高度可以通过惠更斯大气构造探测仪上的压强传感器直接测量[12]，其他变量都不能被直接测量。以上参数可以通过构建全景图像的迭代过程得到，该迭代过程可以改进探测器轨道的地表投影和方位模型，而模型又可以反过来校正探测器飞行轨道，进一步改善全景图像质量等。俯仰和滚动等姿态信息的缺乏，是目前全景图合成及其图像品质，以及后文轨迹的地面投影和风速确定中误差的主要来源。土卫六大气中的雾霾颗粒将使图像变得模糊不清，因此需要进行足够的对比度增强才能够显示更多的细节。对比度增强的操作也会偶尔表现出离散余弦变换数据压缩器的振荡效应，表现为规律性重复出现的明暗相间的条纹，这种效应对于中分辨率图像的影响特别严重。

下一幅拼接图如图2所示，以8 km的高度作为基点作投影，展现了大量的沟渠结构（具体细节如图5所示）。图中的沟渠网络具有两种明显不同的样式：一种是短

features and dendritic features with many branches. The region of the stubby network towards the west is significantly brighter than the dendritic region. The stubby channels are shorter, wider and relatively straight. They are associated with and often begin or end in dark circular areas, which suggest ponds or perhaps pits. The morphology of rectilinear networks with stubby heads is consistent with spring-fed channels or arroyos.

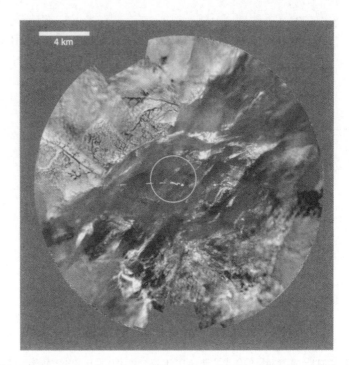

Fig. 2. View from 8 km. Medium-altitude (17 to 8 km) panoramic mosaic projected from 8 km. As in Fig. 1, the preliminary ground-track solution is indicated as points; north is up; scale indicated. At 11 km, the wind direction is at 0° (eastward), reaching −20° (southeastward) at an altitude of 8.5 km. The narrow dark lineaments, interpreted as channels, cut brighter terrain. The circle indicates the outline of the low-altitude pan in Fig. 3.

The dendritic network is consistent with rainfall drainage channels, implying a distributed rather than a localized source of a low-viscosity liquid. Stereo analysis of the dendritic region indicates an elevation of 50–200 m relative to the large darker plain to the south. It suggests that the brighter areas within the darker terrain are higher as well. The topographic differences are evident in Figs 6 and 7, which are three-dimensional renderings of the area just north of the landing site produced from the DISR images. They include the major bright–dark interface seen from above in Fig. 5. Figure 2 depicts many examples of these darker lanes of material between topographically higher, brighter areas. In fact the low contrast of the lowland plane argues that the entire dark region floods, and as the liquid drains the local topography drives flows as seen in the images. If the darker region is interpreted as a dried lakebed, it is too large to have been caused by the creeks and channels visible in the images. It may have been created by other larger river systems or some large-scale catastrophic event, which predates deposition by the rivers seen in these images.

而粗的断株状结构，另一种是具有很多分支的树枝状结构。前者相对更短、更宽和更直。图中指向西侧的断株样网络所在的区域明显比树枝状区域更加明亮。这些断株样沟渠通常起始或终止于深色的圆形区域，这一区域可能是池塘或深坑。以短粗结构为端点的直线网络地貌符合源于泉水的河道或河谷地貌特征。

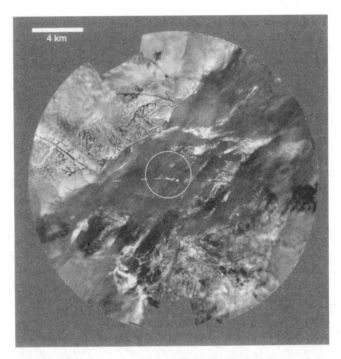

图2. 8 km 高度的视图，是中空(17~8 km)所拍摄照片的全景拼接图。就像图1那样，探测器轨迹地面投影用点表示，图片正上方为北，比例尺标于图中。在11 km高度，探测到的风向位于0°(朝正东方向)；在8.5 km高度，探测到的风向变为-20°(朝东南方向)。图中细窄而暗的线性构造被解译为沟渠，截切了明亮区域。图中圆圈标记出了图3中的全景图像轮廓范围。

树枝状网络的地貌特征与降雨控制的水系河道相符，这意味着存在广泛分布的而非局域性的低黏度的液态物质。关于树枝状区域的立体信息分析显示，该区域相对南部的较暗的大型平原地区抬高了50~200 m；这说明暗色区域内部的亮色区块同样会比周围的暗色区域更高。图6和图7是通过DISR图像生成的着陆点北部区域的三维效果图，它们包含了图5俯视图中主要的明暗分界，显然明暗区域的地貌特征存在很大区别。图2显示了多个存在于地形较高、较明亮的区域之间的较暗的带状分布的物质的实例。事实上，低地平原的较小的起伏说明整个暗色区域被液态物质湮没，而且地势会驱动液态物质的流动，表现出图像中所见的地貌。如果认为较暗的区域是干涸的湖床，显然面积太大，图像中所见到的小溪和河道不可能汇成这样大的湖泊。因此，暗色区域可能是由其他较大的河流系统汇成或者由某种大尺度的灾变造成，但这种灾变要早于图像中所见的河流沉积地貌。

The interpretation of the dark lanes within the brighter highlands as drainage features is so compelling as to dominate subsequent interpretation of other areas of images such as Figs 2 and 3. The prevailing bright–dark boundary of the region becomes a coastline, the bright areas separated from this boundary become islands. Bright streaks running parallel to the albedo boundary may be drift deposits or splays fractured off the bright highlands owing to faulting along the shoreline.

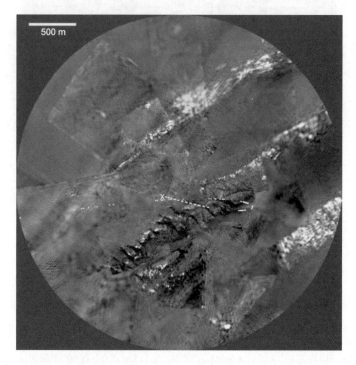

Fig. 3. View from 1.2 km. Low-altitude (7 to 0.5 km) panoramic mosaic projected from 1,200 m. As in Figs 1 and 2, the preliminary ground track is indicated as points; north is up; scale indicated. The probe's steady east-northeast drift halts altogether at an altitude of 7 km and reverses, moving west-northwest for some 1 km during the last 15 min of descent. Note the ridge near the centre, cut by a dozen darker lanes or channels. The projected landing site is marked with an "X" near the continuation of one of the channels, whose direction matches the orientation of the stream-like clearing in the near-foreground of the southward-looking surface image, Fig. 4.

When coupled with Fig. 4, which is an image of a typical offshore dark region, it is clear that the analogy has a limit. At present there is no liquid in the large dark lakebed imaged in Figs 1 to 5. The bright lobate feature, split by an apparently straight dark lane in the western part of the mosaic in Fig. 2, is a possible fissure-fed cryovolcanic flow. However, Fig. 4 also reveals rocks which—whether made of silicates or, more probably hydrocarbon-coated water-ice—appear to be rounded, size-selected and size-layered as though located in the bed of a stream within the large dark lakebed. Rounded stones approximately 15 cm in diameter and probably composed of water-ice, lie on top of a darker, finer-grained surface.

　　将明亮高地区域内的暗色条带看作用于排水的河道是非常可信的一种解释，以至于可以直接主导了图 2 和图 3 等其他图像的解译。这样的话，这一区域内主要的明暗边界可以被认为是海岸线，而被这些边界围限的明亮区域可以解释成岛屿。平行于反照率差异边界延伸的明亮条纹可能是冰川沉积物或者是沿海岸线发生断层活动由明亮高地断裂形成的斜面。

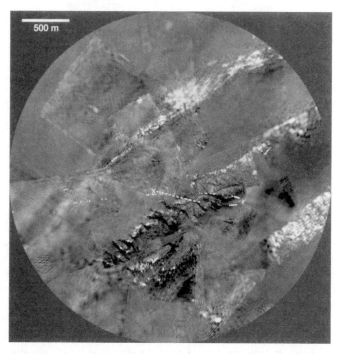

图 3. 1.2 km 高度处的视图。在 7 km ~ 0.5 km 范围内拍摄的图像，被用于制作以 1,200 m 为基点的投影全景图像。如图 1 和图 2，初步推测的探测器轨迹地面投影用点表示，图片正上方为北，比例尺标于图中。探测器首先朝东北偏东方向稳定飘移，到达 7 km 高度时完全停止，后反向运动，在下降阶段的最后 15 分钟探测器向西北偏西移动了大约 1 km。值得注意的是，图中心附近的山脊被若干暗色河道切割。探测器的着陆地点在图中的投影位置用"X"标记出，该位置靠近一条渠道的延长段；渠道与图 4 中显示的视角朝南的地表图像靠近前景的类似溪流的净空地带的延伸方向相一致。

　　图 4 是一片典型的离岸的暗色区域的图像，结合该图可知以上类比显然存在局限性。目前，图 1 至图 5 中的巨型暗色湖床区域并不存在液体。图 2 中西侧被一条直的暗色河道分隔的明亮叶状结构，可能是发源于裂缝的冰岩浆流。然而，图 4 展现的石块就好像是处在大型暗色湖床上的河床内，有较好的磨圆度、分选度和粒序层理，石块的成分可能是硅酸盐，或者更有可能是烃包被的水冰。这些直径大约为 15 cm、可能是由水冰构成的圆形石块位于由更细粒物质构成的更暗的表面之上。

Fig. 4. The view from Titan's surface. Merged MRI and SLI images acquired after the Huygens probe soft-landing. Horizon position implies a pitch of the DISR nose upward by $1.7 \pm 0.2°$ with no measurable roll. "Stones" 10–15 cm in size lie above darker, finer-grained substrate in a variable spatial distribution. Brightening of the upper left side of several rocks suggests solar illumination from that direction, implying a southerly view, which agrees with preliminary evidence from other data sets. A region with a relatively low number of rocks lies between clusters of rocks in the foreground and the background and matches the general orientation of channel-like features in the low-altitude pan of Fig. 3. The bright spot in the lower right corner is the illumination of the DISR surface science lamp.

It is interesting to compare the brightness and colour of the scene shown in Fig. 4 with that of a similar scene on the Earth. The brightness of the surface of the Earth illuminated by full sunlight is about half a million times greater than when illuminated by a full moon. The brightness of the surface of Titan is about a thousand times dimmer than full solar illumination on the Earth (or 500 times brighter than illumination by full moonlight). That is, the illumination level is about that experienced about 10 min after sunset on the Earth. The colour of the sky and the scene on Titan is rather orange due to the much greater attenuation of blue light by Titan's haze relative to red light. If the Sun is high in the sky, it is visible as a small, bright spot, ten times smaller than the solar disk seen from Earth, comparable in size and brightness to a car headlight seen from about 150 m away. The Sun casts sharp shadows, but of low contrast, because some 90% of the illumination comes from the sky. If the Sun is low in the sky, it is not visible.

The sizes of the more than 50 stones in the image in Fig. 4 vary between 3 mm in diameter, the resolution limit of the imager, and 15 cm. No rocks larger than 15 cm are seen. The resolution of the last images taken before landing from a height of 200–300 m would be sufficient to identify metre-sized objects, and none are seen in the 40×35 m field of view. Figure 8 shows the R value, a measure of the fraction of the surface covered by rocks of

636

图 4. 土卫六表面的图像。该图像是由惠更斯探测器软着陆后拍摄的 MRI 与 SLI 图像拼接而成。由地平线的位置可知 DISR 的仰角为 $1.7° \pm 0.2°$，但没有可见的左右倾斜。粒径 $10 \sim 15$ cm 的"石块"不均匀地散布在更暗更细粒的基底上。几块岩石的左上部分被照亮，指示了太阳光的照射位于这一方向，即来自南方，这与其他数据得到的初步结论相符。图中前景石块群和远景石块群之间存在石块数量相对较少的区域，其延伸方向与图 3 低空拼接图中沟渠地貌的大体指向相匹配。图中右下角的亮斑是 DISR 地表科学照明灯的照明。

将图 4 中场景的亮度、颜色与地球上类似的场景进行比较会是非常有趣的。在地球表面上，太阳直射亮度是满月照射亮度的约 50 万倍。而土卫六表面的亮度是地球上太阳直射亮度的约 1,000 分之一，或者约为地球满月照射亮度的 500 倍。即，这一照明水平相当于地球上日落 10 分钟后的情形。土卫六天空和景观呈现橙色，这是由于土卫六上的雾霾对蓝光的衰减作用远大于对红光的作用。因此，如果太阳高悬于天空，看起来就像一个非常小的亮点，是地球上看到的日面的 10 分之一，其尺寸和亮度与在约 150 m 远的地方看一辆小汽车的前灯的情况相当。太阳在土卫六表面投射下清晰的阴影，但是对比度较低，因为大约 90% 的照明来源于天空。如果太阳在天空的位置较低，则无法看到太阳。

对图 4 中 50 多块石头的尺寸进行统计，发现它们的直径尺寸介于 3 mm（成像仪分辨率的极限）到 15 cm 之间，没有发现大于 15 cm 的石块。探测器着陆前的最后一组图像拍摄于 $200 \sim 300$ m 高度，其分辨率足以识别出米量级的物体，然而在 40 m $\times 35$ m 的视场中并没有发现任何这一尺寸的物体。R 值通常被用于表现陨石坑及其周围的溅射物的尺寸−密度分布，在这里表征了土卫六地表被给定尺寸的岩石

a given size frequently used to describe the size distribution of impact craters or crater ejecta. A larger fraction of the surface is covered with rocks greater than 5 cm as opposed to smaller pebbles. The dominance of the cobbles 5–15 cm in size suggests that rocks larger than ~15 cm cannot be transported to the lakebed, while small pebbles (< 5 cm) are quickly removed from the surface. Figure 8 confirms the visual impression given by Fig. 4 that the surface coverage of rocks in the foreground of the image (< 80 cm horizontal distance from the probe) is higher than in the region beyond (about 80–160 cm). However, this trend is not seen for the pebbles less than 5 cm in size.

Elongated dark trails aligned with the general trend of the possible stream-bed visible in the centre of Fig. 4 extend from several of the distant boulders. The direction of the trails agrees with the general northwest–southeast alignment of the stream-like features shown in Fig. 3, because the last upward-looking spectra indicate that the probe settled with DISR facing southward. Images taken from the surface show no traces of the landing of the probe. The viewing direction is probably generally not downwind (the parachute is not visible).

Fig. 5. View of "shoreline" and channels. Panoramic mosaic projected from 6.5 km. showing expanded view of the highlands and bright-dark interface. As in previous figures, north is up; scale indicated. Branching and rectilinear channel networks of dark lanes are shown along an albedo boundary approximately 12 km long.

When coupled with the shapes, size selection and layering of the stones in Fig. 4, the elongated islands and their orientation parallel to the coastline in Fig. 1, the stubby and dendritic channel networks, as well as the ponds in Fig. 2 and Fig. 5, the major elements of the Titan surface albedo variations can be interpreted to be controlled by flow of low-viscosity fluids driven by topographic variation, whether caused by precipitation (the dendritic networks) or spring-fed flows (the stubby networks). We thus interpret the bright–dark albedo difference as follows: irrigation of the bright terrain results in darker material being removed and carried into the channels, which discharge it into the region offshore,

638

覆盖的比例，图 8 给出了 R 值。土卫六表面大部分被尺寸大于 5 cm 的岩石覆盖而不是更小的卵石，5～15 cm 的鹅卵石占主体，表明尺寸大于 ～15 cm 的石块不能够被运移到湖床，而尺寸较小的卵石（＜5 cm）则会很快地被从地表运移走。图 8 中 R 值的分布进一步证实了图 4 给人的视觉印象，即图像前景区域（与探测器的水平距离小于 80 cm）的石块覆盖率高于更远的区域（大约在 80～160 cm）。然而，尺寸小于 5 cm 的石子并没有表现出这种趋势。

从远处几块巨砾延伸出拉长的暗色尾迹，其延伸方向与图 4 中心可辨的疑似河床地形的总体走势平行。同时，尾迹的方向与图 3 中河流状地貌西北–东南的大体走向也相符。因为最后一批朝上拍摄的光谱显示探测器着陆后 DISR 是面向南方的，DISR 在土卫六表面拍摄的图像没有显示探测器着陆的痕迹。视野方向可能不是大体上的下风向（视野内并没有看到降落伞）。

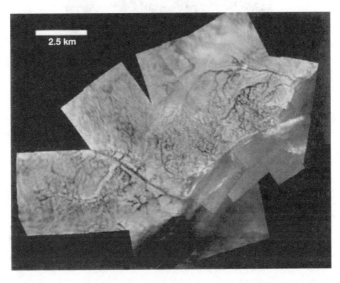

图 5. "海岸线"和河道的图像。6.5 km 高度投影得到的全景拼接图。该图对高地区域和明暗分界进行了放大。与前面的图像一样，其正上方为北，比例尺标于图中。图上展现了沿大约 12 km 长的反照率边界分布的暗色河道的分支和直线网络。

综合图 4 中石块的形状、尺寸筛选以及层状分布，图 1 中和海岸线平行的拉长的岛屿，以及图 2 和图 5 中的断株状、树枝状河道网络及池塘，可以认为形成土卫六表面反照率变化的主要因素是地形起伏驱动的低黏度流体的流动，这些流体可能是源于降雨（对应树枝状河道网络）或源于泉（对应断株状河道网络）。因此，我们对明暗反照率的不同做以下解释：对明亮地带的冲刷将导致较暗物质被带离并被搬运进河道。河流在离岸区域将这些暗色物质卸载，因此这些区域变暗。如阵风等风成

thereby darkening it. Eolian processes such as wind gusts coupled with Titan's low gravity (compared to Earth) may aid this migration.

The dark channels visible in the lowest panorama (Fig. 3) seem to suggest south-easterly fluid flow across the lower plane, depositing or exposing the brighter materials (water ice?) along the upstream faces of the ridges.

Stereographic rendering of the dendritic channels just north of the probe landing site (Figs 6 and 7) shows that the slopes in bright terrain being dissected by the putative methane river channels are extremely rugged; slopes of the order of 30° are common. This suggests relatively rapid erosion by flows in the river beds, resulting in the deeply incised valleys. Erosion by steep landslides on slopes approaching the angle of repose is probably the primary mechanism by which the rugged topography is formed. Figure 7 shows two stereographic views of the shoreline and hillside north of the landing site.

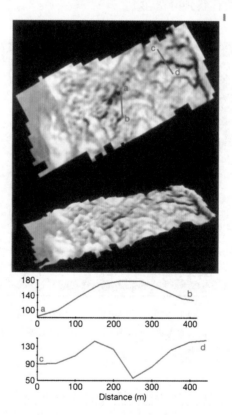

Fig. 6. Topographic model of highland region ~5 km north of the Huygens landing site. The top panel shows an orthorectified HRI image from stereo pair (vertical view). The middle panel shows a perspective view of the topographic model with ~50° tilt angle. No vertical exaggeration was applied (it is 1:1). The bottom panel shows profiles (a–b and c–d from the top panel) that illustrate the extremely steep topography in the region dissected by the drainages. All dimensions are in metres. A DISR stereo pair (using HRI frame 450 and MRI frame 601) was photogrammetrically analysed using a digital stereo workstation. The overlapping area of stereo coverage is about 1 × 3 km; the convergence angle is ~25°. The coincidence of the drainage patterns with the valley floors gives confidence in the reality of the topographic model;

过程和土卫六（相比地球）较低的引力相结合可能有助于这一物质运移过程。

从图 3 全景图像中的暗色河道看，流体是流向东南方向，穿过低地平原，沿山脊面向上游的坡面沉积下或暴露出明亮的物质（水冰?）。

图 6 和图 7 给出了探测器着陆点北侧的树枝状河道的立体影像，显示被假定的甲烷河流分隔的明亮地带中的斜坡是非常崎岖的，其坡度一般在 30° 的量级。这意味着河床遭受了流体相对快速的侵蚀，从而形成了深切河谷。河岸两侧的斜坡经过滑坡作用的侵蚀接近休止角，这种侵蚀过程可能是形成崎岖地形的主要机制。图 7 给出了着陆点北侧海岸线和山坡的立体影像。

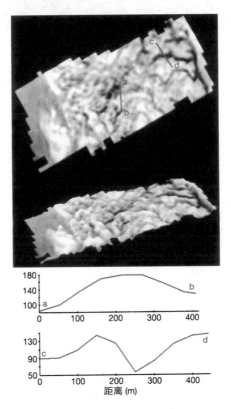

图 6. 惠更斯号着陆点北侧约 5 km 处高地区域的地貌模型。其中，上图表示一对立体图（垂直视图）经过正射校正合成的 HRI 图像；中图表示以约 50° 倾角观察地貌模型的透视图，其垂直方向比例尺没有放大（1:1）；下图给出了上图中从 a 到 b 以及从 c 到 d 点的地形剖面，展示了被河流系统所分割区域的极为陡峭的地形，三维坐标的单位都是米。我们借助数字立体影像工作站，对两幅 DISR 立体图像（HRI 第 450 帧和 MRI 第 601 帧）进行了摄影测绘分析。两幅立体图像的交叠面积大约为 1 km×3 km，会聚角约为 25°。河流系统的分布与谷底位置的相符使我们相信该地貌模型的真实性，其中高度的精确

the height accuracy is ~20 m. This preliminary model has been arbitrarily levelled so that the elevation differences are only relative.

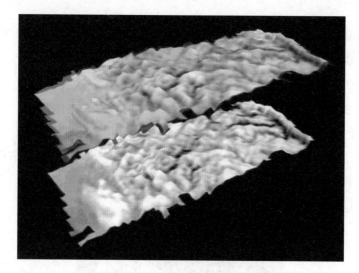

Figure 7. Titan's surface. Perspective view of Titan's surface using a topographic model of the highland region ~5 km north of the Huygens probe landing site derived from the DISR images. The model in greyscale and false colour shows the elevation (pale white highest). The lowland plane or lakebed is to the left side of the display (in blue); the northern highlands (with the dendritic channels) is to the right.

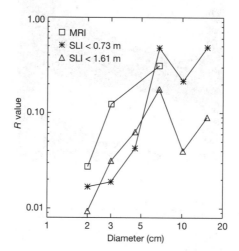

Fig. 8. Distribution of rock on the surface. Rocks larger than 1.63 cm as an R-plot, frequently used to describe size distribution of impact craters or crater ejecta. If N is the number of rocks per centimetre increment of rock size, the fraction of the surface area A covered by rocks with diameters between d and $d+\Delta d$ is approximately $N \times \Delta d \times d^2/A$. By keeping the size bin Δd proportional to the diameter d, the quantity $N \times d^3/A$ (the R value) is also proportional to the surface fraction covered by rocks of diameter d (with a proportionality constant of ~3). The plot shows R values derived from rock counts from the SLI and MRI surface images. For the SLI, R values from counts up to a distance from the probe of 73 cm and up to 161 cm are presented in separate curves. The comparison between the two curves suggests that the count is complete in the displayed size range. The increase of the R value with size corresponds to a higher fraction of the surface covered with large rocks than with smaller ones.

642

度约为 20 m。值得注意的是，这一初步模型的基准面高度是随意选取的，所以高程的变化是相对的。

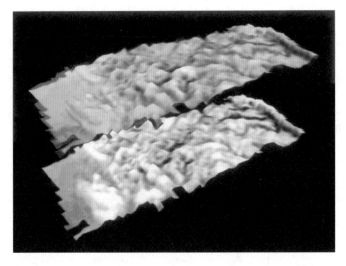

图 7. 土卫六表面的图像。由 DISR 图像得到的、惠更斯号探测器着陆点北侧约 5 km 处高地区域的地貌模型的透视图。其中，模型中的灰度和假彩色表示地形的高度（浅白色代表最高）。图中左侧的蓝色区域代表低地或者湖床，而右侧是布满树枝状沟渠的北方高地。

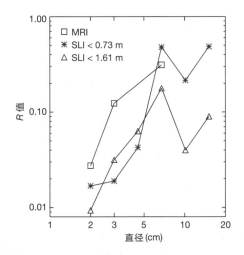

图 8. 地表岩石分布。图中给出了尺寸大于 1.63 cm 的石块的 R 值图，其中 R 通常被用于描述陨石坑及陨石坑周围溅射物的尺寸分布。如果用 N 表示石块尺寸每增加一厘米间隔范围内的石块数量，那么直径介于 d 和 d+Δd 之间的石块在大小为 A 的表面积内的覆盖比例为 $N \times \Delta d \times d^2/A$。如果令尺寸间隔 Δd 正比于石块直径 d，则 R 值的大小 $N \times d^3/A$，也正比于直径为 d 的石块覆盖的面积比例，比例常数约为 3。图中 R 值是通过统计 SLI 和 MRI 表面图像中的石块得到的。对于 SLI，距离探测器 73 cm ~ 161 cm 的石块统计结果由单独的一条曲线表现。通过比较这两条曲线可知，计数统计在所示的尺寸范围内都是完整的。图中 R 值随石块尺寸增加而增加，符合大石块的表面覆盖率高于较小石块的现象。

The Wind Profile

Assembly of the panoramic mosaics leads to the construction of a descent trajectory as part of an iterative process of image reconstruction. The trajectory can be used to derive the probe ground track and extract the implied wind velocity as a function of altitude. Correlation of the roughly 200 usable images acquired by DISR during its descent resulted in longitude and latitude values versus time, displayed in Fig. 9. Fitted by polynomials, these ground tracks were differentiated with respect to time and scaled to derive the horizontal wind speed and direction as functions of altitude.

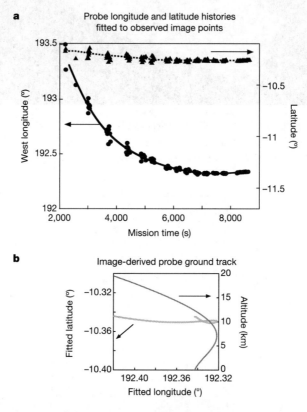

Fig. 9. Probe ground track. **a,** Sub-probe west longitude and latitude histories of the Huygens probe derived from panoramic image reconstructions. Arrows indicate the appropriate vertical axis. The image data points from which the latitude and longitude were derived are shown as triangles and dots respectively. The dotted and solid lines show polynomial fits to the data. Results adjusted to agree with the Descent Trajectory Working Group (DTWG-3) values at 2,200 s after T_0 (mission time). **b,** Probe longitude versus latitude (thicker line) and versus altitude (thinner line). The altitude axis needs to be expanded by a factor of almost six to recover one-to-one correspondence on a linear scale because the total longitudinal variation is less than 4 km. Using the Doppler Wind Experiment's (DWE) high-altitude references[33], the touchdown point (the predicted landing site) was extrapolated to west longitude 192.34°, latitude −10.34°. Using DTWG-3 high-altitude references, the touchdown point was extrapolated to 192.36°, latitude −10.36°.

The results indicate that the probe's steadily increasing eastward drift caused by Titan's

风 况 描 述

　　构建全景拼接图像的同时可以构造出质量较好的探测器降落过程的空间运动轨迹，这也是图像重建迭代过程的重要部分。通过轨迹可以推导出探测器轨迹在地面上的投影，并且可以提取出不同高度的风速信息。我们将 DISR 在下降阶段获取的大约 200 幅有用图像相互关联，以此计算出探测器的经度和纬度随时间的变化，如图 9 所示。我们对轨迹地面投影进行多项式拟合，然后对时间做微分并换算成实际空间距离，可以得出水平的风速和风向随高度变化的函数。

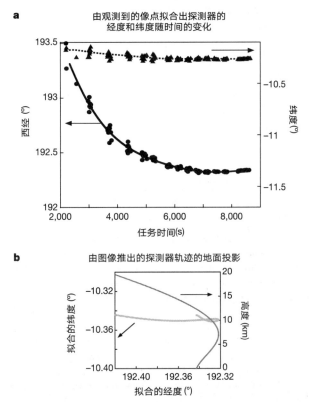

图 9. 探测器轨迹地面投影。a, 根据全景图像重建获得的惠更斯着陆探测器的西经和纬度随时间的变化。图中的箭头指向各自对应的纵轴，三角和圆点分别表示用于推出纬度和经度的图像数据点。图中的点线和实线是对以上数据的拟合曲线。结果已经经过校正以符合下降轨迹工作组 (DTWG-3) 在 T_0 (任务时间) 之后 2,200 s 的值。b, 探测器经度与纬度 (粗线) 以及经度与高度 (细线) 的关系。由于经向距离总变化小于 4 km，与高度对应的轴需要放大接近 6 倍才能实现比例尺上的一一对应。使用多普勒风实验仪 (DWE) 中的高空数据作为参照 [33]，推算出降落点 (预计的着陆点) 为西经 192.34°，纬度为 −10.34°。采用 DTWG-3 的高空数据作为参照，则推算出着地点位于西经 192.36°，纬度 −10.36°。

　　以上结果表明，探测器从 50 km 高度下降至 30 km 高度过程中，由于土卫六风

prograde winds, as shown in Fig. 10, slowed from near 30 to 10 m s^{-1} between altitudes of 50 and 30 km and slowed more rapidly (from 10 to 4 m s^{-1}) between altitudes of 30 and 20 km. The winds drop to zero and reverse at around 7 km, near the expected top of the planetary boundary layer, producing a west-northwestwardly motion extending for about 1 km during the last 15 min of the descent (see Fig. 3). The generally prograde nature of the winds between 50 and 10 km agrees with models of Titan's zonal winds available before the arrival of Cassini or the Huygens probe[7], although the wind speed is somewhat less than the average predicted before entry.

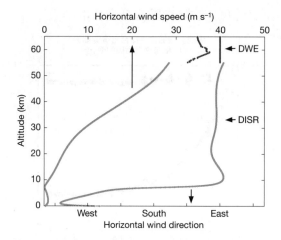

Fig. 10. Observed winds. Horizontal wind speed and direction (counter-clockwise from east) as a function of altitude. The green lines are the DISR data and the blue lines are the high altitude DWE data[33] (showing reasonable consistency between the two). The lines on the left show the wind-speed profile, and on the right is the wind direction. The wind is computed from the combined longitude and latitude reconstructions displayed in Fig. 9. Titan's prograde winds slow from about 28 m s^{-1} at 50 km to 10 m s^{-1} near 30 km altitude, then decrease more rapidly from 30 km (10 m s^{-1}) down to 7 km where they drop to zero. Below 7 km (which is near the expected top of the planetary boundary layer) the winds reverse and become retrograde, and the speed increases to about 1 m s^{-1} around 2–3 km before dropping to almost zero ($\sim$0.3 m s^{-1}) near the surface. The direction begins as due east, and then turns through south (beginning between 9–7 km) to the west-northwest between 7–5 km. The winds are extrapolated to be retrograde at the surface, but the two-sigma error bars (not shown) of 1 m s^{-1} at the surface could include surface prograde winds. The error bars at 55 km altitude (4 m s^{-1}) are consistent with continuity from the DWE measurements.

The planetary boundary layer is calculated to have a thickness of between 4 and 8 km, based on scaling the Earth's near-equatorial planetary-boundary-layer thickness of 1–2 km by the inverse square-rooted ratio of the planetary rotation rates. The minimum horizontal wind speed at 7 km can thus be an indication of entry into the boundary layer[8]. The reversal of wind direction at this altitude is also consistent with the Voyager-derived equatorial temperature profile[9], wherein the temperature gradient changes from dry adiabatic to a sub-adiabatic temperature gradient above 4 km altitude, indicating the top of the boundary layer.

The current ground-track and wind-speed analysis predicts winds of about 0.3 to 1 m s^{-1} near the surface. This velocity can be produced by any number of sources including pressure and temperature gradients and tides[10].

646

的顺行影响平稳地向东飘移，如图 10 所示，但速度从大约 30 m/s 降至 10 m/s；在从 30 km 高度下降至 20 km 高度过程中，速度迅速地从 10 m/s 降至 4 m/s。当探测器降至大约 7 km 高度（靠近预测的行星边界层顶部）时，风速减为零，然后风向改为相反的方向，造成探测器在降落阶段的最后 15 分钟朝西北偏西的方向飘移了约 1 km，如图 3 所示。50 km 到 10 km 高度的风向和土卫六转动的方向相同，这与卡西尼或惠更斯探测器到达前就已提出的土卫六纬向风模型相一致 [7]，但风速比探测器进入前的预测平均值略小。

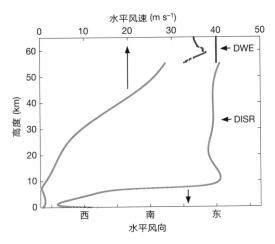

图 10. 观测到的风况。图中给出了水平风速和风向（自东沿逆时针方向）与高度的关系。其中，绿线代表 DISR 数据，而蓝线代表高空 DWE 数据 [33]，二者之间显示出合理的一致性。图中左侧曲线给出了风速剖面，而右侧曲线给出了风向，这两条曲线由图 9 中的经纬度信息重建得到。土卫六上的顺行风的风速由 50 km 高度处大约 28 km/s 下降到 30 km 高度附近约 10 m/s，最后在 7 km 高度迅速降为 0。在 7 km（靠近预期的行星边界层顶部）以下，风向则变为逆行方向，并且风速在 2 ~ 3 km 高度增加到 1 m/s 左右，在到达表面附近时则几乎为 0（~ 0.3 m/s）。风向首先向正东，然后转向南（起始于 9 ~ 7 km），并在 7 ~ 5 km 转为西北偏西。尽管地表附近风向的推断结果是反向风，但地表附近 1 m/s 风速的 2σ 误差范围（未在图中显示）包含了一部分顺向风的区间。55 km 高度处（对应风速 4 m/s）的误差范围与 DWE 测量得到的连续性相一致。

对土卫六和地球的旋转速率比值的平方根取倒数，并乘以地球赤道附近的行星边界层厚度（1 ~ 2 km），就可以计算出土卫六的行星边界层厚度为 4 ~ 8 km。7 km 高度处的最小水平风速，可以被认为是进入行星边界层的信号 [8]。而在 7 km 高度处的风向改变也与旅行者号飞船得出的赤道温度剖面相符 [9]。旅行者号的测量表明，在大于 4 km 的高度，温度梯度从干绝热变化到亚绝热，这意味着边界层的顶部。

目前，通过对探测器轨迹地面投影和风速分析，预测土卫六地表附近的风速大约为 0.3 ~ 1 m/s，这一速度可能是由气压、温度梯度和潮汐等因素引起 [10]。

Migration of Surface Material

The acquisition of visible spectra at known locations in the images allowed correlation of the reflectance spectra with different types of terrain. The Downward-Looking Visible Spectrometer (DLVS) was an imaging spectrometer measuring light between 480 and 960 nm as it projected the image of the slit onto the ground into up to 20 spatial resolution elements for nadir angles from 10° to 50°. Spectra were collected at nominally the same azimuths as the images, though often at slightly different altitudes (on different probe rotations). Interpolation between the times at which the spectral and image data were obtained located the spectra within the images. Determination of the surface reflectivity was hindered by scattering from the haze between the camera and the surface as well as by methane absorption. Correlation of the spectra with images was therefore best performed on measurements during which the altitude changed only slightly.

The centre of the image in Fig. 11 is displayed in true colour (that is, as the human eye would see it under Titan's atmosphere) using actual spectral data from one panorama. The area between the spectra is interpolated in azimuth. The coverage with spectra is similar to that shown in Fig. 12. The orange colour is due mainly to the illumination of the surface. Scattering and absorption (which dominate in the blue) cause the perceived true colour of the surface to change from yellow to orange with decreasing altitude. Note that the passband of the cameras peaked in the near infrared (at 750 nm), and therefore the brightness variations in the images would not necessarily be seen by the human eye.

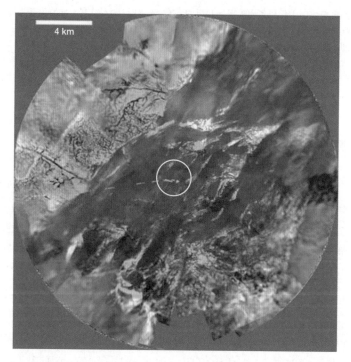

Fig. 11. The surface of Titan displayed in true colour. As seen from an altitude of 8 km. See Fig. 12 for the location of the spectrometer's footprints. Some bright features appear to be overexposed because they are

地表物质的运移

通过采集图像中已知位置的可见光谱，可以将反射光谱与不同类型的地形建立联系。探测器采用俯视可见光谱分析仪（DLVS），是一种用于测量 480～960 nm 波段光线的成像光谱分析仪，它将分析仪狭缝的像投影到地面上，在 10° 到 50° 的天底角最多可以覆盖 20 个空间分辨单元。虽然标称可以测量与图像相同方位角的光谱，但光谱与对应图像的采集高度往往略有不同（探测器未发生明显的偏转）。通过对光谱和图像数据的采集时刻进行内插分析，可以在图像中确定光谱数据对应的覆盖区域。照相机和土卫六地表之间的雾霾的散射作用和甲烷吸收的影响阻碍了地表反射率的确定。因此，测量过程中探测器高度改变较轻微的情况下，光谱和图像之间会有较好的相关性。

借助全景照片的真实光谱数据，图 11 中心附近的图像可以以真实彩色呈现（即和人眼在土卫六大气下实际看到的颜色一样）。光谱之间的区域按方位进行内插。光谱覆盖范围类似于图 12 所示。橙色的色调主要是由于地面被照亮。随着高度的降低，散射和吸收（主要作用在蓝色波段）将使地面的真实颜色从黄色变为橙色。需要注意的是，照相机的滤波通带峰值处在近红外 750 nm，因此，人眼实际观察不一定看到图像中那样的亮度变化。

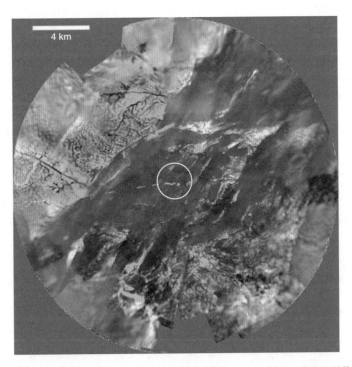

图 11. 真实颜色下的土卫六表面图像。从 8 km 高度看到的图像，图 12 标注了光谱仪覆盖区域的位置。图中亮度主要是根据近红外光谱得出的，部分地貌单元由于颜色过亮而产生过曝。真实彩色表达为红绿

too bright for their colour (the brightness in this image mainly derives from the near-infrared spectrum). True colour is expressed in red-green-blue (RGB) values that are derived by multiplying the spectra with the Commission Internationale de l'Eclairage colour-matching functions (with the 6,500-K correlated colour temperature, the D65 white point). The circle shown is the extent of the lowest panorama (Fig. 3).

Fig. 12. Reflectivity samples of Titan's surface. A panorama of Titan's surface overlaid with DLVS footprints coloured according to the 827 nm/751 nm intensity ratio, coded from red (high) to green (low). Spectral footprints (the small rectangles) selected for analysis in Fig. 13 are outlined in white. The panorama shows an area of 23 by 23 km. Areas A, B and C are referred to in Fig. 13.

In Fig. 12 the images are correlated with the ratio of the intensity in two methane windows (827 nm/751 nm) located in the infrared part of the spectrum where scattering is minimal and the systematic variability with nadir angle can be ignored. Reddening (high 827 nm/ 751 nm ratio) is concentrated at the area covered with drainage channels (north and northwest in the pan of Fig. 12) and to a lesser degree to the lake area adjacent to the coastline. The lake area in the southeast is not reddened. A preliminary analysis of spectra recorded in other panoramic cycles indicates that the land area in the northeast, which is not covered by drainage channels, is only moderately reddened compared to the river area. The reddening is not restricted to these two methane windows. Figure 13 shows that, in fact, it is present over the whole visual range, amounting to about 6% per 100 nm (note that atmospheric backscatter dominates over surface reflection at wavelengths below 600 nm).

蓝（RGB）颜色值，这些值是由光谱乘以国际照明委员会颜色匹配函数（具有 6,500 K 色温，D65 白点）得到的。图中圆圈给出了最低全景图像（如图 3）的范围。

图 12. 土卫六地表的反射率样例。图中将土卫六表面的全景图像与 DLVS 的覆盖区域叠加在一起，其中 DLVS 覆盖区域按照 827 nm/751 nm 两波段强度之比的大小着色，即比值越大则为红色，比值较小则为绿色。图 13 分析的光谱覆盖区用白色矩形框标记出。全景图像的面积为 23 km×23 km。而区域 A、B、C 光谱具体情况可以参考图 13。

在图 12 中，图像与两个甲烷窗口信号（827 nm/751 nm）的强度之比关联起来，这两个窗口位于红外波段，在该波段散射效应最小，并且天底角的系统性变化可以忽略。红化（即高 827 nm/751 nm 比值）集中在排水河道分布的区域（图 12 中的北侧和西北侧）；在靠近海岸线的湖区也存在相对较弱的红化。位于东南侧的湖区没有红化。我们对其他全景拍摄周期记录的光谱数据进行了初步分析，发现东北方向的陆地区域（未分布排水河道）相对于河流区域只是被中等红化。红化现象并不是局限在两个甲烷观测窗口。如图 13 所示，实际上红化存在于整个可见光波段，相当于每 100 nm 大约为 6% 的变化（值得注意的是对于小于 600 nm 的波段，大气背散射的影响显著地强于表面反射）。

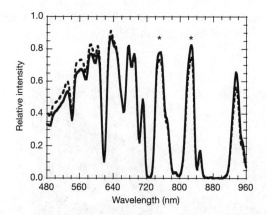

Fig. 13. Spectral comparison of bright highlands and dark lowlands. Spectra of the dendritic river highlands and lakebed lowlands areas are compared. To restrict the influence of the atmosphere, only spectra recorded at the same nadir angle are selected. The solid and dashed curves are the average spectra associated with the two spectral pixels outlined in white near the areas marked "A" (dendritic highlands, solid) and "B" (dark lakebed, dashed), respectively in Fig. 12. The two pixels bordering "C" in Fig. 12 yield a spectrum intermediate to the "A" and "B" spectra. The spectra have been corrected for albedo by dividing by the total intensity to emphasize the difference in slope. Not shown here is that the reflectivity of the dendritic area ("A") is higher than that of the lakebed area ("B") at all wavelengths by roughly a factor of two. The asterisks denote the methane windows taken for the reddening ratio in Fig. 12.

The DLVS data clearly show that the highlands (high-albedo area) are redder than the lakebed (low-albedo area). Spectra of the lakebed just south of the coastline are less red than the highlands but clearly more red than the lakebed further away (that is, to the southeast). The data suggests that the brighter (redder) material of the hilly area may be of local origin, and is corrugated by rivers and drainage channels, and that the darker material (less red) is a substance that seems to be washed from the hills into the lakebed. It could be connected to the alteration of the highland terrain, either by precipitation, wind and/or cryoactivity. Additionally, it could indicate that the surface of the lowland area may be covered by different materials in regions that exhibit diverse morphology.

Surface Reflectivity and Methane Mole Fraction

Spectra taken near Titan's surface allow measurement of its reflectance and determination of the local methane mole fraction. These measurements provide clues as to the make-up of Titan's crust. The Downward-Looking and Upward-Looking Infrared Spectrometers (DLIS and ULIS) cover the region from 840 to 1,700 nm with a resolution of 15 to 20 nm. The ULIS looks up at a half-hemisphere through a diffuser while the DLIS projects its slit into a 3 by 9° field centred at a 20° nadir angle. Below 700 m altitude, a 20-W lamp was turned on to illuminate the surface at wavelengths where solar light had been completely absorbed by methane in Titan's atmosphere. At low altitudes we took repeated DLIS and ULIS spectra at short integration times (1 s). Nine DLIS and seven ULIS spectra were received between 734 and 21 m altitude. The DLIS continued to measure surface spectra free of atmospheric

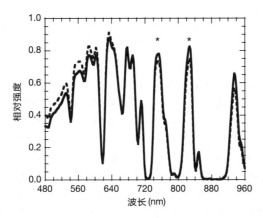

图 13. 较亮高地和较暗低地的光谱对比。我们在图中将分布有树枝状河流的高地的光谱与湖床低地区域的光谱进行了对比。为了限制大气的影响，我们只挑选了以相同天底角采集的光谱。图中实线（虚线）分别表示图 12 中 A(B)区域附近白框标示的两个光谱像素的平均光谱，A 区域是树枝状高地，用实线表示；B 区域是暗的湖床，用虚线表示。而图 12 中 C 区周围的两个像素的平均光谱介于 A 和 B 光谱之间。图中光谱已经通过除以总亮度进行了反照率校正，从而可以突出光谱斜率的不同。图中未显示的信息是，在整个波段范围，树枝状区域（A 区域）的反射率是湖床区域（B 区域）的 2 倍左右。图中星号表示两个甲烷吸收窗波段，这两个波段用于表征图 12 中的红化程度。

DLVS 数据清楚地表明，具有高反照率的高地比低反照率的湖床红化程度更高，紧邻海岸线南侧的湖床的光谱比高地的红化要弱，但比离岸更远些（即东南方向）的湖床红化明显更强。数据说明丘陵地带的更亮（更红）物质可能是在原位形成的，并在河流的作用下形成波状褶皱结构，而较暗（更不红）的物质则看起来像是被从山地冲洗进湖床的。这可能与高原地形的改造有关，可能是降雨、风和（或者）冰冻等机制造成的。此外，低地地区的一些地区展现出地貌的多样性，这些区域的表面可能覆盖着不同的物质。

表面反射率与甲烷摩尔分数

在接近土卫六表面得到的光谱使我们可以测量其反射率，并确定局部的甲烷摩尔分数。这些测量计算可以为确定土卫六地壳组成提供线索。探测器采用的俯视红外光谱分析仪（DLIS）和仰视红外光谱分析仪（ULIS）可以覆盖从 840 nm 到 1,700 nm 的波段，其分辨率为 15～20 nm。ULIS 可以通过漫射器向上观测整个半球，而 DLIS 则将其狭缝投影到以 20° 天底角为中心、大小为 3°×9° 的视场。在小于 700 m 高度，一盏功率为 20 W 的照明灯被开启用来照射土卫六表面，其发光波长恰是太阳光被土卫六大气中的甲烷完全吸收的波段。在低空，我们以较短的积分时间（1 s）不断重复测量 DLIS 和 ULIS 光谱。在 734 m 到 21 m 的下降高度之间，探测器总共

methane absorption after landing. About 20 such identical spectra were acquired from a distance of a few tens of centimetres of the surface.

DLIS spectra at all altitudes clearly showed an additional signal when the lamp was on. However, at the highest altitudes, the lamp reflection from the surface was negligible, so the additional signal was solely due to scattered light from the lamp into the instrument. This scattered light was estimated from the intensity level in the strong methane bands in the highest-altitude spectrum recorded with the lamp on, and removed from all DLIS spectra. After this correction, only spectra at 36 m and especially 21 m showed significant signal from the lamp. This signal dominated the upward intensity due to solar illumination in all regions of strong and moderate methane absorption, while the latter dominated in the methane windows.

This spectrum, which represents the product of the ground reflectivity and the two-way methane transmission, is shown in Fig. 14 and compared to synthetic spectra with methane mole fractions of 3%, 5% and 7%. The ground reflectivity assumed in these model calculations is shown in the inset. Four methane bands are seen in the lamp-only spectrum. The good correlation with the models, notably in the weak structures at 1,140, 1,370 and 1,470 nm, indicates a high signal-to-noise ratio of about 50. The best fit is achieved with a methane mole fraction of 5%, which is in firm agreement with the 4.9% *in situ* measurements made by the Gas Chromatograph Mass Spectrometer[11]. Most structures are well reproduced; notable exceptions are the detailed shape of the 1,000 nm band and the absorption shoulder near 1,320 nm.

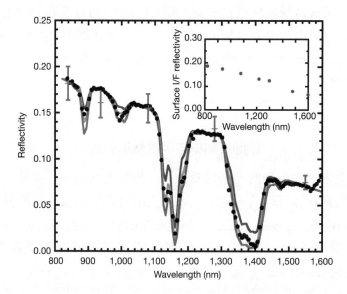

Fig. 14. Derivation of methane mole fraction. Lamp-only downward-looking spectrum from altitude of 21 m (black data points). The red line with three-sigma error bars indicate absolute reflectivity in methane windows estimated from infrared measurements. This spectrum is compared to three models: 3% (blue), 5% (green), and 7% (red) methane mole fractions. These models make use of surface reflectivity at seven

采集了 9 张 DLIS 光谱和 7 张 ULIS 光谱。探测器着陆后，DLIS 在没有大气甲烷吸收影响的情况下继续测量土卫六表面的光谱；在距离表面大约数十厘米的位置获得了大约 20 张这样的光谱。

所有高度的 DLIS 光谱都清楚地显示，当照明灯打开后，出现了一个额外的信号。然而当探测器位于最高的高度时，地面对照明灯光的反射则可以忽略，因此光谱中的额外信号应该仅仅是来自照明灯光的散射。根据照明灯打开后记录的最高高度光谱中的强甲烷吸收带的强度，可以估算出照明灯散射光的强度值，并将其从所有的 DLIS 光谱中剔除掉。经过此校正之后，只有在 36 m，尤其是 21 m 高度处的光谱还存在明显的照明灯信号。该信号决定了向上的光强，这是由于太阳光的所有波段都有强–中等的甲烷吸收，而后者只是在甲烷窗口才起决定作用。

图 14 中给出的光谱是地面反射率与来回两次通过甲烷的穿透率的乘积所给出的结果，并与甲烷摩尔分数分别为 3%、5% 以及 7% 的合成光谱相比较。图 14 右上角的插图给出了计算模型中采用的反射率。在只有照明灯情况下的光谱中可以观察到四个甲烷吸收带。测量数据与模型符合得很好，在 1,140 nm、1,370 nm 和 1,470 nm 位置均出现了较弱的吸收结构，这表明了约为 50 的高信噪比。甲烷摩尔分数为 5% 时，可以实现最佳拟合，这与使用气相色谱–质谱联用仪进行原位测量得到的 4.9% 的甲烷摩尔分数严格一致 [11]。光谱中大多数结构都可以被很好地重现，但 1,000 nm 处吸收带以及 1,320 nm 附近的吸收肩的细节存在明显的差异。

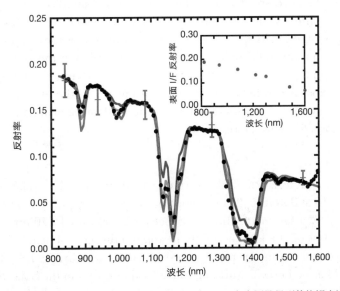

图 14. 甲烷摩尔分数的推导。图中给出了开启照明灯后，在 21 m 高度测量得到的俯视光谱（黑点表示数据点）。红线表示通过红外测量估计得到的甲烷窗口绝对反射率，同时给出了 3σ 误差区间。作为对比，图中也给出了甲烷摩尔分数分别为 3%、5% 和 7% 的三种模型结果，分别用蓝线、绿线和红线表示。这些模型结果通过对 7 个波长处的表面反射率（如图中右上小图中蓝点所示；其中 I/F 表示谱强度与太阳

wavelengths (shown in inset by the blue points; I/F is the ratio of the intensity to the solar flux divided by π) and linearly interpolated between. From the lamp-on infrared spectra, a lamp-only spectrum at 21 m (representing the spectrum observed by DLIS in the absence of solar illumination) was obtained as follows. First, the reflectivity in regions of negligible methane absorption (at 840, 940, 1,070, 1,280 and 1,500–1,600 nm) was estimated by the ratio of mean upward intensity (measured by DLIS) to mean downward intensity (measured by ULIS). The mean upward intensity is the average measured over the seven low-altitude DLIS spectra showing no contribution from the lamp (734 to 53 m). The mean downward intensity was obtained by averaging the strongest intensity with the weakest intensity. This average gives reasonable approximation of the downward flux divided by π. The ratio of the mean upward intensity to the mean downward intensity gives reflectivity. Two corrections were required in this analysis: correction for the spatial response of the ULIS diffuser and correction for the solar illumination in the DLIS 21-m spectrum. The correction for diffuser response ranged from 15% (840 nm) to 25% (1,550 nm), assuming a haze optical depth of ~2 at 938 nm. The contribution of solar illumination in the DLIS spectrum at 21 m was eliminated by subtracting the average of the DLIS spectra at 85 and 109 m where the lamp contribution was negligible. The difference spectrum was then divided by the spectral response of the lamp and scaled by a constant to match the continuum reflectivities inferred previously, producing the lamp-only spectrum at 21 m.

We conclude that the methane abundance is $5 \pm 1\%$ in the atmosphere near the surface. The corresponding (two-way, and including the 20° lamp inclination) methane column abundance in the spectrum is 9.6 m-amagat (or a column 9.6 m high at the standard temperature and pressure of 273 K and 1 atmosphere). With a temperature of 93.8 K and pressure of 1467.6 mbar (ref. 12), the relative humidity of methane is about 50% using Brown and Ziegler's saturation law[13]. Therefore, methane near the surface is not near saturation, and ground fogs caused by methane in the neighbourhood of the landing site are unlikely.

The ratio of the observed spectrum to the methane transmission, restricted to spectral regions where the latter is higher than 90%, is shown in Fig. 15a ("plus" symbols). It is compared to one of the DLIS spectra after landing, divided by the lamp spectral response and rescaled by a constant reflectivity factor. Note that although this spectrum shows signs of methane absorption at 1,140–1,160 nm and 1,300–1,400 nm, no attempt of correction was made, in the absence so far of accurate information on the absorption path lengths. The agreement between the shapes of the two independent determinations of the ground reflectivity adds confidence to the result.

The four major characteristics of the surface spectrum are: (1) a relatively low albedo, peaking around 0.18 at 830 nm; (2) a red slope in the visible range; (3) a quasilinear decrease of the reflectivity by a factor of about two between 830 and 1,420 nm; and (4) a broad absorption, by ~30% of the local continuum, apparently centred near 1,540 nm (although its behaviour beyond 1,600 nm is poorly constrained) as seen in Fig. 15b. This spectrum is very unusual and has no known equivalent on any other object in the Solar System.

656

通量的比值除以 π) 进行线性插值得到。根据开启照明灯后的红外光谱可以推出 21 m 高度处只有照明灯信号的光谱 (代表了没有太阳光照环境下的 DLIS 观测光谱),具体步骤如下。首先,在某些甲烷吸收可以忽略的波段 (如 840 nm、940 nm、1,070 nm、1,280 nm 以及 1,500 ~ 1,600 nm),反射率可以通过向上的平均谱强度 (DLIS 测量得到) 与向下的平均谱强度 (ULIS 测量得到) 的比值估计出来。其中,向上的平均谱强度由不存在照明灯光影响的 7 个低高度 DLIS 光谱测量值 (734 m 到 53 m) 的平均后得到。通过对最大强度和最小强度取平均可以求出向下的平均谱强度;该平均值是向下光通量除以 π 的结果的合理近似。向上的平均谱强度与向下的平均谱强度的比值给出了反射率。在该分析中,需要考虑两种校准:ULIS 漫射器的空间响应的校准和 21 m 高度处 DLIS 光谱的太阳光照的校准。假设雾霾在 938 nm 处的光学深度约为 2,则散射器响应的校准范围介于 15% (对应 840 nm) 和 25% (对应 1,550 nm) 之间。由于在 85 m 和 109 m 高度处照明灯对 DLIS 光谱的贡献可以忽略,因此通过减去这两个高度的 DLIS 光谱平均值,从而消除 21 m 高度处 DLIS 光谱中太阳光照的贡献。然后用相减后得到的光谱除以照明灯的光谱响应,并乘以一个常数,从而与前面推论的连续谱反射率相符;这样就得到了只考虑照明灯情况的 21 m 高度处的反射率光谱。

因此我们得出以下结论:近地表大气中甲烷的丰度为 5% ± 1%。相应的 (双程的且包括了 20 度照明灯倾角的) 谱中甲烷柱丰度为 9.6 米阿马加 (即相当于 273 K 的标准温度和 1 个标准大气压下 9.6 米高度柱)。在实际温度为 93.8 K 和气压为 1467.6 mbar 的条件下 [12],利用布朗和齐格勒的饱和定律 [13],可以计算出甲烷的相对湿度大约为 50%。因此,地表附近的甲烷并未接近饱和,着陆点附近的雾不太可能是由甲烷形成的。

图 15a 给出了光谱区域内观测到的光谱数据与甲烷穿透率的比值 (用加号表示),限制在后者高于 90% 的区域内。图中也从着陆后的 DLIS 光谱中选择了一个与之进行比较,其中该光谱曲线是除以照明灯的光谱响应并利用常量反射率因子重新调整后的结果。值得注意的是,尽管该光谱在 1,140 ~ 1,160 nm 和 1,300 ~ 1,400 nm 波段出现了甲烷吸收的迹象,由于迄今还未获得关于吸收波程长度的精确信息,因此我们并没有尝试对光谱进行校正。由该图可知,通过两种独立方法确定的地面反射率符合得很好,从而进一步提高了结果的可信度。

表面光谱主要有以下四个特征:(1) 相对较低的反照率,在 830 nm 达到峰值约 0.18;(2) 在可见光波段存在红坡;(3) 反射率从 830 nm 到 1,420 nm 准线性地降低了约一半;(4) 如图 15b 所示,存在以 1,540 nm 附近为中心的较宽的吸收峰 (波段宽度约占局域连续谱的 30%),但是在大于 1,600 nm 的情况未得到很好的约束。该光谱特别不同寻常,在太阳系的其他天体上还未发现这样的情况。

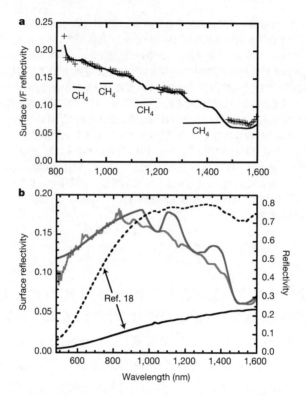

Fig. 15. Reflectance of Titan's surface. **a**, Reflectivity measured from 21 m altitude ("plus" symbols) compared to the reflectivity after landing (solid curve). The methane absorption bands are indicated by the CH₄ symbol. **b**, Surface reflectivity as measured after landing (red line). It is compared with a simulation (blue line) of a mixture of large-grained (750 μm) low-temperature water ice, yellow tholins, and an unknown component with featureless blue slope between 850 and 1,500 nm. Spectra of two different organic tholins: a yellow tholin (dashed line) and a dark tholin (solid black line) from ref. 18 are also shown for comparison (reflectance scale reduced by a factor of 4). We are attempting to identify or synthesize the missing blue material in our laboratory.

Ground-based spectroscopic observations have provided strong evidence, although spectrally restricted to the methane windows, for the presence of water ice on Titan's surface[14], coexisting in variable proportions with a dark component, presumably of organic nature[15,16]. Water ice may explain the 1,540 nm band, as illustrated in Fig. 15b by a simulation of the reflectance spectrum of a mixture of low-temperature water ice[17], yellow tholins[18] and a spectrally neutral dark component. This identification is reasonable in the context of the light-coloured rocks present at the landing site (Fig. 4), but not conclusive, because some organics do show absorption at a similar wavelength. This is the case, notably, for bright yellow-orange tholins produced in laboratory experiments[19,20] (shown in Fig. 15b), which partly contribute to this band in the simulation and which may account for the red slope in the visible range of the surface spectrum. It is probably this material, existing as aerosol particles, that absorbs the blue wavelengths, which would explain the yellow-orange colour of Titan's atmosphere as seen from space or from the surface.

We note the remarkable absence of other absorption features in the surface spectrum along with the 1,540 nm band. This is at odds with predictions that some specific chemical bonds,

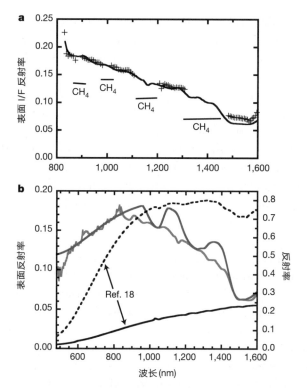

图 15. 土卫六地表的反射率。**a,** 在 21 m 高度测量得到的反射率（加号表示）与着陆后测量的反射率（实线表示）的对比。图中的 CH₄ 标出了甲烷吸收波段的位置。**b,** 探测器着陆后测量得到的表面反射率（红线）与模拟结果（蓝线）的对比。其中，模拟曲线采用的是包含大颗粒（750 μm）的低温水冰、黄色的索林斯以及一种在 850～1,500 nm 之间存在无特征斜坡的未知成分的混合物。同时图中还给出了黄色索林斯（虚线）和暗色索林斯（黑色实线）两种有机物的谱线用于对比（反射强度的纵坐标比例尺缩小为四分之一）[18]。目前，我们正在实验室尝试识别或合成这种缺失的蓝色物质。

尽管基于地表的光谱观测结果限制在几个甲烷窗口内，但该结果也提供了强有力的证据表明土卫六表面存在水冰[14]；这些水冰与某种可能具有有机物特性的较暗的成分以变化的比例共存[15,16]。水冰也许可以解释 1,540 nm 处的吸收峰，图 15b 中给出了低温水冰[17]、黄色索林斯[18]以及一种光谱呈中性的暗色组分的混合物的反射率模拟结果。由于着陆点附近的岩石颜色较淡（见图 4），前文所作的鉴别具有合理性但不是结论性的，因为一些有机物在相似的波段也具有吸收峰。实验室制备的亮橘黄色的索林斯尤其如此[19,20]（如图 15b 所示），它对于模拟中出现的吸收峰具有一定的贡献，同时可能用于解释表面光谱可见光范围内的红端斜坡。或许就是这种物质以气溶胶粒子的形式存在并吸收了蓝色光波段的光，造成了从太空或从土卫六表面观察到的土卫六大气呈现橘黄色。

我们注意到表面光谱中并没有与 1,540 nm 吸收带共存的其他吸收结构，这与以往研究得出的预测不符。根据先前的研究，在表面光谱中，将会出现某些化学键

in particular C–H or C≡N, and possibly the individual bands of atmospherically abundant species, such as ethane (C_2H_6), acetylene (C_2H_2), propane (C_3H_8), ethylene (C_2H_4), hydrogen cyanide (HCN) and their polymers, would show up as signatures in the surface spectrum.

The most intriguing feature in the surface spectrum is its quasilinear featureless "blue slope" between 830 and 1,420 nm. As briefly illustrated in Fig. 15b, a featureless blue slope is not matched by any combination of laboratory spectra of ices and complex organics, including various types of tholins. Depending on their composition and structural state (for example, abundance, extension and/or clustering of sp^2 carbon bonds), organic materials in the near-infrared exhibit either distinct absorption bands (for example, bright yellow-orange tholins, Fig. 15b), or a feature-poor red slope (for medium to low-albedo organics), or a very dark and flat spectrum[18,21].

Assessing the material responsible for the blue slope is a major challenge and also a prerequisite for a secure identification of the 1,540 nm band. If this band is indeed mostly due to water ice, an intimate mixing of this ice with a material displaying a strong "infrared-blue" absorption would explain the absence of the weaker H_2O bands at 1.04 and 1.25 μm in the surface spectrum, as demonstrated for several dark icy satellites, where these bands are hidden by the presence of an organic component (but neutral or reddish). Decreasing the water-ice grain size alone cannot suppress the 1.04- and 1.25-μm bands and at the same time maintain the apparent blue slope that is produced by large-grained water ice (considering only the continuum absorption between the infrared bands). To hide these weak water bands efficiently, the mixture would need to be ice and a material having a stronger and decreasing-with-wavelength infrared absorption.

Haze Particle Size

The haze particles in Titan's atmosphere have long been known to produce both high linear polarization and strong forward scattering. This has been taken to imply that the particles are aggregates of small "monomers" in open structures. The amount of linear polarization constrains the size of the small dimension (monomer radius) while the forward scattering or wavelength dependence of extinction optical depth determines the overall size of the particle or the number of monomers constituting the aggregate.

The DISR instrument measured the degree of linear polarization of scattered sunlight by measuring a vertical strip of sky in two bands centred at 492 and 934 nm. Some 50 measurements of this type were collected during the Titan descent. For the small monomer sizes expected, the direction of polarization would be perpendicular to the scattering plane and reach a maximum near 90° scattering angle at an azimuth opposite to the Sun, and would have a maximum electric field vector in the horizontal direction. We eliminated any polarization measurements made by the DISR that did not have this character, assuming that they were not made at the desired azimuth.

（尤其是 C–H 或 C≡N 键）的特征谱结构，也会出现大气中丰度较高的化合物的特征吸收带，例如乙烷（C_2H_6）、乙炔（C_2H_2）、丙烷（C_3H_8）、乙烯（C_2H_4）、氢氰酸（HCN）以及它们的聚合物。

表面光谱中最有趣的特征是位于 830 nm ~ 1,420 nm 之间的准线性无特征"蓝坡"结构。该斜坡结构还不能与水冰和复杂有机物（包括各种索林斯）的实验室光谱的任何组合相匹配。基于以上物质的不同组成和结构状态（比如，sp^2 碳化学键的丰度、延展和（或）成簇），得到的有机物质在近红外波段或者显示出明显的吸收峰（比如明亮的橘黄色索林斯，如图 15b 所示），或者显示出无特征的红坡（对于中低反照率的有机物），或者显示出非常暗的平坦谱线 [18,21]。

确定"蓝坡"光谱结构产生于何种物质是一项主要的挑战，这也是正确识别1,540 nm 吸收带的前提。如果这一波段确实是主要由水冰造成的，这种冰与具有强"红外–蓝"吸收特征的物质的均匀混合将解释表面光谱 1.04 μm 和 1.25 μm 处更弱的H_2O 吸收峰的缺失。这已经在数颗暗色的冰卫星上得到证实，这是因为这些吸收峰被有机物的谱结构（但是呈现出中性或红色的特征）所掩盖。仅仅减小水冰的晶粒尺寸并不能既抑制 1.04 μm 和 1.25 μm 处的吸收峰，又保持明显的蓝坡，因为蓝坡结构是由较大的水冰晶粒产生的（只考虑了红外波段吸收峰之间的连续吸收结构）。为了能够有效遮盖这些较弱的水的吸收峰，这种混合物质应该包含水冰和另一种具有更强的、但随波长增加而减弱的红外波段吸收特征的物质。

雾霾颗粒的尺寸

人们早就知道土卫六大气中的雾霾颗粒可以产生高度线偏振以及很强的前向散射。这意味着这些颗粒是较小的"单体"通过开放式结构构成的聚集体。线偏振的量约束了颗粒具有小维度的尺寸（单体的半径），而通过前向散射或消光光学深度与波长的关系可以确定颗粒的整体尺寸或构成聚集体的单体数目。

DISR 设备在以 492 nm 和 934 nm 为中心的两个波段，通过测量天空的一段垂直条带，得到太阳散射光的线偏振度。类似的测量在探测器降落过程中大约进行了50 次。对于预期的较小单体尺寸，偏振方向将垂直于散射面并在背向太阳的方向达到最大偏振度（接近 90°），同时，在水平方向上具有最大电场矢量。我们假定不满足以上特征的偏振测量结果是由于 DISR 没有朝向预期的方位上，因此我们将这些结果进行了剔除。

Several polarization measurements showing the expected behaviour in Titan's atmosphere were obtained. A gradual rise to a maximum near a scattering angle of 90° was observed, followed by a decrease on the other side of this peak. The solar aureole camera made several of these measurements at different times through the descent that show a smooth decrease in polarization with increasing optical depth into the atmosphere (Fig. 16). Figure 16a shows a maximum degree of linear polarization of about 60% at altitudes above 120 km in the 934-nm channel. Below, we show that the optical depth at 934 nm is a few tenths at this location in the descent. Comparisons of this degree of polarization with model computations for different-sized fractal aggregate particles produced by binary cluster collision aggregation indicate that the radii of the monomers comprising the aggregate particles is near 0.05 μm, almost independent of the number of monomers in the particle.

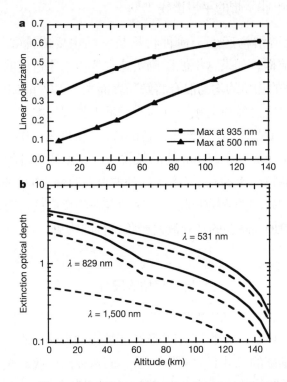

Fig. 16. Haze properties. **a**, The maximum degree of linear polarization measured opposite to the Sun as a function of altitude in our 500-nm channel (triangles) and in the 935-nm channel (dots). **b**, Extinction optical depth versus altitude for three wavelengths, 531 (top), 829 (middle) and 1,500 nm (bottom). The dashed curves correspond to $N = 256$ monomers, and the solid curves correspond to $N = 512$ monomers of the aggregate particles that make up Titan's haze. Note that the 531 (top) curve was constrained above 40 km and extrapolated to the ground. More explicit constraints from the infrared spectrometer will be available after the probe azimuth with time is determined.

Haze Optical Depth and Vertical Distribution

Before the Huygens probe descent, several workers considered the possibility that the haze in Titan's atmosphere clears below an altitude of some 50 or 70 km (ref. 22) owing to condensation of hydrocarbon gases produced at high altitudes that diffuse to lower,

  我们获得了数个与预期的土卫六大气特征相符的偏振测量的结果。我们观察到偏振在 90° 散射角附近逐渐增至最大值，然后在此峰值另一侧逐渐减小。日晕照相机在下降阶段的不同时间对偏振进行了数次测量，结果表明偏振随着大气光学深度的增加而平稳减小（图 16）。其中，由图 16a 可知，934 nm 观测通道的线偏振度在大于 120 km 的高度达到最大值，约为 60%。而图 16b 则表明在下降过程中这一地点 934 nm 通道测量的光深大小仅为十分之几。我们把二元团簇碰撞聚集产生的不同尺寸的分形聚集颗粒带入模型进行计算，并与偏振度的测量结果进行比较，发现构成聚集体颗粒的单体的半径大约为 0.05 µm，且单体的半径几乎与颗粒中单体的数量无关。

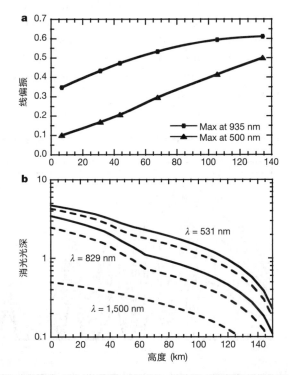

图 16. 雾霾的特性。**a**, 在背向太阳方向，500 nm 和 935 nm 两个通道测量得到的最大线偏振度与高度的关系，分别用三角和圆点表示出。**b**, 图中给出了在三个不同的波长（即 531 nm、829 nm 以及 1,500 nm）处，消光光深与高度的关系，对应曲线分别位于图上部、中部、下部。图中虚线对应组成土卫六雾霾的聚集颗粒的单体数量 $N = 256$ 的情况，实线则对应单体数量 $N = 512$ 的情况。值得注意的是，531 nm（上部）的曲线仅仅在 40 km 以上被约束，40 km 高度到地面的曲线由外推获得。在确定探测器方位随时间变化的关系后，还能进一步从红外光谱分析仪获得更精确的约束。

雾霾的光深和垂直分布

  在惠更斯探测器降落之前，若干研究者认为当低于某一高度（约 50 km 或 70 km）时 [22]，土卫六大气中的雾霾可能会消失，这是因为生成于高层大气的碳氢化

colder, levels of the stratosphere. If such a clearing were to occur in Titan's atmosphere, the intensity seen by the downward-looking DISR spectrometer would be relatively constant below the altitude at which the clearing began.

The brightness looking downward averaged along the DLVS slit and averaged over azimuth increases by a factor of two from the surface to 30 km altitude at a wavelength of 830 nm as shown in Fig. 17a. The increase at 830 nm is due almost solely to scattering by haze between 30 km and the surface. These observations demonstrate that there is significant haze opacity at all altitudes throughout the descent, extending all the way down to the surface.

The brightness of the visible spectra looking upward depends on the azimuth relative to the Sun. Although the probe attitude is not yet well known, it is clear that the minimum intensities are found looking away from the Sun. The upward-looking spectra looking away from the Sun start with low intensities at the highest altitudes and increase in intensity as the altitude decreases from 140 to about 50 km (see Fig. 17b). Below 50 km the intensity decreases at short wavelengths as altitude decreases, while the intensity in continuum regions longer than 700 nm continues to increase, as shown in Fig. 17c.

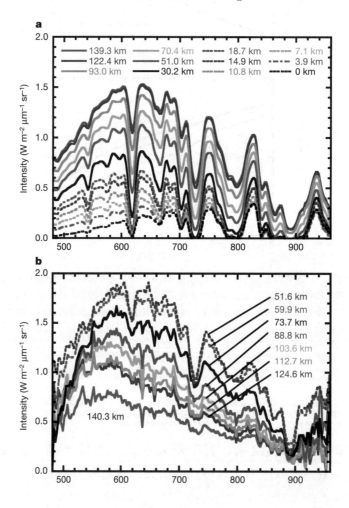

合物气体向下扩散到平流层较冷较低的平面，从而发生凝聚。如果确实如此的话，则当惠更斯探测器降至某一高度后，由于雾霾开始消失，俯视 DISR 光谱分析仪观察到的强度将保持相对固定。

俯视观测到的沿 DLVS 狭缝亮度的以及对方位平均的 830 nm 波长处的平均亮度，从地表到 30 km 高度增加了一倍，如图 17a 所示。这种增加主要是由于 30 km 高度与地表之间的雾霾产生的散射效应。这些观测证实了在整个下降的不同高度，直至土卫六表面都存在显著的由雾霾引起的不透明度。

仰视观测到的可见光谱的亮度则与相对太阳的方位角相关。尽管探测器的姿态尚未完全确定，但探测到的最小强度显然位于背向太阳的方向。在探测器位于最高高度时，背离太阳方向的仰视光谱测量得到的光谱强度较低；随着高度从 140 km 逐渐降低到 50 km，光谱强度逐渐增加，如图 17b 所示。在 50 km 高度以下，短波长波段的强度随着高度降低而减小，但大于 700 nm 的连续波段的强度则继续增加，如图 17c 所示。

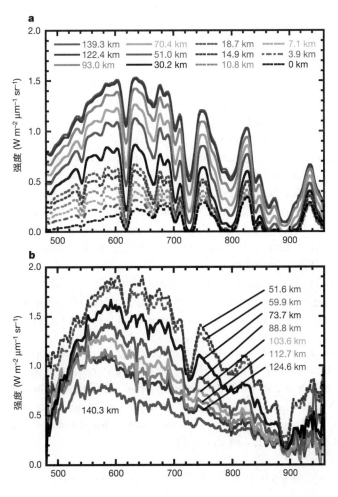

665

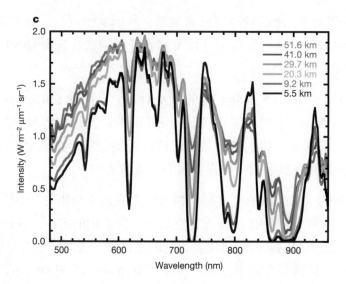

Fig. 17. Atmospheric spectra. **a**, The average intensity looking downward averaged over azimuth and over the length of the slit (10° to 50° nadir angle) as a function of wavelength for several altitudes as labelled. **b**, The intensity measured by the Upward-Looking Visible Spectrometer in the direction opposite the Sun as a function of wavelength for several altitudes as labelled. Note that the brightness begins at a low level at 140 km, and increases as altitude decreases. **c**, Same as **b** but for altitudes below 50 km. Note that the brightness away from the Sun decreases with decreasing altitude at short wavelengths, but increases in continuum regions longward of 700 nm.

The intensity looking upward away from the Sun, and the azimuthally averaged intensity looking downward at each continuum wavelength as functions of altitude, constrain the vertical distribution of aerosol opacity in the atmosphere as well as the aerosol single-scattering albedo. With the monomer radius fixed at 0.05 μm from the polarization measurements, the adjustable parameters include the number of monomers in each aggregate particle, N, as well as the local particle number density, n, in cm^{-3} as a function of altitude. An algorithm developed by (and available from) M.L. was used to determine the single-scattering phase function, the single-scattering albedo, and the extinction cross-section for each aggregate particle as functions of the wavelength, the real and imaginary refractive indices, the monomer radius, and the number of monomers per aggregate particle. This algorithm is based on the discrete dipole approximation and the T-matrix method (M. Lemmon, personal communication) to evaluate the single-scattering properties of the aggregate particles. These computations are most accurate at relatively small particle sizes and depend on extrapolation for N of 256 or larger.

For large particles the wavelength dependence of the extinction optical depth is smaller than for small particles. An N larger than about 100 is required to fit the observations. Hence, models with $N = 256$ or 512 monomers per particle are shown, even though for these values of N the single-scattering algorithm is not as accurate as desired. For these initial models, the real and imaginary refractive indices for the aerosols are taken from the measurements of laboratory tholins in ref. 23.

The radius, R_p, of the circle having the same projected area as an aggregate particle is given

666

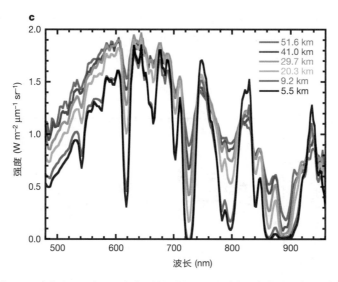

图 17. 大气光谱。**a**, 图中给出了几个不同高度下俯视光谱平均强度与波长的关系（见图中标注），其中平均强度是基于方位角和狭缝长度方向（天底角 10° 到 50° 之间）的平均。**b**, 在背向太阳的方向，仰视可见光波段光谱仪测量得到的不同高度下的强度与波长的关系（见图中标注）。值得注意的是，140 km 高度对应的起始亮度较低，但亮度随着高度的下降而增加。**c** 与 **b** 图表示的关系相同，但曲线对应 50 km 以下的高度。值得注意的是，背向太阳方向的光亮度在短波长波段随着高度的减小而减小，但是在大于 700 nm 的波段则随高度减小而增加。

　　背向太阳方向的仰视测量强度和在各个连续波段得到的对方位平均的俯视强度相对于高度的变化函数，可以用于确定大气中气溶胶不透明度的垂直分布以及气溶胶单散射反照率。根据偏振测量，单体半径固定值为 0.05 μm，那么可调参量包括每个聚集体中颗粒的单体数量 N 以及粒子数局部密度 n（单位为个/cm³）的高度变化函数。利用莱蒙提出的算法（可联系莱蒙获取），可以确定单散射相函数、单散射反照率、每个聚集颗粒的消光截面（波长的函数）、折射指数的实部和虚部、单体的半径以及每个聚集体中的单体数量。该算法基于离散偶极子近似和 T 矩阵方法（莱蒙，个人交流），可以评估聚集颗粒的单散射属性。以上计算对于相对较小的颗粒尺寸具有最高的精确度，而且依托于 $N \geqslant 256$ 的外推。

　　对于较大颗粒而言，较小颗粒的消光光深与波长的关系更加密切。为了拟合观测结果，N 的取值需要大于 100，因此我们在模型中设定每个颗粒中的单体数为 256 或 512。尽管对于这样的 N 取值，单散射算法不如想要的那样精确。对于以上初始模型，气溶胶折射指数的实部和虚部取自实验室中索林斯的测量结果 [23]。

　　设某一圆面积与单一的聚集颗粒的投影面积相同，则该圆的半径 R_p 可以由公式

by $R_p = r \sqrt{(N^{0.925})}$, where r is the monomer radius and N is the number of monomers. Particles with 256 or 512 monomers have the same projected areas (which control their forward scattering properties) as circles with radii 0.65 and 0.9 μm, respectively.

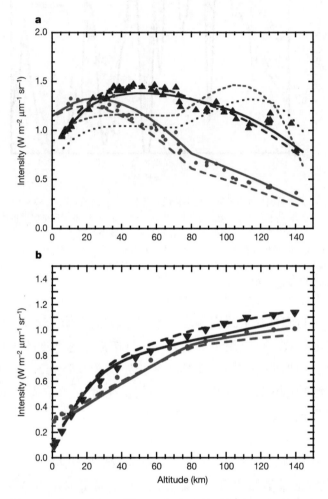

Fig. 18. Haze models versus observations. **a**, Measured upward-looking intensity (points) away from the Sun versus altitude for 531 (blue) and 829 nm (red). Three models are shown compared to the observations at each wavelength. The solid curves are for 512 monomers in each aggregate particle. The model at 531 nm has 12 particles cm^{-3} above 80 km and 18 particles cm^{-3} below that altitude. The corresponding model at 829 nm has 20 particles cm^{-3} above 80 km and 60 cm^{-3} below. The models indicated by long-dashed lines have 256 monomers per particle and at 531 nm the number density is 20 particles cm^{-3} above 80 km and 40 cm^{-3} below. At 829 nm the number density is 30 cm^{-3} above 80 km and 100 cm^{-3} below. The number density of particles differs slightly with wavelength because the model of fractal aggregate particles does not yet reproduce the wavelength dependence of the cross-section to high accuracy. The models indicated by short-dashed lines have 256 monomers per particle, and have the same number of total particles as the models indicated by long-dashed lines, but all the particles are concentrated above 72 km with a clear space below. Such models with clear spaces are clearly not in agreement with the observations. **b**, Downward-looking measured intensities versus altitude (plotted as points) for 531 (blue points) and 829 nm (red points). The two models (plotted as curves) are the same models as those shown by long-dashed lines and solid curves in **a**.

$R_{\mathrm{p}} = r\sqrt{N^{0.925}}$ 计算得到，其中 r 表示单体的半径，N 表示单体的数量。不妨设两种颗粒分别由 256 个单体和 512 个单体组成，则与二者投影面积（决定着前向散射性质）相等的圆的半径分别为 0.65 μm 和 0.9 μm。

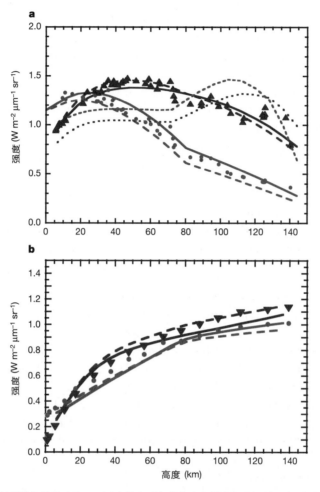

图 18. 雾霾模型与观测结果的对比。**a**, 图中给出了在背向太阳的方向仰视强度（用点表示）与高度的关系，蓝色与红色分别对应 531 nm 和 829 nm 观测通道。同时作为对比，对于每个通道还给出了三个模型的结果。其中，实线代表每个聚集颗粒包含 512 个单体的情况。对于 531 nm 通道，该模型在大于 80 km 高度的粒子数密度为 12 个/cm³，在低于 80 km 的高度则为 18 个/cm³。对于 829 nm 通道，该模型大于 80 km 高度的粒子数密度为 20 个/cm³，在低于 80 km 的高度则为 60 个/cm³。图中用长虚线标示的模型代表每个聚集颗粒包含 256 个单体，对于 531 nm 通道，大于 80 km 高度的粒子数密度为 20 个/cm³，在低于 80 km 的高度则为 40 个/cm³；对于 829 nm 通道，大于 80 km 高度的粒子数密度为 30 个/cm³，在低于 80 km 的高度则为 100 个/cm³。由以上数据可知，粒子数密度随波长变化而有些许变化，这是因为我们根据分形颗粒模型得出的截面积与波长的关系准确度并不高。图中短虚线表示每个颗粒由 256 个单体构成的模型结果，粒子（颗粒）总数与长虚线相同，但是所有的粒子都集中在 72 km 高度以上，低于 72 km 的空间则没有粒子。显然这样一种存在粒子净空层的模型与观测结果不符。**b**, 图中数据点标示了 531 nm 和 829 nm 两个通道测量得到的俯视测量的强度与高度的关系，分别用蓝色点和红色点表示。拟合曲线采用的两种模型与图 **a** 中长虚线和实线所代表的模型一致。

Comparison of the observed downward-streaming intensity looking away from the Sun at wavelengths of 531 and 829 nm with plane-parallel radiative transfer models constrains the vertical distribution of optical depth in Titan's atmosphere. The vertical distribution of particles can be adjusted to fit these curves arbitrarily well. In this preliminary work, only one constant number density above an altitude of 80 km and a second constant number density below 80 km were considered for models with $N = 256$ and $N = 512$, as shown in Fig. 18a. The number densities are larger in the lower half of the atmosphere than the upper half, but only by modest factors of two to three. The number densities are not exactly equal in the models at different wavelengths, but this is probably due to the extrapolation in the wavelength dependence of the cross-sections in the models for these relatively large N values at the shortest wavelengths (and largest size parameters). Average number densities in the entire atmosphere between 30 and 65 cm^{-3} are required if the number of monomers per particle is 256. Average number densities between 15 and 40 cm^{-3} are required if N is 512.

Models with a clear space below an altitude of 72 km are also shown in Fig. 18a. It is apparent that no such clear space exists in the region of the probe's entry. It will be interesting to examine the range of parameters in cloud physics models needed to reproduce the continuous variation of haze opacity throughout Titan's atmosphere.

Haze models must reproduce the upward-streaming intensity observed in the atmosphere as well as the downward-streaming intensity. Models with $N = 256$ and 512 at 531 and 829 nm are compared to the upward intensity averaged in azimuth and along the slit in Fig. 18b. While the fit is not exact, it is clear that the models that fit the downward intensity away from the Sun are also in reasonable agreement with the measured upward intensities. It is interesting to note the ground reflectivity implied by the measurements at 531 and 829 nm. These values include the true shape of the diffuser in the Upward-Looking Visible Spectrometer and produce ground reflectivities of 0.13 at 531 nm and 0.19 at 829 nm. The corrected value at 829 nm is in good agreement with the value measured by the Infrared Spectrometer (0.18) when a correction for the diffuser's non-ideal shape is made to the infrared spectrometer measurements.

Haze Structure and Methane Absorptions

Haze models must fit the observations at all wavelengths. How well do the models derived from the visible spectrometer fit the DISR observations in the infrared? The ULIS spectra in Fig. 19 clearly show absorption by the methane bands around 890, 1,000, 1,160 and 1,380 nm. The depths of these bands increase with decreasing altitude as a result of increasing methane column density. They are correctly reproduced by radiative transfer calculations based on an exponential sum formulation for the methane absorption and a stratospheric methane mole fraction of 1.6% (ref. 24). The agreement is worse at low altitudes in the troposphere, probably owing to inaccuracies in the methane absorption

670

将 531 nm 和 829 nm 两个波长观测到的背向太阳的下行辐射流强度，与平行平面的辐射传输模型进行比较，可以约束土卫六大气中光深的垂直分布。通过调整颗粒的垂直分布，可以对这些曲线进行很好的拟合。在前期工作中，对于 $N = 256$ 和 $N = 512$ 两种模型，我们只假设 80 km 以上的高度具有某一恒定粒子数密度 $N = 256$，而 80 km 以下高度具有另一恒定粒子数密度 $N = 512$，如图 18a 所示。下半层大气层中的粒子数密度比上半层大气层中的粒子数密度大，但只是上半层的 2 到 3 倍。但是在不同波长的模型中，粒子数密度并不是完全相同的，这可能是因为模型引入的截面与波长的关系在短波长和较大 N 取值条件（参数达到最大）下的外推造成的。如果每个颗粒的单体数量为 256，则整个大气层中的平均粒子数密度需要介于 30 个/cm³ 到 65 个/cm³ 之间。如果每颗粒的单体数量为 512，则整个大气层中的平均粒子数密度需要介于 15 个/cm³ 到 40 个/cm³ 之间。

同时，图 18a 还给出了假定 72 km 以下高度不存在雾霾的模型结果。显然，在探测器进入的区域并不存在这样的情况。检验用于重现土卫六大气中薄雾不透明度连续变化的云层物理模型参量的范围将是一项非常有意思的工作。

雾霾模型必须能够重现大气中的仰视方向辐射流强度和俯视方向辐射流强度。我们在 531 nm 和 829 nm 分别将 $N = 256$ 和 $N = 512$ 代入模型，并将模型结果与在不同方位上做平均并沿狭缝做平均后的向上辐射流强度进行比较，如图 18b 所示。尽管拟合还不够精确，但是可以清晰地看出用于拟合背向太阳的向下辐射流强度的模型也可以在一定程度上拟合向上辐射流强度。值得注意的是，在 531 nm 和 829 nm 测量可以得到包括仰视可见光谱仪中漫射器的真实形状等值，并以此获得在 531 nm 和 829 nm 测得的地面反射率（分别对应 0.13 和 0.19）。以上校正后的 829 nm 反射率与经过校正（考虑了红外光谱仪测量过程中漫射器的非理想形状）后的红外光谱仪的测量结果（0.18）相吻合。

雾霾的结构和甲烷的吸收

雾霾模型必须在所有波长都能拟合观测结果。那么根据可见光谱分析仪得到的模型在多大程度上可以拟合 DISR 在红外波段的观测结果呢？图 19 中的 ULIS 光谱清楚地显示，在 890 nm、1,000 nm、1,160 nm 以及 1,380 nm 附近存在甲烷的吸收峰。由于甲烷柱密度随着高度的降低而增加，所以这些吸收峰的深度也随着高度的降低而增加。根据甲烷吸收的指数求和公式以及平流层甲烷摩尔分数 1.6%，以上吸收带可以由辐射传输理论准确地计算出来 [24]。但是在对流层的较低高度，理论计算和实测结果的吻合度较差，这可能是由于采用的甲烷吸收系数不够准确。在甲烷窗

coefficients. In the methane windows the downward average intensity varies by 25% or less between 104 and 10 km, indicating relatively low aerosol absorption in the infrared range.

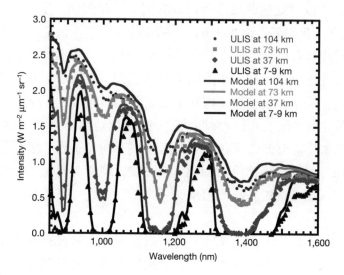

Fig. 19. Vanishing sunlight. ULIS spectra (points) recorded at various altitudes and showing the growth of the methane-band absorption with depth in the atmosphere. The spectra at altitudes greater than 3.5 km have been integrated over several probe rotations and correspond approximately (but not exactly) to azimuth-averaged intensities. The full analysis of these observations must await refinement of the attitude of the probe as a function of time, a task still in progress. Models, with a methane mole fraction of 1.6% in the stratosphere increasing to 5% at the surface, are shown for comparison (lines). At low altitudes (< 20 km), the mismatch between model and observations in the methane windows, specifically around 1,280 and 1,520–1,600 nm, is probably due to errors in the methane absorption coefficients at long path-lengths. The model slightly overestimates the intensity in the 104-km spectrum, because the latter does not correspond to an exact azimuth average. Radiative transfer calculations were based on a 16-term exponential-sum formulation of the methane absorption properties. In the near-infrared (λ > 1,050 nm), these absorption coefficients were calculated from a band model with a modified temperature dependence designed to better match the low-temperature observations. In the visible (λ < 1,050 nm), the absorption coefficients of Karkoschka[34] were used. In practice, for 30 pressure–temperature conditions representative of 30 levels in Titan's atmosphere[12], methane transmissions were calculated for 60 different paths, convolved to the resolution of the DISR spectrometers, and this ensemble of convolved transmissions was fitted each time with an exponential-sum model.

At altitudes less than 3.5 km, we used single short exposures for the infrared spectra rather than long time averages. The 940 nm intensity in the last three ULIS spectra is about four times larger that in the first three spectra, indicating that the Sun is located in their field of view (see Fig. 20). The contrast between the most intense spectrum (with the Sun in the field of view) and the weakest one (with the Sun out of the field of view) increases with wavelength, reaching 17 at 1,550 nm, a consequence of the decreasing haze optical depth. This contrast can be used to constrain the haze optical depth, assuming that the spectra correspond approximately to solar azimuths of 0 and 180°. A satisfactory model, using aggregate particles of 256 monomers, a 0.05-μm monomer radius, and a uniform concentration of 52 particles cm^{-3}, indicates an optical depth of about 2 at 940 nm, decreasing to 0.5 at 1,550 nm. Models with one-half and twice the particle density (and

口，向下平均强度在 104 km 到 10 km 高度之间变化了 25% 或者更小，这说明气溶胶在红外波段的吸收相对较弱。

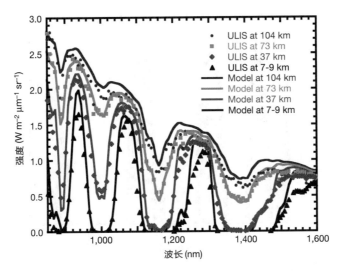

图 19. 逐渐消失的阳光。不同高度处的 ULIS 光谱记录（图中标示为数据点），表明甲烷吸收峰的强度随着进入大气层深度的增加而增加。大于 3.5 km 高度的光谱是对若干个探测器旋转周期进行积分的结果，大体上（但不是完全地）对应了方位平均强度。关于这些观测结果的全面分析尚需精确确定探测器高度与时间的关系，这一工作目前正在进行中。图中的曲线给出了我们采用的模型结果（用实线表示）作为对比，其中采用的甲烷摩尔分数在平流层为 1.6%，然后在表面处增加到 5%。对于较低的高度（<20 km），模型与观测结果在甲烷吸收窗口，特别是在 1,280 nm、1,520 ~ 1,600 nm 附近，存在较大差异；这可能是由于甲烷吸收系数存在误差经过长光程的放大导致的。模型稍稍高估了 104 km 高度处的光谱强度，因为该光谱并不精确地匹配方位平均。辐射传输计算主要基于甲烷吸收特性的 16 项多项式求和公式。在近红外区域（λ > 1,050 nm），通过吸收波段模型可以计算出吸收系数，其中，为更好地拟合低温观测结果，我们对温度的依赖关系进行了修正。在可见光波段（λ < 1,050 nm），则采用了卡尔科施卡计算的吸收系数 [34]。实际操作中，对于代表土卫六大气 30 个高度的 30 个气压–温度条件 [12]，为 60 条不同的路径计算甲烷的穿透率，并与 DISR 光谱分析仪的分辨率作卷积，作卷积后的穿透率每次用一个指数求和模型进行拟合。

在低于 3.5 km 的高度，我们采用单次短时曝光（而不是长时曝光取平均）测量红外光谱。最后的三张 ULIS 光谱中 940 nm 的强度大约是最先三张 ULIS 光谱中 940 nm 强度的 4 倍，这说明太阳位于后三张光谱的拍摄视场中，参见图 20。当太阳位于视野之内时光谱强度显然最高，当太阳位于视野之外时光谱强度则最低；最强与最弱光谱之间对比度随着波长增加而增加，并在 1,550 nm 处达到 17，这是由于雾霾的光深随波长减小。假设以上光谱对应的太阳方位角分别约为 0° 和 180°，则以上对比度可以用来确定雾霾的光深。假设聚集颗粒由 256 个单体组成，每个单体的半径为 0.05 μm，粒子的均匀密度为 52 个/cm³，这一可以被接受的模型可以计算得到：940 nm 处对应的光深大约为 2，在 1,550 nm 处降为 0.5。如果采用粒子数密度

hence optical depth) yield a contrast between spectra with the Sun in and out of the field about twice as large and half as large (respectively) as observed.

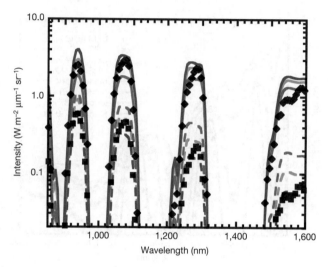

Fig. 20. Determination of total haze optical depth. ULIS spectra (black points) recorded at 734 m (diamonds) and 53 m (squares) above the surface with integration times of 1 s. The one with the highest intensity (734 m) has the Sun in its field of view; the lower one (53 m) does not. The contrast between the two in the methane windows increases with wavelength and is a sensitive function of the haze optical depth. The nominal model, shown for comparison (green line), has an optical depth of 2 at 940 nm decreasing to 0.5 at 1,550 nm. Calculations correspond to intensities averaged over the field of view and azimuths of 0 and 180 degrees with respect to the Sun. Other models show the effect of doubling (red) and halving (blue) the particle concentration. Solid lines show model intensity towards the Sun while the dashed lines show the intensity with the instrument facing away from the Sun.

The methane bands are prominent in the DLIS spectra at all altitudes (see Fig. 21a). The residual intensity in the cores of these bands is due to scattering by aerosol particles between the probe and the surface. Its variation provides a constraint of the vertical profile of the haze particles between approximately 150 and 40 km, as illustrated in Fig. 21a. The method is not sensitive at low altitudes because of absorption of the downward solar flux in the methane bands. A model with a constant particle concentration with altitude provides a good fit of the methane bands. Moderate variations of the particle concentration with height are also acceptable, but a model with a clear space in the lower stratosphere is inconsistent with the data.

The optical depths in the models that fit the visible and infrared spectral observations with $N = 256$ and 512 are shown as functions of altitude in Fig. 16b. The number density was assumed to be 52 cm^{-3}, independent of altitude for the infrared models with $N = 256$. The models computed for comparison with the visible spectrometer contained constant and different number densities above and below 80 km. The average number density above and below 80 km for the models with $N = 256$ (30 and 65) are in reasonable agreement with the single value used in the models derived from the infrared spectrometer. The same particle number densities give the required optical depths from 900 to 1,550 nm, indicating that the algorithm for generating cross-sections from particle sizes is working in a consistent manner. At shorter wave-lengths (531 nm), the size parameter is sufficiently large that the cross-

分别另取为以上粒子数密度（也是光学深度）的 1/2 和两倍的模型，则相应的太阳在视场内外两种光谱的对比度分别为观测值的两倍和一半。

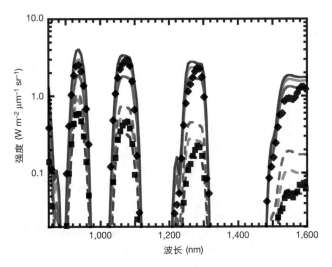

图 20. 雾霾总光深的确定。图中给出了土卫六地表之上 734 m（标示为菱形）和 53 m（标示为方形）处的 ULIS 光谱，积分时间为 1 s。其中，734 m 高度的光谱强度最大，视场中包含了太阳，而强度较弱的 53 m 光谱视场中没有太阳。两个甲烷窗口的对比度随着波长的增加而增加，并且对雾霾光深的变化很敏感。图中给出了标准模型作对比（绿线），其雾霾光深在 940 nm 处为 2，然后逐渐降至 1,550 nm 处的 0.5。计算得到的强度是对视场和相对太阳的 0 和 180° 方位角进行平均后的结果。其他模型表明粒子浓度加倍（红）或减半（蓝）的效应。图中实线表示朝向太阳的光谱强度，而虚线表示仪器背向太阳的光谱强度。

如图 21a 所示，在所有高度的 DLIS 光谱中，甲烷对应的吸收峰都是非常明显的。图中吸收峰核心区内存在强度残差，这是探测器与地表之间的气溶胶粒子散射造成的。这一强度的变化可以用于确定约 150～40 km 高度之间的雾霾颗粒的垂直分布，如图 21a 所示。但该方法在较低高度并不灵敏，因为向下照射的太阳光通量在甲烷吸收峰内被吸收。如果假定粒子浓度不随高度变化，那么相应的模型可以较好地拟合甲烷吸收峰。粒子浓度随高度有适度的变化的模型也是可接受的；但平流层下部有净空层的模型与数据不符。

图 16b 给出了与可见光和红外观测结果吻合的模型中光深与高度的关系，其中 N 的取值分别为 256 和 512。在红外模型（$N = 256$）中假设粒子数密度为 52 个/cm^3，并且与高度变化无关。用于与可见光谱分析仪对比的模型中，以 80 km 高度为界，上下分别采用了恒定但彼此不同的粒子数密度。$N = 256$ 的模型中 80 km 上下的平均粒子数密度（分别为 30 与 65），和基于红外光谱分析仪的模型采用的单值粒子数密度相符。相同的粒子数密度给出了 900～1,500 nm 范围的光深，这说明根据颗粒尺寸推出截面积的算法始终是适用的。在较短的波长（531 nm），粒子尺寸参量过大，

section algorithm is not as accurate, and the number density is decreased slightly to give models that fit the observations. The variation of optical depth with wavelength is modest, decreasing by only about a factor of 2.8 from 500 to 1,000 nm. If 512 monomers are used for the particles, the wavelength dependence is even less steep. The haze optical depth as a function of wavelength is presented in Fig. 21b.

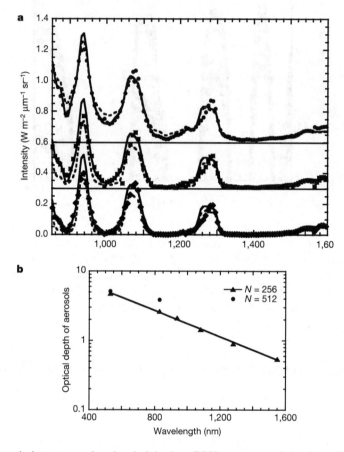

Fig. 21. Haze vertical structure and total optical depth. **a**, DLIS spectra recorded at three altitudes: 104 km (top, dots), 82 km (middle, squares), and 57 km (bottom, diamonds). The intensities for the higher altitudes have been displaced in 0.3 increments for clarity. The points are measured data; the lines are the nominal model; the dashed lines are a modified model with the same optical depth as the nominal model, but with all the haze particles concentrated above 72 km (clearing below). The residual intensity in the core of the methane bands is a sensitive indicator of the presence of scattering particles beneath the probe. The model with the clearing produces too much emission in the core of the CH_4 bands at high altitude and not enough at low altitude. **b**, Total extinction optical depth of the haze alone versus wavelength. The triangles are for models with 256 monomers per particle. The points at the two shortest wavelengths are from models that fit the visible spectrometer measurements. The other four points are from models that fit the infrared spectrometer measurements. The dots are for models with 512 monomers per particle that fit the visible spectrometer measurements.

导致截面算法不够准确，粒子数密度需要略微减小才能使模型与观测结果相符。当波长从 500 nm 增加到 1,000 nm，光深减小为原光深的 2.8 分之一，这一变化并不很大。如果聚集颗粒为 512 个单体构成，则这种变化会更小。图 21b 给出了雾霾光深与波长的关系。

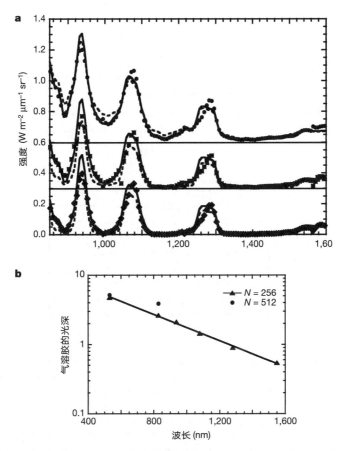

图 21. 雾霾的垂直结构和总光深。a, 图中分别给出了高度为 104 km(上部圆点)、82 km(中部方块)以及 57 km(下部菱形) 的 DLIS 光谱测量数据。为更清楚地显示在一幅图中，较高高度的光谱分别相对较低高度的光谱向上移动 0.3 个单位。图中点表示测得的数据，实线表示标称模型，虚线表示与标称模型具有相同光深的修正模型，其所有的雾霾颗粒都集中在大约 72 km 高度以上(下面没有雾霾颗粒)。图中甲烷吸收峰核心区域存在的强度残差是探测器下面存在散射粒子的很敏感的指示物。对于较高的高度，修正模型在甲烷吸收峰核心附近的辐射强度过大，而对于较低的高度又过小。b, 图中给出了雾霾的总消光光深与波长的关系。其中三角表示每个聚集颗粒由 256 个单体构成的模型，最短的两个波长处的数据点源自可见光谱仪测量结果的拟合模型，而另外四个数据点则源自红外光谱仪测量结果的拟合模型。图中圆点代表了与可见光谱仪测量结果拟合的表示每个聚集颗粒由 512 个单体构成的模型。

A Thin Layer of Haze near 21 km Altitude

Many workers have suggested that hydrocarbons produced at very high altitudes could diffuse downward to cooler levels where they could condense on haze particles. Do our intensity profiles looking towards the horizon detect any thin haze layers at specific altitudes that might be due to this mechanism?

Figure 22 shows the normalized profile of intensity measured by the Side-Looking Imager (SLI) compared to a model. The left-hand side of the plot shows normalized intensity as a function of nadir angle for the observations at altitudes ranging from 20.4 to 22.3 km. The observations at 22.1 km and above and at 20.4 km and below show smooth functions of nadir angle. However, for the measurement at 20.8, and the two measurements at 20.9 km altitude, a dip of about 2% is seen near a nadir angle of 90°.

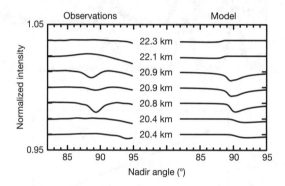

Fig. 22. Thin cloud layer observation at 21 km. In the left-hand side are the intensity profiles as a function of nadir angle divided by the average intensity profile measured by the SLI at the altitudes indicated. The right-hand side of the figure shows the model intensity profiles computed for a cloud layer of absorption optical depth 0.001 which is 1 km thick at an altitude of 21 km. The model is able to reproduce the 2% contrast feature seen in the observations at altitudes of 20.8 and 20.9 km. If the layer is mostly illuminated by diffuse light, the absorption optical depth is equal to total optical depth times the difference between the single-scattering albedos of the material in the layer and the albedo of the background haze. If the layer is primarily illuminated by direct sunlight, the absorption optical depth of the haze is proportional to the total optical depth of the haze times the difference between the phase functions of the material in the layer and the background haze at the scattering angles for any observation.

The curves in the right-hand side of Fig. 22 show the intensities of a model having a thin additional layer of haze at an altitude of 20.9 km. The haze layer of vertical absorption optical depth within a factor of two of 0.001 with a gaussian profile between 1 and 2 km thick can reproduce the depth of the feature. The location of the layer is at 21.0 ± 0.5 km, where the local temperature is 76 K and the pressure is 450 mbar (ref. 12).

This feature at 21 km occurs in the troposphere and may be an indication of methane condensation. It is the only indication of a thin layer seen in the set of SLI images taken from 150 km to the surface. Evidence of condensation of hydrocarbons in the lower stratosphere, where several hydrocarbons might be expected to condense[25], has not yet been

21 km 高度附近的雾霾薄层

许多研究者认为，非常高的大气中产生的碳氢化合物，可以向下扩散到较冷的层位，并在那里凝结成雾霾颗粒。那么我们朝向地平线方向的光谱强度剖面，是否会探测到某一高度存在由于以上机制形成的雾霾薄层呢？

图 22 给出了侧视成像仪（SLI）测量得到的归一化的强度分布曲线，以及与相应模型的比较。图中左侧表示归一化强度与天底角的关系曲线，对应的观测高度从 20.4 km 到 22.3 km。测量曲线在 22.1 km 及以上的区间和 20.4 km 以下的区间内都平滑连续，但是对于位于 20.8 km 高度的一条测量曲线和位于 20.9 km 高度的两条测量曲线，在天底角 90° 附近均存在一个大约 2% 的凹陷。

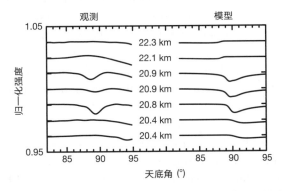

图 22. 21 km 高度处薄云层的观测结果。图中左侧给出了 SLI 在不同高度测量得到的强度与天底角的关系曲线，其中强度是实际强度除以平均强度后的结果。右侧则给出了对厚度为 1 km 的云层的模型计算结果，其中该云层的高度为 21 km，吸收光深为 0.001。该模型可以重现观测结果中 20.8 km 和 20.9 km 曲线 2% 的下陷结构。如果该云层主要由漫射光照亮，则吸收光深等于总光深乘以云层物质的单散射反照率与背景雾霾反照率之差。如果云层主要受直射阳光照亮，对于任何观测，雾霾的吸收光深正比于雾霾的总光深乘以云层物质与散射角方向处的背景雾霾之间的相位函数之差。

图 22 右侧给出了模型的强度曲线，其中该模型在 20.9 km 的高度附加了一层较薄的雾霾。如果这层薄雾霾的垂直吸收光深在 1 ~ 2 km 的厚度范围内为高斯分布，并且其值在 0.001 的两倍范围内变化，则该模型也可以产生类似左图的凹陷结构。该层薄雾霾的高度为 21.0 ± 0.5 km，局域温度为 76 K，压强为 450 mbar[12]。

21 km 处的雾霾薄层位于对流层内，可能意味着甲烷的凝结。SLI 在探测器从 150 km 高度到着陆过程中拍摄的图像里，这是唯一预示雾霾薄层存在的观测结果。但是目前尚未发现人们预期的碳氢化合物在平流层底层的凝结现象（人们预测有几

found, but the search is continuing.

Unravelling Titan's Mysteries

Some of the major questions about Titan concern the nature of the source of methane that replaces the irreversible loss at high altitudes by photochemistry that produces a host of complex organic compounds. Open pools of liquid hydrocarbons on the surface have been suggested, as well as cryovolcanism. Also, if methane photochemistry has been occurring over the lifetime of the Solar System the organic products of these processes should have accumulated to significant depths on the surface and should be seen in images and spectra of the surface.

Although no such liquid bodies were directly imaged by DISR, there is compelling evidence for fluid flow on the surface of Titan, including the dendritic and stubby drainage channel networks, the rounded and size-graded "rocks" at the surface landing site, and the morphology of the shoreline, offshore structures, and the appearance of the darker lakebed region. The stubby networks may imply sapping or spring-fed flows, as the existence of liquid pools on the surface and the frequency of precipitation that could cause the deep dendritic drainage channels are both still unconfirmed. In addition, there are at least a few structures that suggest cryovolcanic flows on the surface.

The ground track derived by the image correlations demonstrates a zonal wind field that is mostly prograde. The general altitude profile and shape agree with predicted average models of the zonal wind flow between 50 and 10 km altitude[7], although with a reduced intensity. Below 10 km, falling wind speeds and an abrupt change of wind direction indicate a planetary boundary layer some 7–8 km thick, scaling nicely from near-equatorial terrestrial boundary layers.

Spatially resolved spectral reflectance measurements of different regions on the surface suggest that the uplands are redder than the lowland lakebed regions. The regions near the mouths of the rivers are also redder than the lake regions. A host of questions about the sequence of flooding and formation of these structures is suggested by these observations.

The reflectivity of the surface at the landing site was measured from 480 nm to 1,600 nm without the interference of methane absorption bands or haze opacity. The peak reflectivity in the dark regions is about 0.18 at 830 nm and decreases towards longer and shorter wavelengths. The red slope in the visible is consistent with organic material, such as tholins, but the blue infrared slope is still unexplained. Between 1,500 and 1,600 nm the reflectivity is low (0.06) and flat, consistent with water ice. Nevertheless, the decrease in reflectivity from 900 to 1,500 nm does not show the expected weak absorption bands of water ice near 1,000 and 1,200 nm, and the identity of the surface component responsible for this blue slope remains unknown.

种碳氢化合物应该会在那里凝结[25]），相关搜寻正在进行中。

揭秘土卫六

关于土卫六的一些主要问题都涉及了甲烷的来源问题。在高层大气中，不可逆光化学反应通过消耗甲烷不断产生大量有机化合物，而源源不断进入大气、填补这一消耗的甲烷来自哪里呢？人们认为土卫六表面存在着液态烃构成的露天水池以及冰火山活动。同时，如果甲烷光化学反应在太阳系整个寿命周期中都不断进行，那么该过程中产生的有机产物就会在土卫六表面积累并达到明显的厚度，从而可以在图像和光谱中被观测到。

然而，DISR 图像并没有直接观测到这样的液体，不过还是有令人信服的证据支持土卫六表面存在液体的流动。这些证据包括树枝状和断株状沟渠网络，着陆点附近有分选性的圆"石"，海岸线地形结构，离岸的各种结构以及较暗的湖床区域的外观。其中，这些断株状网络可能意味着液体下切侵蚀的河道或者源自泉眼的液体流；而表面存在液体湖泊以及形成较深的树枝状排水沟道的频繁降雨，这两者尚未得到证实。此外，至少有几处结构表明地表存在冰火山作用相关的流动构造。

根据图像关联性推出的探测器轨迹地面投影说明了该处的风场是一个总体上与土卫六自转同向的纬向风场。该风场的大体高度剖面和速度曲线的形状与 50～10 km 高度之间的纬向风模型的预测相一致[7]，虽然风的强度相比于模型有所减小。在 10 km 高度以下，风速的降低和风向的骤然改变表明，行星边界层厚度大约为 7～8 km，与地球赤道附近的行星边界层有很好的对应关系。

土卫六表面的空间分辨光谱反射率测量表明，高地比湖床低地区域红化程度更高，河口附近的区域也比湖区更加红化。这些观测结果向我们提出了与洪水活动的顺序以及这些地形结构如何形成等相关的许多问题。

在不受甲烷吸收峰和雾霾不透明度干扰的情况下，探测器在着陆点附近测量了 480～1,600 nm 波段的表面反射率。暗色区域的反射率在 830 nm 处到达峰值，大约为 0.18，然后随着波长的增加和减小都降低。其中，可见光波段的红坡与有机物质（如索林斯）的结果相符，但是红外波段的蓝坡目前尚未得到解释。从 1,500 nm 到 1,600 nm 的范围内，反射率较低（0.06）且比较平坦，与水冰的特征相符。不过，从 900 nm 到 1,500 nm 反射率的下降过程中并未发现 1,000 nm 和 1,200 nm 附近预期存在的较弱的水冰吸收峰。是什么表面物质导致了这一蓝坡结构目前还不得而知。

The nature of the haze aerosols measured by DISR is different in significant ways from the view before the Huygens mission. Before the Huygens probe, cloud physics models with sedimentation and coagulation predicted a strong increase in haze density with decreasing altitude[9]. In addition, measurements of the high degree of linear polarization in light scattered from Titan near a phase angle of 90° by the Pioneer and Voyager spacecraft could only be matched by spherical particles having radii less than or equal to 0.1 μm. Such small particles produced a strong increase in optical depth with decreasing wavelength shortward of 1,000 nm. Fitting the strong methane band at 890 nm constrained the amount of haze at high altitudes. This haze became optically much thicker at the wavelength of the weaker methane band at 619 nm. To fit the observed strength of this band it was necessary to remove the haze permitted by the cloud physics calculations at altitudes below about 70 km by invoking condensation of organic gases produced at very high altitudes as they diffused down to colder levels[26]. The condensation of many organic gases produced by photochemistry at high altitudes on Titan was suggested by Sagan and Thompson in ref. 25, and seemed consistent with this view.

The next development in Titan haze models (pioneered by R.W., P.S., Cabane, M. and M.L.) included the use of fractal aggregate haze particles that had a small component (monomer) with a radius of about 0.06 μm to produce strong linear polarization[26-30]. These monomers stuck together in an aggregation of many tens (or more) monomers. The large size of the aggregation could produce the strong forward scattering required from the Titan haze aerosols while preserving the high degree of linear polarization. However, it was quite laborious to compute the single-scattering properties of such aggregate particles for more than about 100 monomers at visible wavelengths. Particles with an effective radius of about 0.35 μm were required to produce the degree of forward scattering observed by Voyager[31]. This required the number of monomers in an aggregate particle to be about 45, and permitted single-scattering computations of the cross-section and phase function of the particles over the visible range. Of course, even larger numbers of monomers per particle would have matched the observations at high phase angles on Voyager, but these were difficult to perform and have largely gone unexplored. If larger particles had been used, however, the optical depth of the aerosols at shorter wavelengths would not have been nearly so large, and the clear space below 70 km may well not have been necessary.

The new DISR observations give a measurement of the monomer radius of 0.05 μm, in good agreement with previous estimates. Significantly, however, they show that the haze optical depth varies from about 2 at 935 nm to only about 4.5 at 531 nm, and the number of monomers in a haze particle is therefore probably several hundred. A value of 256 for N gives a projected area equal to that of a sphere of radius 0.65 μm, about twice as large as previously assumed. With $N = 512$, the equivalent sphere with the same projected area has a radius of 0.9 μm, nearly three times the size previously used. In any case, it seems that the size of the aggregate particles is several times as large as in some of the older models. A better estimate of the particle size will be available after the analysis of the solar aureole measurements of the variation in brightness near the Sun. In addition, measurements by the DISR violet photometer will extend the optical measurements of the haze to

由 DISR 测量得到的土卫六上的雾霾气溶胶的特性与惠更斯任务之前的观点是非常不同的。在惠更斯探测器之前，考虑了凝固和沉积作用的云层物理模型预测随着高度的降低，雾霾的密度将有较大的增幅[9]。此外，先驱者号和旅行者号飞船在 90°相位角测量了土卫六的散射光，发现存在高度线偏振，两个飞船的测量结果只有在球形颗粒的半径小于或等于 0.1 μm 的情况下才能吻合。这么小的颗粒随着波长（小于 1,000 nm 范围内）的减小，其光深大幅增加。通过拟合 890 nm 的甲烷强吸收峰，可以约束高空雾霾的总量。但在 619 nm 的甲烷弱吸收峰，这些雾霾的光学厚度则要大得多。为了拟合该吸收峰的观测强度，有必要援引产生于高层大气的有机气体扩散到下面较冷的大气层后将发生凝结的观点[26]，以此来移除云层物理模型中允许的 70 km 以下高度的雾霾。萨根和汤普森在参考文献 25 中提出了土卫六高层大气中的许多有机气体通过光化学发生凝结的观点，看似与上述观点相符。

通过引入雾霾颗粒分形聚集，进一步发展了土卫六雾霾模型（由韦斯特、史密斯、卡班和莱蒙率先提出），其采用的小成分（单体）的半径约为 0.06 μm，可以产生很强的线偏振[26-30]。聚集颗粒由数十个（或更多）单体彼此吸附而成。这么大的聚集颗粒不但可以产生土卫六雾霾气溶胶需要的非常强的前向散射，而且能够保持高度线偏振。然而，如果构成颗粒的单体超过 100 个，则计算这么大颗粒在可见光波段的单散射特性是非常艰难的。为了解释旅行者号观测到的前向散射度，聚集颗粒的有效半径需要大约为 0.35 μm[31]，则相应的单体数量大约为 45，这样就可以在可见光范围进行粒子的截面积和相函数的单散射计算。当然，每个颗粒有更大的单体数量也能匹配旅行者号在高相位角的观测结果，但是这是很难计算的，并且在很大程度上也未探索过。然而，如果采用较大的聚集颗粒，这些气溶胶在短波长波段的光深并不一定这么大，而且也没必要假设 70 km 以下存在净空层。

最近的 DISR 观测结果表明，构成雾霾颗粒的单体的半径为 0.05 μm，这与先前的估计相符。但是，明显地，雾霾的光深从 935 nm 处大约为 2 变到 531 nm 处的 4.5，而雾霾颗粒中的单体数量可能高达数百个。如果假设单体数量 $N=256$，则其投影面积相当于一个半径为 0.65 μm 球体，相当于先前假定值的两倍。如果假设单体数量 $N=512$，则相同投影面积的等价球体的半径为 0.9 μm，该尺寸几乎是先前值的三倍。无论哪种情况，聚集颗粒的尺寸看起来都是原先的模型中对应值的数倍。通过对太阳光晕进行太阳周围亮度的变化测量并进行分析，可以更好地估计出粒子的尺寸。此外，DISR 紫外光度计的测量可以将雾霾的光学测量范围扩展到更短的波

wavelengths as short as the band from 350 to 480 nm, also helping to constrain the size of the haze particles.

The number density of the haze particles does not increase with depth nearly as dramatically as predicted by the older cloud physics models. In fact, the number density increases by only a factor of a few over the altitude range from 150 km to the surface. This implies that vertical mixing is much less than had been assumed in the older models where the particles are distributed approximately as the gas is with altitude. In any case, no clear space at low altitudes, which was suggested earlier[32], was seen.

The methane mole fraction of 1.6% measured in the stratosphere by the Composite Infrared Spectrometer (CIRS) and the Gas Chromatograph Mass Spectrometer is consistent with the DISR spectral measurements. At very low altitudes (20 m) DISR measured $5 \pm 1\%$ for the methane mole fraction.

Finally, the entire set of DISR observations gives a new view of Titan, and reinforces the view that processes on Titan's surface are more similar to those on the surface of the Earth than anywhere else in the Solar System.

(**438**, 765-778; 2005)

M. G. Tomasko[1], B. Archinal[2], T. Becker[2], B. Bézard[3], M. Bushroe[1], M. Combes[3], D. Cook[2], A. Coustenis[3],C. de Bergh[3], L. E. Dafoe[1], L. Doose[1], S. Douté[4], A. Eibl[1], S. Engel[1], F. Gliem[5], B. Grieger[6], K. Holso[1], E. Howington-Kraus[2], E. Karkoschka[1], H. U. Keller[6], R. Kirk[2], R. Kramm[6], M. Küppers[6], P. Lanagan[1], E. Lellouch[3], M. Lemmon[7], J. Lunine[1,8], E. McFarlane[1], J. Moores[1], G. M. Prout[1], B. Rizk[1], M. Rosiek[2], P. Rueffer[5], S. E. Schröder[6], B. Schmitt[4], C. See[1], P. Smith[1], L. Soderblom[2], N. Thomas[9] & R. West[10]

[1] Lunar and Planetary Laboratory, University of Arizona, 1629 E. University Blvd, Tucson, Arizona 85721-0092, USA

[2] US Geological Survey, Astrogeology, 2225 N. Gemini Drive, Flagstaff, Arizona 86001, USA

[3] LESIA, Observatoire de Paris, 5 place Janssen, 92195 Meudon, France

[4] Laboratoire de Planétologie de Grenoble, CNRS-UJF, BP 53, 38041 Grenoble, France

[5] Technical University of Braunschweig, Hans-Sommer-Str. 66, D-38106 Braunschweig, Germany

[6] Max Planck Institute for Solar System Research, Max-Planck-Str. 2, D-37191 Katlenburg-Lindau, Germany

[7] Department of Physics, Texas A&M University, College Station, Texas 77843-3150, USA

[8] Istituto Nazionale di Astrofisica — Istituto di Fisica dello Spazio Interplanetario (INAF-IFSI ARTOV), Via del Cavaliere, 100, 00133 Roma, Italia

[9] Department of Physics, University of Bern, Sidlerstr. 5, CH-3012 Bern, Switzerland

[10] Jet Propulsion Laboratory, 4800 Oak Grove Drive, Pasadena, California 91109, USA

Received 26 May; accepted 8 August 2005. Published online 30 November 2005.

References:

1. Coustenis, A. *et al.* Maps of Titan's surface from 1 to 2.5 μm. *Icarus* **177**, 89-105 (2005).

2. Porco, C. C. *et al.* Imaging of Titan from the Cassini spacecraft. *Nature* **434**, 159-168 (2005).

3. Sotin, C. *et al.* Infrared images of Titan. *Nature* **435**, 786-789 (2005).

4. Elachi, C. *et al.* Cassini radar views the surface of Titan. *Science* **308**, 970-974 (2005).

5. Tomasko, M. G. *et al.* The Descent Imager/Spectral Radiometer (DISR) experiment on the Huygens entry probe of Titan. *Space Sci. Rev.* **104**, 469-551 (2002).

6. Ivanov, B. A., Basilevski, A. T. & Neukem, G. Atmospheric entry of large meteoroids: implication to Titan. *Planet. Space Sci.* **45**, 993-1007 (1997).

7. Flasar, F. M., Allison, M. D. & Lunine, J. I. Titan zonal wind model. *ESA Publ.* **SP-1177**, 287-298 (1997).

段（350～480 nm），这也有助于约束雾霾颗粒的尺寸大小。

雾霾颗粒的粒子数密度，并不像早期的云层物理模型预测的那样随着深度的增加而急剧增加。事实上，粒子数密度从 150 km 高度到表面仅仅增加了几倍。这意味着垂直混合程度要比早期模型的假设小得多，后者认为粒子随高度的分布大体和气体的分布是一样的。总之，和先前的研究一样[32]，并没有在较低高度发现雾霾净空层。

合成红外光谱仪（CIRS）和气相色谱–质谱联用分析仪测得的甲烷摩尔分数为 1.6%，这与 DISR 光谱测量结果相一致。在非常低的高度（20 m），DISR 给出的甲烷摩尔分数为 5%±1%。

最后，DISR 的整个观测使我们对土卫六有了全新的认识，使我们进一步意识到土卫六表面上发生的各种过程比太阳系内其他任何天体都更类似于地球表面上发生的过程。

（金世超 翻译；陈含章 审稿）

8. Bond, N. A. Observations of planetary boundary-layer structure in the eastern equatorial Pacific. *J. Atmos. Sci.* **5**, 699-706 (1992).

9. Lindal, G. F., Wood, G. E., Hotz, H. B. & Sweetnam, D. N. The atmosphere of Titan: An analysis of the Voyager 1 radio occultation measurements. *Icarus* **53**, 348-363 (1983).

10. Tokano, T. & Neubauer, F. M. Tidal winds on Titan caused by Saturn. *Icarus* **158**, 499-515 (2002).

11. Niemann, H. B. *et al.* The abundances of constituents of Titan's atmosphere from the GCMS instrument on the Huygens probe. *Nature* doi:10.1038/nature04122 (this issue).

12. Fulchignoni, M. *et al. In situ* measurements of the physical characteristics of Titan's environment. *Nature* doi:10.1038/nature04314 (this issue).

13. Brown, G. N. Jr & Ziegler, W. T. in *Advances in Cryogenetic Engineering* (ed. Timmerhaus, K. D.) Vol. 25, 662-670 (Plenum, New York, 1980).

14. Griffith, C. A., Owen, T., Geballe, T. R., Rayner, J. & Rannou, P. Evidence for the exposure of water ice on Titan's surface. *Science* **300**, 628-630 (2003).

15. Coustenis, A., Lellouch, E., Maillard, J.-P. & McKay, C. P. Titan's surface: composition and variability from the near-infrared albedo. *Icarus* **118**, 87-104 (1995).

16. Lellouch, E., Schmitt, B., Coustenis, A. & Cuby, J.-G. Titan's 5-μm lightcurve. *Icarus* **168**, 204-209 (2004).

17. Grundy, W. & Schmitt, B. The temperature-dependent near-infrared absorption spectrum of hexagonal H_2O ice. *J. Geophys. Res. E* **103**, 25809-25822 (1998).

18. Bernard, J.-M. *et al.* Evidence for chemical variations at the micrometric scale of Titan's tholins: Implications for analysing Cassini-Huygens data. *Icarus* (submitted).

19. Coll, P. *et al.* Experimental laboratory simulation of Titan's atmosphere: aerosols and gas phase. *Planet. Space Sci.* **47**, 1331-1340 (1999).

20. Bernard, J.-M. *et al.* Experimental simulation of Titan's atmosphere: detection of ammonia and ethylene oxide. *Planet. Space Sci.* **51**, 1003-1011 (2003).

21. Moroz, L. V., Arnold, G., Korochantsev, A. V. & Wäsch, R. Natural solid bitumens as possible analogs for cometary and asteroid organics. 1. Reflectance spectroscopy of pure bitumens. *Icarus* **134**, 253-268 (1998).

22. Toon, O. B., McKay, C. P., Griffith, C. A. & Turco, R. P. A physical model of Titan's aerosols. *Icarus* **95**, 24-53 (1992).

23. Khare, B. N. *et al.* Optical constants of organic tholins produced in a simulated Titanian atmosphere—From soft X-ray to microwave frequencies. *Icarus* **60**, 127-137 (1984).

24. Flasar, F. M. *et al.* Titan's atmospheric temperatures, winds, and composition. *Science* **308**, 975-978 (2005).

25. Sagan, C. & Thompson, W. R. Production and condensation of organic gases in the atmosphere of Titan. *Icarus* **59**, 133-161 (1984).

26. Lemmon, M. T. *Properties of Titan's Haze and Surface.* PhD dissertation, (Univ. Arizona, 1994).

27. West, R. A. & Smith, P. H. Evidence for aggregate particles in the atmospheres of Titan and Jupiter. *Icarus* **90**, 330-333 (1991).

28. Cabane, M., Chassefière, E. & Israel, G. Formation and growth of photochemical aerosols in Titan's atmosphere. *Icarus* **96**, 176-189 (1992).

29. Cabane, M., Rannou, P., Chassefière, E. & Israel, G. Fractal aggregates in Titan's atmosphere. *Planet. Space Sci.* **41**, 257-267 (1993).

30. West, R. A. Optical properties of aggregate particles whose outer diameter is comparable to the wavelength. *Appl. Opt.* **30**, 5316-5324 (1991).

31. Rages, K. B. & Pollack, J. Vertical distribution of scattering hazes in Titan's upper atmosphere. *Icarus* **55**, 50-62 (1983).

32. McKay, C. P. *et al.* Physical properties of the organic aerosols and clouds on Titan. *Planet. Space Sci.* **49**, 79-99 (2001).

33. Bird, M. K. *et al.* The vertical profile of winds on Titan. *Nature* doi:10.1038/nature04060 (this issue).

34. Karkoschka, E. Methane, ammonia, and temperature measurements of the Jovian planets and Titan from CCD-spectrophotometry. *Icarus* **133**, 134-146 (1998).

Acknowledgements. We thank the people from the following organizations whose dedication and effort have made this project successful: AETA (Fontenay-aux-Roses, France), Alcatel Space (Cannes, France), Collimated Holes Inc., EADS Deutschland GmbH (formerly Deutsche Aerospace AG, Munich, Germany), ETEL Motion Technology (Mortiers, Switzerland), The European Space Agency's (ESA) European Space and Technology Centre (ESTEC), The European Space Operations Centre (ESOC), The Jet Propulsion Laboratory (JPL), Laboratoire de Planétologie de Grenoble (CNRS-UJF), Loral Fairchild (Tustin, California, USA), Martin Marietta Corporation (Denver, Colorado, USA), Max-Planck-Institut für Sonnensystemforschung (Katlenburg-Lindau, Germany), Observatoire de Paris (Meudon, France), Technische Universität Braunschweig (TUB), Thomson-CSF (Grenoble, France), University of Arizona's Kuiper Lunar and Planetary Laboratory (LPL), and the US Geological Survey (Flagstaff, Arizona, USA).

Author Information. Reprints and permissions information is available at npg.nature.com/reprintsandpermissions. The authors declare no competing financial interests. Correspondence and requests for materials should be addressed to C.S. (csee@lpl.arizona.edu).

Discovery of a Cool Planet of 5.5 Earth Masses through Gravitational Microlensing

J.-P Beaulieu *et al.*

Editor's Note

The search for extrasolar planets continues apace, with nearly 4,000 known in mid-2019. It is difficult from the surface of the Earth to find planets of about the mass of the Earth using the traditional Doppler technique of looking for the wobble in the parent star's position. Here Jean-Philippe Beaulieu and his colleagues report a planet of only 5.5 Earth masses, found using gravitational microlensing (the bending and focusing of light rays by gravity). The planet is about 2.6 astronomical units (the Earth–Sun distance) from its parent star—about where the asteroid belt is in our solar system. The planet orbits the star acting as a lens, which passed between our solar system and a more distant, lensed star.

In the favoured core-accretion model of formation of planetary systems, solid planetesimals accumulate to build up planetary cores, which then accrete nebular gas if they are sufficiently massive. Around M-dwarf stars (the most common stars in our Galaxy), this model favours the formation of Earth-mass ($M_\oplus$) to Neptune-mass planets with orbital radii of 1 to 10 astronomical units (AU), which is consistent with the small number of gas giant planets known to orbit M-dwarf host stars[1-4]. More than 170 extrasolar planets have been discovered with a wide range of masses and orbital periods, but planets of Neptune's mass or less have not hitherto been detected at separations of more than 0.15 AU from normal stars. Here we report the discovery of a $5.5^{+5.5}_{-2.7}M_\oplus$ planetary companion at a separation of $2.6^{+1.5}_{-0.6}$ AU from a $0.22^{+0.21}_{-0.11}M_\odot$ M-dwarf star, where $M_\odot$ refers to a solar mass. (We propose to name it OGLE-2005-BLG-390Lb, indicating a planetary mass companion to the lens star of the microlensing event.) The mass is lower than that of GJ876d (ref. 5), although the error bars overlap. Our detection suggests that such cool, sub-Neptune-mass planets may be more common than gas giant planets, as predicted by the core accretion theory.

GRAVITATIONAL microlensing events can reveal extrasolar planets orbiting the foreground lens stars if the light curves are measured frequently enough to characterize planetary light curve deviations with features lasting a few hours[6-9]. Microlensing is most sensitive to planets in Earth-to-Jupiter-like orbits with semi-major axes in the range 1–5 AU. The sensitivity of the microlensing method to low-mass planets is restricted by the finite angular size of the source stars[10,11], limiting detections to planets of a few $M_\oplus$ for giant

通过微引力透镜发现一颗 5.5 倍地球质量的冷行星

博利厄等

编者按

搜寻系外行星的工作进展迅速,截至 2019 年年中,已有将近 4,000 颗系外行星被发现。在地表利用传统的多普勒技术探测母恒星位置移动,进而发现与地球质量相当的行星是十分困难的。本文中,让-菲利普·博利厄与他的同事报道了通过微引力透镜(光线被引力弯曲和聚焦)找到的一颗仅有地球质量 5.5 倍的行星。这颗行星大致距离其母恒星 2.6 个天文单位(日地距离),大约对应于太阳系中小行星带所处的位置。围绕母星转动的行星与其母星,在经过我们的太阳系与更遥远的、被透镜的恒星之间时,表现为一个"透镜"。

核吸积模型是当前较为流行的行星系统形成模型,该模型认为微行星会聚集形成一个致密的行星核。如果它们的质量足够大,那么它们将逐渐吸积星云气体。M型矮星是银河系中最常见的恒星,根据核吸积模型,在它们附近形成的行星质量通常介于地球质量($M_\oplus$)和海王星质量之间,相应的轨道半径为 1～10 个天文单位(AU)。这个预言与实际观测结果相符合,因为已知围绕 M 型矮星运动的气体巨行星的数目很少[1-4]。目前,人们已经发现了超过 170 颗的系外行星,其质量和轨道周期的范围跨度很大,但是迄今尚未发现与正常恒星相距超过 0.15AU、质量与海王星相当或略小的行星。在本文中,我们发现了一颗质量为 $5.5^{+5.5}_{-2.7} M_\oplus$ 的类行星伴天体,其宿主恒星是质量为 $0.22^{+0.21}_{-0.11} M_\odot$ 的 M 型矮星($M_\odot$表示太阳的质量),二者相距 $2.6^{+1.5}_{-0.6}$ AU。我们建议将这颗星体命名为 OGLE-2005-BLG-390Lb,用以表示是微引力透镜事件中与恒星一起组成透镜系统的行星伴星。相比 GJ8764d,虽然两者的误差棒相互重叠,但是这颗行星的质量更小[5]。我们的探测研究结果表明,同核吸积模型预测的一样,这些温度偏低的亚海王星质量的行星可能比气体巨行星更为普遍。

如果光变曲线的采样频率足够高,以致能够识别时标为几个小时的行星光变曲线偏移特征,微引力透镜事件可以揭示围绕前景透镜恒星转动的系外行星[6-9]。微引力透镜对于探测那些轨道半长轴介于 1～5 AU 之间(对应从地球到木星之间的轨道范围)的行星最为灵敏。该方法在探测低质量行星方面的灵敏度主要受限于背景源恒星有限的角尺度[10,11],因此,对于巨型源恒星,行星质量的探测下限大致为几个

source stars, but allowing the detection of planets as small as $0.1 M_\oplus$ for main-sequence source stars in the Galactic Bulge. The PLANET collaboration[12] maintains the high sampling rate required to detect low-mass planets while monitoring the most promising of the > 500 microlensing events discovered annually by the OGLE collaboration, as well as events discovered by MOA. A decade of pioneering microlensing searches has resulted in the recent detections of two Jupiter-mass extrasolar planets[13,14] with orbital separations of a few AU by the combined observations of the OGLE, MOA, MicroFUN and PLANET collaborations. The absence of perturbations to stellar microlensing events can be used to constrain the presence of planetary lens companions. With large samples of events, upper limits on the frequency of Jupiter-mass planets have been placed over an orbital range of 1–10 AU, down to $M_\oplus$ planets[15-17] for the most common stars of our galaxy.

On 11 July 2005, the OGLE Early Warning System[18] announced the microlensing event OGLE-2005-BLG-390 (right ascension $\alpha = 17$ h 54 min 19.2 s, declination $\delta = -30°\ 22'\ 38''$, J2000) with a relatively bright clump giant as a source star. Subsequently, PLANET, OGLE and MOA monitored it with their different telescopes. After peaking at a maximum magnification of $A_{max} = 3.0$ on 31 July 2005, a short-duration deviation from a single lens light curve was detected on 9 August 2005 by PLANET. As described below, this deviation was due to a low-mass planet orbiting the lens star.

From analysis of colour-magnitude diagrams, we derive the following reddening-corrected colours and magnitudes for the source star: $(V-I)_0 = 0.85$, $I_0 = 14.25$ and $(V-K)_0 = 1.9$. We used the surface brightness relation[20] linking the emerging flux per solid angle of a light-emitting body to its colour, calibrated by interferometric observations, to derive an angular radius of 5.25 ± 0.73 µas, which corresponds to a source radius of $9.6 \pm 1.3 R_\odot$ (where $R_\odot$ is the radius of the Sun) if the source star is at a distance of 8.5 kpc. The source star colours indicate that it is a 5,200 K giant, which corresponds to a G4 III spectral type.

Figure 1 shows our photometric data for microlensing event OGLE-2005-BLG-390 and the best planetary binary lens model. The best-fit model has $\chi^2 = 562.26$ for 650 data points, seven lens parameters, and 12 flux normalization parameters, for a total of 631 degrees of freedom. Model length parameters in Table 1 are expressed in units of the Einstein ring radius R_E (typically ~2 AU for a Galactic Bulge system), the size of the ring image that would be seen in the case of perfect lens–source alignment. In modelling the light curve, we adopted linear limb darkening laws[21] with $\Gamma_I = 0.538$ and $\Gamma_R = 0.626$, appropriate for this G4 III giant source star, to describe the centre-to-limb variation of the intensity profile in the I and R bands. Four different binary lens modelling codes were used to confirm that the model we present is the only acceptable model for the observed light curve. The best alternative model is one with a large-flux-ratio binary source with a single lens, which has gross features that are similar to a planetary microlensing event[22]. However, as shown in Fig. 1, this model fails to account for the PLANET-Perth, PLANET-Danish and OGLE

太阳质量。但对于星系核球中的主序源恒星，则允许探测到 0.1 个太阳质量大小的行星。PLANET 合作组织[12]每年对光学引力透镜实验(OGLE)合作组织以及天文物理微透镜观测(MOA)合作组织发现的超过 500 个最有可能的微引力透镜事件进行监控，其采样频率足够高，可以满足低质量行星探测的要求。结合过去十年 OGLE、MOA、MicroFUN 和 PLANET 合作组织对微引力透镜早期搜寻的观测结果，人们最近发现了两颗质量与木星相当、轨道半径相差几个 AU 的系外行星[13,14]。无扰动的恒星微引力透镜事件可以用来限制透镜中行星伴星的存在。基于大样本的微引力透镜事件，人们不仅对轨道半径在 1～10 AU 范围内木星质量的行星出现频率给出了上限，还对于银河系中最常见的恒星周围低至 $M_\oplus$ 的行星出现频率给出了限制[15-17]。

2005 年 7 月 11 日，OGLE 的预警系统[18]报告发现了微引力透镜事件 OGLE-2005-BLG-390(赤经 $\alpha = 17$ h 54 min 19.2 s，赤纬 $\delta = -30°22'38''$，J2000)，其源恒星为一颗相对明亮的团簇巨星。随后，PLANET、OGLE 和 MOA 分别利用望远镜对其进行了监控观测。在 2005 年 7 月 31 日该引力透镜事件的放大率达到峰值 $A_{max} = 3.0$ 之后，2005 年 8 月 9 日 PLANET 探测到一个短时偏离单透镜光变曲线的现象。如下所述，这种偏离是由一颗围绕透镜恒星运动的低质量行星引起的。

通过分析颜色-星等图，我们可以推知这颗源恒星经过红化校正后的颜色和星等，结果如下：$(V-I)_0 = 0.85$，$I_0 = 14.25$，$(V-K)_0 = 1.9$。我们利用面亮度关系[20]将发光天体单位立体角内的辐射通量与颜色联系起来。经过干涉观测校正，我们推导出源恒星的角半径为 5.25 ± 0.73 μas。如果这颗源恒星的距离为 8.5 kpc，那么上述角半径值对应的物理半径为 $9.6 \pm 1.3 R_\odot$，其中 $R_\odot$ 表示太阳的半径。这颗源恒星的颜色表明，它是一颗温度为 5,200 K 的巨星，对应于 G4 III 光谱类型。

图 1 给出了微引力透镜事件 OGLE-2005-BLG-390 的测光数据以及最佳的行星双透镜模型。最优拟合模型的 $\chi^2 = 562.26$，该拟合具有 650 个数据点、7 个透镜参数、12 个通量归一化参数，总自由度为 631。表 1 给出的模型长度参数以爱因斯坦环半径 R_E(对于一个银河系核球系统而言，其典型值约为 2 AU) 为单位，这个爱因斯坦环的半径对应当透镜、源恒星以及观测者精确排列在一条直线上时所能观测到的环形图像的尺寸。在建模光变曲线时，我们采用线性临边昏暗定律[21]来描述 I 和 R 波段从中心到边缘的发光强度轮廓，参数取值为 $\Gamma_I = 0.538$，$\Gamma_R = 0.626$，适用于 G4 III 巨型源恒星。四套不同的双透镜建模代码被用来验证我们这里所展示的模型，是针对观测到的光变曲线的唯一可接受模型。最好的替代模型包含单一的前景透镜恒星，背景源为具有大流量比的双星，该模型可以粗略地产生类似于行星微引力透镜事件的特征[22]。然而，如图 1 所示，这一模型无法解释由 PLANET-Perth 天文台

measurements near the end of the planetary deviation, and it is formally excluded by $\Delta\chi^2 = 46.25$ with one less model parameter.

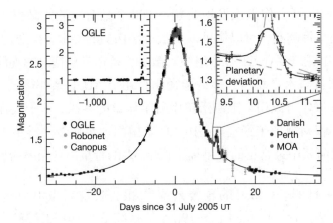

Fig. 1. The observed light curve of the OGLE-2005-BLG-390 microlensing event and best-fit model plotted as a function of time. Error bars are 1σ. The data set consists of 650 data points from PLANET Danish (ESO La Silla, red points), PLANET Perth (blue), PLANET Canopus (Hobart, cyan), RoboNet Faulkes North (Hawaii, green), OGLE (Las Campanas, black), MOA (Mt John Observatory, brown). This photometric monitoring was done in the I band (with the exception of the Faulkes R-band data and the MOA custom red passband) and real-time data reduction was performed with the different OGLE, PLANET and MOA data reduction pipelines. Danish and Perth data were finally reduced by the image subtraction technique[19] with the OGLE pipeline. The top left inset shows the OGLE light curve extending over the previous 4 years, whereas the top right one shows a zoom of the planetary deviation, covering a time interval of 1.5 days. The solid curve is the best binary lens model described in the text with $q = 7.6 \pm 0.7 \times 10^{-5}$, and a projected separation of $d = 1.610 \pm 0.008 R_E$. The dashed grey curve is the best binary source model that is rejected by the data, and the dashed orange line is the best single lens model.

Table 1 Microlensing fit parameters

d	$1.610 \pm 0.008 R_E$
q	$(7.6 \pm 0.7) \times 10^{-5}$
Closest approach	$0.359 \pm 0.005 R_E$
Einstein ring radius crossing time	11.03 ± 0.11 days
Time of closest approach	31.231 ± 0.005 July 2005 UT
Source star radius crossing time	0.282 ± 0.010 days
θ	2.756 ± 0.003 rad

The parameters for the best binary lens model for the OGLE 2005-BLG-390 microlensing event light curve are shown with their 1σ uncertainties. Some of these parameters are scaled to the Einstein ring radius, which is given by $R_E = 2\sqrt{GMD_L(D_S-D_L)/(c^2 D_S)}$, where M is the mass of the lens, G is the newtonian constant of gravitation, c is the speed of light in vacuum, and D_L and D_S are the lens and source distances, respectively.

The planet is designated OGLE-2005-BLG-390Lb, where the "Lb" suffix indicates the secondary component of the lens system with a planetary mass ratio. The microlensing fit

望远镜、PLANET-Danish 天文台望远镜以及 OGLE 观测到的行星偏离末段附近的数据，该模型缺少一个模型参数，因 $\Delta\chi^2 = 46.25$ 从严格意义上被排除。

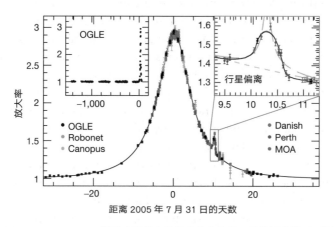

图 1. 观测到的 OGLE-2005-BLG-390 微引力透镜事件的光变曲线以及最佳拟合模型（横轴为时间）。误差棒为 1σ。图中共有 650 个数据点，分别来源于 PLANET-Danish 天文台望远镜（拉西亚欧南台（ESO），红色）、PLANET-Perth 天文台望远镜（蓝色）、PLANET-Canopus 天文台望远镜（霍巴特，青色）、RoboNet 北福克斯望远镜（夏威夷，绿色）、OGLE（拉斯坎帕纳斯，黑色）以及 MOA（约翰山天文台，棕色）。其中，除福克斯望远镜采用 R 波段数据，MOA 采用特制的红色通带外，其他测光监测均在 I 波段进行。我们采用 OGLE、PLANET 以及 MOA 不同的数据处理软件对数据进行实时处理。PLANET-Danish 和 PLANET-Perth 天文台望远镜的数据最终通过 OGLE 数据处理中的图像扣除技术 [19] 进行处理。左上方的插图给出了前四年里的 OGLE 测得的光变曲线，右上方插图给出了行星偏离部分的放大图，其时间跨度为 1.5 天。实线表示文中所述的最佳双透镜模型的理论预言，$q = 7.6 \pm 0.7 \times 10^{-5}$，投影间隔 $d = 1.610 \pm 0.008 R_E$。图中灰色虚线表示不被数据支持的最佳双源模型的理论预言。图中橙色虚线表示最佳单透镜模型的理论预言。

表 1. 微引力透镜的拟合参数

d	$1.610 \pm 0.008 R_E$
q	$(7.6 \pm 0.7) \times 10^{-5}$
最接近点	$0.359 \pm 0.005 R_E$
爱因斯坦环半径的穿越时间	11.03 ± 0.11 天
最接近点时间	31.231 ± 0.005 2005 年 7 月 UT
源恒星半径的穿越时间	0.282 ± 0.010 天
θ	2.756 ± 0.003 rad

上表给出了用于拟合 OGLE 2005-BLG-390 透镜事件光变曲线的最佳双透镜模型的参数值，同时给出了各参数 1σ 不确定度。表中部分参数以爱因斯坦半径 R_E 作为单位，$R_E = 2\sqrt{GMD_L(D_S - D_L)/(c^2 D_S)}$，其中 M 代表透镜的质量，G 表示牛顿万有引力常数，c 表示真空中的光速，D_S 和 D_L 分别表示透镜距离和源距离。

这颗行星被命名为 OGLE-2005-BLG-390Lb，其中后缀"Lb"表示透镜系统包含具有行星质量的次等组分。微引力透镜拟合仅直接确定行星与恒星的质量比

only directly determines the planet–star mass ratio, $q = 7.6 \pm 0.7 \times 10^{-5}$, and the projected planet–star separation, $d = 1.610 \pm 0.008 R_{\mathrm{E}}$. Although the planet and star masses are not directly determined for planetary microlensing events, we can derive their probability densities. We have performed a bayesian analysis[23] employing the Galactic models and mass functions described in refs 11 and 23. We averaged over the distances and velocities of the lens and source stars, subject to the constraints due to the angular diameter of the source and the measured parameters given in Table 1. This analysis gives a 95% probability that the planetary host star is a main-sequence star, a 4% probability that it is a white dwarf, and a probability of < 1% that it is a neutron star or black hole. The host star and planet parameter probability densities for a main sequence lens star are shown in Fig. 2 for the Galactic model used in ref. 23. The medians of the lens parameter probability distributions yield a companion mass of $5.5^{+5.5}_{-2.7} M_{\oplus}$ and an orbital separation of $2.6^{+1.5}_{-0.6}$ AU from the $0.22^{+0.21}_{-0.11} M_{\odot}$ lens star, which is located at a distance of $D_{\mathrm{L}} = 6.6 \pm 1.0$ kpc. These error bars indicate the central 68% confidence interval. These median parameters imply that the planet receives radiation from its host star that is only 0.1% of the radiation that the Earth receives from the Sun, so the probable surface temperature of the planet is ~50 K, similar to the temperatures of Neptune and Pluto.

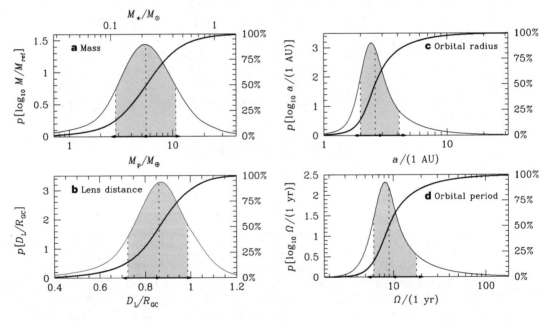

Fig. 2. Bayesian probability densities for the properties of the planet and its host star. **a**, The masses of the lens star and its planet (M_* and M_{p} respectively), **b**, their distance from the observer (D_{L}), **c**, the three-dimensional separation or semi-major axis a of an assumed circular planetary orbit; and **d**, the orbital period Ω of the planet. (In **a**, M_{ref} refers to $M_{\oplus}$ on the upper x axis and $M_{\odot}$ on the lower x axis.) The bold, curved line in each panel is the cumulative distribution, with the percentiles listed on the right. The dashed vertical lines indicate the medians, and the shading indicates the central 68.3% confidence intervals, while dots and arrows on the abscissa mark the expectation value and standard deviation. All estimates follow from a bayesian analysis assuming a standard model for the disk and bulge population of the Milky Way, the stellar mass function of ref. 23, and a gaussian prior distribution for $D_{\mathrm{S}} = 1.05 \pm 0.25 R_{\mathrm{GC}}$ (where $R_{\mathrm{GC}} = 7.62 \pm 0.32$ kpc for the Galactic Centre distance). The medians of these distributions yield a $5.5^{+5.5}_{-2.7} M_{\oplus}$

$q = 7.6 \pm 0.7 \times 10^{-5}$，以及行星与恒星的投影距离，$d = 1.610 \pm 0.008 R_{\mathrm{E}}$。尽管我们不能通过行星微引力透镜事件直接确定恒星和行星的质量，但是却可以推导出它们的概率密度。我们通过引入银河系模型和质量函数[11,23]，进行了贝叶斯分析[23]。基于源的角直径以及表 1 中测量参数的限制，我们对透镜和源恒星在距离和速度上进行了平均。该分析结果表明，该行星宿主恒星是一颗主序恒星的概率为 95%，是一颗白矮星的概率为 4%，是一颗中子星或黑洞的概率小于 1%。图 2 给出了银河系模型[23]下透镜系统中以主序星作为宿主恒星及其行星各参数的概率密度。透镜参数概率分布的中值表明伴星的质量为 $5.5^{+5.5}_{-2.7} M_{\oplus}$，围绕质量为 $0.22^{+0.21}_{-0.11} M_{\odot}$ 的透镜恒星运动，轨道半径为 $2.6^{+1.5}_{-0.6}$ AU，整个透镜系统与我们的距离为 $D_{\mathrm{L}} = 6.6 \pm 1.0$ kpc。图中误差棒给出了中心附近 68% 的置信区间。这些中值参数表明该行星从宿主恒星接收到的辐射，仅为地球从太阳接收到的辐射的 0.1%，因此这颗行星的表面温度很可能仅为 50 K 左右，类似于海王星和冥王星。

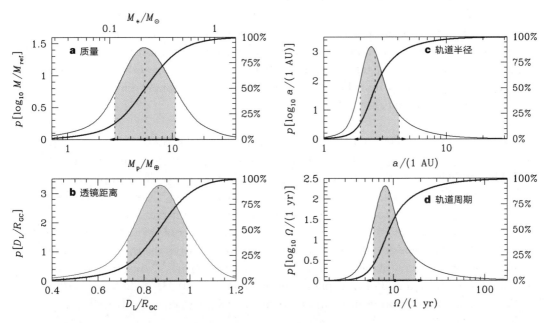

图 2. 行星及其宿主恒星属性的贝叶斯概率密度。**a**，透镜恒星及其行星的质量，分别用 M_* 和 M_{p} 表示；**b**，透镜恒星及其行星与观测者的距离 (D_{L})；**c**，假定圆形行星轨道的三维间隔或半长轴 a；**d**，行星的轨道周期 Ω。（在图 **a** 中，M_{ref} 表示在顶部 x 轴以 $M_{\oplus}$ 为单位，在底部 x 轴以 $M_{\odot}$ 为单位）在每幅图中，加粗曲线表示累计分布，右侧坐标轴给出了相应的百分比值。图中垂直虚线表示中值，阴影部分给出了中心附近 68.3% 的置信区间，横坐标轴上的点和箭头分别标记出期望值和标准差。以上所有估算均是在假定一个标准的银河系盘和核球模型，文献 23 中的恒星质量函数以及 $D_{\mathrm{S}} = 1.05 \pm 0.25 R_{\mathrm{GC}}$ 满足高斯先验分布（其中 R_{GC} 代表银河中心的距离，$R_{\mathrm{GC}} = 7.62 \pm 0.32$ kpc）的情况下利用贝叶斯分析得到的。这些分布的中值表明行星的质量为 $5.5^{+5.5}_{-2.7} M_{\oplus}$，在银河系核球区域的 M 型矮星的质量为 $0.22^{+0.21}_{-0.11} M_{\odot}$，二者相距 $2.6^{+1.5}_{-0.6}$ AU，而整个透镜系统与太阳的距离为 $D_{\mathrm{L}} = 6.6 \pm 1.0$ kpc。行星的中值周期为 9^{+9}_{-3} 年。对概率分布取对数平均（遵循开普勒第三定律），可知这颗行星的轨道周期为 10.4 年，与宿主恒星的距离为 2.9 AU，恒星和行星的质量分别为 $0.22 M_{\odot}$ 和 $5.5 M_{\oplus}$。每幅图中的纵坐标的方括号中列出了该概率密度的独

planetary companion at a separation of $2.6^{+1.5}_{-0.6}$ AU from a $0.22^{+0.21}_{-0.11}M_\odot$ Galactic Bulge M-dwarf at a distance of 6.6 ± 1.0 kpc from the Sun. The median planetary period is 9^{+9}_{-3} years. The logarithmic means of these probability distributions (which obey Kepler's third law) are a separation of 2.9 AU, a period of 10.4 years, and masses of $0.22M_\odot$ and $5.5M_\oplus$ for the star and planet, respectively. In each plot, the independent variable for the probability density is listed within square brackets. The distribution of the planet–star mass ratio was taken to be independent of the stellar mass, and a uniform prior distribution was assumed for the planet–star separation distribution.

The parameters of this event are near the limits of microlensing planet detectability for a giant source star. The separation of $d = 1.61$ is near the outer edge of the so-called lensing zone[7], which has the highest planet detection probability, and the planet's mass is about a factor of two above the detection limit set by the finite size of the source star. Planets with $q > 10^{-3}$ and $d \approx 1$ are much easier to detect, and so it may be that the parameters of OGLE-2005-BLG-390Lb represent a more common type of planet. This can be quantified by simulating planetary light curves with different values of q and θ (where θ is the angle of source motion with respect to the lens axis) but the remaining parameters are fixed to the values for the three known microlensing planets. We find that the probability of detecting a $q \approx 4-7 \times 10^{-3}$ planet, like the first two microlens planets[13,14], is ~50 times larger than the probability of detecting a $q = 7.6 \times 10^{-5}$ planet like OGLE-2005-BLG-390Lb. This suggests that, at the orbital separations probed by microlensing, sub-Neptune-mass planets are significantly more common than large gas giants around the most common stars in our Galaxy. Similarly, the first detection of a sub-Neptune-mass planet at the outer edge of the "lensing zone" provides a hint that these sub-Neptune-mass planets may tend to reside in orbits with semi-major axes $a > 2$ AU.

The core-accretion model of planet formation predicts that rocky/icy $5-15M_\oplus$ planets orbiting their host stars at $1-10$ AU are much more common than Jupiter-mass planets, and this prediction is consistent with the small fraction of M-dwarfs with planets detected by radial velocities[3,5] and with previous limits from microlensing[15]. Our discovery of such a low-mass planet by gravitational microlensing lends further support to this model, but more detections of similar and lower-mass planets over a wide range of orbits are clearly needed. Planets with separations of ~0.1 AU will be detected routinely by the radial velocity method or space observations of planetary transits in the coming years[24-27], but the best chance to increase our understanding of such planets over orbits of $1-10$ AU in the next $5-10$ years is by future interferometer programs[28] and more advanced microlensing surveys[11,29,30].

(**439**, 437-440; 2006)

立变量。这里认为行星−恒星质量比的分布与恒星的质量无关，并假定行星与宿主恒星之间的距离分布服从均匀的先验分布。

这次事件的各参数值接近巨型源恒星的微引力透镜探测极限。恒星、行星相距 $d = 1.61$ 已接近所谓的"透镜区域"的外边界 [7]，从而具有最高的行星探测概率；同时行星的质量是由源恒星的有限尺度所设定的探测极限的 2 倍。显然，那些 $q > 10^{-3}$ 且 $d \approx 1$ 的行星更加容易探测到，因此 OGLE-2005-BLG-390Lb 可能代表了一类更为普遍的行星。这可以在不同 q 和 θ 取值条件下通过对行星光变曲线进行模拟仿真而被量化（其中 θ 表示源恒星运动相对透镜主轴的角度），而剩余参量则对应已知的三个微透镜行星取固定值。我们发现，探测到一颗 $q \approx 4 \sim 7 \times 10^{-3}$ 的行星（类似首先发现的两颗透镜行星）的概率是探测到一颗 $q = 7.6 \times 10^{-5}$ 的行星（类似 OGLE-2005-BLG-390Lb）的概率的约 50 倍。这说明，在微引力透镜可探测的轨道间隔的范围内，对于我们星系中的众多恒星，其四周具有亚海王星质量的行星比大型气态巨行星明显更加普遍。类似地，在"透镜区域"外边缘首次探测到亚海王星质量的行星，暗示着这类行星的轨道半长轴倾向于 >2 AU。

行星形成的核吸积模型预测，在宿主恒星附近普遍存在的是质量为 $5 \sim 15 M_{\oplus}$、轨道半长轴在 $1 \sim 10$ AU 范围内的行星，而不是木星质量大小的行星。这与运用径向速度法很少探测到 M 型矮星附近的行星 [3,5] 的观测结果相吻合，同时也与过去的微引力透镜限制一致 [15]。我们通过微引力透镜方法发现的这样一颗低质量的行星，进一步支持了该模型，但是我们仍需探测更多类似的位于不同轨道上的低质量行星来验证这一模型。在未来几年中，通过径向速度法或空间观测掩食法探测轨道半长轴约为 0.1 AU 的行星 [24-27] 将成为一种常态。然而，未来 $5 \sim 10$ 年里，借助将来的干涉仪探测项目 [29] 以及更为先进的微引力透镜巡天 [11,28,30] 是进一步了解那些轨道介于 $1 \sim 10$ AU 的行星性质的最好的方法。

（金世超 刘项琨 翻译；李然 审稿）

J.-P. Beaulieu[1,4], D. P. Bennett[1,3,5], P. Fouqué[1,6], A. Williams[1,7], M. Dominik[1,8], U. G. Jørgensen[1,9], D. Kubas[1,10], A. Cassan[1,4], C. Coutures[1,11], J. Greenhill[1,12], K. Hill[1,12], J. Menzies[1,13], P. D. Sackett[1,14], M. Albrow[1,15], S. Brillant[1,10], J. A. R. Caldwell[1,16], J. J. Calitz[1,17], K. H. Cook[1,18], E. Corrales[1,4], M. Desort[1,4], S. Dieters[1,12], D. Dominis[1,19], J. Donatowicz[1,20], M. Hoffman[1,19], S. Kane[1,21], J.-B. Marquette[1,4], R. Martin[1,7], P. Meintjes[1,17], K. Pollard[1,15], K. Sahu[1,22], C. Vinter[1,9], J. Wambsganss[1,23], K. Woller[1,9], K. Horne[1,8], I. Steele[1,24], D. M. Bramich[1,8,24], M. Burgdorf[1,24], C. Snodgrass[1,25], M. Bode[1,24], A. Udalski[2,26], M. K. Szymański[2,26], M. Kubiak[2,26], T. Więckowski[2,26], G. Pietrzyński[2,26,27], I. Soszyński[2,26,27], O. Szewczyk[2,26], Ł. Wyrzykowski[2,26,28], B. Paczyński[2,29], F. Abe[3,30], I. A. Bond[3,31], T. R. Britton[3,15,32], A. C. Gilmore[3,15], J. B. Hearnshaw[3,15], Y. Itow[3,30], K. Kamiya[3,30], P. M. Kilmartin[3,15], A. V. Korpela[3,33], K. Masuda[3,30], Y. Matsubara[3,30], M. Motomura[3,30], Y. Muraki[3,30], S. Nakamura[3,30], C. Okada[3,30], K. Ohnishi[3,34], N. J. Rattenbury[3,28], T. Sako[3,30], S. Sato[3,35], M. Sasaki[3,30], T. Sekiguchi[3,30], D. J. Sullivan[3,33], P. J. Tristram[3,32], P. C. M. Yock[3,32] & T. Yoshioka[3,30]

[1] PLANET/RoboNet Collaboration (http://planet.iap.fr and http://www.astro.livjm.ac.uk/RoboNet/)

[2] OGLE Collaboration (http://ogle.astrouw.edu.pl)

[3] MOA Collaboration (http://www.physics.auckland.ac.nz/moa)

[4] Institut d'Astrophysique de Paris, CNRS, Université Pierre et Marie Curie UMR7095, 98bis Boulevard Arago, 75014 Paris, France

[5] University of Notre Dame, Department of Physics, Notre Dame, Indiana 46556-5670, USA

[6] Observatoire Midi-Pyrénées, Laboratoire d'Astrophysique, UMR 5572, Université Paul Sabatier—Toulouse 3, 14 avenue Edouard Belin, 31400 Toulouse, France

[7] Perth Observatory, Walnut Road, Bickley, Perth, WA 6076, Australia

[8] Scottish Universities Physics Alliance, University of St Andrews, School of Physics and Astronomy, North Haugh, St Andrews KY16 9SS, UK

[9] Niels Bohr Institutet, Astronomisk Observatorium, Juliane Maries Vej 30, 2100 København Ø, Denmark

[10] European Southern Observatory, Casilla 19001, Santiago 19, Chile

[11] CEA DAPNIA/SPP Saclay, 91191 Gif-sur-Yvette cedex, France

[12] University of Tasmania, School of Mathematics and Physics, Private Bag 37, Hobart, TAS 7001, Australia

[13] South African Astronomical Observatory, PO Box 9, Observatory 7935, South Africa

[14] Research School of Astronomy and Astrophysics, Australian National University, Mt Stromlo Observatory, Weston Creek, ACT 2611, Australia

[15] University of Canterbury, Department of Physics and Astronomy, Private Bag 4800, Christchurch 8020, New Zealand

[16] McDonald Observatory, 16120 St Hwy Spur 78 #2, Fort Davis, Texas 79734, USA

[17] Boyden Observatory, University of the Free State, Department of Physics, PO Box 339, Bloemfontein 9300, South Africa

[18] Lawrence Livermore National Laboratory, IGPP, PO Box 808, Livermore, California 94551, USA

[19] Universität Potsdam, Institut für Physik, Am Neuen Palais 10, 14469 Potsdam, Astrophysikalisches Institut Potsdam, An der Sternwarte 16, D-14482, Potsdam, Germany

[20] Technische Universität Wien, Wiedner Hauptstrasse 8/020 B.A. 1040 Wien, Austria

[21] Department of Astronomy, University of Florida, 211 Bryant Space Science Center, Gainesville, Florida 32611-2055, USA

[22] Space Telescope Science Institute, 3700 San Martin Drive, Baltimore, Maryland 21218, USA

[23] Astronomisches Rechen-Institut (ARI), Zentrum für Astronomie, Universität Heidelberg, Mönchhofstrasse 12-14, 69120 Heidelberg, Germany

[24] Astrophysics Research Institute, Liverpool John Moores University, Twelve Quays House, Egerton Wharf, Birkenhead CH41 1LD, UK

[25] Astronomy and Planetary Science Division, Department of Physics, Queen's University Belfast, Belfast, UK

[26] Obserwatorium Astronomiczne Uniwersytetu Warszawskiego, Aleje Ujazdowskie 4, 00-478 Warszawa, Poland

[27] Universidad de Concepcion, Departamento de Fisica, Casilla 160-C, Concepcion, Chile

[28] Jodrell Bank Observatory, The University of Manchester, Macclesfield, Cheshire SK11 9DL, UK

[29] Princeton University Observatory, Peyton Hall, Princeton, New Jersey 08544, USA

[30] Solar-Terrestrial Environment Laboratory, Nagoya University, Nagoya 464-860, Japan

[31] Institute for Information and Mathematical Sciences, Massey University, Private Bag 102-904, Auckland, New Zealand

[32] Department of Physics, University of Auckland, Private Bag 92019, Auckland, New Zealand

[33] School of Chemical and Physical Sciences, Victoria University, PO Box 600, Wellington, New Zealand

[34] Nagano National College of Technology, Nagano 381-8550, Japan

[35] Department of Astrophysics, Faculty of Science, Nagoya University, Nagoya 464-860, Japan

Received 28 September; accepted 14 November 2005.

References:

1. Safronov, V. *Evolution of the Protoplanetary Cloud and Formation of the Earth and Planets* (Nauka, Moscow, 1969).

2. Wetherill, G. W. Formation of the terrestrial planets. *Annu. Rev. Astron. Astrophys* **18**, 77-113 (1980).

3. Laughlin, G., Bodenheimer, P. & Adams, F. C. The core accretion model predicts few jovian-mass planets orbiting red dwarfs. *Astrophys. J.* **612**, L73-L76 (2004).

4. Ida, S. & Lin, D. N. C. Toward a deterministic model of planetary formation. II. The formation and retention of gas giant planets around stars with a range of metallicities. *Astrophys. J.* **616**, 567-572 (2004).

5. Rivera, E. *et al.* A ~7.5 Earth-mass planet orbiting the nearby star, GJ 876. *Astrophys. J.* (in the press).

6. Mao, S. & Paczynski, B. Gravitational microlensing by double stars and planetary systems. *Astrophys. J.* **374**, L37-L40 (1991).

7. Gould, A. & Loeb, A. Discovering planetary systems through gravitational Microlenses. *Astrophys. J.* **396**, 104-114 (1992).

8. Wambsganss, J. Discovering Galactic planets by gravitational microlensing: magnification patterns and light curves. *Mon. Not. R. Astron. Soc.* **284**, 172-188 (1997).

9. Griest, K. & Safizadeh, N. The use of high-magnification microlensing events in discovering extrasolar planets. *Astrophys. J.* **500**, 37-50 (1998).

10. Bennett, D. P. & Rhie, S. H. Detecting Earth-mass planets with gravitational microlensing. *Astrophys. J.* **472**, 660-664 (1996).

11. Bennett, D. P. & Rhie, S. H. Simulation of a space-based microlensing survey for terrestrial extrasolar planets. *Astrophys. J.* **574**, 985-1003 (2002).

12. Albrow, M. *et al.* The 1995 pilot campaign of PLANET: searching for microlensing anomalies through precise, rapid, round-the-clock monitoring. *Astrophys. J.* **509**, 687-702 (1998).

13. Bond, I. A. *et al.* OGLE 2003-BLG-235/MOA 2003-BLG-53: A planetary microlensing event. *Astrophys. J.* **606**, L155-L158 (2004).

14. Udalski, A. *et al.* A jovian-mass planet in microlensing event OGLE-2005-BLG- 071. *Astrophys. J.* **628**, L109-L112 (2005).

15. Gaudi, B. S. *et al.* Microlensing constraints on the frequency of Jupiter-mass companions: analysis of 5 years of PLANET photometry. *Astrophys. J.* **566**, 463-499 (2002).

16. Abe, F. *et al.* Search for low-mass exoplanets by gravitational microlensing at high magnification. *Science* **305**, 1264-1267 (2004).

17. Dong, S. *et al.* Planetary detection efficiency of the magnification 3000 microlensing event OGLE-2004-BLG-343. *Astrophys. J.* (submitted); preprint at ⟨http://arXiv.org/astro-ph/0507079⟩ (2005).

18. Udalski, A. The optical gravitational lensing experiment. Real time data analysis systems in the OGLE-III survey. *Acta Astron.* **53**, 291-305 (2003).

19. Alard, C. Image subtraction using a space-varying kernel. *Astron. Astrophys. Suppl.* **144**, 363-370 (2000).

20. Kervella, P. *et al.* Cepheid distances from infrared long-baseline interferometry. III. Calibration of the surface brightness-color relations. *Astron. Astrophys.* **428**, 587-593 (2004).

21. Claret, A., Diaz-Cordoves, J. & Gimenez, A. Linear and non-linear limb-darkening coefficients for the photometric bands R I J H K. *Astron. Astrophys. Suppl.* **114**, 247-252 (1995).

22. Gaudi, B. S. Distinguishing between binary-source and planetary microlensing perturbations. *Astrophys. J.* **506**, 533-539 (1998).

23. Dominik, M. Stochastical distributions of lens and source properties for observed galactic microlensing events. *Mon. Not. R. Astron. Soc.* (submitted); preprint at ⟨http://arXiv.org/astro-ph/0507540⟩ (2005).

24. Vogt, S. S. *et al.* Five new multicomponent planetary systems. *Astrophys. J.* **632**, 638-658 (2005).

25. Mayor, M. *et al.* The CORALIE survey for southern extrasolar planets. XII. Orbital solutions for 16 extrasolar planets discovered with CORALIE. *Astron. Astrophys.* **415**, 391-402 (2004).

26. Borucki, W. *et al.* in *Second Eddington Workshop: Stellar Structure and Habitable Planet Finding* (eds Favata, F., Aigrain, S. & Wilson, A.) 177-182 (ESA SP-538, ESA Publications Division, Noordwijk, 2004).

27. Moutou, C. *et al.* Comparative blind test of five planetary transit detection algorithms on realistic synthetic light curves. *Astron. Astrophys.* **437**, 355-368 (2005).

28. Sozzeti, A. *et al.* Narrow-angle astrometry with the space interferometry mission: the search for extrasolar planets. I. Detection and characterization of single planets. *Pub. Astron. Soc. Pacif.* **114**, 1173-1196 (2002).

29. Bennett, D. P. in *ASP Conf. Ser. on Extrasolar Planets: Today and Tomorrow* (eds Beaulieu, J.-P., Lecavelier des Etangs, A. & Terquem, C.) Vol. 321, 59-68 (ASP, 2004).

30. Beaulieu, J. P. *et al.* PLANET III: searching for Earth-mass planets via microlensing from Dome C? *ESA Publ. Ser.* **14**, 297-302 (2005).

Acknowledgements. PLANET is grateful to the observatories that support our science (the European Southern Observatory, Canopus, Perth; and the South African Astronomical Observatory, Boyden, Faulkes North) and to the ESO team in La Silla for their help in maintaining and operating the Danish telescope. Support for the PLANET project was provided by CNRS, NASA, the NSF, the LLNL/NNSA/DOE, PNP, PICS France-Australia, D. Warren, the DFG, IDA and the SNF. RoboNet is funded by the UK PPARC and the FTN was supported by the Dill Faulkes Educational Trust. Support for the OGLE project, conducted at Las Campanas Observatory (operated by the Carnegie Institution of Washington), was provided by the Polish Ministry of Science, the Foundation for Polish Science, the NSF and NASA. The MOA collaboration is supported by MEXT and JSPS of Japan, and the Marsden Fund of New Zealand.

Author Information. The photometric data set is available at planet.iap.fr and ogle.astrouw.edu.pl. Reprints and permissions information is available at npg.nature.com/reprintsandpermissions. The authors declare no competing financial interests. Correspondence and requests for materials should be addressed to J.P.B. (beaulieu@iap.fr) or D.P.B. (bennett@nd.edu).

Long γ-ray Bursts and Core-collapse Supernovae Have Different Environments

A. S. Fruchter *et al.*

Editor's Note

Long gamma-ray bursts (GRBs) are astrophysical phenomena generally believed to arise in the collapse and subsequent explosion of a very massive star. They had been associated with somewhat anomalous supernovae, but the nature of the link was unclear. Here Andrew Fruchter and colleagues report a careful analysis of galaxies hosting long GRBs and find that the bursts are far more concentrated in the brightest regions of galaxies than are the supernovae. These bright regions have the most massive stars. The authors conclude that bursts are best associated with such stars, and speculate that there is a preference for these stars to occur in environments relatively poor in elements heavier than helium.

When massive stars exhaust their fuel, they collapse and often produce the extraordinarily bright explosions known as core-collapse supernovae. On occasion, this stellar collapse also powers an even more brilliant relativistic explosion known as a long-duration γ-ray burst. One would then expect that these long γ-ray bursts and core-collapse supernovae should be found in similar galactic environments. Here we show that this expectation is wrong. We find that the γ-ray bursts are far more concentrated in the very brightest regions of their host galaxies than are the core-collapse supernovae. Furthermore, the host galaxies of the long γ-ray bursts are significantly fainter and more irregular than the hosts of the core-collapse supernovae. Together these results suggest that long-duration γ-ray bursts are associated with the most extremely massive stars and may be restricted to galaxies of limited chemical evolution. Our results directly imply that long γ-ray bursts are relatively rare in galaxies such as our own Milky Way.

IT is an irony of astrophysics that stellar birth is most spectacularly marked by the deaths of massive stars. Massive stars burn brighter and hotter than smaller stars, and exhaust their fuel far more rapidly. Therefore a region of star formation filled with low mass stars still early in their lives, and in some cases still forming, may also host massive stars already collapsing and producing supernovae. Indeed, with the exception of the now famous type Ia supernovae, which have been so successfully used for cosmological studies[1,2] and which are thought to be formed by the uncontrolled nuclear burning of stellar remnants comparable in mass to the Sun[3], all supernovae are thought to be produced by the collapse of massive stars. The collapse of the most massive stars (tens of solar masses) is thought to leave behind either black holes or neutron stars, depending largely on the state of chemical evolution of the material that formed the star, whereas the demise of stars between approximately 8 and 20 solar masses produces only neutron stars[4].

长伽马射线暴与核坍缩超新星
具有不同的环境

长时标伽马射线暴（简称长暴）一般被认为由大质量恒星的核坍缩及后续爆发所产生。它们与某些超新星成协出现，但成协的本质尚不清楚。本文中，安德鲁·弗鲁赫特与其合作者对长暴和超新星的宿主星系进行了细致的对比分析。他们发现相比于超新星，长暴发生位置更多地集中在星系中最亮的区域，而这些亮区域恰恰存在有超大质量的恒星。作者得出结论，长暴最有可能与星系中的超大质量恒星有着物理关联，并由此推测长暴仅发生在低金属丰度的宿主星系环境中。

当大质量恒星耗尽燃料，它们的内核会发生坍缩，接着星体常常会产生明亮的爆炸，该现象被称为核坍缩超新星。有时这种恒星坍缩会产生更亮的相对论性的爆炸，爆炸会产生持续时标超过两秒的伽马射线辐射，并且辐射能被空间高能卫星探测到。我们简称其为长时标伽马射线暴（即长暴）。一般会认为这些长暴和核坍缩超新星应该处于类似的星系环境中。但我们揭示这一预想是错误的。我们发现长暴比核坍缩超新星更加集中在宿主星系中最亮的区域。而且，长暴的宿主星系明显比核坍缩超新星的宿主星系更暗、更不规则。这些结论表明长暴和超大质量恒星成协，而且可能仅发生在金属丰度较低的星系中。我们的结果直接表明长暴很少会发生在类似银河系样的宿主星系中。

恒星诞生最显著的标志是大质量恒星的死亡，这可以说是天体物理学的一个讽刺。大质量恒星比小质量恒星燃烧时更亮更热，也更快地消耗完自己的燃料。因此在恒星诞生的区域，那里一方面遍布着处于生命期早期的小质量恒星，某些情况下仍有恒星生成；另一方面也可能包含已经坍缩并产生超新星的大质量恒星。诚然，除了最近著名的成功用作宇宙学研究 [1,2] 的 Ia 型超新星，其他类型超新星都被认为是由大质量恒星的核坍缩形成的。Ia 型超新星一般被认为是类似太阳质量大小的恒星遗迹在不可控制的核燃烧 [3] 下形成的。普遍认为，大多数大质量恒星（几十个太阳质量）坍缩后会留有黑洞或中子星，这主要取决于构成此恒星物质的化学演化状态。不过质量约为 8 到 20 倍太阳质量的恒星死亡后仅会产生中子星 [4]。

701

Gamma-ray bursts (GRBs), like supernovae, are a heterogeneous population. GRBs can be divided into two classes: short, hard bursts, which last between milliseconds and about two seconds and have hard high-energy spectra, and long, soft bursts, which last between two and tens of seconds, and have softer high-energy spectra[5]. Only very recently have a few of the short bursts been well localized, and initial studies of their apparent hosts suggest that these bursts may be formed by the binary merger of stellar remnants[6,7]. In contrast, the afterglows of over 80 long GRBs (LGRBs) have been detected in the optical and/or radio parts of the spectrum. And as a result of these detections, it has become clear that LGRBs, like core-collapse supernovae, are related to the deaths of young, massive stars. It is these objects, born of the deaths of massive stars, that we study here.

LGRBs are generally found in extremely blue host galaxies[8-11] that exhibit strong emission lines[12,13], suggesting a significant abundance of young, very massive stars. Furthermore, whereas the light curves of the optical transients associated with LGRBs are often dominated by radiation from the relativistic outflow of the GRB, numerous LGRBs have shown late-time "bumps" in their light curves consistent with the presence of an underlying supernova[14-16]. In several cases spectroscopic evidence has provided confirmation of the light of a supernova superposed on the optical transient[17-20]. Indeed, given the large variations in the brightnesses of optical transients and supernovae, and the limited observations on some GRBs, it seems plausible that all LGRBs have an underlying supernova[21]. And although the energy released in a LGRB often appears to the observer to be orders of magnitude larger than that of a supernova, there is now good evidence suggesting that most LGRBs are highly collimated and often illuminate only a few per cent of the sky[22,23]. When one takes this into account, the energy released in LGRBs more closely resembles that of energetic supernovae. However, not all core-collapse supernovae may be candidates for the production of GRBs. The supernovae with good spectroscopic identifications so far associated with GRBs have been type Ic—that is, core-collapse supernovae that show no evidence of hydrogen or helium in their spectra. (Type Ib supernovae, which are often studied together with type Ic, have spectra that are also largely devoid of hydrogen lines but show strong helium features.) A star may therefore need to lose its outer envelope if a GRB is to be able to burn its way through the stellar atmosphere[24]. Studies that have compared the locations of type Ib/c supernovae with the more numerous type II supernovae (core-collapse supernovae showing hydrogen lines) in local galaxies so far show no differences in either the type of host or the placement of the explosion on the host[25,26]. This result led the authors of ref. 25 to argue that core-collapse supernovae all come from the same mass range of progenitor stars, but that type Ib/c supernovae may have had their envelopes stripped by interaction with a binary stellar companion. Whether type Ic supernovae come from single stars, or binary stars, or both, it is very likely that only a small fraction of these supernovae produce GRBs[27].

Given the common massive stellar origins of core-collapse supernovae and LGRBs, one might expect that their hosts and local environments are quite similar. It has long been argued that core-collapse supernovae should track the blue light in the Universe (the light from massive stars is blue), both in their distribution among galaxies and within their host galaxies themselves. One would expect similar behaviour from LGRBs, and indeed

伽马射线暴，类似于超新星，也分有子类。伽马暴可以分为两类：短硬暴，持续时间大约几毫秒到两秒，具有硬的高能能谱；长软暴，持续时间在两秒到几百秒之间，具有较软的高能能谱[5]。数个短暴的位置直到最近才得到确认。初步研究其宿主星系可知这些短暴可能由两颗恒星各自演化成最后状态的星体然后并合形成[6,7]。相比之下，在光学或射电波段探测到超过80个长暴余辉。这些探测结果显示长暴和核坍缩超新星一样与年轻大质量恒星的死亡有关。我们这里研究的正是这些从大质量恒星死亡中诞生的天体。

长暴一般在呈现强发射线[12,13]的极端蓝宿主星系中被发现[8-11]，这明显说明有大量的年轻的特大质量恒星存在。而且，尽管与长暴成协的光学暂现源通常由来自伽马暴相对论性外向流的辐射主导，还是有很多长暴在它们光变曲线的后期出现"隆起"，与潜在的超新星的出现一致[14-16]。在多个实例中，有光谱证据显示超新星的光线叠加在光学暂现源上[17-20]。确实，考虑到光学暂现源和超新星巨大的亮度变化范围以及对一些伽马暴的有限观测，似乎每颗长暴都具有成协的超新星[21]。尽管长暴释放的各向同性能量通常看来比超新星大几个量级，但是现在有很好的证据显示大多数长暴都高度准直且仅仅照亮几个百分比的天空[22,23]。考虑到这个因素，长暴释放的真实能量更接近能量更强的超新星。不过，不是所有核坍缩超新星都可能是产生伽马暴的候选体。目前与伽马暴成协光谱证认良好的超新星一般都是Ic型超新星——即光谱中没有氢线和氦线的核坍缩超新星(Ib型超新星一般会与Ic型超新星一同被研究，其光谱大多也没有氢线，但表现出很强的氦线特征)。恒星需要失去外包层才能够使伽马暴冲破恒星外表面[24]。目前对邻近星系中Ib/c型超新星和数量更多的II型超新星(具有氢线的核坍缩超新星)位置的比较研究显示二者在宿主星系的类型或在宿主星系爆发的位置均没有差异[25,26]。这一结果使得文献25的作者得出结论，核坍缩超新星全部来自同一质量范围的前身星，但是Ib/c型超新星可能已经通过与双星中的伴星的相互作用被剥离了包层。不管Ic型超新星源自单星、双星抑或二者兼而有之，这些超新星中只有很少一部分会产生伽马暴[27]。

核坍缩超新星和长暴都起源于大质量恒星，我们预计它们的宿主星系与其当地环境都很相似。一直以来都有观点主张核坍缩超新星在星系里的分布和在宿主星系本身里的分布应该都可以示踪宇宙中的蓝光(来自大质量恒星的光是蓝色的)。对长暴也有类似的估计。事实上已经有研究报道了这一相关的粗略证据[28]。这里我们利

rough evidence for such a correlation has been reported[28]. Here we use the high resolution available from Hubble Space Telescope (HST) images, and an analytical technique developed by us that is independent of galaxy morphology, to study the correlation between these objects and the light of their hosts. We also compare the sizes, morphologies and brightnesses of the LGRB hosts with those of the supernovae. Our results reveal surprising and substantial differences between the birthplaces of these cosmic explosions. We find that whereas core-collapse supernovae trace the blue light of their hosts, GRBs are far more concentrated on the brightest regions of their hosts. Furthermore, while the hosts of core-collapse supernovae are approximately equally divided between spiral and irregular galaxies, the overwhelming majority of GRBs are on irregulars, even when we restrict the GRB sample to the same redshift range as the supernova sample. We argue that these results may be best understood if GRBs are formed from the collapse of extremely massive, low-metallicity stars.

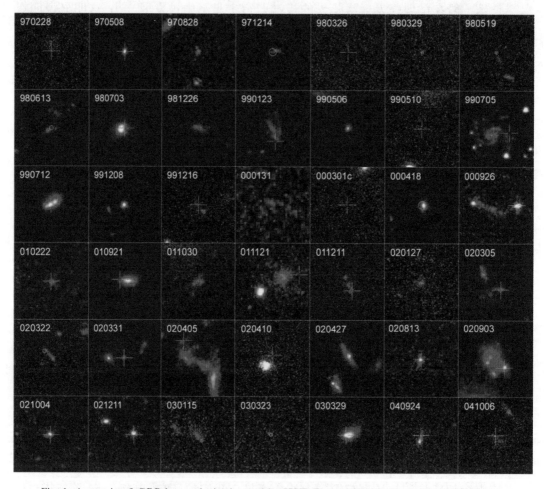

Fig. 1. A mosaic of GRB host galaxies imaged by HST. Each individual image corresponds to a square region on the sky 3.75″ on a side. These images were taken with the Space Telescope Imaging Spectrograph (STIS), the Wide-Field and Planetary Camera 2 (WFPC2) and the Advanced Camera for Surveys (ACS) on HST. In cases where the location of the GRB on the host is known to better than 0.15″ the position of the GRB is shown by a green mark. If the positional error is smaller than the point spread function of the image (0.07″ for STIS and ACS, 0.13″ for WFPC2) the position is marked by a cross-hair;

用哈勃太空望远镜(HST)的高分辨率图像,以及我们提出的独立于星系形态的分析技术来研究这些天体和它们宿主星系光线之间的相关。我们也比较了长暴宿主星系和超新星宿主星系的大小、形态和亮度。结果显示这些宇宙爆炸诞生地存在令人惊讶的重大差异。我们发现尽管核坍缩超新星示踪宿主星系的蓝光,但伽马暴会更加集中于宿主星系的最亮区域。而且,核坍缩超新星的宿主星系中旋涡和不规则星系大致各占一半;甚至将伽马暴样本的选取限定在与超新星样本一样的红移区间时亦是如此。如果伽马暴形成于金属丰度低的超大质量恒星的坍缩,那么也许就能够理解这些结果。

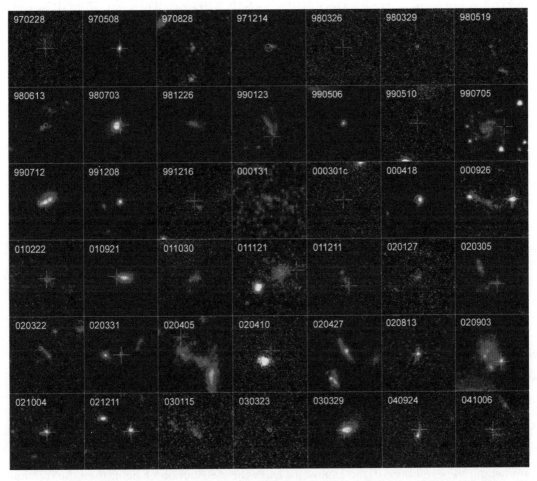

图 1. HST 获得的伽马暴宿主星系拼接组合图。每幅单独的图像对应天球上单边 3.75″ 的正方形天区。这些图像来自 HST 上的太空望远镜成像摄谱仪(STIS)、大视场行星相机 2(WFPC2)和高新巡天相机(ACS)。位置精度优于 0.15″ 的伽马暴用绿色标记出来。位置误差小于图像的点扩散函数(STIS 和 ACS 为 0.07″,WFPC2 的为 0.13″)的伽马暴用十字标记;其他位置误差用圆圈表示。STIS 图像都是白光下拍摄的(没有滤光片),WFPC2 和 ACS 的图像大部分是在 F606W 滤光片下拍摄的(有几种情况此滤光

otherwise the positional error is indicated by a circle. The STIS images were all taken in white light (no filter), and in most cases the WFPC2 and ACS images are in the F606W filter (though in a few cases where images in this filter were not available we have used images in F555W or F775W). The STIS and F606W images can be thought of as broad "V" or visual images, and are, for galaxies exhibiting typical colours of GRB hosts, the single most sensitive settings for these cameras. F555W is close to the ground-based Johnson V band, and F775W corresponds to the ground-based Johnson I band. Owing to the redshifts of the hosts, these images generally correspond to blue or ultraviolet images of the hosts in their rest frame, and thus detect light largely produced by the massive stars in the hosts.

The Sample

Over 40 LGRBs have been observed with the HST at various times after outburst. The HST is unique in its capability easily to resolve the distant hosts of these objects. Shown in Fig. 1 is a mosaic of HST images of the hosts of 42 bursts. These are all LGRBs with public data that had an afterglow detected with better than 3σ significance and a position sufficiently well localized to determine a host galaxy. A list of all the GRBs used in this work can be found in Supplementary Tables 1–3.

The supernovae discussed here were all discovered as part of the Hubble Higher z Supernova Search[29,30], which was done in cooperation with the HST GOODS survey[31]. The GOODS survey observed two ~150 arcmin^2 patches of sky five times each, in epochs separated by 45 days. Supernovae were identified by image subtraction. Here we discuss only the core-collapse supernovae identified in this survey. A list of the supernovae used is presented in Supplementary Table 4, and images of the supernova hosts can be seen in Fig. 2.

Positions of GRBs and Supernovae on Their Hosts

If LGRBs do in fact trace massive star formation, then in the absence of strong extinction we should find a close correlation between their position on their host galaxies and the blue light of those galaxies. However, many of the GRB hosts and quite a few of the supernova hosts are irregular galaxies made up of more than one bright component. As a result, the common astronomical procedure of identifying the centroid of the galaxy's light, and then determining the distance of the object in question from the centroid, is not particularly appropriate for these galaxies—the centroid of light may in fact lie on a rather faint region of the host (examine GRBs 000926 and 020903 in Fig. 1 for excellent illustrations of this effect). We have therefore developed a method that is independent of galaxy morphology. We sort all of the pixels of the host galaxy image from faintest to brightest, and ask what fraction of the total light of the host is contained in pixels fainter than or equal to the pixel containing the explosion. If the explosions track the distribution of light, then the fraction determined by this method should be uniformly distributed between zero and one. (A detailed exposition of this method can be found in Supplementary Information).

片下的图像无法获取，我们用 F555W 或 F775W 中的图像代替）。STIS 和 F606W 的图像可以被认为是宽"V"或者目视图像，而且对于呈现基本色彩的伽马暴宿主，这些相机的设置都是最为敏感的。F555W 接近于地基 Johnson V 波段，F775W 对应于地基 Johnson I 波段。由于宿主星系的红移，这些图像总体上都对应各自静止系的宿主星系的蓝或紫外图像，因此探测到的大部分光线都源自宿主星系里的大质量恒星。

样 本

HST 在爆后不同时间内观测了超过 40 颗长暴。HST 分辨这些天体遥远的宿主星系的能力是独一无二的。图 1 展示的是 HST 测得的 42 颗伽马暴宿主星系的拼接组合图。这些皆是存在公开数据的长暴中测得的余辉，显著性优于 3σ，且具有足以定位到的宿主星系。这一工作中所有伽马暴都在补充表格 1~3 中列出。

本文讨论的超新星是哈勃高红移超新星搜寻项目中 [29,30] 发现的一部分超新星，该研究是与 HST GOODS 巡天合作完成的。GOODS 巡天观测了两块约 150 arcmin² 的天区，每块天区观测 5 次，每次间隔 45 天。通过图像相减发现超新星。这里使用的超新星列在补充表格 4 中，超新星宿主星系的图像见图 2。

伽马暴和超新星在宿主星系的位置

如果长暴真的与大质量恒星密切相关，那么在没有强消光的情况下我们应该能够发现长暴在宿主星系的位置和星系蓝光具有很好的相关。不过，许多伽马暴宿主星系和一些超新星宿主星系是不规则星系，由多于 1 个亮成分组成。因此，尽管天文学普遍的处理步骤是先确定星系光的形心，然后确定要研究的天体离形心的距离，但对于这些星系不是特别适用——光的形心可能位于宿主星系的一个相当暗弱的区域（图 1 中伽马暴 000926 和 020903 很好地描述了这一效应）。我们因此创立了一种独立于星系形态的方法。我们把宿主星系的所有像素从暗到亮排序，然后分别找出像素中包含宿主全部光的部分与包含爆发的部分，再求得前者与后者相比同等亮或更为暗弱的像素的比例。如果爆炸示踪了光的分布，那么这种方法确定的比率应该从 0 到 1 均匀分布（此方法的具体说明见补充资料）。

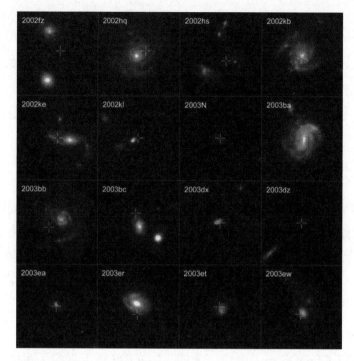

Fig. 2. A mosaic of core-collapse supernova host galaxies imaged with HST as part of the GOODS programme. Each image in the mosaic has a width of 7.5″ on the sky, and thus twice the field of view of each image in the GRB mosaic. The position of each supernova on its host galaxy is marked. In all cases, these positions are known to sub-pixel accuracy. Supernovae in the GOODS sample were identified by ref. 30 as either type Ia or core-collapse supernovae on the basis of their colours, luminosities and light curves, as data allowed (a supernova exploding near the beginning or end of one of the multi-epoch observing runs would have much less data, and sometimes poor colour information). Thus bright type Ib and Ic supernovae, which have colours and luminosities similar to type Ia supernovae, would probably have been classified as type Ia (unless a grism spectrum was taken—however, only a small fraction of objects were observed spectroscopically). On the other hand, fainter type Ib and Ic supernovae ($M_B \gtrsim -18$) could in principle be identified from photometric data; however, in practice the data were rarely sufficient for a clear separation from other core-collapse supernovae. On the basis of surveys of nearby galaxies, one might expect approximately 20% of the core-collapse supernovae to be type Ib or Ic[49,50].

As can be seen in Fig. 3, the core-collapse supernovae do track the light of their hosts as well as could be expected given their small number statistics. A Kolmogorov–Smirnov (KS) test finds that the distribution of the supernovae is indistinguishable from the distribution of the underlying light. The situation is clearly different for LGRBs. As can be seen in Fig. 3, the GRBs do not simply trace the blue light of the hosts; rather, they are far more concentrated on the peaks of light in the hosts than the light itself. A KS test rejects the hypothesis that GRBs are distributed as the light of their hosts with a probability greater than 99.98%. Furthermore, this result is robust: it shows no dependence on GRB host size or magnitude. And in spite of the relatively small number of supernova hosts for which a comparison can be made, the two populations are found by the KS test to be drawn from different distributions with ~99% certainty. In the next section, we show that the surprising differences in the locations of these objects on the underlying light of their hosts may be due not only to the nature of their progenitor stars but also that of their hosts.

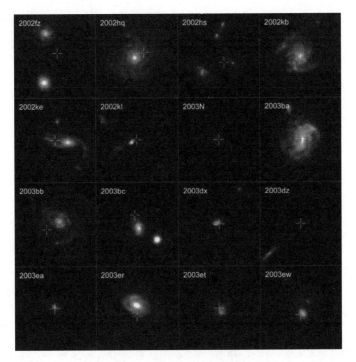

图 2. 作为 GOODS 项目一部分，HST 获得的核坍缩超新星宿主星系拼接组合图。拼接图中的每幅图像对应天球上单边 7.5″ 的天区，因此是伽马暴拼接图中每幅图像视场的两倍。宿主星系里超新星的位置都标记了出来。所有这些位置精度都是在亚像素级别的。GOODS 的超新星样本都在文献 30 中根据颜色、光度和光变曲线数据确定为 Ia 型或核坍缩超新星（如果超新星爆发时接近多次观测中某一次观测的开始或结束，则它的数据会比较少，有时颜色信息也很差）。因此具有类似 Ia 型超新星颜色和光度的亮 Ib 或 Ic 型超新星可能被分类作 Ia 型超新星（除非拍摄获得棱栅光谱——不过，只有对一小部分天体是可以通过光谱观测的）。另一方面，较暗弱的 Ib 或 Ic 型超新星（$M_B \gtrsim -18$）原则上可以依据测光数据分类；不过，实际操作中，缺少充分的数据将其与另外的核坍缩超新星区分。在对附近星系的巡天基础上，估计大约有 20% 的核坍缩超新星是 Ib 或 Ic 型超新星[49,50]。

　　如图 3 所示，在小数目统计下，核坍缩超新星确实如预想的一样很好地示踪了宿主星系光线。柯尔莫戈洛夫 - 斯米尔诺夫（KS）检验发现，超新星的分布和基底光的分布难以区分。而对于长暴，情况则明显不同。如图 3 所示，伽马暴不仅示踪了宿主星系的蓝光，而且它们比光本身更加集中在宿主星系的光的峰值处。KS 检验有大于 99.8% 的概率否定伽马暴与宿主星系光具有相同分布的假设。而且这一结论是很可靠的：它与伽马暴宿主星系的大小或星等无关。尽管超新星宿主星系的数目较少，不足以做出比较，KS 检测发现这两个星族在约 99% 确定度下分布不同。在本文的下一节中，我们将展示这些天体在宿主星系基底光线位置的惊人差异可能不仅源自它们前身星的性质，同时也与其宿主星系的性质有关。

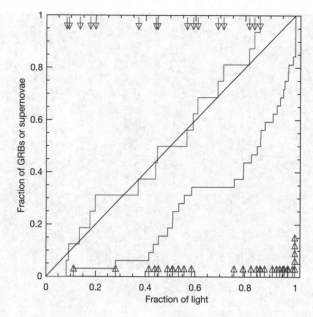

Fig. 3. The locations of the explosions in comparison to the host light. For each object, an arrow indicates the fraction of total host light in pixels fainter than or equal to the light in the pixel at the location of the transient. The cumulative fraction of GRBs or supernovae found at a given fraction of the total light is shown as a histogram. The blue arrows and histogram correspond to the GRBs, and the red arrows and histogram correspond to the supernovae. Were the GRBs and supernovae to track the light identically, their histograms would follow the diagonal line. Whereas the supernova positions do follow the light within the statistical error, the GRBs are far more concentrated on the brightest regions of their hosts. Thus although the probability of a supernova exploding in a particular pixel is roughly proportional to the surface brightness of the galaxy at that pixel, the probability of a GRB at a given location is effectively proportional to a higher power of the local surface brightness.

A Comparison of the Host Populations

An examination of the mosaics of the GRB and supernova hosts (Figs 1 and 2) immediately shows a remarkable contrast—only one GRB host in this set of 42 galaxies is a grand-design spiral, whereas nearly half of the supernova hosts are grand-design spirals. One might wonder if this effect is due to a difference in redshift distribution— the core-collapse supernovae discovered by the GOODS collaboration all lie at redshift $z < 1.2$, whereas LGRBs can be found at much larger redshifts where grand-design spirals are rare to non-existent. Yet if we restrict the GRB population to $z < 1.2$ (and thus produce a population with a nearly identical mean and standard deviation in redshift space compared to the GOODS core-collapse supernovae), the situation remains essentially unchanged: only one out of the eighteen GRB hosts is a grand-design spiral. (For a detailed comparison of GRB hosts to field galaxies, rather than the supernova selected galaxies shown here, see ref. 32.)

Were the difference in spiral fraction the only indication of a difference in the host populations, we could not rule out random chance—given the small number statistics, both populations are barely consistent with each other and a spiral fraction of ~25%. However,

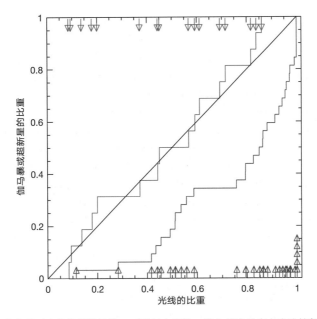

图 3. 爆炸位置和宿主星系光线位置的比较。对于每个天体，箭头代表总宿主光线较暂现位置处光线暗弱或相等的像素的比重。在总光线的给定比例下找到的伽马暴或超新星的累积比重用直方图表示。蓝色箭头和直方图代表伽马暴，红色箭头和直方图代表超新星。如果伽马暴或超新星能同样跟踪光线，那么它们的直方图将沿着对角线分布。超新星位置在统计误差范围的确示踪了光线，而伽马暴的位置则更加集中于宿主星系的极亮区域。因此，尽管超新星在某个像素爆发的概率约正比于在该像素的星系表面亮度，但是在某个位置出现伽马暴的概率实际上正比于能量更高的近域表面亮度。

宿主星族的比较

检查伽马暴和超新星宿主星系的拼接图像（图 1 和图 2）立刻会发现二者的显著差异——在本组 42 个星系中只有一个伽马暴宿主星系是宏大的旋涡结构，而接近一半的超新星宿主星系是宏大旋涡结构。这不禁让人想知道这一效应是否是由于红移分布的不同导致的——GOODS 巡天发现的核坍缩超新星红移全部低于 1.2；长暴可以在更大红移处被发现，但那里却很少甚至没有宏大旋涡星系存在。不过如果我们将伽马暴族限制在 $z < 1.2$（从而产生平均值和标准偏差在红移空间与 GOODS 核坍缩超新星相当的星族），情况基本没有变化：18 个伽马暴宿主星系中只有一个是宏大旋涡星系。（如果相较于此处根据超新星而选择星系进行比较，读者想要获取伽马暴宿主星系和场星系的详细对比，请见文献 32。）

假如旋涡比率的差异是宿主星族的唯一指标，那么我们就不能排除随机因素的影响——因为在小数目统计资料下，星族之间几乎不可能出现彼此一致，旋涡比率约为

the host populations differ strongly in ways other than morphology.

In Fig. 4 we compare the 80% light radius (r_{80}) and absolute magnitude distributions of the GRB and supernova hosts. Included in the comparison are all LGRBs with known redshifts $z < 1.2$ at the time of submission and the 16 core-collapse supernovae of GOODS with spectroscopic or photometric redshifts (see the Supplementary Tables for a complete list of the GRBs, supernovae and associated parameters used in this study). The small minority of GRB hosts in this redshift range without HST imaging are compared only in absolute magnitude and not in size. The absolute magnitudes have been derived from the observed photometry using a cosmology of $\Omega_m = 0.27$, $\Lambda = 0.73$ and $H_0 = 71$ km s^{-1}Mpc^{-1}, and the magnitudes have been corrected for foreground Galactic extinction[33]. For a technical discussion of the determination of the magnitude and size of individual objects, see Supplementary Information.

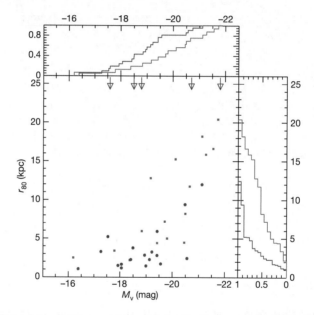

Fig. 4. A comparison of the absolute magnitude and size distributions of the GRB and supernova hosts. In the main panel, the core-collapse supernova hosts are represented as red squares, and the LGRB hosts as blue circles. The absolute magnitudes of the hosts are shown on the x axis, and the lengths of the semi-major axes of the hosts on the y axis. The plot is then projected onto the two side panels where a histogram is displayed for each host population in each of the dimensions—absolute magnitude and semi-major axis. Shown as blue arrows are the absolute magnitudes of GRB hosts with $z < 1.2$ that have been detected from the ground but have not yet been observed by HST. These hosts are only included in the absolute magnitude histogram. The hosts of GRBs are both smaller and fainter than those of supernovae.

As can be readily seen, the two host populations differ substantially both in their intrinsic magnitudes and sizes. The GRB hosts are fainter and smaller than the supernova hosts. Indeed, KS tests reject the hypothesis that these two populations are drawn from the same population with certainties greater than 98.6% and 99.7% for the magnitude and size distributions, respectively.

712

25%的情况。不过，不同的宿主星系族除了形态之外，在其他方面也有很大的差异。

图 4 中我们对伽马暴和超新星宿主星系的 80% 光半径 (r_{80}) 和绝对星等分布进行了比较。包含的比较对象皆为截至投稿时间已知的红移 $z < 1.2$ 的长暴和 16 个具有光谱或测光红移 GOODS 的核坍缩超新星（伽马暴、超新星以及本研究相关参数的完整列表参见补充表格）。在此红移范围的小部分没有 HST 成像的伽马暴宿主星系仅比较绝对星等，不比较大小。绝对星等是在宇宙学参数 $\Omega_m = 0.27$，$\Lambda = 0.73$ 和 $H_0 = 71$ km · s^{-1} · Mpc^{-1} 的条件下，经过银河系前景消光修正后的测光观测数据获得的 [33]。获得单个天体星等和大小的技术讨论见补充资料。

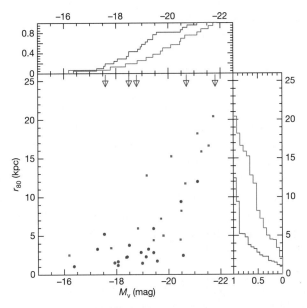

图 4. 伽马暴和超新星宿主星系绝对星等和大小分布的比较。在主图中，核坍缩超新星宿主星系用红色方块表示，长暴宿主星系用蓝色圆圈表示。x 轴表示宿主星系的绝对星等，y 轴表示半长轴的长度。图像投影到面板的两侧得到在每个纬度上——绝对星等和半长轴——各个宿主星系族的直方图。蓝色箭头表示的是在地面探测到，但是没有在 HST 观测到的 $z < 1.2$ 伽马暴宿主星系的绝对星等。这些宿主星系只包含在绝对星等直方图中。伽马暴的宿主星系比超新星的更小更暗弱。

如上所示，两个宿主星系星族在本征星等和大小上都存在巨大差异。伽马暴宿主星系比超新星宿主星系更暗弱且更小。事实上，KS 检验分别在星等和大小分布上以高于 98.6% 和 99.7% 的确定度否定了这两类星系是同一种星系的假设。

Discussion

Although the evidence is now overwhelming that both core-collapse supernovae and LGRBs are formed by the collapse of massive stars, our observations show that the distribution of these objects on their hosts, and the nature of the hosts themselves, are substantially different. How then can this be? We propose here that these surprising findings are the result of the dependence of the probability of GRB formation on the state of the chemical evolution of massive stars in a galaxy.

Even before the association of LGRBs with massive stars had been established, a number of theorists had suggested that these objects could be formed by the collapse of massive stars, which would leave behind rapidly spinning black holes. An accretion disk about the black hole would power the GRB jet. These models, sometimes referred to as "hypernovae" or "collapsar" models, implicitly require very massive stars, as only stars greater than about 18 solar masses form black holes. But in fact it was widely suspected that even more massive stars would be required—if only to provide the required large energies, and to limit the numbers of supernovae progressing to GRBs.

We conclude that LGRBs do indeed form from the most massive stars and that this is the reason that they are even more concentrated on the blue light of their hosts than the light itself. The most massive stars (O stars) are frequently found in large associations. These associations can be extremely bright, and can indeed provide the peak of the light of a galaxy—particularly if that galaxy is a faint, blue irregular, as are the GRB hosts in general. Indeed, a connection of LGRBs with O stars (and perhaps Wolf–Rayet stars) is a natural one— given the strong emission lines (including Ne [III]) seen in many of these hosts[12,13] and the evidence for possible strong winds off the progenitors of the GRBs seen in absorption in some LGRB spectra[34,35].

However, O stars are found in galaxies of all sizes. Indeed, studies of the Magellanic clouds suggest that the initial distribution of masses of stars at formation in these dwarf galaxies is essentially identical to that in our much larger spiral, the Milky Way[36]. Therefore, a difference in the initial mass function of stars is unlikely to be responsible for the differences between the hosts. We propose that the fundamental differences between the LGRB and supernova host populations are not their size or luminosity, but rather their metallicity, or chemical evolution. Some evidence of this already exists. The hosts of seven LGRBs (GRBs 980425 (P.M.V., personal communication), 990712[13], 020903[37], 030323[38], 030329[17], 031203[39] and 050730[40]) have measurements of, or limits on, their metallicity, and in all cases the metallicity is less than one-third solar. The small size and low luminosity of the GRB hosts is then a result of the well known correlation between galaxy mass and metallicity (see ref. 41 and references therein).

But why do LGRBs occur in low-metallicity galaxies? This may be a direct result of the evolution of the most massive stars. It has recently been proposed that metal-rich stars

讨　论

尽管现在有决定性的证据表明核坍缩超新星与长暴是由大质量恒星坍缩而成，但是我们的观测显示这两者在宿主星系里的分布与宿主星系本身的性质上明显不同。这又是为什么呢？我们推测这些令人惊讶的发现是由于伽马暴的形成取决于星系中大质量恒星化学演化状态。

在确定长暴和大质量恒星的成协之前，很多理论已经提出这些天体可能是由大质量恒星坍缩形成，并留下一个快速自转的黑洞。黑洞周围的吸积盘将为伽马暴喷流提供能量。这些模型，即被人们称为"极超新星"或"坍缩星"的模型，内含了形成黑洞需要质量非常大的恒星的要求，例如只有大于 18 个太阳质量的恒星才能形成黑洞。但是实际上人们很大程度上猜测需要更大质量的恒星——如果仅仅是为了提供需要的巨大能量，并限制超新星演化到伽马暴的数目。

我们的结论是，长暴确实由最大质量恒星形成，这就是为什么它们在宿主星系的蓝光区比自身光区更集中。最大质量恒星（O 型星）经常在大的星协里被发现。这些星协极其亮，事实上能够提供星系的光线峰值——特别是如果那个星系是一个微弱的蓝色不规则星系，就像一般的伽马暴宿主星系一样。事实上，长暴和 O 型星（也可能沃尔夫－拉叶星）的关联是很自然的——因为在这些宿主星系里，很多都发现了强发射线（包括 Ne[III]）[12,13]，同时有证据显示在某些长暴光谱吸收中发现可能源自伽马暴前身星的强风 [34,35]。

不过，所有大小的星系中都有 O 型星。事实上，对麦哲伦云的研究显示在这些矮星系中恒星形成时的初始质量分布本质上与我们所生活的更大的旋涡星系——银河系——是一样的 [36]。因此，恒星初始质量分布的差异不太可能导致两类宿主星系的差异。我们认为长暴和超新星宿主星系星族的根本差异不在于其大小或者光度，而是金属丰度或者说化学演化。目前已经有一些这方面的证据。具有测量或极限金属丰度的 7 个长暴的宿主星系（伽马暴 980425（弗雷埃斯维克，个人交流），990712[13]，020903[37]，030323[38]，030329[17]，031203[39] 和 050730[40]），它们的金属丰度全都小于太阳水平的 1/3。伽马暴宿主星系的小尺度和低光度则由著名的星系质量和金属丰度相关导致的（见参考文献 41 及其中的文献）。

但是为什么长暴存在于低金属丰度星系？这可能是最大质量恒星演化直接导致的。最近有研究提出具有几十个太阳质量的富金属恒星的表面风很强（光压作用在富

with masses of tens of solar masses have such large winds off their surfaces (due to the photon pressure on their metal-rich atmospheres) that they lose most of their mass before they collapse and produce supernovae[4]. As a result they leave behind neutron stars, not the black holes necessary for LGRB formation. Ironically, stars of 15–30 solar masses may still form black holes, as they do not possess radiation pressure sufficient to drive off their outer envelopes. Direct evidence for this scenario comes from recent work showing that the Galactic soft γ-ray repeater, SGR 1820–06, is in a cluster of extremely young stars of which the most massive have only started to collapse[42]—yet the progenitor of SGR 1820–06 collapsed to a neutron star, not a black hole. Recent observations of winds from very massive (Wolf–Rayet) stars provide further support for this scenario: outflows from the low-metallicity stars in the Large Magellanic Cloud are substantially smaller than those seen from more metal-rich Galactic stars[43]. The possible importance of metallicity in LGRB formation has therefore not escaped the notice of theorists[44,45].

A preference for low metallicity may also explain one of the most puzzling results of GRB host studies. None of the LGRB hosts is a red, sub-millimetre bright galaxy. These highly dust-enshrouded galaxies at redshifts of ~1–3 are believed to be the sites of a large fraction of the star formation in the distant Universe[46]. And although some LGRB hosts do show sub-millimetre emission, none have the red colours characteristic of the majority of this population. However, it is likely that these red dusty galaxies have substantial metallicities at all redshifts. The low metallicity of hosts may also help explain the fact that a substantial fraction of high-redshift LGRB hosts display strong Lyman-α emission[47].

All well-classified supernovae associated with LGRBs are of type Ic, presumably because the presence of a hydrogen envelope about the collapsing core can block the emergence of a GRB jet[24]. Thus only those supernovae whose progenitors have lost some, but not too much, mass appear to be candidates for the formation of a GRB. Given the large numbers of type Ic supernovae in comparison to the estimated numbers of LGRBs, however, it is likely that only a small fraction of type Ic supernovae produce LGRBs. Indeed, even the number of unusually energetic type Ib/Ic supernovae appears to dwarf the LGRB population[48]. Another process, perhaps the spin-up of the progenitor in a binary[27], may decide which type Ic supernovae produce LGRBs. Interestingly, it was the similar distribution of supernovae on their hosts, and particularly the fact that type Ib/Ic were no more correlated than type II with the UV bright regions of their hosts, that led ref. 25 to the conclusion that type Ib/Ic supernovae form from binaries. LGRBs clearly track light differently from the general type Ic population. However, the samples used by refs 25 and 26 were from supernovae largely discovered on nearby massive galaxies—dwarf irregular hosts are underrepresented in these samples. It will be particularly interesting to see whether large unbiased supernova surveys at present underway produce similar locations for their supernovae.

We do not know, however, what separates the small fraction of low-metallicity type Ic supernovae that turn into LGRBs from the rest of the population. Potentially, the answer is the amount of angular momentum available in the core to form the jet. In this case, the

金属大气上导致），在它们坍缩形成超新星之前已经丢失了大部分质量 [4]。因此它们留下的是中子星而不是形成长暴所需要的黑洞。具有讽刺意味的是，15～30 倍太阳质量的恒星仍然能形成黑洞，因为它们没有足够的辐射压以驱动它们的外包层。这个图景的直接证据来自最近的研究，研究显示银河系的软伽马射线复现源 SGR1820-06 处于由极年轻恒星组成的星团中，其中最大质量的一批恒星刚刚开始坍缩 [42]——不过 SGR1820-06 的前身星已然坍缩成中子星，而不是黑洞。最近对来自超大质量恒星（沃尔夫－拉叶星）星风的观测也进一步支持了这个图景：大麦哲伦云里的低金属度星的外向流比银河系中更为富金属的恒星的外向流明显要小 [43]。在长暴形成过程中可能起重要作用的金属丰度因此也没能逃出理论学家的注意 [44,45]。

低金属丰度也可能用来解释关于伽马暴宿主星系研究中最困难的一个谜题。长暴宿主星系中没有一个是红的、亚毫米波的亮星系。这些红移约为 1～3 处的被尘埃掩盖的星系被认为是遥远宇宙大部分恒星形成之地 [46]。尽管一些长暴宿主星系有亚毫米辐射，但是没有一个具有大部分这类星族的红色特征。不过，很可能这些红色的富尘埃星系在所有红移处都具有充足的金属丰度。宿主星系低金属丰度或许也能帮助解释一部分高红移长暴宿主星系呈现的强莱曼 α 辐射 [47]。

所有与长暴成协的，类型确定的超新星都是 Ic 型，这大概是由于坍缩核外的氢包层阻止了伽马暴喷流出现 [24]。因此只有那些前身星已经丢失一些质量，但又不是很多的超新星才可能是伽马暴的候选。不过，考虑到 Ic 型超新星数量相比长暴的估计数目大得多，可能只有一小部分 Ic 型超新星会形成长暴。事实上，甚至非正常的富能量 Ib/Ic 型超新星的数目看起来都使长暴族相形见绌 [48]。另一个过程，可能是双星前身星的加速 [27]，可能决定了哪一种 Ic 型超新星产生长暴。有意思的是，正是超新星在其宿主星系的相似分布，尤其是 Ib/Ic 型超新星和 II 型超新星与其各自宿主星系的 UV 亮区域之间的相关都不紧密，导致文献 25 得出 Ib/Ic 型超新星来自双星的结论。长暴的清晰示踪的光线和一般 Ic 型超新星不同。不过，文献 25 和 26 使用的样本大部分来自附近大质量宿主星系——矮且不规则的星系中发现的超新星，因而是取样不足的。目前正在进行的大型、无偏超新星巡天发现的超新星是否也具有类似的位置将具有非凡的意义。

不过，我们不知道是什么机制把贫金属 Ic 型超新星从星族分离出来变成长暴。答案有可能是核中角动量是否足够大到形成喷流。在此情形下，贫金属的倾向可能表明双星中单星的演化相比双星相互作用在长暴的形成过程中占主导作用。通过研

preference for low metallicity may indicate that single star evolution dominates over binary interaction in forming LGRBs. Deep, high-spectral-resolution studies of LGRB afterglows may provide insight here, by allowing studies of the winds off the progenitor and any binary companion.

Only a small fraction of LGRBs are found in spiral galaxies, even for LGRBs with redshifts $z < 1$ where spirals are much more common. However, the local metallicity in spirals is known to be anti-correlated with distance from the centre of the galaxy. Thus one might expect LGRBs in spirals to violate the trend that we have seen for the general LGRB population, and avoid the bright central regions of their hosts. The present number of LGRBs known in spirals is still too small to test this prediction. But a sample size a few times larger should begin to allow such a test. Additionally, a survey of the metallicity of the hosts of the GOODS supernovae should find a higher average metallicity than that seen in GRB hosts. Finally, if low metallicity is indeed the primary variable in determining whether LGRBs are produced, then as we observe higher redshifts, where metallicities are lower than in most local galaxies, LGRBs should be more uniformly distributed among star-forming galaxies. Indeed, some evidence of this may already be present in the data[32]. LGRBs, however, are potentially visible to redshifts as high as $z \approx 10$. At significant redshifts, where the metallicities of even relatively large galaxies are expected to be low, we may find that LGRBs do become nearly unbiased tracers of star formation.

(**441**, 463-468; 2006)

A. S. Fruchter[1], A. J. Levan[1,2,3], L. Strolger[1,4], P. M. Vreeswijk[5], S. E. Thorsett[6], D. Bersier[1,7], I. Burud[1,8], J. M. Castro Cerón[1,9], A. J. Castro-Tirado[10], C. Conselice[11,12], T. Dahlen[13], H. C. Ferguson[1], J. P. U. Fynbo[9], P. M. Garnavich[14], R. A. Gibbons[1,15], J. Gorosabel[1,10], T. R. Gull[16], J. Hjorth[9], S. T. Holland[17], C. Kouveliotou[18], Z. Levay[1], M. Livio[1], M. R. Metzger[19], P. E. Nugent[20], L. Petro[1], E. Pian[21], J. E. Rhoads[1], A. G. Riess[1], K. C. Sahu[1], A. Smette[5], N. R. Tanvir[3], R. A. M. J. Wijers[22] & S. E. Woosley[6]

[1] Space Telescope Science Institute, 3700 San Martin Drive, Baltimore, Maryland 21218, USA

[2] Department of Physics and Astronomy, University of Leicester, University Road, Leicester, LE1 7RH, UK

[3] Centre for Astrophysics Research, University of Hertfordshire, College Lane, Hatfield, AL10 9AB, UK

[4] Physics & Astronomy, TCCW 246, Western Kentucky University, 1 Big Red Way, Bowling Green, Kentucky 42101, USA

[5] European Southern Observatory, Alonso de Córdova 3107, Casilla 19001, Santiago, Chile

[6] Department of Astronomy & Astrophysics, University of California, 1156 High Street, Santa Cruz, California 95064, USA

[7] Astrophysics Research Institute, Liverpool John Moores University, Twelve Quays House, Egerton Wharf, Birkenhead, CH41 1LD, UK

[8] Norwegian Meteorological Institute, PO Box 43, Blindern, N-0313 Oslo, Norway

[9] Dark Cosmology Centre, Niels Bohr Institute, University of Copenhagen, DK-2100 Copenhagen, Denmark

[10] Instituto de Astrofísica de Andalucía (CSIC), Camino Bajo de Huétor, 50, 18008 Granada, Spain

[11] California Institute of Technology, Mail Code 105-24, Pasadena, California 91125, USA

[12] School of Physics and Astronomy, University of Nottingham, University Park, Nottingham, NG7 2RD, UK

[13] Department of Physics, Stockholm University, SE-106 91 Stockholm, Sweden

[14] Physics Department, University of Notre Dame, 225 Nieuwland Hall, Notre Dame, Indiana 46556, USA

[15] Vanderbilt University, Department of Physics and Astronomy, 6301 Stevenson Center, Nashville, Tennessee 37235, USA

[16] Code 667, Extraterrestial Planets and Stellar Astrophysics, Exploration of the Universe Division, [17] Code 660.1, Goddard Space Flight Center, Greenbelt, Maryland 20771, USA

[18] NASA/Marshall Space Flight Center, VP-62, National Space Science & Technology Center, 320 Sparkman Drive, Huntsville, Alabama 35805, USA

究前身星风和双子星伴星，长暴余辉的深、高光谱分辨率研究可能会为此问题提供一些思路。

即便在红移 $z<1$ 里十分常见的旋涡星系，也只有小部分长暴被发现存在于其中。不过，旋涡星系的局部金属丰度与离星系中心的距离反相关。因此可以预测旋涡星系中的长暴与我们在一般长暴族观测到的趋势不同，它们将回避其宿主星系中心亮区域。目前已知的旋涡星系中的长暴数量还太少，不足以验证这个预测。等样本数量再增大几倍时就可以进行验证。另外，对 GOODS 超新星宿主星系金属丰度巡天发现的平均金属丰度应该超过伽马暴宿主星系的观测值。最后，如果贫金属丰度确实是决定是否形成长暴的主要参数，那么在更高红移处的金属丰度比大部分邻近星系低，应该看到更多的长暴不均匀地分布于正在形成恒星的星系中。确实，这个趋势可能已经在数据中有所展示[32]。不过，长暴可能在红移 $z \approx 10$ 处都能看到。在显著的红移处，相对较大的星系的金属丰度预期会降低，因而我们可能会发现长暴成为恒星形成的无偏差示踪者。

（肖莉 翻译；徐栋 审稿）

[19] Renaissance Technologies Corporation, 600 Route 25A, East Setauket, New York 11733, USA

[20] Lawrence Berkeley National Laboratory, MS 50F-1650, 1 Cyclotron Road, Berkeley, California 94720, USA

[21] INAF, Osservatorio Astronomico di Trieste, Via G.B. Tiepolo 11, I-34131 Trieste, Italy

[22] Astronomical Institute "Anton Pannekoek", University of Amsterdam, Kruislaan 403, NL-1098 SJ Amsterdam, The Netherlands

Received 22 August 2005; accepted 5 April 2006. Published online 10 May 2006.

References:

1. Riess, A. G. *et al.* Observational evidence from supernovae for an accelerating universe and a cosmological constant. *Astron. J.* **116**, 1009-1038 (1998).

2. Perlmutter, S. *et al.* Measurements of Ω and Λ from 42 high-redshift supernovae. *Astrophys. J.* **517**, 565-586 (1999).

3. Branch, D., Livio, M., Yungelson, L. R., Boffi, F. R. & Baron, E. In search of the progenitors of type IA supernovae. *Publ. Astron. Soc. Pacif.* **107**, 1019-1028 (1995).

4. Heger, A., Fryer, C. L., Woosley, S. E., Langer, N. & Hartmann, D. H. How massive single stars end their life. *Astrophys. J.* **591**, 288-300 (2003).

5. Kouveliotou, C. *et al.* Identification of two classes of gamma-ray bursts. *Astrophys. J.* **413**, 101-104 (1993).

6. Gehrels, N. *et al.* A short γ-ray burst apparently associated with an elliptical galaxy at redshift $z = 0.225$. *Nature* **437**, 851-854 (2005).

7. Prochaska, J. X. *et al.* The galaxy hosts and large-scale environments of short-hard γ-ray bursts. *Astrophys. J. Lett.* (in the press): preprint at ⟨http://arxiv.org/astro-ph/0510022⟩ (2005).

8. Fruchter, A. S. *et al.* HST and Palomar imaging of GRB 990123: Implications for the nature of gamma-ray bursts and their hosts. *Astrophys. J.* **519**, 13-16 (1999).

9. Sokolov, V. V. *et al.* Host galaxies of gamma-ray bursts: Spectral energy distributions and internal extinction. *Astron. Astrophys.* **372**, 438-455 (2001).

10. Le Floc'h, E. *et al.* Are the hosts of gamma-ray bursts sub-luminous and blue galaxies? *Astron. Astrophys.* **400**, 499-510 (2003).

11. Christensen, L., Hjorth, J. & Gorosabel, J. UV star-formation rates of GRB host galaxies. *Astron. Astrophys.* **425**, 913-926 (2004).

12. Bloom, J. S., Djorgovski, S. G., Kulkarni, S. R. & Frail, D. A. The host galaxy of GRB 970508. *Astrophys. J.* **507**, L25-L28 (1998).

13. Vreeswijk, P. M. *et al.* VLT spectroscopy of GRB 990510 and GRB 990712: Probing the faint and bright ends of the gamma-ray burst host galaxy population. *Astrophys. J.* **546**, 672-680 (2001).

14. Bloom, J. S. *et al.* The unusual afterglow of the γ-ray burst of 26 March 1998 as evidence for a supernova connection. *Nature* **401**, 453-456 (1999).

15. Galama, T. J. *et al.* Evidence for a supernova in reanalyzed optical and near-infrared images of GRB 970228. *Astrophys. J.* **536**, 185-194 (2000).

16. Levan, A. *et al.* GRB 020410: A gamma-ray burst afterglow discovered by its supernova light. *Astrophys. J.* **624**, 880-888 (2005).

17. Hjorth, J. *et al.* A very energetic supernova associated with the γ-ray burst of 29 March 2003. *Nature* **423**, 847-850 (2003).

18. Stanek, K. Z. *et al.* Spectroscopic discovery of the supernova 2003dh associated with GRB 030329. *Astrophys. J.* 591, L17-L20 (2003).

19. Della Valle, M. *et al.* Evidence for supernova signatures in the spectrum of the late-time bump of the optical afterglow of GRB 021211. *Astron. Astrophys.* **406**, L33-L37 (2003).

20. Malesani, D. *et al.* SN 2003lw and GRB 031203: A bright supernova for a faint gamma-ray burst. *Astrophys. J.* **609**, L5-L8 (2004).

21. Zeh, A., Klose, S. & Hartmann, D. H. A systematic analysis of supernova light in gamma-ray burst afterglows. *Astrophys. J.* **609**, 952-961 (2004).

22. Panaitescu, A. & Kumar, P. Fundamental physical parameters of collimated gamma-ray burst afterglows. *Astrophys. J.* **560**, L49-L53 (2001).

23. Frail, D. A. *et al.* Beaming in gamma-ray bursts: Evidence for a standard energy reservoir. *Astrophys. J.* **562**, L55-L58 (2001).

24. MacFadyen, A. I., Woosley, S. E. & Heger, A. Supernovae, jets, and collapsars. *Astrophys. J.* **550**, 410-425 (2001).

25. van Dyk, S. D., Hamuy, M. & Filippenko, A. V. Supernovae and massive star formation regions. *Astron. J.* **111**, 2017-2027 (1996).

26. van den Bergh, S., Li, W. & Filippenko, A. V. Classifications of the host galaxies of supernovae, Set III. *Publ. Astron. Soc. Pacif.* **117**, 773-782 (2005).

27. Podsiadlowski, P., Mazzali, P. A., Nomoto, K., Lazzati, D. & Cappellaro, E. The rates of hypernovae and gamma-ray bursts: Implications for their progenitors. *Astrophys. J.* **607**, L17-L20 (2004).

28. Bloom, J. S., Kulkarni, S. R. & Djorgovski, S. G. The observed offset distribution of gamma-ray bursts from their host galaxies: A robust clue to the nature of the progenitors. *Astron. J.* **123**, 1111-1148 (2002).

29. Riess, A. G. *et al.* Identification of Type Ia supernovae at redshift 1.3 and beyond with the Advanced Camera for Surveys on the Hubble Space Telescope. *Astrophys. J.* **600**, L163-L166 (2004).

30. Strolger, L. G. *et al.* The Hubble Higher z Supernova Search: Supernovae to $z \sim 1.6$ and constraints on Type Ia progenitor models. *Astrophys. J.* **613**, 200-223 (2004).

31. Giavalisco, M. *et al.* The Great Observatories Origins Deep Survey: Initial results from optical and near-infrared imaging. *Astrophys. J.* **600**, L93-L98 (2004).

32. Conselice, C. J. *et al.* Gamma-ray burst selected high redshift galaxies: Comparison to field galaxy populations to $z \sim 3$. *Astrophys. J.* **633**, 29-40 (2005).

33. Schlegel, D. J., Finkbeiner, D. P. & Davis, M. Maps of dust infrared emission for use in estimation of reddening and cosmic microwave background radiation foregrounds. *Astrophys. J.* **500**, 525-553 (1998).

34. Mirabal, N. *et al.* GRB 021004: A possible shell nebula around a Wolf-Rayet star gamma-ray burst progenitor. *Astrophys. J.* **595**, 935-949 (2003).

35. Klose, S. *et al.* Probing a gamma-ray burst progenitor at a redshift of $z = 2$: A comprehensive observing campaign of the afterglow of GRB 030226. *Astron. J.* **128**, 1942-1954 (2004).

36. Weidner, C. & Kroupa, P. in *The Initial Mass Function 50 Years Later* (eds Corbelli, E., Plla, F. & Zinnecker, H.) 125-186 (Springer, Dordrecht, 2005).

37. Bersier, D. *et al.* Evidence for a supernova associated with the x-ray flash 020903. *Astrophys. J.* (submitted).

38. Vreeswijk, P. M. *et al.* The host of GRB 030323 at *z* = 3.372: A very high column density DLA system with a low metallicity. *Astron. Astrophys.* **419**, 927-940 (2004).

39. Prochaska, J. X. *et al.* The host galaxy of GRB 031203: Implications of its low metallicity, low redshift, and starburst nature. *Astrophys. J.* **611**, 200-207 (2004).

40. Chen, H.-W., Prochaska, J. X., Bloom, J. S. & Thompson, I. B. Echelle spectroscopy of a GRB afterglow at *z* = 3.969: A new probe of the interstellar and intergalactic media in the young Universe. *Astrophys. J.* **634**, L25-L28 (2005).

41. Kobulnicky, H. A. & Kewley, L. J. Metallicities of galaxies in the GOODS-North Field. *Astrophys. J.* **617**, 240-261 (2004).

42. Figer, D. F., Najarro, F., Geballe, T. R., Blum, R. D. & Kudritzki, R. P. Massive stars in the SGR 1806-20 cluster. *Astrophys. J.* **622**, L49-L52 (2005).

43. Crowther, P. A. & Hadfield, L. J. Reduced Wolf-Rayet line luminosities at low metallicity. *Astron. Astrophys.* **449**, 711-722 (2006).

44. Woosley, S. & Heger, A. The progenitor stars of gamma-ray Bursts. *Astrophys. J.* **637**, 914-921 (2005).

45. Yoon, S.-C. & Langer, N. Evolution of rapidly rotating metal-poor massive stars towards gamma-ray bursts. *Astron. Astrophys.* **443**, 643-648 (2005).

46. Chapman, S. C., Blain, A. W., Smail, I. & Ivison, R. J. A redshift survey of the submillimeter galaxy population. *Astrophys. J.* **622**, 772-796 (2005).

47. Fynbo, J. P. U. *et al.* On the Lyα emission from gamma-ray burst host galaxies: Evidence for low metallicities. *Astron. Astrophys.* **406**, L63-L66 (2003).

48. Soderberg, A. M., Nakar, E., Kulkarni, S. R. & Berger, E. Late-time radio observations of 68 Type Ibc supernovae: Strong constraints on off-axis gamma-ray bursts. *Astrophys. J.* **638**, 930-937 (2006).

49. van den Bergh, S. & Tammann, G. A. Galactic and extragalactic supernova rates. *Annu. Rev. Astron. Astrophys.* **29**, 363-407 (1991).

50. Mannucci, F. *et al.* The supernova rate per unit mass. *Astron. Astrophys.* **433**, 807-814 (2005).

Supplementary Information is linked to the online version of the paper at www.nature.com/nature.

Acknowledgements. Support for this research was provided by NASA through a grant from the Space Telescope Science Institute, which is operated by the Association of Universities for Research in Astronomy, Inc. Observations analysed in this work were taken by the NASA/ESA Hubble Space Telescope under programmes: 7785, 7863, 7966, 8189, 8588, 9074 and 9405 (Principal Investigator, A.S.F.); 7964, 8688, 9180 and 10135 (PI, S. R. Kulkarni); 8640 (PI, S.T.H.). We thank N. Panagia, N. Walborn and A. Soderberg for conversations; A. Filippenko and collaborators for early-time images of GRB 980326; and J. Bloom and collaborators for making public their early observations of GRB 020322.

Author Information. Reprints and permissions information is available at npg.nature.com/reprintsandpermissions. The authors declare no competing financial interests. Correspondence and requests for materials should be addressed to A.S.F. (fruchter@stsci.edu).

The Association of GRB 060218 with a Supernova and the Evolution of the Shock Wave

S. Campana *et al.*

Editor's Note

Although the association between gamma-ray bursts and a rare class of supernovae (exploding stars) is well established, how the jet that characterizes a burst emerges from the exploding star had been unclear. Here Sergio Campana and colleagues report observations of a fairly nearby burst, which they spotted very early and were thus able to study carefully its connection with the associated supernova. They conclude that the properties are best explained by the "break-out" of a shock wave formed when a fast-moving shell of detritus collided with a stellar wind surrounding the supernova's progenitor star. This means that they observed the supernova in the very act of exploding.

Although the link between long γ-ray bursts (GRBs) and supernovae has been established[1-4], hitherto there have been no observations of the beginning of a supernova explosion and its intimate link to a GRB. In particular, we do not know how the jet that defines a γ-ray burst emerges from the star's surface, nor how a GRB progenitor explodes. Here we report observations of the relatively nearby GRB 060218 (ref. 5) and its connection to supernova SN 2006aj (ref. 6). In addition to the classical nonthermal emission, GRB 060218 shows a thermal component in its X-ray spectrum, which cools and shifts into the optical/ultraviolet band as time passes. We interpret these features as arising from the break-out of a shock wave driven by a mildly relativistic shell into the dense wind surrounding the progenitor[7]. We have caught a supernova in the act of exploding, directly observing the shock break-out, which indicates that the GRB progenitor was a Wolf–Rayet star.

G RB 060218 was detected with the Burst Alert Telescope (BAT) instrument[8] onboard the Swift[9] space mission on 18.149 February 2006 Universal Time[5]. The burst profile is unusually long with a T_{90} (the time interval containing 90% of the flux) of $2{,}100 \pm 100$ s (Fig. 1). The flux slowly rose to the peak at 431 ± 60 s (90% containment; times are measured from the BAT trigger time). Swift slewed autonomously to the newly discovered burst. The X-ray Telescope (XRT)[10] found a bright source, which rose smoothly to a peak of ~100 counts s^{-1} (0.3–10 keV) at 985 ± 15 s. The X-ray flux then decayed exponentially with an e-folding time of $2{,}100 \pm 50$ s, followed around 10 ks later by a shallower power-law decay similar to that seen in typical GRB afterglows[11,12] (Fig. 2). The UltraViolet/Optical Telescope (UVOT)[13] found emission steadily brightening by a factor of 5–10 after the first detection, peaking in a broad plateau first in the ultraviolet (31.3 ± 1.8 ks at 188 nm) and later in the optical (39.6 ± 2.5 ks at 439 nm) parts of the spectrum. The light curves reached a minimum at about 200 ks, after which the ultraviolet light curves remained constant while

GRB 060218 与一个超新星成协
以及激波的演化

坎帕纳等

编者按

尽管 γ 射线暴与一类稀有的超新星（即爆炸的恒星）之间的成协已经确立，作为 γ 射线暴特征的喷流如何从爆炸恒星出现却一直不为人所知。本文作者塞尔希奥·坎帕纳与他的同事报道了他们对一个非常近的 γ 射线暴的观测。他们很早便开始观察，因而可以细致地研究这一个 γ 射线暴与成协的超新星之间的联系。他们得出结论，对于所观测到的一系列特征最好的解释是快速运动的爆炸残余物壳层与超新星前身星周围的星风碰撞时形成的激波的"突破"。这意味着他们观测到的超新星正处于爆炸进行之中。

尽管长 γ 射线暴（GRB）和超新星之间的关联已经被确定 [1-4]，迄今尚未观测到超新星爆发的开始时期及其与 GRB 之间的紧密联系。尤其是，我们不知道刻画 γ 射线暴的喷流是如何从恒星的表面突现出来的，也不知道 GRB 的前身星是怎样爆发的。这里我们报道就相对较近的 GRB 060218（参考文献 5）及其与超新星 SN 2006aj（参考文献 6）之间的联系的观测。除了典型的非热辐射，GRB 060218 的 X 射线能谱还显示存在热成分，热成分随着时间推移逐渐变冷并转移到光学/紫外波段。我们将这些特征解释为由轻度相对论性壳层驱动的激波进入前身星周围的致密星风引发的激波暴 [7]。我们正好捕捉到一颗超新星处于爆发之中，并直接观测到激波暴。此激波暴表明该 GRB 的前身星是一颗沃尔夫-拉叶星。

GRB 060218 是于世界时 2006 年 2 月 18.149 日利用载于雨燕 [9] 天文台上的 γ 暴预警望远镜（BAT）[8] 观测到的 [5]。这个 γ 暴的轮廓不寻常的长，T_{90}（包含 90% 流量的时间区间）为 2,100 ± 100 s（图 1）。流量缓慢上升，在 431 ± 60 s 时达到峰值（包含 90%；时间以 BAT 触发时间为起始）。雨燕自动转动到这个最新发现的 γ 暴。X 射线望远镜（XRT）[10] 发现了一颗亮源在平滑上升，并在 985 ± 15 s 时到达峰值，约 100 次计数·s^{-1}（0.3～10 keV）。然后 X 射线流量指数衰减，e 折时间为 2,100 ± 50 s，接下来约 10 ks 后是类似典型 GRB 余辉中看到的那种较缓的幂率衰减 [11,12]（图 2）。紫外/光学望远镜（UVOT）[13] 发现在第一次探测后辐射稳定变亮为原来的 5～10 倍，在一个宽的平台中达到峰值，首次是出现于频谱的紫外波段（31.3 ± 1.8 ks，位于 188 nm），随后出现于光学波段（39.6 ± 2.5 ks，位于 439 nm）。光变曲线在约 200 ks 时达到最低值，之后紫外波段光变曲线一直稳定，光学波段再次变亮，约在

a rebrightening was seen in the optical bands, peaking again at about 700–800 ks (Fig. 2).

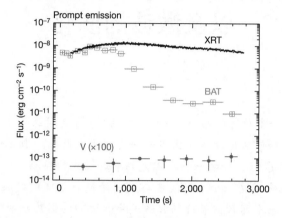

Fig. 1. Early Swift light curve of GRB 060218. GRB 060218 was discovered by the BAT when it came into the BAT field of view during a pre-planned slew. There is no emission at the GRB location up to $-3{,}509$ s. Swift slewed again to the burst position and the XRT and UVOT began observing GRB 060218 159 s later. For each BAT point we converted the observed count rate to flux (15–150 keV band) using the observed spectra. The combined BAT and XRT spectra were fitted with a cut-off power law plus a blackbody, absorbed by interstellar matter in our Galaxy and in the host galaxy at redshift $z = 0.033$. The host galaxy column density is $N_H^z = 5.0 \times 10^{21}$ cm^{-2} and that of our Galaxy is $(0.9–1.1) \times 10^{21}$ cm^{-2}. Errors are at 1σ significance. At a redshift $z = 0.033$ (corresponding to a distance of 145 Mpc with $H_0 = 70$ km s^{-1} Mpc^{-1}) the isotropic equivalent energy, extrapolated to the 1–10,000-keV rest-frame energy band, is $E_{\mathrm{iso}} = (6.2 \pm 0.3) \times 10^{49}$ erg. The peak energy in the GRB spectrum is at $E_{\mathrm{p}} = 4.9^{+0.4}_{-0.3}$ keV. These values are consistent with the Amati correlation, suggesting that GRB 060218 is not an off-axis event[26]. This conclusion is also supported by the lack of achromatic rise behaviour of the light curve in the three Swift observation bands. The BAT fluence is dominated by soft X-ray photons and this burst can be classified as an X-ray flash[27]. A V-band light curve is shown with red filled circles. For clarity the V flux has been multiplied by a factor of 100. Magnitudes have been converted to fluxes using standard UVOT zero points and multiplying the specific flux by the filter Full Width at Half Maximum (FWHM). Gaps in the light curve are due to the automated periodic change of filters during the first observation of the GRB.

Soon after the Swift discovery, low-resolution spectra of the optical afterglow and host galaxy revealed strong emission lines at a redshift of $z = 0.033$ (ref. 14). Spectroscopic indications of the presence of a rising supernova (designated SN 2006aj) were found three days after the burst[6,15] with broad emission features consistent with a type Ic supernova (owing to a lack of hydrogen and helium lines).

The Swift instruments provided valuable spectral information. The high-energy spectra soften with time and can be fitted with (cut-off) power laws. This power-law component can be ascribed to the usual GRB jet and afterglow. The most striking feature, however, is the presence of a soft component in the X-ray spectrum that is present in the XRT data up to $\sim 10{,}000$ s. The blackbody component shows a marginally decreasing temperature ($kT \approx 0.17$ keV, where k is the Boltzmann constant), and a clear increase in luminosity with time, corresponding to an increase in the apparent emission radius from $R_{\mathrm{BB}}^{\mathrm{X}} = (5.2 \pm 0.5) \times 10^{11}$ cm to $R_{\mathrm{BB}}^{\mathrm{X}} = (1.2 \pm 0.1) \times 10^{12}$ cm (Fig.3). During the rapid decay ($t \approx 7{,}000$ s), a blackbody component is still present in the data with a marginally cooler temperature ($kT = 0.10 \pm 0.05$ keV) and a comparable emission radius:

$700 \sim 800 \, \text{ks}$ 再次达到峰值（图 2）。

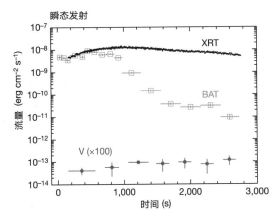

图 1. GRB 060218 的早期雨燕光变曲线。GRB 060218 是在 BAT 按预设回转时进入 BAT 的视野从而被发现的。直到 γ 暴前 3,509 s，在 GRB 的位置还没有辐射。雨燕重新回转到 γ 暴位置，XRT 和 UVOT 在 γ 暴后 159 s 开始观测 GRB 060218。对 BAT 的每个点，我们通过观测的能谱把观测计数率转化为流量（15 ~ 150 keV 波段）。我们使用一个截断幂率谱加一个黑体谱来拟合 BAT 和 XRT 组合能谱，其中考虑了银河系和红移 $z = 0.033$ 的宿主星系中星际介质的吸收。宿主星系的柱密度为 $N_H^{\bar{z}} = 5.0 \times 10^{21} \, \text{cm}^{-2}$，而银河系的柱密度为 $(0.9 \sim 1.1) \times 10^{21} \, \text{cm}^{-2}$。误差为 1σ。在红移 $z = 0.033$ 处（对应距离 145 Mpc，$H_0 = 70 \, \text{km} \cdot \text{s}^{-1} \cdot \text{Mpc}^{-1}$），外推到静止参照系 1 ~ 10,000 keV 能量区间的各向同性等效能量为 $E_{iso} = (6.2 \pm 0.3) \times 10^{49} \, \text{erg}$。GRB 能谱的峰值能量为 $E_p = 4.9^{+0.4}_{-0.3} \, \text{keV}$。这些值和阿马蒂关系一致，表明 GRB 060218 不是一个偏轴的事件[26]。光变曲线在雨燕的三个观测波段没有出现无色的上升行为也支持了这一观点。BAT 的能流主要由软 X 射线光子主导，所以这一 γ 暴被分类为 X 射线闪变[27]。V 波段的光变曲线用红色实心圆表示。为了更清晰地呈现，这里将 V 波段的流量乘以因子 100。星等转化成流量的过程中利用了标准 UVOT 零点，并将此流量乘以了滤波器的半高全宽（FWHM）。光变曲线中的间断是由于第一次观测该 GRB 时滤波器的自动周期性改变所造成的。

在雨燕的发现之后不久，光学余辉和宿主星系的低分辨率光谱显示在红移 $z = 0.033$ 处有很强的发射线（参考文献 14）。GRB 三天后[6,15]分光观测显示出现了一颗增亮中的超新星（命名为 SN 2006aj），宽线辐射特征表明其为 Ic 型超新星（由于缺乏氢线和氦线）。

雨燕搭载的设备提供了丰富的光谱信息。高能光谱随着时间软化，并可以用（截断的）幂率谱进行拟合。这个幂率谱成分通常来自 GRB 的喷流和余辉。不过最显著的特征是，在 XRT 数据中 X 射线谱存在直到约 10,000 s 的软成分。黑体成分呈现温度微弱的降低（$kT \approx 0.17 \, \text{keV}$，此处 k 为玻尔兹曼常数），而光度随时间显著地增大，这对应于视辐射半径从 $R_{BB}^X = (5.2 \pm 0.5) \times 10^{11} \, \text{cm}$ 增大到 $R_{BB}^X = (1.2 \pm 0.1) \times 10^{12} \, \text{cm}$（图 3）。在快速衰减的过程（$t \approx 7,000 \, \text{s}$）中，黑体成分在数据中依然存在，具有一个稍微冷一点的温度（$kT = 0.10 \pm 0.05 \, \text{keV}$）和一个大小相当的辐射半径 $R_{BB}^X =$

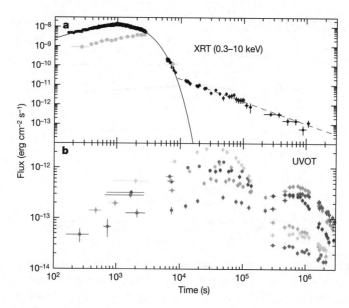

Fig. 2. Long-term Swift light curve of GRB 060218. **a**, The XRT light curve (0.3–10 keV) is shown with open black circles. Count-rate-to-flux conversion factors were derived from time-dependent spectral analysis. We also plotted (filled grey circles) the contribution to the 0.3–10-keV flux by the blackbody component. Its percentage contribution is increasing with time, becoming dominant at the end of the exponential decay. The X-ray light curve has a long, slow power-law rise followed by an exponential (or steep power-law) decay. At about 10,000 s the light curve breaks to a shallower power-law decay (dashed red line) with an index of -1.2 ± 0.1, characteristic of typical GRB afterglows. This classical afterglow can be naturally accounted for by a shock driven into the wind by a shell with kinetic energy $E_{shell} \approx 10^{49}$ erg. The t^{-1} flux decline is valid at the stage where the shell is being decelerated by the wind with the deceleration phase beginning at $t_{dec} \lesssim 10^4$ s for $\dot{M} \gtrsim 10^{-4}(v_{wind}/10^8)M_\odot\,yr^{-1}$ (where v_{wind} is in units of cm s^{-1}), consistent with the mass-loss rate inferred from the thermal X-ray component. Error bars are 1σ. **b**, The UVOT light curve. Filled circles of different colours represent different UVOT filters: red, V (centred at 544 nm); green, B (439 nm); dark blue, U (345 nm); light blue, UVW1 (251 nm); magenta, UVM1 (217 nm); and yellow, UVW2 (188 nm). Specific fluxes have been multiplied by their FWHM widths (75, 98, 88, 70, 51 and 76 nm, respectively). Data have been rebinned to increase the signal-to-noise ratio. The ultraviolet band light curve peaks at about 30 ks owing to the shock break-out from the outer stellar surface and the surrounding dense stellar wind, while the optical band peaks at about 800 ks owing to radioactive heating in the supernova ejecta.

$R_{BB}^{X} = (6.5^{+14}_{-4.4}) \times 10^{11}$ cm. In the optical/ultraviolet band at 9 hours (32 ks) the blackbody peak is still above the UVOT energy range. At 120 ks the peak of the blackbody emission is within the UVOT passband, and the inferred temperature and radius are $kT = 37^{+1.9}_{-0.9}$ eV and $R_{BB}^{UV} = 3.29^{+0.94}_{-0.93} \times 10^{14}$ cm, implying an expansion speed of $(2.7 \pm 0.8) \times 10^9$ cm s^{-1}. This estimate is consistent with what we would expect for a supernova and it is also consistent with the line broadening observed in the optical spectra.

The thermal components are the key to interpreting this anomalous GRB. The high temperature (two million degrees) of the thermal X-ray component suggests that the radiation is emitted by a shock-heated plasma. The characteristic radius of the emitting region, $R_{shell} \approx (E/aT^4)^{1/3} \approx 5 \times 10^{12}$ cm (E is the GRB isotropic energy and a is the radiation density constant), corresponds to the radius of a blue supergiant progenitor. However, the

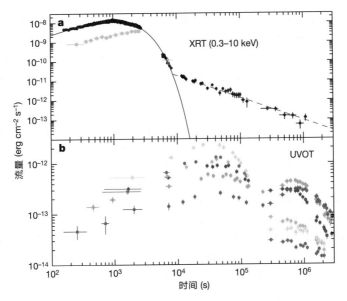

图 2. GRB 060218 的长期雨燕光变曲线。**a**，黑色空心圆代表 XRT(0.3 ~ 10 keV) 的光变曲线。计数率–流量转化因子来自依赖于时间的谱分析。我们也绘制了黑体成分对 0.3 ~ 10 keV 流量的贡献（灰色实心圆）。黑体成分的百分比随着时间而增大，在指数衰减的末期成为主导。X 射线光变曲线具有长而缓慢的幂率上升，然后指数（或以陡的幂率形式）衰减。在约 10,000 s 光变曲线偏折为较浅的幂率衰减（红色虚线），指数为 -1.2 ± 0.1，具有典型的 GRB 余辉的特征。这个典型余辉可以很自然地用被壳层驱动进入星风的激波来描述，壳层动能为 $E_{shell} \approx 10^{49}$ erg。壳层被星风减速，减速相开始于 $t_{dec} \lesssim 10^4$ s，此时 $\dot{M} \gtrsim 10^{-4}(v_{wind}/10^8)M_\odot$ yr^{-1}（这里 v_{wind} 以 cm·s^{-1} 为单位），流量下降符合 t^{-1}。这与从热 X 射线成分推导出的质量损失率一致。误差棒为 1σ。**b**，UVOT 光变曲线。不同颜色的实心圈代表不同的 UVOT 滤光片：红色，V（中心位于 544 nm）；绿色，B（439 nm）；深蓝色，U（345 nm）；浅蓝色，UVW1（251 nm）；品红，UVW1（217 nm）；黄色，UVW2（188 nm）。具体的流量已经乘以了各自的半高全宽（分别为 75 nm，98 nm，88 nm，70 nm，51 nm 和 76 nm）。数据被重新分组以增加信噪比。紫外波段的光变曲线的峰值大约在 30 ks 处，这是由于激波需要从恒星外表面和周围致密星风物质突破出来。然而光学波段的峰值大约在 800 ks 处，源自超新星喷出物中的放射性加热。

$(6.5^{+14}_{-4.4}) \times 10^{11}$ cm。在光学/紫外波段黑体的峰在 9 小时 (32 ks) 时仍然高于 UVOT 的能量范围。在 120 ks 时黑体的峰落入 UVOT 能量范围内，估计的温度和半径分别为 $kT = 3.7^{+1.9}_{-0.9}$ eV 和 $R_{BB}^{UV} = (3.29^{+0.94}_{-0.93}) \times 10^{14}$ cm，这表明膨胀速度为 $(2.7 \pm 0.8) \times 10^9$ cm·s^{-1}。这一估计与我们对超新星的预期一致，也与光谱中观测到的谱线展宽一致。

热成分是解释这一反常 GRB 的关键。X 射线热成分的较高温度（两百万度）表明辐射是由激波加热的等离子体发出。辐射区的特征半径为 $R_{shell} \approx (E/aT^4)^{1/3} \approx 5 \times 10^{12}$ cm（E 是 GRB 各向同性的能量，a 是辐射密度常数），对应于前身星为蓝巨星的半径。不过，超新星光谱缺少氢线表明它是一个更致密的源。大的辐射半径

lack of hydrogen lines in the supernova spectrum suggests a much more compact source. The large emission radius may be explained in this case by the existence of a massive stellar wind surrounding the progenitor, as is common for Wolf–Rayet stars. The thermal radiation is observed once the shock driven into the wind reaches a radius, $\sim R_{shell}$, where the wind becomes optically thin.

The characteristic variability time is $R_{shell}/c \approx 200$ s, consistent with the smoothness of the X-ray pulse and the rapid thermal X-ray flux decrease at the end of the pulse. We interpret this as providing, for the first time, a direct measurement of the shock break-out[16,17] of the stellar envelope and the stellar wind (first investigated by Colgate[18]). The fact that R_{shell} is larger than R_{BB}^{X} suggests that the shock expands in a non-spherical manner, reaching different points on the R_{shell} sphere at different times. This may be due to a non-spherical explosion (such as the presence of a jet), or a non-spherical wind[19,20]. In addition, the shock break-out interpretation provides us with a delay between the supernova explosion and the GRB start of $\lesssim 4$ ks (ref. 21; see Fig. 1).

As the shock propagates into the wind, it compresses the wind plasma into a thin shell. The mass of this shell may be inferred from the requirement that its optical depth be close to unity, $M_{shell} \approx 4\pi R_{shell}^2/\kappa \approx 5 \times 10^{-7} M_{\odot}$ ($\kappa \approx 0.34$ cm^2 g^{-1} is the opacity). This implies that the wind mass-loss rate is $\dot{M} \approx M_{shell} v_{wind}/R_{shell} \approx 3 \times 10^{-4} M_{\odot}$ yr^{-1}, for a wind velocity $v_{wind} = 10^8$ cm s^{-1}, typical for Wolf–Rayet stars. Because the thermal energy density behind a radiation-dominated shock is $aT^4 \approx 3\rho v_s^2$ (ρ is the wind density at R_{shell} and v_s the shock velocity) we have $\rho \approx 10^{-12}$ g cm^{-3}, which implies that the shock must be (mildly) relativistic, $v_s \simeq c$. This is similar to GRB 980425/SN 1998bw, where the ejection of a mildly relativistic shell with energy of $\simeq 5 \times 10^{49}$ erg is believed to have powered radio[22-24] and X-ray emission[7].

The optical–ultraviolet emission observed at an early time of $t \lesssim 10^4$ s may be accounted for as the low-energy tail of the thermal X-ray emission produced by the (radiation) shock driven into the wind. At a later time, the optical–ultraviolet emission is well above that expected from the (collisionless) shock driven into the wind. This emission is most probably due to the expanding envelope of the star, which was heated by the shock passage to a much higher temperature. Initially, this envelope is hidden by the wind. As the star and wind expand, the photosphere propagates inward, revealing shocked stellar plasma. As the star expands, the radiation temperature decreases and the apparent radius increases (Fig. 3). The radius inferred at the peak of the ultraviolet emission, $R_{BB}^{UV} \approx 3 \times 10^{14}$ cm, implies that emission is arising from the outer $\sim 4\pi(R_{BB}^{UV})^2/\kappa \approx 10^{-3} M_{\odot}$ shell at the edge of the shocked star. As the photosphere rapidly cools, this component of the emission fades. The ultraviolet light continues to plummet as cooler temperatures allow elements to recombine and line blanketing to set in, while radioactive decay causes the optical light to begin rising to the primary maximum normally seen in supernova light curves (Fig. 2).

可以用存在于前身星周围的大质量星风来解释，这对于沃尔夫–拉叶星来说很常见。一旦被驱动进入星风的激波到达半径 $\sim R_{shell}$，在此星风变得光薄，热辐射便会被观测到。

特征的变化时标为 $R_{shell}/c \approx 200\ s$，这与平滑的 X 射线脉冲以及在脉冲末端热 X 射线流量的快速减少一致。我们对此的解释是，这是首次直接观测到激波从恒星包层和星风中突破出来的激波暴[16,17]（科尔盖特[18]曾首次进行过研究）。R_{shell} 比 R_{BB}^X 大的事实表明激波以非球形的方式膨胀，在不同的时间到达 R_{shell} 球上的不同点。这可能是由非球形的爆炸（例如存在喷流）或者非球形的星风[19,20]引起。另外，激波暴的解释为我们提供了超新星爆炸和 GRB 开始之间延迟时间的限制，为 $\leqslant 4\ ks$。

激波传播到星风后压缩星风等离子体成为一个薄壳层。这个壳层的质量可以从光深满足接近于 1 这一条件推出，$M_{shell} \approx 4\pi R_{shell}^2/\kappa \approx 5 \times 10^{-7} M_\odot$（$\kappa \approx 0.34\ cm^2 \cdot g^{-1}$ 是不透明度）。这表明对星风速度为 $v_{wind} = 10^8\ cm \cdot s^{-1}$ 的典型沃尔夫–拉叶星而言，星风质量损失为 $\dot{M} \approx M_{shell}v_{wind}/R_{shell} \approx 3 \times 10^{-4} M_\odot\ yr^{-1}$。因为在辐射主导的激波之后的热能量密度为 $aT^4 \approx 3\rho v_s^2$（ρ 为 R_{shell} 处的星风密度，v_s 为激波速度），我们得到 $\rho \approx 10^{-12}\ g \cdot cm^{-3}$，这意味着激波必须是（轻度）相对论性的，$v_s \simeq c$。这与 GRB 980425/SN 1998bw 类似，我们相信其中是一个能量为 $\simeq 5 \times 10^{49}\ erg$ 的轻度相对论性壳层驱动了射电[22-24]和 X 射线辐射[7]。

早期 $t \leqslant 10^4\ s$ 时观测到的光学–紫外辐射可能是来自（辐射）激波进入星风产生的热 X 射线辐射的低能尾巴。之后，光学–紫外的辐射明显亮于对星风中的（无碰撞）激波的预期。这个辐射最有可能来自膨胀的恒星包层，恒星星包由于激波经过而加热到一个大大增高的温度。起初，这个包层隐埋在星风里。当恒星和星风膨胀，光球层向内传播，逐渐显示出激波加速后的恒星等离子体物质。当恒星膨胀时，辐射温度降低，视半径增大（图 3）。紫外辐射达到峰值时推算出的半径 $R_{BB}^{UV} \approx 3 \times 10^{14}\ cm$，这表明辐射来自经激波扫过的恒星外层约 $4\pi(R_{BB}^{UV})^2/\kappa \approx 10^{-3} M_\odot$ 的壳层。当光球快速冷却时，辐射的这一成分衰退。温度降低导致元素发生重合，并开始出现谱线覆盖，因而紫外光继续急速下降。然而放射性衰变而导致可见光开始上升，并达到通常在超新星光变曲线中常见到的主极大值（图 2）。

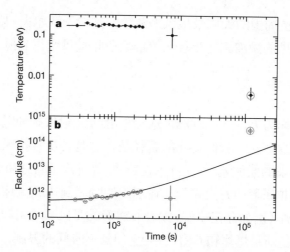

Fig. 3. Evolution of the soft thermal component temperature and radius. **a**, Evolution of the temperature of the soft thermal component. The joint BAT and XRT spectrum has been fitted with a blackbody component plus a (cut-off) power-law in the first ~3,000 s (see also the legend of Fig. 1). The last point (circled in green) comes from a fit to the six UVOT filters, assuming a blackbody model with Galactic reddening, $E(B-V) = 0.14$, and host galaxy reddening. This reddening has been determined by fitting the Rayleigh–Jeans tail of the blackbody emission at 32 ks (9 hours). The data require an intrinsic $E(B-V) = 0.20 \pm 0.03$ (assuming a Small Magellanic Cloud reddening law[28]). Error bars are 1σ. **b**, Evolution of the radius of the soft thermal component. The last point (circled in green) comes from the fitting of UVOT data. The continuous line represents a linear fit to the data.

Because the wind shell is clearly larger than the progenitor radius, we infer that the star radius is definitely smaller than 5×10^{12} cm. Assuming a linear expansion at the beginning (owing to light travel-time effects) we can estimate a star radius of $R_{star} \approx (4 \pm 1) \times 10^{11}$ cm. This is smaller than the radius of the progenitors of type II supernovae, like blue supergiants (4×10^{12} cm for SN 1987A, ref. 25) or red supergiants (3×10^{13} cm). Our results unambiguously indicate that the progenitor of GRB 060218/SN 2006aj was a compact massive star, most probably a Wolf–Rayet star.

(**442**, 1008-1010; 2006)

S. Campana[1], V. Mangano[2], A. J. Blustin[3], P. Brown[4], D. N. Burrows[4], G. Chincarini[1,5], J. R. Cummings[6,7], G. Cusumano[2], M. Della Valle[8,9], D. Malesani[10], P. Mészáros[4,11], J. A. Nousek[4], M. Page[3], T. Sakamoto[6,7], E. Waxman[12], B. Zhang[13], Z. G. Dai[13,14], N. Gehrels[6], S. Immler[6], F. E. Marshall[6], K. O. Mason[15], A. Moretti[1], P. T. O'Brien[16], J. P. Osborne[16], K. L. Page[16], P. Romano[1], P. W. A. Roming[4], G. Tagliaferri[1], L. R. Cominsky[17], P. Giommi[18], O. Godet[16], J. A. Kennea[4], H. Krimm[6,19], L. Angelini[6], S. D. Barthelmy[6], P. T. Boyd[6], D. M. Palmer[20], A. A. Wells[16] & N. E. White[6]

[1] INAF—Osservatorio Astronomico di Brera, via E. Bianchi 46, I-23807 Merate (LC), Italy
[2] INAF—Istituto di Astrofisica Spaziale e Fisica Cosmica di Palermo, via U. La Malfa 153, I-90146 Palermo, Italy
[3] UCL Mullard Space Science Laboratory, Holmbury St. Mary, Dorking, Surrey RH5 6NT, UK
[4] Department of Astronomy and Astrophysics, Pennsylvania State University, University Park, Pennsylvania 16802, USA
[5] Università degli studi di Milano Bicocca, piazza delle Scienze 3, I-20126 Milano, Italy
[6] NASA—Goddard Space Flight Center, Greenbelt, Maryland 20771, USA
[7] National Research Council, 2101 Constitution Avenue NW, Washington DC 20418, USA
[8] INAF—Osservatorio Astrofisico di Arcetri, largo E. Fermi 5, I-50125 Firenze, Italy
[9] Kavli Institute for Theoretical Physics, UC Santa Barbara, California 93106, USA

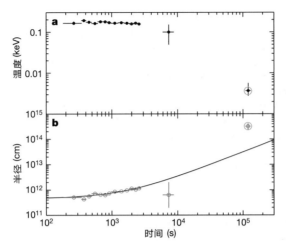

图 3. 软的热成分的温度和半径的演化。**a**，软的热成分温度的演化。BAT 和 XRT 的联合能谱在前约 3,000 s 用一个黑体成分加一个 (截断的) 幂率来拟合 (见图 1 图注)。最后的点 (绿色圆圈内) 来自考虑了黑体模型在银河系红化 $E(B-V) = 0.14$ 和宿主星系红化后对 UVOT 的 6 个滤光片的拟合。红化值的确定来自对在 32 ks (9 小时) 黑体辐射的瑞利-金斯尾的拟合。数据要求本征红化 $E(B-V) = 0.20 \pm 0.03$ (假设为小麦哲伦云的红外定律 [28])。误差棒为 1σ。**b**，软的热成分的半径的演化。最后的点 (绿色圆圈内) 来自对 UVOT 数据的拟合。连续线代表对数据的线性拟合。

因为星风壳层明显比前身星半径大，我们推断恒星半径必然小于 5×10^{12} cm。假设起初星体是线性膨胀的，(由光行时间效应) 我们估计恒星半径为 $R_{star} \approx (4 \pm 1) \times 10^{11}$ cm。这比 II 型超新星的前身星半径，比如蓝巨星 (对 SN 1987A 而言是 4×10^{12} cm，参考文献 25) 或红巨星 (3×10^{13} cm) 要小。我们的结果毫无疑问地证明了 GRB 060218/SN 2006aj 的前身星为致密的大质量恒星，而且最可能是沃尔夫-拉叶星。

(肖莉 翻译；黎卓 审稿)

[10] International School for Advanced Studies (SISSA-ISAS), via Beirut 2-4, I-34014 Trieste, Italy

[11] Department of Physics, Pennsylvania State University, University Park, Pennsylvania 16802, USA

[12] Physics Faculty, Weizmann Institute, Rehovot 76100, Israel

[13] Department of Physics, University of Nevada, Box 454002, Las Vegas, Nevada 89154-4002, USA

[14] Department of Astronomy, Nanjing University, Nanjing, 210093, China

[15] PPARC, Polaris House, North Star Avenue, Swindon SN2 1SZ, UK

[16] Department of Physics and Astronomy, University of Leicester, University Road, Leicester LE1 7RH, UK

[17] Department of Physics and Astronomy, Sonoma State University, Rohnert Park, California 94928-3609, USA

[18] ASI Science Data Center, via G. Galilei, I-00044 Frascati (Roma), Italy

[19] Universities Space Research Association, 10211 Wincopin Circle Suite 500, Columbia, Maryland 21044-3431, USA

[20] Los Alamos National Laboratory, PO Box 1663, Los Alamos, New Mexico 87545, USA

Received 13 March; accepted 10 May 2005.

References:

1. Woosley, S. E. Gamma-ray bursts from stellar mass accretion disks around black holes. *Astrophys. J.* **405**, 273-277 (1993).

2. Paczyński, B. Are gamma-ray bursts in star-forming regions? *Astrophys. J.* **494**, L45-L48 (1993).

3. MacFadyen, A. I. & Woosley, S. E. Collapsars: gamma-ray bursts and explosions in "failed supernovae". *Astrophys. J.* **524**, 262-289 (1999).

4. Galama, T. J. *et al.* An unusual supernova in the error box of the γ-ray burst of 25 April 1998. *Nature* **395**, 670-672 (1998).

5. Cusumano, G. *et al.* GRB060218: Swift-BAT detection of a possible burst. *GCN Circ.* 4775 (2006).

6. Masetti, N. *et al.* GRB060218: VLT spectroscopy. *GCN Circ.* 4803 (2006).

7. Waxman, E. Does the detection of X-ray emission from SN1998bw support its association with GRB980425? *Astrophys. J.* **605**, L97-L100 (2004).

8. Barthelmy, S. D. *et al.* The Burst Alert Telescope (BAT) on the SWIFT Midex Mission. *Space Sci. Rev.* **120**, 143-164 (2005).

9. Gehrels, N. *et al.* The Swift gamma ray burst mission. *Astrophys. J.* **611**, 1005-1020 (2004).

10. Burrows, D. N. *et al.* The Swift X-Ray Telescope. *Space Sci. Rev.* **120**, 165-195 (2005).

11. Tagliaferri, G. *et al.* An unexpectedly rapid decline in the X-ray afterglow emission of long γ-ray bursts. *Nature* **436**, 985-988 (2005).

12. O'Brien, P. T. *et al.* The early X-ray emission from GRBs. *Astrophys. J.* (submitted); preprint at ⟨http://arXiv.org/astro-ph/0601125l⟩ (2006).

13. Roming, P. W. A. *et al.* The Swift Ultra-Violet/Optical Telescope. *Space Sci. Rev.* **120**, 95-142 (2005).

14. Mirabal, N. & Halpern, J. P. GRB060218: MDM Redshift. *GCN Circ.* 4792 (2006).

15. Pian, E. *et al.* An optical supernova associated with the X-ray flash XRF 060218. *Nature* doi:10.1038/nature05082 (this issue).

16. Ensman, L. & Burrows, A. Shock breakout in SN1987A. *Astrophys. J.* **393**, 742-755 (1992).

17. Tan, J. C., Matzner, C. D. & McKee, C. F. Trans-relativistic blast waves in supernovae as gamma-ray burst progenitors. *Astrophys. J.* **551**, 946-972 (2001).

18. Colgate, S. A. Prompt gamma rays and X-rays from supernovae. *Can. J. Phys.* **46**, 476 (1968).

19. Mazzali, P. A. *et al.* An asymmetric, energetic type Ic supernova viewed off-axis and a link to gamma-ray bursts. *Science* **308**, 1284-1287 (2005).

20. Leonard, D. C. *et al.* A non-spherical core in the explosion of supernova SN2004dj. *Nature* **440**, 505-507 (2006).

21. Norris, J. P. & Bonnell, J. T. How can the SN-GRB time delay be measured? *AIP Conf. Proc.* **727**, 412-415 (2004).

22. Kulkarni, S. R. *et al.* Radio emission from the unusual supernova 1998bw and its association with the gamma-ray burst of 25 April 1998. *Nature* **395**, 663-669 (1998).

23. Waxman, E. & Loeb, A. A subrelativistic shock model for the radio emission of SN1998bw. *Astrophys. J.* **515**, 721-725 (1999).

24. Li, Z.-Y. & Chevalier, R. A. Radio supernova SN1998bw and its relation to GRB980425. *Astrophys. J.* **526**, 716-726 (1999).

25. Arnett, W. D. *et al.* Supernova 1987A. *Annu. Rev. Astron. Astrophys.* **27**, 629-700 (1989).

26. Amati, L. *et al.* GRB060218: $E_{p,i}$–E_{iso} correlation. *GCN Circ.* 4846 (2006).

27. Heise, J., *et al.* in *Proceedings of "Gamma-Ray Bursts in the Afterglow Era"* (eds Costa, E., Frontera, F. & Hjorth, J.) 16-21 (Springer, Berlin/Heidelberg, 2001).

28. Pei, Y. C. Interstellar dust from the Milky Way to the Magellanic Clouds. *Astrophys. J.* **395**, 130-139 (1992).

Acknowledgements. We acknowledge support from ASI, NASA and PPARC.

Author Information. Reprints and permissions information is available at www.nature.com/reprints. The authors declare no competing financial interests.

Correspondence and requests for materials should be addressed to S.C. (sergio.campana@brera.inaf.it).